ANIMAL MODELS IN PSYCHIATRY AND NEUROLOGY

Edited by

I. HANIN

Western Psychiatric Institute and Clinic University of Pittsburgh

and

E. USDIN

National Institute of Mental Health, Rockville, Maryland

PERGAMON PRESS

OXFORD · NEW YORK · TORONTO · SYDNEY · PARIS · FRANKFURT

U.K.	Pergamon Press Ltd., Headington Hill Hall, Oxford OX3 0BW, England
U.S.A.	Pergamon Press Inc., Maxwell House, Fairview Park, Elmsford, New York 10523, U.S.A.
CANADA	Pergamon of Canada Ltd., 75 The East Mall, Toronto, Ontario, Canada
AUSTRALIA	Pergamon Press (Aust.) Pty. Ltd., 19a Boundary Street, Rushcutters Bay, N.S.W. 2011, Australia
FRANCE	Pergamon Press SARL, 24 rue des Ecoles, 75240 Paris, Cedex 05, France
WEST GERMANY	Pergamon Press GmbH, 6242 Kronberg-Taunus, Pferdstrasse 1, West Germany

First edition 1977

British Library Cataloguing in Publication Data

Animal models in psychiatry and neurology.
1. Psychology, Physiological - Methodology -
Congresses 2. Animals, Habits and behavior of
- Congresses
I. Hanin, I. II. Usdin, E
612'.8   QP360   77-30306
ISBN 0-08-021556-4

In order to make this volume available as economically and rapidly as possible the authors' typescripts have been reproduced in their original form. This method unfortunately has its typographical limitations but it is hoped that they in no way distract the reader.

Printed in Great Britain by A. Wheaton & Co., Exeter

CONTENTS

Introduction... ix

Foreword   *Israel Hanin and Earl Usdin*............................ xiii

I. Animal Models: An Overview *(Chairmen: Arnold J. Friedhoff and Earl Usdin)*

Animal Models: Promises and Problems *Conan Kornetsky*................ 1

The Physiology of Operant Behavior *Larry Stein, James Belluzzi and C. David Wise*... 9

Significance of Sex and Age Differences *Stata Norton*............... 17

Significance of Species Differences: Rotational Models *Stanley D. Echols and Richard C. Ursillo*.. 27

Minimal Brain Dysfunction Hyperkinesis: Significance of Nutritional Status in Animal Models of Hyperactivity *I Arthur Michaelson, Robert L. Bornschein, Rita K. Loch and Lee S. Rafales*............... 37

II. Models in Psychiatry: An Overview *(Chairman: Julius Axelrod)*

The Psychiatric Clinical Syndrome and Essential Corresponding Qualities of an Animal Model of Receptor Sensitivity Modification *Arnold J. Friedhoff*... 51

III. Models in Psychiatry: Psychoses *(Chairman: Israel Hanin)*

Chronic Stimulant Intoxication Models of Psychosis *Everett H. Ellinwood Jr. and M. Marlyne Kilbey*..................... 61

Animal Models of Human Cognitive Processes *Steven Matthysse*........ 75

Identifying Indoleamine Hallucinogens by Their Preferential Action on Serotonin Autoreceptors *George K. Aghajanian*..................... 83

Central Dopaminergic Systems: Two in vivo Electrophysiological Models for Predicting Therapeutic Efficacy and Neurological Side Effects of Putative Antipsychotic Drugs *Benjamin S. Bunney*................... 91

Effects of Amphetamine on Social Behaviors of Rhesus Macaques: An Animal Model of Paranoia   *Suzanne Haber, Patricia R. Barchas and Jack D. Barchas*.. 107

IV. Models in Psychiatry: Affective Disorders *(Chairman: Julius Axelrod)*

Biobehavioral Models of Depression in Monkeys *William T. McKinney, Jr.* 117

Maternal Separation of Monkey Infants: A Model of Depression *Martin Reite*... 127

The Pharmacology of Kindling *Ronald G. Babington*................... 141

Animal Test Models for Prediction of Clinical Antidepressant Activity *I. S. Sanghvi and S. Gershon*.................................. 157

Nonverbal Communication as an Index of Depression *Robert E. Miller, Candice J. Ranelli and John M. Levine*............................... 171

EEG Sleep Correlates of Depression in Man *David J. Kupfer*.......... 181

The Noradrenergic Cyclic AMP Generating System in the Limbic Fore-Brain: A Functional Postsynaptic Norepinephrine Receptor System and its Modification by Drugs Which Either Precipitate or Alleviate Depression *Fridolin Sulser and Jerzy Vetulani*.................. 189

Approaches to Rapidly Cycling Manic-Depressive Illness *Robert M. Post*.. 201

Animal Models for Mania *Dennis L. Murphy*........................... 211

V. Models in Psychiatry: Stress, Anxiety, Aggression *(Chairman: Conan Kornetsky)*

    Tails of Stress-Related Behavior: A Neuropharmacological Model *Seymour M. Antelman and Anthony R. Caggiula* 227

    Animal Models of Aggressive Behavior *Michael H. Sheard* 247

    Biogenic Amine Metabolism in the Aggressive Mouse-Killing Rat *A. I. Salama and M. E. Goldberg* 259

    The Geller Conflict Test: A Model of Anxiety and a Screening Procedure for Anxiolytics *James L. Howard and Gerald T. Pollard* 269

    The Use of Animal Models for Delineating the Mechanisms of Action of Anxiolytic Agents *Arnold S. Lippa, Eugene N. Greenblatt and Russell W. Pelham* 279

    Alterations in the Function of the Nucleus Locus Coeruleus: A Possible Model for Studies of Anxiety *D. Eugene Redmond, Jr.* 293

VI. Models in Psychiatry: Genetic Factors *(Chairman: Conan Kornetsky)*

    Genetic Models of Behavior Disorders *Benson E. Ginsburg* 307

    A Biochemical Analysis of Strain Differences in Narcotic Action *G. Racagni, F. Cattabeni and R. Paoletti* 315

VII. Extension of Animal Models Studies to Phase I Clinical Pharmacology *(Chairman: Louis Lemberger)*

    Extension of Animal Models in Clinical Evaluation of New Drugs *Louis Lemberger* 321

    The Use of an Animal Model for Parkinsonism and its Extension to Human Parkinsonism *Menek Goldstein, Arthur F. Battista, Abraham Lieberman, Jow Y. Lew and Fumiaki Hata* 331

    Extension of Animal Models to Clinical Evaluation of Antianxiety Agents *Marian W. Fischman, Charles R. Schuster and E. H. Uhlenhuth* .. 339

VIII. Models in Neurology *(Chairman: Harold L. Klawans)*

    Animal Models of Dyskinesia *Harold L. Klawans, Ana Hitri, Paul A. Nausieda and William J. Weiner* 351

    Animal Models of Epilepsy *Peter Lomax and Joseph G. Bajorek* 365

    Animal Models of Hyperactivity *Alan M. Goldberg and Ellen K. Silbergeld* 371

    Animal Models in Hepatic Coma *Josef E. Fischer* 385

    Narcotic Mechanisms and Dyskinesias *Kristin R. Carlson* 391

    Brain Neurotransmitter Receptors and Chronic Antipsychotic Drug Treatment: A Model for Tardive Dyskinesia *R. M. Kobayashi, J. Z. Fields, R. E. Hruska, K. Beaumont and H. I. Yamamura* ·············· 405

    Effect of Striatal Kainic Acid Lesions on Muscarinic Cholinergic Receptor Binding: Correlation with Huntington's Disease *Henry I. Yamamura, Robert E. Hruska, Robert Schwarcz and Joseph Coyle* 411

    Behavioral and Neurochemical Effects of Central Catecholamine Depletion: A Possible Model for "Subclinical" Brain Damage *Michael J. Zigmond and Edward M. Stricker* 415

    An Animal Model of Myoclonus Related to Central Serotonergic Neurons *R. Malcolm Stewart and Ross J. Baldessarini* 431

IX. Models in Studies of Drug Side Effects *(Chairman: Richard C. Ursillo)*

    Neuroleptic-Stimulated Prolactin Secretion in the Rat as an Animal Model for Biological Psychiatry: I. Comparison with Anti-Psychotic Activity *Herbert Y. Meltzer, Victor S. Fang, Richard Fessler, Miljana Simonovic and Dusanka Stanisic* 443

    Tardive Dyskinesia and Antipsychotics *W. E. Fann, J. R. Stafford and Jeanine Wheless* 457

    The Cardiovascular Toxicity of Tricyclic Antidepressants *Duane G. Spiker* 469

Therapeutic and Side Effects of Psychotropic Drugs: The Relevance of
Receptor Binding Methodology *David C. U'Prichard and Solomon H.
Snyder*.. 477

X, Animal Models: Future Directions *(Chairman: Roger J. Kelleher)*........ 497

In the interests of making these
contributions available as rapidly
as possible, no subject index to the
volume has been produced.

INTRODUCTION

Our first intention after agreeing that the time was ripe for a workshop on animal models in psychiatry and neurology was to have an informal one with truly adequate time for discussion among the few participants; we thought in terms of 35 or 40. To keep the workshop simple, we agreed to have it in the Parklawn Building, with the participants sitting around one or several tables. Unfortunately, the time was really ripe and we were pressured into having more and more participants, more and more attendees - until we finally would up with a floating group attending each session of the order of 125. Not only was the pressure to attend applied by university and Government scientists, but it was applied particularly by scientists from the laboratories of pharmaceutical companies. These latter not only were interested in attending and participating, but they also were willing to give financial support. The organizers of the workshop gratefully acknowledge the support of

Burroughs Wellcome Co.
Ciba-Geigy Corporation
Dow Chemical Company
Hoffmann-LaRoche, Inc.
Lederle Laboratories
McNeil Laboratories
Merck Sharp & Dohme Research Laboratories
Merrell-National Laboratories
Sandoz, Inc.
Searle Laboratories
Smith, Kline and French Laboratories
Warner Lambert/Parke Davis
Wyeth Laboratories, Inc.

The workshop could not have been held without the cooperation and support of Pergamon Press and we should like to thank in particular the publisher, Mr. Robert Maxwell for his active assistance. Neither the workshop nor the proceedings would have been possible without the enthusiastic support of our secretaries: Cathy Rupp (at the University of Pittsburgh) and Jean Pierce and Ellen Perella (at the National Institute of Mental Health).

Finally the workshop obviously would not have been possible without the presence and contributions of the active participants:

ACTIVE PARTICIPANTS

George K. Aghajanian
Dept. of Psychiatry
School of Medicine
Yale University
333 Cedar Street
New Haven, CT. 06510

Seymour M. Antelman
Dept. of Psychology
University of Pittsburgh
Crawford Hall
Pittsburgh, PA. 15261

Julius Axelrod
NIMH
Building 10 - Rm 2D-45
Bethesda, Maryland 20014

Ronald G. Babington
Pharmaceutical Research and Development
Sandoz, Inc.
East Hanover, NJ. 07963

Herbert Barry, III
Dept. of Pharmacology
School of Pharmacy
University of Pittsburgh
Salk Hall
Pittsburgh, PA 15261

Lyle Bivens
NIMH
Parklawn Building, Room 10-105
Rockville, MD. 20857

George R. Breese
Dept. of Psychiatry
School of Medicine
University of North Carolina at Chapel Hill
Chapel Hill, NC. 25714

Benjamin S. Bunney
Dept. of Psychiatry and Pharmacology
School of Medicine
Yale University
333 Cedar Street
New Haven, CT. 06510

Kristin R. Carlson
Dept. of Pharmacology
School of Medicine
University of Pittsburgh, Scaife Hall
Pittsburgh, PA. 15261

Neal Castagnoli, Jr.
School of Pharmacy
University of California
San Francisco, California 94143

Thomas N. Chase
NINCDS - NIH
Building 36, Room 5A31
Bethesda, MD. 20014

Leonard Cook
Research Division (Pharmacology)
Hoffmann-La Roche, Inc.
Nutley, NJ. 07110

Barrett R. Cooper
Burroughs Wellcome Co.
Research Triangle Park, NC. 27707

Arnold Davidson
Hoffman-La Roche, Inc.
Roche Park
Nutley, NJ. 07710

Stanley D. Echols
Merrell-National Laboratories
110 E. Amity Road
Cincinnati, OH. 45215

Everett H. Ellinwood, Jr.
Behavioral Neuropharmacology Section
Dept. of Psychiatry
Duke University Medical School
P.O. Box 3870
Durham, NC. 27710

William E. Fann
Dept. of Psychiatry
Baylor College of Medicine
Texas Medical Center
Houston, TX. 77025

Josef E. Fischer
Surgical Physiology Lab.
Massachusetts General Hospital
Harvard Medical School
Boston, MA. 02114

Marian W. Fischman
Dept. of Psychiatry
School of Medicine
University of Chicago
Chicago, IL. 60637

Arnold J. Friedhoff
Millhauser Labs.
School of Medicine
New York University Medical Center
550 First Avenue
New York, NY. 10016

Benson E. Ginsburg
Dept. of Biobehavioral Sciences
College of Liberal Arts and Sciences
The University of Connecticut
Storrs, CT. 06268

Alan M. Goldberg
Dept. of Environmental Health Sciences
School of Hygiene and Public Health
The Johns Hopkins University
615 N. Wolfe Street
Baltimore, MD. 21205

Menek Goldstein
Dept. of Neurochemistry
School of Medicine
New York University Medical Center
550 First Avenue
New York, NY 10016

Suzanne Haber
Dept. of Psychiatry and Behavioral Sciences
School of Medicine
Stanford University
Stanford, CA. 94305

Israel Hanin
Dept. of Psychiatry
Western Psychiatric Institute and Clinic
School of Medicine
University of Pittsburgh
3811 O'Hara Street
Pittsburgh, PA. 15261

James Howard
Burroughs Wellcome Company
3030 Cornwallis Road
Research Triangle Park, NC. 27709

Roger J. Kelleher
NERPRC
Harvard Medical School
One Pine Hall Drive
Southborough, MA. 01772

Marlene Kilbey
Texas Research Institute of Mental Sciences
1300 Moursund
Houston, TX. 77025

Harold L. Klawans
Dept. of Neurological Sciences
Rush-Presbyterian - St. Luke's Medical Center
1753 West Congress Parkway
Chicago, IL. 60612

Arthur Kling
Dept. of Psychiatry
Rutgers University School of Medicine
New Brunswick, NJ. 08903

Ronald M. Kobayashi
Veterans Administration Hospital
3350 La Jolla Village Drive
San Diego, CA. 92161

Conan Kornetsky
Laboratory of Behavioral Pharmacology
Division of Psychiatry
School of Medicine
Boston University
80 E. Concord Street
Boston, MA. 02118

David J. Kupfer
Dept. of Psychiatry
Western Psychiatric Institute and Clinic
School of Medicine
University of Pittsburgh
3811 O'Hara Street
Pittsburgh, PA. 15261

Nancy Leith
Vanderbilt University School of Medicine
    and Tennessee Neuropsychiatric Inst.
1501 Murfreesboro Road
Nashville, TN. 37217

Louis Lemberger
Lilly Laboratories for Clinical Research
William N. Wishard Memorial Hospital
1001 W. Tenth Street
Indianapolis, IN. 46202

Ronald Lipman
NIMH, Parklawn Building, Room 9-101
5600 Fishers Lane
Rockville, MD. 20857

Arnold S. Lippa
Neurophysiology/Psychopharmacology
Central Nervous System Disease Therapy
    Research Section
Lederle Laboratories
Division of American Cyanamid Co.
Pearl River, NY. 10965

Peter Lomax
Dept. of Pharmacology
School of Medicine, UCLA
Los Angeles, CA. 90024

Steven Matthysse
Dept. of Psychiatry
Harvard Medical School
Massachusetts General Hospital
Fruit Street
Boston, MA. 02114

William T. McKinney, Jr.
Dept. of Psychiatry
University of Wisconsin
Madison Center for Health Sciences
1130 University Avenue
Madison, WI. 53706

Herbert Y. Meltzer
Dept. of Psychiatry
Pritzker School of Medicine
University of Chicago
Chicago, IL. 60637

I. Arthur Michaelson
Dept. of Environmental Health
College of Medicine
University of Cincinnati
3223 Eden Avenue
Cincinnati, OH. 45265

Robert E. Miller
Dept. of Psychiatry
Western Psychiatric Institute and Clinic
University of Pittsburgh
3811 O'Hara Street
Pittsburgh, PA. 15261

Dennis L. Murphy
NIMH
Building 10 - Rm 3S229
Bethesda, MD. 20014

Stata Norton
Dept. of Pharmacology
University of Kansas Medical Center
39th Street at Rainbow Blvd.
Kansas City, KS. 66103

Mark Perlow
NIMH
St. Elizabeths Hospital
Washington, D. C. 20032

Robert M. Post
NIMH
Bldg.10 - Room 3S239
Bethesda, MD. 20014

Giorgio Racagni
Institute of Pharmacology and Pharmacognosy
University of Milan
Via A. del Sarto, 21
20129 Milan, Italy

D. Eugene Redmond
Neurobehavioral Laboratory
School of Medicine
Yale University
333 Cedar Street
New Haven, CT. 06510

Martin Reite
Dept. of Psychiatry
University of Colorado Medical Center
4200 E. Ninth Avenue
Denver, CO. 80220

Jeffrey K. Saelens
Ciba-Geigy
556 Morris Avenue
Summit, NJ. 07901

Andre I. Salama
Research Division
Warner Lambert/Parke Davis
2800 Plymouth Rd.
Ann Arbor, MI. 48106

I. S. Sanghvi
USV Pharmaceutical Corporation
1 Scarsdale Rd.
Tucahoe, NY. 10707

Charles R. Schuster, Jr.
Dept. of Psychiatry
School of Medicine
University of Chicago
Chicago, IL. 60637

Norman Share
Merck Frosst Labs.
P.O. Box 1005
Pointe Claire-Dorval
Quebec H9R4P8, Quebec
Canada

Michael H. Sheard
Dept. of Psychiatry
School of Medicine
Yale University
333 Cedar Street
New Haven, CT. 06510

Kathleen Sherman
Dept. of Psychiatry
Western Psychiatric Institute and Clinic
University of Pittsburgh
3811 O'Hara Street
Pittsburgh, PA. 15261

Ellen K. Silbergeld
NIMH
Building 36 - Room 5A10
Bethesda, Maryland 20014

Thomas J. Sobotka
Food and Drug Administration HFF-153
Washington, D. C. 20204

Duane G. Spiker
Dept. of Psychiatry
Western Psychiatric Institute and Clinic
School of Medicine
University of Pittsburgh
3811 O'Hara Street
Pittsburgh, PA. 15261

John R. Stafford
Dept. of Psychiatry
Baylor College of Medicine
Texas Medical Center
Houston, TX. 77025

Larry Stein
Research and Development
Wyeth Laboratories, Inc.
P.O. Box 8299
Philadelphia, PA. 19101

Malcolm R. Stewart
Mailman Research Center
McLain Hospital
115 Mill Street
Belmont, MA.   02178

Edward R. Stricker
Dept. of Psychology
University of Pittsburgh
Crawford Hall
Pittsburgh, PA. 15261

Fridolin Sulser
Vanderbilt University School of Medicine
  and Tennessee Neuropsychiatric Institute
1501 Murfreesboro Road
Nashville, TN. 15261

Ralph E. Tedeschi
Tech. Assistant to Director of Pharma-
  ceutical R and D
Dow Lepetit USA
Indianapolis, IN. 46268

Stanley S. Tenen
Dept. of Biological Research
Searle Laboratories
Division of G. D. Searle & Co.
Box 5110
Chicago, IL. 60680

Earl Thomas
Dept. of Psychology
Bryn Mawr College
Bryn Mawr, PA. 19101

E. H. Uhlenhuth
Dept. of Psychiatry
School of Medicine
University of Chicago
Chicago, IL. 60637

David C. U'Prichard
Dept. of Pharmacology
School of Medicine
The Johns Hopkins University
725 N. Wolfe Street
Baltimore, MD. 21205

Richard C. Ursillo
Merrell-National Laboratories
110 E. Amity Road
Cincinnati, OH. 45215

Earl Usdin
NIMH
Parklawn Building, Room 9-95
5600 Fishers Lane
Rockville, MD. 20857

Richard Vogel
Dept. of Psychiatry
School of Medicine
University of North Carolina at Chapel Hill
Chapel Hill, NC. 25714

Wolfgang H. Vogel
Dept. of Pharmacology
Jefferson Medical College
Thomas Jefferson University
Philadelphia, PA. 19107

Henry I. Yamamura
Dept. of Pharmacology
University of Arizona
Tucson, AZ. 85724

Nancy Zahniser
Dept. of Psychiatry
Western Psychiatric Institute and Clinic
School of Medicine
University of Pittsburgh
3811 O'Hara Street
Pittsburgh, PA. 15261

Michael J. Zigmond
Dept. of Life Sciences
University of Pittsburgh
Crawford Hall
Pittsburgh, PA. 15261

FOREWORD

A good animal model of any particular disease state is one which will simulate and reproduce most accurately the human syndrome which it is designed to represent. Ideally, it should mirror behaviorally the symptoms of the disorder, and should respond to pharmacologic and appropriate therapeutic agents in a manner identical to that observed in the original human state.

Such a concept, of course is a utopian goal. As will be stated repeatedly in the various chapters in this book, one cannot reproduce the exact syndrome in an animal that one is trying to mimic from man. This is particularly the case in the sphere of psychiatry, where emotions, mood, and reactive behavior are all intertwined in a general cluster of symptoms described as a "syndrome". How can one tell, for example, what an animal, particularly a non-primate, is thinking about from our observation of its behavioral patterns? Moreover, at the risk of offending many animal lovers, can one really understand and interpret correctly affective behavioral responses of animals which cannot verbalize their feelings and can only react to specific stimuli or environmental conditions? Add to the above the fact that available diagnostic tools to date in psychiatric practice still are in a state of flux, and that there is still much disagreement among psychiatrists as to what is the best scale to use in characterizing and classifying the type of disorder in their patient population. It is thus not surprising that an animal model fulfilling all the requirements listed in the beginning of this discourse cannot, in reality, be achieved.

A variety of convincing and useful alternatives have, nevertheless, been developed. In some cases, investigators have focused on a specific response, whether it is behavioral, electrophysiological, or biochemical in nature, which is affected in a very specific way by specific psychotropic agents used routinely for the treatment of a particular syndrome. In such a situation, the animal model may or may not resemble at all the illness in question. Using such preparations, the investigators have, nevertheless, access to an animal model for a particular syndrome X, which has been shown to be clinically alleviated by the psychotropic agents in question. Such animal models are particularly useful in terms of screening for potentially effective compounds, and for the investigation of biochemical and neurotransmitter-related correlates of the drugs under investigation. In other cases, there also exists a variety of animal models which imitate not all, but just one or several specific characteristics of the syndrome to be replicated. These characteristics are then considered as representative of the syndrome which is being simulated, since they may be affected pharmacologically, or as a result of a specific physiologic intervention, in a manner similar to that observed in the actual human syndrome. Intermediary approaches to the above two extremes also exist, and examples of all types of approaches are to be found in the various chapters in this book.

Usually, scientific books are edited by authorities on the subject matter of the book itself. In this particular case, we have made an exception to the rule. This book is the result, to paraphrase the playwright, of scientists in search of an animal model, or two or more... Many alternative models exist. But how does one select the best and most adequate animal model for a particular psychiatric or neurologic syndrome? The difficulties are compounded by the lack of agreement in the literature as to which is the choice animal model to be used in each case. Several excellent books on animal models do exist. However, there is not a single comprehensive reference which has attempted to encompass a wide variety of animal models with specific emphasis on psychiatry and neurology.

We consequently discussed the situation and decided to investigate the need for such a reference book. The response was overwhelmingly enthusiastic, and resulted in the "Workshop on Animal Models in Psychiatry and Neurology" which took place in Rockville on June 6-8, 1977. This book is the summary of the above proceedings.

Our intention was to compile a comprehensive collection of chapters representing the "state of the art" in animal models of psychiatry and neurology, as it stands today. By intentional design, participation in the Workshop, and hence in this book, has included individuals from academia, government and industry. The subject matter spans information pertinent to clinical, as well as basic laboratory research interest. The book does not presume to answer all questions pertaining to animal models in general, or to animal models of psychiatric and neurologic disease states in particular. That would be an impossible task. It is hoped, nevertheless, that it will provide an up-to-date index of what types of animal models are available to investigators in this area of interest; also an idea of the problems and limitations associated with the available animal models.

In this sense, the book will serve to educate, as well as stimulate the reader to develop newer and better animal models. Hopefully, these would eventually be useful in aiding and improving clinical care of the psychiatric and in the neurologic patient.

                                                    Israel Hanin
                                                    Pittsburgh, PA.

                                                    Earl Usdin
July, 1977   Rockville, MD.

I. Animal Models: An Overview

ANIMAL MODELS: PROMISES AND PROBLEMS*

Conan Kornetsky+

Laboratory of Behavioral Pharmacology, Division of Psychiatry
Boston University School of Medicine, Boston, Massachusetts

The original title of this paper was, "Animal Models: Essentials and Minimum Requirements." However, after considerable thought and looking at the program of this conference I realized that it would be presumptuous of me to tell this group what the essentials and minimum requirements are for animal models in psychiatry and neurology.

There is a remark attributed to Harry Harlow that goes something like this: you have to be crazy to use animal models for the study of human psychopathology, but you are crazy if you do not because of the possible insights obtained from the animal work. Despite the problems of interpretation posed by animal models they are a necessary approach to the understanding of human pathology.

Animal models have been extensively used in medical research and their utility has been well established. Models for non psychopathological diseases often have the attribute of close similarity between the biological condition in man and what can be created in animals. Many, but far from all of the models in medicine are homologous; that is, there is correspondence in the etiology of the disease and the model. Other models in medicine may be only isomorphic; that is, despite parallelism between the model and the human condition, the cause of the condition in the animal may be quite different than the cause in man. And finally, the model may have no resemblance to the disease but simply be a non homologous or non isomorphic representation that has predictive value of either some aspect of the disease or the therapy used to treat the disease. Models in psychiatry are usually not homologous although in many instances investigators are all too ready to assume that the origin of the behavior in man is identical to the origin in the model.

A model of a disease state may only model one aspect of the biology or behavior we wish to understand. This may be because of the complexity of the total phenomenon or because we believe that we are modeling the most relevant components. Studies of learning in animals are of this type. Certainly when we put a rat in an operant chamber, have it press a lever for a food reinforcement we are not looking at all the variables that lead to active learning in a child in school. The experiment in the animal ignores many organismic and social factors. I do not believe we always know what is reinforcing for the 10 year old child. However, scheduled behavior in an operant chamber does model for us many aspects of learning that are relevant to learning in humans.

We resort to the use of models when certain experiments cannot be carried out in man because of possible insult to the physical and psychological integrity of the subject. However, the simple representation whether it be a single neuron preparation or a frog rectus muscle preparation may be the only way we can understand the mechanisms involved. Baldessarini and Fischer (1) point out that a model is an experimental compromise in that a simple experimental system is used to represent a much more complex and less readily available system, "...the animal to represent the patient, the tissue slice to represent the intact living brain, the isolated nerve ending to represent the intact synapse." Thus we can say that all experiments, even those in human subjects, are models of some aspect of nature. Data obtained while sitting in the bush making observations of baboons do not make a model. Such data might provide a model for human behavior only when hypotheses are made from these observations that concern human social behavior.

---

*Preparation of this manuscript supported in part by NIMH grant MH 12568 and NIDA grant DA 00377.
+NIMH Research Scientist Awardee, MH 1759.
The author would like to acknowledge the help and criticisms of Robert Markowitz in the preparation of this paper.

A model may be nothing more than a series of postulates that are tied together and ultimately are believed to model learning in the human. The work of Clark Hull (2) that was popular when I was a graduate student, was of this type. Thus for Hull the model was really a theory of behavior that was so precisely formulated that the animal experiments to test the theory were clearly mandated.

Psychologists, many years before Hull, attempted to develop a science of human behavior that was rooted in the animal laboratory. The narrower the view of behavior presented the less apparent was relevance to human behavior. For example, Watson the conceded father of "Behaviorism" wrote in 1914, "It is possible to write a psychology, to define it... as a science of behavior and never go back on the definition; never to use the terms consciousness, mental state, mind, content, will, imagery, and the like... It can be done in terms of stimulus and response... (3)". Thus for Watson all behavior could be explained in terms of stimulus and response and he probably would not attempt to model human psychopathology until the abnormal behavior could be redefined in these terms. Skinner, although a behaviorist much in the tradition of Watson, has not felt constrained in interpreting aberrant behavior in an animal as a model for some aspects of abnormal behavior in man (4).

The behaviorists share with Freud a fundamental belief that all behavior is determined. Although Freud's conceptual schema have a systematic logic, the concepts elude the operational definitions necessary for the behaviorists. Thus, although the behaviorists may easily develop animal models and the psychoanalysts can easily develop conceptual models, the former may have little obvious relevance to human behavior and the latter defy experimental verification.

What then makes a good model for human psychopathology? Many scientists would argue that for a model to be viable it must have heuristic value. Let us briefly examine the concept of heuristic value. The dictionary definition is in part, "adj: stimulating interest as a means of furthering investigation (5)". I do not believe that there has been an investigator who has submitted a paper for publication believing that his work did not have heuristic value. The work of Clark Hull and his students (as every graduate student of the period is aware) had a great deal of heuristic value. Each experiment clearly suggested another. Despite this, the system and postulates finally died out. I would be surprised to find a psychology graduate student in the United States who is currently involved in a thesis specifically designed to test a postulate of Clark Hull. Since "heuristic value" is an inflated commodity, it may qualify as a necessary but certainly not a sufficient test of the viability of an animal model. Thus, we must identify some of the other essential components of a good animal model.

Researchers in the field of mental illness are not always as fortunate as those working in other areas of medical science where the defining characteristics of the disorder under study are often easily described. Elevated blood pressure in the monkey is isomorphic with elevated blood pressure in man. Identifying and validating behavioral isomorphisms is much more complex. We can, and daily do, use animal models for the prediction of the effects of drugs on biological systems in man. However, to model a relevant part of a disease whose manifestations are, at least at the present time, only defined in terms of anxiety, depression or a thought disorder is certainly an exercise in either enthusiasm or frustration.

We often make inferences from the antecedent conditions of the ongoing behavior that animals are in pain, hungry, or sexually aroused. These are constructs that cannot always be documented. We can only infer in animals such human states as depression or anxiety. An increase in the number of fecal bolluses in the bottom of a cage may not reflect a state of anxiety but simply a parasympathetic discharge caused by a direct pharmacological action of a drug. As Freud supposedly once said, "a cigar may sometimes only be a good smoke". When attempting to model often ill defined emotional states there is an increased likelihood that the labeling and interpretation may be more in the eye of the beholder than in the brain of the animal. A common source of error is the assumption that similar overt behavior implies similar mood or experience.

Recently Carlton (6) raised the issue of the differences between theories and tautologies. As an example he uses the paradigm of conditioned avoidance response (CAR). In this procedure an animal is presented with some stimulus that signals an impending electric shock. The animal can emit some response that terminates the signal and avoid the impending shock. As is well documented, a wide variety of animals will learn to emit the response and avoid the impending electric shock. The behavior is not isomorphic with any psychopathology, that is, it in no way resembles identifiable aspects of human disease, yet the animal's behavior in the CAR experiment may be altered in a unique and specific way by those drugs that are useful in the treatment of schizophrenia.

The antipsychotic drugs, as well as morphine, all interfere with the animals ability to emit the appropriate response, usually a lever press, and avoid the impending electric shock, at doses that do not interfere with the animals ability to escape from the shock stimulus. The anxiolytic drugs and the barbiturates significantly interfere with the avoidance behavior only at doses that impair the escape behavior.

A hypothetical construct often employed to explain the behavior of the animal is that of fear or anxiety; the animal emits the response because the avoidance of the electric shock reduces its anxiety. Carlton (6) correctly points out that this is not a theory but a tautology. There is no characterization of anxiety that is independent of the avoidance behavior itself. He states that anxiety and avoidance are actually synonyms in this system. Thus in Carlton's schema the inference of anxiety must be independently validated. The failure of drugs used in the treatment of anxiety to block avoidance behavior would certainly argue against anxiety being the relevant intervening variable in the CAR paradigm.

Other constructs have been proferred as explanations of the effects of neuroleptics on CAR. For example, Irwin (7) and Key (8) have attributed the avoidance attenuation to a decrease in afferent stimulation. Posluns (9) has attributed the selective inhibition of avoidance behavior to an inhibition of the motor response, and we have interpreted the effect of neuroleptics on avoidance as a result of a decrease in arousal (10,11). All of these interpretations lack independent validation and thus by Carlton's criteria they are tautologies. Despite this lack of validation the neuroleptic drugs are used in the treatment of anxiety; they do cause some psychomotor retardation; they do decrease afferent stimulation; and they do decrease arousal. However, direct tests of these hypotheses while the animal is performing on the CAR paradigm have not been done and thus we are left with nothing more than hypotheses that remain untested.

As mentioned the CAR is not a model for schizophrenia but it does measure some aspect of the pharmacological action of the neuroleptic drugs that correlates with their antipsychotic activity. However, we cannot prove the null hypothesis; we cannot say that there will never be a drug that does not have the classic effects of neuroleptics on CAR that is not useful for the treatment of schizophrenia. Nor can we say that all drugs that have the classic action will be effective antipsychotic medications. Thus by having too heavy a reliance on the CAR we may discard potentially useful drugs and only find drugs that are similar to those we currently have.

Despite these reservations I have mentioned concerning the CAR, it has proved to be a robust measure that at the present time is probably the best behavioral screening procedure we have for antipsychotic effects. The error is in assuming that the effects of the neuroleptics on CAR are homologous with their therapeutic action in patients.

Even when there is a clear isomorphism between the model and the disease, the cause of the condition in the animal may not completely reflect the cause of the disease in man. An example is the use of tremor induced in animals by the administration of tremorine or oxotremorine as a model of the parkinsonian tremor. Everett (12) demonstrated that the tremorine induced tremor was antagonized by all the anticholinergic antiparkinsonism drugs available at that time. These findings led some to assume a dysfunction in a central cholinergic system was a condition sufficient to account for all the pathologies observed in parkinsonism.

Let us briefly look at one of the current hypotheses concerning a putative neural malfunctioning in the schizophrenic, the "dopamine hypothesis" and the models that support it. Although Synder (13) has pointed out that direct evidence is lacking to support the "dopamine hypothesis" in that it has been conclusively shown that there is anything abnormal about dopamine in the body fluids or brains of the schizophrenic patients, there is indirect evidence that dopamine is significant in the disease process.

Evidence in support of the dopamine hypothesis comes from such findings as; a) the relative potency of various neuroleptic drugs in assays of dopamine blocking activity correlates roughly with milligram potency in the clinic (13,14,15) and, b) the administration of amphetamine or methylphenidate causes an exacerbation of symptoms in acute schizophrenics (16). Thus it is suggested, if not stated as fact, that the neuroleptic drugs exert their antipsychotic action by reducing an overly endowed dopamine system in schizophrenia while amphetamine or methylphenidate exacerbate the symptoms of the disease by producing an additional flooding of the receptors with more dopamine. Although d-amphetamine (17, 18) and apomorphine (19) (which has dopamine-like actions) have been reported to not exacerbate the disease process, most investigators tend to emphasize those experiments that support their hypothesis.

The correlation between the relative potency of dopamine blockade in animals with milligram potency in the clinic predicts the average daily dose used in schizophrenic patients. This correlation is not a model for the relevant neural events taking place in the brain of the schizophrenic nor does it necessarily validate the animal preparations as models. Dopamine activity may be reduced by the neuroleptic drugs but that may not be why the patients show clinical improvement, and this pharmacological action may only be on some secondary manifestation of the disease.

These laboratory models that measure dopamine blocking potency lead to or are based on a number of hypotheses. If only one is incorrect then the theory has difficulty:

1. All antipsychotic drugs have equal efficacy.

2. All drugs that decrease dopamine activity would be good antipsychotic drugs.

3. All antipsychotic drugs will block dopamine.

4. All drugs that increase dopamine activity should cause schizophrenia or at least exacerbate the schizophrenic process.

Since the dopamine hypothesis is dependent to some degree on drug models in animals I would like to briefly discuss some of these models of psychopathology. In medicine drug models have always been popular and this is certainly true in animal models of psychopathology.

A drug model that is currently popular is the amphetamine model (see chapter by Ellinwood). I would like to call this the animal model of a model of schizophrenia in man. It has been stated and often accepted that the chronic administration of high doses of amphetamine to humans will regularly induce a state that is difficult to distinguish from paranoid schizophrenia (20). This observation along with the current theories relating dopamine and schizophrenia, and the actions of amphetamine on the dopamine systems, suggested that amphetamine would provide a model that was biochemically as well as behaviorally isomorphic with paranoid schizophrenia.

The paranoid ideation seen in the schizophrenic patients and the chronic amphetamine user has no direct equivalence in the animal. If the rat or monkey is paranoid we can only infer it. No matter how often you ask the animal he will not be able to tell you that other animals or people are plotting against him. However, there are behaviors seen in animals following amphetamine administration that are loosely isomorphic with those seen in man. These are the stereotypic behaviors. In the rat such behavior is characterized by sniffing, biting and licking. In the cat the dimension of head movements and apparent looking behavior is prominent. The primate adds a further dimension. These are steriotyped hand movements which consist of forefinger probing, pincer-like grasping, palm-clasping and hand examination (21). It is clear that the amphetamine induced behavior in animals seems to be a reasonable model for the amphetamine induced changes seen in man. However, it has its limitations. If the human amphetamine model is biochemically isomorphic with schizophrenia then amphetamine would be expected to exacerbate the schizophrenic syndrome. Such an effect has been described in the acute schizophrenic by Davis & Janowsky (16). However we (18) as well as others (17) have found the opposite to be true in the chronic schizophrenic. These findings are not congruent with the dopamine hypothesis unless one assumes that it applies only to the acute schizophrenic and that in the chronic schizophrenic the dopamine system is burned out or in some other way biochemically different.

A common model that has been employed in the preclinical evaluation of antidepressant drugs is the reserpine-antagonism model. In the application of this model the particular measurement of the reserpine effect that is used is not critical for we equate any reserpine effect with depression. Carlton (6) has referred to this as an equivalence model. In this type of model, since any effect of reserpine is believed to be a potential model for depression, any index that is reliable is chosen, such as sleep time, hypothermia, or ptosis and other drugs are tested in terms of their ability to antagonize the effect. If the assumption of equivalence between reserpine effects and depression were true then it would be a valid model but this is an assumption that few would make.

Many other drug models have been studied. These include LSD and mescaline induced states. Although of interest in their own right they currently are not in favor as models for schizophrenia. This is probably because there is no currently popular serotonin theory of schizophrenia and the behavioral isomorphisms between the LSD or mescaline "psychosis" and schizophrenia are not as compelling as with other drugs.

The difficulty with all these models is that they are all trying to model a human state that may be so uniquely human that no global model can possibly be adequate. Models based on dopamine make the assumption that there is an overabundance of dopamine in some neuronal system, but as mentioned, this overabundance has not been found in the schizophrenic patient. What we may be modeling may be no more than a fantasy of the investigators. The work will tell a great deal about neurotransmitter substances in the brain and the role they may play in behavior but they may tell us nothing about schizophrenia.

Another approach to modeling psychopathology in animals is the approach first made popular by Harlow (22). This model is discussed in the chapters by McKinney, Reite, Miller and Haber.

Ever since Harlow's work in the 1950's reporting the effects of social isolation in monkeys, this procedure has been extensively used to study the role of early environmental effects on monkeys. Social isolation causes severe deficits, in a variety of behaviors (22,23), that are not conducive to survival of the individual or the species. Monkey social isolation models seem to have face validity. This is partially due to the almost human-like quality of the monkey. Certainly the behavior of these monkeys is abnormal but whether or not it is a model of schizophrenia, depression or simply a model of the effects of social isolation of the human infant has not been demonstrated. Children who have been subjected to early isolation are certainly affected by the process, but they do not become schizophrenic, and we do not really know if they become part of our adult depressed population. At least in regard to schizophrenia and depression early isolation is neither a necessary nor sufficient antecedent.

The neonatally isolated monkey does respond to "therapies" that are useful with humans. Suomi, Harlow & McKinney (23) found that placing the isolates with animals they called "Monkey Psychiatrists" useful and McKinney & co-workers (24) found a positive response to chlorpromazine. A better understanding of the model might come from the studies employing antidepressent drugs as well as drugs that are not considered therapeutic in either depression or schizophrenia.

Impaired attention is one of the most commonly reported deficits seen in the schizophrenic patient. We have attempted to model this attentional disturbance in the rat (25). We do not purport to model the disease, for this is probably impossible in the animal, but we do model some significant aspect of schizophrenic behavior. At our current stage of knowledge we define schizophrenia by the behavior of the patient and not by excessive dopamine. Our model is based on the hypothesis that the attentional deficit in the schizophrenic is due to a central state of hyperarousal. Although the hypothesis can be critized on a number of grounds, an attentional deficit was relatively easy to model in animals by the use of a procedure in the rat that was analogous to the Continuous Performance Test (CPT) in man. The rat's performance can be disrupted by either electrical stimulation or chemical stimulation with norepinephrine to the mesencephalic reticular formation. This impairment can be reversed by chlorpromazine but not promethazine or pentobarbital (26). Also the effects of reticular stimulation in animals performing on either a fixed interval or fixed ratio reinforcement schedule are not reversed by chlorpromazine. This model may be open to many of the criticisms I have made in regard to other models. We may have a model for an attentional deficit yet it may be that the attentional deficit may not be the critical behavioral manifestation of the disease process and the drug actions that we observe in the rat may not reflect the relevant action in the patient.

Essential and minimum requirements for a model of schizophrenia were suggested by Matthysee & Haber (27) a few years ago.

1. The aberrant animal behavior ought to be restored to normal, in part, by drugs which are known to be effective in the treatment of schizophrenia.

2. Drugs closely related in chemical structure to the antipsychotic phenothiazines and butyrophenones, but without efficacy in the treatment of psychosis, ought to also be without normalizing effect in the animal model.

3. Tolerance (the tendency, after repeated doses of a drug, to require an increased dose to produce the same effect) should not develop to the behavior-normalizing effect of tranquilizers in the animal model since it does not develop to their antipsychotic action.

4. The normalizing effects of tranquilizers on the abnormal animal behavior should not be blocked by the simultaneous administration of anticholinergic agents.

What is clear to me is that at the present time there are no models that meet all of the criteria suggested by Matthysee and Haber. If you substitute your favorite disease where schizophrenia appears in the above criteria the available models for all psychopathology are all lacking. Even if we successfully meet all the demands of the above criteria, we may still not be modeling the disease process or a relevant aspect of the disease but simply a system for finding more drugs like the ones we already have. Thus, we may have a perfect model but of the wrong thing. This may be the common fault of all models. Thus, I can give no specific essential or minimum requirements, only promises and problems. Certainly the model should be such that it is not contrary to current knowledge, and what I mean by current knowledge is not current theory. But even here we may fall into a trap.

Let us for example assume we have a model for some aspect of schizophrenia in which the behavioral or chemical deficit is not reversed by a known neuroleptic drug. Thus it is contrary to current knowledge. It is possible that the neuroleptic drugs are not reversing primary etiological factors. The model may be immune to the current drugs and yet be more relevant to the primary etiological factors contributing to that disease process. Until we know the cause of the disease we may not be able to adequately model it, yet we may never understand the disease without first modeling it. This brings us back in a full circle to Harlow's statement that you have to be crazy to use animal models and you are crazy if you do not.

REFERENCES

(1) R.J. Baldessarini and J.E. Fischer, Biological models in the study of false neuro-chemical synaptic transmitters. In D.J. Ingle and H.M. Shein (Eds.), (1975) Model Systems in Biological Psychiatry, MIT Press, Cambridge. p 51.

(2) C. Hull (1943) Principles of Behavior, Appleton-Century, N.Y.

(3) J.B. Watson (1914) Behavior: An Introduction to Comparative Psychology, Henry Holt, N.Y. as reviewed in R.S. Woodworth (1948) Contemporary Schools of Psychology, Ronald Press, N.Y. p 68.

(4) B.F. Skinner (1953) Science and Human Behavior, Macmillan, N.Y.

(5) J. Stein and L. Urdang (Eds.) (1967) The Random House Dictionary of the English Language, Random House, N.Y.

(6) P. Carlton, Theories and theoretical models in psychopharmacology. In Psychopharmacology: A Generation of Progress, Raven Press, N.Y., In press.

(7) S. Irwin, Factors influencing acquisition of avoidance behavior and sensitivity to drugs. Fed. Proc. 17, 380 (1958).

(8) B.J. Key, The effects of drugs on discrimination and sensory generalization of auditory stimuli in cats. Psychopharmacologia (Berl.) 2, 352 (1961).

(9) D. Posluns, An analysis of chlorpromazine induced suppression of the avoidance response. Psychopharmacologia (Berl.) 3, 361 (1962).

(10) L.A. Low, M. Eliasson, and C. Kornetsky, Effect of chlorpromazine on avoidance acquisition as a function of CS-US interval length. Psychopharmacologia (Berl.) 10, 148 (1966).

(11) S. Lipper and C. Kornetsky, Effect of chlorpromazine on conditioned avoidance as a function of CS-US interval length. Psychopharmacologia (Berl.) 22, 144 (1971).

(12) G.M. Everett, Tremorine. In J-M Bordeleau (Ed.), (1961) Systeme Extra-Pyramidal et Neuroleptiques, Editions Psychiatrique, Montreal, p 182.

(13) S.H. Snyder, The dopamine hypothesis of schizophrenia: focus on the dopamine receptor. Am. J. Psychiat. 133, 197 (1976).

(14) S. Matthysee, Antipsychotic drug actions: A clue to the neuropathology of schizophrenia? Fed. Proc. 5, 200 (1973).

(15) S. H. Snyder, Amphetamine psychosis: A model schizophrenia mediated by catecholamines. Am. J. Psychiat. 130, 61 (1973).

(16) J.M. Davis and D.S. Janowsky, Amphetamine and methylphenidate psychosis. In E. Usdin and S.H. Snyder (Eds.) (1973) Frontiers in Catecholamine Research, Pergamon Press, N.Y. p 977.

(17) W. Modell, and A.E. Hussar, Failure of dextroamphetamine sulfate to influence eating and sleeping patterns in obese schizophrenic patients. JAMA 193, 275 (1965).

(18) C. Kornetsky, Hyporesponsivity of chronic schizophrenic patients to dextroamphetamine. Arch. Gen. Psychiat. 33, 1426 (1976).

(19) R.C. Smith, C. Tamminga and J.M. Davis, Effect of apomorphine on schizophrenic symptoms. J. Neural Transmission 40, 171 (1977).

(20) J.D. Griffith, J.H. Cavanaugh, J. Held and J.A. Oates, Dextroamphetamine; evaluation of psychotomimetic properties in man. <u>Arch. Gen. Psychiat.</u> 26, 97 (1972).

(21) E.H. Ellinwood Jr., A. Sudilovsky and L.M. Nelson, Evolving behavior in the chemical and experimental amphetamine (Model) Psychosis, <u>Am. J. Psychiat.</u> 130, 1088, (1973).

(22) H.F. Harlow and M.K. Harlow, Social deprivation in monkeys. <u>Sci. Amer</u>. 207, 136 (1962).

(23) S.J. Suomi, H.F. Harlow and W.T. McKinney, Jr., Monkey psychiatrists, <u>Am. J. Psychiat.</u> 128, 41, (1972).

(24) W.T. McKinney, Jr., L.H. Young, S.J. Suomi and J.M. Davis, Chlorpromizine treatment of disturbed monkeys, <u>Arch. Gen. Psychiat.</u> 29, 490 (1973).

(25) C. Kornetsky and M. Eliasson, Reticular stimulation and chlorpromazine: An animal model for schizophrenic overarousal. <u>Science</u> 165, 1273, 1969.

(26) C. Kornetsky and R. Markowitz, Animal models and schizophrenia. In D.J. Ingle and H.M. Shein (Eds.), (1975) Model Systems in Biological Psychiatry, MIT Press, Cambridge. p 26.

(27) S. Matthysee and S. Haber, Animal models of schizophrenia. In D.J. Ingle and H.M. Shein (Eds.), (1975) Model Systems in Biological Psychiatry, MIT Press, Cambridge. P 4.

THE PHYSIOLOGY OF OPERANT BEHAVIOR

Larry Stein, James D. Belluzzi and C. David Wise

Wyeth Laboratories, P. O. Box 8299, Philadelphia, PA 19101

Some theoretical questions of relevance to the physiology and pharmacology of operant behavior will be considered here. First, we will discuss Skinner's (1938) fundamental distinction between the respondent and operant control of behavior--that is, the distinction between the elicitation of behavior by antecedent stimulation (respondent) and its reinforcement by consequent stimulation (operant). Secondly, we will present a classification of operant schedules and attempt to relate the classification to the rudiments of behavioral pharmacology. And, finally, we will describe an approach to the study of the physiology of the operant and briefly review what has been discovered by this method about the anatomy and the chemistry of the positive reinforcement process.

The Respondent-Operant Distinction

All psychiatrists and psychologists are interested in understanding human behavior. To understand so complex a phenomenon it is necessary to break the behavior down into its component processes--to isolate the appropriate units of analysis (for a detailed discussion, see Teitelbaum, 1977). Fifty years ago, behaviorists such as J. B. Watson thought that, in principle at least, such an analysis was already at hand. According to Watson (1924), the goal of psychological investigation was "the ascertaining of such data and laws that given the stimulus, psychology can predict what the response will be; or, given the response, it can specify the nature of the effective stimulus". To explain any behavior, the psychologist merely needs to discover its eliciting stimuli. This view--that the stimulus-response reflex can provide a sufficient basis for behavioral analysis--had its roots in the philosophy of the British Associationists and received nourishment from the successes of Sherringtonian reflexology. Even the problem that much of human behavior is learned, while the reflex is fixed, seemed to have a plausible solution. Pavlov's discovery of the conditioned reflex suggested that, by a simple process of association, any stimulus could acquire the power to elicit any response. Watson's hope thus was sustained that all behavior, learned and unlearned, could be accounted for by the right combination of conditioned and unconditioned reflexes.

We now believe that this hope has failed. Although many behaviors, especially autonomic responses, are under the control of antecedent stimulation, most human behavior of interest to the psychologist does not seem to be reflexly elicited. The problem can be illustrated even in the relatively uncomplicated behavioral repertoire of the rat. For example, how would one elicit a bar-press response? In fact, no combination of conditioned and unconditioned stimuli will reliably cause the rat to bar press. Complex behavior patterns such as bar pressing seem to be emitted by the organism rather than elicited by the environment. Nevertheless, there is an easy way to produce strong bar-pressing behavior. If food is regularly presented to a hungry rat after a bar-press response has "accidentally" been emitted, bar pressing soon comes to dominate the animal's repertoire. Such strengthening of behavior has been termed positive reinforcement and the learning is an example of operant conditioning. Note that the stimulus-response (S-R) correlation characteristic of the reflex has been reversed in the operant case to an R-S correlation. According to Skinner, this R-S correlation constitutes a second irreducible element, the operant, which is as essential as the reflex for behavioral analysis. Respondent and operant relationships are contrasted in Table 1. It is important to note that it is the temporal relationship between S and R, and not the nature of the response itself, that distinguishes respondent and operant behavior. For example, an eye blick elicited by a cinder is respondent, whereas the same eyelid closure, a wink, "voluntarily" emitted to attract the attention of a member of the opposite sex, is operant.

Two-Factor Classification of Operant Schedules

The combination of two factors along the rows and columns of a contingency table permits the classification of all major operant schedules. The first factor may be termed positive or negative contingency--presentation vs. omission of a particular consequence (S) following the occurrence of the response (R). In the case of positive contingency, S is regularly presented after R; in the case of negative contingency, S is regularly omitted after R.

TABLE 1   Temporal Relationships in Respondent and Operant Conditioning

Type of Conditioning	Controlling Stimulus	Effect on Behavior
Respondent (reflex)	Antecedent	Elicitation
Operant	Consequent	Reinforcement

The second factor refers to the hedonic value of S--appetitive (favorable contingency) vs. aversive (unfavorable contingency). When the contingency factor constitutes the columns of a contingency table, and the hedonic factor the rows, as illustrated in Table 2, the cells of the table provide the four basic operant schedules. Response-contingent presentations of an appetitive stimulus yield positive reinforcement schedules (e.g., self-stimulation); response-contingent presentations of an aversive stimulus yield punishment schedules; response-contingent omissions of an aversive stimulus yield negative reinforcement schedules; and response-contingent omissions of an appetitive stimulus yield nonreward schedules. All operant schedules seem to be classified by this table. However, the relationship of the table to behavioral pharmacology is more complex than one originally would have thought. At first, it seemed likely that different drug classes would be differentiated by the rows or columns of the table. For example, it seemed reasonable that there should be drugs that selectively block the effects of aversive schedules; thus, a drug proven to be effective against negative reinforcement should also be effective against punishment. However, such is not the case. As may be seen in Table 3, phenothiazines counteract negative reinforcement schedules but are ineffective against punishment schedules, whereas benzodiazepines, which exert powerful antipunishment effects, have little or no effect against negative reinforcement. Similarly, there seem to be no drugs which selectively block the two appetitive schedules, or which act selectively in either positive or negative contingency schedules. The effects of drugs seem rather to be organized along the diagonals of the contingency table. Thus, phenothiazines are effective in schedules involving either positive or negative reinforcement, whereas benzodiazepines are drugs of choice against schedules involving either punishment or nonreward. Interestingly, the effects of drugs in the different schedules seem to be correlated with the behavioral effects of the schedules. Phenothiazines are active in the two schedules that facilitate behavior, whereas benzodiazepines are active in the two schedules that suppress behavior. The basis of this correlation may be that a common brain mechanism mediates the behavior-facilitating effects of both positive and negative reinforcement, and that a second mechanism mediates in common the inhibitory effects of both punishment and nonreward (Stein, 1964; Stein, Wise and Belluzzi, in press).

TABLE 2   Operant schedules generated by the presentation or omission of a favorable or unfavorable event as a consequence of the behavior. Examples of each schedule are indicated in parentheses.

Consequence	Presentation	Omission
Favorable (appetitive)	Positive reinforcement or reward (self-stimulation)	Frustrative nonreward (extinction)
Unfavorable (aversive)	Punishment or passive avoidance (Geller-Seifter conflict)	Negative reinforcement or active avoidance (Sidman avoidance)

Physiological Basis of Operant Behavior

Confidence in the operant as an elemental behavioral process would be increased if one could identify its biological substrate. The methods of physiology have served well for the identification of the neuronal substrates of reflexes. For example, the method of stimulation is used to map reflex pathways by systematically stimulating various brain points and charting those which elicit the reflex in question. How may the methods of physiology be adapted to chart the brain systems which mediate operant reinforcement?

Consider the following behavioral sequence: a lever-press response delivers food, which then causes rat to salivate. These events are diagrammed in the upper part of Fig. 1.

TABLE 3 Table of operant schedules showing their effects on behavior, drug of choice for pharmacological antagonism, and presumed neurochemical substrates.

Consequence	Presentation	Omission
Favorable	Positive reinforcement (facilitates behavior) Phenothiazines Catecholamines	Frustrative nonreward (inhibits behavior) Benzodiazepines Serotonin, Acetylcholine
Unfavorable	Punishment (inhibits behavior) Benzodiazepines Serotonin, Acetycholine	Negative reinforcement (facilitates behavior) Phenothiazines Catecholamines

Note that the same stimulus, food, stands in operant relationship to an antecedent behavior (lever press) and in respondent relationship to a consequent behavior (salivation). Thus, more or less simultaneously, the same stimulus can exert operant and respondent control over different behaviors. The putative brain events which underlie these behavioral effects are diagrammed in the lower part of Fig. 1. After CNS processing, food stimuli are shown to activate the salivary nuclei to elicit salivation and, by analogy, to activate "reinforcement systems" to reinforce bar pressing. How may one identify these hypothetical reinforcement systems? One approach would be to borrow the method of stimulation from physiology, but with an important change in order to adapt it for operant purposes. The change merely involves giving over operational control of the stimulator to the animal. Use of a Skinner box greatly simplifies the procedure. Each time the animal operates the bar, a specific brain region is stimulated via permanently-indwelling electrodes. Those probes which generate high bar-press rates are presumed to have activated the reinforcement system. It will be obvious to most readers that this experiment has already been performed. It is the famous self-stimulation experiment of Olds and Milner (1954)!

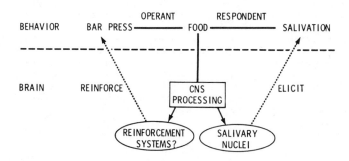

Fig. 1. The same food stimulus causes respondent elicitation of salivation and operant reinforcement of bar pressing. The two behaviors may be mediated by parallel brain events, as shown in the diagram below the broken line.

In the years following the discovery of self-stimulation, Olds and his collaborators prepared extensive maps of brain reinforcement regions (Olds, 1976). The original maps revealed a focus for self-stimulation in the hypothalamus with reward points extending along the medial forebrain bundle into limbic and olfactory forebrain. The discovery of self-stimulation in the hypothalamus was of particular interest, since stimulation of this region also elicited various drive behaviors. Thus eating, drinking, and mating could be obtained from the same hypothalamic electrodes that supported self-stimulation. These observations suggested the possibility that reward neurons might be subdivided into distinctive subgroups. Each subgroup might mediate a particular drive-reward modality and each might make use of a different neurotransmitter. To test the idea that reward neurons are chemically heterogeneous, self-stimulation behavior from a variety of hypothalamic probes were subjected to a pharmacological analysis. The assumption of neurochemical heterogeneity seemed to be quite wrong: indeed, reward behavior maintained by all the probes shared a common pharmacology. In general, it appeared that self-stimulation was controlled by catecholamine neurotransmission. Thus, drugs that enhanced catecholamine activity (for example, amphetamine or cocaine) enhanced self-stimulation, whereas drugs that inhibited catecholamine transmission (for example, chlorpromazine or α-methyl-p-tyrosine) inhibited self-stimulation. No other neurotransmitter system seemed to have so intimate a relationship to self-stimulation (Stein, Wise and Belluzzi, in press).

Even in the case of the catecholamine treatments, however, interpretation of drug effects on self-stimulation was difficult since many factors other than reward can affect response rates. Still, there were hints already in the earliest experiments (Stein, 1962; Stein and Ray, 1960) that amphetamine and chlorpromazine may act specifically on reward thresholds. In one study (Stein, 1962), methamphetamine greatly augmented the very low self-stimulation rates maintained by subthreshold currents. Because methamphetamine's effect closely resembled that produced by a small increase in the brain-stimulation intensity, it was suggested that the drug effectively converts the subthreshold stimulus into a suprathreshold one by lowering the reward threshold. Since a zero-intensity stimulus should not be so benefited, this idea was tested in a second experiment by reducing the current to zero at the time of drug administration. Consistent with the lowered-threshold hypothesis, the rate-enhancing action of methamphetamine was now virtually abolished. In another approach (Stein and Ray, 1960), a current resetting procedure was used to obtain rate-independent estimates of the self-stimulation threshold. Although the confounding effects of nonspecific stimulation or depression are largely excluded by this method, the results supported the suggestion that amphetamine lowers, and chlorpromazine raises, the threshold for brain-stimulation reward.

More recent self-stimulation mapping studies also supported the idea that behavioral reinforcement is mediated by catecholamines (Crow, 1972; Ritter and Stein, 1973). In these studies probes were implanted in the relatively homogeneous noradrenergic and dopaminergic cell concentrations of the locus coeruleus and substantia nigra, respectively (Fig. 2). Good self-stimulation was obtained in both locations. Since noncontingent brain stimulation does not maintain self-stimulation, these results provide further evidence that response-contingent release of catecholamines may mediate positive reinforcement of operant behavior. This idea, which has received critical scrutiny since its inception more than 15 years ago, is supported by a consistent pattern of anatomical and pharmacological data from studies on brain self-stimulation and drug self-administration (Stein, in press). In the absence of definitive evidence favoring one catecholamine over the other, the original question of which catecholamine is the more important has given way to a more detailed inquiry into the precise roles of noradrenaline and dopamine systems. Theories of Crow (1973) and more recently of Herberg, Stephens and Franklin (1976), based partly on catecholamine anatomy and partly on catecholamine psychopharmacology, resemble each other in the attribution of motivational or incentive functions to dopamine systems and reinforcement or memory-generating functions to noradrenaline systems. Abolition of self-stimulation by disruption of either the noradrenaline or dopamine systems suggests that these systems act jointly rather than separately, and that the activity of both systems is required for the successful performance of operant behavior.

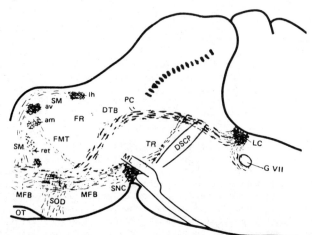

Fig. 2. Diagram of the locus coeruleus (LC) noradrenaline system and the substantia nigra (SNC) dopamine system. Self-stimulation reward is obtained from probes in the cell groups or fiber projections of either system. For other abbreviations, see Lindvall and Bjorklund, 1974.

Opioid Peptides: Mediators of Drive-Reduction Reward?

Common observation provides support both for the view that reward can be identified with events that increase drive (incentives) and for the contrary view that it can be identified with events that reduce drive (satisfiers) (Table 4). Although events of either type may yield positive reinforcement, incentives increase the level of activity and excitement whereas satisfiers cause relaxation and quiescence. Psychologists accordingly have formulated "drive-induction" theories of reward on the one hand and "drive-reduction" theories on the other. In a critical review of this work, Berlyne (1973) has concluded that "the persistence of these apparently opposite theoretical positions through the centuries certainly suggests that both have their elements of validity and that the ultimate answer will be found in some synthesis of them".

TABLE 4 Pharmacological Evidence for Two Types of Reward

Reward Type	Drug of Abuse	Behavioral Effect	Neurochemical Substrate
Incentive	Cocaine or Amphetamine	Increases Drive and Arousal	Catecholamines
Satisfier	Morphine	Reduces Drive and Arousal	Opioid Peptides?

Pharmacological data also reveal that reward may be associated with increase or decrease in the level of arousal. Thus, agents that increase arousal (for example, amphetamine) and those that decrease arousal (for example, morphine) both are avidly self-administered by animals and man. On the basis of this and other evidence, Belluzzi and Stein (1977) recently propoeed that drive-inducing reward functions may be mediated by catecholamines and that drive-reducing reward functions may be mediated by enkephalin or a related opioid peptide.

Consistently with the latter idea, Belluzzi and Stein (1977) found that rats will work for enkephalin injections delivered directly into the ventricles of their own brains. Massive dosages of leucine-enkephalin in particular were taken by the best responders, despite the possibility of tissue damage or other adverse effects that might have been produced by the passage of large volumes of fluid through the brain ventricles.

Avid self-administration of leucine-enkephalin thus supports the idea that a substance similar or identical to enkephalin may serve as a natural euphorigen or reward transmitter. If this conjecture were true, reward or addiction should be produced, not only by administration of exogenous enkephalin, but also by release of endogenous peptide following electrical activation of enkephalin-containing neurons in the brain. Behavioral and immuno-histochemical observations may be consistent with this prediction. Sites that yield high rates of self-stimulation and those that contain dense networks of enkephalin-like immuno-reactivity often overlap in precisely the same brain regions (for example, pontine central gray, zona compacta of the substantia nigra, bed nucleus of the stria terminalis, and nucleus accumbens). According to our hypothesis, self-stimulation of these regions would depend at least in part on the electrically-induced release of enkephalin and the consequent activation of opiate "reward" receptors. If so, the behavior should be suppressed or extinguished following administration of an opiate receptor antagonist such as naloxone.

The central gray region was selected for initial testing because electrical stimulation of this site induces profound analgesia as well as high-rate self-stimulation (Mayer et al., 1971). As predicted, dose-related decreases in the self-stimulation rate were obtained following subcutaneous injections of naloxone. Because the central gray region also contains high concentrations of noradrenaline, the same rats were further tested with different doses of the noradrenaline synthesis inhibitor, diethyldithiocarbamate. Again, response rates were suppressed in a dose-related fashion. These results suggest that central gray self-stimulation may depend on the activation of both noradrenaline-containing and enkephalin-containing neurons; indeed, since all of the enkephalin-rich self-stimulation sites referred to above are also rich in catecholamines, it is possible that the reward process may generally be regulated by the interaction of noradrenaline, dopamine, and enkephalin systems.

Diverse, but complementary, reward functions can also be recognized at the behavioral level. Expected rewards mobilize approach responses and generally increase the readiness to act. Following Crow (1973), we suggest that these motivational or incentive functions may be mediated in large measure by dopamine neurons. Past rewards steer behavior by shaping the response repertoire and elevating the probabilities of previously rewarded responses. This reinforcement or memory-generating function may be mediated in part by noradrenaline neurons (Crow, 1973; Stein et al., 1975). Present rewards bring the behavioral episode to a satisfying termination. We suggest that this gratifying or drive-reducing function may be subserved at least in part by an opioid peptide such as enkephalin (Table 5).

TABLE 5   Different Roles of Reward

Role	Behavioral Effect	Putative Neurotransmitter
Incentive (motivation)	initiate and facilitate pursuit behavior	Dopamine
Reinforcement (learning)	guide response selection via knowledge of response consequences	Noradrenaline
Gratification (drive reduction)	bring behavioral episode to satisfying termination	Enkephalin

REFERENCES

(1) Belluzzi, J. D. and Stein, L., Enkephalin may mediate euphoria and drive-reduction reward, Nature, 266, 556 (1977).

(2) Berlyne, D. E. (1973) The vicissitudes of aplopathematic and thelematoscopic pneumatology (or the hydrography of hedonism), In Berlyne, D. E. and Madsen, K. B. (Eds.) Pleasure, Reward, Preference, Academic Press, New York, p. 1.

(3) Crow, T. J., Catecholamine-containing neurons and electrical self-stimulation, I Review of some data, Psychol. Med., 2, 414 (1972).

(4) Crow, T. J., Catecholamine-containing neurons and electrical self-stimulation, II A theoretical interpretation and some psychiatric implications, Psychol. Med., 3, 66 (1973).

(5) Herberg, L. J., Stephens, D. N. and Franklin, K. B. J., Catecholamines and self-stimulation: evidence suggesting a reinforcing role for noradrenaline and a motivating role for dopamine, Pharmacol. Biochem. Behav., 4, 575 (1976).

(6) Lindvall, O. and Bjorklund, A., The organization of the ascending catecholamine neuron systems in the rat brain as revealed by the glyoxylic acid fluorescence method, Acta Physiol. Scand. Suppl. 412 (1974).

(7) Mayer, D. J., Wolfe, T. L., Akil, H., Carder, B. and Liebeskind, J. C., Analgesia from electrical stimulation in the brainstem of the rat, Science, 174, 1351 (1971).

(8) Olds, J., Reward and drive neurons: 1975, In Wauquier, A. and Rolls, E. T., Brain-stimulation Reward, Amsterdam, North Holland, p. 1.

(9) Olds, J. and Milner, P., Positive reinforcement produced by electrical stimulation of septal area and other regions, J. Comp. Physiol. Psychol., 47, 419 (1954).

(10) Ritter, S. and Stein, L., Self-stimulation of noradrenergic cell group (A6) in locus coeruleus of rats, J. Comp. Physiol. Psychol., 85, 443 (1973).

(11) Skinner, B. F. (1938) The Behavior of Organisms: An Experimental Analysis, Appleton-Century, New York.

(12) Stein, L. (1962) Methods for evaluating stimulant and antidepressant drugs, In Nodine, J. H. and Moyer, J. H., The First Hahnemann Symposium on Psychosomatic Medicine, Lea and Febiger, Philadelphia, p. 297.

(13) Stein, L. (1964) Reciprocal action of reward and punishment, In R. G. Heath (Ed.), The Role of Pleasure in Behavior, Hoeber Medical Division, Harper and Row, New York, p. 113.

(14) Stein, L. (in press) Reward transmitters: catecholamines and opioid peptides, In Lipton, M.A., DiMascio, A. and Killam, K. F. (Eds.), A Review of Psychopharmacology: A Second Decade of Progress, Raven Press, New York.

(15) Stein, L., Belluzzi, J. D. and Wise, C. D., Memory enhancement by central administration of norepinephrine, Brain Res., 84, 329 (1975).

(16) Stein, L. and Ray, O. S., Brain stimulation reward "thresholds" self-determined in rat, Psychopharmacologia (Berl.), 1, 251 (1960).

(17) Stein, L., Wise, C. D. and Belluzzi, J. D. (in press) Neuropharmacology of reward and punishment, In Iverson, L. L., Iverson, S. D. and Snyder, S. H., Handbook of Psychopharmacology, Vol. 8, Plenum Press, New York.

(18) Teitelbaum, P. (1977) Levels of integration of the operant, In W. K. Honig and J. E. R. Staddon (Eds.), Handbook of Operant Behavior, Prentice-Hall, Englewood Cliffs, p. 7.

(19) Watson, J. B. (1924) From the Standpoint of a Behavorist. 2nd Ed., J. B. Lippincott, Philadelphia.

DISCUSSION

Dr. Barry commented that he found in Dr. Stein's four cell table for types of responses and consequences that the upper right-hand cell is especially interesting because its occurrence is so rare: appetitive reinforcement prevented by an active response. Dr. Stein had pointed out that experimental extinction, withholding reinforcement formerly given for an active response is not quite this situation. Dr. Barry then reported that a number of years ago, Donald Blough had published an example which would fit that cell: He trained pigeons to obtain food by holding their heads for several seconds continuously in the intersection between two horizontal photo beams at right angles. He observed that their behavior was agitated, with wings flapping rather than quiescent while they maintained their heads in the required spot. This indicates a conflict between the required behavior and the natural tendency to be active rather than motionless while anticipating food. Since conflict underlies human psychopathology, this might be a useful animal model for investigating therapeutic drug effects. *Dr. Stein* also mentioned that another finding of Blough was that the ability of pigeons to perform this response was improved by chlorpromazine and was impaired by barbiturates.

SIGNIFICANCE OF SEX AND AGE DIFFERENCES

Stata Norton
Department of Pharmacology
University of Kansas Medical Center, Kansas City, Kansas U.S.A.

That there are sex-linked differences in behavior is a truism, but such differences go far beyond behaviors associated with reproduction of the species. Although there may be difficulty in separating the precise contribution of genetics and environment in such differences, the extent of such differences can be investigated without analysis of causes. It is surprising to find scientific papers in which behavioral studies are reported in which the species, but not the sex, of the experimental animals is identified. Not only is the sex of the animal often ignored, some commonly held beliefs concerning sex differences are without scientific foundation, such as the erroneous concept that the adult female albino Norway rat displays an estrus-associated increase in activity in all laboratory environments. The literature contains various well-documented conditions in which male and female behavioral differences are found. These differences may be both quantitative and qualitative and occur in man as well as in other animals. Several examples will be reviewed here.

A second factor which is sometimes ignored in reporting data is the age of the animal. Chronological age and the stage of development of the central nervous system, from embryogenesis to senescence, unquestionably correlate with changes in behavior. The subtlety of behavioral changes which may correlate progressively with the life span needs further investigation. Some of the best documented relationships of age and behavior will be given.

The syndrome of hyperkinesis in humans is an example in which quantitative behavioral differences have been reported to be related to both sex and age. Hyperkinesis is diagnosed primarily in children of school age, with a high incidence of spontaneous improvement with maturation (Hussey and Gendron, 1970). The condition is much more common in males than in females. Ney (1974) has defined four types of hyperkinesis in children with an overall ratio of 3:1 (males:females) and an incidence of 8:1 in the sub-type he calls Minimal Brain Dysfunction. An abnormal sex ratio has been reported in several other conditions in human populations. In infantile autism males predominate with a ratio of 3:1 (King, 1975). Schizophrenia has also been reported to be more frequent in males (Huber et al., 1975). The overall incidence of epilepsy is slightly higher in males, particularly of known etiology (Hauser and Kurland, 1975; Gudmundsson, 1966). Age is clearly a factor in incidence of epilepsy, with peaks early in life, before 10 years of age, and again after 50 years of age (Hauser and Kurland, 1975). Prevalence of migraine has been reported to peak between 30 and 50 years and is almost three times more frequent in females than males (Brewis et al., 1966).

The above examples of age and sex related abnormal brain function in humans are well known. Comparable attention needs to be paid to sex and age related consequences of damage to the central nervous system of experimental animals. Five factors which apply to the assessment of CNS damage are:

1. Developmental stage when damaged.
2. Time elapsed since damage.
3. Sex of animal.
4. Species and strain of animal.
5. Method of measurement.

The vulnerability of the developing brain to many insults is well known. The critical period of gestation occurs when the main structures of the brain are being organized by successive mitoses and migration of neuroblasts from the ependymal layer of the ventricle. Until the brain stem is organized, damage to the brain is likely to result in a non-viable fetus. In the rat and mouse the brain stem nuclei have been laid down by about day 13 of gestation (Berry et al., 1964; Angevine, 1965; Taber Pierce, 1973).

The major development of the diencephalon extends from the end of the formation of the brain stem until about gestational day 16 (Angevine, 1965; Keyser, 1972); the telencephalon extends from the later period of diencephalic development almost until parturition (Berry et al., 1964; Hicks and D'Amato, 1966). These times for peak activity are diagrammed in Fig. 1. The times in Fig. 1 are approximations. Actually, the timing of development of most structures extends over a long period in gestation. The hippocampus is perhaps the most remarkable since dividing granule cells have been found up to 4 months postnatally (Altman, 1963).

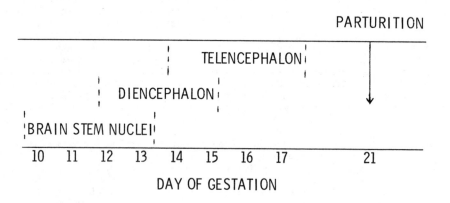

TIME TABLE FOR BRAIN STRUCTURES IN RATS

Fig. 1. Critical periods for development of major brain structures during gestation

Damage to the developing brain may be particularly severe when agents are used which affect mitotic cells. X-irradiation during the fetal period, for example, kills dividing ependymal cells and thereby disrupts not only that mitotic cycle but alters the ordered progression of subsequent migrations of neuroblasts which appear to be dependent on a planned sequence of migrations. Some of these effects have been detailed in a series of papers by Hicks and co-workers (Hicks et al., 1959; Hicks and D'Amato, 1966). However, even after loss of neurons, the capacity of the CNS to recover functionally can be dramatic, and studies on behavioral effects of early insult should take this regenerative capacity into account. In a study on the effects of anoxia produced by exposure to carbon monoxide at 5 days postnatally, young rats developed marked hyperactivity which disappeared by the time they were 3 months old (Culver and Norton, 1976). Adult rats, similarly exposed, became slightly hyperactive but no recovery was found. On the other hand, x-irradiation of fetuses in utero on gestational day 15 resulted in hyperactivity in female rats which became progressively more pronounced with age (Table 1). Comparable irradiation of male rats on day 15 of gestation produced a less hyperactive animal (Norton et al., 1976). The effect of irradiation in Table 1 was most marked in female rats 5 months old, the last age that was tested. Thus these animals showed no recovery, in contrast to the response to carbon monoxide previously mentioned.

TABLE 1 Exploratory Activity of Female Rats Irradiated on Gestational Day 15

Age at testing	Counts / hr (range)	
	Controls	Irradiated
6 weeks	450 (210-800)	630 (380-790)
3 months	480 (320-610)	1680 (1500-1830)
5 months	690 (550-780)	1910 (1750-2310)

Goldman and co-workers (1974) studied the long-term consequences of brain lesions made in neonatal and juvenile monkeys. Behavioral results were considered in relation to both age and sex of these animals. Orbital prefrontal lesions resembled gestational x-irradiation just discussed in that the animals' behavior became progressively more disturbed with increasing age. However, unlike the consequences of gestational x-irradiation which showed greater effect in female rats, male monkeys lesioned between 1 to 8 weeks postnatally showed early impairment on a conditioned reversal test when females lesioned at the same age were like controls. With time the lesioned females' responses in the test became poorer until, 18 months later, they were as impaired as the lesioned males (Goldman et al., 1974).

Since animals may recover from some types of damage inflicted on the immature brain (Culver and Norton, 1976), or may become worse with maturation (Goldman et al., 1974), these factors need to be taken into account in comparing effects in brain damaged experimental animals with the consequences of brain damage in humans. A problem of some importance in following the time-course of recovery of brain function or exacerbation of damage is determination of the stage of maturity of the brain. When, for example, is the rat brain mature? What are the most appropriate criteria for this? Various anatomical, biochemical and behavioral measurements are available but the results, as might be expected, do not all lead to the same conclusions. The rat brain has been proposed to be adult from about 40 days postnatally, at about the time of vaginal opening in the female, to 9 months after birth. Intermediate ages have often been chosen. A casual survey of the literature indicates that 3 to 5 months after birth is commonly regarded as adult age, although failure to define the age is at least as common. Table 2 gives a sampling of the data available on parameters which might be used to determine the adult stage.

TABLE 2   Maturational Times in Development of the Rat Brain

Age in days	Parameter	References
40	acid phosphatase	Friede, 1966
45	dendritic morphology of caudate neurons	Norton and Culver, 1977
50	nicotinamide adenine dinucleotides	Guarneri and Bonavita, 1966
100	cortical acetylcholinesterase	Friede, 1966
120	nocturnal maze activity	Norton et al., 1975b
120	hippocampal granule cell	Altman, 1963
140	capillarity	Craigie, 1974
150	neuronal and glial packing density	Brizzee, 1973
150	DNA	Howard, 1973
180	weight of ventral horn motoneuron	Ford, 1973
360	body weight	Enesco, 1967
700	glia/neuron index	Brizzee et al., 1964
800	myelin	Horrocks, 1973

Several of the parameters listed in Table 2 reach a peak at 4 to 5 months of age and thereafter plateau or decline slowly throughout adult life with a more rapid decline when the rat reaches senescence at 24 to 30 months. An example of this pattern is nocturnal activity in the rat shown in Fig. 2. Activity was measured in a residential maze housing permanent groups of 4 rats (Norton et al., 1975b). Some other parameters do not show a plateau or decline in the adult stage but continue to rise throughout the life of the animal, such as the amount of myelin per brain (Horrocks, 1973). As mentioned above, the albino rat is often considered as an adult by 5 months of age. In view of the number of factors which reach a peak or plateau at about this time, this age may be a reasonable choice for many studies.

When one considers the various kinds of behavioral studies which are performed on adult animals, differences in behavior of the two sexes may be of great interest. As mentioned before, the incidence of several CNS syndromes differs in males and females. Examination of some quantitative differences in behavior in the two sexes follows.

Female rats and female mice are more active than their male counterparts in many testing situations. This applies to exploratory activity, when animals are introduced into a novel environment and their activity is measured from a few minutes up to one or two hours, and also applies to circadian home cage or residential activity. In maze exploration and in two hour tests in running wheels, female mice were found to be more active than males. The activity of both sexes was stable from 4 to 8 months of age and declined by 24 months. Brain acetylcholinesterase and norepinephrine also showed a sex difference and were higher in males than females (Ordy and Schjeide, 1973). A similar decline in activity of mice from 4 to 24 months of age was reported by Goodrick (1967).

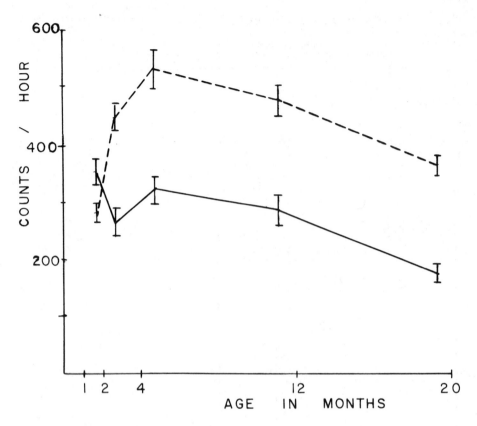

Fig. 2. Nocturnal activity in a residential maze. Groups of 4 female (dotted line) and 4 male (solid line) rats

TABLE 3   Two-hour Exploratory Activity of 3 to 4 Month-old Animals in a Residential Maze

	Counts / Hr
Rats	
Males	336 ± 50
Females	554 ± 46
Gerbils	
Males	289 ± 16
Females	177 ± 42

Two-hour exploratory activity of groups of 4 rats or 4 gerbils, measured in a maze, is shown in Table 3. Female rats were more active than males but the converse was true for gerbils. The difference between sex-related exploratory activity in rats and gerbils extends also to circadian activity (Table 4). Groups of 4 animals of one sex at 3 to 4 months of age were housed in a residential maze of a design previously reported (Norton et al., 1975a). Diurnal activity was measured from 12 noon to 6:00 P.M., on a 6:00 A.M. to 6:00 P.M. light, 6:00 P.M. to 6:00 A.M. dark cycle. Female rats were more active than males during both the light and dark cycle. Male gerbils were more active during the light but the sexes had about the same activity at night. The gerbil shows much less circadian activity than the rat and this holds for both sexes.

Several investigators have noted that male rats habituate faster to novel situations than female rats. This was reported by Hughes (1968) and also can be seen in the data reported by Syme and Syme (1974). The latter investigators analyzed photographs showing the location of groups of male or female rats in an open field. Shortly after introduction of the rats individuals of both sexes were about 100 cm apart. Three hours after introduction, males averaged about 2 cm apart while females were still 100 cm apart. In other studies examining exploratory maze behavior, adult male rats during the second hour in the maze were 20 per cent as active as during the first hour. Female rats of the same age were 72 per cent as active in the second hour as the first hour (Norton et al., 1975b).

TABLE 4 Circadian Activity of Groups of 4 Animals in a Residential Maze

	Counts / hr	
	Diurnal Activity	Nocturnal Activity
Rats		
Males	173 ± 10	400 ± 88
Females	285 ± 31	753 ± 55
Gerbils		
Males	194 ± 8	135 ± 4
Females	121 ± 14	128 ± 25

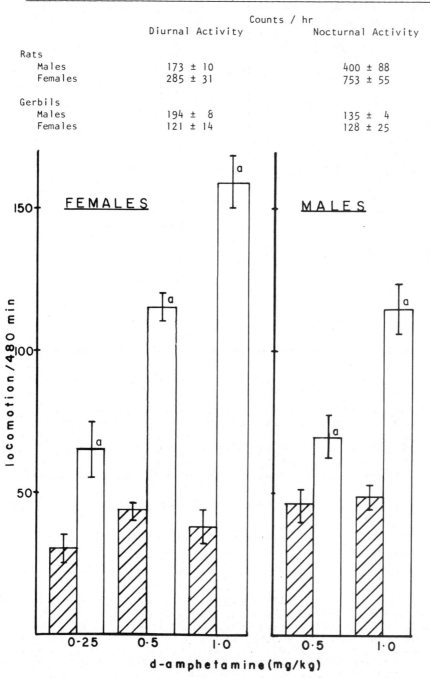

Fig. 3. Movement of groups of 4 adult rats in a residential maze, photographed every 2 minutes for 8 hours after S.C. saline (bars with diagonal lines) or amphetamine (clear bars). Letter "a" indicates drug effect significantly different from saline control. (Taken from Culver, 1975.)

Behavioral effects of brain lesions can differ in the two sexes at least quantitatively, as noted above. Quantitative differences in response to drugs also exist, as might be expected. Female rats respond to amphetamine with a greater increase in activity than males. In Fig. 3 the movement of adult rats in a maze was measured from photographs taken during the light cycle at two-minute intervals. Males changed location as often, or slightly more often, than females after saline injections but they moved much less often than females after receiving amphetamine. While some of the lesser effect of amphetamine in male rats has been attributed to more rapid metabolism (Groppetti and Costa, 1959), this does not appear to be an adequate explanation for the magnitude of the difference in effect (Schneider and Norton, 1977). Furthermore, females also respond to morphine with greater hyperactivity than males. Figure 4 shows the data for morphine, 1 or 2 mg/kg subcutaneously, in an experiment using the same methodology as in Fig. 3. The data for control animals in Figs. 3 and 4 also show that the greater activity of the female rat in some test situations does not apply to all methods of measurement. In this photographic sampling method the lower movement of the females does not correspond to concurrent photo-cell measurement of locomotion which record the females as more active than the males during the same period. Probably the females move in short bursts of activity, returning to the same place to sleep in a huddle while the males walk less but select sleeping areas more randomly.

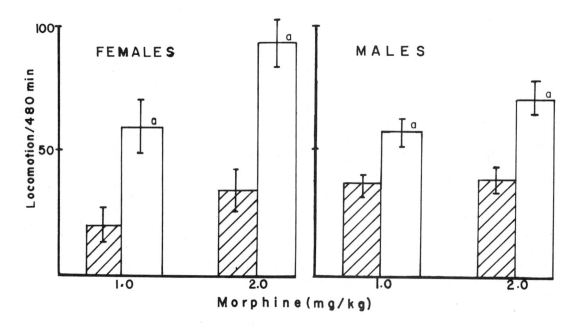

Fig. 4. Movement of groups of 4 adult rats in a residential maze, photographed every 2 minutes for 8 hours after s.c. saline (bars with diagonal lines) or morphine (clear bars). Letter "a" indicates drug effect significantly different from saline control. (Taken from Culver, 1975.)

In addition to the differences already described between the sexes, some subtle differences in behavior exist. Mullenix (1977) has shown that the male exceeds the female in amount of self grooming and this behavior develops in both sexes from weaning to adult age (Tables 5 and 6).

TABLE 5 Time Lapse Photography of Motor Acts of Male Rats as Per Cent of Juvenile Stage (Calculated from Data of Mullenix, 1977)

Age	Grooming Behaviors	Exploratory Behaviors	Attention Behaviors
4 weeks	100	100	100
6 weeks	107	137	90
4 months	141	101	81
5 months	140	83	77

Exploratory behaviors (walking, turning, rearing, pawing, head bobbing and sniffing) decrease in the male by 5 months (Table 5) but the decrease is most marked in comparison with the amount of exploratory behavior in the female (Table 6). Attention behaviors (standing, looking, smelling and head turning) are less altered with age or sex. This detailed analysis was obtained from photographs of paired rats taken at 1 frame per second for 15 minutes. Data obtained with this method confirm the observations of differences in locomotor activity in the sexes but point to the fact that many behaviors, such as grooming, which differ in the sexes are not often recorded. There may be many such differences which have escaped analysis and which may be at least as significant as the more easily measured activity. Examination of the literature on the wild rat (Calhoun, 1962; Barnett, 1963) indicates the richness of the social behavior which the Norway rat can display. Such behavior may as well be applied to models of human behaviors as data obtained from the more conventional laboratory tests.

TABLE 6 Time Lapse Photography of Motor Acts of Male Rats as Per Cent
of Female (Calculated from Data of Mullenix, 1977)

Age	Grooming Behaviors	Exploratory Behaviors	Attention Behaviors
4 weeks	163	81	93
6 weeks	159	89	90
4 months	136	71	90
5 months	181	45	85

SUMMARY AND CONCLUSIONS

It is difficult to generalize about the relationship of age to behavior. One consistent finding, documented in several of the papers discussed here, is the decrease in activity from maturity to senescence. This functional change is accompanied by declines in several, but not all, biochemical and morphological parameters. In the young brain there are no consistent findings. When the brain is damaged during the neonatal period, there may be functional recovery or there may be a progressive deficit. As Goldman and co-workers (1974) have pointed out, the young animal is not always more seriously affected than the adult with the same lesion. The critical factor may be the stage of development of a brain area when the lesion occurs. Thus Goldman and Rosvold (1972) found that neonatal lesions in the caudate nucleus, which develops ahead of the cerebral cortex, were more devastating than comparable lesions in the area of the prefrontal cortex which projects to the caudate. After some types of damage the regenerative capacity of the brain, in functional terms, is quite real. In particular, hyperactivity may occur in brain-damaged animals briefly during the juvenile period. Whether recovery is due to delayed or de novo formation of synaptic connections or to other factors is not known. There is some evidence that synapses in the caudate nucleus in the form of dendritic spines on the interneurons may increase coincident with functional recovery from carbon monoxide exposure (Norton and Culver, 1977).

In gerbils, although diurnal activity is not depressed as it is in rats, the difference in activity of males and females is slight. In rats and mice, the two most common laboratory species, there are marked differences in the activity of the sexes with the female more active in most situations. There may well be species in which the male is distinctly more active. The possible significance of the greater activity of the female rat is that she also seems more prone to developing hyperactivity than the male in response to a variety of brain insults ranging from x-irradiation in utero to drugs causing hyperactivity. The behavioral sensitivity of the female rat has been noted before. Davenport and Hennies (1976) reported that they used female rats to study behavioral consequences of hypothyroidism because both males and females, treated with thiouracil perinatally, became hyperactive but "preliminary observations (indicated) that motivational abnormalities from early thyroid deficiency were more apparent in females." In sex-related differences the female is not always the more sensitive, even in rats. Neonatal male rats have been found to be more sensitive to anoxia than females (Simon and Volicer, 1976) and the earlier response of the male monkey to orbital pre-frontal damage has been noted above (Goldman et al., 1974). There is, in fact, a great range of possibilities in selection of a model in animals for a particular syndrome in humans. A major part of the task is to identify the critical factors and conditions in humans which should be emulated. Then the species with the appropriate characteristics can be rationally searched for and selected.

REFERENCES

Altman, J., Autoradiographic investigation of cell proliferation in the brains of rats and cats. Anat. Rec. 145, 573-591 (1963).

Angevine, J. B., Time of neuron origin in the hippocampal region. Exp. Neurol. Suppl. 2 (1965).

Barnett, S. A., A Study in Behaviour, Methuen, London (1963).

Berry, M., Rogers, A. W., and Eayrs, J. T., The pattern and mechanism of migration of neuroblasts of the developing cerebral cortex. J. Anat. 98, 291 (1964).

Brewis, M., Poskanzer, D. C., Rolland, C., and Miller, H., Neurological disease in an English city. Acta Neurol. Scand. 42, Suppl. 24 (1966).

Brizzee, K. R., Quantitative histological studies on aging changes in cerebral cortex of rhesus monkey and albino rat with notes on effects of prolonged low-dose ionizing radiation in the rat. Prog. Brain Res. 40, 141-160 (1973).

Brizzee, K. R., Vogt, J., and Kharetchke, X., Postnatal changes in glia/neuron index with a comparison of methods of cell enumeration in the white rat. Prog. Brain Res. 4, 136-146 (1964).

Calhoun, J. B., The Ecology and Sociology of the Norway Rat, U.S. Gov. Printing Office, Washington, D.C. (1962).

Craigie, E. H., Changes in the vascularity in the brain stem and cerebellum of the albino rat between birth and maturity. J. Comp. Neurol. 38, 27-48 (1924).

Culver, B. W., Behavioral toxicity produced in developing rats by exposure to carbon monoxide and inorganic lead. Ph.D. Thesis, University of Kansas, No. 76-1292, Xerox University Microfilms, Ann Arbor, Michigan (1975).

Culver, B., and Norton, S., Juvenile hyperactivity in rats after acute exposure to carbon monoxide. Exp. Neurol. 50, 80-98 (1976).

Davenport, J. W., and Hennies, R. S., Perinatal hypothyroidism in rats: persistent motivational and metabolic effects. Dev. Psychobiol. 9, 67-82 (1976).

Enesco, H. E., A cytophotometric analysis of DNA content of rat nuclei in aging. J. Gerontol. 22, 445-448 (1967).

Ford, D. H., Selected maturational changes observed in the postnatal rat brain. Prog. Brain Res. 40, 1-12 (1973).

Friede, R. L., Topographic Brain Chemistry, Academic Press, New York (1966).

Goldman, P. S., Crawford, H. T., Stokes, L. P., Galkin, T. W., and Rosvold, H. E., Sex-dependent behavioral effects of cerebral cortical lesions in the developing rhesus monkey. Science 186, 540-542 (1974).

Goldman, P. S., and Rosvold, H. E., The effects of selective caudate lesions in infant and juvenile Rhesus monkeys. Brain Res. 43, 55-66 (1972).

Goodrick, C. L., Behavioral characteristics of young and senescent inbred female mice of the C57BL/6J strain. J. Gerontol. 22, 459-464 (1967).

Groppetti, A., and Costa, E., Factors affecting the rate of disappearance of amphetamine in rats. Int. J. Neuropharmacol. 8, 209-215 (1969).

Guarneri, R., and Bonavita, U., Nicotinamide adenine dinucleotides in the developing rat brain. Brain Res. 2, 145-150 (1966).

Gudmundsson, G., Epilepsy in Iceland. Acta Neurol. Scand. 43, Suppl. 100 (1966).

Hauser, W. A., and Kurland, L. T., The epidemiology of epilepsy in Rochester, Minnesota, 1935 through 1967, Epilepsia 16, 1-66 (1975).

Hicks, S. P., and D'Amato, C., Effects of ionizing radiation on mammalian development. Advances in Teratology, 196-250 (1966).

Hicks, S. P., D'Amato, C., and Lowe, M. J., The development of the mammalian nervous system. J. Comp. Neurol. 113, 435-458 (1959).

Horrocks, L. A., Composition and metabolism of myelin phosphoglycerides during maturation and aging. Prog. Brain Res. 40, 383-395 (1973).

Howard, E., DNA content of rodent brains during maturation and aging, and autoradiography of postnatal DNA synthesis in monkey brain. Prog. Brain Res. 40, 91-114 (1973).

Huber, G., Gross, J., and Schuttler, R., A long-term follow-up study of schizophrenia: psychiatric course of illness and prognosis. Acta Psychiat. Scand. 52, 49-57 (1975).

Huessy, H. R., and Gendron, R. M., Prevalence of the so-called hyperkinetic syndrome in public school children of Vermont. Acta Paedopsychiatr. 37, 243-248 (1970).

Hughes, R. N., Behaviour of male and female rats with free choice of two environments differing in novelty. Animal Behav. 16, 92-96 (1968).

Keyser, A., The development of the diencephalon of the Chinese hamster. Acta Anat. 83 Suppl. 50 (1972).

King, P. D., Early infantile autism. Relation to schizophrenia. J. Am. Acad. Child Psych. 14, 666-682 (1975).

Mullenix, P., Approaches to studying activity. In: Behavioral Toxicology: an Emerging Discipline, Eds., H. Zenick and L. W. Reiter, U.S. Gov. Printing Office, Washington, D.C. (1977).

Ney, P. G., Four types of hyperkinesis. Canad. Psych. Assoc. J. 19, 543-550 (1974).

Norton, S., and Culver, B., A Golgi analysis of caudate neurons in rats exposed to carbon monoxide. Brain Res. In press (1977).

Norton, S., Culver, B., and Mullenix, P., Measurement of the effects of drugs on activity of permanent groups of rats. Psychopharm. Commun. 1, 131-138 (1975a).

Norton, S., Culver, B., and Mullenix, P., Development of nocturnal behavior in albino rats. Behav. Biol. 15, 317-331 (1975b).

Norton, S., Mullenix, P., and Culver, B., Comparison of the structure of hyperactive behavior in rats after brain damage from x-irradiation, carbon monoxide and pallidal lesions. Brain Res. 116, 49-67 (1976).

Ordy, J. M., and Schjeide, O. A., Univariate and multivariate models for evaluating long-term changes in neurobiological development, maturity and aging. Prog. Brain Res. 40, 25-51 (1973).

Schneider, B. F., and Norton, S., Circadian and sex differences in hyperactivity produced by amphetamine in rats. Fed. Proc. 36, 1046 (1977).

Simon, N., and Volicer, L., Neonatal asphyxia in the rat: greater vulnerability of males and persistent effects on brain monoamine synthesis. J. Neurochem. 26, 893-900 (1976).

Syme, L. A., and Syme, G. J., The role of sex and novelty in determining the social response to lithium chloride. Psychopharmacol. 40, 91-100 (1977).

Taber Pierce, E., Time of origin of neurons in the brain stem of the mouse. Prog. Brain Res. 40, 53-65 (1973).

DISCUSSION

Dr. Sanghvi asked if higher motor activity in females is related to the estrus cycle or to other endocrine hormones. *Dr. Norton* indicated that there may be some such relationships since there is a slight reduction of motor activity in ovariectomized animals. She felt that this was not a complete explanation for the change in motor activity, particularly since hyperactivity associated with the estrus cycle is peculiar to the running wheel and does not occur under most conditions of measuring activity.

Dr. Klawans commented that among sex-related factors in humans are the effects of contraceptive pills on the development of dyskinesias; these effects disappear on withdrawing the pill. He also mentioned the greater incidence of tardive dyskinesias in elderly female schizophrenics. *Dr. Kobayashi* also discussed this phenomenon, pointing out that severe dyskinesias were much more common among women than among men.

SIGNIFICANCE OF SPECIES DIFFERENCES: ROTATIONAL MODELS

Stanley D. Echols and Richard C. Ursillo

Merrell-National Laboratories, Division of Richardson-Merrell Inc.,
Cincinnati, Ohio

INTRODUCTION

Unilateral removal of the caudate of rats by suction was found by Andén et al. (1) to result in assymmetrical motor responses to drugs affecting monoamine transmission. In such animals, peripheral injection of the indirectly acting agent amphetamine and the dopamine receptor stimulant apomorphine caused rotation toward the lesioned side (ipsilateral) which could be blocked by haloperidol (2). A more selective lesioning technique using 6-hydroxy-dopamine (6-OH-DA) was employed by Ungerstedt (3) who demonstrated the loss of dopaminergic terminals of the striatum in rats after microinjection of 6-OH-DA into the striatum or into the zona compacta of the substantia nigra, which contains the cell bodies of the nigro-striatal dopaminergic system. This lesioning technique leaves intact the cell bodies of the striatum and their dopamine receptors. The use of a simple rotometer allowed quantitation of rotational behavior (4). It was found that in rats lesioned by injection of 6-OH-DA into either the striatum or the substantia nigra, peripheral administration of amphetamine caused ipsilateral rotation (4,5) whereas l-dopa or apomorphine caused contralateral rotation (6). The contralateral rotation induced by the direct dopamine receptor stimulants is presumably due to the development of supersensitivity of the dopamine receptors of the striatum on the lesioned side and the consequent more effective stimulation of these receptors than those of the non-lesioned side (6). This model, then, allowed distinction of directly acting dopaminergic agonists from those which are indirect agonists by the direction of the induced rotation. Additionally, it provided a convenient means of quantitating the effect of dopamine receptor antagonists on the striatal receptors.

ANTIPSYCHOTIC DRUGS IN THE RAT ROTATIONAL MODEL

It has been suggested that the antipsychotic drugs exert their major effects on the central nervous system by virtue of a blockade of dopamine receptors (7,8). Functional evidence for this was provided by Ungerstedt who showed that ongoing rotation induced by amphetamine could be halted by the injection of low doses of haloperidol (Fig. 1). Subsequently, a number of investigators, using the rat model, have quantitated the effect of several antipsychotic drugs. Techniques used for lesions of the nigro-striatal system have included electrolytic or 6-OH-DA lesions of the substantia nigra or electrolytic lesioning of the caudate. Rotation has been induced with amphetamine, methamphetamine or apomorphine. Data are shown in Table 1 from ten studies which assessed the ability of pre-treatment with clinically active antipsychotic compounds to decrease the intensity of drug-induced rotational behavior. The ED_{50} values shown (for 50 percent reduction in turning intensity) are those given by the authors or the best estimate from the data presented. In addition to the agonist used, other experimental conditions varied among the different studies. For example, five studies involved 6-OH-DA induced nigral lesions, four used electrolytic nigral lesions and one used electrolytic lesions of the caudate. Doses of amphetamine varied from 2 to 5 mg/kg, i.p.; those of apomorphine, from 0.1 to 0.4 mg/kg, s.c. Three studies used methamphetamine at 5 mg/kg, i.p. The antipsychotic drugs were given at times ranging from 30 minutes to 3 hours prior to injection of the agonist. Despite these differences, the relative potencies and, in most cases, the ED_{50} values of the antipsychotic drugs in inhibiting rotational behavior in the rat model are in remarkable agreement among the different studies. Haloperidol was clearly the most potent of the drugs compared. Chlorpromazine was found to be one one-hundredth the potency of haloperidol; whereas clozapine and thioridazine were quite weak in all studies.

The significance of the relative potencies of the antipsychotic drugs in the rat rotational model has been the subject of much discussion. Crow and Gillbe (13) concluded that dopamine receptor blockade may not underlie clinical antipsychotic activity since thioridazine

27

28

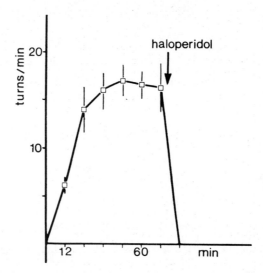

Fig. 1 Inhibition of amphetamine-induced rotation by haloperidol
(1 mg/kg) in rats with unilateral 6-OH-DA lesions of the substantia
nigra. The abcissa represents time after injection of amphetamine.
From Ungerstedt (5).

TABLE 1   Inhibition of Intensity of Drug-Induced Rotation in Rats
with Lesions of the Nigro-Striatal Dopamine System

Agonist	Antagonist ED_{50} mg/kg i.p.[a]					
	Haloperidol	Pimozide	Chlorpromazine	Thioridazine	Clozapine	Ref
Amph	0.05	0.08	2	13.7	25.4	(9)
Methamph			4	>16	>16	(10)
Amph			4	26		(11)
Methamph			<5			(12)
Amph	0.02		3-10			(b)
Methamph			~4	> 8		(13)
Amph	0.05					(14)
Apo				>16	>16	(10)
Apo	0.1					(14)
Apo	< 3 (p.o.)				>20 (p.o.)	(15)

a   Dose necessary for 50% reduction in turning intensity
b   Unpublished data

failed to inhibit rotation, although chlorpromazine and thioridazine are about equally
potent, clinically. However, since thioridazine was shown to be capable of inhibiting
rotation at a higher dose (11), it has been suggested (10,11) that the weak antagonism
of rotational behavior by thioridazine and clozapine may result from a modulation of the
anti-dopaminergic activity of these compounds by their demonstrated central anticholinergic
activity(16,17). This view is in agreement with the demonstrated influence of cholinergic
and anticholinergic drugs on the rotational behavior of rats (10,11,18). On the other
hand, Sayers et. al.(15) and Stawarz et al. (9) have advanced significant exceptions to
this view. For example, unlike other central anticholinergic drugs, clozapine does not
influence the increase in striatal homovanillic acid caused by other antipsychotic compounds; clozapine did not attenuate extrapyramidal side effects during treatment with
fluphenazine or flupenthixol; anticholinergic drugs exacerbate tardive dyskinesias, yet
no cases of these dyskinesias have been reported with clozapine. Bartholini (19) has
pointed out also that the slight increases of acetylcholine output from the striatum by
clozapine is more indicative of a weak dopamine receptor blocking potency than of a strong
anticholinergic effect in this brain area.

Whatever the reason for the weak anti-rotational effects of thioridazine and clozapine in
the rat model, the results given in Table 1 may predict the extrapyramidal liability of
the compounds seen clinically.

ANTIPSYCHOTIC DRUGS IN THE MOUSE ROTATIONAL MODEL

The results in rats were of particular interest to us because of the possibility of predicting the extrapyramidal side effect liability of new antipsychotic compounds. Following
demonstration that the caudate lesioned mouse could also serve as an animal rotational
model(20,21), we prepared such animals in order to examine the effect of some known antipsychotic drugs and to compare these data with those which had been obtained in the rat
model.

We used unilateral 6-OH-DA lesions of the caudate, employing a head mould as described by
Von Voigtlander and Moore (21) and shown in Fig. 2. Correct placement of the guide cannula
in the mould was verified by injection of 0.5 µl of a dye solution in a group of mice before
using the head mould for lesioning. Male Swiss-Webster mice, weighing 20 to 22 g (Charles
River), were used throughout the study. A volume of 4 µl containing 16 µg of 6-OH-DA HBr
and 0.8 µg of ascorbic acid in saline was injected by means of a microsyringe through the

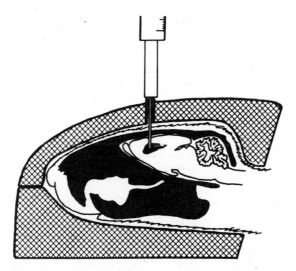

Fig. 2. Saggital view of apparatus used for intra-striatal injections.
The head mould which surrounds and immobilizes the mouse's head is represented by the cross-hatched area. The area at the tip of the needle
represents the corpus striatum. From Von Voigtlander and Moore(21).

guide cannula of the head mould into the left caudate-putamen under methoxyflurane anesthesia. Ten days later, each animal received apomorphine, 0.2 mg/kg s.c., was placed in a spherical glass bowl and the total number of turns during min 2 to 4 were counted. Any animals which did not achieve at least ten contralateral rotations within this period were rejected. The selected animals showed virtually no ipsilateral rotations in response to apomorphine. An apomorphine dose of 0.2 mg/kg s.c. was chosen for all experiments based on a dose-response curve in these animals (Fig. 3). The interval of time between administration of each antipsychotic drug and apomorphine was that yielding maximal blockade

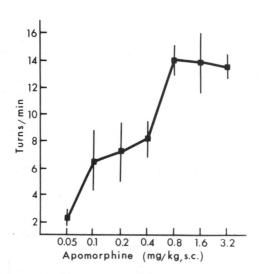

Fig. 3. Apomorphine-induced turning rate in mice with unilateral 6-OH-DA striatal lesions. Each point represents the mean ipsilateral rotation rate of 5 or 10 animals counted during min 2 to 4 following apomorphine. Vertical bars indicate standard error.

of turning as determined by preliminary time-effect studies. The intervals were 2 hours for haloperidol and chlorpromazine, and one hour for clozapine and thioridazine. The antipsychotic drugs were administered intraperitoneally. Following injection of apomorphine, rotation was counted by means of a photocell rotometer. The total number of contralateral turns during min 2 to 10 after apomorphine was used for analysis. Dose response curves were obtained for each drug by regression analysis and the dose necessary to decrease the number of rotations by 50% as compared to saline controls was determined. Ten mice per dose were employed.

The dose-response curves obtained in our study are shown in Fig. 4. Unlike the results in the rat model, our data in the mouse model show chlorpromazine, thioridazine and clozapine to be of nearly equal potency. Haloperidol is weak in comparison to its effect in the rat whereas thioridazine and clozapine are much more potent. The relative potency of haloperidol to the other three agents we examined are not consistent with either relative clinical antipsychotic potency nor with extrapyramidal side effect liability. Two other laboratories have examined these drugs in the mouse model. Lotti (20) removed the striatum by suction. In his study, the interval between antagonist and agonist (apomorphine, 2 mg/kg i.p.) was 1 hr. and measurement was made "10 to 20 min" following agonist. Pycock et al. (22) used 6-OH-DA lesions of the striatum. These authors observed the effect 15 min after apomorphine (2 mg/kg i.p.) and 30 min after amphetamine (5 mg/kg i.p.). The intervals between antagonist and agonist were comparable to those used in our study. It should be pointed out that these two studies used complete inhibition of turning as a measure whereas reduction in turning intensity was used in our study. The ED_{50} values from our study and from these two studies are compared in Table 2. It is obvious that data from our study are not comparable to those from the quantal assays. In the latter, the data are more consistent with those obtained in the rat model.

In considering the possible reasons for the differences between our data and those from other studies, it is of interest to examine another point brought out by Pycock et al. (22, 23). These authors have demonstrated that the rotational inhibitory potency of some antipsychotic drugs appears substantially greater when continuous data are used for analysis

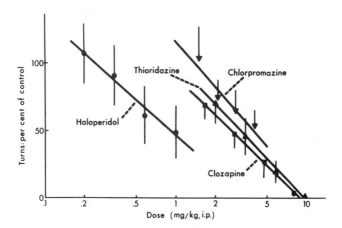

Fig. 4. Inhibition of apomorphine-induced rotation by antipsychotic drugs in mice with unilateral 6-OH-DA striatal lesions. Each point represents the mean total ipsilateral rotations of 10 animals counted during min 2 to 10 after apomorphine, 0.2 mg/kg s.c. Vertical bars indicate standard errors.

TABLE 2 Inhibition of Drug-Induced Rotation in Mice with Striatal Lesions

Agonist	Antagonist ED_{50} mg/kg i.p.[a]					
	Haloperidol	Pimozide	Chlorpromazine	Thioridazine	Clozapine	Ref
Apo	0.91		4.0	3.1	2.6	(b)
Apo	0.15		3.7	16.4		(20)
Apo	0.1	0.31	1.0		17.0	(22)
Amph	0.11	0.19	2.0		13.0	(22)

[a] For present study (b) ED_{50} represents dose decreasing number of turns by 50%; for studies 20, 22 ED_{50} represents dose for complete inhibition of turning in 50% of the mice.

than when using discontinuous (quantal) data. An example of this is shown in Fig. 5. This is expected since complete blockade is a more severe test than partial inhibition. However, inhibition of turning intensity may also be due partially to the sedative effects of compounds, since non-specific sedatives were found to decrease the intensity of turning in a dose-related fashion in the mouse model whereas no complete inhibition of turning was achieved at reasonable doses (22). Quantal analysis of our data for min 8 to 10 (Table 3) yields results still not comparable to those of Lotti (20) or Pycock et al. (22). In addition, re-analysis of our data using inhibition of the peak turning intensity yielded ED_{50} values similar to those derived using reduction of total number of turns during min 2 to 10.

We do not know whether perhaps the dose of apomorphine is an important determinant in the mouse although our dose was near the middle of our dose-response curve and comparable to that used in the rat studies. We are now investigating this possibility.

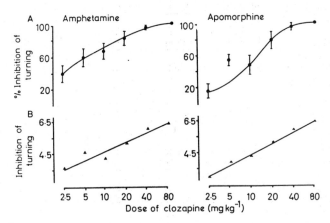

Fig. 5 A.   Dose-response curves for inhibition of intensity of turning 30 min after amphetamine 5 mg/kg i.p.) or 15 min after apomorphine
(2 mg/kg i.p.) by clozapine. Mean ± standard error for 12 observations
at each dose level are shown.

B.   Probit-dose plots for abolition of turning to amphetamine or apomorphine. Each point is the mean of 12 observations. From Pycock et al. (23)

TABLE 3   Inhibition of apomorphine[a]-induced Turning in Mice with
6-OH-DA Striatal Lesions

Haloperidol		Chlorpromazine		Thioridazine		Clozapine	
mg/kg	No. Inhibited[b]	mg/kg	No. Inhibited	mg/kg	No. Inhibited	mg/kg	No. Inhibited
0.2	0/10	1.5	0/10	2.0	0/10	1.65	1/10
0.34	0/10	2.09	0/10	3.4	3/10	2.8	2/10
0.58	2/10	2.89	1/10	5.8	5/10	4.75	5/10
1.0	0/10	4.0	2/10	9.8	10/10	8.0	9/10

[a] 0.2 mg/kg s.c.

[b] Assessed during min 8 to 10 after apomorphine

CONCLUSIONS

The ability of antipsychotic compounds to inhibit apomorphine or amphetamine-induced rotational behavior in rats with lesions of the nigro-striatal dopamine system is well documented.   There is very good agreement among the data derived from a number of different
laboratories despite the variety of experimental conditions. Whether the low potency of
clozapine and thioridazine in rat rotational model and the lack of significant Parkinsonian
effects in man are a result of significant central anticholinergic activity or weak striatal
dopamine receptor blockade is not yet clear. The rotational model in mice yields results
which are not comparable in all cases to those from rats. In contrast to the rat model,
use of inhibition of intensity of turning as a measure in the mouse may be confounded by
effects of antipsychotic compounds other than dopamine receptor blockade and may represent
a true species difference in the response to these drugs.

REFERENCES

(1)   N.-E. Andén, A. Dahlström, K. Fuxe and K. Larsson, Functional role of the nigro-
      neostriatal dopamine neurons, Acta Pharmacol et Toxical. 24, 263 (1966).

(2)   N.-E. Andén, A. Rubenson, K. Fuxe and T. Hökfelt, Evidence for dopamine receptor
      stimulation by apomorphine, J. Pharm. Pharmac. 19, 627 (1967).

(3)   U. Ungerstedt, 6-Hydroxy-dopamine induced degeneration of central monoamine neurons,
      Eur. J. Pharmac. 5, 107 (1968).

(4)   U. Ungerstedt and G. W. Arbuthnott, Quantitative recording of rotational behavior
      in rats after 6-hydroxy-dopamine lesions of the nigrostriatal dopamine system,
      Brain Res. 24, 485 (1970).

(5)   U. Ungerstedt, Striatal dopamine release after amphetamine or nerve degeneration
      revealed by rotational behavior, Acta Physiol. Scand. Suppl. 367, 49 (1971).

(6)   U. Ungerstedt, Postsynaptic supersensitivity after 6-hydroxy-dopamine induced de-
      generation of the nigro-striatal dopamine system, Acta Physiol. Scand.
      Supp. 367, 69 (1971).

(7)   J. M. Van Rossum, The significance of dopamine receptor blockade for the mechanism
      of action of neuroleptic drugs, Arch. Int. Pharmacodyn. 160, 492 (1966).

(8)   R. O'Keefe, D. F. Sharman and M. Vogt, Effect of drugs used in psychoses on cerebral
      dopamine metabolism, Br. J. Pharmac. 38, 287 (1970).

(9)   R. J. Stawarz, H. Hill, S. E. Robinson, P. Setler, J. V. Dingell and F. Sulser,
      On the significance of the increase in homovanillic acid caused by antipsychotic
      drugs in corpus striatum and limbic forebrain, Psychopharmacologia 43, 125 (1975)

(10) P. H. Kelly and R. J. Miller, The interaction of neuroleptic and muscarinic agents
      with central dopaminergic systems, Br. J. Pharmac. 54, 115 (1975).

(11) P. Muller and P. Seeman, Neuroleptics: relation between cataleptic and anti-turning
      actions, and role of the cholinergic system, J. Pharm. Pharmac. 26, 981 (1974).

(12) J. E. Christie and T. J. Crow, Turning behavior as an index of the action of ampheta-
      mine and ephedrines on central dopamine-containing neurones, Br. J. Pharmac.
      43, 658 (1971).

(13) T. J. Crow and C. Gillbe, Dopamine antagonism and antischizophrenic potency of
      neuroleptic drugs, Nature New Biol. 245, 27 (1973).

(14) B. Costall and R. J. Naylor, Stereotyped and circling behaviour induced by dopa-
      minergic agonists after lesions of the midbrain raphe nuclei, Eur. J. Pharmac.
      29, 206 (1974).

(15) A. C. Sayers, H. R. Bürki, W. Ruch and H. Asper, Neuroleptic-induced hypersensitivity
      of striatal dopamine receptors in the rat as a model of tardive dyskinesias.
      Effects of haloperidol, loxapine and chlorpromazine, Psychopharmacologia
      41, 97 (1975).

(16) R. J. Miller and C. R. Hiley, Anti-muscarinic potencies of neuroleptics and drug-
      induced Parkinsonism, Nature 248, 596 (1974).

(17) S. Snyder, D. Greenberg and H. I. Yamamura, Antischizophrenic drugs and brain
      cholinergic receptors, Arch. Gen. Psychiatry 31, 58 (1974).

(18) B. Costall, R. J. Naylor and J. E. Olley, Catalepsy and circling behavior after
      intracerebral injections of neuroleptic, cholinergic and anticholinergic
      agents into the caudate-putamen, globus pallidus and substantia nigra of rat
      brain, Neuropharmac. 11, 645 (1972).

(19) G. Bartholini, Differential effect of neuroleptic drugs on dopamine turnover in the
      extrapyramidal and limbic system, J. Pharm. Pharmac. 28, 429 (1976).

(20) V. J. Lotti, Action of various centrally acting agents in mice with unilateral
      caudate brain lesions, Life Sci. 10, 781 (1971).

(21) P. F. Von Voigtlander and K. E. Moore, Turning behavior of mice with unilateral
      6-hydroxydopamine lesions in the striatum: effects of apomorphine, 1-dopa,
      amantadine, amphetamine and other psychomotor stimulants, Neuropharmac.
      12, 451 (1973).

(22) C. Pycock, D. Tarsy and C. D. Marsden, Inhibition of circling behavior by neuro-
      leptic drugs in mice with unilateral 6-hydroxydopamine lesions of the
      striatum, Psychopharmacologia 45, 211 (1975).

(23) C. Pycock, D. Tarsy and C. D. Marsden, Inhibition of turning behaviour by clozapine
      in mice with unilateral destruction of dopaminergic nerve terminals,
      J. Pharm. Pharmac. 27, 445 (1975).

DISCUSSION

Dr. Cook stated that one of the pharmacological effects that could be responsible for inhibition of the dopamine agonist-stimulated turning in the 6-hydroxydopamine substantia nigra lesioned mouse is simple depression of the motoric response. He wondered to what extent this type of direct motoric depression was responsible for the apparent antagonism of the "turning" phenomenon and how motor depressive potencies compare to potencies in blocking the turning response. He also asked for a comparison of the potency of the compounds discussed by Dr. Ursillo in other tests. *Dr. Ursillo* replied that in amphetamine aggregate toxicity, clozapine and thioridazine are less potent in the turning test but chlorpromazine and haloperidol are more potent.

Dr. Lomax commented that by and large, strain-to-strain differences are not significant in rats but they can be very marked in the case of mice. He wondered if this could account for the differences between Dr. Ursillo's results and those of other investigators. *Dr. Lomax* agreed this was a possibility but felt it would not account for the dissociation of results between rats and mice when comparing several drugs.

Dr. R. Vogel noted that in the current paper, it was reported that apomorphine levels above 0.8 mg/kg produced stereotypies in 6-hydroxydopamine-treated mice unilaterally in the caudate. Since Kelly, Iversen and others have reported that bilateral caudate lesioning with 6-hydroxydopamine blocks amphetamine stereotypy, Dr. Vogel asked for comments on the morphology of this preparation. *Dr. Echols* indicated that what was seen after 0.8 or 1 mg/kg of apomorphine was a mild sniffing stereotypy.

MINIMAL BRAIN DYSFUNCTION HYPERKINESIS: SIGNIFICANCE OF NUTRITIONAL
STATUS IN ANIMAL MODELS OF HYPERACTIVITY

I. Arthur Michaelson, Robert L. Bornschein, Rita K. Loch and
Lee S. Rafales

Department of Environmental Health, Division of Toxicology,
Laboratory Neurobehavioral Toxicology, University of Cincinnati
College of Medicine, Cincinnati, Ohio, 45267

INTRODUCTION - THE HUMAN PROBLEM

History

The syndrome of hyperkinesis in children was succinctly described in a children's book of
verse and droll pictures written more than a century ago by a German physician, Heinrich
Hoffmann (1): "Fidgety Phil, he won't sit still, he wiggles, and giggles, and when admon-
ished, the naughty restless child, grows still more rude and wild." Since then, others have
outlined a syndrome that begins early in life, is more common in boys and is characterized
by overactivity, distractibility, impulsiveness, excitability, difficulty in peer relations,
discipline problems and specific learning problems. Clinicians developed an active interest
in the syndrome during the 1918 epidemic of encephalitis in the United States. The earliest
documented clinically descriptive paper to appear in America was by Hohman (2, 3) and it
described the then newly identified syndrome of "Post-encephalitic Behavioral Disorders in
Children." The concept that abnormal behavioral syndromes in children could be attributed
to brain damage was developed at length in the 1920s when it was noted that children recov-
ering from acute attacks of epidemic *encephalitis lethargica* often manifest bizarre behavioral disturbances. Detailed exploration of these neurologically-activated behavioral,
cognitive and neuromotor disorders had their beginnings in the pioneer research efforts by
Kahn and Cohen (4) and Kramer and Pollnow (5). In the 1940s Strauss and coworkers (6, 7)
propounded the concept that central nervous system dysfunction of non-specific and often
obscure origin was the etiologic referent and corollary to a wide range of behavioral,
emotional, intellectual, and cognitive disorders. Strauss and Lehtinen (6) and Strauss and
Kephart (7) published a series of papers and monographs outlining their approach to the
diagnosis and treatment of "brain-injured" children. They emphasized the cardinal signs and
features of hyperactivity, emotional lability, impulsivity, distractibility, and perseveration that characterized the child with brain damage as distinguished from the child with
mental retardation. Confusion surrounding the etiology of the hyperactive disorder accounts
for the diversity of more than twenty labels attached to the syndrome and reflects theories
of underlying neuropathology. The descriptive phrases delineated essentially the same
symptom complex and served to underline the general lack of consensus. The concept of brain-
damage behavior syndrome has been extended to include many children with relatively borderline disturbances *suggestive* of brain damage.

Gesell and Amatruda (8) developed the concept of "minimal cerebral injury" to explain atypical developmental syndromes in certain infants. Following acceptance of the concept of
"minimal brain damage behavioral syndrome" for diagnostic expression (9, 10) the expression
"minimal brain dysfunction, MBD" was evolved to indicate that this syndrome may be present
in children with "as yet unnamed subtle deviations" of brain function (11). Children considered to have minimal brain dysfunction are a heterogeneous group with regard to the
etiological process as well as the behavioral symptom complexes which they manifest.
Clements (12) described children with minimal brain dysfunction syndrome as being, "....of
near average, average or above average general intelligence with a certain learning or
behavioral disability ranging from mild to severe, which are associated with deviation of
functions of the central nervous system (6, 9). Wender (13) offers the view that there are
two major areas of dysfunction found: A. behavioral problems, and B. perceptual-cognitive
problems. These two areas of dysfunction occur in three possible combinations: a. children
with behavioral difficulties alone, b. children with perceptual-cognitive difficulties, and
c. children with difficulties in both areas. Although Clement's (12) report provides a list
of about one hundred symptoms and signs associated with the child who suffers from MBD there
appears to be five primary features which might be considered the hallmarks of this afflic-
tion and relate to Wender's (13) two major areas of dysfunction: hyperactivity, short
attention span, impulsivity, motor difficulty, i.e., awkwardness and clumsiness, and learning
difficulties. Gross and Wilson (14) point out that the similarities between the signs mani-

fested in the behavior of "brain injured" children (15) and those of "brain dysfunction" with no known organic lesions (11) are striking.

Clinical Symptomatology

Bradley (16) produced one of the earliest descriptions of the behavior of children said to demonstrate organic brain damage. Possibly the single most pathognomonic symptomatic expression of this syndrome is that of a seemingly chronic, irreducible hyperactivity. Hyperkinesis statistically ranks as the most frequently occurring symptom on the minimal brain dysfunction "check-list." The total clinical symptomatology of the minimal brain dysfunction syndrome in children is described by Becker (17). He provides the following description: the child with MBD may initially be viewed as clumsy, awkward, anxious, fidgety, angry or restless on first contact; he is further likely to be either excessively physically active and in constant seemingly purposeless and incessant motion; either to the point of physical exhaustion or behavioral decompensation. He is characteristically prone to unprovoked temper tantrums, overly aggressive acting out and play, and generally has a poor tolerance for dissonance, frustration, attention, commitment, vigilance or concentration of any kind. He responds poorly to requests or demands for his cooperation, involvement or engagement and he can at times find it difficult to move to new tasks or activities (perseveration). He is temperamental, high-strung and sensitive, and is sometimes obstinate, negative, often obstreperous, boisterous, vocal, arbitrary and not infrequently oppositional and alienating. He is seldom viewed as an engaging, related or as a happy child. His sometimes arrogant defiant facade and boldly aggressive and challengingly provocative approach to other children and adults betrays and masks his low esteem, depression, probably constant confusion and inordinate need for unconditional acceptance, approval and love. His flightiness, aimless and often indiscriminately destructive dyssocial inclination and activities in the classroom and at home together with his apparently disorganized approach to perceived structure and his inability to act or to organize his life on the basis of either a goal-concept or principals of expectations, only results in further generating and in fact, insuring a continued negative response for him from all who come in contact with him. He seems only to encounter hostility, and not uncommonly, punitive response from those who in fact, he needs most to understand him. His parents view his behavior as defiant, impulsive, narcissistic, egocentric, manipulative, uneven, demanding, irrational and often inappropriate and without consistency, for which he may be punished, or worse, misunderstood. His misbehaviors and retreat from academic and social or peer related involvement in school leads only to confrontation, crisis and outrage in the classroom and in the playground, and serves to confirm his own self-judgement that he may indeed be a "bad boy," and somehow "different," and "damaged" and really unworthy of his parents' love, understanding and acceptance (17).

Incidence

One of the earliest attempts at estimating the prevalence of the condition in the general child population was by Burks (18). Five hundred twenty-five pupils were rated by teachers of the respective children for characteristics seen in brain injured children. Approximately 8 percent showed a large number of symptoms, from moderate to severe degree. Prechtl and Stemmer (19, 20) reported on the prevalence in the Netherlands using the findings of an involuntary movement disorder ("choreiform" activity) in association with a history of "behavioral problems" and determined this syndrome to be present in 20 percent of elementary school age boys and only 10 percent in girls. Stewart *et al.* (21) found about 4 percent of children between the ages of 5 and 11 had these characteristics. Huessy (22) found the prevalence of "hyperkinesis" in a second grade population to be about 10 percent. Comparable orders of magnitude of involvement (5 to 10 percent) have been reported by Wender (23), Minde *et al.* (24) and Winokur (25). A more conservative figure has been estimated by the U.S. Federal Office of Child Development, 1971, which reports that 3 percent of elementary school children demonstrated enough traits (mild and severe) to be classified as hyperkinetic. On the other extreme there is the study referred to by Wender (23) which investigated the prevalence of behavioral and psychological disturbances in Montgomery County, Maryland. Reporting on a stratified sample of 20 percent of the population of elementary school children (approximately 24,000 out of a total of 120,000 children), teacher ratings indicated that restlessness was a "problem" in approximately 15 percent of the children in grades one to six and that "problems of attention span" were present in approximately 22 percent of the children in those grades. Wender (23) cautions that "problems" is a rather vague referent. It implies that the children were more restless and less attentive than their teachers would have them be. Nonetheless, it indicates that 1:7 to 1:5 of grade school children had minimal brain dysfunction-like problems to some minor degree.

Despite the fact that different investigators have employed different diagnostic criteria, neurological signs (19, 20), teachers' reports (21), and questionnaires (22) studies in different geographic areas have come up with a surprisingly constant prevalence of minimal brain dysfunction which appears to be of the order of magnitude of 5 to 10 percent (22-25).

Sex Linkage

The syndrome, at least when largely defined by the presence of hyperactivity, has a sex linkage being more common in boys than girls. In Stewart *et al.* (21) study of unselected patients the ratio of boys to girls was about 6:1 similar to that found by Chess (26) on brain injured children. Other reports show male-to-female ratios range 9:1 (27) to 3:1 or 4:1 (10). These ratios roughly parallel those of other child psychiatric disorders (23).

It has been estimated that 5 to 10 percent of the child population have MBD. This is comparable to the percentage of adults who suffer severe psychiatric disability, estimated by Srole (28) to be on the order of 15 percent in one psychiatric epidemiological study. In clinics treating psychiatrically disturbed, primary-school age children, minimal brain dysfunction, broadly construed, probably constitutes at least 50 percent of the cases referred (13). This high incidence of minimal brain dysfunction syndrome does not appear to be a unique phenomenon among American children alone. No country, geographical area, or no social class seems to have immunity to the problem of childhood hyperkinesis or other areas of deviant cerebral function. In the area of reading disorders the problem seems to be quite international in scope. UNESCO and WHO studies reveal a similar high incidence of MBD in Great Britain, the Scandinavian countries and other Western Europe countries as well (29).

Etiology

The etiology(ies) of minimal brain dysfunction hyperactivity are unknown. During the course of the past seventy years there has been considerable shift from organic damage to purely psychogenic views and more recently a return to a more biological perspective. There is no one single etiological explanation of the minimal brain dysfunction syndrome. Yet some hypothesized causes seem more probable than others and are more commonplace. Consistent with the complexities of the condition there are some who hold that the minimal brain dysfunction syndrome has several distinct and separate etiologies as well as expressions due to interaction of contributing factors. Many theories have been proposed and in a significant number of cases, the etiology of hyperactivity is thought to be a direct insult to, or dysfunction within, the central nervous system. Theories include: *in utero*-parturition-neonatal factors: organic brain damage following trauma (30), infection (31), pre- peri- and para-complications of pregnancy (32), fetal maldevelopment including minor congenital physical anomalies and stigmata (33, 34), intrauterine random variations in biological development, delayed or maturational lag (35), delayed lateralization of the brain functions (36). Central nervous system alterations: imbalance in cortical and subcortical areas of the brain (37), underarousal of the central nervous system (38), biochemical and/or neurochemical imbalance (23). Psychogenic factors: maternal/oral conflicts (39), deprived mothering (40), psychological determinants (41). Genetic factors: genetic transmission (42), extreme displacement on normal distribution curve (43), biological variations manifest by universal compulsory education (44). Environmental contaminates: food additives (45) low level lead exposure (46-48). Interaction of separate etiological factors: brain damage or pre- peri- and para-birth complications as well as interaction of child rearing factors and biological predisposition (23).

Animal Models in Minimal Brain Dysfunction Hyperactivity

In recent years there has been a growing awareness of the important contributions which animal models can make to the advancement of biomedical sciences. Animal models have become increasingly important as a means by which disease processes occurring in humans can be investigated. It would be ideal if the animal model was analogous to the disease state under study. However, the underlying basis of minimal brain dysfunction is unknown. Thus an animal model which provides symptomatology similar to that described in humans may offer valuable insight into the etiology of this disorder. The role of animal models in developmental research has been well described by Plaut (49). What is an appropriate model for MBD? Sechzer *et al.* (50) and Shaywitz (51) suggest that an appropriate animal model of minimal brain dysfunction-hyperactivity should satisfy the following criteria: (a) Specific symptoms and cardinal features of the syndrome should be replicated in the animal model. Such features may include hyperactivity, cognitive difficulties, attention difficulties, and difficulty habituating to a new environment. The pathogenesis of the syndrome should bear some relationship to the pathogenesis of the disorder in children, being evident shortly after birth, (b) in part in the young developing animal and (c) should appear at comparable ages in children. Shaywitz(51)suggests an additional criterion which may be too stringent and that is that the animal model must follow the same developmental course as found in the human counterpart, i.e., hyperactivity so frequently found in children with MBD decreases in frequency and severity as the children approach adolescence. Since hyperactivity per se is not a common symptom in youngsters with MBD who have matured, he emphasizes that the production of persistent hyperactivity in an adult animal then would not parallel the symptom as seen in children with MBD. (d) The response to medication in the animal model must parallel the response seen clinically in children in that the animals should be managable by amphetamine like compounds, instead of increasing activity should reduce hyperactivity and improve attention, learning and memory.

Not all of the suspect etiological causes are amenable to experimentation with animals; however, analogous aymptomatology could reveal some of the neuroanatomical, neurophysiological or neurochemical substrates identifiable with the symptomatic endpoint. The most outstanding feature in the human is hyperactivity and it is this behavior that is most frequently examined in animal models of hyperactivity.

Examples of some of the types of studies reported include analyses of the following factors as contributors to hyperactivity in experimental animals: brain damage following viral infection (58); brain damage associated with pre- peri- and para-natal complications of pregnancy and birth (53-70); heredity and genetic transmission (71-80); central nervous system lesions (81-97); biochemical lesions (98-107); environmental contaminants and food additives (108-119); and lead exposure (120-129).

Nutrition and the developing nervous system. It has been well documented that the transient period of brain growth, known as the growth spurt (130) is more vulnerable to growth retardation than the periods both before and afterwards. Vulnerability, as Dobbing (130) uses the term, means that mild restriction leads to permanent, irrevocable reduction in the trajectory of bodily growth and to easily detectable distortions and deficits in the adult brain. Retardation during critical periods of brain development has been shown to have effects on neurochemistry, electroneurophysiology and potentially irreversible effects on behavior in adults. These observations have accumulated from a number of different experimental designs, mostly exploiting the postnatal timing of the growth spurt in rodents during their suckling period. One of the more frequently used experimental designs involves retarding growth by increasing the size of the litter. Neonates at weaning are below the normal weight of coetaneous controls. A second experimental design involves underfeeding the mother during the gestational period. Stunted offspring as well as normal control newborns are then cross-fostered at birth to well-fed and undernourished lactating females, the litter size being standardized. Thus there are four groups according to whether the gestational period or the suckling period or both or either was nutritionally restricted. The types of malnutrition which have been most extensively examined in rodents, whether during the first 21 days are either a total caloric and protein deprivation, or a limitation of the protein intake to the lactating dam, which limits the amount rather than the quality of the milk. The final result is a total caloric deprivation to the developing neonate (131). The subject of early undernutrition and brain function encompasses a voluminous literature. Therefore, it would be appropriate to direct the interested reader to two outstanding reviews: 1) CIBA Foundation Symposium: Lipid Malnutrition and the Developing Brain (130, 131) and 2) Nutrition and the Developing Nervous System (132). Dodge *et al.* (132) discusses behavioral consequences of undernutrition (p. 290-303), effects of undernutrition on neurotransmitter and related compounds (p. 269-271) and effects of undernutrition on growth and development of the nervous system (p. 243-265). More specifically, Sereni *et al.* (133) reported decreases in the ontogenic appearance of normal levels of norepinephrine and 5-hydroxytryptamine as well as acetylcholinesterase in undernourished rats. These effects were only demonstrable during the first 14 days after parturition. Likewise, Adlard and Dobbing (134) reported acetylcholinesterase activity in whole brain of 21 day old rats undernourished during fetal life and lactation. In contrast, these same investigators (135) reported elevated acetylcholinesterase activity in adult rats after undernutrition in early life. This picture is even more complex if one considers effects of undernutrition on regional and subcellular neurochemistry and the concept of neurotransmitter balance in integrated normal functioning of the central nervous system.

Undernutrition affects immediate as well as long-term behavioral functioning. The type, duration and timing of undernutrition as well as the age of the animal at testing can influence the observed effect on behavior. Randt and Darby (136) reported behavioral changes in mice undernourished during gestation and lactation and subjected to nutritional rehabilitation. They found alterations in specific types of learning, and in alteration in arousal level reflected by an increase in locomotor activity. Likewise Costellano and Oliverio (137) reported that undernutrition produced by rearing mice in small, intermediate or large litters (respectively, 4, 8 or 16 pups) results in ontogenetic retardation as well as impaired avoidance learning and a two-fold increase in exploratory activity after 45 days of rehabilitation. Enhanced activity during habituation and exploration could be misinterpreted as hyperactivity. It is abundantly clear that nutritional status especially during early development is an important variable influencing the outcome and interpretations of data from animal models.

<u>Undernutrition</u>. Our research group has been studying the effects of early lead exposure on brain development in rodents. Because of the widely cited and potentially significant implications of the Silbergeld and Goldberg(122-125) reports demonstrating a lead-induced hyperactivity in mice suckling milk containing lead from dams drinking lead-containing water and then weaned directly to the same drinking water, we attempted to replicate their findings. These replication attempts have been carried out using the same strain of mouse, the same age mouse, the same exposure levels, the same type of activity monitoring devices and the same testing protocols. While we were unsuccessful in finding increased spontaneous motor activity,we have observed altered pharmacological responses to amphetamine and phenobarbital in mice chronically exposed to lead. These responses take the form of attenuated responses to amphetamine and a transient increase in the level of excitation following phenobarbital. Observation of the animals suggested that the mice exhibiting the greatest alteration in drug response were also those most severely growth retarded. However, this observation was not quantitated.

Since the Silbergeld and Goldberg (122-125)studies did not control for alterations in growth weight and our own observation suggested that growth retardation, and not lead, might be the variable responsible for the altered drug responses, a study was undertaken to investigate the effects of growth retardation on locomotor activity and response to stimulant and depressant drugs employing the same procedure used in our earlier attempts to replicate the results of Silbergeld and Goldberg (122-125).

Mice were raised in either large litters of 16 pups or small litters of 8 pups (controls) in order to produce a large spectrum of growth rates. Six time-pregnant Charles River CD-1 mice served as the initiating pool. The newborn mice were mixed in a common pool and randomly assigned irrespective of sex so that each of two lactating dams suckled 16 pups and each of the two lactating dams suckled 8 pups. Purina mouse chow and tap water were provided <u>ad lib</u> and daily consumption was monitored. Animals were weaned at 21 days of age and weighed once a week throughout the course of the experiment.

Figure 1 shows the growth curve of animals raised in small litters and large litters.

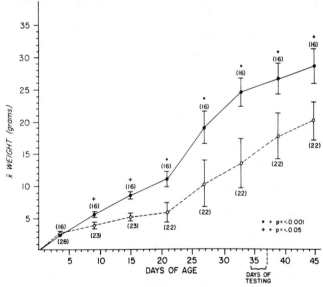

Fig.1. Growth (X+SD), raised in small (●--●, n=8) and large (o--o, n=16) litters. Parentheses indicate total number of animals contributing to each data point.

Growth rates of these two groups are comparable to the control and lead exposed mice as reported by Silbergeld and Goldberg (122-123).

When the mice were 35 to 37 days of age, they were individually tested in activity measuring devices similar to that used by Silbergeld and Goldberg (122-124). Locomotor activity was monitored for a total of 4 hours during the light phase of an animal's 12:12 light-dark cycle. The first hour was designated the habituation hour, and the second hour base line activity. Figure 2 shows the relative activity of mice during the first (habituation) and second hour (base line activity). The body weight of 35 to 37 day old mice from intermediate size litters was 25 ± 2 gms (X̄ ± SD) whereas the body weight of mice from large litters was 17 ± 4. Both the age and weight of these animals correspond to those employed in the lead study of Silbergeld and Goldberg (122-124).

Figure 2 illustrates the effect of d-amphetamine (10 mg/kg i.p.) on locomotor activity of
mice raised in small litters (open bars) and large litters (solid bars). The left panel shows
pre-drug activity of these two groups, and the right panel shows post-drug activity. Multivariate analysis over the 4 hour period showed highly significant differences (p < .001) between the mice raised in large and normal sized litters. Analyses of variance were then
performed to ascertain the effect at each hour.

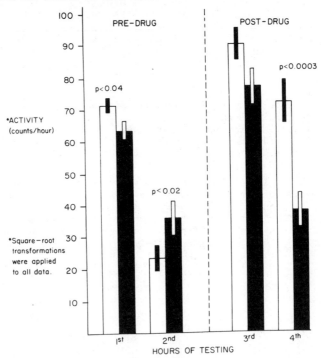

Fig. 2.   Effect of d-amphetamine (10 mg/kg i.p.) on locomotor activity
          (X+SEM) of 35 to 37 day old mice raised in small (▢   n=8)
          and large (▆   n=16) litters.

During the first hour, control mice were significantly more active than the mice reared in
large litters. However, the mice from the large litters, those of the smaller body weights
and retarded growth rates demonstrated significantly greater locomotor activity during the
second "baseline" hour. It should be emphasized that the pattern of habituation differed
for the two groups. The normal weight mice initially demonstrated higher activity levels,
but then habituated more rapidly and to a greater extent than did the growth retarded mice.
Silbergeld and Goldberg (122-124) reported no differences in the habituation rates of normal
and lead exposed mice. Following administration of d-amphetamine, these same growth
retarded mice displayed an attenuated drug response relative to controls, especially during
the second hour. This finding is in full accord with the reports of Silbergeld and
Goldberg (123-124).

When the activity levels of the two groups were analyzed irrespective of initial litter
size it was found that a quadratic relationship between weight and activity could account
for much of the variance during the baseline hours. This relationship took the form of an
inverted U-shaped function (r = 0.48, p < 0.01) (Fig. 3), illustrated on following page.

For the first hour post-drug it was U-shaped. A quadratic relationship was also demonstrated for the *change* in activity following drug administration. This is presented in
Fig. 4. The correlation was highly significant (r = 0.61; p < .0001).

The difference between the second hour post-drug and the baseline activity is plotted against
weight (Fig. 5). The circles represent mice raised in small litters with normal growth
rates, and triangles the growth retarded mice raised in large litters. All the mice raised
in small litters demonstrated increased activity following injection of d-amphetamine.
This is indicated by scores which occur above the dotted line. However, in many of the
large litters, growth retarded mice showed decreases in activity. In other words, these
mice appeared to react "paradoxically" to the drug. Mice in the range of 16 to 20 grams
correspond to the weight range seen in lead exposed mice tested by Silbergeld and Goldberg
(122-125). Figures 4, 5 and 6 are on the following pages.

Figure 6 provides a comparison between the response to methylphenidate obtained in lead-
exposed mice by Silbergeld and Goldberg (123, 124) and the amphetamine response obtained

by us in mice growth retarded to the same extent as the lead-exposed mice of Silbergeld and Goldberg (122).

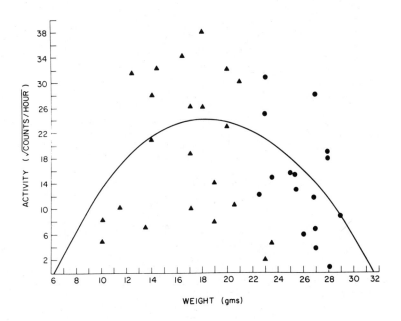

Fig. 3. Curvilinear relationships between body weight (gms) and baseline activity levels of mice raised in large litters (▲ , n=16) and intermediate (● , n=8) litters.

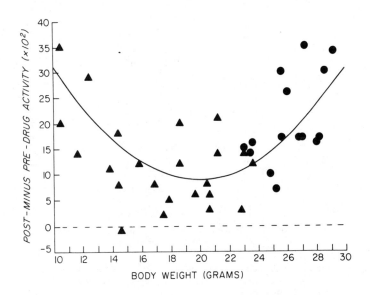

Fig. 4. Relative locomotor activity in mice one hour post-d-amphetamine (10 mg/kg, i.p.) as a function of body weight. Large (▲ , n = 16) and small (● , n = 8) litters.

There are no published data available which permit a direct comparison between the effects of either amphetamine or phenobarbital in lead-exposed and undernourished mice. It is not our intent to suggest that all of the data generated by Silbergeld and Goldberg (122-125)

Murphree *et al.* (52), Shaywitz *et al.* (98-100) can be accounted for by growth retardation. However, the data do suggest that a very important variable, growth rate, has not been controlled. Age-matched controls are not sufficient.

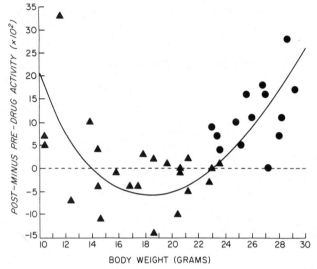

Fig. 5. Relative locomotor activity in mice during the second hour post-d-amphetamine (10 mg/kg, i.p.) as a function of body weight. Large (▲, n = 16) and small (●, n = 8) litters.

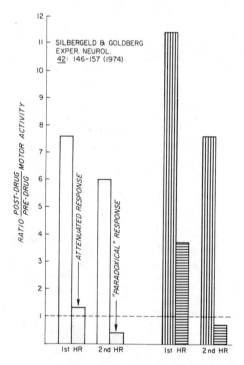

Fig. 6. Effect of methylphenidate (40 mg/kg) in control ☐ and lead exposed ☐ mice (125) and d-amphetamine (10 mg/kg) in normal ▥ and growth retarded ☰ mice.

From these results it would appear that some, but by no means all, of the observations reported by Murphree *et al.* (52) with rat models of post-encephalitic brain damage, Shaywitz and associates (98-100) on 6-OHDA treated neonatal rats, and Silbergeld and Goldberg (122-125) on lead-induced behavioral dysfunction as an animal model of minimal brain dysfunction hyperactivity, may be partially attributed to or confounded by the presence of neonatal growth retardation.

REFERENCES

1. Hoffmann, H. (1845) Der Struwwelpeter: Oder Lustige Geschichten und Drollige Bilder, Insel-Verlag, Leipzig.
2. L. B. Hoffman, Epidemic Encephalitis* (Lethargic Encephalitis) - Its Psychotic Manifestations with a Report of Twenty-Three Cases, Arch. Neurol. and Psychiatry 6, 295 (1921).
3. L. B. Hohman, Post-encephalitic behavior disorders in children, Johns Hopkins Hospital Bulletin, 33, 372 (1922)
4. E. Kahn and L. H. Cohen, Organic Driveness - A Brain-Stem Syndrome and an Experience, New Eng. J. Med. 210, 748 (1934).
5. F. Kramer and H. Pollnow, Uber eine hyperkinetische erkrankung im kindesalter, Monatschr. f. Psychiat. u. Neuro. 82, 1 (1932).
6. Strauss, A. A. and Lehtinen, L. E. (1947) Psychopathology and Education of the Brain-Injured Child, Grune & Stratton, New York.
7. Strauss, A. A. and Kephart, N. (1955) Psychopathology and Education of the Brain-Injured Child: Progress in Clinic and Theory, Vol. 2, Grune & Stratton, New York.
8. Gesell, A. and Amatruda, C. S. (1947) Developmental Diagnosis, Hueber Inc., New York.
9. R. S. Paine, Minimal Chronic Brain Syndromes in Children, Dev. Med. Child. Neurol. 4, 21 (1962).
10. R. Paine, Syndromes of "minimal cerebral damage", Pediatr. Clin. N. Am. 15, 779 (1968).
11. S. D. Clements and J. E. Peters, Minimal brain dysfunctions in the school-age child, Arch. Gen. Psychiat. 6, 183 (1962).
12. S. D. Clements, Minimal brain dysfunction in children, U.S. Department of Health, Education and Welfare, Monograph No. 3, 1 (1966).
13. P. H. Wender, The minimal brain dysfunction syndrome, Annu. Rev. Med. 26, 45 (1975).
14. Gross, M. D. and Wilson, W. C. (1974) Minimal Brain Dysfunction, Brunner-Mazel, New York.
15. C. Ounsted, Hyperkinetic syndrome in epileptic children, Lancet 2, 303 (1955).
16. C. Bradley, Characteristics and management of children with behavioral problems associated with organic brain damage, Pediatric Clinics of North America 4, 1049 (1957).
17. R. D. Becker, Child psychiatry - the continuing controversy. Minimal cerebral (brain) dysfunction - clinical fact or neurological fiction? The syndrome critically re-examined in the light of some hard neurological evidence, Israel Annals of Psychiat. Rel. Disciplines 12, 87 (1947).
18. H. F. Burks, The Hyperkinetic Child, Exceptional Children 27, 18 (1960).
19. H. F. R. Prechtl and J. C. Stemmer, Ein choreatiformes syndrome bei kindern, Wein. Med. Wschr. 22, 461 (1959).
20. H. F. R. Prechtl and J. C. Stemmer, The choreiform syndrome in children, Dev. Med. Child Neuro. 4, 119 (1962).
21. M. A. Stewart, F. N. Pitts, A. G. Craig and W. Dieruf, The hyperactive child syndrome, Am. J. Orthopsychiatry 36, 861 (1966).
22. H. R. Huessy, Study of the prevalence and therapy of the choreatiform syndrome or hyperkinesis in rural Vermont, ACTA Paedopsychiat. 34, 130 (1967).
23. Wender, P. H. (1971) Minimal Brain Dysfunction in Children, Wiley (Interscience), New York.
24. K. Minde, D. Lewin, G. Weiss, H. Laviguevr and E. Sykes, The hyperactive child in elementary school - a five-year controlled followup, Except. Child. 38, 215 (1971).
25. G. Winokur, Depression Spectrum Disease: Description and Family Study, Comp. Psychiat. 13, 3 (1972).
26. S. Chess, Diagnosis and treatment of the hyperactive child, N.Y. State J. Med. 60, 2379 (1960).
27. J. S. Werry, Studies on the hyperactive child. IV. An empirical analysis of the minimal brain dysfunction syndrome, Arch. Gen Psychiat. 19, 9 (1968).
28. Srole, L. (1962) Mental Health in the Metropolis, McGraw, New York.
29. E. G. Ketchum, Reading disorders, Pediat. Clin. N. Am. Aug, 647 (1957).
30. A. Blau, Mental changes following head trauma in children, Arch. Neurol. Psych. 35, 723 (1937).
31. E. Bond, Postencephalitic, ordinary and extraordinary children, J. Pediat. 1, 310 (1932).
32. B. Pasamanick and H. Knobloch, Brain damage and reproductive casualty, Am. J. Orthopsychiat. 30, 298 (1960).
33. M. F. Waldrop and J. D. Goering, Hyperactivity and minor physical anomalies in elementary school children, Amer J. Orthopsychiat. 41, 602 (1971).
34. J. L. Rapoport and P. O. Quinn, Minor physical anomalies (stigmata) and early developmental deviation: a major biologic subgroup of "hyperactive children", Inter. J. Ment. Hlth., 4, 29 (1975).

35. L. B. Silver, A proposed view on the etiology of the neurological learning disability syndrome, J. Learn. Disabil. 4, 6 (1971).

36. Satz, P. and Sparrow, S.S. (1970) Specific Developmental Dyslexia: A Theoretical Formulation, in: Bakker, D.J. and Satz, P. (1970) Specific Reading Disability: Advances in Theory and Method, Rotterdam University Press, Rotterdam.

37. G. B. Baxley and J. M. LeBlanc, The hyperactive child: characteristics, treatment, and evaluation of research design, Adv Child Dev Behav. 11, 1 (1976)

38. J. H. Satterfield, D. P. Cantwell and B. T. Satterfield, Pathophysiology of the Hyperactive child syndrome, Arch. Gen Psychiatry 31, 839 (1974).

39. E. N. Rexford and S. T. Van Amerongen, The influence of unsolved maternal oral conflicts upon impulsive acting "out" in young children, Am. J. Orthopsychiat. 27, 75 (1957).

40. Bereiter, C. and Englemann, S. (1966) Teaching Disadvantaged Children in the Preschool, Prentice Hall, Englewood Cliff.

41. N. M. Lambert, M. Windmiller, J. Sandoval and B. Moore, Hyperactive children and the efficacy of psychoactive drugs as a treatment intervention, Amer J. Orthopsychiat. 46, 335 (1976).

42. D. P. Cantwell, Genetic factors in the hyperkinetic syndrome, J. Am. Acad. Child Psychiat. 15, 214 (1976).

43. M. A. Stewart, Hyperactive children, Scientific Amer. 222, 94 (1970).

44. J. S. Werry, Organic factors in psychopathology of childhood, To appear in: Quay, H. and Werry, J.S. (1977) Psychopathological Disorders of Childhood, Wiley, New York

45. Feingold, B.F. (1975) Why your child is hyperactive, Random House, New York.

46. O. David, J. Clark and K. Voeller, Lead and hyperactivity, Lancet ii, 900 (1972).

47. O. J. David, Association between lower level lead concentrations and hyperactivity in children, Environmental Hlth Pers. 7, 17 (1974).

48. O. J. David, S. P. Hoffman, J. Sverd, J. Clark and K. Voeller, Lead and hyperactivity. Behavioral response to chelation: a pilot study, Am J Psychiatry 133, 1155 (1976).

49. S. M. Plaut, Animal models in developmental research, Pediatric Clinics of N. Am. 22, 619 (1975).

50. J. A. Sechzer, M. D. Faro and W. F. Windle, Studies of monkeys asphyxiated at birth: implications for minimal cerebral dysfunction, Seminars in Psychiatry 5, 19 (1973).

51. B. A. Shaywitz, Minimal brain dysfunction: dopamine depletion?, Science 194, 452 (1976).

52. O. D. Murphree, P. N. Morgan and R. V. Jarman, Learning deficits and activity changes: a partial laboratory model in postencephalitic rats for studies of brain damage, Cond. Reflex 6, 30 (1971).

53. A. W. Brann and R. E. Myers, Central nervous system findings in the newborn monkey following severe in utero partial asphyxia, Neurology 25, 327 (1975).

54. J. A. Sechzer, Memory deficit in monkeys brain damaged by asphyxia neonatorum, Experimental Neurology 24, 497 (1969).

55. J. A. Sechzer, Developmental behaviors: delayed appearance in monkeys asphyxiated at birth, Science 171, 1173 (1971).

56. J. B. Ranck and Wm. F. Windle, Brain damage in the monkey, Macaca mulatta, by asphyxia neonatorum, Experimental Neurology 1, 130 (1959).

57. H. N. Jacobson and W. F. Windle, Responses of foetal and new-born monkeys to asphyxia, J. Physiol. 153, 447 (1960).

58. Wm. F. Windle, Neuropathology of certain forms of mental retardation, Science 140, 1186 (1963).

59. Wm. F. Windle, Brain damage at birth, JAMA 206, 1967 (1968).

60. Wm. F. Windle, Brain damage by asphyxia at birth, Scientific Am. 221, 76 (1969).

61. M. D. Faro and Wm. F. Windle, Transneuronal degeneration in brains of monkeys asphyxiated at birth, Experimental Neurology 24, 38 (1969).

62. R. E. Meyers, Patterns of perinatal brain damage in the monkey, In: Brain Damage in the Fetus and Newborn from Hypoxia or Asphyxia, 57th Ross Conf. on Pediatric Research, James, L.S., Myers, R.E. and Gaull, G.E. (eds.), Ross Laboratories, Columbus (1967).

63. S. V. Saxon and C. G. Ponce, Behavioral defects in monkeys asphyxiated during birth, Experimental Neurology 4, 460 (1961).

64. R. F. Becker and Wm. Donnell, Learning behavior in guinea pigs subjected to asphyxia at birth, J. Comp. Physiol. Psychol. 45, 153 (1952).

65. W. F. Windle and R. F. Becker, Effects of anoxia at birth on central nervous system of the guinea pig, Proc. Soc. Exp. Biol. Med. 51, 213 (1942).

66. W. F. Windle and R. F. Becker, Asphyxia Neonatorum, Am. J. Obstet. Gynecol. 45, 183 (1943).

67. W. F. Windle, R. F. Becker and A. Weil, Alterations in brain structure after asphyxiation at birth: an experimental study in the guinea pig, J. Neuropath. Exp. Neurol 3, 224 (1944).

68. M. L. McCullough and D. E. Blackman, The behavioral effects of prenatal hypoxia in the rat, Develop. Psychobiol. 9, 335 (1976).

69. B. Culver and S. Norton, Juvenile hyperactivity in rats after acute exposure to carbon monoxide, Experimental Neurology 50, 80 (1976).

70. S. Norton, P. Mullenix and B. Culver, Comparison of the structure of hyperactive behavior in rats after brain damage from X-irradiation, carbon monoxide and pallidal lesions, Brain Res. 116, 49 (1976).

71. Hall, C. S. (1951) The Genetics of Behavior, In: Stevens, S. S. (ed.) Handbook of Experimental Psychology, John Wiley and Sons, Inc., New York.

72. G. E. McClearn, The genetics of mouse behavior in Novel stimulations, J. Comp. Physiol. Psychol. 52, 62 (1959).

73. DeFries, J. C. and Hegmann, J. P. (1970) Genetic Analyses of Open-field Behavior, In: Contributions to Behavior-Genetic Analysis: The Mouse as a Prototype. G. L. Lindzey and D. D. Thiesseu (eds.) Appleton-Century-Crofts/Meredith Corp, New York.

74. G. Bignami, Selection for high rates and low rates of avoidance conditioning in the rat, Animal Behavior 13, 221 (1965).

75. G. Bignami, F. Robustelli, I. Janki and D. Boret, Action de l'amphetamine et de quelques agents psychotrapes sur l-acquisition d'un conditionment de faite et d'evitement chez des rats selectionnes en fonction du niveau particulierement bas de leurs performances, C.R. Acad. Sci 260, 4273 (1965).

76. G. S. Omenn, Genetic issues in the syndrome of minimal brain dysfunction, Sems. in Psychiat. 5, 5 (1973).

77. G. S. Omenn, Genetic approaches to the syndrome of minimal brain dysfunction. Pt. IV., Ann. N.Y. Acad. Sci. 285, 212 (1973).

78. S. G. Vandenberg, Possible hereditary factors in minimal brain dysfunction, Ann. N.Y. Acad. Sci. 205, 223 (1973).

79. J. T. Coyle, P. Wender and A. Lipsky, Avoidance Conditioning in Different Strains of Rats: Neurochemical Correlates, Psychopharmacologia 31, 25 (1973).

80. Corson, S. A., Corson, E., O'Leary, Arnold, L. Eugene, and Knopp, W., Animal models of violence and hyperkinesis: interaction of psychopharmacologic and psychosocial therapy in behavior modification, In: Animal Models in Hyman Psychobiology, G. Serban, A. King (eds.), Plenum Press, New York (1974).

81. S. Nurimoto, N. Ogawa and S. Ueki, Hyperemotionality induced by lesions in the olfactory system of the rat, Japan J. Pharmacol. 24, 175 (1974)

82. M. J. Capps and C. W. Stockwell, Lesions in the Midbrain Reticular Formation and the Startle Response in Rats, Physiology and Behav. 3, 661 (1968).

83. J. Cytawa and G. Kutulas, Does chlorpromazine facilitate recovery of emotional behavior in septal forebrain lesioned rats?, Folia Biologica 21, 411 (1973).

84. H. C. Nielson, A. H. McIver and R. S. Boswell, Effect of septal lesions on learning, emotionality, activity and exploratory behavior in rats, Experimental Neurology 11, 147 (1965).

85. B. L. Jacobs, C. Trimbach. E. E. Eubanks and M. Trulson, Hippocampal mediation of raphe lesion- and PCPA-induced hyperactivity in the rat, Brain Res. 94, 253 (1975).

86. G. Lynch, P. Ballantine and B. A. Campbell, Differential rates of recovery following frontal cortical lesions in rats, Physiology and Behav. 7, 737 (1971).

87. J. P. Huston and A. A. Borbely, The thalamic rat: general behavior, operant learning with rewarding hypothalamic stimulation, and effects of amphetamine, Physiology and Behav. 12, 433 (1974).

88. J. N. Papaioannou, Rubral functions in the rat: a lesion study, Neuropsychologia 9, 345 (1971).

89. R. Fog and H. Pakkenberg, Intracerebral lesions causing stereotyped behaviour in rats, ACTA Neurol. Scand. 47, 475 (1971).

90. L. E. Jarrard, Behavior of hippocampal lesioned rats in home cage and novel situations, Physiology and Behav. 3, 65 (1968).

91. D. P. Kimble, The effects of bilateral hippocampal lesions in rats, J. Comparative and Physiological Psych. 56, 273 (1963).

92. H. Teitelbaum and P. Milner, Activity changes following partial hippocampal lesions in rats, J. Comp. and Physiological Psych. 56, 284 (1963).

93. R. Hannon and A. Bader, A comparison of frontal pole, anterior median and caudate nucleus lesions in the rat, Physiology and Behav. 13, 513 (1974).

94. L. P. Lanier, T. LaR. Petit and S. F. Zornetzer, Discrete anterior medial thalamic lesions in the mouse: the production of acute postoperative hyperactivity and death, Brain Res. 91, 133 (1975).

95. F. W. Maire and H. D. Patton, Hyperactivity and pulmonary edema from rostral hypothalamic lesions in rats, Am. J. Physiol. 78, 315 (1954).

96. S. Norton, Hyperactive behavior of rats after lesions of the globus pallidus, Brain Res. Bull. 1, 193 (1976).

97. B. Costall, R. J. Naylor and J. E. Olley, On the involvement of the lesions in caudate-putamen, globus pallidus and substantia nigra with neuroleptic and cholinergic modification of locomotor activity, Neuropharmacology, 11, 317 (1972).

98. B. A. Shaywitz and R. D. Yager, An experimental model of minimal brain dysfunction (MBD) in developing rats-"threshold" brain dopamine concentrations after 6-hydroxydopamine (6-OHDA), Pediatric Res. 9, 385 (1975).

99. B. A. Shaywitz, J. H. Klopper, R. D. Yager and J. W. Gordon, Paradoxical response to amphetamine in developing rats treated with 6-hydroxydopamine, Nature 261, 153 (1976).

100. B. A. Shaywitz, R. D. Yager and J. H. Klopper, Selective brain dopamine depletion in developing rats: an experimental model of minimal brain dysfunction, Science 191, 305 (1976).
101. J. W. Kalat, Minimal brain dysfunction: dopamine depletion?, Science 194, 450 (1976).
102. J. H. McLean, R. M. Kostrezewa and J. G. May, Minimal brain dysfunction: dopamine depletion?, Science 194, 451 (1976).
103. B. A. Pappas, H. B. Ferguson and M. Saari, Minimal brain dysfunction: dopamine depletion?, Science 194, 451 (1976).
104. G. R. Breese and T. D. Traylor, Developmental characteristics of brain catecholamines and tyrosine hydroxylase in the rat: effects of 6-hydroxydopamine, Br. J. Pharmac., 44, 210 (1972).
105. R. D. Smith, B. R. Cooper and G. R. Breese, Growth and behavioral changes in developing rats treated intracisternally with 6-hydroxydopamine: evidence for involvement of brain dopamine, The J. Pharmac. & Exptl. Therap. 185, 609 (1973).
106. A. S. Hollister, G. R. Breese and B. R. Cooper, Comparison of tyrosine hydroxylase and dopamine-β-hydroxylase inhibition with the effects of various 6-hydroxydopamine treatments on d-Amphetamine induced motor activity, Psychopharmacologica 36, 1 (1974).
107. G. R. Breese, R. D. Smith and B. R. Cooper, Effect of various 6-hydroxydopamine treatments during development on growth and ingestive behavior, Pharm. Biochem & Behav. 3, 1097, (1975).
108. W. R. Shannon, Neuropathic manifestations in infants and children as a result of anaphylactic reaction to foods contained in their dietary, Am. J. Diseases of Childhood 24, 89 (1922).
109. Feingold, B. F. (1975) Why Your Child is Hyperactive, Random House, New York.
110. G. W. Brown, Food additives and hyperactivity, J. Learn. Disab. 1, 62 (1974).
111. Anonymous, Diet and hyperactivity: any connection?, Nutr. Rev. 34, 151 (1976).
112. S. Palmer, J. L. Rapoport and P. O. Quinn, Food additives and hyperactivity: a comparison of food additives in the diets of normal and hyperactive boys, Clinical Pediatrics 14, 956 (1975).
113. F. J. Kittler, and D. G. Baldwin, The role of allergic factors in the child with minimal brain dysfunction, Annals of Allergy 28, 203 (1970).
114. L. K. Salzman, Allergy testing, psychological assessment and dietary treatment of the hyperactive child syndrome, Med. J. Australia 2, 248 (1976).
115. M. B. Campbell, Neurologic manifestations of allergic disease, Annals of Allergy 31, 485 (1973).
116. W. G. Crook, Food allergy - the great masquerader, Pediatric Clinics of N. Am. 22, 227 (1975).
117. W. G. Crook, An alternate method of managing the hyperactive child, Pediatrics 54, 656 (1974).
118. P. S. Cook and J. M. Woodhill, The Feingold dietary treatment of the hyperkinetic syndrome, Med. J. Australia 2, 85 (1976).
119. C. K. Conners, C. H. Govette, D. A. Southwick, J. M. Lees and P. A. Andrulonis, Food additives and hyperkinesis: a controlled double-blind experiment, Pediatrics 58, 154 (1976).
120. M. A. Perlstein and R. Attala, Neurological Sequelae of plumbism in children, Clinical Peds. 5, 292 (1966).
121. D. L. Thurston, J. N. Middelkamp and E. Mason, The late effects of lead poisoning, J. Pediatrics 47, 413 (1955).
122. E. K. Silbergeld and A. M. Goldberg, A Lead induced behavioral disorder, Life Sciences 13, 1275 (1973).
123. E. K. Silbergeld and A. M. Goldberg, Hyperactivity: a lead induced behavior disorder, Environmental Hlth Persp. 7, 227 (1974).
124. E. K. Silbergeld and A. M. Goldberg, Lead induced behavioral dysfunction: an animal model of hyperactivity, Experimental Neurology 42, 146 (1974).
125. E. K. Silbergeld and A. M. Goldberg, Pharmacological Neurochemical Investigations of Lead Induced Hyperactivity, Neuropharmacology 14, 431 (1975).
126. G. Bignami, Behavioral pharmacology and toxicology, Am. Rev. Pharmacol. & Tox. 16, 329 (1976).
127. A. H. Neims, M. Warner, P. M. Loughman and J. V. Aranda, Developmental aspects of the hepatic cytochrome P450 Monooxygenase system, Am. Rev. Pharmacol. & Tox. 16, 427 (1976).
128. I. A. Michaelson and M. W. Sauerhoff, An improved model of lead-induced brain dysfunction in the suckling rat, Toxicol. Appl. Pharmacol. 28, 88 (1974).
129. J. Morrison, D. Olton, A. Goldberg and E. Silbergeld, Alterations in consummatory behavior of mice produced by dietary exposure to inorganic lead, Developmental Psychobiol. 8, 389 (1975).
130. Dobbing, J. (1972) Vulnerable Periods of Brain Development, In: CIBA Foundation Symposium: Lipids, Malnutrition and the Developing Brain, K. Elliott and J. Knight (eds.) Elsevier Excepta Medica, Amsterdam.
131. Winick, M., Russo, P. and Brasel, J.A. (1972) Malnutrition and Cellular Growth in the Brain: Existence of Critical Periods, in: CIBA Foundation Symposium: Lipids, Malnutrition and the Developing Brain, K. Elliott and J. Knight (eds.) Elsevier-Excerpta Medica, Amsterdam.

132. Dodge, P.R., Prensky, A.L. and Feigen, R.D. (1975) <u>Nutrition and the Developing Nervous System</u>, C.V. Mosby Company, St. Louis.

133. F. Sereni, N. Principi, L. Perletti and P. Sereni, Undernutrition and the Developing Rat Brain. I. Influence on acetylcholinesterase and successive acid dehydrogenase activities and on norepinephrine and 5-OH-tryptamine tissue concentrations, <u>Biol. Neonate</u> 10, 254 (1966).

134. B.P.F. Adlard and J. Dobbing, Vulnerability of developing brain. III. Development of four enzymes in the brain of normal and undernourished rats, <u>Brain Res</u>. 28, 97 (1971).

135. B. P. F. Adlard and J. Dobbing, Elevated Acetylcholinesterase Activity in Adult Rat Brain after Undernutrition in Early Life, <u>Brain Res</u>. 30, 198 (1971).

136. C. T. Randt and B. Derby, Behavioral and brain correlations in early life nutritional deprivation, <u>Arch. Neurol</u>. 28, 167 (1973).

137. C. Castellano and B. Oliverio, Early malnutrition and postnatal changes in brain and behavior in the mouse, <u>Brain Res</u>. 101, 317 (1976).

DISCUSSION

Dr. Goldberg stated that he was under the impression that there are a number of studies showing that lead-exposed animals are hyperactive in comparison to pair-fed controls. He asked if Dr. Michaelson was proposing undernutrition as a model for hyperkinesis. *Dr. Michaelson* replied that he was trying to indicate that in most studies in his review the results were complicated by undernutrition.

II. Models in Psychiatry: An Overview

THE PSYCHIATRIC CLINICAL SYNDROME AND ESSENTIAL CORRESPONDING
QUALITIES OF AN ANIMAL MODEL OF RECEPTOR SENSITIVITY MODIFICATION

Arnold J. Friedhoff

Millhauser Laboratories of the Department of Psychiatry
NYU School of Medicine, 550 First Avenue, New York, N.Y.   10016

ABSTRACT

An animal model of hyperdopaminergia presumed to be related to striatal
dopamine receptor supersensitivity has been described. After chronic neuroleptic treatment, behavioral response to a dopamine agonist, ^{3}H-dopamine
binding and dopamine sensitive adenylate cyclase were found to be increased. After l-dopa treatment to increase the supply of ligand (dopamine)
the increase in all three parameters was reversed. The model has been used
to develop a paradigm for treatment of tardive dyskinesia, schizophrenia
and Gilles de la Tourette Syndrome.

INTRODUCTION

The development of animal models of clinical syndromes is important for
many reasons. Certain crucial experiments cannot be conducted in man. Also,
conditions can be more closely controlled in non-human species and it may
be more readily possible to uncover basic mechanisms in less complex organisms. However, all models of any aspect of human function must, by definition, be imperfect. Although there have been attempts to develop criteria
as to what constitutes a suitable animal model, no uniform set of criteria
are widely accepted, the reason being that the essential qualities of a
model are dependent on the hypothesis to be tested. In general, a suitable
animal model should correspond to the human state in relation to more than
one system. For example the model may be produced by the same treatment
that produces the human condition, (i.e. amphetamine psychosis), and the
model and the human state may share one or more behavioral, biochemical or
physiological parameters. Often the model and the human state can be
shown to respond to some perturbation in a similar fashion.

In this report I will describe an animal model of the hyperdopaminergic
state. I will present evidence that hyperdopaminergia can be produced
through pharmacologically induced supersensitivity of central dopaminergic
receptors. I will then consider whether this model serves as a model for
studying the treatment of schizophrenia and Gilles de la Tourette Syndrome,
and whether it may serve as an etiological and treatment model of tardive
dyskinesia (Ref. 1). Although considerable evidence of both a behavioral
and biological nature has been accumulated that tends to implicate increased
dopaminergic activity in each of these conditions, the principal evidence is
that each condition responds favorably to the administration of dopaminergic
blocking agents (i.e. antipsychotic drugs).

The utility of the model that I will discuss, that is the supersensitivity
model of hyperdopaminergia, is not dependent on the assumption that any of
the conditions mentioned results from supersensitive dopamine receptors,
although in the case of tardive dyskinesia this may be the case. I will use
the model in the study of a novel way to decrease dopaminergic activity, and
will discuss the behavioral and biochemical effects of this reduction of
sensitivity of central dopamine receptors. I will also consider what we
should expect when this model is tested in humans. The model, in which we
will try to look at essential corresponding qualities of animal and human

states, is one in which we can study the effects of drugs on the animal model and then make predictions about their effect on the three human conditions, schizophrenia, Gilles de la Tourette Syndrome and tardive dyskinesia.

It was definitively shown by Cannon and Rosenblueth (2) that the sensitivity of a neuronal receptor to its specific transmitter is increased when the afferents are interrupted, and they called this phenomenon denervation supersensitivity. By definition, a supersensitive receptor manifests an increased response to a given agonist concentration (Ref. 3). The relationship between the sensitivity of synaptic receptors and the supply of available neurotransmitter has been the subject of many studies (Refs. 4, 5, 6). Hornykiewicz (7) reported that increased activity of the remaining functional dopaminergic neurons may occur in parkinsonism and in syndromes resulting from prolonged administration of dopamine blockers. This increased activity presumably serves to compensate for the decrease in available dopamine at the receptor. Moore and Thornburg (8) proposed that chronic administration of blockers and chronic inhibition of dopamine synthesis could cause conformational changes and/or an increased number or affinity of dopamine receptors, so that greater than normal response would be observed when the dopamine blocking agent was discontinued and neurotransmitter again became available. As has been mentioned dopamine blockade is believed to be important in antipsychotic drug action and the extent of dopamine blockade has been used as a measure of the efficacy of neuroleptics as antipsychotic agents (Ref. 9). It has also been shown that the administration of reserpine, resulting in dopamine depletion and interference with agonist supply to the receptor enhances the response to apomorphine, a dopamine agonist (Ref. 6).

Supersensitivity of central dopaminergic receptors has been defined in terms of several parameters in the rat: at the behavioral level an increase in a stereotyped behavioral response to dopamine agonist has been reported, at the cellular level both increased binding of dopamine and increased activity of dopamine stimulated adenylate cyclase have been found, and at the neurophysiological level an increase in spontaneous electrical activity. In the living human, supersensitivity has been described only in behavioral terms and is believed to be a factor in tardive dyskinesia. I will consider each of these parameters in turn:

In 1974 Tarsy and Baldesserini (10) showed that apomorphine, a potent dopamine agonist, induced stereotyped gnawing behavior in rats following pretreatment with alpha-methyl-p-tyrosine, a synthesis inhibitor, reserpine, a depletor, and after chlorpromazine and haloperidol which are dopamine blockers. All of these agents decrease central dopaminergic activity, although through different mechanisms. Supersensitization did not occur when non-dopamine blockers like phenobarbital were used as a pretreatment, even though behavioral depression was produced as with the antipsychotic drugs. Interestingly, they found that detectable supersensitization did not occur with less than 11 days of pretreatment and this coincides with observations of our own that supersensitivity occurs only after a prolonged period of blockade. Enhanced behavioral supersensitivity has also been demonstrated by Gianutsos et al (11) and Moore and Thornburg (8), following withdrawal of dopamine blocking butyrophenones.

Burt et al (12) and Friedhoff et al (13) have recently found, independently, that persistent blockade of dopamine receptors, followed by washout of the blocking agent, results in increased specific binding of ^{3}H-dopamine to striatal dopamine receptors. Thus we will use increased binding as a second parameter of the supersensitive state.

Kebabian and Greengard (14) demonstrated that both dopamine and apomorphine can activate adenylate cyclase. Kruger et al (15) showed that unilateral nigrostriatal lesions produced apomorphine induced contralateral circling, 93-95% depletion of dopamine and a substantial increase in dopamine stimulated adenylate cyclase on the ipsilateral side. They proposed that the increased adenylate cyclase response was the result of supersensitivity. For these reasons we will use the degree to which adenylate cyclase can be activated by dopamine as a third measure of supersensitivity.

A fourth parameter which appears to be a measure of supersensitivity has been demonstrated by Ungerstedt et al (6). They studied single unit activity in innervated and denervated striatum and were able to show that the number of spontaneously active cells was increased in the dopamine denervated striatum, and that all striatal neurons are slowed by dopamine, apomorphine and cyclic AMP.

In the living human, only behavioral manifestations of presumed dopamine receptor supersensitivity have been observed. The state in humans is believed to be manifested by a condition known as tardive dyskinesia. As in the rat the motoric phenomena became apparent after prolonged treatment with antipsychotic drugs, and particularly at the time the drug is withdrawn. It is widely, but not uniformly, believed that tardive dyskinesia results from supersensitivity of dopamine receptors (Refs. 1, 7). It has been proposed that neuroleptics cause chemical denervation of the dopamine receptor resulting in supersensitivity when the blocker is withdrawn and transmitter is again available (Refs. 4, 10, 11, 16, 17). As in the pharmacologically induced supersensitive state in the rat, symptoms of tardive dyskinesia can be worsened by administering dopaminergic agonists for brief periods (Refs. 1, 18, 19, 20). This strengthens the view that the receptor is supersensitive in tardive dyskinesia.

There is also evidence that receptors cannot only be made supersensitive, but also subsensitive. Katz and Thesleff (21) showed that acetylcholine receptor desensitization can be produced by increasing agonist supply. Thus it appeared to us that it might be possible to tune the sensitivity of some neuronal receptors up or down by manipulating the supply of neurotransmitter at the receptor.

METHODS

In studies carried out in our laboratories by Friedhoff, Bonnet and Rosengarten (13) we decided to determine in our animal model whether we could decrease the sensitivity of striatal dopamine receptors as measured by ^{3}H-dopamine binding and dopamine sensitive adenylate cyclase response, and determine if this decrease produced a corresponding effect on a dopamine stimulated behavioral response.

In order to do this rats were made supersensitive by prolonged administration of a dopamine blocking agent, followed by a washout period. Increased supply of specific agonist, that is dopamine, was produced by administration of l-dopa, followed by a washout period. In addition a group of normal rats was also treated with l-dopa, followed by washout, to determine whether we could reduce sensitivity below normal levels.

We studied three parameters of supersensitivity described previously: 1. behavioral response to apomorphine, 2. dopamine binding and 3. dopamine stimulated adenylate cyclase activity. Results of behavioral studies are only preliminary, but data thus far obtained are consistent with the hypothesis that in the supersensitive animal there is an increased behavioral response to apomorphine which can be reversed after administration of l-dopa followed by a washout period.

RESULTS

Data fron the binding and adenylate cyclase studies are summarized in Table 1.

TABLE 1 Effect of Pharmacological Agents that Modify Ligand Supply on Two Measures of Receptor Sensitivity

(^{3}H-Dopamine Binding. N = 10 Rats/Group)

Phase	Duration	Control	Supersensitive	Reversal	Subsensitive
1	28 days	saline	haloperidol 2 mg/kg/day (i.p.)	haloperidol 2 mg/kg/day (i.p.)	saline
2	10 days	saline	saline	1-dopa 200 mg/kg/day (i.m.)	1-dopa 200 mg/kg/day (i.m.)
3	5 days	No Rx	No Rx	No Rx	No Rx
Specific binding††		34.6 ± 0.8	54.7 ± 1.2**	39.4 ± 2.7	36.4 ± 3.6

(pmoles cAMP Formed/mg/protein/15'. N = 4 Rats/Group)

Phase	Duration	Control	Supersensitive	Reversal	Subsensitive
1	28 days	vehicle	trifluperazine 2.5 mg/kg/day (p.o.)	triflu-perazine 2.5 mg/kg/day (p.o.)	vehicle
2	14 days	vehicle	vehicle	1-dopa 100 mg/kg/day (p.o.)	1-dopa 100 mg/kg/day (p.o.)
3	10 days	No Rx	No Rx	No Rx	No Rx
Net change after dopamine stimulation‡		8.0 ± 1.7	52.3 ± 8.1*	0.63 ± .07*	-2.7 ± 1.7*

*Mean significantly different from mean of control group, $p < .01$; **$p < .001$; students t, two tailed.††Percent total binding displaced by (+) butaclamol; ‡Net change equals stimulated activity minus basal activity for each animal, not the difference of the means. Data for total ^{3}H-dopamine bound and basal adenylate cyclase activity can be found in Ref. 13.

In the animals treated with neuroleptic followed by washout, we found increased specific dopamine binding and increased dopamine specific adenylate cyclase activity. Neuroleptic treatment followed by l-dopa and a washout period, produced a restoration of both binding levels and adenylate cyclase response to normal levels. In the normal, non-neuroleptic treated rat, treated with l-dopa and washout, dopamine stimulated adenylate cyclase response was reduced to subnormal levels but binding was not.

Because of the law of initial values we might expect that it would be more difficult to reduce the sensitivity of normal receptors than it is to reverse the supersensitive state. Indeed we have preliminary data showing that when we administer carbidopa along with l-dopa, in order to raise central dopa levels above that produced by l-dopa alone, we can get a further reduction of adenylate cyclase activity below normal in the normal rat and also a significant decrease in binding.

DISCUSSION AND CONCLUSIONS

From this animal model we have devised a hypothesis for test in the three conditions, schizophrenia, Gilles de la Tourette Syndrome, and tardive dyskinesia (Refs. 23, 24). Our hypothesis predicts that: 1. Each of the conditions should be made worse initially by l-dopa, 2. each of the conditions responding to dopamine blockers should respond to a reduction in dopamine receptor sensitivity after administration of l-dopa, 3. if the condition involved supersensitivity of striatal dopamine receptors it should respond more readily to l-dopa than a hyperdopaminergic state associated with normal receptors, in which hyperdopaminergia is of pre-synaptic origin.

With several collaborators we have initiated a trial of l-dopa treatment in each of the three conditions. The data are very preliminary, but we have found all three conditions are worsened by l-dopa initially, as would be predicted by the model. We have found rather dramatic reversal of symptoms in the few patients with Gilles de la Tourette Syndrome or tardive dyskineaia that we have been able to treat with prolonged l-dopa administration and washout, but have found schizophrenics to improve only slightly with the same dose of l-dopa. We are now studying the cabidopa-l-dopa combination in schizophrenics. It should be borne in mind, however, that all data thus far are from a small number of subjects studied in open trials. The preliminary observations are now being extended in double blind controlled studies. If we consider the corresponding qualities of this animal model of hyperdopaminergia and one of the three clinical states, tardive dyskinesia, it will be seen that:

1. The model and the human condition can be produced in the same way, that is by chronic administration and withdrawal of dopamine blockers.

2. The biochemical and neuropathological state found in the rat, increased dopamine binding and increased adenylate cyclase response, have not yet been convincnngly demonstrated in man.

3. Both the animal behavioral state produced by neuroleptic treatment and withdrawal, and the human state are worsened initially by dopaminergic agents.

4. Both involve disturbances in extrapyramidal motor functions.

5. We are presently determining whether both the animal and human condition respond to treatments as predicted by the hypothesis, and do not respond to other treatments.

Hopefully, findings from ongoing human studies will lead to the development of additional hypotheses to be tested in animals. Thus an animal model should be viewed as a dynamic conceptual model, not a static structure.

The adequacy of a model cannot be determined only by the number of analogous properties of the human and animal state. Rather the success of any animal model should be measured by its usefulness in facilitating the formation of hypotheses that can be verified in humans.

We have shown that receptor sensitivity modification can be demonstrated in the rat by at least three measures. Inasmuch as modification can be affected by drugs that can be administered to humans, it is possible to test the effects of these drugs on human behavior and human motor function. However, the evidence, at present, that receptor sensitivity changes occur in the living human, is only circumstantial. The direct demonstration that human receptor sensitivity can be modified would constitute an important step in further establishing the validity of the current conceptual model.

REFERENCES

(1) H. L. Klawans, The pharmacology of tardive dyskinesias, _Amer. J. Psychiat._ 130, 82 (1973).

(2) Cannon, W. B. and Rosenbleuth, A. (1949) _The Supersensitivity of Denervated Structures_, Macmillan, New York.

(3) L. L. Iversen, in _Advances in Neurology_ (Eds. D. Calne, T. N. Chase, and A. Barbeau) vol. 9, p. 415. Raven Press, New York (1975).

(4) P. F. Von Voigtlander, S. Y. Boukma and G. A. Johnson, Dopaminergic denervation supersensitivity and dopamine stimulated adenyl cyclase activity, _Neuropharmacology_ 12, 1081 (1973).

(5) P. F. Von Voigtlander and K. E. Moore, Turning behavior of mice with unilateral 6-hydroxydopamine lesions in the striatum: effects of apomorphine, l-dopa, amanthadine, amphetamine and other psycho-motor stimulants, _Neuropharmacology_ 12, 451 (1973).

(6) U. Ungerstedt, T. Ljunberg, B. Hoffer and G. Siggens, in _Advances in Neurology_ (Eds. D. Calne, T. N. Chase and A. Barbeau) vol. 9, p. 57. Raven Press, New York (1975).

(7) O. Hornykiewicz, in _Advances in Neurology_ (Eds. D. Calne, T. N. Chase and A. Barbeau) vol. 9, p. 155. Raven Press, New York (1975).

(8) K. E. Moore and J. E. Thornburg, in _Advances in Neurology_ (Eds. D. Calne, T. N. Chase and A. Barbeau) vol. 9, p. 93. Raven Press, New York (1975).

(9) I. Creese, D. R. Burt, and S. H. Snyder, Dopamine receptor binding: differentiation of agonist and antagonist states with ^{3}H-dopamine and ^{3}H-haloperidol, _Life Sci._ 17, 993 (1975).

(10) D. Tarsy and R. J. Baldessarini, Behavioral supersensitivity to apomorphine following chronic treatment with drugs which interfere with the synaptic function of catecholamines, _Neuropharmacology_ 13, 927 (1974).

(11) G. Gianutsos, R. B. Drawbaugh, M. D. Hynes and H. Lal, Behavioral evidence for dopaminergic supersensitivity after chronic haloperidol, _Life Sci._ 14, 887 (1974).

(12) D. R. Burt, I. Creese and S. H. Snyder, Antischizophrenic drugs: chronic treatment elevates dopamine receptor binding in brain, Science 196, 326 (1977).

(13) A. J. Friedhoff, K. Bonnet and H. Rosengarten, Reversal of two manifestations of dopamine receptor supersensitivity by administration of l-dopa, Res. Comm. Chem. Pathol. Pharm. 16, 411 (1977).

(14) J. W. Kebabian and P. Greengard, Dopamine-sensitive adenyl cyclase: possible role in synaptic transmission, Science 174, 1346 (1971).

(15) B. K. Krueger, J. Forn, J. R. Walters, R. H. Roth and P. Greengard, Stimulation by dopamine of adenosine cyclic 3',5'-monophosphate formation in rat caudate nucleus: effect of lesions of the nigroneostriatal pathway, Mol. Pharmacol. 12, 639 (1976).

(16) P. F. Von Voigtlander, Behavioral and biochemical investigation of dopamine supersensitivity induced by chronic neuroleptic treatment, Fed. Proc. 33, 578 (1974).

(17) A. Carlsson, in L-Dopa and Parkinsonism (Eds. A. Barbeau and F. H. McDowell) F. A. Davis, Philadelphia (1970).

(18) A. Barbeau, L-dopa therapy in Parkinson's disease: critical review of nine years experience, Can. Med. Assoc. J. 101, 59 (1969).

(19) H. L. Klawans, P. Crossett, and N. Dana, in Advances in Neurology (Eds. D. Calne, T. N. Chase and A. Barbeau) vol. 9, p. 105. Raven Press, New York (1975).

(20) C. H. Markham, The choreathetoid movement disorder induced by levodopa, Clin. Pharmacol. Ther. 12, 340 (1971).

(21) B. Katz and S. Thesleff, A study of the "desensitization" produced by acetylcholine at the motor end plate, J. Physiol. 138, 63 (1957).

(22) A. J. Friedhoff, H. Rosengarten and K. Bonnet, in preparation (1977).

(23) A. J. Friedhoff and M. Alpert, in Psychopharmacology: A Generation of Progress (Eds. M. Lipton, A. DiMascio and K. Killam) Raven Press, New York (1977).

(24) A. J. Friedhoff, in Psychopathology and Brain Dysfunction (Eds. C. Shagass, S. Gershon and A. J. Friedhoff) p. 139. Raven Press, New York (1977).

(25) T. Arnfred and A. Randrup, Cholinergic mechanism in brain inhibiting amphetamine induced sterotyped behavior, Acta Pharmacol. et Toxicol. 26, 384 (1968).

(26) J. W. Kebabian, G. L. Petzold and P. Greengard, Dopamine-sensitive adenylate cyclase in caudate nucleus of rat brain and its similarity to the "dopamine receptor", Proc. Nat. Acad. Sci. 69, 2145 (1972).

(27) J. W. Kebabian, Y. C. Clement-Cormier, G. L. Petzold and P. Greengard, in Advances in Neurology (Eds. D. Calne, T. N. Chase and A. Barbeau) vol. 9, p. 1. Raven Press, New York (1975).

(28)   J. W. Kebabian, M. Zatz, J. A. Romero and J. Axelrod, Rapid changes
       in rat pineal β-adrenergic receptor: alterations in l-(^{3}H)
       alprenolol binding and adenylate cyclase, <u>Proc. Nat. Acad. Sci.</u>
       72, 3735 (1975).

(29)   B. S. Bunney, J. R. Walters, R. H. Roth and G. K. Aghajanian,
       Dopaminergic neurons: effect of antipsychotic drugs and amphet-
       amine on single cell activity, <u>J. Pharmacol. Exp. Ther</u>. 185, 560
       (1973).

DISCUSSION

Dr. Lomax asked how much of a contribution could a change in endogenous transmitter play in the competition for the binding site of labelled dopamine? If there were more, or less, dopamine present, could this affect the binding data? *Dr. Friedhoff* did not think the concentrations are adequate to have much effect on the system he is using for assaying specific dopamine binding. He agreed that he was probably ignoring important adjustments on the presynaptic side. He has also side-stepped the problem of supersensitivity by defining it as being increased behavioral response, increased binding, increased adenylate cyclase response, i.e., using an operational definition.

Dr. Zigmond commented that he had been studying supersensitivity after 6-hydroxydopamine and found that supersensitivity was observed only after very large lesions (85–90% depletion of dopamine). Further, supersensitivity continues to develop for at least 60 days after the lesion. *Dr. Friedhoff* indicated that he thought there is probably a point at which a qualitative change occurs after blockade becomes X% effective, but he does not know what % that is. Prior to this, nothing much happens. In his system, there is some reversal in regard to binding after about three weeks. At that point, adenylate cyclase is still up, but there is decrease in the increased binding.

CHRONIC STIMULANT INTOXICATION MODELS OF PSYCHOSIS

Everett H. Ellinwood, Jr. and M. Marlyne Kilbey

Behavioral Neuropharmacology Section
Department of Psychiatry
Duke University Medical Center, Durham, North Carolina 27710   USA

INTRODUCTION

Animal models of human neuropsychopathological conditions may be examined from numerous viewpoints. For example, the utility of models, per se, as a strategy in scientific investigations (Serban et al., 1976) or the requirements of useful models (Kornetsky, Friedhoff, this volume; Hinde, 1976) may be considered. More commonly, though, animal models of various pathologies, e.g., dyskinesias, epilepsy, hyperactivity, depression, and mania, are specified on the basis of similarities between the animal characteristics and the human pathology on some minimum number of biochemical, physiological, or behavioral parameters. In spite of large areas of non-concordance between the animal model and the human condition, use of models, specified in this way, is deemed worthwhile if these simpler systems allow identification and evaluation of possible mechanisms underlying phenomena of interest. The work described by Klawans, McKinney, Murphy (this volume), and Domino (1976) is illustrative of this application of animal models.

The use of stimulant-induced abnormal behavior in animals as a model of psychosis is unique in that the animal model interfaces the human pathology via a human model psychosis, i.e., amphetamine-induced psychosis. Interest in the animal model originated in the observation of the resemblance amphetamine-induced psychosis in man bears to paranoid schizophrenia (Young and Scoville, 1938; Connell, 1958; Ellinwood, 1969), a resemblance which has occasionally resulted in misdiagnosis of the amphetamine abuser (Beamish and Kiloh, 1960). Over the past decade, stimulant administration in man and experimental animals has allowed a more comprehensive description of the acute, chronic, and residual effects of these drugs and formulation of hypotheses of the basic mechanisms underlying these phenomena which, hopefully, will broaden our understanding of psychoses. In this paper, we intend first to review work which has described stimulant-induced psychosis in man. We shall consider the hypotheses these descriptions have generated and the intriguing, although infrequent, clinical-experimental investigations of these hypotheses.

In addition, we shall review the descriptions of acute, chronic, and residual effects of stimulants in experimental animals (but especially in cats), the hypotheses currently being investigated, and the utility of this particular model in investigations of human neuropsychopathology. We have chosen to review recent work with cats for several reasons. An extensive review of the topography of stimulant-induced abnormal behavior in man, monkey, cat, and rat is available (Ellinwood and Kilbey, in press). The review indicates that, across laboratories, stimulant-induced abnormal behaviors in the rat are fairly homogeneous. Investigations of these phenomena in monkeys are completed on small numbers of animals due to practical considerations; hence, detection of typical behavior patterns is difficult. Using cats, we have identified some typical response patterns as well as wider individual differences in initial response to stimulants. Similar to man, the initial behavior once enhanced appears to contribute to the expression of behavioral patterns developing with chronic drug administration.

Furthermore, we shall discuss preliminary results of experiments in which the availability of various neurotransmitters was manipulated and stimulant-induced behaviors of interest were observed. These data indicate that, in the cat, stimulant-induced abnormal behaviors are sensitive to transmitter functions in addition to those involving dopamine (DA).

THE STIMULANT MODEL PSYCHOSIS IN MAN

Acute Effects of Stimulants

Most clinical psychopharmacologists agree that drug effects in man reflect the interaction of drug, subject, and environmental factors. One of the most consistent characteristics of acute administration of stimulants is the energizing, mood-elevating effect which is sometimes described as euphorigenic. It is not invariably found, however (Bahnsen et al., 1938; Flory and Gilbert, 1943; Tecce and Cole, 1974). The relationship between disparate effects (such as energizing vs. non-energizing, or pleasant vs. unpleasant consequences) following initial use, and disparate responses (such as emergent psychosis vs. non-psychosis) following sustained use, is a fascinating, but uninvestigated, question. At least to a point, pleasant subjective effects increase with dose, as dose-response data for acute administration of cocaine in man have yielded significant dose-related augmentation in terms of self-perception of drug "highs," in addition to increasing autonomic functions, such as systolic blood pressure (Resnick et al., 1977) and heart rate (Fischman et al., 1977). Acute high doses of amphetamine may induce toxic, hallucinatory states (Connell, 1958; Rickman et al., 1961); however, it is not clear that psychosis can be induced with an acute dose of amphetamine in the absence of predisposing factors such as a previous history of drug-induced psychosis, critical personality characteristics, and/or borderline psychotic states (Ellinwood and Kilbey, 1975).

Chronic Effects of Stimulants

The ability of chronic stimulant administration in man to produce psychosis has provided a human model psychosis. Development of the psychosis is associated with increases in daily drug intake which sometimes reaches 3000 mg/day (Ellinwood and Kilbey, in press). There appear to be individual differences in susceptibility to amphetamine psychosis (Ellinwood, 1967). Paranoid symptoms are generally the hallmark of the psychosis; these include delusions of persecution; ideas of reference; visual, auditory, and olfactory hallucinations; and changes in body image.

Descriptions of the effects of chronic amphetamine administration based upon standardized interviews and tests have established drug effects which appear to be common to users who do--and those who do not--develop psychoses, and effects which predominate in the former group (Ellinwood, 1967, 1972). Effects common to both groups include induction of stereotyped touching and picking of the face and extremities, and changes in perception of time and space. Effects which predominate in those who develop psychoses include assembling and/or disassembling objects, suspiciousness, olfactory and tactile hallucinations, and extreme reactivity (Ellinwood, 1967, 1972).

Dyskinetic and dystonic reactions to chronic stimulant use have been described (Schiorring, 1977; Ellinwood, 1967; Kramer, et al., 1967; Rylander, 1969). Observed are frozen, bizarre, catatonic-like postures (Tatetsu et al., 1956) and an aphasic, paralyzed condition--"overamping"--which may last for hours following i.v. use of high doses of amphetamine (Kramer, 1969). Chronic users of high doses may manifest tics, buccolingual dyskinesias, and choreiform movements (Ellinwood, 1973). Dyskinesia may develop even after use of therapeutic doses of amphetamine (Mattson and Calverley, 1968). These motor disorders are not dependent upon development of an amphetamine psychosis, as they were seen in almost equal numbers among persons who did not develop psychosis (63%) and those who did (70%) (Ellinwood, 1967).

Residual Effects of Stimulants

Once a stimulant-induced psychosis is manifested, certain of the symptoms persist, and the individual has a lower threshold for precipitation of psychotic effects with subsequent amphetamine use. Psychological residua include delusions and mannerisms that persist in some individuals long after the chronic stimulant abuse has ceased (Ellinwood, 1971a, 1973). Following development of stimulant-induced psychosis, various aspects of the syndrome may reappear after only moderately high doses of amphetamine, despite long periods of intervening abstinence (Kramer, 1969; Bell, 1973; Ellinwood, 1973). Utena (1974) has described re-initiation of psychosis following stress in persons no longer using drug. Bell (1973) describes 14 patients in whom psychosis was re-created experimentally with only moderately large doses of amphetamine, and his cases are excellent descriptive examples of the re-triggering of amphetamine psychosis following subsequent administration in currently abstinent individuals who have previously developed the psychotic syndromes during chronic abuse. One case is highly illustrative:

After receiving an accumulative total dose of 120 mg of amphetamine, and 7 minutes after receiving the last 60 mg, the patient said, "Oh, not this again. I'm starting to get the wrong ideas about people. I feel they are all against me. All the questions in my mind -- going through my mind. Is it night? Is it night? This is it, all over again." (Bell, 1973). Shortly thereafter, questions about sexual intercourse and children were fantacized. Forty-nine minutes after the last dose, "mental telepathy machines" were discussed and the patient observed with some stress the similarity between her present state and her original state upon admission.

Tests of Stimulant-Induced Effects in Schizophrenia

There are surprisingly few tests of stimulant effects in schizophrenic patients. Small doses of methylphenidate or amphetamine have been reported to exacerbate a variety of pre-existing schizophrenic symptoms in actively psychotic patients (Janowsky et al., 1973), but not in remitted patients. Janowsky et al. (1977) have argued, on the basis of projective test data, that methylphenidate affects thought processes of schizophrenics, causing loosening of associations. Neuroleptics ameliorate amphetamine psychosis (Angrist and Gershon, 1974), as they do endogenous schizophrenia.

Hypotheses Based on Observations and Manipulations of Stimulant-Induced Psychoses

As the major acute pharmacological effect of amphetamine is believed to be catecholamine release (Carlsson, 1970; Rutledge et al., 1973), this property is postulated to be integral to stimulant-induced psychosis. Because the stereotypic effects of amphetamine and apomorphine, a DA agonist, are similarly blocked by neuroleptics, DA release has been considered to be the primary pharmacological property of interest. The "dopamine hypothesis," stated simply, holds that schizophrenia may relate to excessive DA neuronal activity. Associated with this hypothesis is the proposition that chronic depletion of DA may underlie some of the effects seen after chronic high-dose stimulant administration (Ellinwood and Duarte-Escalante, 1970a; Seiden et al., 1977), and that effects seen when drug is readministered after a period of abstinence may reflect DA receptor supersensitivity (Ellinwood et al., 1977).

Evidence bearing on these hypotheses, based predominantly on animal model systems, has been reviewed previously (Ellinwood, 1967, 1969, 1972; Klawans et al., 1972; Snyder, 1972; Snyder et al., 1972, 1974; Ellinwood et al., 1972, 1977; Matthysse, 1974, 1977; Ellinwood and Kilbey, 1975; Meltzer and Stahl, 1976). The evidence supporting DA hyperactivity as a possible mechanism in neuropsychiatric disorders has been based primarily upon three neurochemical test systems: 1) accumulation of homovanillic acid (HVA) as an estimation of DA turnover; 2) binding characteristics of radioactive DA or neuroleptics as an estimate of the number of DA receptors; and 3) the inhibition of DA-sensitive adenyl cyclase as an estimate of DA antagonist properties; three physiological test systems: 1) release of serum prolactin; 2) release of DA from striatal tissue; and 3) amphetamine and DA agonist inhibition of DA neuronal firing; and three behavioral test systems: 1) stimulant-induced stereotypy; 2) hyperactivity in nucleus accumbens-lesioned animals; and 3) rotational behavior in striatal-lesioned animals. However, as Matthysse (1977) points out, there exists no direct evidence supporting the DA hypothesis, insofar as schizophrenic patients have not been shown to differ on any parameter of DA activity implicated by the test systems outlined above. Furthermore, most of the experimental evidence has been based on acute administration of drug and, consequently, whether or not the relationships being tested hold in the case of chronic treatment is largely unevaluated.

STIMULANT-INDUCED ABNORMAL BEHAVIOR IN THE CAT AS A MODEL PSYCHOSIS

While the models outlined above have generated evidence supportive of the DA hypothesis, the work that has been completed in tissue systems and in rats has not provided a completely satisfactory explanation of psychosis. For example, intensification of abnormal behaviors is seen following chronic stimulant administration in both monkeys and rats. Yet, while chronic stimulant administration in monkeys establishes long-term DA depletion (Seiden et al., 1977), a prerequisite for development of DA receptor supersensitivity, similar effects for rats have not been established. Furthermore, Burt et al. (1977) have recently shown that DA receptor binding characteristics in rats are not altered by chronic amphetamine treatment. Thus, it is questionable whether the augmented stereotyped responses seen after chronic stimulant treatment in this species can be attributed, wholly or partially, to DA receptor supersensitivity. Likewise, motor abnormalities analogous to dyskinesias, tics, and choreiform movements seen in schizophrenia (Bleuler, 1924) are difficult to document in the rat.

TABLE 1   Characteristic Acute, Chronic, and Residual Behavioral Effects of High-Dose Stimulant Administration Suggested by Clinical Reports in Man and Long-Term Investigation in Experimental Animals

	MAN	MONKEY	CAT	RAT
Acute Peak and Early Chronic Effects	-exploratory, repetitious behaviors or perceptuomotor compulsions -visual scanning -beginning states of suspiciousness	-stereotyped: eye-hand examination; visual scanning, tracking, checking -investigatory attitude	-stereotyped perceptuomotor movements of head -investigatory attitude -dysjunctive movements -postural fixation	-restricted locomotion and exploration -stereotyped sniffing, licking, gnawing
Progressive Effects	-paranoid schizophrenic-like psychoses with delusional experiences; auditory, visual, tactile hallucinations -delusions of parasitosis prominent with dermatological excoriations -reactive behavior -oral dyskinesias and bruxism; facial tics -choreiform movements -frozen, immobile, catatonic-like behavior (overamping)	-more constricted bizarre stereotyped components -dyskinesias, especially oral -dystonic movements -reactive attitude -excoriating self-grooming behavior -failure to maintain normal social, sexual, and maternal behavior	-reactive attitude with hyperstartle and/or paw or head shaking and recoil -akathisia -pivoting movements -dystonic postures; e.g., camel-back -ataxia -dysjunctive movements increasingly isolated from original sequence	-increased restrictiveness or intensity of stereotypy with same dose of stimulant or apomorphine -dyskinetic jerking movements -backward walking and jumping -more reactive
Residual Effects	-low dose threshold for induction of psychosis or dyskinesias	-1 to 2 doses reinitiate chronic end-stage behavior	-conditioned behavior to injection and/or setting -low dose of stimulants produce end-stage behavior	-persistent augmented response to same stimulant dose or to apomorphine

We have recently reviewed the effects of chronic stimulant administration in four species (Ellinwood and Kilbey, in press) and noted certain similarities across species in acute, chronic, and residual effects. As can be seen from Table 1, an acute effect common to all species is the initiation of exploratory behavior, which typically develops into the well-known stereotyped behavior characteristic of moderately high-dose amphetamine. In all species this can be expressed as motor behavior, e.g., sniffing, licking, and gnawing in the rat. Additionally, in cats, monkeys, and man it is often expressed as visual exploratory behavior. Chronic effects in all species involve a more complex behavioral syndrome (than the better-known stereotypy) which includes reactivity and postural abnormalities, although the latter are not easily quantified by current techniques in the rat. Inspection of Table 1 indicates that the stimulant-associated behavior of the cat is more varied than that of the rat and, like the monkey, more analogous to that seen in man. In our chronic studies, we used the less expensive cat rather than the monkey. A brief outline of the acute, chronic, and residual effects of stimulants in cats must consider the following information.

Acute Effects

Acute effects of stimulants in cats include a dose-dependent induction of stereotyped head turning or sniffing (Wallach and Gershon, 1972) and motor dysjunction (Ellinwood et al., 1972; Sudilovsky et al., 1974). These and other effects have been measured using the Behavior Rating Inventory for Drug Generated Effects (BRIDGE), which provides a quantifiable method for documenting acute, chronic, and residual stimulant effects in cats (Ellinwood et al., 1972; Sudilovsky et al., 1975).

Chronic Effects

Using the BRIDGE scale, the following behaviors have been found to increase with chronic administration of amphetamine and cocaine: speed of head movement; dystonia; hyperreactive attitude or hyperstartle responses; dyskinesias and akathisia-like pivoting movements; head shakes and limb flicks (Ellinwood et al., 1972; Sudilovsky et al., 1974; Castellani et al., submitted for publication). Ataxia and dysjunctive movements are also seen. (See Table 2 for definitions of these chronic end-stage behaviors.) These plus the original acute intoxi-cation manifestations (e.g., stereotypy) make up the chronic behavior syndrome.

Residual Effects

Two major residual effects that have been identified are conditioned autonomic and behavioral responses (Ellinwood and Duarte-Escalante, 1970a; Ellinwood, 1971c) and reinstitution of behavior(s) characteristic of the chronic drug state, following a period of drug abstinence, using a dose which would not elicit the behavior in a drug-naive cat. These responses include bizarre constricted stereotypies, hyperreactive startle responses, dysjunctive postures, limb shakes or flicks, dyskinesias, and obstinate progression.

CHRONIC STIMULANT-INDUCED BEHAVIORS IN THE CAT FOLLOWING NEUROTRANSMITTER MANIPULATION

Over the past year, we have investigated the biologenic bases of specific behaviors which evolve with daily cocaine administration (10 mg/kg). Although general observations were made, behaviors more specifically investigated include hyperreactive behaviors (e.g., startle responses, recoil, flycatching, "hallucinatory" attention transients, limb flicks, and head shakes) as well as obstinate progression, dysjunctive behaviors, and dystonic postures. When we administered cocaine (10 mg/kg i.m., daily) to cats for three weeks, three of them developed hyperstartle, reactive behaviors. In these animals, apomorphine (5 mg/kg i.v.) elicited strong reactive hyperstartle responses even more intense than those seen after cocaine, but the response was transient (lasting only from 5 to 15 min post-injection). The four other cats which had not developed reactive behaviors following cocaine did not show a reactive response to apomorphine and, of course, none of the cats had shown this behavior when administered apomorphine prior to the chronic cocaine period. Cocaine-induced hyperreactivity which developed (4 cats) during chronic (2 months) treatment was blocked in these four animals with 5.0 mg/kg clozapine, i.m., while 2.5 mg/kg blocked the response in three of the four cats. Pimozide, i.m. (0.8 mg/kg), had no effect on this response, whereas 2.0 mg/kg blocked the hyperreactive response in two of four cats. At three months of chronic intoxication, apomorphine challenges were again introduced at four-day intervals, with cocaine in-between. Apomorphine (0.5 mg/kg) induced a weak reactivity in three of four animals; 0.25 mg/kg had no response except transiently in one animal; 0.75 induced the response in four animals. Gradually the cats became less

TABLE 2 Definitions of Chronic Stimulant End-Stage Behaviors in the Cat

I. Hyperreactivity

    (Ellinwood, 1971a, 1974; Ellinwood and Sudilovsky, 1973):

      1) Increase in startle response.

      2) Sudden reaction to--or withdrawal from--apparently non-existent stimuli, hallucinatory-like; e.g., sudden recoil or flycatching behavior involving tongue.

      3) Specific increase in head and paw shake (included also under hallucinatory-like).

      4) Fearful hyperreactivity; e.g., crouching to stimuli or presence of experimenter.

II. Dysjunctive Behavior*

    (Ellinwood and Duarte-Escalante, 1970a; Ellinwood, 1971a; Ellinwood et al., 1972):

    Defined as lack of proper relationship among postures or movements in two or more body regions, resulting from fragmentation of the static or kinetic configuration of the body; often appears as isolated remnant or abortive behaviors (e.g., grooming).

III. Dystonia

    (Ellinwood et al., 1972; Ellinwood and Sudilovsky, 1973; Sudilovsky et al., 1974):

    Awkward movement or posture characterized by hyperextension or hyperflexion.

IV. Twitches and Dyskinesias

    (Ellinwood and Duarte-Escalante, 1970b; Ellinwood, 1971b; Sudilovsky et al., 1974):

    e.g., torsion of tongue, blepharospasm, facial twitches.

V. Obstinate Progression

    (Ellinwood et al., 1972; Ellinwood and Sudilovsky, 1973):

    Persistent propulsion pushing against an immovable obstacle.

VI. Akathisia

    (Ellinwood and Sudilovsky, 1973; Sudilovsky et al., 1974):

    Restlessness and/or repeated weight-shifting movements with or without changing place or orientation.

VII. Ataxia

    (Ellinwood et al., 1972; Ellinwood and Sudilovsky, 1973; Sudilovsky et al., 1974):

    Inadequate control of range and precision of movement in static or kinetic state; often refers to staggering or swaying movements.

*Also noted as a late phenomenon in acute intoxication.

responsive to apomorphine. At four months, 0.5 mg/kg was ineffective, and 1.0 mg/kg induced hyperreactivity in only three cats. The reactive response to cocaine remained vigorous, often lasting over an hour.

Since hyperreactivity has been reported to respond to serotonin (5HT) manipulation (Trulson, 1976; Cools, 1974), we examined the effect of para-chlorophenylalanine (PCPA) and LSD in the chronic cocaine cats. Six cats with a daily dose of cocaine (10 mg/kg i.m.) for over eight months were administered in sequence the following drugs: Study day 1--saline; days 2-6--50, 100, 150, 150, 150 mg/kg PCPA, respectively; day 7--saline injection followed by cocaine (10 mg/kg i.m.); day 8--cocaine (10 mg/kg); day 9--LSD (100 µg/kg i.p.). Behavior was rated for one hour. In order to compare our results with those reported by Trulson (1976), occurrences of head and limb shakes, abortive grooming, and hallucinatory-like behavior were also counted for one hour after injection. The results of cocaine and serotonergic effects in chronic animals are tabulated in Table 3. It can be noted that head shake and forelimb flicks are induced by cocaine alone, and the response is not enhanced by PCPA. LSD plus PCPA induced the maximum overall response. Hindlimb flicks, abortive grooming, and hallucinatory behavior are apparently very sensitive to serotonergic manipulation, in that PCPA plus LSD induced by far the greatest rates. PCPA alone was not striking except for a large increase in hindlimb flicks in two cats. PCPA pretreatment facilitated a cocaine induction of marked dysjunctive postures and severe hindlimb dystonia. Following LSD and PCPA, two cats also aggressively attacked inanimate objects.

TABLE 3   Number of Animals Showing Serotonergic Behavioral Syndrome and the Mean* Number of Occurrences per hour Following Several Manipulations in Chronic Cocaine Cats

BEHAVIOR	TREATMENT									
	Saline		Saline + PCPA		LSD + PCPA		Cocaine		Cocaine + PCPA	
	#	$\overline{X}$	#	$\overline{X}$	#	$\overline{X}$	#	$\overline{X}$	#	$\overline{X}$
Headshakes	2/6	1	4/6	2	6/6	34	6/6	25	6/6	19
Forelimb Flicks	0/6	--	0/6	--	6/6	16	6/6	18	5/6	14
Hindlimb Flicks	2/6	2	2/6	6	6/6	29	2/6	3	6/6	3
Abortive Grooming	0/6	--	0/6	--	6/6	13	6/6	7	6/6	5
Hallucinatory Behavior (e.g., recoil)	0/6	--	0/6	--	6/6	30	3/6	1	5/6	3

* $\overline{X}$ calculated for those subjects that showed the behavior

To What Extent Are Monoamines Involved in Maintaining End-Stage Behaviors?

At the onset, the issue should be raised as to whether neurotransmitter alterations involved in the ontogeny of the end-stage behaviors are necessary for their maintenance. As we have discussed previously (Ellinwood et al., 1977), mechanisms other than neurotransmitter alterations may be involved, including conditioned or learned responses and/or electrophysiological reorganization; e.g., kindling. However, even once reorganized, manipulating neurotransmitters remains our most robust means of altering the behaviors noted in the chronic stimulant syndromes. The finding reported here of initially strong, but transient, apomorphine activation of hyperreactivity is consistent with a DA receptor supersensitivity response; however, apomorphine, when tested over a period of time, loses its activating properties. The explanation for these results may reside in a change of the balance between the post-synaptic vs. pre-synaptic efficacy of apomorphine. Initially, apomorphine may be more effective at the post-synaptic sites, gradually being overcome by the pre-synaptic inhibition of tyrosine hydroxylase. The greater efficacy of clozapine (rather than pimozide) in blocking the hyperreactivity is also difficult to explain. At the very least, neither set of data supports a simple DA receptor supersensitivity model for explaining the end-stage hyperreactive behavior maintenance, even though in the developing phase apomorphine induction of these behaviors would be consistent with this type of model. The doses of clozapine effectively blocking hyperreactivity are within the human therapeutic range; pimozide, a potent stereotypy blocker in acute stimulant intoxication, was relatively ineffective, even at very high doses. If these effects are replicable in subsequent studies, they may generate hypotheses leading to the elucidation of the mechanisms of action of the non-cataleptogenic neuroleptic. For example, in our cat model, hyperreactivity may be more sensitive to DA activity in the mesolimbic system, in that clozapine has been touted to be more effective in blocking DA receptors in this region than in the neostriatum (Anden and Stock, 1973; Zivkovic et al., 1975; Stawarz et al., 1975).

The other results of this report indicate that the main chronic cocaine reactive behaviors do not appear to be blocked or potentiated by 5HT depletion, although the spectrum of behaviors noted is the same as with LSD plus PCPA, and--furthermore--when a behavior appears under both cocaine alone or LSD plus PCPA in the same animal, it is quite similar, often with identical sequences of behavioral components. Thus, we have an intriguing similarity between behavioral manifestations, yet 5HT does not appear to have a very direct effect upon the cocaine-induced syndrome.

These results were quite surprising to us, in that there is reasonable background literature (as listed in Table 4) indicating that hyperreactivity, including startle responses and hallucinatory-like reactivity, can be manipulated by alterations of primarily 5HT and secondarily, DA and NE. Also indicated in Table 4 are other end-stage behaviors which can be augmented with neurotransmitter alterations. Insofar as hyperstartle reactivity, work in the rat has indicated 5HT, NE, and/or DA in these responses (e.g., when raphe neurons are dysfunctional following lesions or inhibition by LSD administration, startle response is augmented) (Davis and Sheard, 1974a, b; Aghajanian et al., 1970). Apomorphine produces a biphasic response in which augmentation of startle occurs immediately after administration followed by a marked inhibition (Davis and Aghajanian, 1976). This response may be very similar to the transient reactivity in our chronic cocaine animals when apomorphine is administered. NE has been demonstrated to be involved in inhibiting startle reactivity in that amine-depleting doses of reserpine enhance the startle reaction and the NE receptor stimulating agent, clonodine, produces a marked reduction in startle reactivity in the reserpinized subjects, even below the normal control levels (Fechter, 1974). In contrast, apomorphine in a moderate dose does not reverse the response.

Hyperreactivity is an overriding phenomenon involved in, or related to, several of the behavioral patterns, and is an important aspect of the final expression of the chronic stimulant intoxication syndrome. Reactivity has been indicated in several areas of neuropharmacology, including LSD or other hallucinogenic induction of aggressiveness which many investigators agree is secondary to the enhancement of reactivity or responsiveness, rather than from a specific effect on aggressive patterned behaviors (Miczek and Barry, 1976). Both 5HT and DA appear to play a central role in modulating reactivity. There are paradoxes involved, in that DA stimulation in the acute situation leads to stereotyped behavior during which reactivity to environmental stimuli is markedly reduced; however, in the chronic stimulant syndrome, it would appear that both DA and 5HT (with a contribution from NE) are involved in its modulation (see Table 4). Increasingly, experiments in neuropharmacology have begun to take into consideration the interrelationships between neurotransmitters; the functional link between 5HT and DA, including the potential circuits involved, has been addressed several times recently (Cools and Janssen, 1974; Grabowska et al., 1976; Wang and Aghajanian, 1977).

From the literature, serotonin and dopamine would appear to have reasonably robust interaction. Imipramine pretreatment intensifies the apomorphine-induced gnawing stereotypy, and this response is inhibited by methysergide or PCPA pretreatment (Dadkar et al., 1976). Cools and Janssen (1974) have also demonstrated that reversible raphe lesions induced by local application of procaine block apomorphine-induced stereotypy, both when apomorphine is instilled into the caudate as well as when given systemically. In an analagous vein, raphe lesions are capable of blocking neuroleptic actions; PCPA treatment or raphe lesions block the cataleptic effects of haloperidol treatment in rats (Gumulka et al., 1973). These behavioral findings are all in general agreement with McKenzie's (1973) earlier observation that electrolytic lesions in the raphe block apomorphine-induced aggression and gnawing.

Westerink and Korf (1975) have demonstrated that PCPA and raphe lesions inhibit the haloperidol-induced HVA elevation. Whereas compounds which inhibit tyrosine hydroxylase, thus DA synthesis, increase neostriatal 5HT content (Kitsikis and Roberge, 1973), compounds which inhibit dopamine-beta-hydroxylase involved in the synthesis of NE, decrease 5HT turnover. The neuroleptic-induced HVA elevation has been suggested to be modulated via a striatal-pallido-nigral gabaminergic circuit (Lahti and Losey, 1974). On the basis of their cannula studies, Cools and Janssen (1974, 1976) have postulated that DA in the extrapyramidal system is functionally linked in circuits (perhaps in series) involving both gamma-amino-butyric acid (GABA) and 5HT, and that these interfaces are essential for the

TABLE 4 Manipulations Enhancing Chronic Stimulant End-Stage Behaviors

I.   HYPERREACTIVITY

    A. Startle Response

        1) Raphe lesions (Davis and Sheard, 1974a)

        2) Low doses of LSD or dimethyltryptamine (DMT) which selectively inhibit 5HT neurons (Davis and Sheard, 1974b)

        3) Functional DA increase or NE decrease (Davis and Aghajanian, 1976; Fechter, 1974)

    B. Hallucinatory-like

      Recoil:

        1) Chronic depletion of 5HT (Dement et al., 1969)

        2) Chronic amphetamine effect increased by NE depletion (Ellinwood and Sudilovsky, 1973)

      Limb shake; head shake:

        1) 5HT depletion, LSD, psilocybin (Trulson, 1976)

        2) 5HT instilled into anterioventral--or DA into rostromedial--caudate (Cools, 1972, 1974)

        3) 5HT depletion enhances only hind limb flicks, and hallucinatory behavior with chronic cocaine (This report)

II.   DYSJUNCTIVE BEHAVIOR

    A. Abortive Grooming

        1) Chronic PCPA depletion of 5HT, low doses of LSD (Trulson, 1976)

        2) 5HT into anterioventral caudate (Cools, 1974)

        3) Superior colliculus lesion with 5HT depletion (Randall and Trulson, 1974)

    B. Repeated Behavioral Remnant

        1) 5HT into anterioventral caudate (Cools, 1973)

III. DYSTONIC POSTURE AND OBSTINATE MOVEMENT

    A. Camel-back Posture

        1) LSD (Adey, 1962)

        2) Mescaline (DeJong, 1945)

        3) Functional NE decrease plus amphetamine (Ellinwood and Sudilovsky, 1973)

    B. Obstinate Progression

        1) Instillation of 5HT into anterioventral caudate (Cools, 1973)

IV.   TWITCHES AND DYSKINESIAS

        1) Instillation of DA into rostromedial--or 5HT into anterioventral--or acetylcholine (ACh) into anterioventral--caudate (Cools, 1974)

        2) Suprahyoidal muscle twitch; central administration of DA or 5HT (Bierger, 1972)

behavioral expression of DA-mediated responses. In keeping with this view are reports by Grabowska et al. (1973, 1976) demonstrating that not only neuroleptics but also mesencephalic-diencephalic transection blocks the augmentation of 5HT and 5-hydroxyindoleacetic acid (5-HIAA) in the mesencephalon induced by apomorphine, indicating that a functional circuit is critical rather than a direct action of apomorphine on the 5HT neurons. More recently, Wang and Aghajanian (1977) have described a forebrain circuit via the habenula to the raphe serotonergic cells. They have presented evidence that GABA is the linking inhibitory neurotransmitter in the habenula.

Thus, in summary, we think it is important to consider that the end-stage behaviors produced by chronic stimulant intoxication provide a range of behaviors that demonstrate functional interaction between several neurotransmitters. With chronic intoxication, the neurotransmitter-induced behavioral expressions are not simply determined (e.g., DA induction of stereotypy). One must consider receptor supersensitivity, including 5HT receptor supersensitivity (Klawans et al., 1975), pre- and post-synaptic receptor balances, as well as the functional feedback between neurotransmitter systems. Examination of the time course of drug actions--both short- and long-term--thus becomes doubly important when one is dealing with feedback metabolic inhibition or issues of supersensitivity, and may in part explain some of the paradoxical results in the literature. Consideration of these chronic stimulant end-stage behaviors on a base of interacting neurotransmitters over time certainly makes categorization of specific effects more difficult, but the complexity represented may be more analagous to the psychotic syndromes in humans than the simple one-neurotransmitter hypotheses that were more recently in vogue.

SUMMARY

Among experimental animal models of human psychopathology, stimulant-induced model psychosis is unique in that there exists an analogous robust human model (i.e., amphetamine-induced psychosis). Thus, investigators have in recent years regarded the observation and comparison of effects of stimulant administration in man and experimental animals as a heuristically worthwhile endeavor. Work with acute stimulant administration induction of stereotypy in animals has led to formulation of the current "DA hypothesis;" associated hypotheses point to DA depletion and DA receptor supersensitivity as possible factors in the effects seen after chronic high-dose administration and after readministration following a period of abstinence, respectively. There are no human data which directly support these hypotheses, however.

Practical and other considerations (e.g., the homogeneity of stimulant effects in rats) suggest that the more complex chronic stimulant-induced abnormal behavioral syndrome in higher animals--e.g., in the cat--may provide a particularly useful animal model of psychosis. In the authors' laboratory, cats were employed in recent experiments which were designed to assess the effect of various neurotransmitter-availability manipulations on stimulant-induced behaviors. The preliminary indications are that, in addition to dopamine, other neurotransmitter interactions may be involved in the abnormal behaviors induced by chronic administration of stimulant drugs.

ACKNOWLEDGEMENTS

Animals were cared for in accordance with federal guidelines and regulations. This work was supported by grant DA 00057 from the National Institute on Drug Abuse.

REFERENCES

Adey, W. R., Bell, B. R., and Dennis, B. J., Effects of LSD-25, psilocybin, and psilocin on temporal lobe EEG patterns and learned behavior in the cat, Neurology 12, 591 (1962).

Aghajanian, G. K., Foote, W. E., and Sheard, M. H., Action of psychotogenic drugs on single midbrain raphe neurons, J. Pharmacol. Exp. Therap. 171, 178 (1970).

Anden, N.-E., and Stock, G., Effect of clozapine on the turnover of dopamine in the corpus striatum and in the limbic system, J. Pharm. Pharmacol. 25, 346 (1973).

Angrist, B., and Gershon, S., Dopamine and psychotic states: Preliminary remarks, in: Usdin, E., Ed. (1974) Neuropsychopharmacology of Monoamines and Their Regulatory Enzymes, Raven Press, New York, p. 211.

Bahnsen, P., Jacobsen, E., and Thesliff, H., The subjective effect of beta-phenylisopropylaminsulfate on normal adults, Acta Med. Scand. 97, 89 (1938).

Beamish, P., and Kiloh, L. G., Psychosis due to amphetamine consumption, J. Ment. Sci. 106, 337 (1960).

Bell, D. S., The experimental reproduction of amphetamine psychosis, Arch. Gen. Psychiat. 29, 35 (1973).

Bieger, D., LaRochelle, L., and Hornykiewicz, O., A model for the quantitative study of central dopaminergic and serotonergic activity, Eur. J. Pharmacol. 18, 128 (1972).

Bleuler, E. (1924) Textbook of Psychiatry, MacMillan, New York.

Burt, D. R., Creese, I., and Snyder, S. H., Antischizophrenic drugs: Chronic treatment elevates dopamine receptor binding in brain, Science 196, 326 (1977).

Carlsson, A., Amphetamine and brain catecholamines, in: Costa, A., and Garatini, S., Eds. (1970) Amphetamines and Related Compounds, Raven Press, New York, p. 289.

Castellani, S., Ellinwood, E. H., Jr., and Kilbey, M. M., Behavioral analysis of chronic cocaine intoxication in the cat (submitted for publication).

Connell, P. H. (1958) Amphetamine Psychosis, Chapman and Hill, London.

Cools, A. R., Athetoid and choreiform hyperkinesias produced by caudate application of dopamine in cats, Psychopharmacologia 25, 229 (1972).

Cools, A. R., Serotonin: A behaviourally active compound in the caudate nucleus of cats, Israel J. Med. Sci. 9, 5 (1973).

Cools, A. R., The transsynaptic relationship between dopamine and serotonin in the caudate nucleus of cats, Psychopharmacologia 36, 17 (1974).

Cools, A. R., and Janssen, H. J., The nucleus linearis intermedius raphe and behavior evoked by direct and indirect stimulation of dopamine sensitive sites within the caudate nucleus of cats, Eur. J. Pharmacol. 28, 266 (1974).

Cools, A. R., and Janssen, H. J., Gamma-amino-butyric acid: The essential mediator of behaviour triggered by neostriatially applied apomorphine and haloperidol, J. Pharm. Pharmacol. 28, 70 (1976).

Dadkar, N. K., Dohadwalla, A. N., and Bhattacharya, B. J., The involvement of serotonergic and noradrenergic systems in the compulsive gnawing in mice induced by imipramine and apomorphine, J. Pharm. Pharmacol. 28, 68 (1976).

Davis, M., and Aghajanian, G. K., Effect of apomorphine and haloperidol on the acoustic startle response in rats, Psychopharmacology 47, 217 (1976).

Davis, M., and Sheard, M. H., Effects of lysergic acid diethylamide (LSD) on habituation and sensitization of the startle response in the rat, Pharmacol. Biochem. Behav. 2, 675 (1974a).

Davis, M., and Sheard, M. H., Biphasic dose response effects of N,N-dimethyltryptamine on the rat startle reflex, Pharmacol. Biochem. Behav. 2, 827 (1974b).

DeJong, H. (1945) Experimental Catatonia, William and Wilkins, Baltimore.

Dement, W., Zarcone, V., Ferguson, J., Cohen, H., Pivik, T., and Barchas, J., Some parallel findings in schizophrenic patients and serotonin-depleted cats, in: Sankar, D. V. S., Ed. (1969) Schizophrenia: Current Concepts and Research, P.J.D., Hicksville, N.Y.

Domino, E. F., Indole hallucinogens as animal models of schizophrenia, in: Serban, G., and Kling, A., Eds. (1976) Animal Models in Human Psychology, Plenum Press, New York, p. 187.

Ellinwood, E. H., Jr., Amphetamine psychosis I: Description of the individuals and process, J. Nerv. Ment. Dis. 144, 273 (1967).

Ellinwood, E. H., Jr., Amphetamine psychosis: A multi-dimensional process, Seminars in Psychiat. 1, 208 (1969).

Ellinwood, E. H., Jr., Effects of chronic methamphetamine intoxication in rhesus monkeys, Biol. Psychiat. 3, 25 (1971a).

Ellinwood, E. H., Jr., Chronic amphetamine intoxication in several experimental animals, Psychopharmakologie 4, 351 (1971b).

Ellinwood, E. H., Jr., Accidental conditioning with chronic methamphetamine intoxication: Implications for a theory of drug habituation, Psychopharmacologia 21, 131 (1971c).

Ellinwood, E. H., Jr., Amphetamine psychosis: Individuals, settings, and sequences, in: Ellinwood, E. H., Jr., and Cohen, S., Eds. (1972) Current Concepts on Amphetamine Abuse, U.S. Government Printing Office, Washington, p. 143.

Ellinwood, E. H., Jr., Amphetamine and stimulant drugs, in: (1973) Drug Use in America: Problem in Perspective (Second Report of the National Commission on Marihuana and Drug Abuse), U.S. Government Printing Office, Washington, p. 140.

Ellinwood, E. H., Jr., and Duarte-Escalante, O., Chronic amphetamine effect on the olfactory forebrain, Biol. Psychiat. 2, 189 (1970a).

Ellinwood, E. H., Jr., and Duarte-Escalante, O., Behavioral and histopathological findings during chronic methedrine intoxication, Biol. Psychiat. 2, 27 (1970b).

Ellinwood, E. H., Jr., and Kilbey, M. M., Species differences in response to amphetamine, in: Eleftheriou, B., Ed. (1975) Psychopharmacogenetics, Plenum Press, New York.

Ellinwood, E. H., Jr., and Kilbey, M. M., Stimulant abuse in man: The use of observational animal models to assess and predict behavioral toxicity, in: Thompson, T., and Unna, K., Eds. (in press) Predictions of Abuse Liability of Stimulants and Depressant Drugs, Academic Press, New York.

Ellinwood, E. H., Jr., and Sudilovsky, A., Chronic amphetamine intoxication: Behavioral model of psychoses, in: Cole, J. O., Freedman, A. M., and Friedhoff, A. J., Eds. (1973) Psychopathology and Psychopharmacology, Johns Hopkins University Press, Baltimore, p. 51.

Ellinwood, E. H., Jr., Sudilovsky, A., and Nelson, L., Behavioral analysis of chronic amphetamine intoxication, Biol. Psychiat. 4, 215 (1972).

Ellinwood, E. H., Jr., Stripling, J. S., and Kilbey, M. M., Chronic changes with amphetamine intoxication: Underlying processes, in: Usdin, E., Hamburg, D. A., and Barchas, J. D., Eds. (1977) Neuro-Regulators and Psychiatric Disorders, Oxford University Press, New York, p. 578.

Fechter, L. D., The effects of L-DOPA, clonodine, and apomorphine on the acoustic startle reaction in rats, Psychopharmacologia 39, 331 (1974).

Fischman, M. W., Schuster, C. R., and Krasnegor, N. A., Physiological and behavioral effects of intravenous cocaine in man, in: Ellinwood, E. H., Jr., and Kilbey, M. M., Eds. (1977) Cocaine and Other Stimulants, Plenum Press, New York, p. 647.

Flory, C. D., and Gilbert, J., The effects of benzedrine sulfate and caffeine citrate on the efficiency of college students, J. Appl. Psychol. 27, 121 (1943).

Grabowska, M., Antiewiez, J., May, J., and Michaluk, J., Apomorphine and central serotonin neurons, J. Pharm. Pharmacol. 25, 29 (1973).

Grabowska, M., Przewlocki, R., and Smialowska, M., On the direct or indirect influence of apomorphine in central serotonin neurons, J. Pharm. Pharmacol. 28, 64 (1976).

Gumulka, W. W., Kostowski, W., and Czlonkowski, A., Role of 5HT in the action of some drugs affecting extrapyramidal system, Pharmacology 10, 363 (1973).

Hinde, R. A., The use of differences and similarities in comparative psychopathology, in: Serban, G., and Kling, A., Eds. (1976) Animal Models in Human Psychology, Plenum Press, New York, p. 187.

Janowsky, D. S., El-Yousef, M. K., and Davis, J. M., Provocation of schizophrenic symptoms by intravenous administration of methylphenidate, Arch. Gen. Psychiat. 28, 185 (1973).

Janowsky, D. S., Huey, L., and Storms, L., Psychologic test responses to methylphenidate, in: Ellinwood, E. H., Jr., and Kilbey, M. M., Eds. (1977) Cocaine and Other Stimulants, Plenum Press, New York, p. 675.

Kitsikis, A., and Roberge, A. G., Behavioral and biochemical effects of alpha-methyltyrosine in cats, Psychopharmacologia 31, 143 (1973).

Klawans, H. L., Jr., Goetz, C., and Westheimer, R., Pathophysiology of schizophrenia and the striatum, Dis. Nerv. Syst. 33, 711 (1972).

Klawans, H. L., D'Amico, B. J., and Patel, D. C., Behavioral supersensitivity to 5-hydroxy-tryptophan induced by chronic methysurgide pretreatment, Psychopharmacologia 44, 297 (1975).

Kramer, J. C., Introduction to amphetamine abuse, J. Psychedelic Drugs 2, 1 (1969).

Kramer, J. C., Fischman, V. S., and Littlefield, D. C., Amphetamine abuse--pattern and effect of high doses taken intravenously, J.A.M.A. 201, 305 (1967).

Lahti, R. A., and Losey, E. G., Antagonism of the effect of chlorpromazine and morphine on dopamine metabolism by GABA, Res. Comm. Pathol. Pharmacol. 7, 31 (1974).

Mattson, R. H., and Calverley, J. R., Dextroamphetamine-sulfate-induced dyskinesias, J.A.M.A. 204, 400 (1968).

Matthyse, S., Dopamine and the pharmacology of schizophrenia: The state of the evidence, J. Psychiat. Res. 11, 107 (1974).

Matthyse, S., The role of dopamine in schizophrenia, in: Usdin, E., Hamburg, D. A., and Barchas, J. D., Eds. (1977) Neuro-Regulators and Psychiatric Disorders, Oxford University Press, New York, p. 3.

Meltzer, H. Y., and Stahl, S. M., The dopamine hypothesis of schizophrenia: A review, Schizophrenia Bull. 2, 19 (1976).

Miczek, C. A., and Barry, H., Pharmacology of sexual aggression, in: Glick, S., and Goldfarb, J., Eds. (1976) Behavioral Pharmacology, C.R. Mosley, St. Louis, p. 176.

Morrison, J. R., Catatonic, retarded and excited types, Arch. Gen. Psychiat. 28, 39 (1973).

Randall, W., and Trulson, M., A serotonergic system involved in the grooming behavior of cats with pontill lesions, Pharmacol. Biochem. Behav. 2, 355 (1974).

Resnick, R. B., Kestenbaum, R. S., and Schwartz, L. K., Acute systemic effects of cocaine in man: A controlled study of intranasal and intravenous routes of administration, in: Ellinwood, E. H., Jr., and Kilbey, M. M., Eds. (1977) Cocaine and Other Stimulants, Plenum Press, New York, p. 615.

Rickman, E. E., Williams, E. Y., and Brown, R. K., Acute toxic psychiatric reaction related to amphetamine medication, Med. Ann. D.C. 30, 209 (1961).

Rutledge, C. O., Azzaro, A. J., and Ziance, R. J., Dissociation of amphetamine-induced release of norepinephrine from inhibition of neuronal uptake in isolated brain tissue, in: Usdin, E., and Snyder, S. H., Eds. (1973) Frontiers in Catecholamine Research, Pergamon Press, New York, p. 973.

Rylander, G., Clinical and medical criminal aspects of addiction to central stimulating drugs, in: Sjoqvist, F., and Tottie, M., Eds. (1969) Abuse of Central Stimulants, Raven Press, New York, p. 251.

Schiorring, E., Changes in individual and social behavior induced by amphetamine and related compounds in monkey and man, in: Ellinwood, E. H., Jr., and Kilbey, M. M., Eds. (1977) Cocaine and Other Stimulants, Plenum Press, New York, p. 481.

Seiden, L. S., Fischman, M. W., and Schuster, C. R., Changes in brain catecholamine induced by long-term methamphetamine administration in rhesus monkeys, in: Ellinwood, E. H., Jr., and Kilbey, M. M., Eds. (1977) Cocaine and Other Stimulants, Plenum Press, New York, p. 179.

Serban, G., Pichot, P., Freedman, A. F., and Kittay, S., New perspectives in psychiatry: Relevance of the psychopathological animal model to the human, in: Serban, G., and Kling, A., Eds. (1976) Animal Models in Human Psychobiology, Plenum Press, New York, p. 1.

Snyder, S. H., Catecholamines in the brain as mediators of amphetamine psychosis, Arch. Gen. Psychiat. 27, 169 (1972).

Snyder, S. H., Aghajanian, G. K., and Matthysse, S., Drug-induced psychoses, Neurosci. Res. Prog. Bull. 10, 430 (1972).

Snyder, S. H., Banerjee, S. P., Yamamura, H. I., and Greenburg, D., Drugs, neurotransmitters, and schizophrenia, Science 184, 1243 (1974).

Stawarz, R. J., Hill, H., Robinson, S. E., Sittler, P., Dingell, J. V., and Sulser, F., On the significance of the increase in homovannilic acid (HVA) caused by antipsychotic drugs in corpus striatum and limbic forebrain, Psychopharmacologia 43, 125 (1975).

Sudilovsky, A., Nelson, L., and Ellinwood, E. H., Jr., Amphetamine dyskinesia: A direct observational study in cats, in: Singh, J. M., and Lal, H., Eds. (1974) Drug Addiction (Vol. 4), Stratton Intercontinental Medical Book Corp., New York, p. 17.

Sudilovsky, A., Ellinwood, E. H., Jr., Dorsey, F., and Nelson, L., Evaluation of the Duke University Behavior Rating Inventory for Drug-Generated Effects (BRIDGE), in: Sudilovsky, A., Gershon, S., and Bier, B., Eds. (1975) Predictability in Psychopharmacology: Preclinical and Clinical Correlations, Raven Press, New York, p. 189.

Tatetsu, S., Goto, A., and Fujiwara, T. (1956) The Awakening Drug Intoxication, Igaku Shoin, Tokyo.

Tecce, J. J., and Cole, J. O., Amphetamine effects in man: Paradoxical drowsiness and lowered electrical brain activity (CNV), Science 185, 451 (1974).

Trulson, M. E., Biological bases for the integration of appetitive and consummatory grooming behaviors in the cat: A review, Pharmacol. Biochem. Behav. 4, 329 (1976).

Utena, H., On relapse-liability; Schizophrenia, amphetamine psychosis and animal model, in: Mitsuda, H., and Fukuda, T., Eds. (1974) Biological Mechanisms of Schizophrenia and Schizophrenia-Like Psychoses, Igaku Shoin, Tokyo, p. 285.

Wallach, M. B., and Gershon, S., The induction and antagonism of central nervous system stimulant-induced stereotyped behavior in the cat, Eur. J. Pharmacol. 18, 22 (1972).

Wang, R. Y.., and Aghajanian, G. K., Physiological evidence for habenula as major link between forebrain and midbrain raphe, Science 197, 89 (1977).

Westerink, B. H. C., and Korf, J., Influence of drugs on striatal and limbic homovannilic acid concentration in the rat brain, Eur. J. Pharmacol. 33, 31 (1975).

Young, D., and Scoville, W. B., Paranoid psychosis in narcolepsy and the possible dangers of benzedrine treatment, Med. Clin. D.A. (Boston) 22, 637 (1938).

Zivkovic, B., Guidotti, A., Revvelta, A., and Costa, E., Effect of thioridazine, clozapine and other antipsychotics on the kinetic state of tyrosine hydroxylase and on the turnover rate of dopamine in striatum and nucleus accumbens, J. Pharmacol. Exp. Therap. 194, 37 (1975).

ANIMAL MODELS OF HUMAN COGNITIVE PROCESSES

Steven Matthysse

Laboratories for Psychiatric Research
Mailman Research Center, McLean Hospital, Belmont, Massachusetts

The tests used to screen antipsychotic drugs bear remarkably little resemblance to the phenomena of psychosis. Drugs are selected which induce cataleptic states; impair the ability to stay on a rotating rod or a tight rope; lower rectal temperature; cause drooping eyelids; potentiate barbiturate-induced sleeping time; inhibit locomotor exploration (1). The complete lack of relationship between these effects and human mental states makes them unconvincing as screening tests, even though they do correlate in a general way with the potency of established antipsychotic drugs. A number of tests are based on catecholamine, especially dopamine antagonism: reduction of catecholamine toxicity, inhibition of apomorphine-induced emesis, blockade of stereotypy and agitation produced by amphetamine or apomorphine. Even stereotypy blockade, which is perhaps the most successful of these screening tests, has substantial limitations. Thioridazine and clozapine are effective antipsychotic drugs with only weak stereotypy blocking effects. These two drugs also have relatively less tendency than the other antipsychotics to produce Parkinsonian side effects. Curiously, the stereotypy test seems to be a better predictor of motor side effects than of antipsychotic actions! Reliance on it may encourage the production of still more drugs with extrapyramidal effects. Nevertheless, there is a grain of truth in the stereotypy concept, since schizophrenic patients do show stereotypy in behavior, ideation and attention. Mental stereotypy, like motor stereotypy, may be related to dopamine (2). One wonders: could tests be designed for mental, rather than motor stereotypy which might differentiate therapeutic effects from side effects?

My general thesis is that it is possible to study many of the cognitive processes we customarily ascribe only to humans, in animals, at least in primates. Even some of the processes which are characteristically disturbed in schizophrenia can be observed in lower primates. Thus if we are willing to broaden our attention beyond purely linguistic phenomena, a practically unexplored world of schizophrenia-related cognitive processes accessible in primates becomes available to us. The solution of the drug screening problem might lie in this direction.

Kornetsky's Continuous Performance Test is a very successful prototype for cross-species tests of cognition. The test strikingly differentiates schizophrenics from normals, and also has a simple animal version. Reaction time is a fruitful source of paradigms for which animal analogues have not yet been constructed. Perhaps the best-established psychological "marker" for schizophrenia is the crossover effect discovered by Rodnick and Shakow (Fig. 1) (3). When the preparatory interval between the warning stimulus and the cue which triggers the reaction is held constant on successive trials ("regular series"), schizophrenics actually perform worse than on an irregular series, whereas normals improve. This paradoxical decrement can be demonstrated especially clearly by comparing successive trials within a regular series (Fig. 2) (4). Primates ought to be capable of mastering reaction time

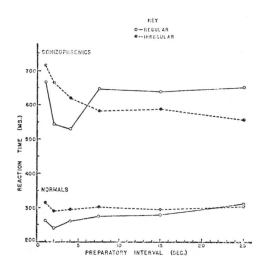

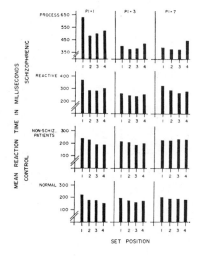

Fig. 1.   Fig. 2.

paradigms of this type; the first attempt in this direction is now being made by Carlson and Werner. A closely related deficit in schizophrenic performance can be seen in the <u>previous preparatory interval</u> paradigm. If a long preparatory interval is followed by a short one, the reaction of schizophrenic subjects is delayed, as if they were still concentrating on the previous preparatory interval (Fig. 3). Conversely, if a short preparatory interval is followed by a long one, schizophrenics make more anticipatory responses (5). A third variant is the <u>cross-modal reaction time</u> paradigm, in which the cue which triggers the response can be presented in either of two different sensory channels. After a visual cue, the response of schizophrenics to a tone is delayed, more than the response of normal subjects (Fig. 4) (6). An elegant method of studying <u>attention switching</u> has been developed by Kristofferson and Allan

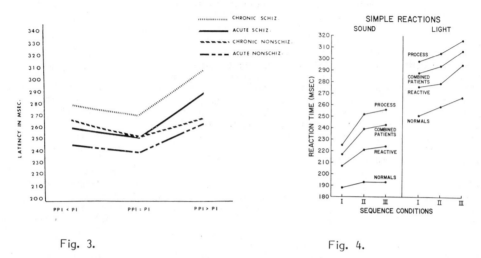

Fig. 3.   Fig. 4.

using two stimuli simultaneously presented in different channels. The subject's task is to determine which stimulus terminated first. In order to make this discrimination, the subject must switch attention from one stimulus to the other and back again, to verify whether or not the stimulus is still on. Kristofferson and Allan find that a "quantum" or minimum switching time must elapse between the termination times in order for discrimination accuracy to be better than chance (Fig. 5) (7). There is some evidence that schizophrenics have a longer "quantum" than normals (8). With careful training, monkeys ought to be capable of learning such a discrimination.

One of the most widely discussed "markers" of schizophrenic dysfunction is the defect in smooth pursuit <u>visual tracking</u> which has recently been investigated intensively by Holzman and his colleagues in schizophrenic patients and their relatives. Patients lag behind the stimulus and make saccadic movements to catch up (Fig. 6). Holzman ascribes this disturbance to "failure to maintain

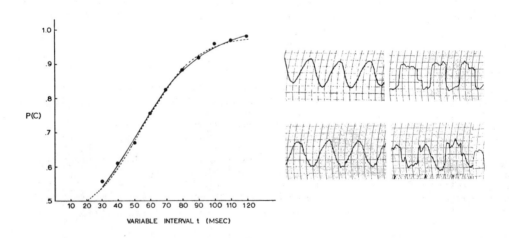

Fig. 5.   Fig. 6.

a visual-attentive focus that is continually locked onto the moving pendulum" (9). This experiment is an excellent example of how paradigms that differentiate schizophrenics from normals can be successfully adapted to primates. Schuster has been able to condition Rhesus monkeys to track a moving spot of light by making a reward contingent on rapidly responding to the dimming of the spot. The response to the dimming spot is learned with the target stationary, and the monkey then tracks the moving target without any specific training for visual pursuit. Another visual attention paradigm in which schizophrenics show deviations, although less thoroughly studied than smooth pursuit, is the scanning of a visual image. Some schizophrenics have an abnormally restricted range of fixation, narrowing their attention to one small portion of the presented stimulus (Fig. 7) (10). Since Rhesus monkeys are capable of responding appropriately to pictures, it should be possible to investigate whether visual attention is restricted by stereotypy-producing drugs such as amphetamine.

Schizophrenics are known to deviate from normals on a number of paradigms which may involve levels of information processing still more elementary than reaction time, attention switching, and visual scanning. Animal versions of these paradigms may be especially useful for neurophysiological and neuroanatomical experimentation. In the backward masking experiment, a visual stimulus is briefly exhibited, e.g., either a T or an A; after an interval a "mask" (a bright letter W) is presented in the visual field. The "mask" is thought to obliterate the visual stimulus trace, making it unavailable for subsequent encoding in verbal form. There is an interval beyond which the mask no longer impairs discrimination of the T and A, so encoding is thought to be complete at that time. The interval for complete encoding appears to be longer in schizophrenics than in normal control subjects (Fig. 8) (11).

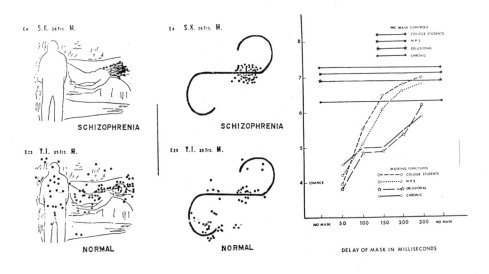

Fig. 7.   Fig. 8.

It may be that the prolonged encoding process is related to the increased minimum attention switching time ("quantum"). The backward masking paradigm is actually not quite "elementary", because performance depends on at least two factors, the persistence of the visual trace (iconic image) and the rate of verbal encoding. An elegant experiment has been designed by Knight to study iconic imagery apart from encoding, and his paradigm is particularly adaptable to primates. A fragmented stimulus is presented tachistoscopically, which subjects are unable to recognize. After an interval another fragmented stimulus is presented, which is exactly complementary in the sense that when the two stimuli are superimposed a complete, readily recognizable object appears. If, therefore, the iconic image persists for the interstimulus interval, the subject will succeed in naming the object. Stimulus pairs are chosen (Fig. 9) which are recognized by nearly all normal subjects when the inter-stimulus in-

Fig. 9.

terval is sufficiently short, but rarely when the interval is long. In this illustration the objects are flag, coat, and whale. Knight is presently comparing the persistence time of iconic images in schizophrenics and normals using this paradigm. A monkey could be conditioned to respond selectively to one of several stimulus objects, which would then be presented as pairs of fragmented images. Another "elementary" aspect of information processing which has been studied in schizophrenics is reaction time to single and double pulses of light of equal energy. Ordinarily, the response of observers to short pulses of light depends only on their energy (integrated intensity). Collins, Kietzman and Sutton have noted that schizophrenics, unlike normals, seem to react differentially to a double pulse (two 2-msec. pulses separated by a 2-msec. dark period) and a single 4-msec. pulse of equal energy. They have a longer reaction time to the double pulse (Fig. 10) (12). One attractive feature of this paradigm is that it is very hard to account for the results with schizophrenics in terms of a deficit in motivation or responsiveness, since the patients are actually making a discrimination the normals do not make. This paradigm, too, is readily adaptable to animals.

A search for other simple perceptual and cognitive tasks capable of being tested in animals, which differentiate schizophrenics from normal subjects, would be worthwhile. Although the task may not seem very closely related to the content of the psychosis, it may utilize synaptic mechanisms (specific neurotransmitters, local inhibitory circuits, etc.) which are also involved in the psychotic process. For example, local inhibitory circuits may be involved in generating the selective human visual response to certain spatial frequencies. In Fig. 11, the spatial frequency of the grid varies from left to

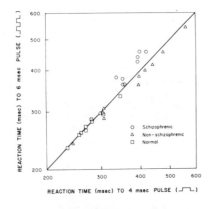

Fig. 10.   Fig. 11.

right, while along the vertical axis the contrast is uniformly decreasing. The discriminability of the grid when the contrast is low depends on spatial frequency (13). Binocular disparity as a depth cue may also have some interesting applications to schizophrenia and to animal models. The two random-dot patterns shown in Fig. 12 are identical except for a center square, which is shifted to the left by six dots in the

Fig. 12.

pattern on the right. Julesz has shown that when viewed stereoscopically, such patterns give rise to three-dimensional patterns. In this case, the square seems to float above the remainder of the pattern. Certain random-dot stereograms are ambiguous, having an equal probability of being perceived as advancing in front of or as receding behind the background. Julesz has coined the term "attention time" to refer to the tendency of an unambiguous pattern to bias the perception of a subsequent ambiguous pattern, if presented within a certain interval (14). Times measured in this way might be compared with the Kristofferson-Allan "quantum", the backward masking encoding time, or the iconic imagery decay time. Julesz has also noted the phenomenon of "hysteresis": once fused, binocular images seem to remain fused even after they have been gradually pulled apart (Fig. 13) (15); the "breakaway" disparity could be compared in schizophrenic and normal subjects.

While we are imagining experiments that could be done both by schizophrenic patients and monkeys, it is interesting to consider multistable objects. These are objects which can be perceived in several ways, and reverse spontaneously or because of the context. For example, in Fig. 14 the drawing on the right in the top row has an equal probability, when seen alone, of being perceived as a man or a girl; the actual interpretation of the figure depends on the sequence (16). Subjects who make the switch later than normal in the sequence might be said to exhibit a kind of perceptual rigidity. Primates could be trained to respond differentially to the two stimuli representing the end points of the sequence, and their switch point could similarly be determined. One wonders, for example, whether amphetamine stereotypy, at a mental level, would be manifested by persisting longer in the original interpretation of the stimulus. Paranoia could be studied in primates using a similar design. The animal would be trained to respond to pictures of threatening monkeys by pressing lever A, and to pictures of friendly monkeys by pressing lever B. After a drug or psychological manipulation, one would present neutral pictures (judged by an equal probability of A or B response by untreated animals) and look for a skewing of the response in the direction of lever A. Another variant of the procedure might be used to study cerebral dominance. In Fig. 15 one of the two faces is usually seen as happy and one as sad. Fixate on the nose of each one. Which is which? Actually they are mirror images. Apparently the affective interpretation depends on which cues are seen by the right hemisphere (17). It is not known how schizophrenics respond to these ambiguous faces. Primates could be trained to make an affective discrimination with unambiguous faces, and then tested with the asymmetric stimuli.

In conclusion, it would seem too narrow to rely exclusively on the screening tests now in common use, when there are many paradigms for assessing human mental function which can be extended to primates. Some of these paradigms show clear deviances in schizophrenia, and others would be worth testing in patient groups. Actually the usefulness of cross-species paradigms goes beyond screening tests. Insofar as schizophrenia is not a purely linguistic disorder, but involves abnormalities in more elementary perceptual, attentional and cognitive processes, such paradigms can be used to investigate the anatomical, physiological and neurochemical mechanisms underlying the disease.

80

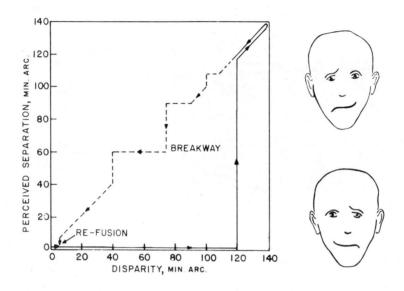

Fig. 13.   Fig. 15.

Fig. 14.

REFERENCES

1. Fielding, S. and Lal, H., Eds. (1974) Industrial Pharmacology: Neuroleptics. Futura, New York,
    pp. 53-129.

2. S. Matthysse, The role of dopamine in schizophrenia. In: Neuroregulators and Psychiatric Dis-
    orders (Usdin, E., Hamburg, D.A. and Barchas, J.D., Eds.) Oxford, New York, pp. 3-13
    (1977).

3. E. Rodnick and D. Shakow, Set in the schizophrenic as measured by a composite reaction time
    index. Am. J. Psychiat. 97, 214-225 (1940).

4. A. Bellisimo and R.A. Steffy, Redundancy-associated deficit in schizophrenic reaction time
    performance, J. Abn. Psychol. 80, 299-307 (1972).

5. R.M. Nideffer, J.M. Neale, J.H. Kopfstein, and R.L. Cromwell, The effects of previous pre-
    paratory intervals upon anticipatory responses in the reaction time of schizophrenic
    and nonschizophrenic patients, J. Nerv. Ment. Dis. 153, 360-365 (1971).

6. J. Zubin and S. Sutton, Assessment of physiological, sensory, perceptual, psychomotor, and
    conceptual functioning in schizophrenic patients, Acta Psychiat. Scand. 46, 247-263
    (1970).

7. A.B. Kristofferson and L.G. Allan, Successiveness and duration discrimination, In: Attention
    and Performance, Vol. 4 (Kornblum, S., Ed.) Academic Press, New York, pp. 737-749
    (1973).

8. L.G. Allan, The attention switching model: implications for research in schizophrenia.
    J. Psychiat. Res. (in press, 1978).

9. P.S. Holzman, D.L. Levy and L.R. Proctor, Smooth pursuit eye movements, attention and
    schizophrenia, Arch. Gen. Psychiat. 33, 1415-1420 (1976).

10. H. Moriya et al., Eye movements during perception of pictures in chronic schizophrenia,
    Folia Psychiat. et. Neurol. 26, 189-199 (1972).

11. D.P. Saccuzzo, M. Hirt and T.J. Spencer, Backward masking as a measure of attention in schizo-
    phrenia, J. Abn. Psych. 83, 512-522 (1974).

12. P.J. Collins, M.L. Kietzman, S. Sutton and E. Shapiro, Visual temporal integration in psychi-
    atric patients, J. Psychiat. Res. (in press, 1978).

13. F. Ratliff, Contour and contrast, Scientific American 226, 90-101 (1972).

14. B. Julesz, Texture and visual perception, Scientific American 212, 38-48 (1965).

15. Szentagothai, J. and Arbib, M.A., Eds. (1975) Conceptual Models of Neural Organization,
    MIT Press, Cambridge, p. 103.

16. F. Attneave, Multistability in perception, Scientific American 225, 63-71 (1971).

17. Jaynes, J. (1976) The Origin of Consciousness in the Breakdown of the Bicameral Mind.
    Houghton Mifflin, Boston, p. 120.

DISCUSSION

Dr. Hanin asked if Dr. Matthysse had considered showing monkeys pictures of the unpleasant monkey to whom they were conditioned to react and, if so, whether they recognized pictures of their enemies and friends. *Dr. Matthysse* indicated that Dr. Haber had done this and the monkeys did show such recognition.

IDENTIFYING INDOLEAMINE HALLUCINOGENS BY THEIR PREFERENTIAL ACTION
ON SEROTONIN AUTORECEPTORS

George K. Aghajanian
Yale University School of Medicine and the
Connecticut Mental Health Center, New Haven CT 06508

INTRODUCTION

Because of their structural similarity to the endogenous indoleamine serotonin (5-hydroxy-tryptamine; 5-HT), it has long been thought that indoleamine hallucinogens might act upon 5-HT receptors in the CNS (1,2,3). More recently it has been found by means of single-cell recording and microiontophoretic techniques that there are several different types of central 5-HT receptors (see Ref. 4 for review). Furthermore, there is evidence that indoleamine hallucinogens have a preferential action upon particular types of 5-HT receptors which have been termed "5-HT autoreceptors" (see below). The delineation of a selective pattern of receptor action for known indoleamine hallucinogens provides a model for the evaluation of unknowns with potential hallucinogenic activity. To lay the ground-work for such a model, we will first present evidence for the existence of 5-HT auto-receptors. Next, procedures for determining the preferential actions of known indoleamine hallucinogens on 5-HT autoreceptors (as compared to identified postsynaptic 5-HT receptors) will be described. Finally, we will describe the application of these test procedures to suspected endogenous hallucinogenic substances.

5-HT AUTORECEPTORS

The existence of 5-HT autoreceptors was first suggested by studies in which microionto-phoretically applied 5-HT was found to have a powerful direct inhibitory action on 5-HT neurons of the dorsal raphe nucleus (5,6,7). The receptors mediating this response may be termed "autoreceptors" since they mediate the response of a neuron to its own transmitter. 5-HT neurons are also extremely sensitive to the inhibitory action of LSD and other hallucinogenic indoleamines (7,8,9). In fact, 5-HT neurons are more sensitive to these compounds than are neurons receiving a 5-HT input (6,10). The preferential action of hallucinogenic drugs at presynaptic 5-HT receptors (i.e., 5-HT autoreceptors) is a distinguish-ing feature of this class of compounds (see below). It should be noted that by directly inhibiting 5-HT neurons, LSD or other hallucinogens would impair 5-HT release and therefore transmission (11).

Two possible physiological functions can be suggested for 5-HT autoreceptors. The auto-receptor could reflect the presence of 5-HT axon collaterals within or between the various raphe nuclei. Another possibility is that the entire presynaptic membrane, including that of raphe terminals, is sensitive to 5-HT and structurally related indoleamines such as LSD. This sensitivity could serve as part of a local feedback mechanism at the terminal. The block by LSD in 5-HT release from isolated brain slices (12,13) supports the possibility of the existence of such a receptor-mediated feedback mechanism for the regulation of 5-HT release.

Recently both anatomical and physiological evidence has been presented in favor of the concept that 5-HT neurons receive a 5-HT synaptic input. After the intraventricular in-jection of large doses of 5,7-DHT, degenerating 5-HT terminals have been found over 5-HT cell bodies and dendrites (14). These presumably represent the terminals of 5-HT axon collaterals. In addition, apparent connections between different groups of raphe cells have also been detected by means of retrograde tracing methods (15, 16). Direct physiolog-ical evidence for recurrent or collateral inhibition in the raphe system has been provided by studies in which the major ascending 5-HT pathway at the level of ventromedial tegmentum (VMT) was stimulated electrically during the recording of 5-HT cells in the raphe (17,18). Raphe cell with a characteristic slow, regular firing pattern (5,6,7,8,9,10,11,19,20) can be antidromically driven by stimulating the 5-HT pathway at the level of VMT. The anti-dromic activation of 5-HT neurons is associated with a period of post stimulus suppression in spontaneous activity. There are several lines of evidence which indicate that this inhibition probably is mediated through 5-HT-axon collaterals (18). The evidence may be summarized as follows: 1) after VMT-stimulation, in addition to the expected antidromic responses, 5-HT neurons uniformly show a period of post stimulus depression in firing; the duration of the depression is linearly related to the stimulus intensity; 2) the latency of the latter depression is approximately the same as the antidromic action potentials; 3) iontophoresis of 5-HT directly onto these identified 5-HT-containing cells invariably produces inhibition; and 4) the post-stimulus depression is potentiated by iontophoretic application of a 5-HT uptake blocker (chlorimiphramine) but not by a norepinephrine uptake blocker (desipramine). Taken together, the above results suggest that 5-HT autoreceptors serve the physiological role of mediating collateral inhibition between 5-HT neurons.

PREFERENTIAL ACTION OF KNOWN INDOLEAMINE HALLUCINOGENS ON 5-HT AUTORECEPTORS

A series of hallucinogenic indoleamines and related substances have been tested for possible differential actions on 5-HT autoreceptors (i.e., presynaptic receptors) and post-synaptic 5-HT receptors. As previously described, small intravenous doses of d-lysergic acid diethylamide (LSD) produce an inhibition of firing of 5-HT neurons in the dorsal midbrain and other brain stem raphe nuclei. Subsequent studies have shown that LSD has a powerful direct inhibitory action upon 5-HT neurons in the raphe when applied by micro-iontophoresis (5,6,7). The inhibition of 5-HT neurons by LSD is more pronounced than its inhibitory effect upon neurons receiving an identified serotonergic input in several visual and limbic areas (6). In contrast, 5-HT itself shows no such preferential activity. By inhibiting 5-HT neurons directly, LSD could release postsynaptic neurons from a tonic inhibitory 5-HT influence. Such a release might account for the sensory, cognitive and other disturbances caused by LSD. Three simple indoleamine hallucinogenic drugs, psilocin (4-hydroxy-N,N-dimethyltryptamine), DMT (N,N-dimethyltryptamine) and bufotenine (5-hydroxy-N,N-dimethyltryptamine) have also tested by microiontophoretic techniques for their relative potencies in affecting 5-HT neurons in the dorsal raphe nucleus and neurons in postsynaptic areas (10). In addition, the primary amine analogues of each drug (i.e., 4-hydroxytryptamine (4-HT), tryptamine (T), and 5-HT) were tested to evaluate possible differential effects of N-methyl and ring hydroxyl substitutions on pre- and postsynaptic potency. Psilocin and DMT were found to have a preferential inhibitory action upon 5-HT neurons as compared to postsynaptic neurons in the VLG and amygdala. In this respect, these two drugs are similar to LSD. On the other hand, there is only a slight difference between the pre- and post-synaptic actions of bufotenine. In general, these results correspond well with the known hallucinogenic (or psychotogenic) actions of these compounds. Except for LSD and some of its congeners, psilocin is the most potent of the indoleamine hallucinogens that have been tested in human subjects (21). The effects of another powerful indoleamine hallucinogen, psilocybin (4-phosophoryl-N,N-dimethyltryptamine) are most probably mediated through its rapid metabolic conversion to psilocin (22). The psychological effects of DMT are similar to those of LSD and psilocin except for its lesser potency (23,24). Bufotenine, on the other hand, has little hallucinogenic action even at high doses (25,26). Perhaps part of the reason for bufotenine's low hallucinogenic activity is the fact that it penetrates the brain poorly (27). On the other hand, it is also possible that bufotenine has low activity as an hallucinogen because it differentiates less well between pre- and postsynaptic serotonergic sites than do psilocin and DMT.

From a structural standpoint it is of interest that, next to LSD, psilocin best discriminates between 5-HT autoreceptors and postsynaptic 5-HT receptors. Psilocin has a ring substitution on the "4" position of the indole ring, the same position in which LSD is substituted. Furthermore, conformational and molecular orbital analysis indicates that psilocin can approximate the "A," "B "and "C" rings of LSD (28,29). Thus, the 4-substitution in psilocin and the analogous substitution in LSD may be of significance in accounting for both hallucinogenic activity and preferential action at 5-HT autoreceptors. DMT, which lacks a ring hydroxylation, also shows a preferential action at 5-HT autoreceptors. T lacks this preferential effect, suggesting that the presence of N-methyl groups account for DMT's greater potency at the presynaptic site. However, in view of bufotenine's minimal differential action, it appears that hydroxylation at the "5" position can in part cancel out the effect of N-methylation. Interestingly, 4-hydroxylation even in the absence of N-methylation confers some preferential presynaptic action in the case of 4-HT. With the addition of N,N-dimethylation in psilocin, the action upon 5-HT autoreceptors is intensified. Thus, it would appear that both the position of the indole ring hydroxylation as well as presence of side chain N-methylation can influence preferential activity at pre- versus postsynaptic 5-HT receptors.

In conclusion, these results are consistent with the hypothesis that LSD and indoleamine hallucinogens such as psilocin and DMT produce their hallucinogenic effects by acting preferentially upon 5-HT autoreceptors located on or near the soma of 5-HT neurons in the raphe nuclei. The hallucinogens appear to mimic 5-HT effectively at 5-HT autoreceptors but are relatively ineffective in mimicing 5-HT at postsynaptic receptors.

ENDOGENOUS INDOLEAMINE HALLUCINOGENS AND 5-HT AUTORECEPTORS

The finding that known hallucinogenic indoleamines can be identified by their preferential action on 5-HT autoreceptors suggests that the same identification procedure can be used to screen endogenous substances for potential hallucinogenic activity.

For many years, there has been speculation about the possibility that psychotogenic (hallucinogenic) indoleamines can be formed in the body (see Ref. 30 for review). DMT is the compound which has attracted most of the attention of workers in the field. In 1962, Axelrod (31) found an enzyme in the lung of rabbits that is capable of N-dimethylating tryptamine to form DMT. The psychotomimetic potency of DMT has been amply confirmed in clinical studies (26,32,33,34). However, doubt has been cast on the proposal that DMT is formed enzymatically by mammalian brain (35,36). Furthermore, it remains uncertain whether its metabolism differs in psychiatric patients as compared to normals (see Refs. 34, 37 for review).

5-Methoxytryptamine (5-MT) and 5-methoxy-N,N-dimethyltryptamine (5-MDMT) are also candidates as endogenous psychotogens. There is some anecdotal evidence that 5-MDMT has powerful psychotomimetic effects in humans (38) but controlled studies remain to be done (39). The presence of 5-MDMT has been reported in blood and urine of acute schizophrenic patients (40,41). Moreover, it has been reported that 5-MDMT can be formed in vitro by human pineal gland (42). 5-MT has not been tested in humans but it has been shown to disrupt behavior in all animal species where it has been tested (43,44,45,56,47). Its effects are greatly enhanced by monoamine oxidase inhibitors and at least some of these effects appear to be mediated centrally. 5-MT has been identified with gas chromatography - mass spectrometry in the rat hypothalamus (48,49). Its concentration was not significantly altered by chronic pinealectomy suggesting that its presence in the hypothalamus is not dependent on its synthesis in the pineal gland. Elevated levels of 5-MT in the CSF of manic and acute schizophrenic patients have been reported (50,51).

In our most recent single-cell studies we have investigated the relative potencies of 5-MT and 5-MDMT on 5-HT autoreceptors and post-synaptic 5-HT receptors (52). As before, these compounds were applied microiontophoretically to 5-HT neurons of the midbrain raphe and to neurons of the VLG and amygdala. Both 5-MT and 5-MDMT were found to exert a more powerful depressive effect on presynaptic than postsynaptic neurons; in this respect these compounds resemble LSD and other hallucinogenic indoleamines. The preferential presynaptic action of 5-MDMT was more pronounced than that of 5-MT: their respective pre- to postsynaptic ratios were 4.3 and 1.8 (ratio for LSD was 5.6). Since the two compounds seemed to be approximately equipotent when applied to the presynaptic 5-HT-containing neurons, the greater than two-fold difference in their ratios is attributable to the greater effect of 5-MT on postsynaptic units. In the case of 5-MDMT, the very high ratio found in the present study is consistent with its reported hallucinogenic properties (38). Although the ratio of 5-MT pre- and postsynaptic potencies is lower than that of LSD and 5-MDMT, it is still greater than that of 5-HT. On this basis, it might be predicted that this compound should also be an hallucinogen, although its rapid metabolism and poor blood brain barrier penetration could be expected to reduce its potency. The fact that low doses of 5-MT disrupt behavior only in the presence of a monoamine oxidase inhibitor (46) tends to support this assumption. However, a definite conclusion cannot be drawn in the absence of clinical studies.

After systemic administration, we found a much greater depressant effect of 5-MDMT on the firing frequency of 5-HT neurons than on postsynaptic neurons. This result is consistent with the above microiontophoretic studies which showed a preferential action of 5-MDMT on 5-HT autoreceptors. The inability of low doses of 5-MT administered intravenously to alter the activity of 5-HT neurons is consistent with the report of Green et al. (46) who failed to detect any elevation of 5-MT concentration in the brain following parenteral administration of the drug in the absence of a monoamine oxidase inhibitor. This can be explained, at least partially, by its rapid oxidation and its relatively poor penetration into brain, since following pretreatment with a monoamine oxidase inhibitor the same authors were able to show a relatively small but dose-related penetration of the drug into the brain.

SUMMARY AND CONCLUSIONS

All indoleamine hallucinogens tested to date (i.e., LSD, DMT, psilocin, bufotenine, 5-MDMT) have exhibited a preferential action upon 5-HT autoreceptors (i.e., receptors responsive to 5-HT which are located on 5-HT neurons _per se_). In contrast, 5-HT itself shows no such preferential actions. Thus, the hallucinogens differ in this fundamental way from the natural transmitter to which they are structurally related. As depicted in Fig. 1, these findings are of theoretical interest for distinguishing between different classes of 5-HT receptors in the CNS.

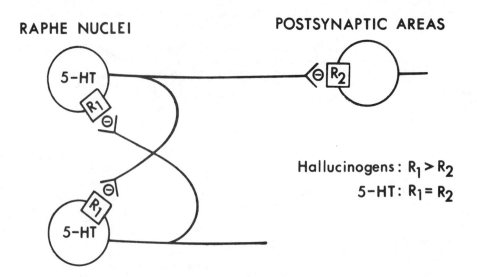

RAPHE NUCLEI   POSTSYNAPTIC AREAS

Hallucinogens: $R_1 > R_2$

5-HT: $R_1 = R_2$

Fig. 1. A schematic representation of the differential localization
of 5-HT autoreceptors (R_1) and postsynaptic 5-HT receptors (R_2). 5-HT
autoreceptors are shown in relation to inhibitory 5-HT axon collaterals
within the raphe nuclei; postsynaptic 5-HT receptors are shown in relation to a prototypical postsynaptic neuron.

Since known indoleamine hallucinogens can be identified by their preferential action on 5-HT
autoreceptors, potentially hallucinogenic compounds might be detected by testing them for a
similar preferential action. In terms of this model, we have found that the endogenous indoleamine 5-MT is similar to known hallucinogenic drugs in having a preferential action upon
presynaptic 5-HT receptors. 5-MT differs in its chemical structure from the established
hallucinogenic drugs in its lack of N-methyl substitutions. This structural difference may
account for 5-MT's poor penetration into brain. On the other hand, 5-MT appears to have a
fundamental action on central 5-HT receptors in common with the known hallucinogenic drugs.
These findings point up an advantage of the proposed model: a direct, microiontophoretic
test for patterns of receptor sensitivity circumvents extraneous factors such as peripheral
metabolism and uptake into brain. Thus, one novel implication of the results with 5-MT is
that the N-methylation of indoleamines may be a requirement for efficient entry into a brain
but not for an hallucinogenic pattern of receptor activity in the CNS. A further implication
of this work is that a search for possible endogenous indoleamine psychotogens need not be
restricted to N-methylated derivatives since O-methylation and conceivably other types of
substitutions may result in preferential activity upon 5-HT autoreceptors.

ACKNOWLEDGEMENTS

Supported by PHS. Grants MH-17871, MH-25642 and the State of Connecticut.

REFERENCES

(1) J. H. Gaddum, Antagonism between lysergic acid diethylamide and 5-hydroxy-tryptamine,
J. Physiol. 121, 15P (1953).

(2) A. W. Woolley and E. Shaw, A biochemical and pharmacological suggestion about certain
mental disorders, Proc. Natl. Acad. Sci. 40, 228 (1954).

(3) D. X. Freedman, Effects of LSD-25 on brain serotonin, J. Pharmacol. Exp. Ther. 134,
160 (1961).

(4) G. K. Aghajanian, H. J. Haigler, and J. L. Bennett, Amine receptors in brain-III-5-hydroxy-tryptamine, in Iversen, L. L. and Snyder, S. H., eds. (1975) Handbook of Psychopharmacology, Plenum Press, New York.

(5) G. K. Aghajanian, H. J. Haigler, and F. E. Bloom, Lysergic acid diethylamide and serotonin: direct actions on serotonin-containing neurons, Life Sci. 11,615 (1972).

(6) H. J. Haigler and G. K. Aghajanian, Lysergic acid diethylamide and serotonin: a comparison of effects on serotonergic neurons and neurons receiving a serotonergic input, J. Pharmacol. Exp. Ther. 188,688 (1974).

(7) G. J. Bramwell and T. Gonye, Responses of midbrain neurones to microiontophoretically applied 5-hydroxytryptamine: comparison with the response to intravenously administered lysergic acid diethylamide, Neuropharmacology 15,457 (1976).

(8) G. K. Aghajanian, W. E. Foote, and M. H. Sheard, Lysergic acid diethylamide: sensitive neuronal units in the midbrain raphe, Science 161,706 (1968).

(9) G. K. Aghajanian, W. E. Foote, and M. H. Sheard, Action of psychotogenic drugs on single midbrain raphe neurons, J. Pharmacol. Exp. Ther. 171, 178 (1970).

(10) G. K. Aghajanian and H. J. Haigler, Hallucinogenic indoleamines: preferential action upon presynaptic serotonin receptors, Psychopharmacol. Comm. 1,619 (1975).

(11) D. W. Gallager and G. K. Aghajanian, Effects of chlorimiphramine and lysergic acid diethylamide on efflux of precursor-formed ^{3}H-serotonin: correlations with serotonergic impulse flow, J. Pharmacol. Exp. Ther. 193,785 (1975).

(12) T. N. Chase, G.R. Breese, and I. J. Kopin, Serotonin release from brain slices by electrical stimulation: regional differences and effect of LSD, Science 157,1471 (1967).

(13) L. O. Farnebo and B. Hamberger, Drug-induced changes in the release of ^{3}H-monoamines from field stimulated rat brain slices, Acta Physiol. Scand. 371,35 (1971).

(14) R. Y. Wang and G. K. Aghajanian, in preparation.

(15) G. K. Aghajanian and R. Y. Wang, Habenular and other midbrain raphe afferents demonstrated by a modified retrograde tracing technique, Brain Res. 122,229 (1977).

(16) S. S. Mosko, D. Haubrick, and B. L. Jacobs, Serotonergic afferents to the dorsal raphe nucleus: evidence from HRP and synaptosomal uptake studies, Brain Res.119,267 (1977).

(17) R. Y. Wang and G. K. Aghajanian, Inhibition of neurons in the amygdala by dorsal raphe stimulation: mediation through a direct serotonergic pathway, Brain Res. 120,85 (1977).

(18) R. Y. Wang and G. K. Aghajanian, Antidromically identified serotonergic neurons in the rat midbrain: evidence for collateral inhibition, Brain Res., in press (1977).

(19) G. J. Bramwell, Factors affecting the activity of 5-HT-containing neurons, Brain Res. 79,515 (1974).

(20) S. S. Mosko and B. L. Jacobs, Midbrain raphe neurons: spontaneous activity and response to light, Physiol. Behav. 13, 589 (1974).

(21) A. B. Wolbach, Jr., E. J. Miner, and H. Isbell, Comparison of psilocin with psilocybin, mescaline and LSD-25, Psychopharmacologia 3, 219 (1962).

(22) A. Horita, Some biochemical studies on pilocybin and psilocin, J. Neuropsychiat. 4, 270 (1963).

(23) S. Szara, Dimethyltryptamine: its metabolism in man; the relation of its psychotic effect to the serotonin metabolism, Experientia 12, 441 (1956).

(24) D. E. Rosenberg, H. Isbell, E. J. Miner, and C. R. Logan, The effect of N,N-dimethyltryptamine in human subjects tolerant to lysergic acid diethylamide, Psychopharmacologia 5, 217 (1964).

(25) H. D. Fabing and R. J. Hawkins, Intravenous bufotenine injection in the human being, Science 123,886 (1956).

(26) W. J. Turner and S. Merlis, Effect of some indolealkylamines on man, Arch. Neurol. Psychiat. 81, 121 (1959).

(27) E. Sanders and M. T. Bush, Distribution, metabolism and excretion of bufotenine in the rat with preliminary studies on its o-methyl derivative, J. Pharmacol. Exp. Ther. 148, 340 (1967).

(28) S. H. Snyder and E. Richelson, Psychedelic drugs: steric factors that predict psychotropic activity, Proc. Nat. Acad. Sci. 60, 206 (1968).

(29) S. Kang, C. L. Johnson, and J. P. Green, Theoretical studies on the conformations of psilocin and mescaline, Molec. Pharmacol. 9, 640 (1973).

(30) R. J. Wyatt, B. Termini, and J. M. Davis, Biochemical and sleep studies of schizophrenia: a review of the literature - 1960 - 1970, parts 1 and 11. Schizophrenia Bull. 4, 10 (1972).

(31) J. Axelrod, Enzymatic formation of psychotomimetic metabolites from normally occurring compounds, Science 134, 343 (1961).

(32) S. Szara, L. H. Rockland, D. Rosenthal, and J. H. Handlon, Arch. Gen. Psychiat. 15, 320 (1966).

(33) P. Bickel, A. Dittrich, and J. Schoepf, An experimental study on altered state of consciousness induced by N,N-dimethyltryptamine (DMT), Pharmakopsychiat. 9, 220 (1976).

(34) J. C. Gillin, J. Kaplan, R. Stillman, and R. J. Wyatt, The psychedelic model of schizophrenia: the case of N,N-dimethyltryptamine, Am. J. Psychiat. 133, 203 (1976).

(35) U. R. Gomes, A. L. Neethling, and B. C. Shanley, Enzymatic N-methylation of indoleamines by mammalian brain: fact or artefact, J. Neurochem. 27, 901 (1976).

(36) M. R. Boarder and R. Rodnight, Tryptamine-N-methylatransferase activity in brain tissue: a re-examination, Brain Res. 114, 359 (1976).

(37) E. F. Domino, Indole alkyl amines as psychotogen precursors - possible neurotransmitter imbalance, in Domino, E. F. and Davis, J. M., eds. (1975) Neurotransmitter Balances Regulating Behavior, Ann Arbor.

(38) H. Shulgin, Pharmacological studies of 5-methoxy-N,N-dimethyltryptamine, LSD and other hallucinogens, in the discussion of P. K. Gessner, in Efron, D. H., ed. (1970) Psychotomimetic Drugs, Raven Press, New York.

(39) J. C. Gillin, J. Tinklenberg, D. M. Stott, R. Stillman, J. S. Shertlidge, and R. J. Wyatt, 5-methoxy-N,N-dimethyltryptamine: behavioral and toxicological effects in animals. Biol. Psychiat. 11, 355 (1976).

(40) N. Narasimhachari, B. Heller, J. Spaide, L. Haskovec, M. Fujimori, K. Tabushi, and H. E. Himwich, Urinary studies of schizophrenics and controls, Biol. Psychiat. 3, 9 (1971).

(41) N. Narasimhachari, B. Heller, J. Spaide, L. Haskovec, H. Meltzor, M. Strahilritz, and H. E. Himwich, N,N-dimethylated indoleamines in blood, Biol. Psychiat. 3, 21 (1971).

(42) R. B. Guchhait, Biosynthesis of 5-methoxy-N,N-dimethyltryptamine in human pineal gland, J. Neurochem. 26, 187 (1976).

(43) P. K. Gessner, W. M. McIssac, and I. H. Page, Pharmacological actions of some methoxy-indole alkylamines, Nature 190, 178 (1961).

(44) M. D. Mashovskii and A. S. Arntyunyan, Pharmacology of 5-methoxytryptamine hydrochloride (Mexamine), Fed. Proc. 23, 125 (1964).

(45) M. Vasko, M. P. Lutz, and E. F. Domino. Structure activity relations of some indole-alkylamines in comparison to phenethylamines on motor activity and acquisition of avoidance behavior, Psychopharmacologia 36, 49 (1974).

(46) A. R. Green, J. P. Hughes, and A. F. C. Tordoff, The concentration of 5-methoxy-tryptamine in rat brain and its effects on behavior following its peripheral injection, Neuropharmcology, 14, 601 (1975).

(47) D. J. Boullin and A. R. Green, 5-methoxytryptamine: stimulation of 5-HT receptors mediating the rat hyperactivity syndrome and blood platelet aggression, Adv. Biochem. Psychopharmacol. 15, 127 (1976).

(48) A. R. Green, S. H. Koslow, and E. Costa, Identification and quantitation of a new indolealkylamine in rat hypothalamus, Brain Res. 51, 371 (1973).

(49) S. H. Koslow, 5-methoxytryptamine: a possible central nervous system transmitter, Adv. Biochem. Pharmacol. 11, 95 (1974).

(50) S. H. Koslow, R. Post, F. Goodwin, and C. Gillin, Mass fragmentographic identification and quantification of 5-methoxytraptamine in human cerebrospinal fluid, Neurosci. Abst. 1, 557 (1975).

(51) S. H. Koslow, The biochemical and biobehavioral profile of 5-methoxytryptamine, in Usdin, E. and Sandler, M., Eds. (1976) Trace Amines and the Brain, Dekker, New York.

(52) C. de Montigny and G. K. Aghajanian, Preferential action of 5-methoxytryptamine and 5-methoxydimethyltryptamine on presynaptic serotonin receptors: a comparative iontophoretic study with LSD and serotonin, Neuropharmacology, in press (1977).

DISCUSSION

Dr. Hanin suggested that since both 4- and 5-hydroxytryptamine are inactive, one might be able to get an indication as to the site of action: with hydrogen bonding, the 4- and 5- might interact between the N- and the -OH, but the 6 might not. Thus, there could be the necessity to have a transfer across the membrane and interaction on the inside of some receptor; i.e., there may be a lipophilic barrier which has to be crossed before action is observed. *Dr. Aghajanian* agreed that the lipophilic factor is important for uptake into brain. He also thinks N-methyl groups and O-methylation are important for brain uptake. He found it interesting that the 4-hydroxy compound has ring substitution in the same position as in LSD.

Dr. R. Vogel asked if iontophoretic application of 5-HTP had been tested. *Dr. Aghajanian* replied that 5-HTP has no direct effect on the receptor. In a normal animal, 5-HTP slowly produces an inhibition if injected for a long time. However, if decarboxylation is inhibited, 5-HTP has no effect at all, even though 5HT continues to be effective.

Dr. Castagnoli queried about the specificity in terms of indole hallucinogens as opposed to phenethylamine hallucinogens. *Dr. Aghajanian* indicated that mescaline-type compounds are inactive on these receptors; they may have a similar net effect postsynaptically, but through a different mechanism.

Dr. Breese asked if chlorpromazine blocked effects of LSD clinically; *Dr. Aghajanian* indicated that chlorpromazine is useful in the later stages as the LSD effects are wearing off, but not in the early stages.

In response to a question of *Dr. Shares's*, *Dr. Aghajanian* stated that the effects of LSD and serotonin are indistinguishable on the raphe cell itself, but that an additive effect is obtained if they are given together. LSD seems to be a partial agonist on the postsynaptic cell.

In reply to *Dr. Meltzer*, *Dr. Aghajanian* replied that GABA can be antagonized with picrotoxin in the raphe, but the serotonin response is unaffected by picrotoxin. He also indicated that beta-carbolines are inactive in his system.

CENTRAL DOPAMINERGIC SYSTEMS: TWO IN VIVO ELECTROPHYSIOLOGICAL MODELS FOR PREDICTING THERAPEUTIC EFFICACY AND NEUROLOGICAL SIDE EFFECTS OF PUTATIVE ANTIPSYCHOTIC DRUGS

Benjamin S. Bunney

Departments of Psychiatry and Pharmacology, Yale University School of Medicine
New Haven, Connecticut   06510

INTRODUCTION

The neurobiological basis of psychiatric disorders has greatly expanded over the last 15 years. A major part of this expansion has been due to a marked increase in our understanding, at a cellular level, of the mechanism and site of action of drugs used to treat mental illness, especially the antipsychotic and antidepressant drugs. This heavy investment in pharmacological research is, of course, based on the assumption that knowledge of where and how therapeutically effective drugs act in the central nervous system might give us significant clues as to the pathology underlying the various mental disorders these drugs are used to treat. So far, this approach has fallen short of its goal. One reason for this is the wide gap between our knowledge of cellular mechanisms and our knowledge of how such cellular effects finally lead to behavioral change. However, discovering organic factors involved in the etiology of schizophrenia or depression is not the only possible benefit from this pharmacological approach. By studying the cellular mechanism of action of a drug information is obtained which may make it possible to design better drugs with fewer side effects. Described below are two in vivo electrophysiological models based on studies of the mechanism of action of antipsychotic drugs which we feel may allow us to predict, prior to clinical trials, whether a putative neuroleptic will have antipsychotic efficacy and neurological side effects. Both models involve the effects of drugs on a particular neuronal system in the brain--the dopaminergic (DA) system. In the first model, drug-induced changes in the activity of the DA cells themselves are used as indicators of the presence or absence in putative neuroleptics of therapeutic effects and adverse neurological side effects. The second model, which is only partially developed at this time, is based on drug-induced changes in the ability of microiontophoretically applied DA to induce inhibition of cells in various areas innervated by central DA systems.

NEUROANATOMY OF THE DOPAMINE SYSTEM

Fluorescence histochemical studies first performed by Dahlström and Fuxe in 1964 (1) and later extended by Ungerstedt (2), Lindvall et al. (3) and Berger et al. (4,5), have demonstrated that the great majority of DA cells are located in the zona compacta (ZC) of the substantia nigra and the adjacent midbrain ventral tegmental (VT) area (Fig. 1). The ZC DA neurons project mainly to the caudate nucleus whereas the VT cells have terminals in parts of the limbic system (e.g. accumbens nucleus, olfactory tubercles, lateral septal nucleus, central nucleus of the amygdala) and the cortex (prefrontal cortex, entorhinal cortex and cingulate gyrus). This precise identification of central DA systems has made it possible to study intensively their function in the brain as well as the effects of drugs upon them. The combined data from these studies form the rationale for choosing the DA systems around which to build a model for predicting putative neuroleptic antipsychotic efficacy and extrapyramidal side effect incidence. Some of these data are reviewed below.

REASONS FOR CHOOSING CENTRAL DOPAMINE SYSTEMS FOR PREDICTIVE MODELS

1. Biochemical Evidence for a selective action of neuroleptics on central DA systems

Although the first antipsychotic drug, chlorpromazine, was discovered in 1953, it was not until 10 years later that the first hint of its mechanism of action in the central nervous system was obtained. Thus, in 1963 biochemical studies were performed which suggested that the antipsychotic drugs chlorpromazine and haloperidol increased DA turnover (6,7). Promethazine, a phenothiazine lacking antipsychotic efficacy and extrapyramidal side effects, was found to have no effect. Based on these findings, Carlsson and Lindqvist (6) hypothesized that the increase in dopamine turnover might be due to the ability of antipsychotic drugs to block DA receptors on cells which form part of a neuronal feedback circuit (see Fig. 2) modulating DA cell activity. They suggested that the blockade would lead to a com-

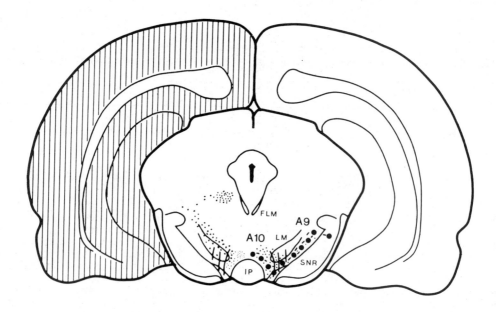

Fig. 1. Representative transverse section of the rat midbrain at the level of the interpeduncular nucleus, showing location of dopamine-containing cell bodies. Black dots represent dopamine cells in the ZC of the substantia nigra and in the VT area. FLM, fasciculus longitudinalis medialis; IP, nucleus interpeduncularis; LM, lemniscus medialis, SNR, substantia nigra, zona reticulata. (Modified, from ref. 2.)

pensatory increase in the firing rate of DA neurons and thus to an increase in turnover. A great deal of biochemical evidence now exists supporting this mechanism as being one of the major ways antipsychotic drugs act in the brain. For example, both Andén et al. (8) and Nyback and Sedvall (9) have shown that an anatomically intact DA system is necessary for antipsychotic drugs to cause changes in DA turnover. A DA sensitive adenylate cyclase has been found in several areas of the brain which is blocked by antipsychotic drugs in rank order of potency approximating that found for their therapeutic potency (10,11,12). In vitro binding studies have shown that tritiated haloperidol is displaced by other antipsychotic drugs, also in rank order of potency corresponding to their therapeutic effects (13, 14).

Thus, there is convincing biochemical evidence that many antipsychotic drugs have a specific effect on central DA systems. Although phenothiazines have many other effects on biological systems, their effect on DA systems appears to correlate best with antipsychotic efficacy and therapeutic potency (15,16).

2. Evidence for a dopaminergic site of action for neuroleptic extrapyramidal side effects

There is evidence suggesting that at least some of the Parkinson-like side effects of the antipsychotic drugs are attributable to their effect on the DA system. We know that the primary anatomical defect in Parkinson's disease is the destruction of the substantia nigra, ZC, dopaminergic neurons (17). This destruction leads to decreased DA in the neostriatum which in turn is thought to be responsible for at least some of the symptoms of Parkinson's disease (18). Schizophrenic patients treated with an antipsychotic drug often develop symptoms similar to those seen in Parkinson's disease: akinesia, rigidity and tremor (19). These side effects have been linked by biochemical and behavioral evidence to an action on the nigro-striatal DA system (20, 21). These neurological side effects are thought to be mediated through the ability of antipsychotic drugs to block DA receptors. Receptor blockade, it is suggested, leads to a functional deficiency of DA at postsynaptic receptors just as destruction of DA neurons leads to such an actual deficiency in Parkinson's disease. It is this line of reasoning which has led investigators to suggest that the antipsychotic drugs may cause their extrapyramidal side effects through the blockade of DA receptors in the neostriatum (20,21).

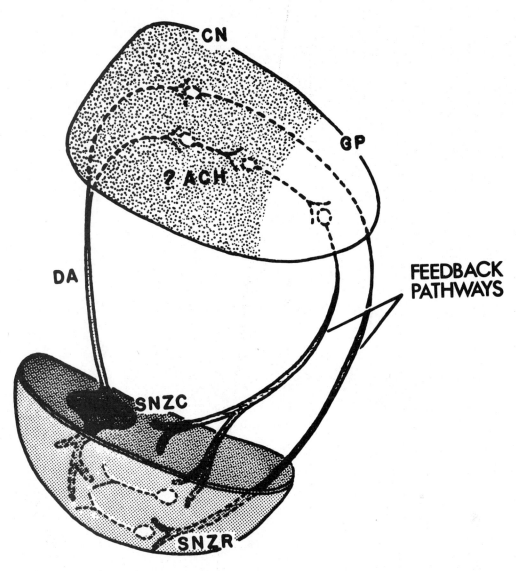

Fig. 2. Schematic representation of possible feedback pathways from the basal ganglia to the substantia nigra. A dopaminergic (DA) neuron is shown leaving the substantia nigra zona compacta (SNZC) and projecting to the caudate nucleus (CN). Two feedback pathways from the basal ganglia to the substantia nigra are depicted. Neurons in both the caudate nucleus and the globus pallidus (GP) have been shown to project to the substantia nigra. Some of the possible configurations for other links in these pathways are also shown. ACH, acetylcholine; SNZR, substantia nigra zona reticulata. (Modified, from ref. 43.)

3. Some evidence for a dopaminergic site of action for neuroleptic therapeutic effects

The fact that antipsychotic drugs have an effect on the DA system in the CNS is well established (see above). It is argued that the antipsychotic action of neuroleptics must be mediated through their effect on some part of this system. The nigro-striatal DA system appears to be mainly involved in somatic motor mechanisms (22,23). The tuberoinfundibular DA system is mainly concerned with hormonal mechanisms (24,25). The remaining DA systems, therefore, seem by far the most likely candidates for the antipsychotic action of neuroleptics in terms of what little is known about the function of the areas they innervate: e.g. accumbens nucleus, amygdaloid complex and cerebral cortex. For example, various parts of the limbic system have been shown to be involved in the production or control of sexual and aggressive behavior, and implicated in the mediation of punishment and reward. The cortical areas innervated by the DA system have been implicated in the phenomena of "behavioral anticipation" and "temporal stability" in behavioral programs (26). Many psychotic patients show deficits in one or more of these areas. For these reasons, the limbic and cortical areas innervated by the DA system have been proposed as the most likely

94

site of action for the antipsychotic effects of neuroleptics (15,16,20,27,28). It should
be emphasized, however that here the evidence is much less conclusive than the evidence for
the site of action of antipsychotic drug neurological side effects.

Thus we feel that there is substantial evidence linking the site of action of the antipsychotic drugs to the DA system--extrapyramidal side effects to the ZC nigrostriatal system and antipsychotic properties perhaps to the VT mesolimbic and mesocortical systems.
In addition, we have been provided with a detailed map of its anatomical location. It is
for these reasons that we choose the DA system for our models.

MODEL 1: DRUG INDUCED CHANGES IN THE ACTIVITY OF DOPAMINERGIC NEURONS

This model involves single unit recording from DA-containing cells in the ZC and VT
areas (29,30). With this technique we are able to investigate in vivo the effect of drugs
on the activity of DA neurons and have obtained the following results. We have confirmed
Carlsson's hypothesis concerning the effect of antipsychotic drugs on DA cell firing rate.
It was found that the systemic administration of many antipsychotic compounds, including
chlorpromazine (Fig. 3) and haloperidol, increases the firing rate of DA-containing ZC and

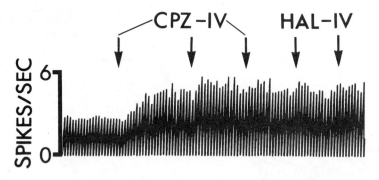

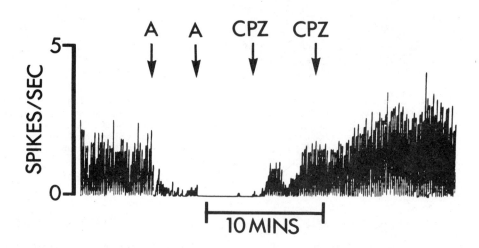

Fig. 3. (Upper) Effect of chlorpromazine (CPZ) on the firing rate of a ZC dopaminergic
cell. Intravenous administration of CPZ in a series of small doses (0.5, 0.5, and 1.0 mg/kg)
maximally increased firing rate to double its baseline rate. Once a maximal rate of firing
was obtained with CPZ, haloperidol (HAL; 0.1 and 0.5 mg/kg) had no further effect, suggesting that these antipsychotic agents act on the same DA receptors.

(Lower) Antagonism by CPZ of d-amphetamine (A) induced depression of ZC dopaminertic cell
activity. d-Amphetamine given intravenously in a total dose of 0.75 mg/kg stopped the cell
completely. After CPZ (0.25 and 0.50 mg/kg) was administered intravenously cell firing re-
sumed and increased to above baseline levels. (From ref. 30).

VT neurons in very small doses (31). This effect of antipsychotic drugs was specific to DA neurons in that non-DA cells did not respond in a similar manner even at high doses. In addition non-neuroleptics such as LSD and imipramine had no effect on DA cell activity.

Another drug whose mechanism of action has been linked to the DA system is amphetamine. It produces a paranoid psychosis many feel is indistinguishable, in most cases, from paranoid schizophrenia (32). Antipsychotic agents useful in the treatment of schizophrenia rapidly reverse the psychosis induced by amphetamine (32). Amphetamine is thought to stimulate DA receptors indirectly through increasing DA release and blocking its reuptake (33,34,35). More DA is thus available in the synaptic cleft for stimulation of DA receptors. Corrodi and coworkers (36) hypothesized that increased DA receptor stimulation would cause a decrease in DA cell firing rate mediated via a postsynaptic neuronal feedback pathway. Recently this hypothesis was confirmed, again using single unit recording techniques (31,37). Amphetamine in small doses (0.25-1.6 mg/kg, i.v.) markedly inhibits these cells (Fig. 3). DA receptor blockers such as the antipsychotic drugs, chlorpromazine and haloperidol, reverse the d-amphetamine-induced depression of firing rate (Fig. 3) (31,37), thus providing a direct neurophysiological parallel with the clinical interaction of these two drugs in producing and reversing a paranoid psychosis in man.

In evaluating the significance of these drug effects on DA neurons it is essential to determine whether the ability of chlorpromazine to reverse the d-amphetamine-induced depression of DA cell firing is a property of all phenothiazines or just those with antipsychotic properties and/or extrapyramidal side effects. To begin to answer this question, promethazine, a drug with no antipsychotic efficacy or extrapyramidal side effects (19) and no effect on DA metabolism (38), was tested for its ability to reverse the d-amphetamine-induced depression of DA cell activity in both the ZC and VT areas. It was found to have no effect in either area (Fig. 4), suggesting that the antipsychotic properties and/or the extrapyramidal side effects of the phenothiazines may, indeed, correlate with their ability to reverse the effects of d-amphetamine (31). However, from the data presented so far one cannot distinguish between these possibilities. In order to evaluate further this problem we needed to test a phenothiazine with no antipsychotic efficacy but one that does produce extrapyramidal side effects. Such a drug is mepazine.

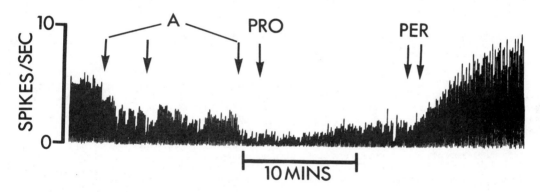

Fig. 4. Effect of promethazine (PRO) on dopamine cell activity subsequent to d-amphetamine (A) depression. d-Amphetamine (1.25 mg/kg, i.v., total dose) depressed unit activity. Promethazine (10 mg/kg, i.p.), a phenothiazine lacking antipsychotic efficacy or extrapyramidal side effects, failed to elicit the usual increase in firing rate induced by antipsychotic phenothiazines. The moderate increase in firing rate seen following promethazine administration is no greater than the spontaneous recovery from amphetamine seen with these cells. However, perphenazine (PER: 0.2 mg/kg, total dose), a clinically active phenothiazine, produced a rapid increase in rate to above baseline levels. (From ref. 31.)

Mepazine was found to reverse d-amphetamine-induced depression of DA cells in the ZC group but not in the VT area (29). In subsequent studies we have found a 100 percent correlation between the ability of a drug to reverse d-amphetamine-induced inhibition of VT cells and that drug's therapeutic efficacy. That is, all drugs tested that were known to be good antipsychotic agents (e.g. chlorpromazine, perphenazine, thioridazine, trifluoperazine, clozapine, haloperidol, molindone) reversed and blocked the inhibitory effects of d-amphetamine on VT neurons, whereas drugs lacking antipsychotic properties (e.g. mepazine, promethazine, diethazine, tricyclic antidepressants, anesthetics, LSD, etc.) had no effect. In the latter class of drugs (non-neuroleptic) we found two exceptions which will be discussed separately (see below). These findings suggest that we may have a model which can be used to differentiate putative neuroleptics with antipsychotic properties from those lacking clinical efficacy by testing their ability to reverse d-amphetamine-induced depression of VT dopaminergic neurons (29,30).

Since antipsychotic efficacy in a given drug is often accompanied by a moderate to high incidence of extrapyramidal side effects, it would be useful if our model could distinguish between neuroleptics with a high incidence versus those with a low incidence of these side effects. To test the ability of our model system to distinguish between these two groups of drugs, we determined the effect on ZC dopaminergic cell activity of representatives from each of these two groups. We chose the nigrostriatal DA system for this part of the model because it is the one most implicated in the pathogenesis of these symptoms (see above).

Antipsychotic drugs with a moderate to high incidence of extrapyramidal side effects, such as chlorpromazine, perphenazine, trifluoperazine, and haloperidol (19), were found to increase the activity of ZC cells 20 to 100 percent above baseline rate at low doses (29). In addition they reversed d-amphetamine-induced depression of these cells to above baseline levels (Fig. 3). However, when neuroleptics with a low incidence of extrapyramidal side effects (e.g., thioridazine and clozapine) were tested, they never increased the firing rate of ZC dopaminergic cells above baseline even when given in doses 10 times that needed for chlorpromazine to cause an increase in rate (Fig. 5) (29,30). In addition, although they did reverse amphetamine-induced depression of these cells, they did so only back to baseline and never above (Fig. 5). Thus antipsychotic drugs with a moderate to high incidence of extrapyramidal side effects increase ZC cell firing rate above baseline whether given alone or after d-amphetamine. Conversely, antipsychotic drugs with a low incidence of these side effects do not increase ZC cell activity. Therefore, there is a significant difference between drugs with a high incidence versus those with a low incidence of extrapyramidal side effects in terms of their effect on the firing rate of ZC cells. It is important to note, however, that there is no difference between these two groups of drugs in terms of their antipsychotic efficacy or effect on the firing rate of VT DA-containing cells.

The difference in response of the ZC dopaminergic neurons to antipsychotic drugs with differing incidence of extrapyramidal side effects suggests that this model system may provide a means for predicting the incidence of extrapyramidal side effects of putative neuroleptics prior to clinical trials in man (29,30).

MODEL 2: DRUG-INDUCED CHANGES IN DOPAMINE INHIBITION OF DOPAMINE INNERVATED NEURONS

Model 1 has at least one major drawback. If we assume that DA receptor blockade is the sine qua non of antipsychotic drug therapeutic efficacy then we are studying the ability of drugs to block DA receptors at least one step removed from this effect. Thus, in model 1 we are using changes in DA cell activity as a reflection of the effect of drugs on DA receptors located on cells which form part of a neuronal feedback pathway responsible for mediating the observed changes (Fig. 2). This is further complicated by the fact that there are at least two feedback pathways (39,40) and each pathway is probably polysynaptic (Fig. 2). Although two of the transmitters used by neurons in these pathways are most likely gamma-aminobutyric acid (GABA) and substance P (41,42), there are probably other neurons involved whose neurotransmitters are totally unknown. Model 1 is, therefore, susceptible to "false positives". Thus, a drug could reverse d-amphetamine induced inhibition of DA cells or cause an increase in their firing rate through interfering with neural transmission in one of the feedback pathways rather than by blocking DA receptors. As mentioned earlier two "false positives" have already been observed--both picrotoxin (43) and diazepam (44) reverse d-amphetamine-induced inhibition of DA cells (Fig. 6). It is for these reasons that we have begun to develop a second in vivo model system in which the ability of drugs to block DA receptors on cells in areas innervated by midbrain DA systems (see Fig. 7) is studied directly.

Previously it has been shown that DA, applied iontophoretically, has a predominantly depressant effect on the activity of non-dopaminergic cells in the caudate nucleus, accumbens nucleus and frontal cortex (45,46,47), all of which are innervated by the DA system. By recording from cells in these "postsynaptic" areas while administering antipsychotic drugs i.v. or iontophoretically, it is possible to determine the effect of these drugs on DA-induced changes in neuronal activity in various areas innervated by the DA system, including the caudate nucleus, accumbens nucleus and prefrontal cortex.

DA receptor blockers should block the depressant effect of DA applied iontophoretically. In addition, since it appears possible to categorize drugs according to their differential effects on the firing rate of ZC cells vs. VT cells, it would be of interest to compare these drugs in terms of their effects on the activity of postsynaptic cells in different regions of the brain which are innervated by the midbrain DA systems (i.e. nigrostriatal vs. meso-limbic and meso-cortical). Thus, if our hypotheses are correct, drugs with antipsychotic properties should block DA-induced depression of postsynaptic cells in

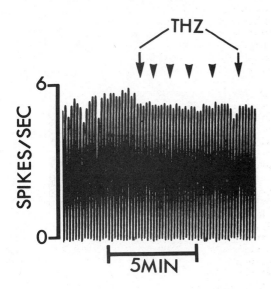

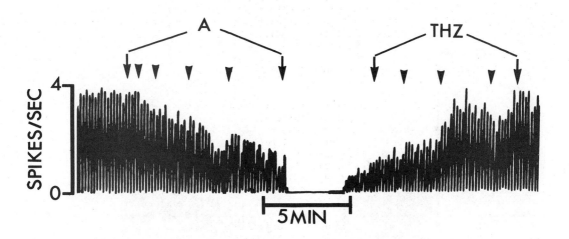

Fig. 5.   Effects of thioridazine (THZ) on the firing rate of dopaminergic neurons in the
          substantia nigra zona compacta.

(Upper) Thioridazine administered intravenously in a total dose of 32 mg/kg had no effect
on baseline firing rate. These results are in marked contrast to those with chlorpromazine
where the average dose needed to increase baseline firing rate 50 to 100 percent was approximately 2.0 mg/kg.

(Lower) d-Amphetamine (A) (1.6 mg/kg, i.v.) stopped the firing of this dopamine cell.
Thioridazine given intravenously in a sequence of doses (0.25, 0.25, 0.5, 1.0) increased
unit activity back to baseline rate. However, an additional 8 mg/kg (single dose) did not
increase activity above baseline level. Chlorpromazine administered after a comparable
dose of d-amphetamine would have markedly increased firing rate above baseline in a dose
range of 1.5 to 2.0 mg/kg.

98

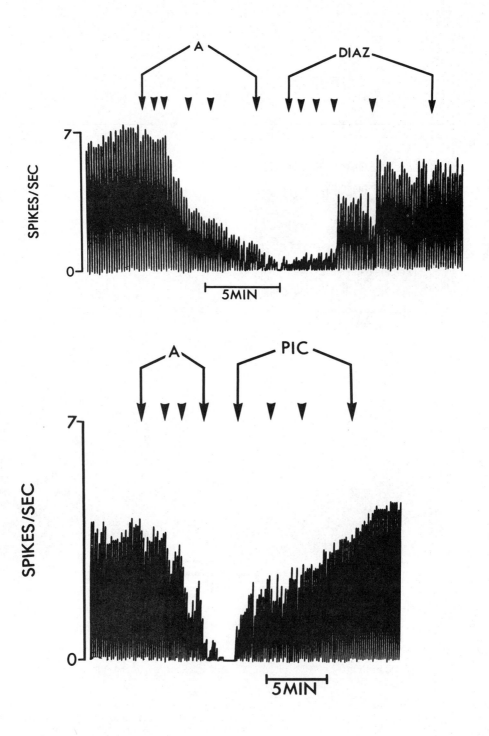

Fig. 6. (Upper) Antagonism by diazepam (DIAZ) of d-amphetamine (A)-induced slowing of a VT dopaminergic neuron. d-Amphetamine (3.2 mg/kg) markedly slowed this cell. Diazepam maximally reversed this depression at a total dose of 3.2 mg/kg. An additional 6.4 mg/kg had no further effect. All drugs were administered intravenously. (From ref. 44.)

(Lower) Interaction between the effects of picrotoxin (PIC) and those of d-amphetamine (A) on the firing rate of a dopaminergic neuron in the substantia nigra zona compacta. A, in a total dose of 3.2 mg/kg, temporarily stopped the cell. PIC, in a cumulative dose of 0.8 mg/kg (0.2, 0.2, 0.4), reversed the A-induced depression back to base line. An additional 0.8 mg/kg increased activity above base line rate. Both drugs were administered intravenously. (From ref. 43.)

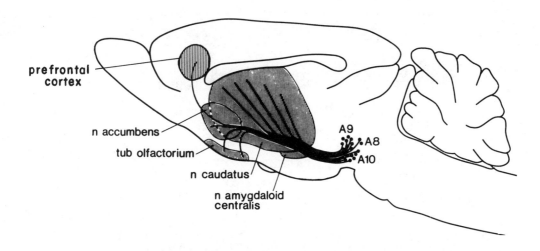

Fig. 7. Representative sagittal section of the rat brain illustrating projection areas of midbrain dopaminergic cells. Black dots represent cell bodies, striped regions represent areas innervated. A9 designates dopaminergic neurons in the zona compacta (ZC) of the substantia nigra. A10 designates dopamine-containing cells in the ventral tegmental (VT) area. (Modified, from ref. 2.)

the accumbens nucleus and cerebral cortex, while phenothiazines lacking clinical efficacy should have no effect. Antipsychotic drugs with a high or moderate incidence of extrapyramidal side effects should block DA-induced depression of caudate nucleus cells when they are applied iontophoretically. Those with a low incidence should have little effect.

Results to date include the following:

1. Effect of antipsychotic drugs on DA-induced inhibition of DA-innervated cells in the cortical and limbic systems.

Microiontophoretically applied trifluoperazine (TFP) blocks DA-induced inhibition of cells in layers V and VI of the prefrontal cortex (Fig. 8) (48). These cells receive predominantly a DA innervation (3,4,5). In contrast, TFP does not block norepinephrine (NE)-induced inhibition of cells in two areas which receive an NE innervation--the superficial layers of the prefrontal cortex and the pyramidal cells in the hippocampus (Fig. 8). TFP applied microiontophoretically (48) and haloperidol given i.v. (44) block DA-induced inhibition of nucleus accumbens cells. TFP applied microiontophoretically blocks DA inhibition of cells in the lateral septal nucleus and in the central nucleus of the amygdala (Bunney, unpublished data).

2. Effect of antipsychotic drugs on DA-induced inhibition of DA-innervated cells in the caudate nucleus.

Haloperidol (Bunney and Aghajanian, unpublished data) and chlorpromazine (49) administered i.v. block DA-induced inhibition of neurons in the caudate nucleus. Microiontophoretic TFP also blocks DA-induced depression of these cells (Bunney unpublished data).

3. Effect of an antipsychotic drug lacking extrapyramidal side effects on DA-induced inhibition of DA innervated cells in the accumbens nucleus and caudate nucleus.

Clozapine (CLZ), in preliminary studies, has been found to block DA-induced inhibition of nucleus accumbens cells (44) but not caudate nucleus neurons (Bunney and Aghajanian, unpublished data).

Obviously, with only the above results available so far, it is impossible to tell whether this second model will work. However, the finding that TFP appears to be a selective blocker of DA receptors and that CLZ blocks nucleus accumbens but not caudate nucleus DA receptors suggests that its continuing development is justified.

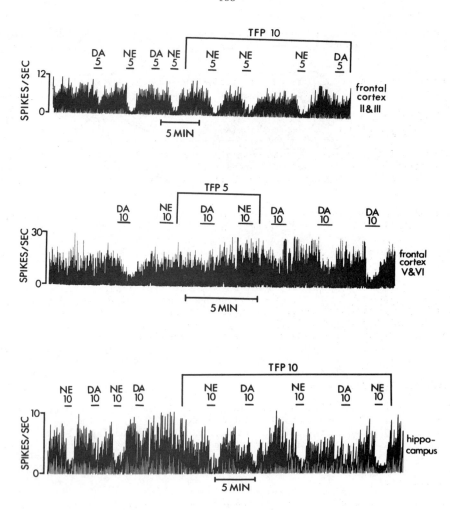

Fig. 8 (Upper) Typical interaction of trifluoperazine (TFP) with dopamine (DA) and norepinephrine (NE) applied microiontophoretically to a cell in layer II or III of the prefrontal
cortex (primary innervation--NE). This cell was inhibited to a greater extent by NE than
by DA. TFP did not block the inhibition of this cell by NE. TFP failed to block the effects
of NE whether NE was applied alternately with DA or sequentially as illustrated here.

(Middle) Typical interaction of trifluoperazine (TFP) with dopamine (DA) and norepinephrine (NE) applied microiontophoretically to a cell in layer V or VI of the prefrontal cortex
(primary innervation--DA). This cell was inhibited to a greater extent by DA than by NE.
TFP totally blocked the inhibitory effect of DA. Following cessation of TFP ejection the
inhibitory effect of DA gradually returned.

(Lower) Typical interaction of trifluoperazine (TFP) with dopamine (DA) and norepinephrine
(NE) applied microiontophoretically to norepinephrine-innervated pyramidal cells in the dorsal hippocampus. This cell was inhibited to a greater degree by NE than by DA. Paralleling
the effect of TFP when its interaction with DA and NE were tested in layers II and III of
the prefrontal cortex, the inhibitory effect of NE and DA were not blocked. Horizontal bars
indicate duration of ejection. Numbers above bars specify ejection current in nA. (From
ref. 48.)

SUMMARY AND CONCLUSIONS

We have presented two in vivo models which use drug induced changes in the activity
of central DA-containing and DA-innervated neurons, respectively, to predict therapeutic
and adversive neurological effects of putative antipyschotic drugs. In the first model
specific changes in the firing rate of DA neurons appear to correlate with the clinical
actions of known neuroleptics. These correlations are:

(a) Antipsychotic drugs administered intravenously reverse the amphetamine-induced depression of DA-containing neurons in the midbrain VT area. Most drugs lacking antipsychotic effects are inactive in this regard.

(b) Drugs possessing a high to moderate incidence of extrapyramidal side effects increase the activity of substantia nigra ZC dopaminergic neurons above baseline and reverse d-amphetamine-induced depression of these cells to above baseline levels. Drugs with a low incidence of extrapyramidal side effects do not increase the firing rate of these cells even in high doses and do not reverse amphetamine-induced depression above baseline.

In the second model drug-induced changes in the ability of microiontophoretic DA to depress the activity of neurons in various areas receiving DA innervation appear to correlate with the clinical effect of known neuroleptics. Much of the data presented for this model is preliminary and a great deal of further work is necessary.

It is hoped that the development of these models will not only provide a predictive tool but will contribute to our further understanding of how and where antipsychotic drugs act to produce their clinical effects. In addition, these models may contribute to the knowledge needed to synthesize drugs with optimal therapeutic efficacy and minimal adverse side effects.

Finally, as a note of caution, it should be pointed out that both models are based on acute drug effects, whereas treatment with antipsychotic drugs takes days, and sometimes weeks, to produce a maximal therapeutic response. For this reason, it may be that similar models examining the chronic effects of neuroleptics will yield the most relevant and useful information. The development of such models is in progress.

ACKNOWLEDGEMENTS

Supported by NIMH Grants (MH-1781; MH-14459; MH-25642; MH-28849), The Benevolent Foundation of Scottish Rite Freemasonry, Northern Jurisdiction, U.S.A. and the State of Connecticut.

REFERENCES

(1) A. Dahlström and K. Fuxe, Evidence for the existence of monoamine-containing neurons in the central nervous system, Acta Physiol. Scand. 62, suppl. 232, 1-55 (1964).

(2) U. Ungerstedt, Stereotaxic mapping of the monoamine pathways in the rat brain, Acta Physiol. Scand., Suppl. 367, 1-48 (1971).

(3) O. Lindvall, A. Björklund, R. Y. Moore, and U. Stenevi, Mesencephalic dopamine neurons projecting to neocortex, Brain Res. 81, 325-331 (1974).

(4) B. Berger, J. P. Tassin, G. Blanc, M. A. Moyne, and A. M. Thierry, Histochemical confirmation for dopaminergic innervation of the rat cerebral cortex after destruction of the noradrenergic ascending pathways, Brain Res. 81, 332-337 (1974).

(5) B. Berger, A. M. Thierry, J. P. Tassin and M. A. Moyne, Dopaminergic innervation of the rat prefrontal cortex: A fluorescence histochemical study, Brain Res. 106, 133-145 (1976).

(6) A. Carlsson and M. Lindqvist, Effect of chlorpromazine and haloperidol on formation of 3-methoxytyramine and normetanephrine in mouse brain, Acta Pharmacol. Toxicol. 20, 140-144 (1963).

(7) N. E. Andén, B. E. Roos, and B. Werdinius, Effects of chlorpromazine, haloperidol and reserpine on the levels of phenolic acids in rabbit corpus striatum, Life Sci. 3, 149-158 (1964).

(8) N. E. Andén, H. Corrodi, K. Fuxe, and U. Ungerstedt, Importance of nervous impulse flow for the neuroleptic induced increase in amine turnover in central dopamine neurons, Eur. J. Pharmacol. 15, 193-199 (1971).

(9) H. Nyback and G. Sedvall, Effect of nigral lesion on chlorpromazine-induced acceleration of dopamine synthesis from ^{14}C-tyrosine, J. Pharm. Pharmacol. 23, 322-326 (1971).

(10) H. S. Ahn, R. K. Mishra, C. Demirjian, and M. H. Makman, Catecholamine-sensitive adenylate cyclase in frontal cortex of primate brain, Brain Res. 116(3), 437-454 (1976).

(11) Y. C. Clement-Cormier, J. W. Kebabian, and P. Greengard, Dopamine-sensitive adenylate cyclase in mammalian brain: A possible site of action of antipsychotic drugs, Proc. Nat. Acad. Sci. U.S.A. 71(4), 113-117 (1974).

(12) R. J. Miller, Comparison of the inhibitory effects of neuroleptic drugs on adenylate cyclase in rat tissue stimulated by dopamine, noradrenaline and glucogon, Biochem. Pharmacol. 25, 537-541 (1976).

(13) I. Creese, D. R. Burt, and S. H. Snyder, Dopamine receptor binding predicts clinical and pharmacological potencies of antipsychotic drugs, Science 192, 481-483 (1976).

(14) P. Seeman, M. Chau-Wong, J. Tedesco, and K. Wong, Brain receptors for antipsychotic drugs and dopamine: Direct binding assays, Proc. Nat. Acad. Sci. U.S.A. 72(11), 4376-4380 (1975).

(15) S. Matthysse, Antipsychotic drug actions: A clue to the neuropathology of schizo-phrenia? Fed. Proc. 32, 200-205 (1973).

(16) S. H. Snyder, S. P. Banerjee, H. I. Yamamura, and D. Greenberg, Drugs, neurotrans-mitters and schizophrenia, Science 184, 1243-1253 (1974).

(17) R. Hassler, Extrapyramidal-motorische syndrome und erkrankungen, In Hb. Inn. Med. 5, edited by G. V. Bergmann, W. Frey, H. Schwiegk and R. Jung, Springer, Berlin, Güttingen, Heidelberg (1953).

(18) O. Hornykiewicz, Die topische lokalisation und das Verhalten von noradrenalin und dopamin(3-hydroxytryamin) in der substantia nigra des normalin und Parkinsonkraken, Menschen, Wien. Klin. Wschr. 75, 309 (1963).

(19) D. E. Klein and J. M. Davis (1969) Diagnosis and Drug Treatment of Psychiatric Dis-orders, Williams and Wilkins, Baltimore.

(20) H. Y. Meltzer and S. M. Stahl, The dopamine hypothesis of schizophrenia: A review, Schizophrenia Bull. 2, 19-76 (1976).

(21) G. L. Stimmel, Neuroleptics and the corpus striatum: clinical implications, Dis. Nerv. Syst. 37, 219-224 (1976).

(22) W. J. H. Nauta and W. R. Mehler, Projections of the lentiform nucleus in the monkey, Brain Res. 1, 3-42 (1966).

(23) M. R. De Long, Activity of pallidal neurons during movement, J. Neurophysiol. 34, 414-427 (1971).

(24) H. Y. Meltzer and V. S. Fang, Effect of apomorphine plus 5-hydroxytryptophan on plasma prolactin levels in male rats, Psychopharmacol. Comm. 2(3), 189-198 (1976).

(25) W. Lichtensteiger and P. J. Keller, Tubero-infundibular dopamine neurons and the secretion of lutenizing hormone and prolactin: Extrahypothalamic influences, inter-action with cholinergic systems and the effect of urethane anesthesia, Brain Res. 74, 279-303 (1974).

(26) W. J. H. Nauta, The problem of the frontal lobe. A reinterpretation, J. Psychiatr. Res. 8, 167-187 (1971).

(27) J. R. Stevens, An anatomy of schizophrenia? Arch. Gen Psychiat. 29, 177 (1973).

(28) T. Hökfelt, Å. Ljungdahl, K. Fuxe, and O. Johansson, Dopamine nerve terminals in the rat limbic cortex: Aspects of the dopamine hypothesis of schizophrenia, Science 184, 177-179 (1974).

(29) B. S. Bunney and G. K. Aghajanian, Central dopaminergic neurons: A model for predict-ing the efficacy of putative antipsychotic drugs? In Use of Model Systems in Biological Psychiatry, edited by D. Ingle, MIT Press, Cambridge, 97-112 (1975).

(30) B. S. Bunney and G. K. Aghajanian, Antipsychotic drugs and central dopaminergic neurons: A model for predicting therapeutic efficacy and incidence of extrapyrami-dal side effects, In Predictability in Psychopharmacology: Preclinical and Clinical

Correlations, edited by A. Sudilovsky, S. Gershon, and B. Beers, Raven Press, New York, 225-245 (1975).

(31) B. S. Bunney, J. R. Walters, R. H. Roth and G. K. Aghajanian, Dopaminergic neurons: Effect of antipsychotic drugs and amphetamine on single cell activity, J. Pharm. Exp. Ther. 185, 560-571 (1973).

(32) S. H. Snyder, Catecholamines in the brain as mediators of amphetamine psychosis, Arch. Gen. Psychiat. 27, 343 (1972).

(33) A. Carlsson, K. Fuxe, B. Hamberger, and M. Lindqvist, Biochemical and histochemical studies on the effects of imipramine-like drugs and (+)-amphetamine on central and peripheral catecholamine neurons, Acta Physiol. Scand. 67, 481-497 (1966).

(34) M. Besson, A. Cheramy, P. Feltz, and J. Glowinski, Dopamine: Spontaneous and drug-induced release from the caudate nucleus in the cat, Brain Res. 32, 407 (1971).

(35) M. Besson, A. Cheramy, and J. Glowinski, Effects of some psychotropic drugs on dopamine synthesis in the rat striatum, J. Pharm. Exp. Ther. 177, 196-205 (1971).

(36) H. Corrodi, K. Fuxe, and T. Hökfelt, The effect of some psychoactive drugs on central monoamine neurons, Eur. J. Pharmacol. 1, 363-368 (1967).

(37) B. S. Bunney, G. K. Aghajanian, and R. H. Roth, Comparison of effects of L-DOPA, amphetamine and apomorphine on the firing rate of rat dopaminergic neurons, Nature New Biol. 245, 123-125 (1973).

(38) H. Nyback, A. Borzecki and G. Sedvall, Accumulation and disappearance of catecholamines formed from tyrosine-^{14}C in mouse brain: effect of some psychiatric drugs, Eur. J. Pharmacol. 4, 395 (1968).

(39) P. L. McGeer, H. C. Fibiger, T. Hattori, V. K. Singh, E. G. McGeer, and L. Maler, Biochemical neuroanatomy of the basal ganglia, in Advances in Behavioral Biology, Vol. 10, edited by R. D. Myers and R. R. Drucker-Colin, Plenum Press, New York, pp. 27-47 (1974).

(40) B. S. Bunney and G. K. Aghajanian, The precise localization of nigral afferents in the rat as determined by a retrograde tracing technique, Brain Res. 117, 423-435 (1976).

(41) I. Kanazawa, P. C. Emson, and A. C. Cuello, Evidence for the existence of substance P-containing fibres in striato-nigral and pallido-nigral pathways in rat brain, Brain Res. 119, 447-453 (1977).

(42) J. S. Hong, H.-Y.T. Yang, G. Racagni, and E. Costa, Projections of substance P containing neurons from neostriatum to substantia nigra, Brain Res. 122, 541-544 (1977).

(43) B. S. Bunney and G. K. Aghajanian, Dopaminergic influence in the basal ganglia: Evidence for striatonigral feedback regulation, In The Basal Ganglia, edited by M. D. Yahr, Raven Press, New York, pp. 249-267 (1976).

(44) B. S. Bunney and G. K. Aghajanian, The effect of antipsychotic drugs on the firing of dopaminergic neurons: a Reappraisal, In Antipsychotic Drugs, Pharmacodynamics and Pharmacokinetics, Wenner-Gren Center International Symposium Series, Pergamon Press, New York, pp. 305-318 (1975).

(45) F. E. Bloom, E. Costa and G. Salmoiraghi, Anesthesia and the responsiveness of individual neurons of the caudate nucleus of the cat to acetylcholine, norepinephrine and dopamine administered by microelectrophoresis, J. Pharm. Exp. Ther. 150, 244-252 (1965).

(46) D. H. York, The inhibitory action of dopamine on neurones of the caudate nucleus, Brain Res. 5, 263-266 (1967).

(47) B. S. Bunney and G. K. Aghajanian, Electrophysiological effects of amphetamine on dopaminergic neurons, In Frontiers in Catecholamine Research, edited by S. Snyder and E. Usdin, Pergamon Press, New York, pp. 957-962 (1973).

(48) B. S. Bunney and G. K. Aghajanian, Dopamine and norepinephrine innervated cells in the rat prefrontal cortex: Pharmacological differentiation using microiontophoretic techniques, Life Sci. 19, 1783-1792 (1976).

(49) D. H. York, Dopamine receptor blockade--a central action of chlorpromazine on striatal neurons, Brain Res. 37, 91-99 (1971).

DISCUSSION

Dr. Zigmond wondered if Dr. Bunney thought that some of his systemically administered drugs might be acting on the cell body autoreceptor rather than on the terminal postsynaptic cell. *Dr. Bunney* replied that he thought some of them may be acting on the autoreceptor. He then mentioned the current controversy on the mechanism of action of amphetamine, and that he feels his own data indicate such action is almost solely from a feedback pathway effect. He agreed that antipsychotics may act on autoreceptors but pointed out that that was different from the concept of dopamine receptors interacting with each other.

Dr. Cook asked if Dr. Bunney had seen similar dissociation with thioridazine in his two systems as he had described with clozapine; he also asked if Dr. Bunney had done interaction studies with a centrally active anticholinergic in both systems. *Dr. Bunney* emphasized that his second model was only very preliminary and little work has yet been done with it. He had compared clozapine and scopolamine with regard to their effects on the firing rate of dopamine cells when given i.v.; there is no similarity: scopolamine does not reverse amphetamine-induced depression, clozapine does.

Dr. Sulser asked if Dr. Bunney had tested the very crucial drug metaclopromide - a very potent postsynaptic dopamine blocker, but not an antipsychotic. *Dr. Bunney* replied that he had not yet looked at that drug.

Dr. Antelman commented that he and Alan Eichler had been looking at the effects of chronic neuroleptic administration with self-stimulation in different dopamine pathways. He found that in A10 cells, there is no tolerance to the effects of neuroleptics and that anticholinergics do have a modifying effect. He also had found a failure of tolerance to develop in nucleus accumbens, but there anticholinergics do not modify the neuroleptic effect.

EFFECTS OF AMPHETAMINE ON SOCIAL BEHAVIORS OF RHESUS MACAQUES:
AN ANIMAL MODEL OF PARANOIA

Suzanne Haber, Patricia R. Barchas[*], and Jack D. Barchas

Nancy Pritzker Laboratory of Behavioral Neurochemistry
Department of Psychiatry and Behavioral Sciences
Stanford University School of Medicine
and
Department of Sociology[*]
Stanford University
Stanford, California

A good animal model of paranoia should sharpen our inquiry into the disorder
and its underlying mechanisms. Such a model must be dependent upon behaviors
such as observable social interaction and locomotor activity and not dependent
upon inferences about the cognitive or emotional state of the subject. In our
model, we assume that one basic feature of paranoia is an increase in perception of threats coming from the environment. We derive this assumption from
classical descriptions of the clinical state of paranoia. These include hypervigilance, guardedness, hostility, suspicion, mistrust, isolation, and
rigidity. In humans, these characteristics are reliably produced, in humans,
by amphetamine, a finding we utilize in our model. We asked whether an animal
will show an increase in species-specific responses to danger when in a relatively neutral or unchanged environment and when the normal perceptual system
is altered by amphetamine. Or, more directly, does amphetamine produce an
animal that behaves as if there was a threat in his environment.

"If we want to compare human and animal behavior, the first thing we have to
agree on is the use of a common language." (1) If such a language could be
found to describe any of the human psychopathologies, it would be of great
heuristic value in creating animal models. To achieve this goal, it will be
crucial first to consider a clearly defined human condition in which fundamental features appear reliably. We have chosen paranoia. Because of
complexities that human speech presents to the development of reliable behavioral models in animals, it seems important to construct models based on
nonverbal behaviors. Utilizing ethologic methods of observation, we then
create a "common language" which can describe and compare human and animal
behavior. Such a language should be adaptable across species.

Clearly, the survival of both animals and man has been dependent on effective
recognition of danger; the ability to perceive real danger or threat is by no
means limited to the human species. Viewing paranoia as an alteration in this
ability has several advantages. It provides a logical way to proceed in
creating an animal model. It does not limit the model to any particular
species, since any animal having the ability to perceive threat, presumably
may be made to demonstrate some malfunction. Finally, such abnormalities need
not be expressed in the same overt fashion, but will be consistent with the
animal's normal behavioral patterns.

HUMAN PARANOIA

Despite the continuing debate on the definition, classification, and etiology
of human paranoid syndromes, most clinicians agree on general behavioral
characteristics (2,3). Paranoia reflects an attitude, which Sullivan
describes as a "paranoid slant on life," permeating everything the individual
does (4). The paranoid person perceives serious threat to himself; explaining
and coping with that threat dominates his entire life.

The paranoid searches for hidden meanings and motives in even the simplest
interactions, often misperceiving neutral and even friendly gestures as
threatening and hostile. Like someone in constant danger, the paranoid is
hypervigilant and guarded, orienting to the slightest sound or movement.
Shapiro refers to a "fugitive life-style," in which the behaviors are maintained for the purpose of avoiding danger (3). The patient accepts nothing at

face value and scrutinizes everything. Interpersonal relationships are marked by a sense of mistrust and betrayal, thus the paranoid becomes socially isolated. He does not recognize that the hostility, uncertainty, and threat he perceives may come from within rather than from without. Finally, as the anxiety and fear in his world become unendurable, explanation is sought and usually found in the "crystallization" of the delusion (5). It is striking that, in the general characteristcs used to describe paranoia, words such as suspicion, mistrust, betrayal, hypervigilence, guardedness, isolation, and rigidity are used over and over. Such symptomology suggests a fundamental condition of paranoia may involve a distorted perception of threat to the individual.

Who is susceptible to paranoia? Are only certain individuals predisposed towards the disorder? Or is it possible that a malfunction in normal perceptual processes could provide each of us with the potential for paranoia? Evidence for the latter is suggested from the mental deterioration which often accompanies certain diseases and drug abuses. Paranoid psychosis and ideation have been documented in patients without a predisposition towards paranoia. "Organic paranoid syndromes" have been described in the following disease states (3): infectious disease, as in syphilis; diseases producing general brain atrophy or degeneration, as in Huntington's Chorea; various endocrine disorders; and certain drug abuse conditions.

The difficulties associated with using many of these conditions to study paranoia are three-fold. First, with some notable exceptions, the syndrome is not reliably elicited. Evident paranoid ideation does not appear with consistent regularity, even when considering only the subpopulation of affected persons who have mental or psychotic reactions. Second, paranoia is often only one of many symptoms which include delirium, confusion, disorientation, and impaired consciousness. Finally, many of the illnesses affect the brain in undifferentiated ways. For an effective modeling system, the minimum requirement is to find a reliable, consistent, mechanism for the induction of paranoid symptoms in clear consciousness and in the absence of other mental and physical defects.

The effects of chronic amphetamine use in humans appear to come close to meeting these criteria. In hospital emergency rooms, amphetamine psychosis has often been mistaken for paranoid schizophrenia. For example, in 1958, Connell reported a study in some London hospitals which indicated 42 patients with amphetamine psychosis misdiagnosed as schizophrenia (6). Controlled hospital studies, reported that 12 out of 14 patients were diagnosed as having a paranoid psychosis after treatment with large doses of methamphetamine HCL (7). Griffith found similar results in 5 out of 6 subjects receiving D-amphetamine sulfate (8). In addition, connections between amphetamine and paranoia are supported by reports from drug abusers as well as "street literature" describing the syndrome (9). Taken together, the evidence is compelling that amphetamine is capable of producing a paranoid syndrome in clear consciousness, and that vulnerability to amphetamine-induced paranoid ideation is not limited to particular individuals.

AMPHETAMINE BEHAVIOR IN HUMANS AND PRIMATES

Amphetamine psychosis in humans usually results from chronic use of the drug. Common features include: a) unusual motor behavior such as grinding teeth; b) general hyperactivity which gradually subsides as the psychosis emerges; c) heightened curiosity and interest in novel objects; d) compulsion to analyze details; e) a tendency towards focusing attention on objects for extended periods of time; f) vigilance; and g) an attribution of an emotional quality, especially hostility to neutral and even pleasant surroundings. The amphetamine user initially perseverates on various stimuli without assigning them negative emotional labels and, in fact, usually finds them pleasureable. At some point, however, perceptions take on a hostile and emotional quality; vigilance and guardedness replace curiosity, and the perseveration becomes negatively focused (6,10,11).

There is remarkable similarity between amphetamine behaviors reported in humans and in animals. Nonhuman primate reactions to the drug include: an initial increase in activity levels which then gradually decreases despite elevation of dose; biting and chewing movements; an increase in searching and examining behavior; and the appearance of a repetitive orienting or "checking" behavior (12,13). Experiments with chimpanzees (14), rhesus (12), and Japanese macaques (15) have described searching and examining behavior, such as picking, self-grooming and probing body parts and inanimate objects, dramatically parallel to that seen in humans. The animals fixate on manipulating objects, a behavior deemed reminiscent of the "hung up" attitude of amphetamine abusers. Garver et al. (13) interprets the "checking behavior" as similar to the hypervigilance seen in the humans. Finally, as amphetamine intoxication progresses, primates tend to become hyperresponsive to stimuli and may appear fearful (12,15).

AMPHETAMINE-INDUCED PRIMATE MODEL OF PARANOIA

The ability of amphetamine to produce paranoid psychosis in humans, and to elicit similar behaviors in both man and primates suggests a basis for an animal model of paranoia. If our concept is valid, that a perceptual system which distorts danger and threat underlies paranoid behaviors, then chronic doses of amphetamine would be expected to elicit species-specific behaviors in the primate which are consistent with perception of a threat in his environment.

The rhesus macaques were chosen because of their aggressivity and rigid hierarchical social system. They, as all nonhuman primates, exhibit a rich repertoire of behavior which enables them to express affiliation, submission, and aggression in a number of ways (16). We felt that a perception of threat, real or imaginary, by an indiviudal in a social colony would likely manifest itself in some overt behavioral change. Maintenance of the hierarchy in a group of rhesus mokeys is critical for the stability of the group (17). When this system is threatened there is an increase in aggression and general disorder until the hierarchy is reestablished. There is strong support suggesting that the hierarchy is maintained not by the dominant animals, but rather by the submissive behavior of subordinants who show submission without obvious cues (18). However, a dominant animal will increase his aggressive behaviors to defend his position in the colony when threatened either by a stranger or by a challenging subordinate. Thus, in these experiments, the amphetmine treatment was expected to elicit an increase in agonistic relation- ships and in normal social distance between the experimental animal and the remainder of the colony. Individual behaviors consistent with perception of threat as, "sit tense and orient," should also increase. Furthermore, behav- iors not directly related to coping with threat, for example, "self-groom" and "explore," should not change in any consistent way across animals.

Our studies have been carried out on two colonies of rhesus monkeys that have been living together for at least six months prior to the start of the experiment. One colony consisted of four adult monkeys (two males and two females) and one infant. They were housed in an 8' x 8' x 7' room. Five adult animals comprised the other colony (one male and four females). These animals were housed in a colony room 8' x 12' x 14'. In both colonies, food and water was available ad libitum. Each animal had been trained to come out of the enclosure each morning to receive fruit, vitamins, and an intramuscular injection of either amphetamine or placebo. Amphetamine was initially given to all animals in a concentration of 0.1 mg/kg. This dose was increased 0.1 mg/kg daily the first week. For the following two weeks, further increments were made only if stereotypy was not observed.

Before any drug was administered, baseline data was collected for three weeks (3). At this point, one animal was chosen randomly from the group and injected with amphetamine once a day for three weeks, while all other animals received saline injections. At the conclusion of any three-week drug period and before any other member of the colony was treated with amphetamine, all animals received saline injections for three weeks. Thus, a typical experimental design would be as follows:

             Period 1 - Baseline, all animals on saline
             Period 2 - A on amphetamine, rest of the colony on saline
             Period 3 - Baseline II, all animals on saline
             Period 4 - B on amphetamine, rest of the colony on saline
             Period 5 - Baseline III, all animals on saline
             Period 6 - C on amphetamine, rest of the colony on saline
             Period 7 - Baseline IV, all animals on saline
             Period 8 - D on amphetamine, rest of the colony on saline
             Period 9 - Baseline V, all animals on saline

Behavioral data on each colony was collected twice daily at 10:00 AM and at 2:00 PM. Data was collected for 15 minutes on each animal and included absolute frequencies of such behaviors as mounts, presents, and agonistic

behaviors. Duration of behaviors such as groom, sit idle, and others were
recorded as the number of half-minute episodes. Table I gives a complete list
of the behaviors and their definitions. In addition the distance between
animals was recorded every 30 seconds. The relative frequency with which each
behavior for each animal occurred during both drug and baseline periods was
computed. Analysis of variance was used to isolate drug effects. The F test
used time X subjects as the error mean square.

The results indicate that amphetamine administration caused a significant
decrease in the time spent "eating" and huddling/sleeping" for each animal
during the three weeks of drug administration only. There were no consistent
changes across animals in "explore," self-groom," or "pace." The frequency of
the behaviors increased for some animals, but was decreased or unchanged for
others. There were dramatic decreases in "sit idle" and "watching other
calmly" while "sit tense and orient" categories sharply increased (Fig. 1).

The average distance between an animal receiving amphetamine and the other
colony members did not increase as we had expected. However, when we looked
at the individual relationships, it became evident that the amphetamine-
treated animal maintained closer contact with one individual, while moving
further away or having less contact with others in the group. In each case we
have examined, the increased contact was directed towards that member of the
colony with whom the test animal, during baseline periods, spent the most time
in affiliative social behaviors as grooming and huddling (Fig. 2).

The most dramatic result, however, was the effect of amphetamine on agonistic
encounters. Agonistic responses dramatically increased during drug periods (P
< 0.01). The dominant animals of a colony, increased their threatening during
drug periods. However, the frequency of submissive responses by other animals
did not increase accordingly. On the other hand, subordinate members of the
group did not increase their threatening behaviors when on the drug, but did
significantly increase their submissive responses. However, the number of
threats or approaches by the dominant animals did not increase (Fig. 3). It
is clear from these results that amphetamine administration increased the
frequency of agonistic behavior in the treated animals. Furthermore, the type
of agonistic behavior was dependent on, and appropriate to, the normal social
position of the individual animal within the colony. However, these increases
in agonistic behaviors, whether aggressive or submissive, were not accompanied
by either increases in eliciting cues or increases in responses by other
members of the colony.

Three major amphetamine-induced alterations in social and individual behaviors
in our primate colonies are clearly compatible with those reported in humans.
A monkey treated with amphetamines appears more tense in its posturing and
shows an increased orientation to noises and movements of other animals. The
animal increases his association with one member of the colony and isolates
himself from the remaining members. Although unexpected, this finding is not
inconsistent with our hypothesis, since, in conflict or threatening
situations, rhesus monkeys do form coalitions and alliances with close
associates (19). Finally, consistent with his behavior patterns but
inappropriate to the cues of his environment, the animal displays marked
increases in agonistic behaviors. Each of these changes may be interpreted as
consistent with a distorted perception of threat or danger in a normally
neutral or unchanged environment. At the descriptive level, then, the model
captures the behaviors associated with amphetamine-induced paranoia in humans.

In conclusion, the evidence suggests that amphetamine administration coupled
with ethological methods provides a means to elicit and observe an animal
analogue of a fundamental feature of human paranoia. It promises a productive
line of research. Although the linguistic subtleties of human paranoia are
not captured in the model, many of the behavioral analogues are. However, the
strength of such a model must reside not only in the behavioral analogues
described here, but also in the convergence of data from pharmacological
studies. We want to establish that: anti-anxiety drugs such as diazapam, do
not decrease amphetamine-induced behaviors associated with threat; drugs which
exacerbate symptoms in humans, such as L-DOPA, also exacerbate behaviors
associated with threat in the primate model; and anti-psychotic agents which
ameliorate symptoms in humans do the same in the primate model.

An important advantage of this model is that we can study the interaction of
social and pharmacological effects. For example, what happens if we
manipulate the environment by removing the target animal's close associated or

TABLE 1   Behavioral Code Definitions

Absolute Frequency

A   Inspect. Peering closely at another animal (particularly at
     genital areas, but can occur before grooming), touching,
     sniffing.
B   Approach. Animal approaches another animal within two feet.
C   Leave. Animal moves out of two-feet range.
D   Social Present. Animal presents to another for grooming or
     mounting, but not in an agonistic context.
E   Mount. To include attempted mounts.
F   Orient. Quick, jerky head movement towards some real (or
     imagined) stimuli, i.e., loud noise, sudden movement,
     interrupting an activity.
G   Toy. Manipulating, biting, playing with toys (Plexiglas).

Agonistic Behaviors

H   Submit I. Turn away, crouch, look away, fear grimace,
     submissive present - but occupying same general area.
I   Submit II. Avoid, moves out of the way of another animal,
     usually a rapid getaway.
J   Threat I. Threat yawn, stare, open-mouth, scalp flick.
K   Threat II. Open mouth, head thrust forward, lunge, but animal
     occupies approximately same space.
L   Threat III. Chase, reaching out to hit, but does not make
     contact.
M   Threat IV. Attack, contact is made.
N   Lip Smack.

Duration

A' Idle Sit. Animal is engaging in no other activity and is
     sitting quietly.
B' Tense Sit. Animal looking around, checking other animals, no
     other activity taking place. Usually looking from animal to
     animal, very ready to move at the slightest cue. Character-
     ized by fidgeting, rapid head and eye movement towards
     others.
C' Social Groom.
D' Self-Groom.
E' Eat/Drink. Chewing food, drinking, not merely picking at food.
F' Watching Others. Not as in Tense Sit (B'). No darting eyes,
     not fidgeting, calm gaze directed toward other animals.
G' Individual Idiosyncratic Behavior. Animals' nondrug peculiar
     behavior.
H' Pace. Walking back and forth covering the same area at least
     four times.
I' Huddle/Sleep.
J' Explore - Manual (sitting, fiddling). Picking at inanimate
     objects, licking or biting objects, fingering (parts of
     cage, tire, food [not eating]). NOT TOYS (see G).

Individual Amphetamine or Drug Behaviors

V-Y Categories to depend on individual drug differences, i.e.,
     rocking, staring: particular stereotypy if not in above
     categories.

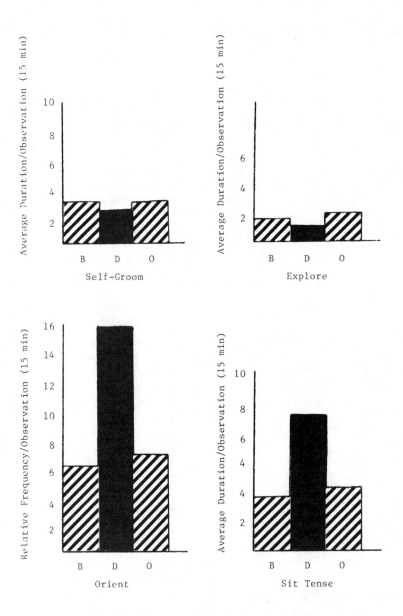

Fig. 1 Differential response of behaviors to amphetamine. Data
for each animal was grouped in the following way: B –
baseline period, no animal on amphetamine; D – amphetamine
period for that animal; O – amphetamine periods of other
members in the colony.
Each graph represents the mean across animals for each
period.

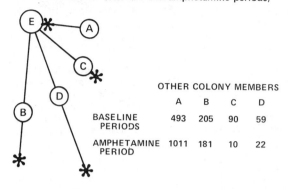

PROXIMITY (baseline and amphetamine periods)

	OTHER COLONY MEMBERS			
	A	B	C	D
BASELINE PERIODS	493	205	90	59
AMPHETAMINE PERIOD	1011	181	10	22

Fig. 2 Amphetamine-induced changes in proximity. Number of 30-second intervals that the target animal E spends in proximity to other members of the colony (ABCD). The length of each line is inversely proportional to the amount of time spent in proximity. The circles represent the mean for the (non-drug) baseline periods; the asterisks represent the value when E was treated with amphetamine.

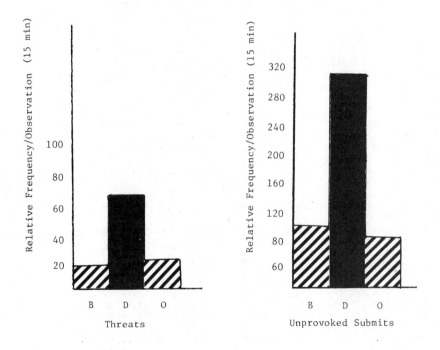

Fig. 3 Agonistic interaction by rank of animal. A. Relative frequency of threats of dominant animals to subordinate animals (data taken from dominant animals only). B. Relative frequency of unprovoked submits to dominant animals (data taken from subordinate animals only).

Data for each animal was grouped in the following way: B - baseline period, no animal on amphetamine; D - amphetamine period for that animal; O - amphetamine periods of other members in the colony.
Each graph represents the mean across animals for each period.

introducing a stranger to the group? How would other species of primates not so dependent on aggression for their social stability respond? As experiments provide increased information about the anatomical and neurochemical specificity of amphetamine, various techniques could be tested toward understanding possible underlying mechanisms, and for better social and pharmacological treatment of the paranoid patient.

ACKNOWLEDGMENTS

This work was supported in part by NIDA grant, DA 01207; Scottish Rite Schizophrenia Research Program; Harry Frank Guggenheim Foundation; and the Office of Naval Research. SH holds a fellowship from Biosciences Training Grant MH 8304. JDB is recipient of Research Scientist Development Award MH 24161. The authors would like to thank R. Bruce Holman and Richard Castillo for their valuable assistance in the preparation of this paper.

REFERENCES

1. N. Tinbergen, The Animal In Its World, Vol. II - Laboratory Experiments and General Papers, Harvard University Press, Cambridge, 164, 1972.
2. P. Polatin, The paranoid states. In: Comprehensive Textbook of Psychiatry, A. M. Freedman, H. I. Kaplan and B. J. Sadlock (eds.), Williams & Wilkins Co., 1975.
3. D. W. Swanson, P. J. Bohnert and J. A. Smith. The Paranoid, Little, Brown, Boston, 1970.
4. H. S. Sullivan, Clinical studies in psychiatry. In: The Collected Works of Harry Stack Sullivan, Vol. 2, H. S. Perry, M. L. Gawel and M. Gibbon (eds.), Norton, New York, 1965.
5. N. Cameron, Behavior Pathology, Houghton-Mifflin, Boston, 1951.
6. P. H. Connell, Amphetamine Psychosis, Chapman and Hall, London, 1958.
7. D. S. Bell, The experimental reproduction of amphetamine psychosis. Arch. Gen. Psychiatry 29:35, 1973.
8. J. D. Griffith, H. Cavanaugh, J. Held and J. A. Oates, Experimental psychosis induced by the administration of D-amphetamine. In: Amphetamines and Related Compounds - International Symposium, E. Costa and S. Garattini (eds), Raven Press, New York, 1970.
9. A. Ellinwood and L. N. Sudilovsky, Evolving behavior in the clinical and experimental amphetamine (model) psychosis. Am. J. Psychiatry 130:1088, 1973.
10. E. H. Ellinwood, Amphetamine psychosis: I Description of the individuals and process. J. Nerv. Ment. Dis. 144:273, 1967.
11. D. S. Bell, The experimental reproduction of amphetamine psychosis. Arch. Gen. Psychiatry 29:35, 1973.
12. E. H. Ellinwood, Effect of chronic methamphetamine intoxications in rhesus monkeys. Biol. Psychiatry 3:25, 1971.
13. D. Garver, F. Schlommer, J. Maas and J. Davis, A schizophreniform behavioral psychosis mediated by dopamine. Am. J. Psychiatry 132:1, 1975.
14. F. Fitzgerald, Effects of d-amphetamine upon behavior of young chimpanzees reared under different conditions. Neuropsychopharmacology V:23, 1967.
15. Y. Machiyama, H. Utena and M. Kikuchi, Behavioral disorders in Japanese monkeys produced by long-term administration of methamphetamine. Proc. Jap. Acad. 46:738, 1970.
16. R. A. Hinde and T. E. Rowell, Communication by postures and facial expressions in the rhesus monkey (Macaca mulatta). Proc. Zool. Soc. London 138:1, 1962.
17. I. S. Bernstein, T. P. Gordon and R. M. Rose, Aggression and social controls in rhesus monkey (Macaca mulatta) groups revealed in group formation studies. Folia Primat. 21:81, 1974.
18. T. Rowell, The Social Behavior of Monkeys, Penguin Books, Ltd., Middlesex, England, 162, 1972.
19. M. E. Agar and G. Mitchell, Behavior of free-ranging adult rhesus macaques: a review. In: The Rhesus Monkey, Vol. I. Geoffrey H. Bourne (ed), Academic Press, New York, 1975.

DISCUSSION

Dr. Echols suggested that one might conclude from Dr. Haber's studies on amphetamines that the previously dominant monkeys dominated more and the previously submissive monkeys submitted more. He wondered if instead of considering this as a model of paranoia, it might be that the animals were performing inappropriately the same sorts of behavior they had performed beforehand. *Dr. Haber* disagreed because of the individual variations of the dominant and subordinate behaviors: they were directed mostly towards the animal in the group with whom that animal had the most ambivalent relationship. For example, a very subordinate animal did not submit any more to the top animal in the group; their relationship was clear and there wasn't much change. However, there was a tremendous increase in submission to the animal directly above her. Thus, the study shows that the individual animals respond most to the animal with whom they had the most ambivalent relationship.

IV. Models in Psychiatry: Affective Disorders

BIOBEHAVIORAL MODELS OF DEPRESSION IN MONKEYS

By

William T. McKinney, Jr., M.D.
Professor and Chairman
Department of Psychiatry
University of Wisconsin School of Medicine
Madison, Wisconsin 53706

INTRODUCTION

An understanding of the appropriate and inappropriate uses of animal models is basic to this workshop. Previously published papers provide more detail on this topic (1-5). However, recent interactions with a variety of professional groups leads me sadly to the conclusion that animal models are poorly, if at all, understood. Indeed, I think we are at a stage where we are confronted with two dangers: 1) Animal models are ignored; 2) The expectations from animal models research are too high. Historically, the animal modeling field has been plagued by these two dangers, that is, by obscure methodology leading to its being ignored, and premature clinical labeling, leading to unrealistic expectations.

Research involving nonhuman primates has come to occupy a special role in the animal modeling field, though equally interesting animal models have also been created in other species. It is true that primates provide investigators with subjects whose physical, cognitive and behavioral characteristics may be quite similar to humans in some regards. However, it should be kept in mind that nonhuman primates are not furry little men with tails. Indeed, I need not tell this group that there are proper scientific restraints that limit cross-species reasoning and ethical restraints that correctly (and incorrectly in some cases) limit our degrees of freedom in both human and nonhuman primate research.

CRITERIA

It has been pointed out by Seligman (6) and by McKinney and Bunney (1-5) that the difficulty in moving from a dramatic analogue to animal models has been due in large part to the lack of ground rules or criteria that might validate the model. Four such criteria have been proposed:

1) Similarity of inducing conditions

2) Similarity of behavioral states produced

3) Similar underlying neurobiological mechanisms

4) Reversal by clinical effective treatment techniques

A number of you will immediately recognize the issue of circularity. One could argue with considerable justification that these parameters are incompletely understood in most forms of human psychopathology. However, this fact does not preclude the use of these criteria as guidelines in the approach to the creation and study of animal models. As a matter of fact, one of the purposes of animal models is to help clarify in a more systematic way some parameters of the human syndromes.

No one of the above four areas can be viewed in isolation from the others. An important implication of the above conceptualization is that models must be viewed as open systems for study, rather than replicas or static endpoints. If you will, a model is a model is a model .... By definition, it is not the real thing. One can only engage in successive approximations. A model can never be proved to be the real thing, but that's not the major reason to develop animal models. Models are created as open systems for study to help understand mechanisms.

A common fallacy is to think that animal models are "proven" on the basis of apparent behavioral similarities between species. For example, someone maintaining this position would say that mechanism studies of separation in primates are justified only if it can be proven beforehand that separation represents a "valid" model of depression. We strongly disagree with this viewpoint. To extend the example further: one does not have to be able to label separated monkeys as "depressed" prior to doing neurobiological and/or rehabilitative studies. We do not know for sure if monkeys separated from their mothers or from peers are depressed, grieving, or "bored." We think it is not the latter and in this presentation will present evidence in this regard. However, the point is that one can get a striking pattern of behavioral alteration following the event of separation which is worth studying in its own right to better understand the underlying mechanisms associated with separation reactions. It is not necessary and even dangerous, perhaps, to make a priori assumptions about clinical labels that might apply to separation, to social isolation or other situations. This point is especially important with the increased emphasis on applied research and the reluctance of some study sections to look favorably on this kind of work unless it can be securely anchored to a specific human clinical syndrome. We feel this is dangerous and shortsighted reasoning.

CLASSICAL SEPARATION STUDIES IN PRIMATES

The role of separation or exit type events as frequent antecedents or precipitants of certain forms of human depression is reasonably well established (7, 8). The literature is not unanimous in this regard (9), however. Certainly the vast majority of the work suggests that for certain predisposed individuals separation is a high risk event in terms of affective disorders. This applies to adults, of course, and a number of workers in particular have studied the occurrence of anaclitic depression in children as a consequence of maternal separation (9-15). Such a syndrome does not always occur following maternal separation, but when it does it is exceedingly dramatic. There is considerable question regarding the relationship between the disorder of anaclitic depression in children and adult depression. Rie (16) has presented an extensive review of some of the relevant literature in this regard and has concluded that such generalization may not be valid. Rie's assertion, if correct, would of course limit the generalizability of anaclitic depression models in humans and animals in relationship to adult depressions.

There have been numerous reports, some of which have been documented and some of which are anecdotal, which deal with the consequences of naturally occurring separation of nonhuman infant primates from their mother. This usually occurs in the wild when the mother has died. For example, Hamburg, Hamburg and Barchas (17) have pointed out that depression in young chimpanzees is a common consequence of loss of the mother. They report a period of active protest, followed by a period of despairing inactivity, social withdrawal and apathy. Most of the chimpanzee infants who lost their mothers died within a few months unless adopted by older siblings. There have been scattered reports of other species which are similar in nature.

The first of the laboratory studies concerned with mother-infant separation in rhesus macaque was performed at Wisconsin by Seay, Hanson and Harlow in 1962 (18). In the original study, four infants and four mothers served as subjects. Two of the infants underwent maternal separation at 169 and 170 days of age, and two at 206 and 207 days of age. The experiment was divided into three time blocks of three weeks each: pre-separation period, separation period, and post-separation period. In this particular study, plexiglass dividers were used, and thus the infants could continue to see and hear all mothers. Detailed behavioral observations were made each day. The initial reaction of all mothers and all infants after separation indicated a high degree of emotional disturbance. Immediately after separation, the infants' behavior included disoriented scampering, high-pitched screeching, including vocalizations, and huddling against the barrier in close proximity to their mother. The mothers displayed an increase in barking vocalizations and in threats directed toward the experimenter. The mother's emotional response appeared to be both less intense and of shorter duration than that of the infants. Later during the separation period, most complex infant-to-infant social behaviors exhibited a drastic decrease in frequency of occurrence. For instance, non-contact play was for all practical purposes obliterated. Threats, approaches and withdrawals also decreased. During the post-separation period, infant-mother clinging rose significantly. This initial investigation demonstrated that separating a rhesus monkey from its mother had striking behavioral effects, but with individual variation in terms of the nature of the response. In general, the sequence of stages seen paralleled the protest and despair stages of Bowlby and suggested a powerful animal analogue for anaclitic depression.

Another study by Seay and Harlow (19) used eight rhesus mother-infant pairs separated at about 207 days of age. In contrast to the above study masonite paneling was used during the separation phase and the infants were not able to see any of the mothers. Behavioral changes were similar to those already described.

Jenson and Tolman (20) in one of the early separation studies examined the short-term effects of separation of mother-infant pairs (Macaca Nemestrina) for less than one hour and found that the infants screamed almost continuously during the separation. The mothers attacked their own cages and tried to escape. On return of the infant, there was a striking increase in the intensity of the mother-infant relationship.

Kaufman and Rosenblum (21, 22) have studied the separation reactions of infant pigtail monkeys (Macaca Nemestrina) and infant-body (Macaca Radiata). Their separation studies differed in several key ways from those previously described. Their subjects lived in groups consisting of several adult females, at least one adult male, and some adolescents. Separation was accomplished by removing the biological mother from the group with her infant being left with remaining members of the group. Separation from the mother in an infant-pigtail study was done at age 4-6 months, and the length of separation was four weeks, after which the mother was brought back into the group. Reunion data were taken for three months. Before separation, bonnet groups were observed to spend long periods of time in passive contact with each other in contrast to pigtails, which did not make much physical contact, except to engage in active social interactions like grooming, aggression, etc. Bonnet females even maintained their high degree of passive contact with other adult bonnets, despite the birth and continued presence of their developing infant. Pigtail mothers, by contrast, were very reluctant to engage socially with others immediately following the birth of their offspring. As a result, bonnet infants spent time during the first months interacting with other adults in addition to their own mother, whereas pigtail infants spent almost all their time with their own mother. During the separation phase, three of the four pigtail infants exhibited the reactions which Kaufman and Rosemblum term "agitation", "depression" and "recovery." The bonnet infants did not exhibit this kind of reaction.

Hinde's group has focused their attention on the short and long-term effects of short separations of rhesus monkeys (23-26). The animals they studied lived in groups which consisted of a male, two to three females, and their young. Separation was accomplished by removing the mother from the infant's sight at approximately 30-32 weeks of age. The length of separation was six days. The infants responded to removal of the mother with behaviors similar to those described above for the pigtail infant. They initially exhibited "whoo" calls and were hyperactive. Then there followed a period of "depression" including hunched posture, decreased activity and social withdrawal. Upon reunion, three of the four infants returned to the pre-separation level of activity within a week, but the fourth showed less activity up to four weeks afterwards. This study also noted that the less an infant had been in contact with his mother before separation, the less clinging he displayed on return, and the more quickly he returned to behaviors similar to those seen before separation, thus demonstrating for the first time that the nature of the pre-separation relationship affects the severity of reactions to mother removal.

Several other aspects of early separation were studied by Hinde, including the effects of the length of separation in which he found that infants who had had a 13 day separation were affected more severely than those that had only a 6 day separation. He also retested infants who had been separated earlier in life when they became 12 and 30 months old, that is, five months and two years after the original separation, to determine if there were long-term consequences of the early separation experience. He found that previously separated infants, when confronted with strange objects in a strange cage, were less likely to approach them than control animals who had not had an early separation experience. Thus, the effects of a mere 6 day absence from the mother were clearly discernable five months later. When tested two years later, the differences were less marked but still present.

Mitchell et al. (27) have also studied the effects of 48 hour separations of Macaca Mulatta (rhesus monkeys) from their mother at three different age periods, two months, three and a half months and five months. At all ages, the infants exhibited signs of protest and despair, even with such a short separation period. This same group also reported that six of twenty-four infants used in their studies showed signs of "detachment" when reunited with their mothers. That is, the mothers would try to retrieve the infants as usual, but the infants would screech and run away before finally establishing ventral contact despite persistent efforts of the mothers to do so.

Preston et al. (28) have extended the research on mother-infant separation to a non-macaque species (Erythrocebus Patas). The subjects were six infant patas monkeys which underwent separation from their mothers when they were approximately seven months of age. During the three week separation, each was individually housed, except for one hour per day when the six of them were permitted access to each other. Thus, the infants underwent maternal separation, but had the opportunity to interact with each other during the separation period. The immediate reaction to removal from their mother was reported to include frequent and intense cooing, frantic searching about, and wide-eyed scanning of the room. The reaction was most intense during only the first half-hour. The infants also stayed in close proximity to each other, and their usual behaviors, for example play, fell to low levels immediately after removal from the mother, though the infants were reported to have remained alert. Thus, in this species the initial protest stage was observed, but little indication of macaque like despair.

120

In summary, anaclitic depression is hardly an invariant consequence of all mother-infant separations which occur among nonhuman primates. In some species, investigators have as yet been unable to produce a depressive-like phenomenon via maternal separation, though this may reflect the fact that species specific techniques have not yet been developed. Studies involving nonmacaque primate species have revealed a similar lack of depressive response to separation from the mother. Like bonnet macacs, patas infants intensified their interactions with remaining group members following departure of their mothers. Reunion revealed detachment-like behaviors. Kaplan (29) and Jones and Clark (30) studied the reactions of squirrel monkeys, a species of new world monkeys to separation from mother. Both groups reported that squirrel monkeys exhibited well defined agitation immediately after separation, but prolonged despair behavior was not seen in these subjects.

Indeed, recent research suggests that there exists as much variation in the reaction of rhesus infants to separation from mother as exist across the several primate species whose reaction to maternal separation have been studied empirically. In rhesus monkeys, as in humans, not all separations from mother yield anaclitic depression. For example, Young, Lewis and McKinney (31) reported a series of five studies involving physical separation of mother and infant monkeys. Parameters of age were varied, along with housing conditions during separation. Considerable variation in the reaction of the infants to separation of their mothers was reported. Some subjects responded with the classical protest-despair sequence, whereas other infants exhibited considerable agitation, but little or no subsequent despair. Finally, a few subjects showed little, if any, behavioral disruptions following separation from their mother. The diverse findings in response to separation did not reflect methodological or procedural differences in the studies. There appeared to be almost as much variation among subjects separated under similar or identical conditions as there was across the various studies.

These kinds of data would lead one to the current conclusion that the best view of maternal separation in macaque infants is that it might provide an appropriate and probably an excellent model of maternal separation in human infants. Clearly not all of human mother-infant separations yield depression, and then again not all mother-infant separations in primates yield depressive-like behavior. There appears to exist enormous variation, both in humans and in monkeys with regard to maternal separation. Studies currently underway at our lab and others are attempting to study some of the parameters which influence the nature of the response following maternal separation. When the response does occur, it is dramatic and sometimes leads to death. A picture of an infant showing the despair response is shown in Fig 1.

Fig. 1. A typical rhesus monkey during "despair" stage following separation from mother.

For example, the age at which the infant monkey is separated from his mother has been systematically varied (32). The general finding from these studies has been that age of the infant at time of separation seems to be relatively unimportant if the subjects are under a year of age, but at least sixty days old, and all other variables are held constant. The one exception to this rule occurs among subjects three months old at the time of separation. Such infants exhibit a quantitatively more severe reaction to separation from mother to those either older or younger.

Sex of the separated infant has been examined as a relevant variable, but the findings to date have not been definitive. There is little evidence that male infants are a priori more prone to depressive reaction following separation from mother than young females or vice versa.

However, several studies, including some cited above have indicated that the nature of an infant's attachment relationship may affect the nature of his response to removal from the mother (32). Thus, the nature of the attachment prior to separation apparently is a variable of import in determining the reaction to maternal separation in both monkey and man.

Another set of variables that have been found to influence the nature of an infant's reaction to separation from mother involves parameters of the separation environment. For example, duration of separation has been shown to be an important variable in determining both the course of separation reaction and the extent of long-term consequences on the infant's behavioral development. Generally speaking, the longer the separation between mother and the infant is maintained, the more severe are both the immediate and long-term consequences (32).

The form of separation may also be a relevant variable, and in some of the earlier studies it was found that subjects who could still see their mothers throughout the period of separation generally exhibited more disturbance than those who could not see their mothers (32).

There is data to indicate that variation in the nature of the post-separation social environment can profoundly influence the infant's reaction to loss of mother. Such an environment can either facilitate or ameliorate the response to separation. For example, those infants whose separation social environment contains socially interested adult female caretakers seem least likely to suffer severe consequences from the separation.

One other set of variables that can influence the long-term consequences of separation from mother include those which characterize the nature of the infant's environment following reunion with mother. Generally speaking, it has been found that the more similar the reunion environment is to the pre-separation environment, both with respect to its physical characteristics and to the social behavior of the individuals within it, the more rapid and complete will be the infant's return to his previous pattern of social development (32).

PEER-PEER SEPARATION IN RHESUS MONKEYS

Recent research has indicated that the mother-infant bond is not the only affectional bond whose disruption may lead to depressive-like behaviors in rhesus monkeys. It has been found in studies by Suomi, Harlow and Domek (33), and by Bowden and McKinney (34) that peer-peer separation of rhesus macaque monkeys has powerful effects that in many ways mirror the effects following maternal separation. It is clear that infant monkeys reared with peers and then separated will under certain circumstances exhibit the biphasic protest-despair reaction described for mother-infant separation. It has recently been demonstrated that most of the same variables which determine the nature and course of mother-infant separation have similar effects for peer-peer separation. If this is so, then the peer-peer separation model has many practical advantages. Importantly, peer bonds are more susceptible to experimental manipulation than are mother-infant bonds. One can do separations and reunions on a repetitive basis, thus allowing drug trials with cross-over and blind designs. Such niceties would be virtually impossible with mother-infant separation paradigms.

The most striking protest-despair reaction to peer-peer separation has occurred in monkeys under a year of age or in older animals. Efforts to produce depression in 3-year old monkeys, vis a vis peer separations, have not been very fruitful. For example, Bowden and McKinney (34) housed feral-born adolescent rhesus monkeys in pairs for six to eight months, a period of time felt to be sufficient for the development of mutual attachment bonds. Pair members were then physically and visually separated from each other for a period of two weeks, and then reunited. All subjects responded to peer separation with active protest during the first few days, but there was little indication of subsequent despair. Instead, the subjects returned to pre-separation levels of most nonsocial activities, except for a gradual but slight increase in self-directed activity and occasional stereotypic rocking. In another study, McKinney, Suomi and Harlow (35) studied the responses of well socialized adolescent monkeys to repetitive peer separation. The subjects were three-year-old male monkeys that had been reared with their mothers and peers for their first two years of life, and then housed individually for the third year. Two groups of four monkeys were formed, and members of each group were permitted unlimited interaction with each other for four weeks. When subjected to repetitive separations, they showed a clear-cut protest stage following separation, but little evidence of despair during the two weeks. A third study involving peer-peer separation of juvenile age rhesus monkeys by Erwin et al. (36) reported essentially the same findings.

However, when one moves to peer-peer separation in even older animals, different results are obtained. Peer separation reactions have been described by Suomi et al. (37) in five-year-old rhesus macaques reared in enriched nuclear family environments and separated from their families. Those housed individually during the separation did show the protest-despair sequence, and of even greater importance when reintroduced into their families where they clearly had the opportunities for social interaction in a familiar setting, were passive and nonresponsive to social initiation. In other words, just because most of the separation work has been done with young subjects just under a year of age, they should not be dismissed as relevant only to childhood depression. Although one does not see the biphasic reaction in juvenile age rhesus monkeys, when one studies older animals one again detects a biphasic reaction to peer-peer separation.

Although I have spoken of the cautions one must exercise in assessing the significance of separation work in relation to human depression, I am so often asked the question how the separation syndrome in rhesus macaques compares to human depression, that the following table has been organized.

TABLE 1 Comparison of Separation Syndromes in Rhesus Macaques with Human Depression

Criteria for Animal Model	Human Depression	Separation Syndrome(s) in Rhesus Macaque
1. Inducing condition	Can be separation	Separation
2. Behavioral similarities	Activity↓ Intake↓ Interest in surroundings↓ Social Activities↓ Sleep Disturbances Self-Directed Behaviors↑ Death	↓Locomotion ↓Food and water intake ↓Environmental Exploration ↓Play ↓Social Explore Probable ↑Self clasp, Huddle, Rocking Yes
3. Neurobiological Mechanisms	Extensive literature Suggest variety of disturbances	Protest Stage - yes Despair Stage - not studied
4. Rehabilitation	Social Tricyclics MAOI ECT Lithium Other agents	Yes - needs further study Yes (Imipramine) No Data - needs study Prevented death in 3 of 4 cases No Data - needs study Needs study

ARE THE RESULTS OF SEPARATION "ARTIFACTUAL" OR "REAL"

Serious questions have been raised about whether the changes seen following either kind of separation are artifactual or real. The above data, plus the following considerations indicate one is witnessing more than artifact, i.e., fewer behaviors being possible during separation, being housed alone, not giving the subjects ample opportunity for normal behaviors following attachment bond disruption, etc.:

1) The reaction to peer separation can be obtained even with other unfamiliar subjects present during the separation.

2) Not all subjects show the reaction, even though all have the same artifactual opportunities (for a model to be valid, one would expect individual differences not homogeneity).

3) Virtually all the parameters that influence the occurrence and severity of the response to maternal separation have been shown to have similar effects in peer separation studied.

4) The fact that the syndrome is responsive to imipramine treatment and not placebo. An important study to compliment this finding would be one showing the lack of effectiveness of an active, but not antidepressant neuroleptic.

5) Peer separation reactions that have been described in younger animals have also been shown to occur in 5-year-old rhesus macaques reared in an enriched nuclear family environment and separated from their families.

Let me reiterate something that was said above, that separation reactions can be likened to human depression in a number of aspects which have been listed. If one carefully examines the experimental situations in which depressive type reactions occur as a consequence of the separation, certain consistencies appear as outlined by Suomi (38). In each case: 1) the monkey loses a critical portion of its environment, i.e., it no longer has available part of its social environment towards whom a major portion of its interactions were previously directed; 2) there is nothing in its separation environment which can replace what it lost via separation; and 3) it has no power or ability to change its current social situation. We do not argue, of course, that separation is the only social event which can precipitate depressive reaction, either in individuals or monkeys. It is merely one which has been studied more than other possibilities. In conclusion, we contend that separation studies in nonhuman primates offer a useable animal model for understanding certain aspects of human depression especially in relation to separation, an event frequently associated with human depression. At a minimum, they offer an experimental system in which the various parameters which influence the response to separation can be systematically studied, including underlying neurobiological mechanisms. In many ways this is the most important use of separation models. Rather than endless debates about whether or not they model human depression, we should study the behaviors produced as a consequence of separation in terms of possible underlying neurobiological mechanisms and try various agents that might reverse these reactions.

RECENT DIRECTIONS IN SEPARATION RESEARCH

There are a number of new and exciting avenues being pursued by investigators interested in studying separation in primates. These include, but are certainly not limited to, the following: 1) the study of the various behavioral parameters influencing the nature of the response to maternal and to peer-peer separation; 2) the examination of how subtle alteration of amines by the use of enzyme inhibitors might influence the response to peer-peer separation; and 3) how treatment drugs might alter the response to peer separation. The first of these has already been discussed. Perhaps a quick overview of the last two would be helpful.

How Subtle Alterations Of Amines By The Use Of Enzyme Inhibitors Might Influence The Response To Peer Separation.

The protest-despair response to separation from a peer group is usually compared to or described in terms of its similarity to the more widely studied phenomenon of protest and despair following mother-infant separation in monkeys and humans. As has been found to be the case with mother-infant separation, all groups and individuals do not show the same severity of response. With regard to peer separation, the identification of groups which either do or do not respond adversely to peer separation presents the possibility of examining the pharmacological interactions of drugs which may exacerbate or ameliorate some of the effects of separation. In this regard, we have been studying the possible interactive effects of various biogenic amine enzyme inhibitors with the response to peer separation. That is, can we exacerbate the response to peer separation by the use of enzyme inhibitors which alter amine metabolism. An interactive effect has been defined according to the following criteria:

1) Drug treatment should have no effect on group social behavior.

2) Drug treatment should have no effect on chronically separated, singly housed monkeys, even though the lack of stimulation from the environment is identical to that of recently separated monkeys.

3) Drug treatment should alter the response to separation from the peer group as compared to the separation response when receiving placebo.

Three inhibitors have been studied thus far - alphamethylparatyrosine (AMPT), parachlorylphenylalanine (PCPA), and fusaric acid. In a series of studies, peer groups of rhesus monkeys were repeatedly separated and reunited while receiving on a daily basis placebo or active drug via nasogastric intubation. Initiation of drug treatment was always one week before separation while the animals were housed in a peer group. Therefore, we were able to observe whether the drug had any effect on social behavior, and the onset of drug effect was never coincident with peer separation. Separations on AMPT, fusaric acid, PCPA, or PCPA and AMPT in combination, as compared to appropriate placebo separations revealed: 1) that AMPT potentiated the despair response to separation in doses as low as 25 mg/Kgm per day (39); 2) fusaric acid ameliorated the despair response in doses as low as 5 mg/Kgm per day; 3) PCPA and PCPA and AMPT in combination had no interactive effect in peer separation in doses as high as 200 mg/Kgm per day (PCPA alone or 50:50 of the combination). AMPT and fusaric acid did not have effects on group behavior or on the behavior of chronically singly caged animals at dose levels that interacted with separation. Of significance is the finding that this interaction occurs at a dose of 25 mg/Kgm per day, whereas a previous study reported effects on primate social behavior at a dose of 250-500 mg/Kgm per day. The fact that a 10-20 fold dose response difference might be observed between these two paradigms suggest that effects of the high dose could be due to sedative or toxic effects of AMPT; whereas, the fact that behavioral effects of 25 mg/Kgm per day could only be observed in the separation context suggests that sedation is not a factor at this dose.

These data suggests that alterations in catecholamine metabolism which may have no observable effect in social groups or in chronically single housed monkeys may significantly alter the response to separation. The fact that chronically single housed monkeys do not show effects at this dose suggested it is not the deprived stimulus conditions of the single housed situation which interact with drugs, but features the separation event itself.

How Treatment Type Drugs Might Alter The Response To Peer Separation.

One of the criteria which has been proposed with regard to validating animal models concerns their reversibility by clinically effective treatment techniques. One study has been done designed to examine the effects of the tricyclic antidepressant imipramine in rhesus monkeys subjected to repetitive peer separations (40). Imipramine was chosen because its effectiveness as an antidepressant is well established. We used repetitive peer separation because previous work with this technique had indicated that this paradigm had numerous methodological advantages in a drug study as opposed to the usual mother-infant separations.

The subjects were eight rhesus monkeys, each removed from their mothers at birth and reared for the first thirty days of life according to standard laboratory nursery procedures. They were then divided into two groups: a placebo group, and an imipramine treatment group. Once all animals were at least thirty days of age, a group was removed from the laboratory nursery and placed in a living cage in which all group members had unlimited physical access to each other 24 hours per day. This group housing early in life was for the purpose of developing strong mutual attachments in the infant monkeys. Thus, each group remained together for eight weeks prior to the initial separation (pre-separation phase). When all members were at least ninety days of age, the separation phase of the study was initiated. Each member of the group was removed from the living cage and housed individually in a bare wire-mesh cage in another room, where it could see and hear, but not physically contact the other group members. After nineteen days of such separation, all subjects were returned to their living cage for a nine-day period of reunion. Immediately after the ninth day of reunion, the subjects were separated again for a second nineteen-day period, during which time they were housed exactly as they had been during the first separation. They were then reunited for nine days as before. A third nineteen-day separation and nine-day reunion followed. After the final day of the third reunion the group members were permitted unlimited mutual interaction 24 hours per day without interruption, as was the case prior to the first separation. They were maintained in this manner for an additional twelve-week period (post-separation phase).

On the first day of the first reunion, placebo intubation was started for all subjects in both groups. The placebo intubation consisted of administration of 1 cc. per kilogram of body weight of tap water via nasogastric intubation. The placebo administration continued thereafter for the placebo group throughout the study. On the eighth day of the second separation after 16 days of placebo intubation, actual drug treatment was begun for members of the experimental (imipramine) group, and it continued on a daily basis until the end of the third reunion. Dosage was set at 10 mg/Kgm of body weight. The dosage volume was the same as for the control group. To summarize, during the first separation, no intubation or drug administration took place. Placebo intubation began on the first day of the first reunion, and actual administration of imipramine began on the eighth day of the second separation and continued through the final day of the third reunion. There was no intubation or drug administration during the post-separation phase of the experiment. Data were collected by blind testers according to thirteen predefined behavioral categories.

The results of this study are extremely interesting. First of all, analysis of behavioral scores recorded during the pre-separation phase revealed no statistically significant difference between the two groups. That is, prior to separation, subjects in both groups demonstrated age appropriate social behaviors and did not differ from each other. Both the experimental and control groups responded to the first separation in a manner consistent with that reported in previous studies of peer separation in infant rhesus monkeys. This includes significant increases of pre-separation levels of self-clasping and huddling and decreased exploration. Separation too was marked by several differences between the groups in which the animals on imipramine actually showed some slightly increased levels of self-clasp and huddling, and slightly decreased levels of locomotion during the separation. If anything, the imipramine monkeys showed somewhat more pronounced depressive-like behaviors during the second separation. Remember, they had been started on imipramine half-way through the separation. However, the group differences observed through the second separation were largely reversed during the third separation. By this time, the imipramine animals had been on drugs approximately fourteen days. The data suggest that chronic imipramine treatment was associated with less self-directed behaviors such as self-clasping and huddling and more interest in environment in comparison with the same sex placebo treated controls. In other words, monkeys after chronic imipramine treatment reacted to separation with generally less self-directed behaviors and more environmentally directed behaviors than they did at the start of treatment. After chronic placebo treatment, monkeys reacted to separation with more self-directed behavior and less environmentally directed behavior than they did when placebo treatment was begun. During the post-separation and the first two weeks of the post-separation period, it was apparent that having been on imipramine had prevented the experimental group from showing the maturational arrest that is normally seen as a result of repetitive peer separations. However, during the last two weeks of follow-up, that is the experimental group had been off of imipramine 11-12 weeks, there was basically no difference between the groups.

Thus, the results of the present study suggest that imipramine treatment greatly influences the behavior of monkeys during and immediately following peer separation, but that these effects are largely transitory once treatment is ceased.

The repetitive peer separation paradigm could lend itself to the testing of a variety of agents, some of which have been listed in a previous table.

By the separate and combined use of the above two approaches, that is the use of agents which enhance the response to peer separation, as well as those which might reverse it, it's hoped that studies in animal models will enable us to do at least two things: 1) help us in understanding some of the neurobiological mechanisms associated with separation syndromes, and 2) provide a useful social system for preclinical psychopharmacological evaluations.

CONCLUSION AND SUMMARY

I have tried to outline some of the models that my collaborators and I are working with at Wisconsin. There are other animal model systems being studied that are relevant to this workshop that I did not discuss. For example, the learned helplessness models, the social isolation models, etc. The latter probably does not relate to depression per se, but provides a useful system for studying: 1) the effects of early social deprivation and psychobiological functioning, and 2) another system for preclinical psychopharmacological type evaluations.

I view the animal modeling field as already in a new and exciting era. Older, naive simplistic conceptualizations about animal behavior in relation to human psychopathology are being discarded. I think one of the unique contributions of animal models is to provide systems in which the complex interactions between social and biological variables can be studied and more than just correlations obtained.

REFERENCES

1. McKinney, W.T. and Bunney, W.E., Animal model of depression, I. Review of evidence: implications for research, Arch. Gen. Psych. 21, 240 (1969).
2. McKinney, W.T., Animal models in psychiatry, Persp. in Bio. & Med. 17, 529 (1974).
3. McKinney, W.T., Suomi, S.J. and Harlow, H.F., Methods and models in primate personality research. (1973) Individual Differences in Children. Wiley, New York.
4. Morrison, H.L. and McKinney, W.T., Environments of dysfunction: the relevance of primate animal models, (1976) Environments as Therapy for Brain Dysfunction. Plenum, New York.
5. Morrison, H.L. and McKinney, W.T., Models of human psychopathology: Experimental Approaches in primates, Introduction to Clinical Neuropharmacology. Spectrum, New York, (in press).
6. Seligman, M.E.P., Lein, D.C. and Miller, W.R. Depression, Handbook of Behavior Therapy. Appleton-Centery Crafts, New York, (in press).
7. Leff, M., Rvatchz, J., and Bunney, W.E. Environmental factors preceding the onset of severe depressions, Psychiatry. 33, 293 (1970).
8. Paykel, E., Myers, J., Dievelt, M., et al., Life events and depression, Arch. Gen. Psychiatry. 21, 753 (1970).

9. Hudgens, R., Morrison, J., Barchha, R., Life events and onset of primary affective disorders, Arch. Gen. Psychiatry. 16, 134 (1967).
10. Robertson, J., Some responses of young children to loss of maternal care, Nurs. Times. 49, 382 (1955).
11. Robertson, J. and Bowlby, J., Responses of young children to separation from their mothers, Cour du Centre International de l'Enfance. 2, 131 (1952).
12. Bowlby, J. Grief and mourning in infancy and early childhood, Psychoanal. Study Child. 15, 9 (1960).
13. Bowlby, J., Separation anxiety, Int. J. Psychoanal. 41, 89 (1960).
14. Bowlby, J., Childhood mourning and its implications for psychiatry, Am. J. Psychiatry. 118, 481 (1961).
15. Spitz, R.A., Anaclitic depression: An inquiry into the genesis of psychiatric conditions in early childhood, II., Psychoanal. Study Child. 2, 313 (1946).
16. Rie, H.E., Depression in childhood, J. Am. Acad. Child Psychiatry. 5, 653 (1966).
17. Hamburg, D.A., Hamburg, B.A. and Barchas, J.D., Anger and Depression of Behavioral Biology, (1974) Parameters of Emotion, Raven Press, New York.
18. Seay, B., Hansen, E. and Harlow, H., Mother infant separation in monkeys, J. Child. Psychol. Psychiatry. 3, 123 (1962).
19. Seay, B. and Harlow, H.F., Maternal separation in the rhesus monkey, J. Nerv. Ment. Dis. 140, 434 (1965).
20. Jensen, G.D. and Tolman, C.W., Mother-infant relationship in the monkey, macaca nemestrina: the effect of brief separation and mother-infant specificity, J. Comp. Physiol. Psychol. 55, 131 (1962).
21. Kaufman, I.C. and Rosenblum, L.A., The reaction to separation in infant monkeys: Anaclitic depression and conservation-withdrawal, Psychosom. Med. 29, 648 (1967).
22. Kaufman, I.C. and Rosenblum, L.A., Depression in infant monkeys separated from their mothers, Science. 155, 1030 (1967).
23. Hinde, R.A. and Davies, L.M., Changes in mother-infant relationship after separation in rhesus monkeys, Nature. 239, 41 (1972).
24. Hinde, R.A. and Davies, L., Removing infant rhesus from mother for 13 days compared with removing mother from infant, J. Child. Psychol. Psychiatry. 13, 227 (1972).
25. Hinde, R.A., Spencer-Booth, Y. and Bruce, M., Effects of 6-day maternal deprivation on rhesus monkey infants, Nature. 1021 (1966).
26 Hinde, R.A. and Spencer-Booth, Y. Effects of brief separation from mother on rhesus monkeys, Science. 111 (1971).
27. Mitchell, G.D., Abnormal Behavior in Primates, (1970) Primate Behavior, Academic Press, New York.
28. Preston, D.G., Baker, R.P. and Seay, B., Mother-infant separation in the patas monkey, Dev. Psycho. 3, 298 (1970).
29. Kaplan, J., The effects of separation and reunion on the behavior of mother and infant squirrel monkeys, Dev. Psychobiol. 3, 43 (1970).
30. Jones, B.C. and Clark, D.L., Mother-infant separation in squirrel monkeys living in a group, Dev. Psychobiol. 6, 259 (1973).
31. Lewis, J.K., McKinney, W.T. and Young, L.D., Mother-infant separation in rhesus monkeys as a model of human depression: a reconsideration, Arch. Gen. Psychiatry. 6, 699 (1976).
32. Suomi, S.J., Factors Affecting Responses to Social Separation in Rhesus Monkeys (1976) Animal Models in Human Psychobiology, Plenum Press, New York.
33. Suomi, S.J., Harlow, H.F. and Domek, C.J., Effect of repetitive infant-infant separation of young monkeys, J. Abnorm. Psychol. 76, 161 (1970).
34. Bowden, D.M. and McKinney, W.T., Behavioral effects of peer separation, isolation and reunion on adolescent male rhesus monkeys, Dev. Psychobio. 5, 353 (1972).
35. McKinney, W.T., Suomi, S.J. and Harlow, H.F., Repetitive peer separations of juvenile-age rhesus monkeys, Arch. Gen. Psychiatry. 27, 200 (1972).
36. Erwin, J., Mobaldi, J. and Mitchell, G., Separation of rhesus monkey juveniles of the same sex, J. Abnorm. Psycho. 78, 134 (1971).
37. Suomi, S.J., Eisele, C.D., Grady, S.A. and Harlow, H.F., Depressive behavior in adult monkeys following separation from family environment, J. Abn. Psycho. 84, 576 (1975).
38. Suomi, S.J. and Harlow, H.F., Production and Alleviation of Depressive Behaviors in Monkeys, Psychopathology: Experimental Models, Freeman, San Francisco. (in press)
39. Morrison, H.L., Kraemer, G.W. and McKinney, W.T., Protest-despair response to peer separation in rhesus monkeys and its potentiation by alteration of catecholamine metabolism, presentation Am. Psychoso. Soc., Pittsburg, 1976.
40. Suomi, S.J., Seaman, S.F., Lewis, J.K., Delizio, R.D. and McKinney, W.T., Effects of imipramine treatment on separation-induced social disorders in rhesus monkeys, Arc. Gen. Psychiatry. (in press)

MATERNAL SEPARATION IN MONKEY INFANTS: A MODEL OF DEPRESSION

Martin Reite

University of Colorado Medical Center, Denver, Colorado 80262

INTRODUCTION

With the goal of developing a useful animal model of depression, our laboratory has been studying the physiological correlates of the "agitation-depression" or "protest-despair" reaction seen in infant monkeys who have been separated from their mothers. In group living pigtail (M. nemestrina) monkey infants, the period immediately following maternal separation (the mother is removed, leaving the infant in a familiar social group) is characterized by increased locomotor activity, often of a searching nature, and frequent cooing, the distress call of the young macaque. This period of behavioral "agitation" is followed in 18-24 hours by a constellation of behaviors characterizing the "depressive" reaction, which include a decrease in locomotion, impaired coordination, decreased play, a characteristic slouched posture, and a sad facial expression quite similar to that which Darwin described as "universally and instantly recognized as that of grief. . ." (ref. 1). The depressive component of the agitation-depression reaction has been thought by several investigators to represent an animal model of certain loss-related human affective disorders (ref. 2,3,4,5).

My goals in this paper are four-fold and include: (1) an examination of certain of the conceptual problems involved in using animal models in psychiatry; (2) an outline of the overall systems approach we have developed in our laboratory for studying physiology and behavior in unrestrained group living monkey infants; (3) a summary of the data we have obtained to date on the physiology of the agitation-depression reaction in infant pigtail monkeys; and, (4) a brief discussion of how this animal model might relate to human affective disturbances.

Animal Models in Psychiatry: Some General Considerations

The development of more comprehensive animal models of human mental illness may well presage a major step forward for psychiatric research. While advances in basic medical research of the last half century have been both dramatic and related in a very direct manner to the intelligent utilization of animal models, advances in psychiatric research by comparison have been less impressive, a situation that has been due in part to our inability in psychiatry to utilize animal models in a similar fashion.

The utilization of animal models in other areas of medicine stems from the fulfillment of the basic requirement that the phenomena in question must share biological structures and/or mechanisms common to man and the experimental animal. This requirement is more easily fulfilled when we are dealing with the structure and function of basic physiological systems and pathophysiological processes, many of which are common to the mammalian class, and perhaps even animals generally.

Animal model utilization is equally important to psychiatry if we are to see research advances commensurate with those in other areas of medicine, but considerably more difficult. We might propose a general rule: the degree of possible extrapolation from animal models of behavioral disorders to man is directly related to the extent to which the phenomena in question share common biological and psychosocial structures and mechanisms. Most importantly, behavioral similarity is not sufficient in and of itself. This point cannot be overstressed, for the psychiatric literature is replete with nonsensical analogies suggesting aberrant behavior in all manner of creatures as models of mental illness in man.

Similar behaviors, normal and abnormal, need not imply similar mechanisms. An example of this, as has been pointed out by Bowlby (ref. 6), is seen in the area of mother-infant attachment behavior. Well-documented attachment behavior is seen in birds, and attachment behavior is prominently manifested by young mammals, especially primates. Yet the common ancestor of present day birds and mammals was an early reptile, and the evolutionary lines have been distinct since that time. Thus, it appears almost certain that attachment behavior has evolved independently in both groups, an example of parallel evolution. This, in addition to the fact that brain structure (and the relationship of brain structure to behavior), is very different in birds and mammals, makes it almost certain that the mechanisms underlying attachment behavior are very different in birds and mammals. The fact that the behavior is present in both groups tells us that the behavior in question has significant survival value, and it may help us evaluate those factors relating to the evolutionary pressures that select the similar-appearing behaviors in both groups, but clearly, if we are interested in the ability to extrapolate as to mechanisms, one is not a proper model of the other--in either direction.

The advances in nonhuman primate behavioral biology of the last several decades, considered in the light of the ever-increasing evidence that certain of the major mental illnesses in man are strongly rooted in biological disturbances, offer the promise of developing animal model systems of value to scientists concerned with certain areas of human psychopathology.

When considering the role of animal models in the study of human mental illness, the higher primates possess certain unique advantages. For example, there is little doubt that high primates more closely resemble man in terms of both CNS morphology and behavioral plasticity and richness than do other existing nonprimate species. There is also little doubt that man shares both a greater and more recent phylogenetic heritage with the neighboring primates than with other species (ref. 7).

The model concept as we use it in our paradigm, relating to the behavioral depressive response following separation, accentuates two areas, early developmental behavioral-physiological correlations and affective behavioral disturbances, which we believe will maximize the probability of obtaining useful results.

The general problem of extrapolation from primate model to the human condition appears less complex when dealing with early development. It seems probable that the immature nonhuman primate much more closely resembles the immature human in terms of physiological-behavioral interactions and development than would be true of the mature adult. Clearly, developmental rates vary significantly, but most obviously with this configuration we are not forced to deal with the complexities of adult human CNS functions (e.g. language) and social influences.

Likewise, it is reasonable to believe that there is far more similarity in the CNS activation patterns underlying emotional and affective behavioral states in man and neighboring primates than those underlying the most complex cognitive activity available to each, as when dealing with affective behaviors we are dealing with phenomena which would appear to be based in significant part on the activity of CNS systems (e.g. limbic and mesodiencephalic) shared by the higher primates. The assumption that homology in function and ultimately behavior appears reasonable, although we cannot speak to the issue with certain knowledge. This concept is supported by Bowlby (ref. 6), who states ". . .whatever (attachment) behavior is found in subhuman primates is likely to be truly homologous with what obtains in man".

The foregoing limitations implicitly suggest that there may be areas of human behavior and human psychopathology for which good animal models cannot exist. When dealing with human phenomena which depend primarily upon activity in CNS systems essentially unique to man, e.g. those involving the very highest level of cognitive function, our limitation would suggest that animals lacking those CNS structures or functions probably would not be usable as models of the human condition. Failing to take such limitations into account would lead to the continued inappropriate use of animal models based upon behavioral similarity alone.

A SYSTEMS APPROACH TO STUDYING PHYSIOLOGY AND BEHAVIOR IN MONKEY INFANTS

It was early apparent that if we were going to derive maximum benefit from the maternal separation paradigm as an animal model, we would have to know more about what was happening to the behaviorally depressed monkey infant; specifically, what was the physiological status of the infant. Since our experimental paradigm involved the study of mother-reared, group living infants, collecting the physiological data initially presented some obstacles, as we could not catch, restrain, tie-down, or connect wires to our young subjects. The only alternative was to collect our physiological data utilizing radio telemetry.

Commercially available biotelemetry systems did not meet our needs, which included small size, very low power consumption, and total implantability. I stress the importance of implantability for two reasons. First, it avoids the necessity of penetrating the skin barrier with the resultant chronic infections and tissue reactions, an important consideration in any long term study. Secondly, by avoiding any externally mounted devices in our group living subjects, we avoid a source of artifact that may further complicate an already very complex social situation.

Largely through the efforts of Dr. J. Donald Pauley in our laboratory, we developed a totally implantable, seven channel biotelemetry system that would allow us to transmit body temperature, heart rate, eye movement, muscle activity, and three channels of EEG from an unrestrained monkey infant (ref. 8). A block diagram of the telemetry transmitter is illustrated in Fig. 1.

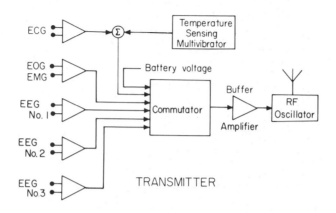

Fig. 1. Block diagram of telemetry transmitter.

Seven channels of physiological information are encoded on five transmitted data channels. The electrocardiogram (ECG) and body temperature are multiplexed on one channel, as are eye movement and muscle activity. Body temperature is multiplexed on the ECG waveform by means of a temperature sensitive oscillator which frequency modulates a nominal 150 Hz sine wave, which is superimposed on the ECG waveform and later separated by band pass filtering. In a like manner, eye movement (EOG) and muscle activity (EMG) are combined on a single channel by placing one electrode of a differential input amplifier next to the eye and the other electrode in the posterior nuchal musculature, again to be later separated by band pass filtering. Each of the three EEG channels has its own differential input amplifier with a frequency response essentially flat from 1 to 100 Hz. The five data channels are then commutated along with a battery voltage sample at a 1.7 KHz rate, producing a pulse amplitude modulated subcarrier used to modulate an FM carrier in the 88-108 MHz band.

The telemetry transmitter is constructed by hand on a 1/32 inch thick PC board using discrete components and commercially available IC's. The completed circuit board is coated with a waterproofing compound,[1] potted in epoxy, coated with a layer of a 50-50 mixture of beeswax and paraffin and finally coated with medical grade silastic prior to implantation. The transmitter is powered by a 2.8 volt lithium cell, which provides a nominal 1000 hours of continuous operation and a range of 10 M. The system can be turned off and on inside the animal using either a magnetically operated bistable latching reed switch or an RF-activated command receiver attached to the lithium cell (ref. 9).

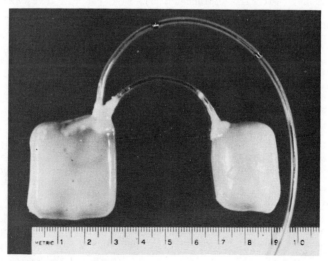

Fig. 2. Telemetry system prior to implantation.

---

[1]Humiseal Type 1B66, Humiseal Div., Columbia Technical Corp., Woodside, New York 11377.

A photograph of a completed telemetry transmitter is illustrated in Fig. 2. The larger square package contains the signal conditioning electronics and the RF oscillator; the smaller package contains the lithium cell and the command receiver and/or latching reed switch. The cable exiting the top of the larger package contains the electrode lead wires, as well as a pair of attached EKG electrodes. All wires used in the electrode lead wires, EKG electrodes, and the cable connecting the battery to the transmitter package, are multi-stranded, teflon coated stainless steel.[2]

The completed telemetry system is sterilized with ethylene oxide gas and then, under general (pentobarbital) anesthesia, surgically implanted in the anterior abdominal walls, with the electrode leads being tunneled subcutaneously to their various recording sites (ref. 10). The infants tolerate this surgical procedure quite well. In over 30 implantations to date we have yet to experience a surgical or anesthetic fatality. The implanted infants are returned to their mothers as they recover from the surgical anesthesia. By the time the sutures are removed 10 days later, their behavior has returned to normal.

A block diagram of the receiver and demodulator is illustrated in Fig. 3.

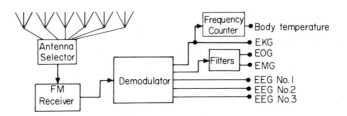

RECEIVING EQUIPMENT

Fig. 3. Block diagram of telemetry receiver and demodulator.

The signal transmitted from the animal is received on one of six dipole antennas located around the group pen. The antenna selector automatically selects an antenna with a good quality signal and ignores those with poor signals (ref. 11). The receiver is a slightly modified Dynaco Model FM-5A commercial FM receiver. The PAM subcarrier is demodulated and decoded into the original five transmitted data channels. The combined body temperature and EKG waveform is separated into low frequency components (the EKG waveform) and high frequency components (body temperature). EOG and EMG are separated by similar band pass filtering.

A block diagram of our overall data collection and analytic system as presently constituted is illustrated in Fig. 4. With the implanted subject on the far left, there are three major components to the system, separated by the vertical dashed lines. The left hand component, entitled "Data Collection" on the bottom, constitutes the various types of data input and the primary recording media. The central component is made up of the PDP-12 laboratory computer, including its input software, output software, and various output devices. The far right-hand component is a CDC-6400 computer located at the University of Colorado at Boulder, some 25 miles distant. Beginning with the left-hand input column, a master time code generator reader generates an IRIG-B BCD timecode that is used to time-lock the FM tape recorder and the audio cassette recorder as well as provide a direct input through an IO-BUS to the laboratory computer. A simultaneous slow code is recorded on the polygraph. Again under the input column, still photographs are taken of the experimental infants as appropriate, and as are illustrated in Fig. 5 and 6 below. The telemetered signal is picked-up by the FM tuner, demodulated, and the physiological data is recorded simultaneously on FM tape and a 16-channel Grass Model 78 polygraph. Additionally, a stripchart recorder and digital counter are used to record the body temperature. A microphone, located inside the group pen, picks up the various animal and other sounds, which are then recorded on the audio track of the videotape recorder. A TV camera with remote pan, tilt, and zoom controls and infrared recording capability is available for monitoring the infant as appropriate; this data is also recorded on videotape. A video time/date generator, synchronized to the master time code generator-reader, places date and time (hour, minute, second) on videotape. Finally, the observer, who watches the animals through oneway windows, dictates ongoing

---

[2]Medwire part No. 31655 7/44T; Medwire Corp., Mt. Vernon, New York 10553.

behavior as it occurs, utilizing a taxonomy similar to that developed by Kaufman and Rosen-blum (ref. 12) onto one track of a stereo audio cassette, the other track of which contains the IRIG-B time code.

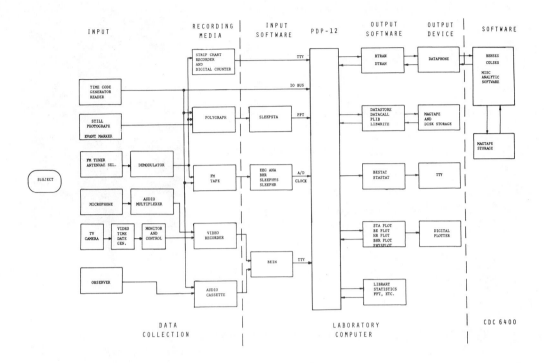

Fig. 4. Block diagram of overall data collection and analysis system.

Thus the behavioral and physiological data are recorded and stored in a variety of forms, including videotape, FM tape, audio tape, and paper records prior to computer entry. A number of different software routines have been developed in the laboratory to facilitate entry of the various data into the PDP-12 computer. Under the input software column, data from the strip chart recorder and the digital counter can be directly entered via teletype to permit statistical analysis of temperature rhythms. As mentioned above, the IRIG-B time code is directly available to the PDP-12 computer via an in-house designed hardware inter-face. We use the SLEEPSTA program to score the all-night paper sleep records (ref. 13). The paper record is scored conventionally by hand, and for each page (20 sec epoch) of the paper record, a specific ASR 33 teletype key is punched corresponding to that stage which consti-tutes 50% or more of the page. The punched paper tape so generated is later read by the computer, which provides a statistical summary of the night's sleep, as well as an all-night sleep histogram (output software programs STASTAT and STAPLOT). Data from the FM tape, in-cluding the several EEG channels and the EKG data, is entered via the A/D convertors or KW12 clock utilizing a variety of general and special purpose software. The BEIN program allows the behavioral observer to replay either the audio cassette or the video tape and enter behavior into the computer keyboard, providing a visual display on the computer screen of both the observation time and the various behaviors that are ongoing. The computer then can generate both plots and statistical summaries of the observation periods using the BESTAT or BEPLOT programs (under output software). The output software includes the BTRAN and DTRAN programs which send behavioral and physiological data to the CDC-6400 computer via a data-phone link. The DATASTOR, DATACALL, PLIB, and LIBWRITE programs constitute a general pur-pose data handling system developed inhouse for storing multivariable physiological and be-havioral data (ref. 14). Mass storage devices include magnetic tape (LINCTAPE) and/or RK8E cartridge disks. The plot routines illustrated permit the plotting of various physiological parameters, e.g. all-night heart rate, body temperature, sleep patterns, and behavioral sessions, on an incremental digital plotter. Additionally, we utilize a library of statisti-cal routines, FFT algorithms, etc., for general purpose processing.

As an example of the physiological data, Fig. 5 illustrates an implanted infant being en-closed by its mother with the simultaneously transmitted physiological data on the left. The power spectrum illustrated on the lower right of Fig. 5 was computed on the illustrated epoch of central temporal EEG and illustrates power (on the Y axis) as a function of frequency from 0 to 32 Hz (on the X axis). The dominant frequency of the CT EEG in this power spectrum is 8.25 Hz.

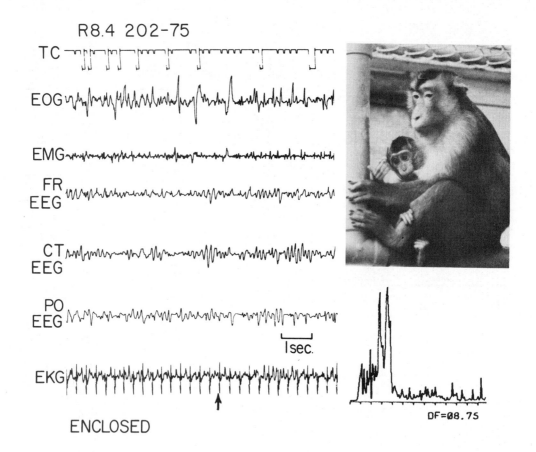

Fig. 5. Implanted pigtail infant enclosed by its mother.
TC=time code. EOG=eye movement. EMG=muscle
activity. FR EEG=frontal EEG. CT EEG=central
temporal EEG. PO EEG=parieto occipital EEG.
EKG=electrocardiogram. The photograph was taken
at the time indicated by the arrow beneath the
EKG channel.

The obvious advantage of implantable biotelemetry in our application is that it permits physiological data to be obtained without in any way interfering with the normal behavior or social environment of the animal. The types of physiological-behavioral correlations made possible by these techniques are quite considerable. For example, long term changes in various circadian and ultradian biological rhythms can be correlated with long term quantified alterations in the subject's behavior. At the other extreme, a "microdissection" of physiology and behavior can be performed in ways similar to that illustrated in Fig. 6. In this illustration a 30 second epoch of physiological data, including three channels of EEG and beat-to-beat heart rate, is accompanied by five still photographs of an infant enclosed by its mother, taken at the time indicated by the circled arrows. At the beginning of the sequence, the infant is in light sleep, as indicated by EEG activity patterns. With no apparent accompanying behavioral change, the illustrated cardiac acceleration of some 50 beats per minute occurred prior to any noticeable EEG changes and prior to the infant's subsequent arousal. In photographs 1 and 2, the infant remains asleep, although evidence of EEG arousal is now seen; the infant opens its eyes in the third photograph, with continued EEG arousal. The mother's eyes are not opened until the fourth photograph, possibly in response to her infant's arousal, and both mother and infant fall back to sleep in the fifth panel. Thus, in sum, we have evidence of an arousal on the part of the infant, preceded by some 10 seconds by a marked cardiac acceleration, not accompanied by overt EEG or behavioral changes. Such a "microdissection" of physiology and behavior, especially when simultaneous behavior is videotaped permitting frame-to-frame correlation of changes in behavior with changes in physiological status, appears to represent a powerful research strategy.

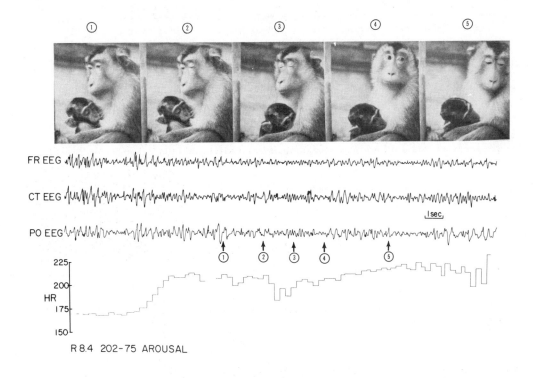

Fig. 6. Behavioral arousal with associated physiological changes.

We are presently engaged in the development of a biobehavioral developmental profile (BDP) for the pigtail monkey infant. Using behavioral data collected from birth onward in a large number of infants, and normative physiological data and physiological-behavioral correlations collected with the system just described, we will construct a family of curves representing the normal or expected development of the various behavioral and physiological parameters so obtained. This BDP will then permit any future infants' biobehavioral development to be assessed in comparison to these norms. We might then expect to find infants exhibiting either biobehavioral acceleration and/or retardation in one or several areas and we might expect such infants to show differences in their behavioral and/or physiological maturation, in their reactions to various stressful situations, and in their assumption of social roles as adults. The next step would be the investigation of the mechanisms underlying such differences. It is our hope that with the construction of the BDP, the usefulness of the monkey infant as a more general model of several aspects of human development will be considerably enhanced.

THE PHYSIOLOGY OF THE AGITATION-DEPRESSION REACTION

It is not my purpose to detail our findings to date (which will be reported in full), but rather to outline what appear at this time to be the major physiological changes accompanying the agitation-depression reaction, so that the entire reaction can be pictured in a more comprehensive manner. We have studied nearly 20 infants to date, with about half having four-day maternal separations; the other half, 10-day separations.

The animals are housed in sizeable group pens that contain six to eight females, several with infants, and one adult male. Timer-controlled lights are off between 2000 and 0700 daily. In most cases, the infants have been observed behaviorally from birth onward, so that we have quantified measurements of their behavioral development and their infant-mother interaction patterns. The infants are normally implanted between five and six months of age, and after they have recovered from surgery and their behavior in the group has returned to normal, the telemetry units are turned on and recording continues 24 hours a day thereafter. We obtain baseline (pre-separation) recordings for four to five consecutive days, at which point the mother is separated from the infant and from the group, while the infant remains in the group in which it was raised. The mother is kept out of sight and sound while behavioral observations and physiological recording continues, and the infant goes through the agitation-depression reaction. The mother is then returned to the group, and the recording continues for at least another four days following reunion. Those physiological variables for which data are presently available include sleep patterns, heart rate and body temperature rhythms, and EEG activity patterns. In this discussion I will concentrate on how infants react as a group. There is marked inter-individual variability in the reaction to separation, both behaviorally and physiologically, and while this indeed is most important and one of the real advantages of this model, it is a sizeable and complex topic in its own right and best considered in detail elsewhere.

Sleep Patterns

Nocturnal sleep in normal infant pigtail monkeys exhibits a considerable degree of night-to-night intra-individual variability seemingly unrelated to various measures of the preceding day's behavior and in the absence of major environmental changes (ref. 15). However, when sleep patterns are averaged over several nights, the animals are quite similar to each other. As with other primates, most deep Stage 3-4 sleep occurs early in the night, while REM tends to occur more often during the latter half of the night. The inter-REM interval (time interval between REM periods), is about 60 minutes in the infant pigtail, as opposed to about 90 minutes in adult man, and 51 minutes in adult Rhesus (ref. 16). Pigtail infants have about 100 minutes of REM sleep every night divided up among 9-10 REM periods.

Sleep patterns the night following separation from the mother (the night following the agitation reaction) are characterized by marked increased latency to sleep onset, more time awake, more frequent arousals, greatly increased REM latency, and a marked decrease in REM time (ref. 17). In terms of actual minutes, however, slow wave sleep (both Stage 3-4 and Stage 2) is not changed from baseline. Subsequently, during the depressive behavioral reaction, the major alteration in sleep patterns is the change in REM sleep. REM time remains depressed, REM latency remains increased, and the number of REM periods are diminished from baseline. All of these changes tend to return towards the normal as separation continues, although there are residual changes in REM sleep that persist throughout 10 days of separation. Slow wave sleep has not been significantly diminished during separation in the infants we have studied to date. Sleep patterns return to normal following reunion with the mother.

These sleep pattern changes are similar to those described in man as constituting the so-called "first night effect", which results from a change in sleeping environment (ref. 18), and this indeed may contribute in part to our findings. However, it is not likely the total explanation. In another study we raised several pigtail monkey infants from birth on cloth surrogates (ref. 19). When they were later separated from their surrogates, there was no evidence of a prominent agitation-depression reaction either behaviorally or physiologically, with the exception of sleep pattern changes which were similar, but not as prominent, as those found in the group living infants described above. We believe that the magnitude of the sleep pattern changes found in the surrogate-reared infants were due to the environmental disruption, or the so-called "first night effect", and the sleep pattern changes above and beyond that as seen in the above-described group living infants are the result of some other additional effect related to the loss of the mother.

Heart Rate

Accompanying the agitation reaction immediately following separation (which usually takes place in the early afternoon), there is a marked increase in heart rate, which tends to remain somewhat higher than normal throughout the remainder of the first day of separation. Subsequently, most separated infants exhibit significant decreases in both daytime and nighttime heart rate mean values for the remainder of the separation period, with again a tendency for heart rate values to return towards normal in most animals as separation continues (up to the maximum period of 10 days). There is marked inter-individual variability in terms of the degree and the persistance of the decreases in heart rate. Increased variability in heart rate persists throughout reunion with the mother, with some infants showing increases to levels higher than baseline and others showing marked day-to-day variability. A most interesting observation is that occasional infants exhibit persistant separation-induced bradycardia that lasts up to at least several weeks following reunion with the mother.

Normal (unseparated) pigtail infants exhibit varying degrees of cardiac arrhythmias during the baseline period. There is a significant negative correlation between mean heart rate and mean number of arrhythmias in normal infants; that is, those animals which have the slowest heart rates also tend to have a greater number of arrhythmias. Most of the arrhythmias appear to be sinus in origin. Cardiac arrhythmias are increased during the depressive reaction following separation, but we cannot yet say whether this can be explained solely by the decreased heart rate, or whether it is a function of other influences as well.

There is, of course, a considerable literature on stress-related alterations in heart rate and cardiac rhythms for various species, including man (ref. 20,21,22,23,24). And there is considerable evidence that the meaning or symbolic significance of a situation may be a most important determinant of the physiological reaction to that situation. The mechanism underlying the bradycardia we have observed is as of yet unknown, although we might anticipate that it is a vagally-mediated phenomenon. However, the elaboration of underlying mechanisms awaits further study.

Body Temperature

As with heart rate, body temperature also increases following the actual separation experience and remains somewhat higher than normal during the remainder of the agitated behavior that accompanies the first afternoon and early evening of separation. Usually beginning during the first night of separation, and accompanying the subsequent depression reaction, there are significant drops in body temperature, both day and night. The 24-hour circadian rhythm is preserved, but it is shifted to a lower level. Again, as with heart rate, group values for mean body temperature tend to return towards normal baseline levels as separation continues. In all animals studied to date, body temperature values returned to baseline levels following reunion with the mother.

Not all infants exhibit the lowest body temperature of the first separation night; some have the lowest values later. Two infants have exhibited a sudden transient nocturnal drop in body temperature ($\approx$ 1-2° C) later in separation. These drops occurred in Stage 3-4 sleep in one case, and during an awake period in another case. Neither of these sudden nocturnal drops in body temperature were associated with marked changes in heart rate. These observations have suggested to us that there may be a qualitative impairment of thermoregulatory mechanisms that accompanies the depressive behavioral reaction.

Body temperature in pigtail infants is a complex function of several variables, one of which is the nature of the contact with the mother (infants who are tightly held tend to have slight increases in body temperature). Since separation from the mother does entail loss of a heat source, this in itself may have some influence on the infant's body temperature. However, a single pigtail infant raised from birth on a heated cloth surrogate did not experience a decrease in body temperature when separated from the heated surrogate, and if we extrapolate from these data to the group living situation, it appears unlikely that the loss of maternal body heat is the sole explanation for the decrease in the infant's body temperature, with some additional mechanism being implicated.

EEG Changes

While most of the EEG data we have collected to date is not yet completely analyzed, we do have some interesting preliminary findings. There is a characteristic high central rhythmicity seen in most infants we have studied to date, and which is illustrated in the central-temporal lead in Fig. 5. In this case the infant is enclosed by its mother, and at rest, and the dominant frequency of the central-temporal EEG is 8.75 Hz. This central activity is quite different from the normal alpha activity seen in infant pigtail monkeys which is generally limited to the parietal-occipital channel and is seen only during lights-out (ref. 25). We think this activity characterizes a state of phasic motor inhibition, and may be homologous to the rhythme en arceau described in man (ref. 26). Two changes have been noted in this activity during the behavioral depressive reaction period: first, it tends to become slower, and secondly, it occurs more frequently in behaviors in which it is not prominent in baseline.

Figure 7 illustrates an infant exhibiting the slouched posture of depression (the same animal as in Fig. 5). Here the power spectrum of the illustrated epoch of CT EEG shows the dominant frequency to be 6.00 Hz, a decrease of 2.75 Hz. The mean decrease in frequency of this activity for a group of 10 infants was 1.0 Hz (ref. 27). These changes have suggested to us that there may be an increase in the activity of a central inhibitory system accompanying the depressive behavior, as well as a decrease in its major cortical EEG frequency.

By comparison, posterior alpha activity does not show similar changes during the depressive reaction and in fact, if anything, tends to increase slightly (a mean increase of 0.3 Hz for a group of 10 infants) during the depressive reaction. Thus, alpha generators and those underlying the characteristic central-temporal rhythms in pigtail monkey infants appear to be physiologically dissociated and differentially influenced by the stress of separation (ref. 28).

Overview of the Physiological Changes

The mechanisms underlying the various behavioral and physiological components of the

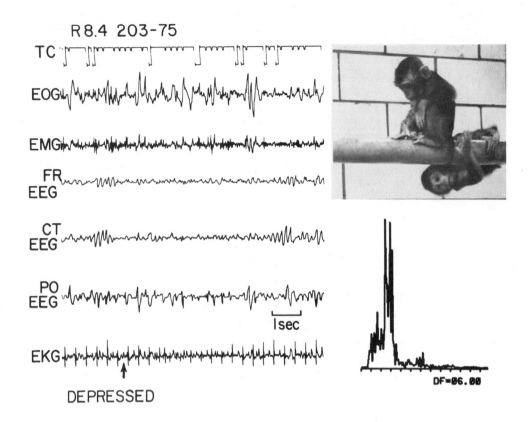

Fig. 7. Behaviorally depressed infant exhibiting characteristic slouched posture.

agitation-depression reaction have yet to be identified. Although we have no direct evidence implicating neurochemical systems, it is tempting to speculate as regards to their possible role. Ordy and his collaborators (ref. 29) found a significant decrease in hypothalamic norepinephrine levels in squirrel monkeys after 12 hours of social isolation. And Redmond et al. (ref. 30) found changes in social behavior (decreases in grooming and play) in Macaca speciosa monkeys subjected to inescapable shock that they thought to be attributable to decreases in central norepinephrine known to accompany inescapable shock. In several species decreases in central norepinephrine have been thought to be associated with motoric slowing (ref. 31,32), and the dopamine-beta-hydroxylase inhibitor FLA-63 (which depletes central norepinephrine) has been reported to both lower body temperature and decrease motility in rats, with both effects thought to result from the central norepinephrine depletion (ref. 33). Weitzman et al. (ref. 34) found that alpha-methyl-para-tyrosine, which inhibits the synthesis of dopamine and norepinephrine, selectively decreased REM sleep in Rhesus monkeys. Thus, certain of our behavioral and physiological findings might be compatible with a disturbance in central catecholinergic (e.g. noradrenergic) systems. Of course, a comprehensive explanatory mechanism would have to account for the other behavioral and EEG changes known to accompany the agitation-depression reaction as well. Although attractive from the standpoint of simplicity, it seems unlikely that the functional disturbance of a single neurochemical system would account for the totality of this complex syndrome.

While the physiological changes tend to be maximal early in the separation period, the depressive behavioral changes usually do not reach their maximum until 4 to 6 days after separation. There is not a complete dissociation, however, as we have found that those infants who have had the most severe sleep disturbances the first separation night (i.e. more time awake, less total sleep, and less total REM time) also appear to become the most depressed behaviorally later in separation (ref. 17). Evidence for other correlations between physiology and behavior is currently being evaluated.

IMPLICATIONS FOR A MODEL OF DEPRESSION

The emergence of these data now permits a more comprehensive view of the organismic nature of the reaction to maternal separation in monkey infants. In addition to the previously well-documented behavioral changes, we see wide-ranging disturbances in many areas of physiological functioning, with alterations in heart rate, body temperature, sleep patterns and EEG rhythms. An important and obvious question is to exactly what extent these findings in separated monkey infants can be extrapolated to affective disturbances in man. From the standpoint of the physiological findings, a major problem resides in the absence of comparable physiological data in man. Although Kraepelin (ref. 35) described melancholia as being accompanied by lower than normal heart rate and body temperature, he did not present quantified data. Unfortunately, long term physiological recordings have yet to be obtained in human affective illnesses. It is most germane, however, to consider the question of the relationship of loss or separation to affective illness.

Although attachment behavior and separation and its sequellae in various species have been the subject of extensive recent discussions (ref. 6,36,37), the question of exactly how separation experiences may be related to the various clinical depressions remains a moot one. The early formulations of Spitz (ref. 38) and the syndrome of anaclitic depression, and Robertson and Bowlby (ref. 39) and the protest-despair reaction seen in human youngsters separated from their mothers are by now common knowledge. Although Rutter (ref. 40) has tended to discount the importance of separation in childhood in terms of its influence in the development of antisocial behavior, he leaves open the question of whether early separations might be related to the development of depression in later life. However, both Sethi (ref. 41) and Cardoret et al. (ref. 42) found a positive relationship between depression and antecedant losses or separations. Heinicke (ref. 43) felt that there was a significant relationship between parent loss in childhood and depression in adult life. Stenback (ref. 44) found a significant relationship between object loss and depression in the aged; Paykel et al. (ref. 45), studying 105 depressed patients with matched controls, found the experimental subjects had experienced more unpleasant life events, especially losses and separations, prior to the onset of their illnesses than had controls, although other investigators have not been able to confirm these findings (ref. 46). In a recent review of the relationship between recent events and the onset of depressive illness, Paykel (ref. 47) states:

> Life events bear a causal relationship to the onset of most depressions and probably to a greater extent than with most other psychiatric disorders. A variety of types of event are implicated. Most prominent among these are separations and interpersonal losses.

It is only relatively recently that significant attention has been paid to depression in childhood, but here there appears to be an even clearer relationship between separation and/or loss to subsequent depression (ref. 48). This position is clearly stated by McKnew and Cytryn (ref. 49), who studied 50 children with affective disorders, and stated that "In summary, our investigation suggests that the sine qua non in acute depressive reactions is the sudden loss of a love object".

While the necessary role of separation and loss in human depression is moot, it is hardly inconsequential. Object loss may well be but one of several etiologically predisposing factors which contribute to depression as a "final common pathway" (ref. 50,51). Thus, although separation is certainly not the whole question when it comes to depression, and we would in no way imply this is the case, it does appear to be one contributory aspect of a complex clinical picture, and to the extent that we can obtain basic information about the effects of separation on the psychobiological functioning of the organism, we will know that much more about the entire picture.

In a very real sense the mother-infant bond is the biological prototype of all future attachment bonds. In monkeys future bonds may extend to peers, companions, offspring, other animals, and perhaps in ways we are not yet aware of, to certain surroundings or other objects. In man future bonds may extend to include family, job or profession, material possessions, and more nebulous but real concepts relating to aspects of self-image. But the mother-infant bond remains for both man and monkey the primordial attachment bond. To exactly what extent observations on the effects of its rupture can be extended to other more elaborated or derivative bonds is as of yet unclear, but we expect that basic processes may be quite generally relevant. It may be that future work will disclose that, indeed, the reaction to mother loss in infant monkeys is the prototype of a more general psychobiological reaction to object loss in the high primates.

REFERENCES

(1) Darwin, C. (1955) The Expression of Emotions in Man and Animals (1872), Philosophical Library, New York.
(2) I. C. Kaufman and L. Rosenblum, The reaction to separation in infant monkeys: Anaclitic depression and conservation withdrawal, Psychosom. Med. 29, 649-675 (1967).
(3) W. R. McKinney and W. E. Bunney, Animal model of depression, Arch. Gen. Psychiat. 21, 240-248 (1969).

(4)   H. Harlow, J. P. Gluck, and S. J. Suomi, Generalization of behavioral data between non-human and human animals, Am. Psychol. 27, 709-716 (1972).

(5)   M. Reite, I. C. Kaufman, J. D. Pauley, et al., Depression in infant monkeys: Physiological correlates, Psychosom. Med. 36, 363-367 (1974).

(6)   Bowlby, J. (1969) Attachment and Loss (Vol. I: Attachment), Basic Books, New York.

(7)   A. C. Wilson and V. M. Sarich, A molecular time scale for human evolution, Proc. Nat. Acad. Sci. U.S.A. 63, 1088-1093 (1969).

(8)   J. D. Pauley, M. Reite, and S. D. Walker, An implantable multi-channel biotelemetry system, Electroenceph. Clin. Neurophysiol. 37, 153-160 (1974).

(9)   J. D. Pauley and M. Reite, A command receiver for use in low power implantable biotelemetry systems, in: Fryer, T. B., Miller, H. A., and Sandler, H., Eds. (1977) Biotelemetry III, Academic Press, New York.

(10) M. Reite, S. D. Walker, and J. D. Pauley, Implantation surgery in infant monkeys, Lab. Primate Newslet. 4, 1-6 (1973).

(11) J. D. Pauley and M. Reite, Automatic antenna selector for use in biotelemetry applications, Physiol. and Behav. 18, 169-170 (1977).

(12) I. C. Kaufman and L. Rosenblum, A behavioral taxonomy for Macaca nemestrina and Macaca radiata: Based on longitudinal observation of family groups in the laboratory, Primates 7, 205-258 (1966).

(13) S. D. Walker and M. Reite, SLEEPSTA: Sleep stage data handler, Psychophysiol. 13, 159 (1976).

(14) S. D. Walker and M. Reite, A multivariable data handling system, in press: Behav. Res. Meth. and Inst. (1977).

(15) M. Reite, A. J. Stynes, L. Vaughn, et al., Sleep in infant monkeys: Normal values and behavioral correlates, Physiol. and Behav. 16, 245-251 (1976).

(16) D. F. Kripke, M. Reite, G. V. Pegram, et al., Nocturnal sleep in rhesus monkeys, Electroenceph. Clin. Neurophysiol. 24, 582-586 (1968).

(17) M. Reite and R. Short, Nocturnal sleep in separated monkey infants, in press: Arch. Gen. Psychiat. (1977).

(18) H. W. Agnew, W. B. Webb, and R. L. Williams, The first night effect: An EEG study, Psychophysiol. 2, 263-266 (1966).

(19) M. Reite, R. Short, and C. Seiler, Physiological correlates of separation in surrogate reared infants: A study in altered attachment bonds, in press: Dev. Psychobiol.

(20) W. D. Cannon, "Voodoo" death, Am. Anthrop. 44, 169 (1942).

(21) M. A. Hofer, Cardiac and respiratory function during sudden prolonged immobility in wild rodents, Psychosom. Med. 32, 633-647 (1970).

(22) C. P. Richter, On the phenomenon of sudden death in animals and man, Psychosom. Med. 19, 191-198 (1957).

(23) S. Wolf, Cardiovascular reactions to symbolic stimuli, Circulation 18, 287-292 (1958).

(24) S. Wolf, The end of the rope: The role of the brain in cardiac death, Canad. Med. Assn. J. 97, 1022-1025 (1967).

(25) M. Reite, J. D. Pauley, I. C. Kaufman, et al., Normal physiological patterns and physiological-behavioral correlations in unrestrained monkey infants, Physiol. and Behav. 12, 1021-1033 (1974).

(26) H. Gastaut, H. Terzian, and Y. Gastaut, Etude d'une activité électroencéphalographique méconnue: "le rhythme rolandique en arceau", Marseille Méd. 89, 296 (1952).

(27) R. Short, S. Iwata, and M. Reite, EEG changes during the depressive reaction following maternal separation, Psychophysiol. 14, 120 (1977).

(28) R. Short and M. Reite, Alpha activity in the infant Macaca nemestrina following maternal separation, presented at the Inaugural Meeting of the American Society of Primatologists, Seattle, Washington (April, 1977).

(29) J. M. Ordy, T. Samorajski, and D. Schroeder, Concurrent changes in hypothalamic and cardiac catecholamine levels after anesthetics, tranquilizers and stress in a subhuman primate, J. Pharmacol. Exp. Ther. 152, 445-475 (1966).

(30) D. E. Redmond, Jr., J. W. Maas, H. Dekirmenjian, et al., Changes in social behavior in monkeys after inescapable shock, Psychosom. Med. 35, 448-449 (1973).

(31) D. S. Segal and A. J. Mandell, Behavioral activation of rats during intraventricular infusion of norepinephrine, Proc. Nat. Acad. Sci. U.S.A. 66, 289-293 (1970).

(32) E. A. Stone and S. Medlinger, Effect of intraventricular amines in motor activity in hypothermic rats, Res. Comm. in Chem. Path. and Pharm. 7, 549-556 (1974).

(33) T. H. Svensson and B. Waldeck, On the significance of central noradrenaline for motor activity: Experiments with a new dopamine Beta-hydroxylase inhibitor, Europ. J. Pharmacol. 7, 278-282 (1969).

(34) E. D. Weitzman, P. McGregor, C. Moore, et al., The effect of alpha-methyl-para-tyrosine on sleep patterns of the monkey, Life Sci. 8, 751-755 (1969).

(35) Kraepelin, E. (1921) Lehrbuch der Psychiatrie, as abstracted and adapted by A. R. Diefendorf, Clinical Psychiatry, MacMillan, New York.

(36) Bowlby, J. (1973) Attachment and Loss (Vol. II: Separation), Basic Books, New York.

(37) Scott, J. P. and Senay, E. C. (1973) Separation and Depression, AAAS, Washington, D.C.

(38) R. A. Spitz, Anaclitic depression, Psychoanal. Study Child 2, 313-342 (1946).

(39) J. Robertson and J. Bowlby, Responses of young children to separation from their mothers, Courrier Centre Internat. Enfouce 2, 131 (1952).

(40) M. Rutter, Parent-child separation: Psychological effects on the children, J. Child
        Psychol. Psychiat. 12, 233-260 (1971).
(41) B. B. Sethi, Relationship of separation to depression, Arch. Gen. Psychiat. 10, 486-
        496 (1964).
(42) R. J. Cadoret, G. Winokur, J. Dorzab, et al., Depressive disease: Life events and
        onset of illness, Arch. Gen. Psychiat. 26, 133-136 (1972).
(43) C. Heinicke, Parental deprivation in early childhood, in: Scott, J. and Senay, E., Eds.
        (1973) Separation and Depression: Clinical and Research Aspects, AAAS, Washing-
        ton, D.C.
(44) A. Stenback, Object loss and depression, Arch. Gen. Psychiat. 12, 144-151 (1965).
(45) E. S. Paykel, J. K. Myers, M. N. Dienelt, et al., Life events and depression, Arch.
        Gen. Psychiat. 21, 753-760 (1969).
(46) R. W. Hudgens, J. R. Morrison, and R. G. Barchha, Life events and onset of primary
        affective disorders, Arch. Gen. Psychiat. 16, 134-145 (1967).
(47) E. S. Paykel, Environmental variables in the etiology of depression, in: Flach, F. F.
        and Draghi, S. C., Eds. (1975) The Nature and Treatment of Depression, Wiley,
        New York.
(48) C. P. Malmquist, Depression in childhood, in: Flach, F. F. and Draghi, S. C., Eds.
        (1975) The Nature and Treatment of Depression, Wiley, New York.
(49) D. H. McKnew and L. Cytryn, Historical background in children with affective disorders,
        Am. J. Psychiat. 130, 1278-1280 (1973).
(50) H. S. Akiskal and W. T. McKinney, Depressive disorders: Toward a unified hypothesis,
        Science 182, 20-29 (1973).
(51) H. Akiskal and W. T. McKinney, Overview of recent research in depression, Arch. Gen.
        Psychiat. 32, 285-305 (1975).

DISCUSSION

In response to several questions on methodology, *Dr. Reite* indicated that since the thermistors are located inside the telemetry package, the temperature is recorded from the abdominal wall beneath the musculature, immediately external to the peritoneum. This is as close as he could get to the core temperature. The animal that had shown a prolonged change in heart rate did not show a corresponding drop in body temperature.

Following separation from mothers, the animals have a tachycardia lasting from half an hour to several hours. During this time, the animals are behaviorally agitated, hyperactive and excited. Dr. Reite emphasized that he was not claiming that this was homologous to clinical psychotic depression; only what happens when infant monkeys are separated from their mothers.

As far as sleep studies are concerned, Dr. Reite has seen a number of studies which suggest that if the right clinical subgroup of depressed patients is chosen one can find almost any type of sleep disturbance – from REM suppression to increased REM, from increased REM latency to decreased REM latency. These differences may even be useful for determining clinical subgroups. He feels his model has the greatest homology to an anaclitic depression in an infant, but there is not yet available physiological data on behavioral changes in human infants.

THE PHARMACOLOGY OF KINDLING

Ronald G. Babington
Sandoz Pharmaceuticals, East Hanover, New Jersey 07936

Historically, most of the important advances in the pharmacotherapy of mental and neurologic diseases were the result of fortuitous clinical observations. Thus, the development of animal models to predict therapeutic efficacy is a particular challenge to investigators working on the central nervous system (CNS). A basic shortcoming is the lack of pathologic models directly related to the various disease states in man, necessitating the use of observation profiles in normal animals or drug interaction procedures to screen for CNS activity. Despite the obvious predicament that synthetic compounds might possess a highly selective mechanism of action not revealed by the usual laboratory models, the battery-of-tests approach continues as the modus operandi for determining the therapeutic potential of experimental drugs. Unfortunately, the available models continue to identify compounds with a mechanism of action similar to the standard drugs.

In contrast, there are several preclinical models for testing antiepileptic compounds that can predict efficacy with a substantial degree of reliability. Still, when Goddard et al. (1) published details of the kindling phenomenon, it appeared to be a particularly applicable model for studying pharmacologic effects on the gradual development of seizures. As described by the Goddard group, daily low-intensity electrical stimulation of selective brain sites eventually leads to the appearance of overt behavioral seizures; and the seizure is preceded in time by electrical afterdischarges at the stimulus site. Several neural structures are susceptible to kindling, especially the limbic system; but sensitivity from site to site can be quite variable.

To familiarize those not acquainted with kindling, a brief description of the methodology and behavioral consequences is in order. Rats are implanted with a stainless steel bipolar electrode (Plastic Products MS 303) and allowed to recover from the effects of the surgery. Beginning 10 days postoperatively, each animal is stimulated electrically once daily for one minute with a 50 μA, 60-Hz constant current sine wave. During the first few stimulation sessions, the subjects show no overt signs; but after eight to ten days, facial movements and bilateral, clonic seizures are seen. Once the sensitization process is initiated, stimulation is continued only until motor movements appear, whereupon the current is shut off and the duration of the seizure is measured. Within four to five weeks, sensitization has progressed such that according to individual animals, five to seven seconds of stimulation elicits seizures that last from 45-75 seconds. Although the above procedure is specific for rats, kindling has been produced in several species and the methodology is essentially the same.

The pattern of the kindled seizures in rats has been described in detail by Goddard et al. (1). Ambulatory activity ceases 5-7 seconds after the start of stimulation and precedes the onset of gross seizures. Next, the animals rear back on their hind limbs and tail, and bilateral, clonic forelimb seizures begin. Facial contractions and foaming at the mouth occur in most rats. There is a frequent loss of balance during the seizure episode, but the rats invariably return to the rearing position. Under the prescribed conditions, hind limb involvement rarely occurs and tonic seizures never. There is a sudden cessation of the seizure with the animal assuming a normal body position and proceeding with searching behavior. For several minutes after termination of the seizure, the subjects are extremely hyper-irritable, reacting violently to tactile or auditory stimuli.

Repeated testing at 15-minute intervals over extended periods of time has shown the phenomenon to be extremely stable. For each animal, both the length of stimulation required to trigger the seizure and the duration of each successive seizure is nearly identical trial after trial; and that consistency is maintained for months even though the animals are only tested once a week after drug experimentation is started.

142

The obvious assumption that kindled seizures should be blocked by antiepileptic drugs has been verified in several laboratories in several
species; but the results have been less conclusive than expected. Babington
and Wedeking (2) reported that in rats, diphenylhydantoin and phenacemide
(Fig. 1) block kindled seizures, but only at doses that depress motor
activity or cause ataxia. Wise and Chinerman (3) studied the effect of

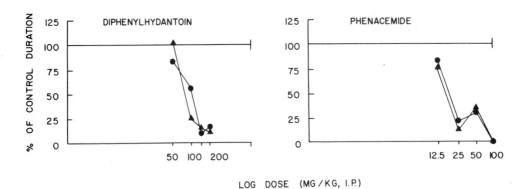

LOG DOSE (MG/KG, I.P.)

Fig. 1. Effects of two antiepileptic drugs on the duration
of kindled seizures. Seizures were elicited from
the amygdala (●) or sensorimotor cortex (▲).

diphenylhydantoin on several parameters of the kindling process in rats,
including genesis and threshold; but at the dose level tested, the drug was
completely inactive. Racine et al. (4) obtained seizure blockade with
diphenylhydantoin in rats when the anterior neocortex was the site of stimulation; but in contrast, amygdaloid-elicited convulsions were enhanced.
Tanaka (5), using rabbits, found that diphenylhydantoin had little or no
effect on kindled seizures. In a study carried out by Wada et al. (6) in
cats, diphenylhydantoin was ineffective, even at toxic plasma levels.

Aside from the fact that seizures per se were involved, kindling was of
particular interest as a test system because of the seizure pattern. Unlike
most rodent convulsive models, the seizures are not generalized but are
restricted to the head and forelimbs; and tonic extension is not involved.
The possibility that the kindling models might serve specifically for agents
useful in the treatment of temporal lobe epilepsy led to the testing of
carbamazepine, a drug with documented therapeutic effectiveness in temporal
lobe as well as grand mal and psychomotor epilepsies (8-12). Babington
and Horovitz (13) reported that in rats carbamazepine was not only efficacious, it was much more active than the other antiepileptic drugs.
Moreover, unlike other antiepileptic agents it appeared to exert a more
selective effect on the amygdala than the neocortex. Wada et al. (6) also
reported that carbamazepine effectively suppressed kindling in cats.

Because the limbic system is acknowledged to be closely involved with
emotionality (14), it is a prime substrate for examining the effects of
psychotherapeutic drugs and several authors (15-18) have reported that
antidepressive agents suppress limbic afterdischarges. The fact that
Hernandez-Péon (19) found that carbamazepine blocked limbic afterdischarges,
the fact that several clinical investigators (9, 11, 12) realized that
carbamazepine produced an antidepressive effect in epileptic patients that
could be dissociated from an improvement in their seizure status, and the
fact that carbamazepine behaved so differently than the standard antiepileptic
drugs in the kindling procedure led to the speculation that activity
responsible for suppressing kindled seizures was related to an antidepressive rather than an antiseizure component. Thus, several antidepressive
agents were tested in the kindling model (2, 7).

The effects of three standard antidepressive agents on the duration of
amygdaloid- and cortical-induced seizures are illustrated in Fig. 2. As
can be seen, amitriptyline was quite potent in inhibiting kindled seizures
elicited from the amygdala. Cortically-induced seizures were also suppressed, but at higher dose levels. Similarly, Fig. 2 shows that nortriptyline and imipramine were more active in suppressing amygdaloid, as opposed
to cortical seizures. However, neither was as potent against amygdaloid-

induced seizures as amitriptyline; and the separation of the activity
curves was least manifest with imipramine. With all three antidepressives,

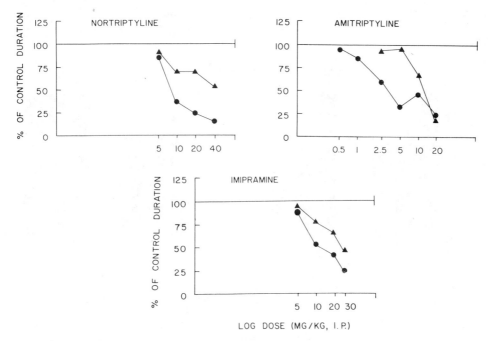

Fig. 2. Effects of three antidepressive drugs on the duration of kindled seizures elicited from the amygdala
(●) or sensorimotor cortex (▲).

the difference in the slope of the amygdaloid and cortical dose-response
curves was significantly different ($P < 0.05$). Although not illustrated in
Fig. 2, the dose-response curves of all three antidepressives against
seizures originating from kindled sites in the septum fall slightly to the
right of the cortical dose-response curves (20). At the dose levels represented, blockade of the seizures was not complete; and even at higher doses
where overt signs of CNS depression were observed, there was never a 100
percent suppression of the seizures.

To date, the only antidepressive with clinically-proven activity that has
been ineffective against kindled seizures has been iprindole, a compound
that does not behave like the other tricyclic antidepressives in the standard laboratory tests (21). Thyrotropin-releasing hormone has also been
reported to relieve depression in man (22) but was inactive over a wide
range of doses against kindled seizures. Another type of antidepressive,
the monoamine oxidase inhibitors, had activity but the effect was one of
slight potentiation rather than blockade of seizure duration (Babington,
unpublished data).

Figure 3 presents the results of an experiment designed to determine the
effects of chronic dosing with an antidepressive agent. Two groups of eight
rats with amygdaloid seizures were dosed with one mg amitriptyline/kg, a
dose that has little or no effect on seizure duration. Two other groups of
eight rats received ten mg amitriptyline/kg, a dose that effectively blocks
seizures. One group from each dose level was administered drug daily for
14 days while the remaining two groups received amitriptyline only on the
seventh and fourteenth days. On test days, all animals were electrically
stimulated 30 minutes after drug dosing with the same length of stimulation
required to produce seizures on day 0.

In both groups treated with ten mg amitriptyline/kg, seizures were suppressed when the animals were challenged with amygdaloid stimulation.
Similarly, seizure duration was blocked to nearly the same extent in the
group receiving a 1-mg/kg daily dose. However, 1 mg amitriptyline/kg was
not sufficient for suppressing kindled seizures unless administered chronically. Note on the other hand, that chronic treatment with the higher

dose of amitriptyline was no more effective than a single dose administered
30 minutes prior to amygdaloid stimulation.

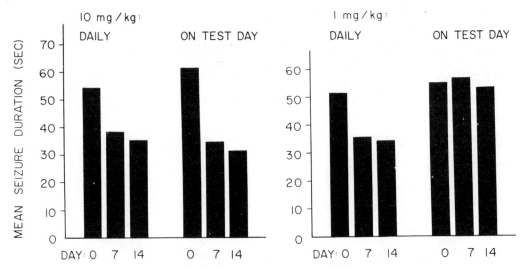

Fig. 3. Effect of acute or chronic treatment with amitrip-
         tyline on the duration of seizures kindled in the
         amygdala. Two dose levels were studied - acutely
         effective (10 mg/kg) and acutely subeffective (1 mg/
         kg) - and each dose level was administered either
         daily for 14 days or only on test days (days 7 and
         14). On test days, each animal was challenged 30
         min after dosing with the same length of stimulation
         required to produce seizures on day 0 (Babington,
         unpublished data).

Electroconvulsive therapy is a widely used and effective form of psychiatric
treatment, particularly for endogenous depression. Unfortunately, most
animal models used to screen compounds for antidepressive activity are not
suitable for testing electroconvulsive shock (ECS) for comparative purposes.
As Babington and Wedeking (23) reported, however, the kindling model readily
accomodates ECS testing. Figure 4 illustrates the effect of a single ECS
treatment on the duration of seizures kindled from the amygdala, septum, or
cortex. In this experiment, each animal received a 0.25 sec transpinnal
train of 60-Hz current at 50 ± 2 ma. This current was sufficient to produce
tonic limb extension. Groups of 8 rats were challenged at the various time
intervals after ECS with each rat receiving the same length of stimulation
used to produce the control kindled seizures.

Note that like the antidepressive drugs, ECS suppressed the duration of
elicited seizures, and the most profound reduction was against the amygda-
loid-induced response. Not only was the magnitude of the suppression
greater, but the duration of the effect on the amygdaloid tardive seizures
was much longer, lasting five to six hours after ECS treatment.

In a second experiment, only rats with electrodes implanted in the amygdala
were used, and seizures were not established prior to ECS treatment. The
parameters for both kindled seizures and ECS were the same as in the first
experiment. The procedure was to administer ECS at: (a) 30 min prior to
electrical stimulation of the amygdala; (b) 5 min after cessation of the
1-min stimulation period (or in those cases where short periods of seizures
were evoked, 5 min after cessation of the seizures); or (c) 4 h after
amygdaloid stimulation. On day 20, ECS treatment was terminated.

Figure 5 illustrates the results of administration of ECS during the sensitization process. ECS treatment 30 min prior to stimulation of the
amygdala completely blocked establishment of seizures, but as soon as ECS
was terminated, the sensitization process proceeded in normal fashion.
The results with ECS cannot merely be ascribed to the postictal depression
associated with electroconvulsions. Administration of ECS approximately
24 h prior to sensitization, although not as effective as the other two

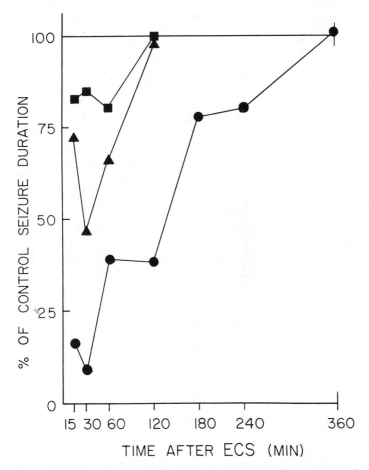

Fig. 4. Effect of a single electroconvulsive shock (ECS)
on the duration of tardive seizures elicited from
the amygdala (●), neocortex (▲) or septum (■).
Groups of 8 rats were tested at the various time
intervals after ECS.

ECS schedules, certainly had a profound effect on seizure development; and
ECS 4 h after (or 20 h prior to) the kindling experience was as effective as
ECS 30 min prior to kindling stimulation. Again, sensitization proceeded
rapidly once ECS was discontinued.

Success with the antidepressive class of drugs prompted us (2) to investigate the effects of a variety of other CNS-active drugs in kindled rats.
Figure 6 shows the results with four antianxiety agents: chlordiazepoxide,
diazepam, oxazepam, and meprobamate. A common characteristic for the
anxiolytic-type drugs was the marked potency as well as efficacy. Wise and
Chinerman (3) also found that diazepam would block amygdaloid-kindled
seizures in rats. In addition, the drug interfered with seizure development,
elevated seizure threshold in kindled animals and attenuated the contralateral propagation of the electrographic afterdischarge. Also in rats,
Racine et al. (4) noted that diazepam blocked or suppressed amygdaloid-
elicited seizures but had little or no effect on area 2 neocortical seizures.
In rabbits, Tanaka (5) found that diazepam abolished the behavioral
phenomenon but the amygdaloid electrographic afterdischarge was unaffected.

Among the drugs that Babington and Wedeking (2), Tanaka (5), Wise and
Chinerman (4) and Wada et al. (6) found to suppress kindled seizures were
the barbiturates, phenobarbital and pentobarbital. As shown in Fig. 7,
both agents are quite potent in the suppression of established seizures
in rats, with no motor deficits evident at efficacious dose levels.

Two drugs that exert stimulatory effects on behavior in rats, d-amphetamine
and methylphenidate both prolong seizure duration (Fig. 8). Unexpectedly,

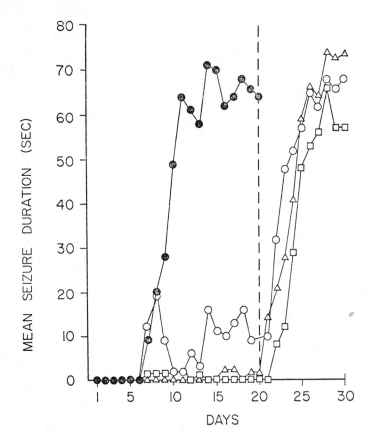

Fig. 5. Prevention by electroconvulsive shock of development of tardive seizures in the amygdala. ECS was
administered 30 min (□), 20 h (△) or 24 h (o) prior
to amygdaloid stimulation with the same stimulus
parameters as control (●). ECS treatment was terminated at day 20 (dashed line). Each plot represents a groups of 8 animals.

the enhancement reached a plateau and although higher doses of either drug
caused enhanced behavioral excitation, seizure duration did not increase.
In rabbits, Tanaka (5) reported that methamphetamine blocked all aspects of
the kindling response.

Other drugs that have been reported to be effective antiseizure agents in
the kindling model include Δ^9-tetrahydrocannabinol (24), lidocaine (5) and
atropine (25); reserpine (25), procaine HCl (4) and 6-hydroxydopamine (25,
26) were reported to enhance the kindling phenomenon; inactive CNS agents
include the neuroleptics (2, 5) antihistamines (2), narcotic analgetics (2),
and atropine (2, 27); and chemical kindling has been reported with cocaine
(28).

As a neuropharmacologic test system, the tardive seizure procedure possesses
numerous desirable features, including the fact that the response is a
pathologic entity, an attribute lacking in most test models. Other
attributes include: (a) the seizure duration of each animal is remarkably
consistent from trial to trial; (b) the response is easy to measure and has
a clear endpoint; (c) the apparatus is simple and inexpensive; (d) the rats
can be used repeatedly; (e) both increases and decreases in the duration
of seizures can be measured; (f) various types of centrally-active agents
have differential effects; (g) with experience, the testing can be carried
out quickly; and (h) once a colony of kindled rats is established, the
procedure can be utilized as a limited screening test.

To date, the procedure has reliably distinguished several classes of CNS-
active agents and has proved to be a valuable preclinical test model.
Whether the model is capable of identifying psychoactive drugs with unique
mechanisms of action remains to be seen.

147

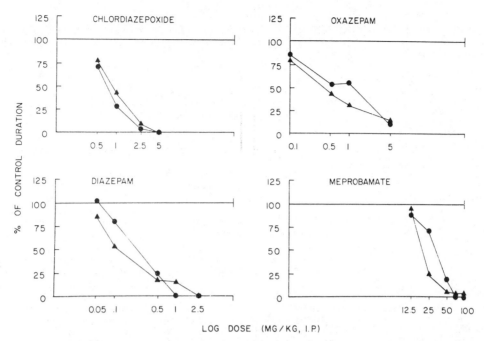

Fig. 6. Effects of four antianxiety drugs on the duration of
kindled seizures from the amygdala (●) or sensorimotor cortex (▲).

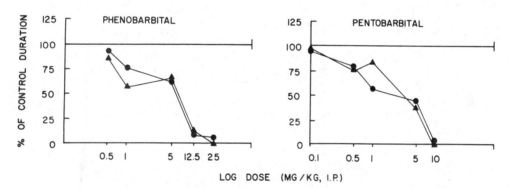

Fig. 7. Effects of two barbiturates on the duration of
kindled seizures from the amygdala (●) or sensorimotor cortex (▲).

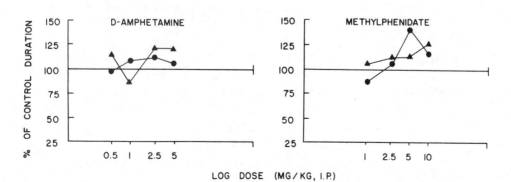

Fig. 8. Effects of two CNS-stimulating drugs on the duration
of kindled seizures from the amygdala (●) or sensorimotor cortex (▲).

148

Acknowledgements

I wish to credit Paul W. Wedeking who collaborated on all of the studies in which I was involved. I also wish to thank Ms. Jeanne Oster who prepared the manuscript.

1. Goddard, G.V., D.C. McIntyre, and C.K. Leech, A permanent change in brain function resulting from daily electrical stimulation, Expl. Neurol., 25, 295 (1969).
2. Babington, R.G. and P.W. Wedeking, The pharmacology of seizures induced by sensitization with low intensity brain stimulation, Pharmac. Biochem. Behav., 1, 461 (1973).
3. Wise, R.A. and J. Chinerman, Effects of diazepam and phenobarbital on electrically-induced amygdaloid seizures and seizure development, Expl. Neurol., 45, 355 (1974).
4. Racine, R., K. Livingston, and A. Joaquin, Effects of procaine hydrochloride, diazepam, and diphenylhydantoin on seizure development in cortical and subcortical structures in rats, Electroenceph. Clin. Neurophysiol., 38, 355 (1975).
5. Tanaka, A., Progressive changes of behavioral and electroencephalographic responses to daily stimulation of the amygdala in rabbit, Fukuoka Acta Med., 63, 152 (1972).
6. Wada, J.A., M. Sato, A. Wake, J.R. Green, and A.S. Troupin, Prophylactic effects of phenytoin, phenobarbital, and carbamazepine examined in kindled cat preparations, Arch. Neurol., 33, 426 (1976).
7. Babington, R.G. (1975) Antidepressants, Futura, Mt. Kisco, New York.
8. Jongmans, J.W., Report on the anti-epileptic action of Tegretol, Epilepsia, 5, 74 (1964).
9. Arieff, A.J. and M. Mier, Anticonvulsant and psychotropic action of Tegretol, Neurology, 16, 107 (1966).
10. Bird, C.A.K., B.P. Griffin, J.M. Miklaszewska, and A.W. Galbraith, Tegretol (carbamazepine): A controlled trial of a new anticonvulsant, Brit. J. Psychiat, 112, 737 (1966).
11. Livingston, S., C. Villamater, Y. Sakata, and L.L. Pauli, Use of carbamazepine in epilepsy, J. Amer. Med. Ass., 200, 204 (1967).
12. Marjerrison, G., S.M. Jedlicki, R.P. Keogh, W. Hrychuk, and G.M. Poulakakis, Carbamazepine: Behavioral, anticonvulsant and EEG effects in chronically-hospitalized epileptics, Dis. Nerv. Syst., 29, 133 (1968).
13. Babington, R.G. and Z.P. Horovitz, Neuropharmacology of SQ10, 996, a compound with several therapeutic indications, Arch. Int. Pharmacodyn. Ther., 202, 106 (1973).
14. MacLean, P.D., Contrasting functions of limbic and neocortical systems of the brain and their relevance to psychophysiological aspects of medicine, Am. J. Med., 25, 611 (1958).
15. Horovitz, Z.P. (1966) Antidepressant Drugs, Exerpta Medica Fdn., Amsterdam.
16. Penaloza-Rojas, J.H., G. Bach-Y-Rita, H.F. Rubio-Chevannier, and R. Hernandez-Péon, Effects of imipramine upon hypothalamic and amygdaloid excitability, Expl. Neurol., 4, 205 (1961).
17. Schmitt, H. and H. Schmitt, Action de l'imipramine, de l'amitriptyline et de leurs dérivés monodéméthyles sur les postdécharges provoquéss par l'excitation de certaines structures rhinencéphaliques chez le lapin, Thérapie, 21, 675 (1966).
18. Berger, F.M., M. Kletzkin, and S. Margolin (1966) Antidepressant Drugs, Exerpta Medica Fdn., Amsterdam.
19. Hernandez-Peon, R. (1963) Neuropsychopharmacology, Vol. 3, Elsevier, Amsterdam.
20. Wedeking, P.W. and R.G. Babington, Effects of neurological agents on seizures elicited by low-intensity brain stimulation in rats, Pharmacologist, 11, 279 (1969).
21. Gluckman, M.I. and T. Baum, The pharmacology of iprindole, a new antidepressant, Psychopharmacologia, 15, 169 (1969).
22. Prange, A.J., I.C. Wilson, P.P. Lara, L.B. Alltop, and G.R. Breese, Effects of thyrotrophin-releasing hormone in depression, Lancet Nov. 11, 999 (1972).
23. Babington, R.G. and P.W. Wedeking, Blockade of tardive seizures in rats by electroconvulsive shock, Brain Res., 88, 141 (1975).

24. Corcoran, M.E., J.A. McCaughran, Jr., and J.A. Wada, Acute antiepileptic effects of Δ^9-tetrahydrocannabinol in rats with kindled seizures, Expl. Neurol., 40, 471 (1973).
25. Arnold, P.S., R.J. Racine, and R.A. Wise, Effects of atropine, reserpine, 6-hydroxydopamine, and handling on seizure development in the rat, Expl. Neurol., 40, 457 (1973).
26. Corcoran, M.E., H.C. Fibiger, J.A. McCaughran, Jr., and J.A. Wada, Potentiation of amygdaloid kindling and metrazol-induced seizures by 6-hydroxydopamine in rats, Expl. Neurol., 45, 118 (1974).
27. Corcoran, M.E., J.A. Wada, A. Wake, and H. Urstad, Failure of atropine to retard amygdaloid kindling, Expl. Neurol., 51, 271 (1976).
28. Post, R.M. and R.T. Kopanda, Cocaine, kindling, and psychosis, Am. J. Psychiat., 133, 627 (1976).

DISCUSSION

Dr. Axelrod inquired about neurochemical studies done by Dr. Babington in the amygdala. *Dr. Babington* replied that the site had been too small, that he had not been able to observe changes in any of the major transmitters. He had electrically kindled animals using a chematrode; after removing the electrical stimulus, he injected possible transmitters down the cannula to see if seizures could be elicited. Acetylcholine and some other cholinergics did elicit seizures but none of the other neurotransmitters did. Although atropine had no effect on either electrically-elicited seizures or most chemically-elicited seizures, it will block the seizures elicited from acetylcholine.

Dr. R. Vogel commented on the intercranial self-stimulation technique, used by Dr. Stein and others in screening for antidepressants. In his laboratory, he usually finds at the beginning of a session (during the "warm-up" period), kindling in a tardive seizure at the outset of the intracranial self-stimulation period which soon disappears. *Dr. Babington* replied that this had been reported several times, not always gross motor seizures, but after discharges which precede the actual appearance of motor seizures.

ANIMAL TEST MODELS FOR PREDICTION OF CLINICAL
ANTIDEPRESSANT ACTIVITY

I.S. Sanghvi and S. Gershon
U.S.V. Pharmaceutical Corp. and N.Y.U. Medical Center
Neuropsychopharmacology Research Unit

Many animal experimental model systems are employed to screen and predict
clinical antidepressant activity. Predictability of drug activity for a
clinical condition is more reliable the more closely the model system in the
animal resembles the condition to be treated in man. Thus, muscle relaxant
or hypotensive activity is more predictable than antidepressant activity,
because valid animal models of mental depression do not yet exist. In the
absence of an endogenously depressed animal model, psychopharmacologists had
to rely on drug interactions to characterize potential antidepressants. It
is important to note that the techniques for pharmacological evaluation of
these compounds and for predictive assays have all been developed retrospectively after these drugs had been shown to exert antidepressant activity in
man suffering from depression. Most animal tests employed in the evaluation
of potential antidepressants are based on the catecholamine hypothesis for
mood, suggested by Schildkraut and Kety (1967). Furthermore, the validity
of the tests is based on the empirical finding that two clinically-proven
classes of antidepressants, the monoamine oxidase (MAO) inhibitors and the
tricyclic compounds, show selective activity. In this chapter the tests
most commonly employed are outlined and critically reviewed to indicate specificity and objectivity of the measurements utilized to evaluate antidepressant activity. The tests are simple to perform, require simple equipment
and most importantly, are relatively economical.

ANTAGONISM OF THE EFFECTS OF RESERPINE-LIKE DRUGS

Reversal of Hypothermia

Reserpine-produced hypothermia in rodents can be measured objectively
and can be reversed by antidepressant drugs. Tricyclics can prevent and reverse the effect of reserpine, whereas MAO inhibitors have only a preventive
action. According to the method of Askew (1963), mice are given oral or
i.p. doses of test drug 18 hours after an i.p. dose of 5 mg/kg of reserpine.
Oral or rectal temperatures are taken at 30-min intervals for two hours.
Post-drug temperatures are then compared to either those of the vehicle control group or their non-predrug values. The experiment should be carried
out at $20^{\circ} \pm 1^{\circ}$ C.

One criticism of the test is that drugs other than antidepressants also reverse or prevent reserpine-induced hypothermia. For example, amphetamine
(Spencer, 1966), chlorpromazine and aspirin (Whittle, 1967), and diethyldithiocarbamate (DDTC) (Barnett and Taber, 1968).

Antagonism of Ptosis

Reserpine and tetrabenazine or RO 4-1284 (2-hydroxy-2-ethyl-3-isobutyl 9,10-
dimethoxy-1,2,3,4,6,7-hexahydro(r)quinolizine) produce ptosis in rodents.
Tetrabenazine is preferred in this test because of its rapid onset and its
selective action in the central nervous system. The ptosis induced by reserpine or tetrabenazine is rated from 0-4 (Rubin et al., 1957). The dose-
response effect of pretreatment with a test compound is determined, and
PD_{50}, a dose at which 50% of the animals have eyelid opening scores of 3 or
more, can be determined by the method of Litchfield and Wilcoxon (1949).

The shortcomings of the method include the non-specificity for antidepressant
drugs. Drugs like amphetamines and antihistamines also prevent ptosis.

---

*Authors' work reported in this paper was supported by USPHS Grant MH 12383.

Antagonism of Locomotor Depression

This test, first described by Vernier et al. (1962), is usually carried out simultaneously with the previous two tests. Reserpine or tetrabenazine causes locomotor depression in mice and rats. Tricyclics prevent this depression. MAO inhibitors not only prevent tetrabenazine-induced depression, but cause amphetamine-like stimulation. Another drawback of the test is that repeated administration of tricyclics is necessary to produce excitation in rats.

POTENTIATION OF PHENETHYLAMINES

Potentiation of Amphetamine Effects in Rats

This test measures amphetamine-induced stereotyped behavior characterized by head-searching, licking, and gnawing (Quinton and Halliwell, 1963). The ability of a test drug to potentiate the stereotyped behavior induced by a standard dose of amphetamine is measured according to a rating scale from 0-6. Imipramine-like drugs enhance amphetamine-induced stereotypy, probably by interfering with the metabolism of amphetamine. Drugs like chlorpromazine and chlorprothixene also enhance amphetamine stereotypy in rats (Halliwell et al., 1964). Thus, the specificity of this test for antidepressants is less clear-cut.

Methamphetamine Potentiation in Self-stimulation Tests for Rats

Antidepressants, although ineffective in increasing self-stimulation rates, can potentiate the effects of amphetamine-like drugs on this measure (Stein, 1962). Rats are stereotaxically implanted with electrodes in the medial forebrain bundle or in the lateral hypothalamus, and trained to press a lever to receive electrical stimulation in these "reward" areas. After a control period to establish threshold current for eliciting stable self-stimulation rate, a 0.25 mg/kg i.p. dose of methamphetamine is given 15 minutes after treatment with a test drug, and the self-stimulation rate is determined. A comparison of rates before and after test drug is made. Antidepressants produce a marked increase in self-stimulation rate. Again the specificity of the test is lost since various antihistamines also enhance the self-stimulation rate (Stein, 1962).

Potentiation of Amphetamine-induced Locomotor Activity in Mice

This is another test which measures the ability of potential antidepressants to potentiate the locomotor activity of mice induced by amphetamine. The activity is measured in photocell activity cage systems.

Potentiation of the Effects of DOPA in Mice

l-Dopa, precursor of dopamine and norepinephrine (NE), when given in the presence of the MAO inhibitor, pargyline, to mice, produces adrenergic signs such as piloerection, salivation, increased motor activity, irritability, vocalization, and aggression (Everett and Wiegand, 1962; Everett, 1966). Test drugs are administered 4 or 24 hours prior to l-dopa (100 mg/kg i.p.) injection. The behavior of the animals is measured by an arbitrary rating scale from 1 to 3. Scores obtained by groups treated with test drugs are compared with those of both vehicle control groups and of those treated with a standard dose of pargyline or amitriptyline. Dr. Everett also suggests that compounds found positive in this test should also be tested for possible MAO inhibiting activity. This test also lacks specificity, since antidepressants are not the only group of compounds which gives positive tests in this method. Antihistamine, chlorpheniramine, and a quarternary anticholinergic, methyl atropine, to which the blood-brain barrier is highly resistant, are also found to be positive in this test (Sigg and Hill, 1967).

Effects on the Autonomic Nervous System

This test, first suggested by Dr. Sigg (1959), is done in anesthetized cats or dogs. The nicitating membrane-superior cervical ganglion preparation is quite efficient in showing that established antidepressants potentiate the NE response. However, in this test, all drugs which facilitate NE release and/or block NE uptake, such as amphetamines, cocaine, and similar agents, also potentiat the NE response, thus lacking in specificity for antidepressants. Acute studies using this preparation may be misleading with regard to potency relationships between cat and man. Chronic imipramine administration for 4 to 5 days increase the sensitivity of nictitating membrane to NE manifold. This observation has important practical implications. In man, imipramine-like compounds do not produce therapeutic effects until almost two weeks after the initiation of treatment. It is quite plausible that desipramine, a major metabolite of imipramine (Bickel, 1966), is responsible for enhanced sensitivity of the nicitating membrane to NE and for a therapeutic effect in depression.

A similar method, but using blood pressure of "ganglion-blocked" anesthetized dog, is also used to test antidepressant agents. Phenethylamine (300 mg/kg i.v.) and norepinephrine (NE) produce a pressor response in this preparation. Tricyclic antidepressants antagonize the pressor effect of phenethylamine and potentiate NE effect (Stone, 1967). However, phenothiazines, antihistamines and CNS stimulants also antagonize the pressor effect of phenethylamine (Stone, 1967), suggesting non-specificity of the test for antidepressants.

MONOAMINE OXIDASE INHIBITION

Pharmacological Tests

Potentiation of 5-hydroxytryptophan-induced head twitch. MAO inhibitors potentiate 5-hydroxytryptophan (5-HTP)-induced head twitch in mice (Corne et al., 1963). The test drug is given orally to groups of mice, and four hours later 50 mg/kg 5-HTP is given intraperitoneally. The number of head twitches displayed by each group is counted during three 2-min periods: 19-21, 23-25, 27-29 min, after injecting 5-HTP. During this observation period, 5-HTP alone produces little or no head twitching, so potentiation by MAO inhibitors is easily detectable. Criticism of this test is that drugs with different modes of action may be found negative in this test.

DOPA potentiation. This test was already outlined earlier.

SIDMAN CONTINUOUS AVOIDANCE PROCEDURE

In this procedure is tested the ability of a compound to: 1) potentiate amphetamine, 2) potentiate cocaine, and 3) "stimulate" with a non-depressant dose of tetrabenazine (Scheckel and Boff, 1964). The doses of test compound employed are such that they do not produce any behavioral effects. Various classes of drugs have been tested in this procedure. Antidepressants of tricyclic group and cocaine produced potentiation or stimulation in all three drug-interaction tests. However, the MAO inhibitor, iproniazid, potentiates amphetamine and cocaine responses only. It fails to stimulate with tetrabenazine. Anticholinergics, antihistamines and tranquilizers have no significant effects. This is a useful test and indicates that an antidepressant agent should cause potentiation and stimulation in all three drug interactions. Failure of an MAO inhibitor to stimulate with tetrabenazine may indicate a deficiency of the test, as this group of compounds is known to be antidepressant. Its usefulness lies in its being an objective test and it should be included as an important unit in a battery of tests for evaluation of antidepressant activity.

A reference should be made to a test which is reported to be a behavioral model of clinical depression (Seltzer and Tonge, 1975). Amphetamine or methamphetamine is chronically administered to rats for two weeks. After two weeks water is substituted for amphetamine. Twenty-four hours later, these animals exhibit signs of "depression" such as hypothermia, ptosis and reduced motor activity. Imipramine given 23 hours after amphetamine withdrawal prevents the behavioral "depression" and the animals look like normal controls. This test needs verification. Quite apart from its pragmatic significance, this post-amphetamine behavior may not justifiably be considered as an analog of psychiatric depression.

BEHAVIORAL TESTS

Antagonism of Muricide Behavior in Rats

All preceding tests for antidepressants employ at least one other chemical agent besides the test substances. One test which does not involve drug interaction is prevention of mouse-killing (muricide) behavior in rats, described by Horovits et al. (1966). Certain strains of rats will promptly kill a mouse that is placed in the same cage, by biting the cervical vertebrae. Pretreatment of killer rats with antidepressants prevents this muricide behavior. The mechanism of muricide behavior probably involves a discrete area of the brain, the amygdala, since bilateral lesions of amygdala abolish muricide behavior (Karli, 1955; Horovitz et al., 1966). Furthermore, antidepressants have been shown to depress amygdaloid afterdischarge (Penalozo-Rojas et al., 1961; Plas and Nequet, 1961); thus it is conceivable that the effect of antidepressants on amygdala may be related to their antimuricide actions. Another test which is done along with this test is the ability of rats to maintain equilibrium on a rotorod, since a compound which affects the muricide behavior may interfere with motor activity of the animal. A drug is considered positive if the ratio of its ED_{50} for muricide and for impairment of rotorod performance is greater than one. Antidepressant imipramine, amitriptyline, and desipramine, as well as MAO inhibitors, were positive in this test. However, other classes of drugs were also positive in the test, namely, certain CNS stimulants, anticholinergics, and antihistamines.

Action on the "Grasping Reflex"

This is another behavioral test to evaluate antidepressant acitivity. The test measures the duration of time a rat will remain suspended from a horizontal rod (Molinengo and Orsetti, 1976). Established antidepressants, amphetamine, and neuroleptics prolong the "grasping" time, whereas hypnotics decrease it. To differentiate CNS stimulants from the thymoleptics, another test is run at the same time. In the latter test, the duration of time a rat will swim in water is determined. The test drug is given at a dose which prolongs the "grasping" time. CNS stimulants prolonged the swimming time as well whereas thymoleptics and hypnotics or depressants did not have significant effect on swimming time. This test, like other tests, lacks specificity and objectivity but may be useful in conjunction with other tests.

Effect on Immobility in Water Induced by Forced Swimming

Recently a new test has been proposed as an animal model of depression (Porsolt et al., 1977). The method is based on the observation that a rat, when forced to swim without recourse to escape, will eventually cease to move altogether, making only the movements necessary to keep its head above water. This behavioral immobility is compared to a state of despair from which escape is impossible and the animal resigns itself to the experimental conditions. In essence, the test measures the duration of immobility in water and the ability of antidepressants to reduce the period of immobility. For this measure, the drugs are administered in three doses at various times before the test. All established thymoleptics of the tricyclic group and nialamide reduced the period of immobility. In addition, some newer compounds such as iprindole, mianserin and viloxazine also reduced the immobility. The test is able to distinguish CNS stimulants, anxiolytics and neuroleptics such as chlorpromazine and reserpine from antidepressants. Anxiolytics have no effects while neuroleptics increas the period of immobility. CNS stimulants also reduce the period of immobility but in addition produce hyperactivity and stereotypy. It should be pointed out that one of the compounds, mianserin, which failed to show any activity in common antidepressant tests, did show an activity similar to other antidepressants. Mianserin has been shown to be effective in the treatment of depression (Murphy, 1975; Wheatley, 1975). However, these latter studies did not include a placebo group; a proper double blind control study, including placebo group, is necessary to establish its clinical effectiveness. It is claimed to be as effective as imipramine or amitriptyline. This test needs verification and can be useful in conjunction with other tests. It remains to be established if a single dose administration, instead of three doses, will produce similar effects.

Behavior of Bulbectomized Rat

Removal of olfactory bulbs results in diverse behavioral changes in rodents. One of the effects of bilateral ablation of olfactory bulbs of rats is increased thirst in such animals. Rigter and coworkers (1977) have shown that this behavioral change, which they call "anxiosoif" test, can be used to predict clinical antidepressant activity. The test is essentially a passive avoidance paradigm in which thirsty rats learn to avoid an electrified water spout. Bulbectomized animals show deficient avoidance behavior. Antidepressants like amitriptyline, imipramine, chlorimipramine and mianserin selectively reverse this behavioral deficiency. Anxiolytics and neuroleptics such as chlordiazepoxide, diazepam and chlorpromazine aggravate this deficiency. This method appears to be useful in distinguishing antidepressants on one hand and anxiolytics and neuroleptics on the other. The effects of CNS stimulants have not been reported in this test and it would be useful to see if CNS stimulants will show different effects than antidepressants. One feature of this test different from other tests outlined above is the pretreatment schedule. The test drug was administered for seven days before the experiment. Actually, this procedure may have certain relevance since clinically, the antidepressant activity is evident only after several days of treatment. On the other hand, the increased duration of time necessary to perform the test may be a discouraging factor compared to other acute tests.

POTENTIATION OF YOHIMBINE-INDUCED BEHAVIORAL AND CARDIOVASCULAR CHANGES IN CONSCIOUS DOG

Most of the animal tests for prediction of antidepressant activity were developed retrospectively following their therapeutic effectiveness. Up until the early 1960's most common tests for antidepressant activity consisted of reversal of reserpine or tetrabenazine-induced sedation and hypothermia and potentiation of norepinephrine response of nictitating membrane of anesthetized cat. Lang and Gershon (1963) first developed and proposed an empirical test in conscious dog for prediction and evaluation of clinically useful antidepressant agents; since then, it has been successfully explored for preliminary evaluation of potential antidepressant agents (Sanghvi et al., 1969; Sanghvi and Gershon, 1969; Sanghvi et al., 1971; Geyer et al., 1973). Compounds with divergent chemical structures predicted as potential antidepressants by means of many of the previously-described tests, have been evaluated in this test system with positive correlation with clinical studies.

TABLE 1

BEHAVIORAL RATING SCALE

		0	1	2
1.	Stance	standing erect little weight on harness	leaning heavily on harness	collapsed in harness with feet spread apart
2.	Back position	level	slightly arched	severe arching
3.	Head position	elevated	lowered or held in an unusual position	lowered to table
4.	Tail position	elevated or lowered with distinct outward curl	lowered parallel with hind legs	curled inward between legs
5.	Movement	wagging tail or calm movements of head	agitated movements of head and lifting of legs	forceful attempt to escape harness - needs restraint
6.	Respond to stimuli	calm, investigative movements toward mild random stimuli	no response to mild stimuli, and slow, delayed movement to specific stimuli, i.e. whistle, clapping	no response to any stimuli, only movement of eyes OR exaggerated response, i.e. jumping or arching of the back
7.	Response to handling	positive response presses body toward hand	no response	withdrawal, arching of back
8.	Pain response	strong, quick withdrawal	slow, weak, or delayed withdrawal	no response
9.	vocalization	none	occasional whining, moaning	constant whining, moaning or distinct barking
10.	Tremors	none	mild or seldom	severe or constant
11.	Penile erection	none	mild	extreme
12.	Piloerection	none	mild	extreme
13.	Defecation	none	formed boli	diarrhea
14.	Salivation	none	mild, licking of lips	pouring from mouth
15.	Pupil size	normal or restricted	mild dilatation	extreme dilatation
16.	Pupil response (to light)	normal	moderate	severe
17.	Vascularization of the conjunctiva	normal	moderate	severe
18.	Heart rate	normal	+50%	+100%

The test involves changes in behavioral and cardiovascular effects produced
by a standard intravenous dose of yohimbine in the conscious dog. Mongrel
dogs of either sex are trained to stand in a Pavlov-type stand and harness.
The behavioral changes induced by yohimbine and its interaction with a test
drug are scored on a three-point rating scale (Geyer et al., 1973), which
includes features such as body position, restlessness measured as movement,
responses to stimuli, body tremor, and autonomic features such as salivation,
piloerection, pupil size, and pupil response to light and heart rate. Systolic blood pressure is continuously measured by direct cannulation of ex-
teriorized carotid artery (Gershon and Lang, 1962; Brown and Korol, 1968).
The observations are made "blind" by the experienced raters. Experiments are
carried out at weekly intervals and are designed such that each dog acts as
its own control. In each test for a new antidepressant, imipramine is used
as a standard antidepressant for comparison. The test proposed that a potential antidepressant should significantly enhance the behavioral and cardiovascular effects of yohimbine. In an actual test, the test compound or imipramine (1.5 mg/kg) is administered intravenously or intraperitoneally 15-30
minutes before a challenge dose of yohimbine (0.5 mg/kg i.v.). The animals
are rated according to the rating scale, at 5, 10, and 30 minutes following
yohimbine administration. The peak responses are compared with those obtained following a control yohimbine injection. Changes in behavioral scores
and systemic blood pressure are analyzed statistically, using Student's "t"
test. The cardiovascular measures are completely objective, and therefore,
in case of doubt, are utilized exclusively (Table 1).

To date, several compounds with wide variations in their chemical structures
have been tested in this animal test system, and also clinically to verify
its validity. In addition to imipramine, amitriptyline, and the MAO inhibitor nialamide (Lang and Gershon, 1963), the following compounds were tested:

EXP 461 (4-phenyl-bicyclo (2,2,2) octan-1-amine hydrochloride monohydrate).
This compound was found to be more potent than imipramine in several animal
tests. In our test system, it failed to modify behavioral and cardiovascular
responses to yohimbine. Clinically, it exhibited stimulant properties, but
could not be used as a therapeutic agent in the treatment of depression
(Gershon et al., 1968).

BLKR-140 (N-acetonyl-N,N-dimethylbenzylammonium chloride). Its preclinical
pharmacology indicated its potential antidepressant activity, but clinically
it did not exhibit any such activity (Hekimian et al., 1968). In our test
system too, it failed to modify yohimbine response (Fig. 1) (Sanghvi et al.,
1969).

Amphetamine. It is believed in some clinical circles that amphetamine is a
good antidepressant. However, clinically depressed patients feel initial
stimulation, followed by a greater degree of depression. In order to verify
the specificity of the yohimbine test for the prediction of clinical antidepressant agents, it is essential to test the effects of CNS stimulants such
as amphetamine on the yohimbine response in conscious dogs. The preclinical profile of amphetamine would suggest that it should be a good antidepressant.
In our test system, amphetamine, although producing stimulation in the dog,
failed to potentiate yohimbine responses.

MJ-1986 (N,N-dimethyl-1-phenylindene-1-ethylamine). This compound, also
known as Indriline, was also found to be positive in animal tests for antidepressants. In fact, it was about five times as potent as imipramine or amitriptyline in anti-reserpine effect on ptosis. Like imipramine, it potentiated endogenous norepinephrine stimulation of Sidman avoidance behavior in
rats (Kissel, 1967). In clinical tests, this compound caused worsening of
symptoms or no change in 8 of 10 patients. Five of the 10 patients exhibited
mild stimulation resulting in restlessness, tremor, and agitation, and it
failed to produce any antidepressant effect (Hekimian et al., 1970). In our
test system, MJ-1986 exhibited transient central nervous system stimulation,
but failed to potentiate yohimbine responses.

Cocaine. Cocaine is well known as a central nervous system stimulant and has many pharmacological actions similar to imipramine, viz., potentiation of norepinephrine response on nictitating membrane, reversal of
reserpine-induced ptosis, and antagonism of benzoquinoline-induced suppression
of behavioral responses in rats (Ryall, 1961; Haefley et al., 1964; Kissel,
1967). Clinically, cocaine produces pronounced, but short-lasting, effects
on mood and behavior. In our test system, cocaine did produce stimulant ef-

157

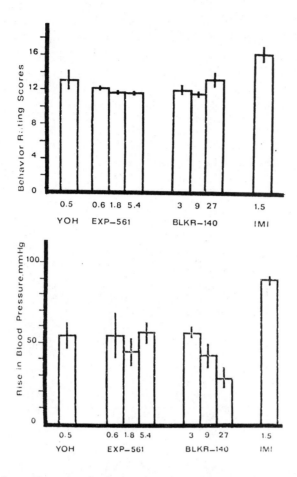

Fig. 1. Changes in behavioral rating scores (upper)
and rise in arterial pressure (lower) in conscious
dogs caused by yohimbine (YOH) before and after treatment with EXP-561, BL-KR 140 and imipramine (IMI).
Each column represents mean effect of 406 experiments.
Vertical bars indicate S.E. All drugs were administered intravenously. Number under each column refers
to mg/kg.

fects, but failed to modify yohimbine responses (Sanghvi and Gershon, 1969)
(Fig. 2).

5-Hydroxytryptophan (5-HTP). Although 5-HTP, precursor of serotonin, was not
evaluated in usual animal tests for antidepressants, clinically it was found
beneficial in certain depressives (Coppen et al., 1967). However, this contention is controversial, and opposite claims have also been made. While
our test system was being developed, a report by Bogdanski and co-workers
(1958) was found, in which they reported the effects of 5-HTP in various species, including dog. The effects of moderate dose (25-30 mg/kg) of 5-HTP, as
they reported, appeared similar to yohimbine effects in our study. So, we
compared the effects of 5-HTP with yohimbine in our laboratory, and found
that 5-HTP markedly potentiated yohimbine effects in dogs (Sanghvi and Gershon, 1970) (Fig. 3). Clinically, the value of 5-HTP in treatment of depression remains to be resolved.

1-Dopa. 1-Dopa, precursor of dopamine and norepinephrine, is claimed to
have beneficial effects in certain depressive patients.. It was, therefore,
felt that it would be useful to test 1-dopa in our test system to verify the specificity
of the test for antidepressants. We found that 1-dopa not only failed to potentiate yo-

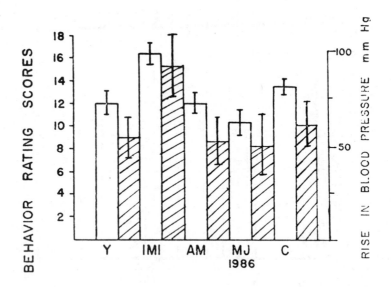

Fig. 2. Changes in behavioral rating scores (open) and rise in arterial pressure (hatched) in conscious dogs caused by yohimbine (Y) before and after imipramine (IMI)(1.5 mg/kg), amphetamine (AM)(1.0 mg/kg), MJ 1986 (1.5-210 mg/kg) and cocaine (C)(1.5 mg/kg). All drugs were administered intravenously.

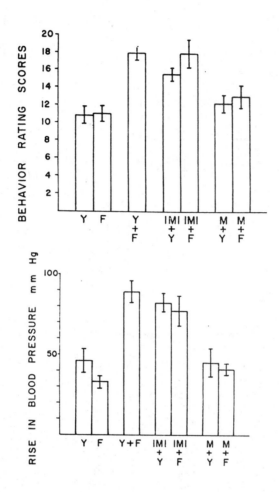

Fig. 3. Changes in behavioral rating scores (upper)
and rise in arterial pressure (lower) in conscious
dogs caused by yohimbine (Y) and 5-hydroxytryptophan
(F) and their combination with imipramine (IMI) and
methysergide (M). Doses: Y, 0.5; 5-HTP, 25; IMI,
1.5; and M, 0.5. All doses in mg/kg, I.V.

160

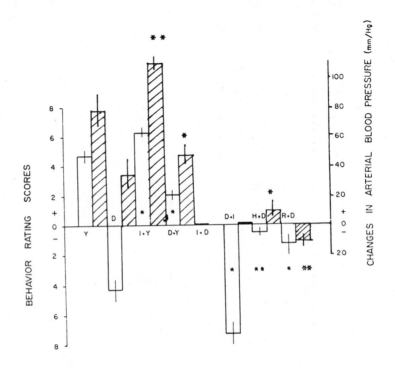

Fig. 4. Changes in behavioral rating scores (open)
and rise in arterial pressure (hatched) in conscious
dogs caused by yohimbine (Y) and l-dopa (D) before
and after various treatments. I, imipramine (1.5); H,
haloperidol (0.1-0.2); R, R04-4602 (50); l-dopa (25-
30). All doses refer to mg/kg, I.V., except l-dopa,
which was administered I.P.

himbine effects, but partially antagonized them. At most, it may be classified as a motor mobilizer (Sanghvi et al., 1971) (Fig. 4). Subsequent
clinical studies have also concluded that l-dopa produces motor mobilization,
and have confirmed its lack of thymoleptic activity.

Lithium. Lithium is a therapeutic agent for manic phase of manic-depressive
psychosis (Baastrup and Schou, 1967). Several clinical reports have indicated that lithium is beneficial in the treatment of depression (Adreani et
al., 1968; Fieve et al., 1968; Mendels et al., 1972). In view of these reports, we anticipated that lithium might potentiate yohimbine response in the
dog. Our results following acute and chronic lithium treatments indicated
that lithium in blood concentrations found therapeutic in the treatment of
mania did not potentiate yohimbine response in the dog (Geyer et al., 1973)
(Figs. 5 and 6). Based on the results in this test, lithium does not appear
to exhibit antidepressant activity. The issue of thymoleptic activity of
lithium in treatment of endogenous depression remains to be resolved.

Thyrotropin-Releasing Hormone (TRH). TRH is reported to be beneficial in
the treatment of unipolar depression (Prange et al., 1972) and the depressant phase of manic depressive psychosis (Kastin et al., 1972). Furthermore
this beneficial effect of TRH is evident after a single dose administration.
In view of these reports, TRH was tested in our animal model (Hine et al.,
1973). TRH at a dose of 50 and 100 µg per kg failed to potentiate yohimbine
effects in dog (Fig. 7). These doses of TRH were at least 25 times greater
than those used clinically on a mg per kg basis. Lower doses of TRH were
ineffective. TRH itself produced signs of sympathetic activation; thus,
mild salivation, a moderate degree of restlessness, and an increase in
arterial blood pressure and heart rate occurred. According to the criteria
for a potential antidepressant agent, TRH thus can not be considered to
exhibit antidepressant activity.

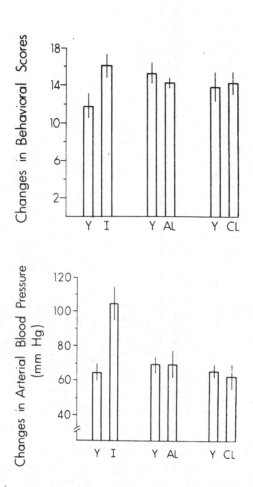

Fig. 5. Changes in behavioral rating scores (upper)
and rise in arterial pressure (lower) in conscious
dogs caused by yohimbine (Y) before and after imipramine (I), acute lithium (AL) and chronic lithium
(CL) treatments. Doses: I, 1.5, lithium chloride,
50, as mg/kg, I.V. and I.P. respectively. Chronic
lithium was administered as lithium carbonate in a
total daily dose of 600 mg/kg, p.o., for four to five
days.

162

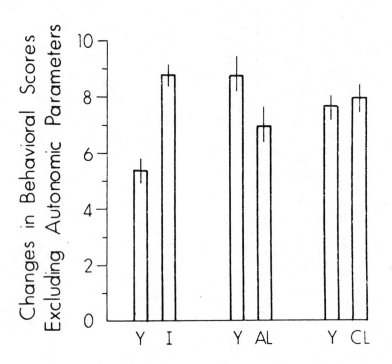

Fig. 6. Changes in behavioral rating scores excluding autonomic parameters in conscious dogs produced by yohimbine (Y) before and after imipramine (I), acute lithium (AL) and chronic lithium (CL).

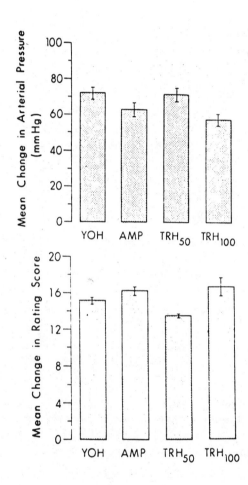

Fig. 7. Changes in behavioral rating scores (upper)
and rise in arterial pressure (lower) in conscious
dogs produced by yohimbine (YOH) before and after
amphetamine (AMP) and thyrotropin-releasing hormone
(TRH). Doses: AMP, 0.5 mg/kg and TRH, 50 and 100
ug/kg. All drugs were administered intravenously.

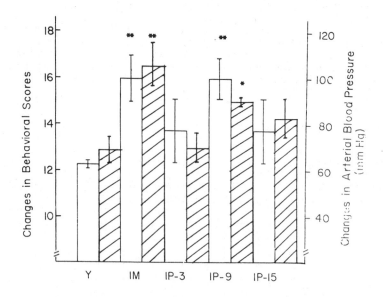

Fig. 8. Changes in behavioral rating scores (open) and rise in arterial pressure (hatched) in conscious dogs produced by yohimbine (Y) before and after imipramine (IM) and iprindole (IP). Numbers following IP refer to the dose as mg/kg. All drugs were administered intravenously. *, $p < 0.02$ and **, $p < 0.01$, two-tailed.

Iprindole. This is a tricyclic indole compound with an eight-membered saturated third ring. Clinically it is reported to be an antidepressant agent (Daneman, 1967; Sterlin et al., 1968; Ayd, 1969). In our test model iprindole potentiated the effects of yohimbine, although not as markedly as imipramine (Fig. 8) (Sanghvi et al., 1976). According to the criteria of the test, therefore, iprindole can be predicted to possess clinical antidepressant activity.

Present animal tests for prediction of antidepressant activity leave much to be desired. Species differences are an important factor in drug evaluation in treatment for human illnesses. In addition, most tests involve drug interaction, and it would be preferable to develop animal models of behavior in which depression is induced by environmental or conditioning factors. Some attempts have been made to develop such a model behavioral state in animals. Irwin and co-workers (1965) suggested a rat swimming test as a model of "hopelessness," and antidepressants were found to have antifatigue activity in such a swimming test (Latz et al., 1966). However, this test is highly variable and is difficult to standardize. Harlow and co-workers (1970) have attempted to develop a model "depression" in monkeys by separation of infants from mother to cause emotional stress. They have also shown that antidepressants reverse the "depressive" state in these animals. This test is promising, but it is very elaborate and not subject to easy application in a routine manner. Furthermore, it would be presumptuous to describe the separation syndrome as clinical depression. This model is discussed in more detail elsewhere in this volume. Finally, the available animal tests evaluate acute effects of test drug, whereas, clinically, antidepressant drugs of MAO inhibitor and tricyclic groups do not produce their optimal responses for several days or weeks. More experiments following chronic administration of test compounds on desired parameters should be developed. It is possible that compounds found negative in acute tests may show positive effects in chronic experiments or give a different profile of its pharmacological effects. One such compound is chlorpromazine. Acutely, chlorpromazine produces the well-established effects, such as lowering of blood pressure, adrenergic blockade, and serotonin antagonism in conscious dogs. However, after continued oral administration of chlorpromazine for several weeks, the adrenergic responses to noradrenaline and yohimbine are significantly potentiated (Lang et al., 1966). This effect of chronic chlorpromazine administration is similar to the effects of the antidepressant, imipramine, and its analogues. Clinically, beneficial effects of chlorpromazine in treatment of depression have been reported by several investigators (Cohen, 1964; Denber, 1957; Fink et al., 1965; Overall et al., 1964). It is conceivable that changes in pharmacological activity observed in the conscious dog may form the basis of beneficial effect of chlorpromazine in depression.

SUMMARY

Review of the commonly used and some newer animal test models for prediction of clinical antidepressant activity indicates that no single test is reliable across the wide range of compounds (Table 2). Many variables such as dose, duration of administration or pretreatment time, species, objectivity of observations, etc., can contribute to this lack of reliability. A selective use of some of the tests outlined in this review may serve as preliminary tests for prediction of probable antidepressant activity.

TABLE 2

PROFILES OF VARIOUS CHEMICAL COMPOUNDS IN SOME OF THE ANIMAL MODELS
AND THEIR CLINICAL ANTIDEPRESSANT ACTIVITY

+ = POSITIVE
– = NEGATIVE
BLANK OR ? = NOT KNOWN

DRUGS	MODELS							
	RESERPINE	TBZ	CAR	SWIM	MURICIDE	GRASPING TEST	YOH	CLINICAL
IMIPRAMINE	+	+	+	+	+	+	+	+
AMITRIPTYLINE	+	+	+	+	+	+	+	+
IPRINDOLE	+			+			+	+
MIANSERIN	+			+				+
MAO I.	+	+	–	+	+	+	+	+
AMPHETAMINE	+	+	+	+	+	+	–	(–)
COCAINE	+	+	+				(–)	(–)
L-DOPA	–						–	
5-HTP	–						+	(+)
TRH							–	–
LITHIUM							–	(±)?
EXP-561	+	+					–	–
BLKR-140	+						–	–
MJ-1986	+		+				–	–
AHR-1118							+	?

REFERENCES

G. Adreani, G. Caselli and G. Martelli, Rilieve clinici et elettroencefalo Grafici durante il tetramento con sli di litio in malati psichiatrici, Arch. Psichiat. Neuropat. 86, 273-328 (1968).

B.M. Askew, A simple screening procedure for imipramine-like antidepressant accents, Life Sci. 1, 725-730 (1963).

F.J. Ayd, Jr., Clinical evaluation of a new tricyclic antidepressant, iprindole, Dis. Nerv. Syst. 30, 818-824 (1969).

P.C. Baastrup and M. Schou, Lithium as prophylactic agent, Arch. Gen. Psychiat. 16, 162-172 (1967).

A. Barnett and R.I. Taber, The effects of diethyldithiocarbonate and L-Dopa on body temperature in mice, J. Pharm. Pharmac. 20, 600-604 (1968).

M.H. Bickel, Metabolism and structure-activity relationships of thymoleptic drugs, In: Antidepressant Drugs (S. Garattini and M.N.G. Dukes, Eds.), pp. 3-9, Excerpta Medica Foundation, Amsterdam (1966).

D.F. Bogdansi, H. Weissbach and S. Udenfriend, Pharmacological studies with the serotonin precursor, 5-hydroxytryptophan, J. Pharmacol. Exp. Ther. 122, 184-194 (1958).

M.L. Brown and B. Korol, Surgical preparation of externalized carotid artery loops in dogs, Physiol. Behav. 3, 207-208 (1968).

S. Cohen, Is thioridazine (Mellaril) an antidepressant, West. Medicine 5, 359-361 (1964).

A.D. Copin, M. Shaw, B. Herzberg and R. Maggs, Tryptophan in the treatment of depression, Lancet II, 1178 (1967).

S.J. Corne, R.W. Pickering and B. Warner, A method for assessing the effects of drugs on the central action of 5-hydroxytryptamine, Brit. J. Pharmacol. 20, 106-120 (1963).

E.A. Daneman, Treatment of depressed patients with iprindole, Psychosomatics 8, 216-221 (1967).

H.C.B. Denber, Some aspects of chlorpromazine treatment with reference to the problem of depression, Dis. Nerv. Syst. 18, 35-36 (1957).

G.M. Everett, The dopa response potentiation test and its use in screening for antidepressant drugs, In: Antidepressant Drugs (S. Garattini and M.N.G. Dukes, Eds.), pp. 164-171, Excerpta Medica Foundation, Amsterdam (1966).

G.M. Everett and R.G. Wiegand, Central amines and behavioral states: A critique and new data, In: Proc. First Int. Pharmacol. Mtg. (W.D.M. Paton, Ed.) Vol. 8, pp. 85-92, MacMillan, New York (1962).

R.R. Fieve, S.R. Platman and R.R. Plutchik, The use of lithium in affective disorders, I. acute endogenous depression, Amer. J. Psychiat. 125, 487-491 (1968).

M. Fink, D.F. Klein and J.C. Kramer, Clinical efficacy of chlorpromazine-procyclidine combination imipramine and placebo in depressive disorders, Psychopharmacologia 7, 27-36 (1965).

S. Gershon and W. Lang, A psycho-pharmacological study of some indole alkaloids, Arch. Int. Pharmacology 135, 31-56 (1962).

S. Gershon, L.J. Hekimian and A. Floyd, Non-correlation of preclinical-clinical evaluation of a proposed antidepressant 4-phenyl-bicyclic (222) octan-1-amine hydrochloride monohydrate (exp-5), Arzneim. Forsch. 18, 243-245 (1968).

167

H. Geyer, III, I. Sanghvi and S. Gershon, Exploration of the anti-depressant potential of lithium, Psychopharmacologia (Berlin) 28, 107-113 (1973).

W.A. Haefely, H. Hurlimann and H. Thoenen, A quantitative study of the effect of cocaine on the response of the cat nictitating membrane to nerve stimulation and to injected noradrenaline, Brit. J. Pharmacol. 22, 5-21 (1964).

G. Halliwell, R.M. Quinton and F.E. Williams, A comparison of imipramine, chlorpromazine and related drugs in various tests involving autonomic functions and antagonism of reserpine, Brit. J. Pharmacol. 23, 330-350 (1964).

H.F. Harlow, S.J. Suomi and W.T. McKinney, Experimental production of depression in monkeys, Mainly Monkeys 1, 6-12 (1970).

L.J. Hekimian, S. Gershon and A. Floyd, Discrepancies in the evaluation of an antidepressant, Curr. Ther. Res. 10, 282-287 (1968).

L.J. Hekimian, S. Gershon and A. Floyd, The clinical evaluation of four proposed antidepressants: Relationship to their animal pharmacology, Int. Pharmacopsychiat. 3, 65-76 (1970).

B. Hine, I. Sanghvi and S. Gershon, Evaluation of Thryotropin-releasing hormone as a potential antidepressant agent in the conscious dog, Life Sci. 13, 1789-1797 (1973).

Z.P. Horovitz, J.J. Piala, N.P. High, J.C. Burke and R.C. Leaf, Effects of drugs on the mouse-killing (muricided) test and its relationship to amygdalo function, Int. J. Neuropharmacol. 5, 405-411 (1966).

S. Irwin, D. Buxbaum and A. Feinberg, Pharmacologist 7, 153 (1965).

G. Johnson, S. Gershon and L.J. Hekimian, Controlled evaluation of lithium and chlorpromazine in the treatment of manic states: an interim report, Comp. Psychiat. 9, 563-573 (1968).

P. Karli, The Norway rat's killing response to the white mouse: An experimental analysis, Behavior (Leiden) 10, 81-103 (1956).

A.J. Kastin, D.S. Schalch, R.H. Ehrensing and M.S. Anderson, Improvement in mental depression with decreased thyrotropin response after administration of thyrotropin-releasing hormone, Lancet II, 740-742 (1972).

J.W. Kissel, The pharmacology of MJ 1986, phenylindene with properties of antidepressant psychostimula significance. In: Antidepressant Drugs (S. Garattini and M.N.G. Dukes, Eds.), pp. 233-240, Excerpta Medica Foundation, Amsterdam (1966).

W. Lang and S. Gershon, Effects of psychoactive drugs on yohimbine induced response in conscious dogs, Arch. Int. Pharmacodyn. 142, 457-472 (1963).

W. Lange, M.L. Brown, I. Sletten, B. Korol and S. Gershon, Effects of chronic chlorpromazine administration upon the blood pressure responses to autonomic drugs in conscious dogs, Arch. Int. Pharmacodyn. 162, 330-344 (1966).

A. Latz, A. Bain, M. Goldman and C. Kornetsky, Swimming performance of mice as affected by antidepressant drugs and baseline levels, Psychopharmacologia (Berlin) 10, 67-88 (1966).

J.T. Litchfield, Jr., and F. Wilcoxon, Simplified method of evaluating dose-effect experiments, J. Pharmacol. Exp. Ther. 96, 99-113 (1949).

W. Lovenberg, R.J. Lavine and A. Sjoerdsma, A sensitive assay of monoamine oxidase activity in intro application to heart and sympathetic ganglia, J. Pharmacol. Exp. Ther. 135, 7-10 (1962).

J. Mandels, S.K. Secunda and W.L. Dyson, A controlled study of the antidepressant effects of lithium carbonate, Arch. Gen. Psychiat. 26, 154-157 (1972).

168

L. Molinengo and M. Orsetti, Drug action on the "grasping" reflex and on swimming endurance: An attempt to characterize experimentally antidepressant drugs, Neuropharmacology 15, 257-260 (1976).

J.E. Murphy, A comparative clinical trial of OR & GB 94 and imipramine in the treatment of depression in general practice, J. Int. Med. Res. 3, 251-260 (1975).

J.E. Overall, L.E. Hollister, F. Meyer, I. Kimbell and J. Shelton, Imipramine and thioridazine in depressed and schizophrenic patients. Are there specific antidepressant drugs? J. Amer. Med. Assoc. 189, 605-608 (1964).

J.H. Penaloza, G. Backy Rita, H.F. Rubio-Chevannier and R. Hernandez-Peon, Effects of imipramine upon hypothalamic and amygdaloid excitability, Exp. Neurol. 4, 205-213 (1961).

R. Plas and R. Naquet, Contribution to the Neurophysiological study of imipramine, C.R. Soc. Biol. 155, 840-843 (1961).

R.D. Porsolt, M. Le Pichon and M. Jalfre, Depression: A new animal model sensitive to antidepressant treatments, Nature (London) 266, 730-732 (1977).

A.J. Prange, I.C. Wilson, P.P. Lara, L.B. Alltop and G.R. Breese, Effects of thyrotropin-releasing hormone in depression, Lancet II, 999-1002 (1972).

R.M. Quinton and G. Halliwell, Effects of alpha-methyl dopa and dopa on the amphetamine excitatory response in reserpinized rats, Nature (London) 200, 178-179 (1963).

H. Rigter, H. van Riezen and A. Wren, Pharmacological validation of a new test for the detection of antidepressant activity of drugs, Br. J. Pharmac. 59, 451P-452P (1977).

B. Rubin, M. Malone, M. Waugh and J.C. Burke, Bioassay of rauwolfia roots and alkaloids, J. Pharmacol. Exp. Ther. 120, 125-136 (1957).

R.W. Ryall, Effects of cocaine and antidepressant drugs on the nictitating membrane of the cat, Brit. J. Pharmacol. 17, 339-357 (1961).

I. Sanghvi, E. Bindler and S. Gershon, The evaluation of a new animal method for the prediction of clinical antidepressant activity, Life Sci. 8, 99-106 (1969).

I. Sanghvi and S. Gershon, The evaluation of central nervous system stimulants in a new laboratory test for antidepressants, Life Sci. 8, 449-457 (1969).

I. Sanghvi and S. Gershon, Similarities between behavioral and pharmacological actions of yohimbine and 5-hydroxy-tryptophan in the conscious dog, Eur. J. Pharmacol. 11, 125-129 (1970).

I. Sanghvi, X. Urquiaga and S. Gershon, Exploration of the antidepressant potential of L-dopa, Psychopharmacologia (Berlin) 20, 118-127 (1971).

I. Sanghvi, H. Geyer and S. Gershon, Exploration of the antidepressant potential of iprindole, Life Sci. 18, 569-574 (1976).

C.L. Scheckel and E. Boff, Behavioral effects of interacting imipramine and other drugs with D-amphetamine, cocaine and tetrabenazine, Psychopharmacologia (Berlin) 5, 198-208 (1964).

J. Schildkraut and S. Kety, Biogenic amines and emotion, Science 156, 21-37 (1967).

V. Seltzer and S.R. Tonge, Methylamphetamine withdrawal as a model for the depressive state: Antagonism of post-amphetamine depression by imipramine, J. Pharm. Pharmac. 27 (suppl.) 16P (1975).

E.B. Sigg, Pharmacological studies with tofranil, <u>Canad. Psychiat. Assoc. J.</u>
4, 75 (1959).

E.B. Sigg and R.T. Hill, The effect of imipramine on central adrenergic mechanisms, In: <u>Neuro-Psycho-Pharmacology</u> (H. Brill, Ed.), p. 367, Excerpta Medica Foundation, Amsterdam (1967).

P.S.J. Spencer, Antagonism of Hypothermia in the mouse by antidepressants, <u>Antidepressant Drugs</u> (S. Garattini and M.N.G. Dukes, Eds.), pp. 194-204, Excerpta Medica Foundation, Amsterdam (1966).

L. Stein, Effects and interactions of imipramine, chlorpromazine, reserpine and amphetamine on self-stimulation possible neurophysiological basis of depression, <u>Recent Advances in Biological Psychiatry</u> (J. Wortis, Ed.), Vol. IV, pp. 288-309, Plenum Press, New York (1961).

C. Sterlin, H.E. Lehmann, R.F. Oliveros, T.A. Ban and B.M. Saxena, A preliminary investigation of WY-3263 versus amitriptyline in depression, <u>Curr. Ther. Res.</u> 10, 576-582 (1968).

C.A. Stone, Relationship of some autonomic actions to potential antidepressant activity among antidepressant drugs, <u>Antidepressant Drugs</u> (S. Garattini and M.N.G. Dukes, Eds.), pp. 158-163, Excerpta Medica Foundation, Amsterdam (1966).

V.G. Vernier, H.M. Hanson and C.A. Stone, The pharmacodynamics of amitriptyline, <u>Psychosomatic Medicine</u> (J.H. Nodine and J.H. Moyer, Eds.), p. 683, Lea and Febiger, Philadelphia (1962).

B.A. Whittle, Reversal of Reserpine-induced hypotyermia by pharmacological agents other than antidepressants, <u>Nature</u> (London) 216, 579-580 (1967).

NONVERBAL COMMUNICATION AS AN INDEX OF DEPRESSION

Robert E. Miller, Candice J. Ranelli and John M. Levine

Department of Psychiatry, University of Pittsburgh Medical School,
3811 O'Hara Street, Pittsburgh, PA. 15261.

In a discussion of developments in the study of animal behavior, the British ethologist, Robert Hinde (1972), enumerated four ways that the study of various animal species may contribute to the study of human behavior: (1) through the development of methods and procedures which can be extended and adapted for investigations in man, (2) through the use of the animal model to study specific behavioral aberrations similar to those found in human disorders, (3) in establishing principles of behavior which can be assessed throughout various phyla and species, including man, (4) to make a determination of those features of behavior which man shares with his evolutionary kin and those which are emergent only in the human species, thus describing the unique qualities of man. John Bowlby (1967) added to this list two additional ways that animal studies have facilitated our understanding of man: (1) the study of certain typical features in animal behavior may suggest that these same behaviors, perhaps in modified form, are also typical of human behavior, and (2) the study of the ontogeny of behavior in nonhuman forms may provide clues to developmental staging in the human infant.

The specified purpose of this Workshop is to consider and evaluate the use of animal models in psychiatry and neurology. This is a most worthy topic, and the papers and discussions have provided convincing evidence that animal models have much to offer in terms of their relative simplicity, both conceptually and procedurally; their ability to be induced acutely and, often, to be reversed easily so that variables affecting their course can be studied; and, very importantly, their avoidance of some of the ethical issues which preclude many studies in man. As is true of biology and medical science generally, the careful use of animal models of human disorder in psychiatry will continue to enlighten and enrich our understanding of the mechanisms of human health and disease.

There are, however, two self-evident constraints on the use of animal models in psychiatry. First, just as the identification of abnormal behavior in man requires a thorough and detailed understanding of normal behavior, one cannot employ an animal model of a psychiatric symptom without extensive experience with the normal animal, so that a rather complete knowledge of the behavioral repertoire and the relative frequencies of behavioral acts is acquired. Had Harry Harlow not had years of experience in the observation of the rhesus monkey in various laboratory and social situations, it is conceivable that his important studies on the effects of social isolation might have been interpreted simply in terms of the learning capability of the infant monkey, rather than opening up a whole field of study on affectional systems and their importance to social development. The careful observation and description of the behavioral responses of the normal animal are essential to the use of animal models in psychiatric and neurologic research. The investigation of animal behaviors, their adaptive value in the survival of the individual and the species, and the organization and coordination of individual behaviors to form and maintain stable social groups informs us about the variety and the consistency of mechanisms which have evolved to permit the individual to cope with the demands and stresses of life. Man, no less than his more humble relatives, must face the anxieties and indignities of life and has his own repertoire of behaviors which allow him to cope with the pains and stresses of life. That these adaptations sometimes fail and the individual succumbs to a physical or emotional illness is not surprising. Yet, most animals and most men do, indeed, somehow adapt to their conditions and do survive to live another day.

In regard to the second constraint, we should not forget in our use of the animal model that, in many cases, we have pushed the animal subject beyond his adaptive limits and that he survives only through the diligent ministrations of his human caretaker. The Harlow social isolate infant would not exist in nature; it would simply perish within a few hours. Likewise, there is, to my knowledge, no natural analog to the chronic amphetamine-treated monkey. The brain lesioned primate not only requires competent post-operative care following surgery but, even after recovery, may not survive in the natural habitat (e.g., Dicks, Myers, &

171

Kling, 1969; Myers, Swett, & Miller, 1973). Thus, even in the animal model it may prove impossible to determine the full impact of an experimental procedure on the natural behavior of the subject in its own ecological niche. Of course, it is equally true of man that the more severe behavioral and emotional disturbances often require sequestering and special caretaking from others to safeguard the individual and the larger society to which he belongs.

In my own field of interest, nonverbal communication in animals and man, animal studies have made tremendous contributions in both theory and methodology. In 1872 Darwin wrote his treatise, "The Expression of the Emotions in Man and Animals", which, although predated by other significant work, is generally recognized as the first major scientific treatment of the subject of nonverbal communication. For the next several decades there was a rather considerable amount of experimental investigation of the nonverbal capabilities of man, concentrating almost exclusively on the facial expression of emotion. This work was conducted by some of the major figures in the experimental psychology of that time: Robert Woodworth (1938), Carney Landis (1924, 1929), Norman Munn (1940), and Harold Schlosberg (1952), among others. Due, in part, to the absence of some of the more sophisticated technology which is available today and because these investigators seem to have asked the basic question in an inefficient way, the results of these studies were quite meager and seemed to suggest that man was not able to perceive and identify the subtle nuances of behavioral expression in others. The judgments of facial expression were reliable only in very broad categories (Woodworth, 1938) or when placed on a very limited number of dimensions (Schlosberg, 1952). In the face of a series of underwhelming experiments on the facial expression of emotion, this field was largely abandoned for almost 25 years.

Meanwhile, however, the field of ethology was really beginning to blossom in Europe and later here in the United States. Detailed descriptions of the intricate and elegant interplay of movements and postures in the greylag goose, the fiddler crab, the stickleback, and many other diverse forms provided incontrovertible evidence of the vital necessity of intraspecific communication in the coordination of social and sexual functions of these animals. It was convincingly demonstrated that social processes could not be culminated successfully if even very subtle nuances of behavior were missed in the chain of interaction. The description of natural interactions observed in the habitat led to clever and systematic experimental interventions, which permitted a close specification of the necessary and sufficient components, both structural and functional, which elicited the appropriate behavioral response in a member of the species. Although the theories arising from ethology have not been without controversy (e.g., Lehrman, 1953), the contribution of these scientists to our understanding and appreciation of the complex and species specific communication patterns in insects, birds, fish, and mammals is simply enormous.

If studies reveal that every form of animal, from the very primitive to our closest primate relatives, is so exquisitely sensitive to and dependent upon the exchange of nonverbal cues through movements, postures, facial expressions, vocalization, and/or olfactory cues, is it not probable that human social behavior is also affected by similar nonverbal cues? In the late 1950s, investigators began to explore the field of nonverbal communication in man, using some of the observational and descriptive methods which had proven to be such powerful techniques in the hands of the ethologists. These first efforts seemed to arise at about the same time in the laboratories of scientists from several different scientific disciplines: sociologists, psychologists, anthropologists, and linguists. Each tended to bring his own methods and theoretical orientation to the field, and the result has been that the study of human nonverbal communication enjoys a richness and diversity of approach and interpretation that has proven to be uniquely productive. In a somewhat negative way, the price of success in these investigations can be gauged by the popularity and commercial appeal of a succession of paperback books purporting to explain "body language" to the layperson. However, it is also true that a sound and scientifically reliable literature is increasing at an exponential rate.

Our work in nonverbal communication is, I believe, a mirror image of the theme of this Workshop; i.e., rather than developing an animal analog of a human condition, we have extended our work with primates to the study of man. It is, of course, to the advantage of science for experimental techniques and observations to flow in both directions: from man to the animal model, and from animal studies to man. For example, the finding that the seal has reflex bradycardia on diving beneath the surface has been checked in man and found to occur with such regularity that it has been invoked as a probable cause of death in young, healthy adults who die suddenly while swimming. Similar instances of the extension of research from animal to man are numerous. Therefore, our attempts to study human behavior, even the clinical disorders of psychiatric disturbances, in the same ways that we would examine a new species of monkey is simply a phase shift of 180 degrees from the topic of the Workshop. There may even be a significant advantage to looking at man from the vantage point of a primatologist, in that fewer of the biases concerning psychodynamic causality or any other theoretical etiology will intrude upon the observations.

Our study of nonverbal communication in the rhesus monkey began in the mid-1950s. We were attempting to investigate the effects on discrimination learning of pairing a naive monkey with a more sophisticated partner during the learning sessions to see whether the observation of an accurate performance would enable the untrained monkey to solve the task at a more rapid rate than if he had to go through the whole trial-and-error process alone. While the results of these experiments indicated that there was a social facilitation of learning, it was also observed that the social relationships of the partners seemed to play a very significant role in the performance. As discrimination accuracy improved, the dominant animal, through a very subtle series of cues, asserted his prerogative of preempting the correct choice, and the submissive partner was relegated to the unrewarded choice or, as frequently happened, to hanging back and making no choice at all (Miller & Murphy, 1956). It was apparent that some of the more interesting social exchanges were occurring right before our eyes, but that we were not equipped to record or measure them, so we embarked on a series of studies designed to examine and describe the exchange of social information via the postures, gestures, and vocalizations of our laboratory animals.

A series of experiments was initiated in which social relationships in groups of monkeys were systematically altered experimentally by conditioning a dominant male to perform an instrumental avoidance response each time that he was exposed to the sight of a submissive partner. In tests outside the conditioning situation, it was found that the status roles were reversed and, moreover, that the formerly submissive monkey had adopted the behavioral movements and postures appropriate to social dominance (Murphy, Miller, & Mirsky, 1955; Miller, Murphy, & Mirsky, 1955; Murphy & Miller, 1956).

Then experiments were initiated to determine exactly which cues were significant in the communication between pairs of monkeys. The first studies were rather crude but did indicate that facial expressions of emotion were readily detected and responded to by the rhesus monkey (Mirsky, Miller, & Murphy, 1958; Miller, Murphy, & Mirsky, 1959). A more sensitive technique was devised to isolate and identify nonverbal behaviors, specifically facial expressive cues. This method, cooperative conditioning, has been used extensively to investigate communicative phenomena in both normal and "abnormal" animals. Monkeys are first trained to perform instrumental avoidance responses (Miller, Banks, & Ogawa, 1962, 1963) or instrumental rewarded responses (Miller, Banks, & Kuwahara, 1966). The conditioned stimulus is an end-lighted lucite plaque located at the end of a darkened wooden tunnel which also contains a videocamera. Thus, when the monkey looks at the CS, a full face video picture is obtained. The CS-US interval which we have employed is six seconds. When the monkeys have learned to perform the requisite instrumental response within six seconds of stimulus presentation and to withhold responses in the absence of the stimulus, the cooperative conditioning phase is begun. One animal, the stimulus monkey, is placed before the stimulus tunnel but is not provided with the bar necessary to perform a conditioned response. A second monkey, the responder, is placed in another test room where it is equipped with the appropriate bar to make the instrumental response but is not given the conditioned stimulus plaque which would signal that a trial is being presented. Instead, the responder is permitted to view the head and face of the stimulus monkey via closed circuit television. The premise of the experiment is that, if a presentation of the conditioned stimulus elicits an affective response in the stimulus monkey which is reflected in a facial expression and if that expressive change can be perceived and properly interpreted by the responder within six seconds, the responder will then make the appropriate instrumental response to either avoid shock or, in the case of instrumental reward, to obtain a pellet of food. The experimental arrangement has also enabled us to take concurrent psychophysiological recordings from the pair of animals to determine if both the stimulus and responder monkeys display autonomic responses during the trial periods.

A number of studies were performed on untreated rhesus monkeys. These were animals which had been captured in the wild after, hopefully, normal rearing experiences and which had not been subjects in other experimental studies of social behavior in our laboratory. These experiments have been detailed in reviews (Miller, 1967, 1971) and will not be covered here. In brief, they showed that monkeys performed immediately and at a high level of accuracy on the cooperative conditioning task; that the specific facial expressive cues of stimulus animals which triggered responses in their partners could be isolated and identified; and that autonomic responses of the responder were consistent with the notion that the animals reacted affectively to the facial expressions displayed by their partners. Rhesus monkeys were found to utilize nonverbal cues with great accuracy and efficiency; indeed, they performed as well with social signals as they did with the original stimuli used in conditioning (i.e., the light used as the CS).

The next ventures were concerned with the examination of what kinds of social disruptions in both cooperative conditioning and group social behavior might occur when non-normal subjects were tested. Our initial study was made possible through the generous cooperation of Dr. Harry Harlow, who made available to us three total social isolates from his laboratory in Wisconsin. These monkeys, ranging between 4 and 5 years of age when they arrived at our laboratory, had been kept in total isolation for the first year of life (Rowland, 1964) and had later been subjected to an extensive series of observations of social deficit at Wisconsin (Mitchell, 1966; Mitchell, et.al., 1966; Sackett, 1965). Three age-matched feral control animals were selected from our colony for the cooperative conditioning experiment. The animals were tested in all combinations of role: isolate stimulus monkeys sending to normal responders; isolate stimulus and isolate responders; normal stimulus and isolate responders; and normal stimulus and normal responders. The results of this experiment revealed that, although monkeys reared in total social isolation were perfectly capable of acquiring a conditioned response to a physical stimulus, they were completely unable to process social cues from the facial expression of another monkey, either isolate or normal. Further, the isolate was found to be deficient in expressing clear and discriminable facial expressions which could be "read" by normals or other isolates. The heart rate responses of the isolates indicated that, while they displayed a significant tachycardia to the light which served as the CS, there was absolutely no change when they were required to receive social messages (Miller, Caul, & Mirsky, 1967).

Psychoactive drugs were also found to affect both nonverbal communication and social performance in the group. Amphetamine sulfate (0.25 mg/kg,IM) heightened the display of facial expressions in nonverbal communication so that undrugged responders discriminated even more clearly the expressions of treated stimulus monkeys. Amphetamine-treated responders also performed well in the receipt of expressive signals from undrugged stimulus monkeys, but they did tend to make sightly more errors in that they made instrumental responses when the stimulus monkey was not receiving a trial. When one of a group of three monkeys was treated with amphetamine, the amount of social interaction increased markedly, particularly that directed by undrugged animals toward the treated subject, and the quality of interaction was extremely benign--agnostic responses fell to 10% of that seen when none of of the subjects had received the drug. Phencyclidine hydrochloride (0.25 mg/kg,IM) impaired the expression of affect in the treated stimulus monkey but did not affect the performance of responders. This hallucinogen did, however, have a very drastic effect on group behavior. Undrugged subjects avoided each other as well as the treated animal, and the proportion of agnostic responses increased some 360%. Chlorpromazine has an interesting effect. It significantly reduced both the sending and the receiving of nonverbal cues in cooperative conditioning (0.25 mg/kg,IM) and almost eliminated the social responses of undrugged subjects toward the treated subject. It was almost as though the chlorpromazine-treated monkey was treated as a "nonperson" by his partners (Miller, 1974; Miller, Levine & Mirsky, 1973). In another study, delta-9-THC (1.0 mg/kg,orally) impaired the discriminability of facial expressions of treated stimulus monkeys but improved the performance of responders in the receipt of social cues from either undrugged or drugged stimulus monkeys (Miller & Deets, 1976). In freely behaving social groups, if all the subjects received THC, there was a marked reemergence of juvenile forms of play activities, but, when drugged and non-drugged subjects were placed together in a social group, the untreated monkeys attacked their drugged partners (Deets & Miller, unpublished).

It was our interpretation from all of these data that the exchange of nonverbal expressive cues is an integral and essential feature of primate social behavior. Systematic deficits in sending of these cues, or sending defective and bizarre cues, or failing to recognize and discriminate such expressions in other conspecifics can produce the most drastic social disruptions in group behaviors. Individuals who do not monitor the behaviors of others in an appropriate way and who do not respond promptly with a proper nonverbal return are ignored, avoided, or attacked by others in the social group and are not admitted to full social participation in the group.

If nonverbal communication plays a crucial role in the integration of a primate social group, is the same true of man? And what changes in behavior are potential mediators of the social deficit and social isolation of those who are failing to sustain their position in the social milieu? We have attempted to extend our studies of the rhesus monkey, using modifications of the methods which worked with that species, to examine the behavior of a group of human subjects whose social deficit is a prominent feature of their condition--clinical depression. There is a great deal of evidence and consensus that paucity of social relationships and increased social friction is characteristic of the depressed patient (Salzman, 1975; Klerman, 1974; Ferster, 1974; Ruesch & Brodsky, 1968; Paykel & Weissman, 1973; Gruenberg, 1967; Weissman, Paykel, Seigel & Klerman, 1971). The depressed individual presents a formidable obstacle to a social initiation by healthy partners and, through his constant complaints, his insatiable demands for comfort and assurance, and his sadness and self-pity, he angers and drives away even those who love him most (Salzman, 1975,pp.50-51). There is some evidence from a factor-analytic study of social deficit in depressed women (Paykel & Weissman, 1973) that, while some areas of social deficit return to normal levels upon remission, the factors of inhibited communication and interpersonal friction remain impaired as trait rather than state variables. This implies that some social deficits may be associated in an etiological way with the development and recurrence of depressive episodes. Our studies are designed to identify those specific behavioral acts which may be associated with the depressive illness and which may impair or prevent a normally productive interpersonal relationship by disrupting the nonverbal communications among the partners.

Our first efforts were directed to adapting the cooperative conditioning task for use with human subjects. The essential features were to be the same: one subject was videotaped while in the performance of a task involving a discrimination, and the second subject, the responder, was required to process facial expressive cues of the stimulus subject to make a discriminated response. The experimental task also had to meet the criteria of being sufficiently interesting to engage the attention and cooperation of the subjects, require no learning trials for either of the subjects, be suitable for subjects of various ages and mental/emotional conditions, and be totally nonaversive in terms of the delivery of reinforcement. The first experiments used a series of stimulus pictures of five different types: sexual content; mothers and their babies; scenic landscapes; ambiguous, or difficult to interpret, pictures; and unpleasant views of burn or surgery patients. These pictures were shown to normal college students, while other students watched their facial expression on a closed circuit television and tried (a) to guess which category of slide was being shown to the stimulus subject and (b) to estimate the degree of affect on a pleasant-unpleasant dimension. These studies revealed that there were both personality and physiological variables associated with the sending and receiving of facial expressive cues in man (Buck, Savin, Miller, & Caul, 1972; Buck, Miller, & Caul, 1974). In further studies using the same method, Buck (1975, 1977) has studied the differentiation of facial expressive cues in the nonverbal communication of preschool children with the intent of finding the age at which sex differences in this important behavior emerge.

In our own laboratories, we made further refinements in the cooperative conditioning task by devising a putative problem-solving task in which the subjects were attempting to win coins. A "slot-machine" which delivers quarters, pennies, or jackpots was fabricated. The stimulus subject is told that signal lights which indicate a particular denomination of reinforcement availability will be illuminated for eight seconds, during which the subject must make a response on one of three unmarked bars. If the choice has been "correct", the appropriate coin/coins is automatically dispensed at the termination of the signal light. The instructions are designed to suggest to the subject that the solution of the slot-machine task requires a fairly complex set of sequential moves in order to continue to obtain coins. In actual fact, the stimulus subject's responses are not instrumental to reward, but reinforcements are programmed by the experimenters in a learning curve fashion, i.e., few reinforcements are given early in the series, but many are given in the last ten trials. The subjects are informed that a videotape is being made during the test and are asked to sign a release to permit the use of the videotape for research purposes. Subjects are also instrumented for the recording of heart rate and skin potential throughout the stimulus sessions. The videotaped sequences of the face and head of stimulus subjects are then shown to responders to determine how well they can predict the correct denomination of coin for each of the trials, simply on the basis of the anticipatory facial expressions of the stimulus subjects. Responders make their choices by pressing an appropriately labeled bar within six seconds of the initiation of a trial on the videotape. This task was extensively tested with normal college students and medical students and found to be effective in assessing cooperative conditioning in humans (Miller, Giannini, & Levine, 1977).

For the past year we have been studying the cooperative conditioning performance and the behavior exhibited in an interview situation by hospitalized depressed patients. These patients are diagnosed as primary depressives and have met the admission criteria for an amitriptyline-placebo study which involves a 35-day protocol (MH 5R01-24652). We see the patients on the first day of protocol and weekly thereafter for a total of six sessions, ending on the final day of the protocol. Since the entire study is double-blind, none of the research personnel know which of the patients receive active medication and which receive placebo.

In the interview, a psychologist, Candice Ranelli, asks the patient 25 open-ended questions which deal with present mood, current somatic complaints, perceptions of the hospital regimen, social and family affairs, and attitudes of the patient towards himself and his future. These inverviews will be videotaped in future studies, but, up to this point, two behavioral raters have recorded on-line the behaviors of the patient in some 30 categories of behavioral acts. The categorization is accomplished with a counter-timer mechanism which was originally constructed to facilitate behavioral observations in the rhesus monkey. Since the observers perform the categorizations behind a one-way vision mirror, they cannot hear the verbal conversation between the patient and the interviewer, which is recorded on audio tape for subsequent analysis of paralinguistic features of speech. Since the interview length is dependent upon the responses of the patient, the frequency and duration of behavioral measures are adjusted to a per minute basis.

When the interview has been completed, the patient is escorted to the laboratory for a cooperative conditioning session. Although it will be of interest to test patients' nonverbal capabilities both as expressors and as receivers of nonverbal expressions, it is not feasible to test both of these roles at the same time and in the same patients, so our preliminary work has been devoted exclusively to their behavior as stimulus subjects. Since both the interviews and cooperative conditioning are obtained weekly throughout the protocol period, there are six videotaped samples of facial expression generated on the slot-machine task. These samples are then shown in random sequence (weeks are shown in various orders so that judges cannot determine which week will be shown) to a panel of nondepressed, normal judges who are required to utilize facial expressive changes to predict which denomination of coin was available on each trial. In addition, judges are asked to guess which of the six sessions each of the videotapes represent (whether the patient was near admission, at the middle of treatment, or near the end of protocol) and to rate, on a scale of 1-7, the severity of illness represented in each sequence of videotape.

Before presenting some of the preliminary results of these studies, let us make some disclaimers and caveats. The data represent results from only 17 patients and, more importantly, we do not have data from the control groups which will be essential to place the results in context. We have, to this time, insufficient resources to be able to recruit matched nondepressed control subjects and to bring them into our laboratories weekly for six consecutive weeks. I would also like to emphasize that we have taken the attitude that we should have no preconceptions about the nature and course of depression but should simply try to look at behavioral differences in depression just as one would characterize a new species of primate in the field. Therefore, some of the results seem to be readily interpretable, but others are not. We intend to let the data lead us, hopefully, to new descriptions and new understandings about the phenomena of depressive illness.

The behavioral observations from the interview situation were first subjected to repeated measures analyses of variance with two groups (amitriptyline and placebo) and six sessions. The results of these analyses revealed that there was a very marked effect on the behavior of placebo subjects when they were informed on the final day of protocol that they had not received the active medication during the previous 35 days. Even though they had been questioned weekly during the interview about the research protocol, including their feelings about a placebo group, and had been asked to predict each week to which group they felt they had been assigned, the actual revelation by the attending psychiatrist that they had been in the placebo group had a very significant behavioral effect (Ranelli & Miller, 1977). The frequency of frowns, head nods, shoulder shrugs, and arm tosses increased dramatically during the last interview, which followed the revelation of group assignment by no more than 24 hours. These data would lead one to suspect that the discovery that they had received placebo treatment was not received enthusiastically by the patients. Other findings from these analyses indicated that the frequency of illustrators (those hand movements associated with speech) increased over weeks in both groups of subjects and that, for patients treated with amitritpyline, adaptors (hand movements associated with self-manipulation, e.g. face rubbing, stroking of one's body) also increased over weeks. Adaptors decreased in frequency over the first five sessions for placebo patients but increased in the sixth week after disclosure of group assignment. The frequency of eye contact increased linearly over sessions for the drug group but decreased over the first five sessions for placebo patients.

The next set of analyses were conducted on a three group split: drug subjects who showed a favorable response to treatment (final Hamilton score of less than 10), drug subjects who did not show favorable response to the drug (Hamilton score at the end of protocol of greater than 10), and placebo subjects. Both frequency and duration measures were subjected to repeated measures ANOVAs, making a total of 50 separate analyses. Eighteen of these analyses yielded statistically significant results beyond the .05 level, indicating that there were reliable behavioral differences between groups, or over sessions within groups which discriminated among depressed subjects who improved over the course of treatment and those who did not. It is, of course quite likely that with appropriate nondepressed control groups, even more interesting and significant behavioral discriminations will be achieved. While space does not permit the full exposition of these results in this paper, let me just mention a couple of the more intriguing findings. The results indicate that those depressed patients who subsequently fail to make a clinical improvement with amitriptyline display markedly elevated speech latencies from the very beginning of the protocol and show an increasing frequency of foot sliding over weeks. Another foot behavior, foot tapping, showed a major increase over sessions for the placebo subjects. Duration of speech was affected by the drug in both responders and nonresponders early in the treatment period, but the drug responders late in treatment showed a significant increase in the duration of their replies during the interview.

While the cooperative conditioning data are not completely analyzed at this date, it is clear that some depressed subjects were very difficult for our judges to "read" during early sessions and then displayed much more discriminable expressions in later sessions. Other subjects showed the reverse pattern: they were easy to read at the beginning of treatment but their expressions became more ambiguous as they neared the end of protocol. Until the judges have completed their task of responding to all of the patients' videotapes, the relationship between facial expression and clinical improvement is undetermined. It is encouraging to note, however, that the task clearly differentiates among individual patients and also the same patient over time.

Our preliminary findings and the previously published resprts of Ekman and Friesen (1968, 1969, 1974) on the coding of behaviors in depressed women suggest that this is a productive line of research. If one were able to characterize with some precision those elements of behavior which seem to be specific to depressive illness, it would provide a powerful tool for the diagnosis of the disorder (perhaps discriminating among the several sub-classifications of depression) and for the assessment of treatment effects in the individual patient. In addition, some clues to etiology might become more apparent from observing the constellation of behavior change which accompanies the mood disorder. Most significantly, one might be able to focus on the incidence and duration of objective behavioral acts so that early signs of relapse might be detected in time to institute early preventive measures. In accord with our earlier thesis, the objective study of behavior in "naturalistic" settings, which was largely rediscovered by the ethologists, is one example of the benefits which animal investigations have made to the study of man. In this case, the animal does not serve as the model of a human condition. Rather, the careful methodology and the obvious success of the techniques in providing new insights and understanding of the role that behaviors play in the developemnt and survival of the many species of animals which have been studied encourage us to believe that these methods might also profoundly assist us in understanding ourselves.

REFERENCES

J. Bowlby, Human personality in an ethological light, In: E. Serban and A. Kling (eds.), Animal Models in Human Psychobiology (1976). Plenum Press, New York, pp.27-36.

R.Buck, Nonverbal communication of affect in children, J. Personal. Soc. Psychol. 31, 644 (1975).

R. Buck. Nonverbal communication of affect in preschool children: Relationships with personality and skin conductance, J. Personal. Soc. Psychol. 35, 225 (1977).

R. Buck, V. Savin, R. Miller, and W. Caul, Communication of affect through facial expression in humans, J. Personal. Soc. Psychol. 23, 362 (1972).

R. Buck, R. Miller, and W. Caul, Sex, personality, and physiological variables in the communication of emotion via facial expression, J. Personal. Soc. Psychol. 30, 587 (1974).

C. Darwin (1872) The Expression of the Emotions in Man and Animals, Murray, London.

D. Dicks, R.E. Myers, and A. Kling, Uncus and amygdala lesions: Effects on social behavior in the free-ranging rhesus monkey, Science, 165, 69 (1969).

P. Ekman and W. Friesen, Nonverbal behavior in psychotherapy research, In: J. Shlien (ed.), Research in Psychotherapy, Vol. 3, American Psychological Association, Washington, D.C. (1968), pp. 179-216.

P. Ekman and W. Friesen, Nonverbal leakage and clues to deception, Psychiatry, 32, 88,(1969)

P. Ekman and W. Friesen, Nonverbal behavior and psychopathology, In: R. Friedman and M. Katz (eds.), The Psychology of Depression: Contemporary Theory and Research, Winston & Sons, Washington, D.C. (1974), pp. 203-224.

C.B. Ferster, A functional analysis of depression, Amer. Psychol. 28, 857 (1973).

E.M. Gruenberg, The social breakdown syndrome--some origins, Amer. J. Psychiat. 123, 1481 (1967).

R.A. Hinde (1972) Social Behavior and its Development in Subhuman Primates, Oregon State System of Higher Education, Eugene, Oregon.

G.L. Klerman, Depression and adaptation, In: R. Friedman and M. Katz (eds.),The Psychology of Depression: Contemporary Theory and Research, Winston & Sons, Washington, D.C. (1974), pp. 127-145.

C. Landis, Studies of emotional reactions: General behavior and facial expression, J. Comp. Psychol. 4, 447 (1924).

C. Landis, The interpretation of facial expression in emotion, J. Gen. Psychol. 2, 59(1929).

D.S. Lehrman, A critique on Konrad Lorenz'stheory of instinctive behavior, Quart. Rev. Biol. 28, 337 (1953).

R.E. Miller, Experimental approaches to the autonomic and behavioral aspects of affective communication in rhesus monkeys, In: S. Altmann (ed.), Social Communication Among Primates, Univ. of Chicago Press, Chicago, Ill. (1967), pp. 125-134.

R.E. Miller, Experimental studies of communication in the monkey, In: R. Rosenblum (ed.), Primate Behavior: Developments in Field and Laboratory Research, Vol. 2, Academic Press, New York (1971), pp. 139-175/

R.E. Miller, Social and pharmacological influences on the nonverbal communication of monkeys and man, In: L. Krames, P. Pliner, and T. Alloway (eds.), Nonverbal Communication: Advances in the Study of Communication and Affect, Plenum Press, New York (1974), pp. 77-102.

R.E. Miller, J.V. Murphy and I.A. Mirsky, The modification of social dominance in a group of monkeys by inter-animal conditioning, J. Comp. Physiol. Psychol. 48, 392 (1955).

R.E. Miller and J.V. Murphy, Social interactions of rhesus monkeys: II. Effects of social interaction on the learning of discrimination tasks, J. Comp. Physiol. Psychol. 49, 207 (1956).

R.E. Miller, J.V. Murphy and I.A. Mirsky, The relevance of facial expression and posture as cues in the communication of affect between monkeys, Arch. Gen. Psychiat. 1, 480 (1959).

R.E. Miller, J.H. Banks, and N. Ogawa, Communication of affect in "cooperative conditioning" of rhesus monkeys, J. Abn. Soc. Psychol. 64, 343 (1962).

R.E. Miller, J.H. Banks, and N. Ogawa, The role of facial expression in "cooperative avoidance" conditioning in monkeys. J. Abn. Soc. Psychol. 67, 24 (1963).

R.E. Miller, J.H. Banks, and H. Kuwahara, The communication of affects in monkeys: Cooperative reward conditioning, J. Genet. Psychol. 108, 121 (1966).

R.E. Miller, W.F. Caul, and I.A. Mirsky, Communication of affects between feral and socially isolated monkeys, J. Personal. Soc. Psychol. 7, 231, (1967).

R.E. Miller, J.M. Levine, and I.A. Mirsky, The effect of psychoactive drugs on nonverbal communication and group social behavior of monkeys, J. Personal. Soc. Psychol. 28, 396 (1973).

R.E. Miller and A.C. Deets, Delta-9-THC and nonverbal communication in monkeys, Psychopharmacology, 48, 53 (1976).

R.E. Miller, J. Giannini, and J. Levine, Nonverbal communication in man with a cooperative conditioning task, J. Soc. Psychol. in press (1977).

I.A. Mirsky, R.E. Miller, and J.V. Murphy, The communication of affect in rhesus monkeys: I. An experimental method, J. Amer. Psychoanal. Assoc. 6, 433 (1958).

G.D. Mitchell (1966) A Follow-up Study of Total Social Isolation in the Rhesus Monkey, Unpublished Doctoral Dissertation, Univ. of Wisconsin, Madison.

G.D. Mitchell, E.J. Raymond, G.C. Ruppenthal, and H.F. Harlow, Long-term effects of total social isolation upon the behavior of rhesus monkeys, Psychol. Rep. 18, 567 (1966).

N.L. Munn, The effect of knowledge of the situation upon judgment of emotions from facial expressions, J. Abn. Soc. Psychol. 35, 324 (1940).

J.V. Murphy, R.E. Miller, and I.A. Mirsky, Inter-animal conditioning in the monkey, J. Comp. Physiol. Psychol. 48, 211 (1955).

J.V. Murphy and R.E. Miller, The manipulation of dominance in monkeys with conditioned fear, J. Abn. Soc. Psychol. 53, 244 (1956).

R.E. Myers, C. Swett, and M. Miller, Loss of social group affinity following prefrontal lesions in free-ranging macaques, Brain Res. 64, 257 (1973).

E.S. Paykel, and M.M. Weissman, Social adjustment and depression: A longitudinal study, Arch. Gen. Psychiat. 28, 659 (1973).

C. Ranelli and R. Miller, Nonverbal communication in clinical depression, Program of the Eastern Psychological Assn. Convention, Boston, Mass (1977), p. 91 (abstract).

G.L. Rowland (1964) The Effects of Total Social Isolation upon Learning and Social Behavior in Rhesus Monkeys, Unpublished Doctoral Dissertation, Univ. of Wisconsin, Madison.

J. Ruesch and C.M. Brodsky, The concept of social disability, Arch. Gen. Psychiat. 19, 394 (1968).

G.P. Sackett, Response of rhesus monkeys to social stimulation presented by means of colored slides, Percept.Motor Skills, 20, 1027 (1965).

G.P. Sackett, Interpersonal factors in depression, In: F. Flach and S. Draghi (eds.), The Nature and Treatment of Depression, Wiley, New York (1975), pp.43-56.

H. Schlosberg, The description of facial expressions in terms of two dimensions, <u>J. Exp. Psychol</u>. 44, 229 (1952).

M. Weissman, E. Paykel, R. Siegel, and G. Klerman, The social role performance of depressed women: Comparisons with a normal group, <u>Amer. J. Orthopsychiat</u>. 41, 390 (1971).

R.S. Woodworth (1938) <u>Experimental Psychology</u>, Holt, New York.

EEG SLEEP CORRELATES OF DEPRESSION IN MAN

David J. Kupfer

Department of Psychiatry, University of Pittsburgh School of Medicine
Pittsburgh, Pennsylvania 15261

Although there has been a long-standing interest in developing an adequate animal model for
depression, the bridge between man who presents with heterogeneous types of depression, on
the one hand, and animals appropriate for investigation as models for depression, on the
other hand, has been a difficult one to cross. In fact, an argument could be made that one
major difficulty in the continuing search for refined, more effective antidepressant treatment is the lack of adequate animal models of depression. For indeed, the animal models
need to both resemble depressive illness and also be selectively sensitive to those antidepressant treatments which are clinically effective. Despite these requirements, there
has been considerable activity and progress in the development of animal models as aids in
the investigation of psychiatric states.

In developing animal models, particularly for affective illness, attention has been paid to
certain regulatory functions such as motor activity, appetite, sexual interest, and sleep.
Yet, relatively little research has been conducted using EEG sleep or sleep changes as a
major component of an animal model in depression. Selected EEG sleep studies have been performed in primate models of depression based on separation and loss and it appears that certain sleep phenomena such as increased wakefulness and decreased REM sleep are characteristics of young primates who are suffering loss or separation (1). Given the extensive EEG
sleep research conducted during the last two decades in humans, and the relatively small
amount of data available from human models, it becomes clear that both the need for and the
type of EEG sleep procedures that might be used in animal studies should be reexamined.
However, prior to our suggestions for future animal experiments, advances in EEG sleep correlates of depression should be reviewed.

The major findings uncovered in the last decade in EEG sleep research in man have indicated
the following four sleep features to be characteristic of depressive states occurring in
adult life. First, depressed patients frequently show changes in sleep continuity or inability to remain asleep throughout the night. Most frequently this alteration is manifested by
intermittent wakefulness during the night as well as by early morning awakening (a final
awakening). It should be noted, however, that in 15-20% of depressed patients (usually bipolar patients) rather than decreased sleep, the sleep continuity "disturbance" is in the
direction of increased sleep (2). These changes in sleep continuity commonly found in depressive states, may reflect either transient alterations of mood, or acute responses to
stress. Regardless of the propensity to have increased sleep or decreased sleep under
stress, or during a depressive episode, patients who ultimately develop a psychotic depression show considerable decreases in overall sleep time with a loss of even 50% of their normal sleep time (3)(Table 1).

Table 1   EEG Sleep Features in Primary Depression

1.   Shortened REM latency

2.   Sleep continuity disturbance (80-85% of all cases)

3.   Reduced delta (SWS) sleep

4.   Increased REM activity

A second characteristic feature of EEG sleep is the reduced delta (SWS) sleep (Stages 3 and
4). Although delta sleep decreases with age, the reduction of Stage 3 and 4 during episodes
of depression are independent of age effects (4). It should be noted that decreases in delta sleep alone are considered nonspecific for affective states since these delta sleep changes are often found in other neuropsychiatric and medical disease states, especially those
with a chronic course (3).

To date, the most prominent finding with respect to primary depressive states appears to be shortened REM latency (REM latency is the interval from the onset of sleep until the onset of the first REM period). Shortened REM latency is a feature present in virtually all drug-free patients suffering from primary depression but absent in most patients with secondary depression (5). REM latency is shortened in primary affective disease regardless of whether it is unipolar or bipolar type. We have postulated that with the exception of drug withdrawal states (such as CNS depressant or amphetamine withdrawal) and narcolepsy, shortened REM latency points to a strong affective component in the patient's illness. Shortened REM latency has also been observed in patients suffering from mania and schizoaffective illness as well as in certain schizophrenic patients who require tricyclic antidepressants in their management (2). In depression associated with primary unipolar or bipolar affective disease, this psychobiologic marker is a persistent, rather than a transient, phenomenon and can be observed over a period of several weeks to several months unless a patient's condition is improved through clinical intervention or spontaneous remission. REM latency appears to be a dependable, measurable marker for the diagnosis of primary depression, and one which our studies show to be independent of age and changes in other sleep parameters. This finding has been confirmed in patients requiring hospitalization, as well as depressed outpatients (6,7,8, Vogel personal communication). The short REM latency found in these states is refractory to the effects of adaptation to the sleep laboratory (8).

In addition to the shortened REM latency, depressed patients often show increased REM activity (an increased number of rapid eye movements without necessarily an increase in the number of minutes of REM sleep time). More recently it has been argued that this phasic change in REM sleep (change in REM activity) may be more pronounced in the first half of the night than the latter half of the night and may best be measured by examining the REM intensity of the first REM period (9). As will be discussed later, some of the changes we have described, especially the short REM latency and the intense first REM period, may represent a so-called "defect" in the cholinergic programming of REM sleep cycles. In fact, Vogel (personal communication) has recently argued that the presence of increased REM activity during the first REM period might be considered a "REM saturation state" reflecting the patient's reduced capacity to postpone REM sleep after sleep onset. In our own sleep laboratory, the examination of the REM sleep of thirty-five inpatients with primary depression by means of an automated REM analyzer emphasizing the initial level of phasic and tonic activity, showed a higher REM frequency (that is, increased number of rapid eye movements) in the first REM period than would have been expected in normals and higher than the frequency of eye movements present in subsequent REM periods. The REM frequency decreased significantly from REM period 1 to REM period 2 and then remained constant. This finding was in contrast to the actual REM density (average size of each REM movement) which increased in both normals and depressed patients as the night progressed. Average REM size increased by a significant degree from REM periods 2 to 3 and from 3 to 4 with REM size being significantly greater at the end of the night (REMP4) than at the beginning (REMP1)(Fig. 1).

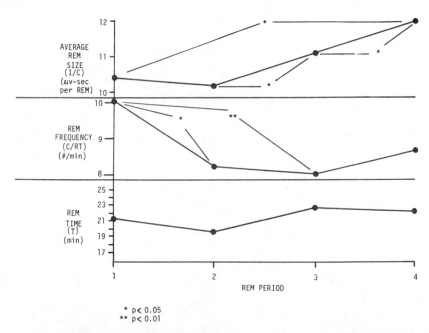

* $p < 0.05$
** $p < 0.01$

Fig. 1. Average orthogonal automated REM period measures vs. the rank of the REM period derived from 7 nights of sleep from each of 35 primary depressed hospitalized patients.

Previous explanations for this type of so-called primary REM change in the sleep of depressed patients have included: increased REM pressure; reduced REM sleep inhibition; and a rotation of ultradian rhythms. Regardless of the etiology, this high initial REM frequency
coupled with the short REM latency characteristic of depression suggests that depressives
have a high phasic REM pressure at sleep onset which is rapidly dissipated by the first REM
period leading to lower values of REM frequency and normal REM cycle lengths. That severity
of depression is associated with the initial phasic REM pressure is supported by the significant correlations of REM latency to severity of depression (10) and to the REM frequency
of the first REM period.

Even though there is a great need for better controlled studies, it is clear that we can
classify certain subtypes of depressive disorders using EEG sleep. For example, the EEG
sleep pattern in the typical, agitated episode of unipolar depression, in addition to showing changes in delta sleep and sleep continuity, is marked by normal to high levels of REM
activity, reflecting the intensity of the REM sleep, as well as decreased delta (Stages 3
and 4) (3). Measurements of motor activity indicate increased levels in agitated depressives
compared to normals of the same age (2). In contrast, certain recurrent depressives (15-20%)
and most bipolar patients who are experiencing a depressive episode have shortened REM latencies and reduced delta sleep, but show a minimal sleep continuity disturbance and reduced
motor activity levels. Some patients even have increased sleep, often approaching hypersom-
nia (2). In summary, our current understanding suggests that sleep criteria discriminating
primary and secondary depression exist in the form of shortened REM latency and changes in
REM activity, that psychotic and nonpsychotic primary depression can be separated on the basis of sleep continuity measures, and that a "medical-depressive" group show profound reductions in REM activity as contrasted to other depressive states (11,12). Despite this extensive set of studies available on untreated primary depressives, research data on EEG sleep
during drug treatment of depression is sparse and largely uncontrolled. This is unfortunate
since these later studies could represent an approach for obtaining biological correlates
during the treatment phase of depression, as well as developing predictors for treatment
response.

In order to appreciate the possible application of the EEG sleep findings based on studies
of depressed patients to the development of more useful animal models which could incorpor-
ate techniques for testing new treatment strategies, we now need to update our understanding
of the role of the neurotransmitters in sleep.

The current consensus in terms of the neurotransmitter mechanisms in sleep presumes that both
the raphe complex and the locus coeruleus play major roles in the onset and alteration of
various stages of sleep. Available scientific evidence suggests that both serotonergic and
cholinergic systems in the raphe complex have a prime role both in the sleep-wakefulness
cycle and the regulation of REM sleep onset and the length of each REM period (13). In contrast, the norepinephrine (NE) system in the locus coeruleus is thought to be more concerned
with the executive function of REM sleep which would imply that the locus coculeus plays a
role in regulating the phasic intensity of REM sleep periods(Fig. 2).

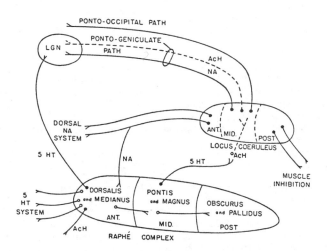

Fig. 2. Schematic representation of possible routes of interaction in neuro-
         transmitter mechanisms of sleep. Adapted from Chemical anatomy of
         brain circuits in relation to sleep and wakefulness. (Adapted from
         P.J. Morgane and W.C. Stern, Advances in Sleep Research, Vol. I, 1974)

It has been established that, in the rat, serotonin (5HT) is primarily related to the initiation and duration of slow wave sleep (SWS) and that lesions in the rostral raphe system (which depletes 5HT) or administration of parachlorophenyalanine (pCPA and inhibitor of 5HT synthesis) leau to insomnia and loss of SWS (14). While pCPA also affects REM sleep, the drug's major effect is to reduce SWS; interestingly, this SWS "deprivation" is not followed by SWS rebound or compensation when pCPA administration is discontinued.

As indicated above, the cholinergic system perhaps via the ascending reticular formation, is related to both the raphe system and locus coeruleus. Evidence from recent investigations (15) in the cat provide data showing that cholinergic neurons rostral to the nesencephalon are primarily involved in arousal mechanisms, while cholinergic neurons in the pons are involved in REM sleep. It is thought that the initiation of REM sleep may be induced by cholinergic stimulation, a notion supported by experiments both in man and in animals (15). In fact, it is suggested that 5HT mechanisms may act on a cholinergic relay which in turn "triggers" the executive mechanism of REM sleep (such as phasic REM activity). Evidence for this so-called cholinergic connection is that a direct transition has been shown between waking and REM sleep via direct stimulation of cholinergic neurons located in the pons, thus avoiding the necessary "priming " function of 5HT stimulation and SWS. It can be suggested that cholinergic neurons in the raphe complex may be responsible for controlling the initiation of REM sleep (e.g., the REM latency) as well as the cycle lengths between subsequent REM periods during the night (16). However, these same authors have recently argued that this "system" appears to modulate the timing of sleep rather than its maintenance. Physostigmine (0.5 mg IV) more readily induces REM sleep when given 35 minutes after sleep onset than when given 5 minutes after sleep onset. The same dose given at REM onset and during the second non-REM period is associated with different intensities of arousal. A lower dose (0.25 mg), however, is able to produce REM without awakening when infused during the second non-REM period (17). Furthermore, it appears that physostigmine accelerates the time of onset without changing the duration of REM periods. These cholinergic neurons then interact with the locus coeruleus where the actual intensity and perhaps length of each REM period is controlled. Although the locus coeruleus is considered primarily a NE center, biochemical support for this cholinergic interaction is derived from the ACh concentrations which have recently been noted to occur in this region (18).

The work of Hobson, McCarley and coworkers on sleep cycle oscillation involving reciprocal discharge by two brainstem groups is directly pertinent to this discussion on sleep, neurotransmitters, and depression. Hobson et al. (19) have demonstrated the presence of giant reticular neurons (FTG) in the so-called "giant-cellular" tegmental field located near the locus coeruleus using single cell recording techniques in cats. These cells appear to be involved in a reciprocal relationship with the locus coeruleus neurons during REM sleep. It has been proposed that activity of the locus coeruleus neurons via NE release is present at the onset of a REM period, but during the REM period, the FTG neurons increase firing as the locus coeruleus neurons decrease firing. This increased FTG firing is accompanied by increase in pontine-genticulate-occipital (PGO) spiking. More recently, this same group has reported that the giant cells of the FTG best satisfy the correlational criteria for eye movement generation during REM sleep, adding further support to the hypothesis that the FTG is part of a generator system for REM sleep (20). Both measurements of ACh content in the pontine tegmentum and activities of FTG neurons by carbachol (a cholinergic agonist) point to the cholinoceptive nature of the cells. If further evidence is accumulated in this direction, the role of FTG neurons is more important in the regulation of REM sleep, via cholinergic mechanism, than previously thought.

Although PGO spiking (considered the correlate of REMs in man) may be controlled from the locus coeruleus, continuous PGO activity occurs both after lesions in the locus coeruleus or raphe, and after pretreatment with reserpine and pCPA. Therefore, it could be hypothesized that the inhibition mechanisms for the PGO generator reside within the locus coeruleus. The cholinergic "signal" from the raphe (which in turn is "primed" by 5HT stimulation) could normally trigger the onset of REM sleep, which would then confine PGO spiking primarily to the REM period. Since it has been shown that PGO spiking is the major correlate of so-called REM rebound, rather than the tonic notion of the length of the REM period, in a sense, then, REM sleep deprivation is manifested primarily as phasic REM deprivation of PGO spiking.

With regard to the role of NE mechanisms of REM sleep, it should be mentioned that L-dopamine (L-DOPA), a NE precursor, reduces the amount of REM sleep and SWS and increases REM latency (16). However, it has been shown that L-DOPA administration reduces 5HT activity, which might be responsible for its effect on SWS. In man, but not in animals, the blocker, α-methyl-para-tyrosine (AMPT), appears to increase the amount of REM sleep by lengthening each REM period rather than changing the REM cycle or the number of REM periods (16).

Thus, as our understanding of the relationships among sleep changes, neurotransmitter changes, and depression improves, it is clear that cholinergic mechansims, especially in relation to the timing of REM sleep, need to be emphasized in the development of research strategies. As mentioned earlier, one explanation for the type and extent of REM changes seen in depression could involve a phase shift or rotation in ultradian rhythms. Since it has now been established that acetylcholine levels demonstrate circadian rhythms (21,22), one might hypothesize that the 24 hour rhythm of normal ACh levels is altered in primary depression and furthermore that tricyclic antidepressants may alter this circadian rhythm. The increasing body of data suggesting that the tricyclic antidepressants show similar phenomena to anticholinergic agents has been augmented by a recent report arguing the similarity of atropine blocked transitions into REM sleep to that of imipramine blocked transitions (23). This result coupled with the finding that physostigmine antagonizes the imipramine induced REM abnormalities in cats led the authors to suggest that the affect of tricyclic antidepressants in sleep is primarily mediated by a decrease in available acetylcholine in the brain. Furthermore, it appears that the MAO inhibitors do not operate via cholinergic mechanisms in effecting changes in REM sleep. This data is particularly useful since an examination of the effect of tricyclics and MAOI on EEG sleep in man suggests similar mechanisms (Table 3).

Table 3. Antidepressants and EEG Sleep

	Sleep continuity	REM latency	REM sleep time
Tricyclics	↑ ↑	↑ ↑	↓ ↓
MAOI	--	↑ ↑	↓ ↓ ↓

If one accepts the hypothesis that the priming of REM sleep occurs in the anterior region of the raphe complex, we could assess the effect of drugs on REM cycle length in depressed patients via raphe "effect" on cholinergic systems. The locus coeruleus, which is primarily responsible for the phasic component of REM sleep together with the FTG neurons, has been shown to operate primarily via NE mechanisms. If we hypothesize that the mechanism of action of tricyclics on sleep occurs via its effect on the raphe system then perhaps a drug such as amitriptyline acts directly on the 5HT system, producing its sedative action at the rostral portion of the raphe system. It is likely that tricyclics also affect the cholinergic system by increasing the REM cycle length between subsequent REM periods (Table 3). Amitriptyline may not act very powerfully at the locus coeruleus thus accounting for its transient effect on the intensity of the phasic eye movements. This hypothesis might explain the particular REM findings with regard to the administration of the tricyclic drugs.

These speculations are consistent with the data on the timing of REM sleep with physostigmine, as well as data recently presented by Vogel arguing that the important factor in EEG sleep changes in relation to the clinical course of depression is the rate of increase in REM change rather than the absolute total REM change during the process of a recovery from a depression. Vogel (APSS, 1977) reported that REM sleep deprivation increases REM % during the latter portion of the night (late REM %) in control subjects and that when depressives are treated by REM sleep deprivation the amount of improvement is proportional to the increase in late REM %. In addition, REM sleep deprivation increases late REM time (REM time after the first REM period) in controls while increasing REM time for the first REM period in depressives. These data have been interpreted as indicating an inability of depressives in general to delay, until the latter half of the night, excess REM sleep (in terms of REM time) caused by REM sleep deprivation and that the worse this inability the less likelihood of successful clinical response to REM sleep deprivation treatment (Vogel, APSS, 1977). In essence, unimproved patients appear unable to sustain their higher levels of late REM %, whereas improved patients reached a peak late REM percent faster than the unimproved. Furthermore, this data would be consistent with the notions of REM inhibitors and REM generators as advocated by Hobson and McCarley. In fact, one might argue that the so-called REM inhibitor is what is primarily affected in depression and it needs to be strengthened in terms of its relationship or association with clinical remission in depression.

Another strategy used in understanding the drug effect on EEG sleep in depressed patients has been an effort to establish EEG sleep correlates of clinical response in depressed patients receiving tricyclic antidepressants. For several years, we have been interested in sleep EEG changes as potential predictors of clinical response in depression. Since drug effects on EEG sleep have been noted within several nights after the initiation of drug treatment while clinical changes take a minimum of 10 days, such a strategy seemed feasible. In our first study of 18 patients who were treated with amitriptyline, it was noted that although the sedative characteristics of amitriptyline did not differentiate clear-cut good responders (n=7) from poor responders (n=11) until the third week of treatment, the good responders showed significant increases in REM latency, decreases in REM sleep percent and REM activity after only two nights of drug treatment (24) (Fig. 3). These early differences in REM sleep variables in good responders were more intensive and pronounced than could be explained on the basis of drug effect alone. Thus, our own work has suggested that the more rapid the suppression of REM sleep which is best reflected in the increase in REM latency, the more likely that the patient will respond to tricyclic antidepressants. In fact, we have argued repeatedly that there should be a period of sustained increase in REM latency associated with the clinical remission. Continued work in this area with a new sample of seventeen depressed patients receiving stepwise increases in amitriptyline (Kupfer et al., unpublished observations) is confirming the earlier results. At 150mg amitriptyline per night (between the 8th night of drug administration and the 13th night), the drug responder group showed a significantly more prolonged REM latency than the nonresponder group (Fig.4).

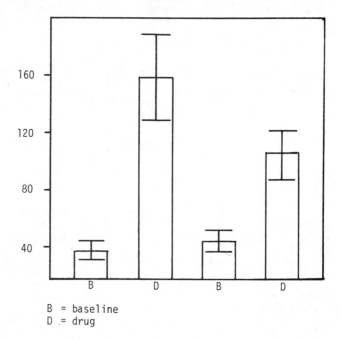

B = baseline
D = drug

Fig. 3. REM latency changes on amitriptyline

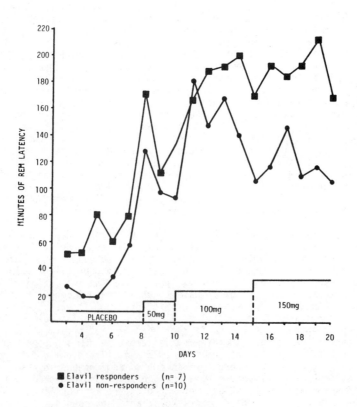

Fig. 4. REM latencies of responders and nonresponders

These results support our notion that clinical response is associated with a period of sustained elevation of REM latency. While there were no sleep continuity discriminators between the two groups in the baseline period, within four days of drug administration sleep latency increased in the nonresponder group and decreased in the responder group and remained significant for the next week (during days 15-20). Thus, despite the lack of significant differences in clinical improvement between the two groups in the early period of the treatment phase, by the time the 150mg level was reached, there were both REM and sleep continuity measures that discriminated responders from nonresponders. In fact, during the 150 mg dosage period, the following significant relationships between final Hamilton score and EEG sleep variables were noted: Sleep latency $r = .57$ ($p<.001$); REM latency $r = -.74$; ($p < .001$); sleep efficiency $r = .63$ ($p<.01$). These significant relationships persisted through the first week of 200mg amitriptyline administration (drug days 14-21): Sleep latency $r = .61$ ($p<.001$); REM latency $r = -.45$ ($p<.01$); sleep efficiency $r = -.55$ ($p<.01$). Of further interest is that the percentage of stage 2-REM (which may represent a displacement of phasic activity during sleep) is increased in nonresponders. The REM latency, as well as other REM sleep variables, at the end of three weeks of drug administration is not necessarily different in clinical responders and nonresponders, but the rate of change of REM latency lengthening in the early stages of treatment appears to relate clearly to eventual treatment response. These human studies contain the ingredients and stimulus for well-designed animal studies, especially in relation to tricyclic antidepressants and EEG sleep.

The intriguing findings of shortened REM latency and increased REM intensity suggest the existence of abnormal cholinergic actions in the sleep of depressed patients. Cholinergic neurons have been implicated in both arousal and REM sleep mechanisms. On the basis of recent reports indicating that physostigmine infusions can hasten the initiation of REM sleep without altering the duration of individual REM periods, it would appear that cholinergic mechanisms may modulate the "timing" rather than the "maintenance" of REM. Another body of evidence points to a sleep oscillator, perhaps under cholinergic control, involving the reciprocal interaction of two groups of neurons - a REM generator group and a REM inhibitor group. Both the short REM latency and "overactive" REM density, especially in the first REM period, may represent an excessive cholinergic outpouring or at the very minimum, a multi-neurotransmitter imbalance which, in effect, yields a "hyperactive" cholinergic discharge. The tricyclic antidepressants, with their anticholinergic action, would be effective in reestablishing the oscillation balance, as well as prolonging the onset of REM sleep (lengthening of REM latency). This anticholinergic action might be related to the actual antidepressant mechanism of the drug, or perhaps reflective of the "triggering" action preceding the antidepressant effects. Recent work on tricyclic antidepressants suggests that the rate of clinical improvement may be correlated with changes in EEG sleep variables which in turn are associated with increasing plasma levels. While the testing of further hypotheses can be best pursued in animal models, interestingly enough, the basic observations have been derived from the "animal" model called man.

REFERENCES

(1) M. Reite, I.C. Kaufman, J.D. Pauley, and A.J. Stynes. Depression in infant monkeys: physiological correlates, Psychosom. Med. 36, 363 (1974).
(2) D.J. Kupfer, F.G. Foster, T.P. Detre, and J. Himmelhoch. Sleep EEG and motor activity as indicators in affective states, Neuropsychobiology 1, 296 (1975).
(3) D.J. Kupfer and F.G. Foster. The sleep of psychotic patients: does it all look alike? In Biology of the Major Psychoses, D.X. Freedman (Ed.) pp 143-164 Raven Press, New York.
(4) J. Mendels and D.A. Chernik. Sleep changes and affective illness, In The Nature and Treatment of Depression, F.F. Flack, S.C. Dragh (Eds.), pp 309-334, Wiley, New York.
(5) D.J. Kupfer. REM latency: a psychobiologic marker for primary depressive disease, Biol. Psychiat. 11, 159 (1976).
(6) D.A. Chernik and J. Mendels. Sleep in bipolar and unipolar depressed patients, Sleep Research 3, 123 (1974).
(7) J.C. Gillin, W.E. Bunney and R. Buchbinder. Sleep changes in unipolar and bipolar depressed patients as compared with normals, Presented, 2nd Int'l Congress of Sleep Res.(1975).
(8) P. Coble, F.G. Foster, and D.J. Kupfer. EEG sleep diagnosis of primary depression. Arch. Gen. Psychiat. 33, 1124 (1976).
(9) R.J. McPartland, D.J. Kupfer, P. Coble, D. Spiker and G. Matthews. The REM analyzer: automated REM measures in depression, Sleep Research, 6, in press.
(10) D.G. Spiker, P. Coble, J. Cofsky, F.G. Foster and D.J. Kupfer. Sleep studies and severity of depression, Sleep Research 6, in press.
(11) F.G. Foster, D.J. Kupfer, P. Coble, R.J. McPartland. Rapid eye movement sleep density: an objective indicator in severe medical-depressive syndromes, Arch. Gen. Psychiat. 33, 1119 (1976).
(12) D.J. Kupfer, F.G. Foster, P. Coble, R.J. McPartland, and R.F. Ulrich. The application of EEG sleep for the differential diagnosis of affective disorders, in press.
(13) P.J. Morgane and W.C. Stern. Chemical anatomy of brain circuits in relation to sleep and wakefulness. In Advances in Sleep Research, I, E. Weitzman (Ed.) pp 1-125, (1974) Spectrum, New York.

188

(14) J.F Pujol, A. Buguet, J.L. Froment, B. Jones and M. Jouvet. The central metabolism of serotonin in the cat during insomnia: a neurophysiological and biochemical study after p-chlorophenylalanine or destruction of the raphe system, Brain Res. 29, 195 (1971).
(15) M. Jouvet. Cholinergic mechanisms and sleep. In Cholinergic Mechanisms, P. Waser (Ed.) pp 456-476. Raven Press, New York (1975).
(16) N. Sitaram, R.J. Wyatt, S. Dawson, and J.C. Gillin. REM sleep induction by physostigmine infusion during sleep, Science 191, 1281 (1976).
(17) N. Sitaram, W. Mendelson, R.J. Wyatt and J.C. Gillin. The time-dependent induction of REM sleep and arousal by physostigmine infusion during normal human sleep, Brain Res. 122, 562 (1977).
(18) D.L. Cheney, G. Racagni and E. Costa. Appendix II: Distribution of acetylcholine and choline acetyltransferase in specific nuclei and tracts of rat brain. In, Biology of Cholinergic Function, A.M. Goldberg and I. Hanin (Eds.) pp 655-659, Raven Press, New York (1976).
(19) J.A. Hobson, R.W. McCarley and P.W. Wyzinski. Sleep cycle oscillation: reciprocal discharge by two brainstem neuronal groups, Science, 189, 55 (1975).
(20) R.T. Pivik, R.W. McCarley and J.A. Hobson. Eye movement-associated discharge in brain stem neurons during desynchronized sleep, Brain Res. 121, 59 (1977).
(21) I. Hanin, R. Massarelli, E. Costa. Acetylcholine concentrations in rat brain: diurnal oscillation, Science, 170, 341 (1970).
(22) Y. Saito, I. Yamashita, K. Yamazaki, F. Okada, R. Satomi, and T. Fujieda. Circadian fluctuation of brain acetylcholine in rats, Life Sci. 16, 281 (1975).
(23) L.L. Glenn, R.A. Mancina and W.C. Dement. Antidepressants and REM sleep: II. Neurochemical substrates, Sleep Research 6, in press.
(24) D.J. Kupfer, F.G. Foster, L. Reich, K.S. Thompson and B. Weiss. EEG sleep changes as as predictors in depression, Am. J. Psychiat. 133, 622 (1976).

ACKNOWLEDGEMENT

This study was supported in part by NIMH 24652. These research studies have been aided immensely by Dr. F. Gordon Foster, Dr. Duane Spiker, Ms. Patricia Coble, Mr. Richard McPartland, Mr. Richard Ulrich, and the nursing staff of the Clinical Research Unit at WPIC.

THE NORADRENERGIC CYCLIC AMP GENERATING SYSTEM IN THE LIMBIC FORE-
BRAIN: A FUNCTIONAL POSTSYNAPTIC NOREPINEPHRINE RECEPTOR SYSTEM AND
ITS MODIFICATION BY DRUGS WHICH EITHER PRECIPITATE OR ALLEVIATE
DEPRESSION.

Fridolin Sulser and Jerzy Vetulani[*]
Department of Pharmacology, Vanderbilt University
School of Medicine and Tennessee Neuropsychiatric
Institute, Nashville, Tennessee 37217

INTRODUCTION

The action of antidepressant drugs on animal behavior as well as their therapeutic effects
in man have been mostly related to their acute modifications of presynaptic neuronal events
involving such processes as storage, reuptake, metabolism and/or regulation of synthesis
of biogenic amines in brain. Two major facts have, however, prompted a search for possible
alternate and/or complimentary mechanisms to explain the therapeutic action of antidepressants:

1) The existing discrepancy in the time course between presynaptic biochemical or pharmacological effects elicited by tricyclic antidepressants within minutes and their clinical
therapeutic action which requires treatment for several weeks

2) The clinical efficacy of tricyclic drugs which do not influence either uptake or turnover
of biogenic amines, e.g. the tricyclic antidepressant iprindole.

Recent studies from our laboratories have been concerned with norepinephrine (NE) - receptor
interactions in slices of the limbic forebrain. We have chosen as a functional model the
noradrenergic cyclic AMP generating system because this particular cyclic AMP generating
system displays a number of characteristics which are compatible with those of a central
NE receptor. Neurophysiologically, the limbic forebrain provides a key integrating system
related to selective modulation of emotion and sensory mechanisms (1) and the anatomical
situation of the limbic system within the forebrain is such as to enable it to correlate
and integrate every form of internal and external perception (2,3).

It is the aim of this paper to briefly characterize the noradrenergic cyclic AMP generating
system in the rat limbic forebrain and to discuss recent results on the modification of
this system by drugs which can either precipitate or alleviate depression in man.

Characterization of the Noradrenergic Cyclic AMP Generating System in Slices of the Limbic
Forebrain: Agonist and Antagonist Properties.
The cyclic AMP generating system in slices of the limbic forebrain is stimulated by low
concentrations of NE but responds poorly if at all to dopamine or serotonin (Fig. 1).
The system is also stimulated by the β-agonist isoproterenol and by adenosine (5). The rise
in the nucleotide caused by isoproterenol to maximum levels is, however, more rapid than
following NE though the maximal response is less than half of that elicited by NE. Studies
on structure-activity relationships and steric requirements for phenylethylamines as agonists
have shown that agonist activity of β-phenylethylamines requires a β-3,4-dihydroxyphenylethy-
lamine with a β-hydroxyl group in the R configuration (6). The α-adrenergic blocking agent
phentolamine and the β-receptor antagonist propranolol blocked in a dose dependent manner
the increase of the nucleotide caused by NE while not significantly affecting the basal level
of cyclic AMP (5). The cyclic AMP response to isoproterenol is blocked by very low concentrations of propranolol but unlike the NE response, the isoproterenol induced stimulation
is not affected by phentolamine even at concentrations 10 times higher than those of the
agonists (Table 1). It is noteworthy that neither phentolamine or propranolol - at molar
concentrations which maximally inhibited the cyclic AMP response to NE - inhibited significantly the increase of the nucleotide caused by adenosine. These data suggest at first
glance that the NE induced rise in cyclic AMP may be mediated by both α- and β-adrenoceptors,
as classically defined. While the presence of classical β-receptors appears to be well

[*]Present address: Institute of Pharmacology, Polish Academy of Sciences,
Krakow, Poland.

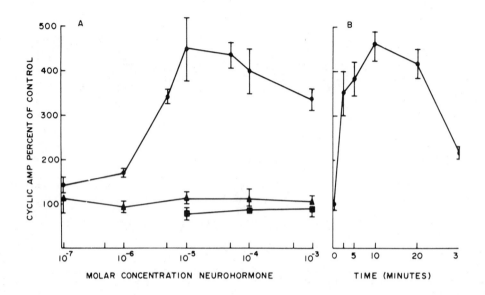

Fig. 1. (A) Effect of various concentrations of (ℓ)-norepinephrine
($\bullet$————$\bullet$) dopamine ($\blacktriangle$————$\blacktriangle$) and 5HT-($\blacksquare$————$\blacksquare$) on the
accumulation of cyclic AMP in slices from the limbic forebrain of rats.
Basal control values were 26.0 ± 2.0 p moles cyclic AMP/mg protein.
(B) Time course of the effect of (ℓ)-norepinephrine (50 μM) on the
accumulation of cyclic AMP in tissue slices from the limbic forebrain.
All values are expressed as the mean percentage of control values ±
S.E.M.. N = 4. From Blumberg et al. (4).

established, the blockade of the NE response by phentolamine does not necessarily imply the
presence of classical α-receptors. Thus, the classical peripheral α-stimulants phenylephrine,
metaraminol and methoxamine, did not elevate the level of cyclic AMP in slices of the limbic
forebrain in concentrations up to 10^{-4} M (6). As a matter of fact, the α-stimulants phenylephrine and clonidine did not only not elevate the level of the nucleotide in the limbic
forebrain, but significantly inhibited the response to NE, but not to isoproteronol (7).
Similar unexpected results with α-agonists have been obtained in slices of the rat cerebral
cortex (8). The data clearly indicate that an extrapolation from peripheral α-agonist activity to that in brain is not valid and that the terms α- and β-adrenergic activity, as
classically defined, have to be used with caution if applied to neuronal functions in the
central nervous system.

While not altering the basal level of the nucleotide, clinically effective antipsychotic
drugs cause a dose dependent inhibition of the limbic noradrenergic cyclic AMP response
with clozapine being particularly potent (5, 9). Recent studies conducted with the antipsychotic drug butaclamol indicate that its blocking effect of the specific noradrenergic
cyclic AMP generating system resides in the (+) enantiomer, thus also demonstrating stereo-
selectivity for central noradrenergic receptor blockade (10).

Recently, a NE sensitive adenylate cyclase system has been described in homogenates of the
rat limbic forebrain using Krebs-Ringer buffer as the homogenizing medium (11). By and
large, this system displays properties somewhat similar to those of slice preparations with
regard to agonist activity though differences are apparent with regard to antagonist-pro-
perties of antipsychotic drugs (9,11). It remains to be seen whether or not this system
in homogenates will display receptor characteristics such as changes in the sensitivity
to NE as a consequence of a changed availability of the neurohormone as the intregrity of
the cell appears to be a prerequisite for adaptive changes to occur in the reactivity of
receptors to catecholamines (12, 13, unpublished observations from this laboratory).

TABLE 1 Effect of Adrenergic Blocking Drugs on the Cyclic AMP Response to Isoproterenol and Adenosine in Slices of the Limbic Forebrain

	Isoproterenol 5 µM		Adenosine 100 µM	
	Cyclic AMP pmoles/mg protein±SEM	% of Basal Level	Cyclic AMP pmoles/mg protein±SEM	% of Basal Level
Agonist	87.9 ± 10.1 (6)	160	379.8 ± 54.3 (5)	1075
Agonist + Phentolamine[a]	85.9 ± 4.7 (4)	156	437.0 ± 77.2 (5)	1236
Agonist + Propranolol[b]	54.8 ± 3.5 (6)	100	321.3 ± 23.9 (5)	908
Basal Level	54.8 ± 4.3 (5)	100	35.3 ± 7.2 (5)	100

The adrenergic blocking drugs were added to the incubation medium 14 minutes before the agonists. The reaction was terminated 10 minutes after addition of the agonists. Numbers in parentheses indicate the number of samples. From Blumberg et al. (5).

[a]50 µM

[b]0.5 µM in experiment with isoproterenol
50 µM in experiment with adenosine

Functional Modification of the Limbic NE Receptor Coupled Adenylate Cyclase System.
So far, we have demonstrated several characteristics of the noradrenergic cyclic AMP generating system in slices of the limbic forebrain that are compatible with those of a central NE receptor system: high sensitivity for agonists, stringent structural requirements and stereospecificity for agonists and stereoselective blockade by antagonists. Another characteristic of a receptor is its ability to develop supersensitivity following prolonged reduction in the neurohormone-receptor interaction (Denervation or disuse supersensitivity). Several behavioral studies (14,15) have indicated enhanced receptor reactivity to intraventricular NE following reserpinization or chemical sympathectomy with 6-hydroxydopamine (6-OHDA). An increased responsiveness of the cyclic AMP generating system to NE following 6-OHDA in cortical slices (16,17,18,19) or in slices from the hypothalamus and brain stem (20) have also been reported. More recent studies from our laboratories now show that the sensitivity to NE of the specific noradrenergic cyclic AMP generating system in the limbic forebrain is markedly increased following destruction of noradrenergic nerve terminals with 6-OHDA. The response to the β-agonist isoproteronol is also enhanced while the response to adenosine is not significantly affected (Table 2).

TABLE 2 Effect of Chemosympathectomy with 6-Hydroxydopamine on the Cyclic AMP Response to Norepinephrine, Isoproterenol and Adenosine

	Cyclic AMP pmoles/mg protein±SEM		Reactivity
	Control	6-Hydroxydopamine	Index[1]
Basal Level	28.93±1.68 (6)	47.90±8.13 (4)	----
Norepinephrine, 5 µM	53.62±5.38 (9)	119.06±6.39 (9)†	288
Isoproterenol, 5 µM	44.77±2.56 (9)	69.23±8.30 (9)*	135
Adenosine, 100 µM	175.24±17.36 (6)	216.69±26.84 (6)	115

[1]Calculated with the formula:

$$\frac{\text{Stimulated level (Drug) - Basal level (Drug)}}{\text{Stimulated level (Control) - Basal level (Control)}} \times 100$$

*p < 0.05   †p < 0.001   From Vetulani et al. (21).

The changes in the reactivity of the cyclic AMP generating system appear to be related to changes in the availability of NE and not to that of other neurohormones as protection by desipramine (DMI) of noradrenergic nerve terminals against the neurotoxic action of 6-OHDA prevents the development of hypersensitivity of the system to NE (Fig. 2).

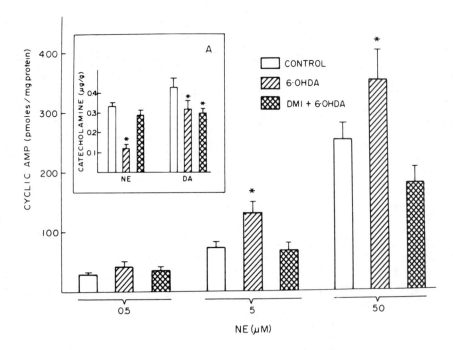

Fig. 2. Effect on 6-OHDA with and without desipramine (DMI) on the cyclic AMP response to NE in slices of the limbic forebrain and on the levels of brain catecholamines (insert A). Each bar represents the mean values of cyclic AMP in pmoles/mg protein±SEM. (N=8) and of catecholamines in µg/g±S.E.M. (N=11 to 16).
DA: Dopamine. From Vetulani et al. (21).
*p < 0.01

The enhanced responsiveness of the cyclic AMP generating system to NE following chemosympathectomy with 6-OHDA appears in all probability to be the consequence of changes occurring at postsynaptic noradrenergic receptor sites. Thus, the sensitivity to isoproteronol, which is not taken up by presynaptic nerve terminals (22), is also enhanced following chemosympathectomy with 6-OHDA. It does not seem likely that the enhanced responsiveness to NE following destruction of nerve terminals by 6-OHDA is related to changes in the activity of cyclic nucleotide phosphodiesterase (18,19).

Subchronic treatment with reserpine also enhances the response to NE in slice preparations from the limbic forebrain while not changing the basal level of the nucleotide. It is pertinent that supersensitivity caused by either 6-OHDA or reserpine is not associated with a change in the apparent affinity of NE to the receptor moiety of the system as the EC_{50} values for NE (concentration that causes half maximal stimulation) are identical to those obtained in slice preparations from control animals. It is tempting to speculate that prolonged NE-deprivation of the noradrenergic receptor could change the actual number of receptors. The increased sensitivity of adrenergic β-receptors in rat cerebral cortex following 6-OHDA has indeed been shown to be associated with an increase in the density of β-adrenergic receptors (23).

Surgical lesions of the medial forebrain bundle which cause a denervation of limbic forebrain structures, elicit also supersensitivity of the system to NE (unpublished results from this laboratory). It is noteworthy that enhanced cyclic AMP responses to NE following lesions of the medial forebrain bundle have also been observed in rat cortical slices (24).

Effect of Various Antidepressant Treatments on the Sensitivity of the Noradrenergic Receptor Coupled Adenylate Cyclase System in the Limbic Forebrain.

Since the above described changes in noradrenergic receptor sensitivity can be interpreted as being the consequence of a prolonged reduction in the NE receptor interaction, it was of interest to study the biochemical consequences following the administration of various antidepressant drugs after a single dose or after administration of the drugs on a clinically more relevant time basis.

Inhibition of monoamineoxidase (MAO) has been shown to be associated with an increased availability of NE at presumptive catecholaminergic receptor sites in the brain (25). The relative cyclic AMP responses to NE in slices of the limbic forebrain of rats following the acute and chronic administration of either pargyline or nialamid are shown in Fig. 3.

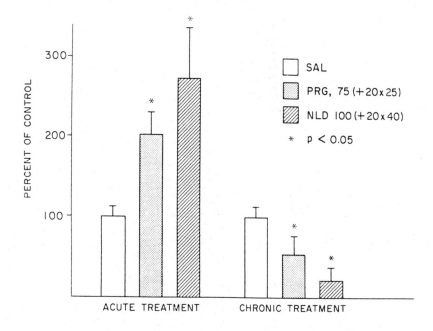

Fig. 3. Relative cyclic AMP responses to 5 μM NE in slices of the limbic forebrain of rats following acute and chronic treatment with MAO inhibitors. Animals were killed 18 hrs. after the last injection. Pargyline (PRG) was administered at a dose of 75 mg/kg I.P. followed by daily doses of 25 mg/kg for 20 days; nialamide (NLD) was given at a dose of 100 mg/kg i.p. followed by daily doses of 40 mg/kg for 20 days. SAL=Saline. The data are expressed as a percentage of the control response to NE ±SEM.

While a single dose of the MAO inhibitors results in an enhanced response of the system to NE, treatment for 3 days or for up to two weeks did not cause significant changes in the responsiveness to NE despite the fact that the level of NE was maximally increased in brain thus already indicating some type of postsynaptic adaptation (21). The administration of the MAO inhibitors for a period of three weeks caused a marked decrease in the reactivity of the noradrenergic receptor coupled adenylate cyclase system (Fig. 3). Preliminary results have shown that withdrawal of the MAO inhibitors for 9 days after treatment for 3 weeks resulted in normalization of the neurohormonal responses. These data indicate the ability of "normal" postsynaptic noradrenergic receptors to adapt to a prolonged increase in the availability of NE and provide evidence for an important regulatory mechanism involving the noradrenergic receptor coupled adenylate cyclase system in the limbic forebrain.

In the next series of experiments, our research group has investigated the action of two clinically effective tricyclic antidepressants on this system: desipramine, a potent blocker of presynaptic NE reuptake, and iprindole, a drug which does not affect uptake, metabolism or turnover of NE (26,27,28). While a single dose of desipramine or iprindole or short term treatment with the tricyclic antidepressants did not alter either the basal level of the cyclic nucleotide or the neurohormonal response to NE, the administration of the drugs for 3 to 6 weeks caused a marked reduction in the sensitivity of the noradrenergic receter coupled adenylate cyclase system (Table 3).

TABLE 3   Effect of Long-Term (4-8 weeks) Treatment with Desipramine or Iprindole on the Response of the Cyclic AMP generating System in the Rat Limbic Forebrain to NE.

	Time of sacrifice[a] (h)	Basal level of cyclic AMP pmoles/mg protein ± SEM	Cyclic AMP response to NE[b] pmoles/mg Protein ± SEM	% of Control
Control	1 or 24	17.8 ± 2.6 (15)	20.4 ± 2.7 (15)	100
Desipramine	1	20.5 ± 2.7 (12)	9.9 ± 3.5* (12)	49
Desipramine	24	16.6 ± 1.6 (14)	6.9 ± 2.1*** (14)	34
Iprindole	1	22.3 ± 3.6 (13)	9.4 ± 4.7* (13)	46
Iprindole	24	16.9 ± 1.5 (15)	7.9 ± 2.4** (15)	38

[a]Time after last injection

[b]Difference in the level of cyclic AMP between the preparation exposed to 5 µM NE and that of the control preparation (corresponding hemisection). Numbers in parenthesis indicate number of samples. From Vetulani et al. (29)

*$p < 0.05$
**$p < 0.01$
***$p < 0.001$ (differences from control responses; Student t-test).

The decreased hormonal sensitivity to NE appeared not to be due to a direct effect of the drugs as no correlation exists between the concentration of the drugs in brain tissue and the change in the cyclic AMP response to the catecholamine (29). Tricyclic antidepressants thus share this delayed action on the noradrenergic receptor coupled adenylate cyclase system in the limbic forebrain with MAO inhibitor type antidepressants. The development of delayed subsensitivity following the administration of tricyclic antidepressants has recently been confirmed by Schultz (30). He has shown that chronic but not acute administration of imipramine causes subsensitivity of the NE stimulated formation of cyclic AMP in rat brain cortex. The finding that this effect was shared by chlorpromazine is not surprising as chlorpromazine also increases the availability of NE at postsynaptic sites (presynaptic α-blockade, blockade of reuptake of NE) thus clearly sharing pharmacological properties with classical tricyclic antidepressants.

Electro-shock treatment (ECT) has been reported to increase the turnover of NE in brain (31) and to lower the high affinity uptake for NE (32). Decreased affinity for NE uptake could imply an increased availability of NE at its receptor sites. It is thus of interest that ECT also reduced the cyclic AMP response to NE (Fig. 4). It is noteworthy that concomitant treatment for 4 days with ECT was capable to significantly prevent the development of the reserpine induced hypersensitivity of the noradrenergic receptor coupled cyclic AMP generating system in the limbic forebrain (Fig. 5). In the next series of experiments, we tested the effect of ECT on the already developed postsynaptic hypersensitivity to NE. As previously shown, destruction of noradrenergic nerve terminals with intraventricular 6-OHDA markedly increased the response of the cyclic AMP generating system to NE. While ECT, applied for eight days, did not significantly alter the basal level of cyclic AMP in slices prepared from animals treated with 6-OHDA, it significantly reduced the enhanced cyclic AMP response to NE (Fig. 5). The findings that ECT is capable of preventing as well as reducing the already developed hypersensitivity of the NE receptor coupled adenylate cyclase system, have heuristic value for the formulation of a catecholamine hypothesis of affective disorders which takes into consideration postsynaptic receptor mediated events. It is tempting to speculate that depression or subtypes of depression might indeed be associated with increased postsynaptic sensitivity of the NE receptor coupled adenylate cyclase system in the limbic forebrain (and conceivably in other areas of the brain) and that the therapeutic action of antidepressants might be related to desensitization of hypersensitive NE receptors - a process that requires time - rather than to acute presynaptic events which occur within minutes or hours.

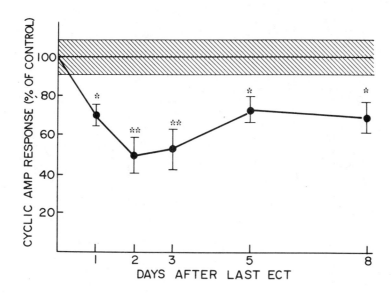

Fig. 4. Effect of ECT on the response of the cyclic AMP generating system to NE in the rat limbic forebrain. Super maximal electro-shock (100 mA; 300 mS) was applied for 8 consecutive days through ear-clip electrodes. The animals were sacrificed at various times after the last shock. Control animals were handled in the same manner but no current was passed. The values are expressed as a percentage of the corresponding drug control value (N = 5 to 20).
*p < 0.05   **p < 0.001

CONCLUSIONS

Generally, the activity of adrenergic neurons appears to be regulated by drugs and non-pharmacological means by intraneuronal or interneuronal feedback mechanisms mostly involving regulation of catecholamine biosynthesis. Results from our laboratories, summarized in this paper, now provide evidence for a postsynaptic regulatory mechanism in the central nervous system involving the noradrenergic receptor coupled adenylate cyclase system that can adapt its sensitivity to NE in a manner inversely related to the degree of its stimulation by the catecholamine.

The findings that psychotropic drugs which can either precipitate or alleviate mental depression cause predictable opposite changes in the reactivity of this system, shift the emphasis on mechanism of action of antidepressant drugs from presynaptic to postsynaptic receptor mediated events.

By analogy to the regulation by cyclic AMP of skeletal muscle glycogenolysis (where at present the cyclic nucleotide action is best understood) the mechanism of cyclic AMP modulation of other physiological events has been proposed to occur by a system of protein phosphorylation - dephosphorylation (33). It is tempting to speculate that the NE - receptor coupled adenylate cyclase-protein kinase system in the limbic forebrain (and in other areas of the brain) is also functioning as a highly effective kinetic amplificational mechanism. If so, studies aimed at the elucidation of both the molecular basis and the neurobiological consequences of changed noradrenergic receptor activity in brain structures - β-adrenergic activity and NE receptor activity that is not β in nature - should be highly rewarding.

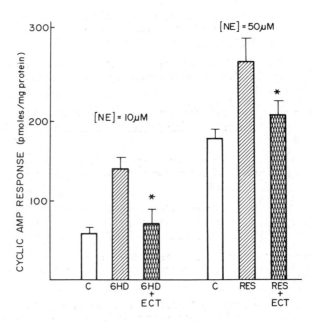

Fig. 5. Effect of ECT on enhanced cyclic AMP responses to NE.
Left: Hypersensitivity was developed following 6-hydroxydopamine
(6 HD; 250 μg/ventricle). Supermaximal shock was applied daily
(90 mA 300 ms) 10 days following 6 HD for a period of 8 days. Controls were handled in the same manner but no current was passed.
The animals were sacrificed 1 day after the last ECT. The data are
expressed as pmoles cyclic AMP/mg protein ± SEM. (N = 5 to 8).
Right: Prevention by ECT of hypersensitivity elicited by reserpine
(2.5 mg/kg i.p. for 4 days). ECT was administered concomitantly
(100 mA; 300 ms) through ear-clip electrodes. The animals were
sacrificed 18 hours after the last ECT treatment.

*$p < 0.05$

Acknowledgements
The original investigations from this laboratory have been supported by USPHS grants MH-
11468, MH-29228 and by the Tennessee Department of Mental Health and Mental Retardation.

REFERENCES

1. E.W. Powell and G. Hines, The limbic system: An interface, Behav. Biol. 12, 149 (1974).

2. P.D. MacLean, Psychosomatic disease and the "visceral brain", Psychosom. Med. 11, 338,
    (1949).

3. P.D. MacLean, Some psychiatric implications of physiological studies on fronto temporal
    position of limbic system (visceral brain), Electroencephalogr. Clin. Neurophysiol. 4,
    407 (1952).

4. J.B. Blumberg, R.E. Taylor and F. Sulser, Blockade by pimozide of a norepinephrine-sen-
    sitive adenylate cyclase in the limbic forebrain: Possible role of limbic noradrenergic
    mechanisms in the mode of action of antipsychotics. J. Pharm. Pharmacol. 27, 125 (1975).

5. J.B. Blumberg, J. Vetulani, R.J. Stawarz and F. Sulser, The noradrenergic cyclic AMP
    generating system in the limbic forebrain: Pharmacological characterization and possible
    role of limbic noradrenergic mechanisms in the mode of action of antipsychotic drugs.
    Europ. J. Pharmacol. 37, 357 (1976).

6. H.E. Smith, E.P. Burrows, P.L. Mobley, S.E. Robinson and F. Sulser, Agonist effects of β-phenethylamines on the noradrenergic cyclic AMP generating system in rat limbic forebrain. Stereoisomers of p-hydroxynorephedrine. J. Med. Chem., in press (1977).

7. J. Vetulani, N.J. Leith, R.J. Stawarz and F. Sulser, Effect of clonidine on the noradrenergic cyclic AMP generating system in the limbic forebrain and on medial forebrain self-stimulation behavior. Experientia, in press (1977).

8. P. Skolnick and J.W. Daly, Stimulation of adenosine 3', 5' monophosphate formation by α and β adrenergic agonists in rat cerebral cortical slices: Effect of clonidine. Molec. Pharmacol. 11, 545 (1975).

9. F. Sulser and S.E. Robinson, Clinical implications of pharmacological differences among antipsychotics. Psychopharmacology: A Generation of Progress, Raven Press, N.Y. in press (1977).

10. S.E. Robinson and F. Sulser, The noradrenergic cyclic AMP generating system of the rat limbic forebrain and its stereospecificity for butaclamol. J. Pharm. Pharmacol. 28, 645 (1976).

11. A.S. Horn and O.T. Phillipson, A noradrenaline sensitive adenylate cyclase in the rat limbic forebrain: preparation, properties and the effects of agonists, adrenolytics and neuroleptic drugs. Europ. J. Pharmacol. 37, 1 (1976).

12. M. Baudry, M.P. Martres and J.C. Schwartz, Modulation in the sensitivity of noradrenergic receptors in the CNS studied by the responsiveness of the cyclic AMP system, Brain Res. 116, 111 (1976).

13. B.K. Krueger, J. Forn, J.R. Walters, R.H. Roth and P. Greengard, Stimulation by dopamine of adenosine cyclic 3', 5' - monophosphate formation in rat caudate nucleus: Effect of lesions of the nigroneostriatal pathway. Mol. Pharmacol. 12, 639 (1976).

14. M.A. Geyer and D.S. Segal, Differential effects of reserpine and α-methyl-p-tyrosine on norepinephrine and dopamine induced behavioral activity, Psychopharmacologia 19, 131 (1973).

15. A.J. Mandell, The role of adaptive regulation in the pathophysiology of psychiatric disease, J. Psychiat. Res. 11, 173 (1974).

16. G.C. Palmer, Increased cyclic AMP response to norepinephrine in the rat brain following 6-hydroxydopamine, Neuropharmacology 11, 145 (1972).

17. B. Weiss and S. Strada, Neuroendocrine control of the cyclic AMP system of brain and pineal gland. Adv. Cyclic Nucl. Res. 1, 357 (1972).

18. M. Huang, A.K.S. Ho and J.W. Daly, Accumulation of adenosine 3', 5' monophosphate in rat cerebral cortical slices. Molec. Pharmacol. 9, 711 (1973).

19. A. Kalisker, C.H.O. Rutledge and J.P. Perkins, Effect of nerve degeneration by 6-hydroxydopamine on catecholamine-stimulated adenosine 3', 5' - monophosphate formation in rat cerebral cortex, Molec. Pharmacol. 9, 619 (1973).

20. G.C. Palmer, F. Sulser and G.A. Robison, Effect of neurohumoral and adrenergic agents on cyclic AMP levels in various areas of the rat brain in vitro. Neuropharmacology 12, 327 (1973).

21. J. Vetulani, R.J. Stawarz and F. Sulser, Adaptive mechanisms of the noradrenergic cyclic AMP generating system in the limbic forebrain of the rat: Adaptation to persistent changes in the availability of norepinephrine (NE). J. Neurochem. 27, 661 (1976).

22. B.A. Callingham and A.S.W. Burgen, The uptake of isoprenaline by the perfused rat liver. Molec. Pharmacol. 2, 37 (1966).

23. J.R. Sporn, T.K. Harden, B.B. Wolfe and P.B. Molinoff, β-adrenergic receptor involvement in 6-hydroxy-dopamine induced supersensitivity in rat cerebral cortex, Science 194, 624 (1975).

24. R.K. Dismukes, P. Ghosh, C.R. Creveling and J.W. Daly, Altered responsiveness of adenosine 3', 5' monophosphate generating systems in rat cortical slices after lesions of the medial forebrain bundle, Exp. Neurol. 49, 725 (1975).

25. S.J. Strada and F. Sulser, Effect of monoamine oxidase inhibitors on metabolism and <u>in</u> <u>vivo</u> release of ^{3}H-norepinephrine from the hypothalamus. <u>Europ</u>. <u>J</u>. <u>Pharmacol</u>. 18, 303 <u>(1972)</u>.

26. R.A. Lahti and R.P. Maickel, The tricyclic antidepressants - inhibition of norepinephrine uptake as related to potentiation of norepinephrine and clinical efficacy, <u>Biochem</u>. <u>Pharmacol</u>. 20, 482 (1971).

27. J.J. Freeman and F. Sulser, Iprindole-amphetamine interactions in the rat: The role of aromatic hydroxylation of amphetamine in its mode of action. <u>J</u>. <u>Pharmacol</u>. <u>Exp</u>. <u>Ther</u>. 183, 307 (1972).

28 B.N. Rosloff and J.M. Davis, Effect of iprindole on norepinephrine turnover and transport, <u>Psychopharmacologia</u> 40, 53 (1974).

29. J. Vetulani, R.J. Stawarz, J.V. Dingell and F. Sulser, A possible common mechanism of action of antidepressant treatments. <u>Naunyn-Schmiedeberg's</u> <u>Arch</u>. <u>Pharmacol</u>. 293, 109 (1976).

30. J. Schultz, Psychoactive drug effects on a system which generates cyclic AMP in brain, <u>Nature</u> 261, 417 (1976).

31. A.M. Thierry, F. Javoy, J. Glowinski and S.S. Kety, Effects of stress on the metabolism of norepinephrine, dopamine and serotonin in the central nervous system of the rat. I. Modifications of norepinephrine turnover. <u>J</u>. <u>Pharmacol</u>. <u>Exp</u>. <u>Therap</u>. 163, 163 (1968).

32. E.D. Hendley and B.L. Welch, Electroconvulsive shock: Sustained decrease in norepinephrine uptake in a reserpine model of depression, <u>Life</u> <u>Sci</u>. 16, 45 (1975).

33. D.A. Walsh and C.D. Ashby, Protein kinases: Aspects of their regulation and diversity. <u>Recent</u> <u>Prog</u>. <u>Horm</u>. <u>Res</u>. 29, 329 (1973).

DISCUSSION

Dr. Stricker inquired about transmitter specificity in the hypersensitive receptor. *Dr. Sulser* replied that he had tested only R-norepinephrine and R-isopranolol and both showed supersensitivity; he had not, e.g., tested dopamine. He went on to state that iso-proterenol also works in the normal animal, and that the maximal response which can be elicited from isoproterenol is only about 1/3 to 1/4 of the norepinephrine response. The two responses are not additive. Dr. Sulser thinks that part, but not all, of the norepine-phrine response is a beta-response.

Dr. Salama commented that he used the same limbic system as Dr. Sulser but found dopamine sensitivity rather than norepinephrine sensitivity. He wondered if the difference could be the result of his adenine prelabeling. *Dr. Sulser* indicated that he was not using adenine prelabeling; that he was measuring endogenous cyclic AMP.

APPROACHES TO RAPIDLY CYCLING MANIC-DEPRESSIVE ILLNESS

Robert M. Post
National Institute of Mental Health, Bethesda, Maryland

INTRODUCTION

This paper focuses on possible models for two temporal characteristics of manic-depressive illness - the tendency for rapid cyclicity and for recurrences to occur with increased frequency over time. The cyclic aspects of manic-depressive illness are well-known with some patients showing irregular patterns of manic-depressive mood swings and others showing patterns with such precise regularity that the presence of an underlying or a rhythmic biological substrate is difficult to reject. Recently, Grof, Angst, and Haines (1) reviewed histories of 593 patients with recurrent depressive illness (of the unipolar variety, i.e. those without a prior history of mania) and reported that at all ages, the average interval between successive episodes decreased. This second apparent temporal characteristic of sensitization or increased proclivity for recurrence over time will be reviewed from the perspective of several animal models. Thus, the emphasis in this discussion is on selective characteristics of affective syndromes; the models suggested are offered not as close parallels to the illness, but as approaches which may offer new tools and perspectives to dissect sensitization and oscillatory phenomena in the central nervous system. Considering different electrophysiological and pharmacological paradigms may be useful in considering general principles relevant to the development of oscillation in manic illness.

The psychomotor stimulants such as amphetamine and cocaine produce prominent effects on mood, cognition, alertness, and arousal that parallel in many respects clinical phenomena observed during mania (2). The psychomotor activating effects are well documented in anecdotal reports and in the clinical literature as well as in the laboratory in a variety of experimental animal species. Thus, at lower doses or following less chronic administration, the psycho-motor stimulants may offer possible models for mood and activity disturbances (2,3) in addition to their previously discussed roles as possible models for a paranoid psychosis (3,4). A wide variety of evidence now suggests that intermittent, repetitive administration of psychomotor stimulants, as well as the dopaminergic active drugs bromocriptine and L-DOPA, may produce behavioral sensitization to the motor activating and stereotypic effects of these compounds (4-6).

Drug Sensitization

Following the early reports of Downs and Eddy (7,8), and Tatum and Seevers (9), we demon-strated that repeated administration of cocaine to rats produced increases in intensity of hyperactivity and stereotypy at low doses (10 mg/kg i.p.) (10) while higher doses (60 mg/kg i.p.), which were initially subconvulsive, eventually led to convulsions and death (11). Lidocaine, a local anesthetic of equal potency to cocaine but without its psychomotor stimulant effects, also produced increased susceptibility to convulsions following repetitive administration and an increased frequency of omniphagic behavior (12). In the monkey, repetitive administration of cocaine for periods up to six months eventually produced an inhibitory syndrome characterized by motor inhibition, catalepsy, abnormal visual searching and staring behavior, and in four of 13 animals, prominent oral-buccal-lingual dyskinesias (13). This inhibitory syndrome, as well as the dyskinesias, appeared to develop with increasing severity over time. Similar changes have recently been reported following chronic amphetamine or cocaine by investigators in several other laboratories (4,14-21)

Dopaminergic Mechanisms

Various mechanisms have been suggested to account for this reverse tolerance or sensitization phenomenon including agonist supersensitivity (5,19), denervation, or as discussed below, kindling (4,6). Recently, Seeman and collaborators (22) have reported a paradoxical increase in the number of apomorphine receptor binding sites demonstrated after in vitro administration of dopamine. Klawans and collaborators (5) have also reported increased number of dopamine receptors following chronic once daily L-DOPA administration in animals. Thus, it is possible that chronic treatment with the psychomotor stimulants or direct or indirectly acting dopaminergic agonists may be associated with not only a behavioral sensitization but one documented on a receptor level as well.

The possible involvement of dopaminergic mechanisms in both phases of manic-depressive illness had been reviewed extensively elsewhere (23-28). In one patient amphetamine administration on two separate occasions provoked manic episodes lasting three and four days duration. Later treatment with the dopamine receptor agonist piribedil (ET 495) also produced recurrent cycles of manic behavior (23). Pimozide, a dopamine receptor antagonist, was partially successful in suppressing this recurrent manic behavior in this patient, as well as in a larger series of manic patients (29). Similarly, L-DOPA administration may be associated with the precipitation of manic episodes in bipolar patients (28).

Thus, if dopaminergic systems were being activated during manic episodes in a fashion similar to that observed in animal studies during repetitive psychomotor stimulant administration, it is possible that some of the mechanisms mediating sensitization to the psychomotor stimulants could be evoked during the clinical episode. An agonist supersensitivity or kindling model of mania might lead to several new clinical and experimental approaches, and predictions could be directly tested. Do patients with recurrent manic episodes demonstrate an increased proclivity for recurrences as do unipolar depressives? Does the severity of an episode increase with repetition and duration of illness? Does lithium block agonist supersensitivity as preliminary reports suggest it blocks supersensitivity following withdrawal of neuroleptics (5)? Would clinical lithium prophylaxis prevent the development of increased susceptibility?

L-DOPA Induced On-Off Effect

Chronic treatment of Parkinsonian patients with L-DOPA not only leads to sensitization phenomena in terms of increasing frequency of L-DOPA induced dyskinesias (30) and psychotogenic effects (31) with duration of treatment, but also is associated with a progressively increased frequency of the "on-off effect" over time (32). During this phase of treatment immobilizing akinesia rapidly alternates with severe dyskinesias. The on-off effect may initially be mild but in its progressively developing later stages, oscillations occur more rapidly, with larger amplitudes, and may become totally incapacitating for the patient.

Thus, chronic treatment with dopamine agonists may not only be of behavioral relevance in relation to sensitization phenomena but, under some conditions, may be associated with the development of rapidly oscillating behavioral manifestations. The on-off effect in Parkinson's disease and rapid mood cycling in manic-depressive illness appear to have a related time course of development - both occur relatively late in the illness and may emerge with increased frequency of cycles, increased rapidity of "switches", increased severity of symptoms, and may be relatively treatment resistant. Study of the pathophysiology of the on-off effect may help define some of the important clinical characteristics and mechanisms involved in other rapidly cycling disorders.

Kindling

Another aspect of the cocaine sensitization phenomena led us to explore the interaction between electrical sensitization phenomena (kindling) and that produced by drugs. That is, repetitive treatment with subconvulsive doses of local anesthetics eventually evoked convulsive phenomena (Fig. 1). The temporal characteristics appeared to be parallel to those of kindling in which repetitive subthreshold stimulation of a variety of subcortical sites, particularly in the limbic system, eventually produces increasingly long after-discharge activity, increasing spread of this activity outside of the limbic system, and finally is associated with the development of full-blown seizures (33-36). This altered responsivity in neuronal excitability appears to be relatively permanent. The relevance of this model for epileptogenesis (37-39) and its pharmacological dissection has been reviewed in detail in this session by Babington (40). Its possible use as a model for learning and memory have also been suggested by Goddard and Douglas (41). The kindling paradigm may likewise be useful in examining some of the temporal characteristics of clinical phenomenology in the endogenous psychoses (6).

Oscillations in Amygdala Excitability

In the course of studies designed to examine the interaction between drug and electrical sensitization paradigms (42), we observed that a subgroup of animals receiving once daily electrical stimulation of the amygdala developed regular patterns of oscillations in amygdala excitability as documented by alterations in after-discharge duration (43). In three separate studies, rats were implanted with unilateral amygdala electrodes and were stimulated

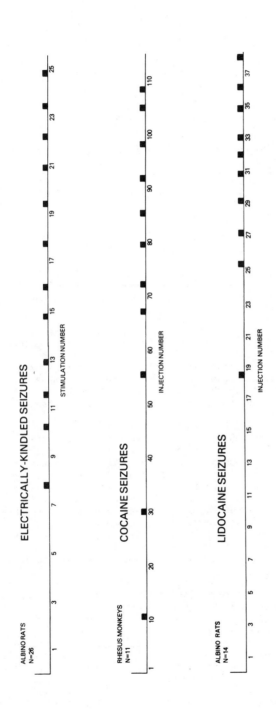

Fig. 1

Electrically kindled rats received once daily stimulation of the left amygdala with 200 µamps, 50 Hz, for 500 milliseconds. Eventually animals developed seizures to the previously subthreshold stimulations. Rhesus monkeys received the same dose of cocaine, i.p., five days per week for periods up to six months (13). Lidocaine (60 mg/kg, i.p.) was administered once daily, five times per week to male Sprague-Dawley rats (12). Seizures following repetitive local anesthetic administration also appeared with increasing frequency over time.

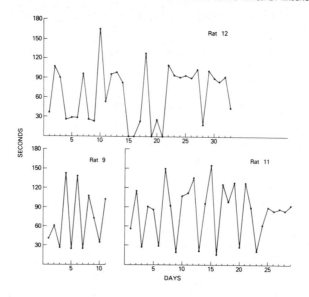

RHYTHMIC OSCILLATIONS IN AFTER-DISCHARGE DURATION DURING AMYGDALA KINDLING

Fig. 2

In individual rats, after-discharge durations recorded from the amygdala following once daily
electrical stimulation showed different oscillatory
patterns.

once daily with 200 microamp, 50 Hz current for 500 milliseconds with a constant current
stimulator. Approximately 40% of the animals demonstrated rhythmic oscillations in after-
discharge duration, similar to that illustrated in Fig. 2. Animals showed a typical
kindling-like increase in after-discharge duration over the first three to seven days of
stimulation and then developed regular rhythmic oscillations in after-discharge threshold;
several animals appeared to develop the oscillations following the onset of the first after-
discharge.

Individual animals showed different but stable rhythmic patterns of oscillation and after-
discharge duration. Some animals showed a tendency for 48-hour cycling (Fig. 3),
demonstrating long after-discharges one day, short the next, long on the following day, and
so on. Other animals developed other patterns (three or five day cycles) which were maintained fairly stably over a period of several weeks of stimulation. The presence or absence
of generalized seizures following a given stimulation also showed a similar pattern of
rhythmicity with seizures usually occurring on days with long after-discharges and the
absence of seizures when after-discharge durations were short (Fig. 3). When after-
discharge thresholds were assessed these also appeared to show some regular and cyclic
variation from day to day in some animals. These oscillations in excitability were not
related to histologically verified location of the electrode within the amygdala.

Wake and Wada (44) have reported similar marked oscillations in after-discharge duration and
seizure stages in cats following long-term kindling in various cortical areas. Thus, the
appearance of marked oscillations in after-discharge duration following repetitive intermittent stimulation may be reflective of the propensity of several neural substrates
to show this phenomenon. It appears unlikely that specific characteristics of the kindling
stimulus such as the development of a relative refractory period or a subsequent excitatory
period account for the appearance of rhythmic oscillations. If the stimulation characteristics
were producing the oscillations in electrical activity it would be expected that all animals
would show a similar periodicity of cycle length.

Kindling and Relation to Interval Between Stimulation

Goddard et al., (33) reported that the interval between kindling stimulations is important in
the rate of development of kindling itself. That is, animals stimulated once every 24 hours
required the shortest number of stimulations before the onset of the first seizure, while
animals stimulated more frequently (5-10 minutes) require many more stimulations to complete
kindling. Continuous stimulations may prevent kindling. Similarly, Mucha and Pinel (45)

have recently documented that stimulation of the amygdala once every 1.5 hours is associated
with after-discharge suppression rather than the facilitation seen following stimulation once
every 24 hours. While the interval between electrical stimulations is thus clearly important
in the development of sensitization phenomena such as kindling, it would not appear to account
for the oscillations in amygdala excitability documented in our studies and those of Wake and
Wada in the cortex (44) since a constant interval of one stimulation per 24 hours was ad-

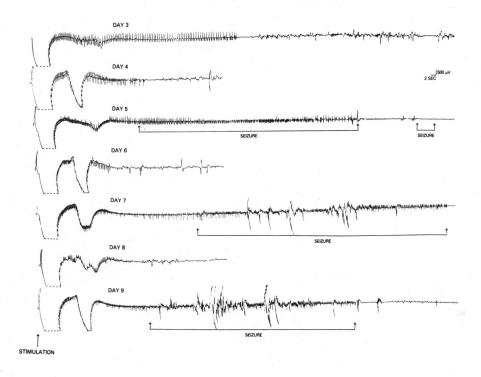

Fig. 3

A 48-hour pattern of oscillation in amygdala after-discharge durations
and seizure occurrence is illustrated during electrical kindling in
one rat.

ministered in each case. Consideration of the duration of the interval between drug application and whether continuous versus intermittent drug levels are achieved may help resolve
some of the controversy in the literature regarding the development of tolerance versus
sensitization. In this way the electrical kindling model helps to focus attention on drug
administration characteristics and suggests direct clinical and laboratory tests of important
temporal variables such as continuous versus intermittent application.

Behavioral Effects of Kindling

The behavioral relevance of the kindling paradigm is particularly highlighted by Adamec (46,
47). He first demonstrated that cats with differences in predatory aggression had differences
in amygdala after-discharge thresholds. That is, cats which were rat killers showed higher
amygdala after-discharge thresholds than non-killers. However, when the killers were
subjected to repetitive electrical stimulation of the amygdala (kindling), in order to
lower their after-discharge threshold, they converted into non-killer animals (46).
McIntyre and collaborators (48,49) have also emphasized the possible behavioral relevance
of the electrical kindling processes in their studies of memory and learning. They demonstrated that kindled animals have deficits in active avoidance and conditioned emotionality
studies. Woodruff (50) suggests that experimental epileptic foci may exert more profound
and disruptive behavioral effects than ablative lesions and that the deficits are specific
to the pathways involved. Numerous investigators including Papez (51), MacLean (52),
Gellhorn (53), Heath (54), and Stevens (55) have emphasized the role of the limbic system

in the modulation of emotional behavior. The kindling paradigm helps refocus attention on the limbic system and especially the amygdala because of its particular seizure susceptibility and behavioral relevance. Moreover, kindling appears to provide a new tool for producing long-lasting alterations in excitability of selective neural substrates in a relatively specific manner as demonstrated in the work of Adamec (46).

Acute stimulation and lesion experiments in the amygdala have produced marked alterations in emotional behavior, including placidity or rage, the effects depending on the experimental animal, the precise location of the stimulation or lesion or other characterisitics of the experiment (56-58). Moreover, the functional differentiation of the amygdala into a basolateral and corticomedial grouping appears to provide some neuroanatomical and electrophysiological evidence for opposing mechanisms within the amygdala nucleus for modulation of affective, aggressive, sexual, endocrine, and autonomic reactivity (56-58). Thus, it is tempting to speculate that oscillations in the excitability of the amygdala or one of its functional subdivisions as revealed by the kindling studies reported here could be associated with marked alterations in emotional behavior, motor, and neuroendocrine reactivity like that seen during the cyclic mood disorders.

Drug-Kindling Interactions

In our recent experiments it appears that chronic electrical stimulation of the amygdala alters animals' subsequent responsivity to a variety of pharmacological challenges (42). Amygdala kindled animals were less sensitive to the effect of high dose cocaine (40 mg/kg) on vertical hyperactivity and showed markedly increased effects following the challenge with lidocaine (60 mg/kg i.p.). Fourteen of 16 amygdala kindled animals had lidocaine induced convulsions compared to only four of 17 implanted, sham stimulated controls ($x^2 = 13.59$, p <.001). Conversely, differences in prior handling of animals also appear to influence the rate of kindling. As previously reported by Arnold et al. (59) we found that handling and saline injecting animals (presumedly increasing stress) retards kindling (the time to the development of first after-discharges) compared to non-handling controls. Although, pretreatment with low doses of cocaine (10 mg/kg) had no effect on the rate of kindling, experiments are currently in progress to determine whether drug sensitization with lidocaine or high dose cocaine alters the rate of amygdala kindling.

It is thus suggested that the emergence of oscillations in amygdala excitability during kindling may represent a useful model system for examining and manipulating various sub-cortical substrates thought to be intimately involved in the modulation of emotional behavior. If these oscillations were in some way related to the behavioral oscillations observed during mania and depression, one might expect lithium carbonate to likewise exert an ameliorative effect. Preliminary experiments in collaboration with Agu Pert and Kathleen Squillace suggest that lithium carbonate pretreatment may not block the development of oscillations in amygdala excitability, however. Nineteen animals were treated with oral lithium (0.65 milliequivalents/15 grams of lab chow) beginning seven days prior to amygdala kindling and continuing during 21 days of electrical stimulation. Both the lithium pretreated group and 18 non-lithium treated kindled animals showed a similar rate of onset of kindling as measured by the number of stimulations to the first seizure. Approximately 40% of the animals in both groups developed marked oscillations in after-discharge duration following chronic kindling. The lithium levels achieved in these studies however, only bordered on the low therapeutic ranges observed in humans (lithium levels averaged 0.5 meq/liter). Whether lithium in higher doses would affect these characteristics of the kindling paradigm or, even in the current doses, affect the periodicity or cycle frequency remains to be examined in further studies.

CONCLUSIONS

The experiments reported help to emphasize the possible interactions between electro-physiological and pharmacological paradigms and their possible effects on a final common pathway affecting behavior. The kindling and drug sensitization paradigms emphasize important temporal characteristics of the behavioral effects and the progressive and long-term alterations produced by these chronic experimental manipulations. They highlight the possibility that environmental events and severe stresses acting through similar pharmacological and neuronal final common pathways may be capable of activating similar pathways important in the modulation of behavior. During the course of both the drug sensitization paradigm and the electrical kindling paradigm marked alterations in behavioral responsivity or in electrical excitability have been observed. These procedures may be useful in exploring long-term facilitation or disruption with an "active lesion" of selective neural pathways.

The appearance of the pathological on-off effect with increasing duration of L-DOPA treatment of Parkinson's disease highlights the possibility that chronic dopaminergic stimulation may interact with diseased or supersensitive receptors to produce rapidly alterating neurological and behavioral pathology. The oscillations in amygdala after-discharge duration following electrical kindling may provide a useful tool for examining the effects of pharmacological manipulations and lesions of this cyclic phenomena. Since only a portion of the kindled animals develop oscillations in after-discharge duration, the model might also help in the exploration of the underlying genetic, biological, and experiential variables that predispose to the rapid cycling. Thus, careful definition of the biochemical and electrophysiological parameters of these oscillating systems may lead to new directions in affective illness research, therapy, and prophylaxis. Kindling and drug sensitization paradigms may help elucidate normal and inherent capacities for oscillation in the central nervous system even though they represent only partial and incomplete models for the mood disorders.

REFERENCES

1. P. Grof, J. Angst, T. Haines, The clinical course of depression: Practical Issues. In Symposia Medica, Vol. 8, F. K. Schattauer, ed., (1973), p. 141, Hoechst, Stuietegaard, Germany.

2. D. Murphy, Animal models of mania. In Proceedings of the Workshop on Animal Models in Psychiatry and Neurology, Rockville, Maryland, June 6-8, 1977, Pergamon Press, New York, (this volume).

3. R. M. Post, Cocaine psychoses: A continuum model, Am. J. Psychiatry 132, 225 (1975).

4. E. H. Ellinwood, Stimulant models of psychosis. In Proceedings of the Workshop on Animal Models in Psychiatry and Neurology, Rockville, Maryland, June 6-8, 1977, Pergamon Press, New York, (this volume).

5. H. L. Klawans, Animal models of dyskinesias. In Proceedings of the Workshop on Animal Models in Psychiatry and Neurology, Rockville, Maryland, June 6-8, 1977, Pergamon Press, New York, (this volume).

6. R. M. Post, R. T. Kopanda, Cocaine, kindling, and psychosis, Am. J. Psychiatry 133, 627 (1976).

7. A. W. Downs, N. B. Eddy, The effect of repeated doses of cocaine on the rat, J. Pharmacol. Exp. Ther. 46, 199 (1932).

8. A. W. Downs, N. B. Eddy, The effect of repeated doses of cocaine on the dog, J. Pharmacol. Exp. Ther. 46, 195 (1932).

9. A. L. Tatum, M. H. Seevers, Experimental cocaine addiction, J. Pharmacol. Exp. Ther. 36, 401 (1929).

10. R. M. Post, H. Rose, Increasing effects of repetitive cocaine administration in the rat, Nature 260, 731 (1976).

11. R. M. Post, Progressive changes in behavior and seizures following chronic cocaine administration: Relationship to kindling and psychosis. In Cocaine and Other Stimulants (Advances in Behavioral Biology, Volume 21), E. H. Everett and M. M. Kilbey, eds., (1977), p. 353, Plenum Press, New York.

12. R. M. Post, R. T. Kopanda, A. Lee, Progressive behavioral changes during chronic lidocaine administration: Relationship to kindling, Life Sci. 17, 943 (1975).

13. R. M. Post, R. T. Kopanda, K. E. Black, Progressive effects of cocaine on behavior and central amine metabolism in rhesus monkeys: Relationship to kindling and psychoses, Biol. Psychiatry 11, 403 (1976).

14. M. M. Kilbey, E. H. Ellinwood, Reverse tolerance to stimulant-induced abnormal behavior, Life Sci. 20, 1063 (1977).

15. J. S. Stripling, E. H. Ellinwood, Sensitization to cocaine following chronic administration in the rat. In Cocaine and Other Stimulants (Advances in Behavioral Biology, Volume 21), E. H. Everett and M. M. Kilbey, eds., (1977), p. 327, Plenum Press, New York.

16. P. H. Short, L. Shuster, Changes in brain norepinephrine associated with sensitization to d-amphetamine, Psychopharmacol. 48, 59 (1976).

208

17. L. Shuster, Y. Grace, A. Bates, Sensitization to cocaine stimulation in mice, Psychopharmacol. 52, 185 (1977).

18. D. S. Segal, A. J. Mandell, Long-term administration of d-amphetamine: Progressive augmentation of motor activity and stereotypy, Pharmacol. Biochem. Behav. 2, 249 (1974).

19. H. L. Klawans, D. I. Margolin, Amphetamine-induced dopaminergic sensitivity in guinea pigs, Arch. Gen. Psychiatry 32, 725 (1975).

20. P. M. Groves, G. V. Rebec, Changes in neuronal activity in the neostriatum and reticular formation following acute or long-term amphetamine administration. In Cocaine and Other Stimulants (Advances in Behavioral Biology, Volume 21), E. H. Everett and M. M. Kilbey, eds., (1977) p. 269, Plenum Press, New York.

21. B. T. Ho, D. L. Taylor, V. S. Estevez, Behavioral effects of cocaine--metabolic and neurochemical approach. In Cocaine and Other Stimulants (Advances in Behavioral Biology, Volume 21), E. H. Everett and M. M. Kilbey, eds., (1977) p. 229, Plenum Press, New York.

22. T. Lee, P. Seeman, Dopamine receptors in human and calf brains. In Abstracts of the Thirty-Second Annual Convention of the Society of Biological Psychiatry, April 28 - May 1, 1977, Abstract #25, p. 36.

23. R. H. Gerner, R. M. Post, W. E. Bunney, Jr., A dopaminergic mechanism in mania, Amer. J. Psychiatry 133, 1177 (1976).

24. A. Randrup, I. Munkvad, R. Fog, J. Gerlach, L. Molander, B. Kjellberg, J. Scheel-Kruger, Mania, depression, and brain dopamine, Curr. Dev. Psychopharmacol. 2, 207 (1975).

25. D. L. Murphy, L-DOPA, behavioral activation, and psychopathology. In Neurotransmitters (Research Publications, Association for Research in Nervous and Mental Disorders, Volume 50), I. J. Kopin, ed., (1972) p. 472, Raven Press, New York.

26. R. M. Post, Frontiers in affective disorder research: New pharmacological agents and methodologies. In Psychopharmacology - A Generation of Progress, K. Killam, A. DiMascio, M. Liptom, eds., Raven Press, New York, (in press).

27. R. M. Post, R. H. Gerner, J. S. Carmen, W. E. Bunney, Jr., A dopamine receptor stimulator in depression. In Scientific Proceedings of the 128th Annual Meeting of the American Psychiatric Association, Anaheim, California, May, 1975.

28. D. L. Murphy, H. K. H. Brodie, F. K. Goodwin, W. E. Bunney, Jr., L-DOPA: Regular induction of hypomania in bipolar manic-depressive patients, Nature 229, 135 (1971).

29. R. M. Post, D. C. Jimerson, F. K. Goodwin, Time-related biochemical effects on neuroleptics. In Scientific Proceedings of the 129th Annual Meeting of the American Psychiatric Association, Miami Beach, Florida, May, 1976.

30. H. L. Klawans, P. Crossett, N. Dana, Effect of chronic amphetamine exposure on stereotyped behavior: Implications for pathogenesis of L-DOPA-induced dyskinesias. In Advances in Neurology, Volume 9, D. B. Calne, T. N. Chase, and A. Barbeau, eds., (1975) p. 105, Raven Press, New York.

31. H. L. Klawans, C. Moskovitz, H. Moses, Levodopa-induced psychosis: A kindling phenomenon. In Scientific Proceedings of the One-hundred and Thirthieth Annual Meeting of the American Psychiatric Association, Toronto, May, 1977, Abstract #344, p. 206.

32. F. McDowell, A. Barbeau, eds., Advances in Neurology, Volume 5, Raven Press, New York, (1974).

33. G. V. Goddard, D. C. McIntyre, C. K. Leech, A permanent change in brain function resulting from daily electrical stimulation, Exp. Neurol. 25, 295 (1969).

34. R. J. Racine, Modification of seizure activity by electrical stimulation: I Afterdischarge threshold, Electroencephalogr. Clin. Neurophysiol. 32, 269 (1972).

35. R. J. Racine, Modification of seizure activity by electrical stimulation: II Motor seizure, Electroencephalogr. Clin. Neurophysiol. 32, 281 (1972).

36. J. A. Wada, M. S. Sato, Generalized convulsive seizures induced by daily electrical stimulation of the amygdala in cats, Neurology 24, 565 (1974).

37. J. A. Wada, M. S. Sato, M. E. Corcoran, Persistent seizure susceptibility and recurrent spontaneous seizure in kindled cats, Epilepsia 15, 465 (1974).

38. F. Morrell, Goddard's kindling phenomenon: A new model of the "mirror focus". In Chemical Modulation of Brain Function, H. C. Sabelli, ed., (1973) p. 207, Raven Press, New York.

39. J. P. Pinel, P. H. Van Oot, Generality of the kindling phenomenon: Some clinical implications, Can. J. Neurol. Sci. 2, 467 (1975).

40. R. G. Babington, The pharmacology of kindling in rats. In Proceedings of the Workshop on Animal Models in Psychiatry and Neurology, Rockville, Maryland, June 6-8, 1977, Pergamon Press, New York, (this volume).

41. G. V. Goddard, R. M. Douglas, Does the engram of kindling model the engram of normal long-term memory? Can. J. Neurol. Sci. 2, 385 (1975).

42. R. M. Post, K. M. Squillace, W. Sass, A. Pert, Drug sensitization and electrical kindling. In Abstracts of the Society for Neuroscience, Anaheim, California, November, 1977.

43. R. M. Post, K. M. Squillace, A. Pert, Kindling and oscillation in amygdala excitability. In Proceedings of the Annual Meeting of the Society of Biological Psychiatry, Toronto, May, 1977.

44. A. Wake, J. Wada, Frontal cortical kindling in cats, Can. J. Neurol. Sci. 2, 493 (1975).

45. R. F. Mucha, J. P. Pinel, Postseizure inhibition of kindled seizures, Exp. Neurol. 54, 266 (1977).

46. R. Adamec, Behavioral and epileptic determinants of predatory attack behavior in the cat, Can. J. Neurol. Sci. 2, 457 (1975).

47. R. Adamec, Normal and epileptic limbic system mechanisms for prolonged emotive biasing. In Proceedings of the Symposium on the Continuing Evolution of the Limbic System Concept, November 5-6, 1976, K. Livingston, O. Hornykiewicz, eds., (1977), Plenum Press, New York.

48. D. C. McIntrye, H. Reichert, State-dependent learning in rats induced by kindled convulsions, Physiol. Behav. 7, 15 (1971).

49. D. C. McIntrye, Kindling and memory: The adrenal system and the bisected brain. In Proceedings of the Symposium on the Continuing Evolution of the Limbic System Concept, November 5-6, 1976, K. Livingston, O. Hornykiewicz, eds., (1976) Plenum Press, New York, (in press).

50. M. L. Woodruff, Subconvulsive epileptiform discharge and behavioral impairment, Behav. Biol. 11, 431 (1974).

51. J. W. Papez, A proposed mechanism of emotion, Arch. Neurol. Psychiatry 38, 725 (1937).

52. P. D. MacLean, A triune concept of the brain and behavior. In The Clarence M. Hicks Memorial Lectures, 1969, T. J. Boag and D. Campbell, eds., (1973), p. 4, University of Toronto Press, Toronto.

53. E. Gellhorn, Prolegomena to a theory of the emotions, Perspect. Biol. Med. 4, 403, (1961).

54. R. G. Heath, Correlation of brain function with emotional behavior, Biol. Psychiatry 11, 463 (1976).

55. J. R. Stevens, An anatomy of schizophrenia? Arch. Gen. Psychiatry 29, 177 (1972).

56. P. Gloor, Temporal lobe epilepsy: Its possible contribution to the understanding of the functional significance of the amygdala and of its interaction with neocortical-temporal mechanisms. In The Neurology of the Amygdala (Advances in Behavioral Biology, Volume 2), B. E. Eleftheriou, ed., (1972) Plenum Press, New York.

57. M. D. Egger, J. P. Flynn, Further studies on the effects of amygdaloid stimulation and ablation in hypothalamically elicited attack behavior in cats. In Structure and Function of the Limbic System (Progress in Brain Research, Volume 27), W. R. Adey and T. Tokizane, eds., (1967) p. 165, Elsevier, Amsterdam.

210

58. B. R. Kaada, Stimulation and regional ablation of the amygdaloid complex with reference to functional representations. In The Neurobiology of the Amygdala (Advances in Behavioral Biology, Volume 2), B. E. Eleftheriou, ed., (1972), p. 205, Plenum Press, New York.

59. P. S. Arnold, R. J. Racine, R. A. Wise, Effects of atrophine, reserpine, 6-hydroxy-dopamine, and handling on seizure development in the rat, Exp. Neurol. 40, 457 (1973).

ANIMAL MODELS FOR MANIA

Dennis L. Murphy
Clinical Neuropharmacology Branch, NIMH
NIH Clinical Center 10-3S229, Bethesda, MD 20014

ABSTRACT

Animal models proposed for mania are mostly based upon hyperactivity induced in animals by drugs such as amphetamine (given alone in low doses or in combination with chlordiazepoxide), morphine, or desmethylimipramine combined with a monoamine releasing drug. The major evidence supporting the relevance of such drug-induced hyperactivity to clinical mania is its reversal by pretreatment with lithium. This agent has minimal antianxiety or neuroleptic efficacy, but possesses definite antimanic effects as well as apparent prophylactic efficacy in patients with recurrent manic-depressive cycles. Lithium also attenuates amphetamine-induced activation, euphoria and hypomanic symptomatology in man, and may block L-dopa-related hypomania in man. In animal studies, increased exploratory activity and increased activity in response to novel environments seem preferentially sensitive to lithium; similarly, psychological data suggests that lithium diminishes interest in novel stimuli and physical task initiation in man. In contrast, lithium does not antagonize various kinds of generalized hyperactivity or the stereotyped hyperactivity produced by high doses of amphetamine in rodents, nor does it appear to antagonize morphine- or alcohol-induced euphoria in man. Mania seems an especially promising disorder for the exploration of mediating mechanisms and possibly better treatments using behavioral-biological modeling approaches in animals and man because of the availability of cross-species as well as within-species models (which include similarities in behavioral changes, treatment responses and postulated neurochemical mechanisms) and because mania is a psychopathologic entity more completely expressed in objectively-assessable behavior than other clinical syndromes.

INTRODUCTION

The difficulties in postulating animal models for human psychopathology have been much discussed. In addition to the large gaps in behavioral repertoires between man and his nearest non-human primate relatives which have resulted from three million years of evolution, the dimensions of speech, thought and human socio-cultural interactions found in clinical psychopathologic syndromes and missing in primate species markedly limit the potential applicability of animal models to these complex disorders. The phylogenetic distances from man to rats, mice and other species commonly utilized in psychoactive drug studies is even greater. Furthermore, the potential for finding simple, general models for psychiatric disorders is limited by the likelihood that most human behavioral disorders are actually syndromes of heterogeneous etiology, and may depend upon interactions between one or more predisposing genetic factors and various environmental contributions.

For all of these reasons, an animal model for a human behavioral disorder seems most likely to apply only to some component, ideally a major feature, of the disorder. Most animal models can be classified according to whether they appear homologous to the human disorder in (a) symptoms, (b) postulated etiology, (c) mediating mechanisms and (d) treatment responses. Thus, although it is unlikely that any particular animal model could be demonstrated to be an exact analog of the human disorder, it is possible to evaluate the partial validity of models in terms of specific components of human psychopathology.

MANIA

Clinical features. This "affective disorder" seems more clearly expressed in overt, characteristic behavioral features than any other major psychopathologic syndrome. For example, experimental cluster analyses using symptom features of psychiatric patients have been relatively more successful in separating out patients clinically diagnosed as manic than patients with clinical diagnoses of depression or schizophrenia (1). Similarly, rater agreement on behaviorally based scales used in psychiatric patient populations is consistently highest for mania-related items compared to other factors such as depression, anxiety, psychosis or anger (2). In contrast to some depressed or schizophrenic patients who do not clearly or directly communicate their despair or delusions to the observer, the manic patient not only seeks out others to involve in his activities but also characteristically sheds inhibitions and some elements of judgment in drawing attention to his symptoms (3). Because of its direct, behaviorally-manifested symptoms, mania would thus seem to offer special opportunities to find and test with confidence animal behavioral analogs for the spontaneous disorder.

Mania is an episodic behavioral disorder which usually occurs in irregular cycles which alternate with depression in patients with bipolar affective disorders. Its symptoms include: (a) increased motor and verbal activity, (b) distractability, (c) lack of need for sleep, (d) a labile affective state sometimes characterized by elation, but often including bursts of depression, irritability and anger, (3) increased social contact, (f) increased aggressive and sexual behavior and (g) impaired judgment and impulse control. In some instances hallucinations, delusional thinking and confusion may develop, particularly as the course of the manic state peaks or becomes prolonged (3).

There is a strong familial pattern of occurrence of bipolar symptomatology. In some families, but clearly not in all, the disorder seems to be inherited as a dominant, sex-linked single gene trait (4). Acute manic episodes may be precipitated by various drugs including the tricyclic antidepressants, monoamine oxidase inhibitors and corticosteroids (5). Briefer and less severe (hypomanic) episodes may follow L-dopa or d-amphetamine administration (6,7). Individuals with previous spontaneous episodes of mania are markedly more susceptible to the development of manic or hypomanic symptoms upon exposure to these drugs (5-8).

Etiology, biological factors and drug treatment of mania. The etiology of mania remains unknown. On the basis of knowledge about the mechanisms of action of drugs which can precipitate manic episodes, a functional excess of brain dopamine and norepinephrine, perhaps together with functional deficits in brain serotonergic and cholinergic pathways, has been postulated in mania (9). The evidence is most compelling for dopamine mediating the enhanced activity, arousal and affective expression observed during the manic episode (9,10). An enhanced vulnerability to develop hypomanic or manic symptoms following increased catecholamine synthesis has been suggested in bipolar patients (5) and may be related to altered amine receptor sensitivity (11), altered biogenic amine uptake function (5) or reduced degradation of dopamine and other amines via monoamine oxidase (12,13) in these patients. Other suggestions that an "endogenous amphetamine" (phenylethylamine) or other less well studied neuromodulator substances including tryptamine, certain "false neurotransmitters," GABA and peptides such as the endorphins might contribute to human psychopathology, including mania, have only recently begun to be evaluated.

For the most part, the biochemical effects of drugs which are effective antimanic agents are in accord with these suggestions. Antagonists of catecholamine function including α-methyl-p-tyrosine, phenothiazines and, to some extent, reserpine have antimanic activity. Lithium carbonate and L-tryptophan enhance serotonergic function and have antimanic activity. Lithium carbonate also has been suggested to act as an antagonist of some dopamine- and norepinephrine-mediated cell functions. Evidence that physostigmine possesses partial antimanic efficacy has been used to suggest that reduced cholinergic function could contribute to manic symptomatology. The effects of lithium are of special note since the phenothiazines and butyrophenones have a broad spectrum of efficacy across schizophrenic, depressed and anxious patients. Lithium seems less useful in schizophrenia and anxiety, while its most striking actions are in the prevention and stabilization of individuals with recurrent manic episodes or manic-depressive cycles, and in the treatment of acute mania.

These hypotheses on biological contributions to mania and on the mechanism of action of lithium are reviewed in greater detail elsewhere (3,9,15,16). Observations and speculations on other contributions to mania from the viewpoints of clinical characteristics, psychodynamics, psychophysiologic studies and genetic investigations have also been recently reviewed (3,4,17,18).

ANIMAL MODELS FOR MANIA

Most of the suggested animal models for mania are based upon behavioral changes such as hyperactivity produced by psychoactive drugs like amphetamine, which are capable of being blocked by antimanic agents, especially lithium. Because of the postulated relative specificity of lithium as an antagonist of mania and of manic-depressive cycles in man, lithium responsivity appears to have been adopted as a major criterion for evaluating animal behaviors induced by drugs or environmental stimuli as potentially valid models for mania. Such models thus include apparently homologous symptoms and treatment responses, and may imply similar mediating mechanisms and postulated etiologic factors (e.g., the well-studied enhancement of norepinephrine and dopamine system function by amphetamine). These animal model studies have been primarily carried out in rodents. Recently, there have been some attempts to evaluate whether, in fact, lithium is capable of antagonizing in man the stimulant effects of amphetamine and other agents studied in the proposed animal models for mania.

Amphetamine-Induced Hyperactivity as a Model for Mania

Features. Hyperactivity in mice and rats following low doses of d-amphetamine, usually given together with chlordiazepoxide, has been proposed as an animal model for mania (19). Amphetamine given together with chlordiazepoxide or barbiturates markedly increases spontaneous motor activity in rats and mice (20-23). This hyperactivity is described as "fast and coordinated but characteristically repetitive and apparently compulsive walking, as well as other activities, depending on the species of animal and the kind of test environment used" (19). It has been interpreted as "activity for activity's sake" (24) and as having little in common with the stereotyped activity induced by larger doses of amphetamine which is characterized by sniffing, gnawing, head shaking and other small movements (19,25).

Antagonism by lithium and other drugs. In several early studies, lithium pretreatment was found to have no antagonistic effects on hyperactivity and stereotyped behavior induced by amphetamine used in doses of 3 to 10 mg/kg; in fact, in some instances the stereotyped behavior was prolonged in the lithium-treated animals (26-29). In contrast, increased locomotor activity induced by 0.5 mg/kg d-amphetamine in rats (30) and increased activity in mice produced by 1.18 mg/kg d-amphetamine in a novel situation (19) were both significantly attenuated by lithium pretreatment. Lithium was effective when administered either immediately before the amphetamine injection (19,30) or for one to eight days prior to the test day (30).

In the first of a series of studies of the effects of lithium on amphetamine-chlordiazepoxide hyperactivity, a single dose of 2 mEq/kg lithium chloride was found to markedly diminish the increased activity induced by d-amphetamine (1.18 mg/kg) used together with chlordiazepoxide (12.5 mg/kg) in rats (31). Activity was measured on first exposure to a Y-shaped runway. Lithium treatment did not alter the activity of saline-treated control animals, nor did it significantly reduce the smaller activity changes in the runway produced by d-amphetamine or chlordiazepoxide given alone.

In a second study in mice (32), activity in photocell-monitored cages was increased only when higher doses of the amphetamine-chlordiazepoxide mixture or of amphetamine alone were used. This increased activity was not affected by lithium chloride (3 mEq/kg) pretreatment. However, using a different behavioral assessment procedure, mixtures of 0.5 or 1.18 mg/kg d-amphetamine combined with 7.5 or 12.5 mg/kg chlordiazepoxide reliably increased activity, and this enhancement was completely prevented by lithium pretreatment. The behavioral assessment procedure utilized the hole board originally described by Boissier and Simon (33); the number of times the animals dipped their heads into the holes in three minutes was counted. The differences in results between cage and hole board activity, both in respect to the doses of drugs used to elicit increased activity and the antagonism by lithium treatment, were interpreted as resulting from an interaction of the novel hole board environment with the drugs, and the inability of the photocell monitors to discriminate different types of activity.

In a more comprehensive series of studies of amphetamine-chlordiazepoxide-related hyperactivity, Davies et al. (19) replicated earlier results demonstrating lithium antagonism of hole board head dips in mice treated with this drug combination. Photocell-monitored cage activity was not altered by lithium. Lithium, however, did antagonize enhanced Y-maze running activity in a subgroup of animals who had been exposed for eight days to open field testing and who received chlordiazepoxide (12.5 mg/kg) alone. Non-open field tested animals given chlordiazepoxide alone did not exhibit the greater activity increase observed in the "experienced" animals, nor was their behavior changed by lithium. These experiments were interpreted as indicating that lithium seemed only to affect certain types of increased activity, and did not have a suppressive effect on normal, baseline activity levels. Furthermore, enhanced exploratory-type activities, as in the open field, Y-maze, and especially the hole board situations, seemed most sensitive to the antagonist effects of lithium.

To evaluate the neurochemical basis of hyperactivity in animals treated with amphetamine and chlordiazepoxide, the catecholamine synthesis inhibitor α-methyl-p-tyrosine was studied by Davies et al. (19). Pretreatment with this agent (100 mg/kg X 3) was as effective as lithium in blocking the hole board hyperactivity. In subsequent studies by Poitou et al. (34), hole board hyperactivity induced by amphetamine and chlordiazepoxide was also found to be reduced by the serotonin precursor L-tryptophan (300 mg/kg) and by two dopamine β-hydroxylase inhibitors, fusaric acid (75 mg/kg) and FLA 63 (50 mg/kg). The serotonin synthesis inhibitor p-chlorophenylalanine was without effect. These results were interpreted within the framework of the hypotheses suggesting that there exists a balance in the central catecholamine and indoleamine neurotransmitter systems capable of modifying certain forms of hyperactivity. Lithium is also known to affect both catecholamine and indoleamine systems in ways thought to be compatible with the effects of L-tryptophan and possibly also α-methyl-p-tyrosine and the dopamine β-hydroxylase inhibitors (9,14,19,34).

Hyperactivity Induced by Other Drugs as Models for Mania

Morphine. The behavioral activation ("intense stereotyped hyperactivity") induced in mice by
morphine sulfate (25 mg/kg) was reported by Carroll and Sharp (35) to be significantly reduced by lithium chloride (5 mEq/kg X 4 days). On the basis of this observation plus indications that morphine hyperactivity in mice was reduced by other drugs said to be effective
in treating mania (chlorpromazine, haloperidol, cinanserin, methylsergide, α-methyl-p-tyrosine
and p-chlorophenylalanine), it was proposed that morphine activation might provide an animal
model for mania (35). An attempted replication of the lithium antagonism of morphine hyperactivity using identical doses of both drugs did not indicate any evidence of a lithium
effect, however (36). While the question of a morphine-related behavioral model for mania
does not appear to have been pursued further using different experimental environments,
different activity measurements or a range of drug doses, there are other indications that
lithium may interact with morphine in reducing other behaviors, including morphine self-
administration (37) and morphine-enhanced self-stimulation using electrodes placed in the
substantia nigra of rats (38).

DMI--RO4-1284: The motor hyperactivity induced in rats by combined treatment with the tricyclic antidepressant desmethylimipramine (DMI) and a benzoquinolizine, RO4-1284, was found
to be antagonized by lithium pretreatment by Matussek and Linsmayer (26). A longer duration
of lithium pretreatment tended to be more effective than a shorter period, although acute
pretreatment was markedly antagonistic. This hyperactivity was also reversed by α-methyl-p-
tyrosine and chlorpromazine. In a similar study using the same dose of DMI (20 mg/kg), but
tetrabenazine in place of RO4-1284, the drug combination-induced hyperactivity was unaffected
by lithium pretreatment (27). This possible model for mania has not been pursued in more
detailed studies, although there is one report of the nociceptive effects of another tricyclic
antidepressant, protriptyline, being reversed by lithium pretreatment (39).

Other drugs. A variety of agents which produce mood changes, including euphoria, in man, or
which produce behavioral changes other than pronounced hyperactivity in animals have been
evaluated for interactions with lithium treatment. Lithium has been reported to antagonize
alcohol drinking in rats (40). It also antagonized p-chlorophenylalanine-induced aggressive
and sexual behavior, but not the increased open-field activity produced by this drug in rats
(41). It did not affect the behavioral changes which follow yohimbine administration to
dogs (42) or 5-hydroxytryptophan to mice (43). The hyperactivity in rodents produced by MAO-
inhibiting antidepressants such as tranylcypromine (44,45) and nialamide (46) is enhanced
rather than antagonized by lithium. The enhancement of open-field activity produced by the
MAO B-inhibiting drug, pargyline, was not influenced by lithium (47). None of these behaviors
have been specifically proposed as mania analogs nor comprehensively evaluated for their
possible relevance to manic hyperactivity.

Non-Drug-Induced Behaviors as Animal Models for Mania

In the experiments described above, lithium treatment alone in doses which blocked various
drug-induced behaviors was uniformly noted to be ineffective in altering basal levels of
spontaneous activity and was said not to produce any apparent sedation or ataxia. Nonetheless,
following Cade's (48) original observation that lithium-treated guinea pigs appeared lethargic,
which led to the suggestion that lithium might possess antimanic efficacy, there have been
occasional reports suggesting that lithium reduced some forms of locomotor activity in sub-
toxic doses. While these reports are exceptions and stand in contrast to the frequently-
expressed suggestions of minimal or no effect of lithium on most objectively-evaluated
spontaneous behavior [ranging from activity wheels to self-stimulating behavior (24,49)],
these reports are of some interest.

Johnson and Wormington (50), for example, observed a reduction in the frequency of rearing
behavior in rats given 2 mEq/kg lithium chloride. Rearing (standing on two legs) has been
interpreted as a form of exploratory behavior which occurs in response to environmental
stimulation, and the effect of lithium was suggested to represent reduced stimulus control
of behavior. Lithium's effects in this situation did not seem explainable on the basis of
any general impairment in muscular function, as rearing height was unaffected in the lithium-
treated animals.

There exist a number of suggestions that lithium treatment rendered animals apparently less
reactive to test situations (50,51). Lithium has also been found to differentially affect
animals tested for activity in pairs versus alone, suggesting that social interaction is a
variable which needs to be considered in all of these studies (52).

Open-field activity behavior was found to be reduced by lithium pretreatment in two studies
(47,53). This change was interpreted as congruent with suggestions from the studies of rearing behavior in rats that lithium might more specifically antagonize exploratory behaviors.

Lithium has also been reported to antagonize aggressive behavior in several species (54).

Territorial aggression, foot shock-induced fighting behavior, and drug-induced aggressiveness were all diminished by lithium pretreatment (41,51,55,56). In the foot shock experiments, lithium produced no apparent changes in thresholds for flinch and jump responses (51,56). These studies of aggressive behavior are of interest because of the large contribution of aggression-related behaviors to the clinical syndrome of mania.

There do not appear to be any explicit suggestions for models for mania from non-human primate studies. In fact, it is of some interest that separation-induced behavioral changes in non-human primates, which some have considered models for depression and other forms of psycho-pathology, have not included manic-like phenomena. Mania, according to one line of theo-retical reasoning, has been postulated to be a more severe form of the same basic process as depression in a manic-depressive affective disorder continuum (57); if separation was a key element in this process, manic-like phenomena might have been predicted to occur. Although increased vocalization and agitated behavior occurs as part of the initial, "protest stage" response to maternal and peer separation, other manic-like behaviors apparently are not present (58). Further evaluation of the separation model in non-human primates with this and other hypotheses for mania (3) in mind might be of interest, however.

EVALUATION OF PROPOSED ANIMAL MODELS FOR MANIA BY STUDIES IN MAN

While the questions of how closely the proposed animal models for mania approach the spontane-ous disorder in man has not been explicitly examined, there exists some information of rele-vance from drug interaction studies in man. The largest amount of such data is from investiga-tions of the effects of lithium pretreatment on responses to amphetamine, morphine and some other drugs, primarily accomplished in psychiatric patient groups and normals. In addition, there is a small amount of information on the effects of lithium on behaviors other than those found in manic patients which is of interest in regard to the non-drug-induced animal behaviors which are affected by lithium treatment.

Amphetamine-lithium interactions in man. Flemembaum (59) provided case reports of several amphetamine users who noted diminished stimulant effects of the drug while they were receiv-ing lithium carbonate. A double-blind examination of the effects of lithium pretreatment on amphetamine-induced subjective and behavioral changes in patients hospitalized for depression was reported by van Kammen and Murphy (7). Moderate to marked (3-4 fold) elevations in clusters of self-rated items reflecting activation and euphoria, and reductions in items re-flecting depression, were observed to peak approximately two hours following treatment with d-amphetamine (30 mg). These changes were markedly attenuated when these same individuals were restudied during lithium carbonate treatment (0.9 - 2.1 g/day) for ten days or longer. Smaller self-rated changes occurred with l-amphetamine (30 mg) administration, and these were similarly reduced by lithium. Behavioral ratings made by staff observers were less sensitive than self-ratings to the amphetamine effects, but statistically significant correlations be-tween the patient's self-rating and the staff's behavioral observations were obtained. The amphetamine responses were examined on two separate days both before and after lithium, and proved to be highly replicable within the same individual, although large differences were observed between individuals. Brief periods of increased talkativeness, intrusiveness, hyper-activity and emotional lability considered typical of hypomanic behavior occurred in response to d-amphetamine in two individuals, who were two of the three individuals included in the study with histories of previous hypomanic or manic episodes. These overt hypomanic symptoms were markedly reduced by lithium in these individuals.

In addition to reductions in subjective and behavioral responses to d- and l-amphetamine, lithium pretreatment also led to an attenuation of cortical average evoked response changes produced in these individuals by amphetamine (60). Lithium pretreatment also modified changes in the urinary excretion of the catecholamine metabolite, 3-methoxy 4-hydroxyphenyl glycol produced by d-amphetamine (61). Insomnia, including a reduction in REM sleep, pro-duced by amphetamine was not affected by lithium, however (62). Peak plasma levels of amphetamine and amphetamine half-lives in plasma were not altered by lithium in this study, indicating that the effects of lithium were not due to a direct effect on amphetamine metabolism (63).

These studies indicate that despite known differences in amphetamine and lithium metabolism found in rats, mice and man, some attenuation of the behavioral effects of low doses of am-phetamine (approximately 0.5 mg/kg for both man and rodents) can be observed in both species. The provocation of mild and brief but nonetheless typical symptoms of mania in two individuals receiving amphetamine, and the blockade or reduction of these hypomanic symptoms by lithium (7) offers additional support for proposing that this drug interaction may be relevant to the spontaneous disorder. Furthermore, evidence from both the rat (19) and man (64) that the catecholamine synthesis inhibiting drug, α-methyl-p-tyrosine, is also capable of reducing amphetamine-induced stimulant effects suggests that the well-studied effects of amphetamine on the dopamine and norepinephrine transmitter systems may mediate these effects in both species. The demonstrated partial efficacy of α-methyl-p-tyrosine as an antimanic agent (65) adds further weight to these postulated interrelationships between the amphetamine model and spontaneous mania.

Morphine-lithium interactions in man. Jasinski and coworkers (66) examined the effects of lithium pretreatment (0.9 - 1.5 g/d) on the subjective, behavioral and pupillary responses to morphine (7.5 and 15 mg) in narcotic-abusing prisoners. No significant antagonism by lithium of the euphoric, subjective or miotic effects of morphine occurred. The only difference observed was a small increase in the subjects' self-rating of liking the morphine-lithium combination better than morphine alone. The results were interpreted as arguing against a common mechanism for drug-induced euphorias.

It should also be noted that there are large differences between the morphine doses used to produce hyperactivity in rodents (25 mg/kg) and those required to provide antagonism of pain responses in animals (3-5 mg/kg) and euphoria in man (0.2 mg/kg), and that morphine within the dosage ranges tolerated in man does not lead to increased motor activity or any other manic-like symptoms. Nonetheless, interest in morphine euphoria has been rekindled by the discovery of endorphins, peptides with morphine-like activity found in mammalian brain tissue, which possess a range of behavioral effects in animals from tranquilization (α endorphin) and catatonia (β endorphin) to irritability and agitation (γ endorphin) (67-69). The opiate antagonist naloxone is currently being studied in schizophrenic, manic and other psychiatric patient groups for possible therapeutic effects. In addition, it has been hypothesized that lithium may act through modification of the affinity of opiate receptors for an endorphin (70), a suggestion based on the sodium and lithium modifications of the configuration of this receptor in brain (71).

Interactions between other drugs and lithium in man. There do not appear to be systematic studies of desmethylimipramine or other tricyclic antidepressant drugs studied in the same patients with and without lithium pretreatment, nor of MAO-inhibiting antidepressants and lithium, although combinations of these drugs have been used in depressed patient populations. It seems clear that lithium treatment does not block the occasional precipitation of a manic episode during tricyclic antidepressant treatment (5). Lithium did not appear to attenuate the "high" induced by 95% ethanol (1.3 ml/kg) in normal subjects (72). There is a case report indicating that persistent hypomania resulting from L-dopa treatment of a patient with Parkinson's disease responded to lithium carbonate treatment (73). A comparison of the effects of lithium on various drug-induced states in animals and man is presented in Table 1.

TABLE 1   Antagonism by Lithium of Drug-Induced States in Animals and Man

State	Lithium Effect	
	Animals	Man
Amphetamine (low dose) Hyperactivity, Activation	Blocked	Reduced
Amphetamine (high dose) Stereotypic Behavior	Unchanged or Prolonged	--
Morphine Excitement, Euphoria	Reduced or Unchanged	Unchanged
DMI + Tetrabenazine or RO4-1284 Hyperactivity	Reduced or Unchanged	--
MAO-Inhibitor Hyperactivity	Enhanced	--
L-Dopa Hyperactivity, Hypomania	--	Possibly Reduced
Ethanol "High"	--	Unchanged

Interactions between specific behaviors and lithium in man: In animals, lithium does not antagonize all forms of drug-induced hyperactivity (such as that measured in photocell-monitored cages) but rather seems to have preferential effects on exploratory behavior or behavior in novel settings. In patients or the occasional normal subjects given lithium in a non-systematic fashion in moderate doses, the drug has most commonly been reported to have minimal psychological, sedative or other effects, in contrast to its marked effects in patients with mania and some other forms of psychopathology. Some tiredness and muscular heaviness was noted in one study using moderate lithium doses (925 mg/day) (74). However, in large doses of 1850 mg/day, lithium treatment led to greater number of physiological and psychological symptoms including malaise, increased mental effort required to initiate physical tasks, passivity, irritability and decreased responses to environmental stimuli (74). These subjective impressions were obtained from three researchers who received lithium carbonate under non-blind conditions.

In the only comprehensive, double-blind study of the effects of lithium carbonate on the psychological state of normals, Judd and coworkers (75) reported significant self-rated increases in lethargy, dysphoria, and muddled, less-clearheaded cognition. Lithium serum levels averaged 0.91 mEq/L (range 0.7 - 1.37 mEq/L); the lithium dosage used was not reported. Of particular interest to the animal findings, their subjects also indicated that "they did not want to be bothered by environmental demands, nor to have to deal with other people, nor to have to respond to new stimuli." Furthermore, on an inkblot test, the subjects selected significantly more "popular" and commonplace responses and had longer delays in responding to the inkblots than they did during the placebo periods. The discrepancy between these observations and those from patient populations may be in part due to the use of self-ratings in this study, since observer's ratings have most commonly been used in the patient studies and these detected no lithium-placebo differences in this study.

A Genetic Model for Vulnerability to Mania in Man and Its Relation to Animal Behavior

Reduced activity of a major enzyme in the degradatory pathways for biogenic amines, monoamine oxidase (MAO) has been reported in patients with bipolar affective disorder (12,13). The lack of change in this reduced enzyme activity in patients studied while manic, depressed or well (76), together with the results of twin studies in normals and patients indicating that most of the inter-individual variability in platelet MAO activity seems attributable to genetic factors (77), suggests that these enzyme differences may represent stable characteristics of bipolar patients. Since the well first degree relatives of bipolar patients also have reduced platelet MAO activity (13), it has been hypothesized that reduced activity of this enzyme might represent a familial vulnerability factor for bipolar affective disorder (and consequently for mania, since mania is the differentiating feature for the bipolar-unipolar affective disorder dichotomy).

Because of the well-known difficulties in evaluating biological factors in patient groups due to drug treatment, hospitalization and other possible bases for artifacts, these studies of the significance of MAO activity differences for psychopathology have been pursued recently in normal populations. Individuals selected from a large normal populations on the sole basis of their having platelet MAO activities in the lowest decile were found to have a number of indications of significantly greater psychosocial difficulties, including a higher incidence of consulting a psychiatrist or psychologist and an eight-fold higher familial incidence of suicide (78). Moreover, individuals with low platelet MAO activity were found to have higher MMPI mania subscale scores and higher Zuckerman Sensation-Seeking Scale scores (79) -- the latter test including many items indicative of hypomanic behavior (80). Further work has identified associations between low platelet MAO activity, suicide attempts and augmenting characteristics on cortical average evoked response (AER) measures (81). AER augmenting has previously been reported to be a characteristic of patients with bipolar affective disorders, to correlate positively with Zuckerman Sensation-Seeking scores and to be diminished during treatment with lithium carbonate (18,82).

These results in patients with bipolar affective disorders and normal individuals with reduced platelet MAO activity are of interest because an earlier study examining rhesus monkey behaviors over a four-month period (to obtain an estimate of relatively stable behavioral characteristics of the individual animals) reported a number of significant correlations between individual monkey behavior profiles and their platelet MAO activities (83). Social contact behavior, ambulatory movement as well as specific behaviors such as play in males were negatively correlated with platelet MAO activity. In contrast, opposite behaviors such as inactivity, time spent alone and self-grooming were positively correlated with MAO activity. While by no means reflective of "psychopathology" in these primates, the behavioral profiles associated with differences in platelet MAO activities somewhat resemble the findings in man, and seem worthy of further study.

DISCUSSION

To some extent, the brief periods of psychomotor activation, euphoria, reduced sleep and apparent hypomania following low doses of amphetamine and L-dopa in man can be considered within-species models for some components of the syndrome of mania. These within-species behavioral homologs have not yet been comprehensively evaluated for their validity as mania models. For example, it would be of definite interest to utilize some of the suggestions from animal studies regarding lithium's dampening effects on the external stimulus control of behavior (84) to ask whether the clinically evident and objectively ratable "distractibility" component of mania (2,3) was evident in the course of amphetamine or L-dopa associated behavioral changes, using recently devised techniques for evaluating human information processing and storage mechanisms (85,86). This approach seems especially cogent in regard to the concepts that vulnerability to bipolar affective disorders as well as some of their symptomatology may coincide with sensation (or stimulus) seeking behaviors. These within-species models also help bridge the gap between rodents and man in evaluating the proposed amphetamine-lithium interactional models for mania which are most readily studied for basic mechanisms in lower species. They offer further support for the conceptions emphasizing the role of dopamine in the psychomotor activation observed in both mania and these drug-induced hyperactivity states (3,6,9,10,77), and also support the relevance of lithium's antagonistic effects on dopamine and norepinephrine processes via both direct actions as well as via serotonergic system interactions (14,30).

The study of increased exploratory activities as models for mania in rodents has primarily been approached through the use of stimulant drugs like amphetamine, and the antagonism of this increased activity by lithium. However, many potentially informative drug combinations have yet to be studied using suitably complex environmental situations and taking into account such features as novelty effects and social interactional contributions to the elicited behaviors. Sex and age differences have seldom been evaluated or even noted in many reports. In particular, strain differences have not yet been comparatively studied. As there are now several well-studied examples of rodent strains differing in (a) behavioral features such as spontaneous activity levels; (b) pharmacologic response differences, including response differences to amphetamine; and (c) brain monoamine enzyme and metabolism differences, there are many rich opportunities for the development of a variety of animal models for mania which may further the understanding of some features of altered behavior and its treatment responsivity in animals and man (87).

Almost all of the emphasis in the literature and in this review has been on the manic syndrome itself, and its precipitation by drugs. The fact that mania occurs in bipolar patients who generally exhibit periodic cycles in psychomotor activity and mood cannot be neglected. Although man has evolved beyond mating periods and has engineered environments which no longer make hibernation inactivity useful, he clearly maintains many cyclic phenomena. It is not inconceivable that some manic episodes might be hypothesized to represent vestigial activity cycle surges akin to the several week mating periods prominent in many species which are characterized by continuous sexual activity as well as general hyperactivity, heightened social contact, diminished sleep and aggressive behavior. It would be of interest to evaluate whether lithium might affect other cyclic behavioral, hormonal and biological phenomena in animals and man (88). The effects of lithium on neurohormonal responses in man have not yet been comprehensively examined, although slight effects on the amplitude and circadian rhythm of corticosteroid secretion and on thyroid function have been reported (15,16). Lithium has also been found to have a slight but significant slowing effect on the activity cycle of the desert rat and to alter the period length of the rhythmic petal movements of Kalanchoe flowers (89). The possibilities of exploring mania and manic-depressive phenomena from the viewpoint of cyclic phenomena in animals seem wide open. Even in regard to the amphetamine model, it is of note that the post-amphetamine withdrawal period in animals, which is similar to the "down" or depressed state that follows amphetamine in man, has been suggested as a pharmacological model for depression, with some advantages over the reserpine depression model (90). In our study of amphetamine-lithium interactions in man, it had been observed that lithium not only attenuated the amphetamine antidepressant response, but also tended to diminish the elevation in afternoon depression ratings observed on the days amphetamine (but not placebo) was given (7). This post-amphetamine withdrawal has apparently not yet been assessed in animal studies of lithium-amphetamine interactions. The amphetamine "down" may be of pertinence to the questions of the relationship of mania to manic-depressive cycles, because it has frequently been noted that depressive periods commonly follow directly after manic episodes, and it has been hypothesized that lithium may act to prevent depression in patients with manic-depressive cycles chiefly by suppressing mania (91).

REFERENCES

1. B.S. Everitt, A.J. Gourlay and R.E. Kendall, An attempt at validation of traditional psychiatric syndromes by cluster analysis, Br. J. Psychiatry, 119, 399 (1971).

2. D.L. Murphy, A. Beigel, H. Weingartner and W.E. Bunney, Jr., The quantitation of manic behavior. In Modern Problems in Pharmacopsychiatry, P. Pichot, ed. (1974), 7, 202, S. Karger, Basel.

3. D.L. Murphy, F.K. Goodwin and W.E. Bunney, Jr., The psychobiology of mania. In American Handbook of Psychiatry, S. Arieti, ed. (in press), Basic Books, New York.

4. R.R. Fieve, J. Mendelewicz, J.D. Rainer and J.L. Fleiss, A dominant X-linked factor in manic-depressive illness: Studies with color blindness. In Genetic Research in Psychiatry, D. Rosenthal and H. Brill, eds. (1975), p. 241, Johns Hopkins Univ. Press, Baltimore.

5. W.E. Bunney, Jr., F.K. Goodwin, D.L. Murphy, K.M. House and E.K. Gordon, The "switch process" in manic-depressive illness. II. Relationship to catecholamines, REM sleep and drugs. Arch. Gen. Psychiatry 27, 304 (1972).

6. D.L. Murphy, H.K.H. Brodie, F.K. Goodwin and W.E. Bunney, Jr., L-dopa: Regular induction of hypomania in bipolar manic-depressive patients. Nature 229, 135 (1971).

7. D.P. van Kammen and D.L. Murphy, Attenuation of the euphoriant and activating effects of d- and l-amphetamine by lithium carbonate treatment. Psychopharmacologia 44, 215 (1975).

8. D.L. Murphy, The behavioral toxicity of monoamine oxidase inhibiting antidepressants. In Advances in Pharmacology and Chemotherapy, R.J. Schnitzer, ed. (1976), 81, 178, Academic Press, New York.

9. D.L. Murphy and D.E. Redmond, The catecholamines: Possible role in affect, mood and emotional behavior in man and animals. In Catecholamines and Behavior, A.J. Friedhoff, ed. (1975), p. 73, Plenum Press, New York.

10. D.L. Murphy, L-dopa, behavioral activation and psychopathology. In Neurotransmitters, I.J. Kopin, ed. (1972), Res. Publ. A.R.N.M.D., 50, 472.

11. W.E. Bunney, Jr. and D.L. Murphy, Strategies for the study of neurotransmitter receptor function in man. In Pre- and Post-Synaptic Receptors, E. Usdin and W.E. Bunney, Jr., eds. (1975), p. 283, Marcel Dekker, New York.

12. D.L. Murphy and R. Weiss, Reduced monoamine oxidase activity in blood platelets from bipolar depressed patients. Am. J. Psychiatry 128, 1351 (1972).

13. J.F. Leckman, E.S. Gershon, A.S. Nichols and D.L. Murphy, Reduced platelet monoamine oxidase activity in first degree relatives of individuals with bipolar affective disorders: A preliminary report. Arch. Gen. Psychiatry (1977).

14. W.E. Bunney, Jr. and D.L. Murphy (1976) The Neurobiology of Lithium, Neurosciences Res. Prog. Bull. 14, 111.

15. M. Schou, Possible mechanisms of action of lithium: Approaches and perspectives. Biochem. Soc. Trans. 1, 81 (1973).

16. S. Gershon and B. Shopsin (1973) Lithium, Plenum Press, New York.

17. G. Winokur, P.J. Clayton and T. Reich (1969) Manic Depressive Illness, C.V. Mosby Co., St. Louis.

18. M. Buchsbaum, F.K. Goodwin, D.L. Murphy and G. Borge, Average evoked responses in affective disorders. Am. J. Psychiatry 128, 19 (1971).

19. C. Davies, D.J. Sanger, H. Steinberg, M. Tomkiewicz and D.C. U'Prichard, Lithium and α-methyl-p-tyrosine prevent "manic" activity in rodents. Psychopharmacologia 36, 263 (1974).

20. R. Rushton and H. Steinberg, Mutual potentiation of amphetamine and amylobarbitone measured by activity in rats. Br. J. Pharmacol. 21, 295 (1963).

21. R. Rushton and H. Steinberg, Combined effects of chlordiazepoxide and dexamphetamine on activity of rats in an unfamiliar environment. Nature 211, 1312 (1966).

22. H. Steinberg and M. Tomkiewicz, Animal behaviour models in psychopharmacology. In Chemical Influences on Behaviours, R. Porter and J. Birch, eds. (1970), CIBA Foundation Study Group, 35, p. 199, Churchill, London.

23. M. Dorr, D. Joyce, R.D. Porsold, H. Steinberg, A. Summerfield and M. Tomkiewicz, Persistence of dose related behaviour in mice. Nature 228, 469 (1971).

24. H. Steinberg, Animal models for behavioural and biochemical studies on the effects of lithium salts. Biochem. Soc. Trans. 1, 38 (1973).

25. A. Randrup and I. Munkvad, Stereotyped activities produced by amphetamine in several animal species and man. Psychopharmacologia 11, 300 (1967).

26. N. Matussek and M. Linsmayer, The effect of lithium and amphetamine on desmethyl-imipramine-RO 4-1284 induced motor hyperactivity. Life Sci. 7, 371 (1968).

27. P.S. D'Encarnacao and K. Anderson, Effects of lithium pretreatment on amphetamine and DMI tetrabenazine produced psychomotor behavior. Dis. Nerv. Syst. 31, 494 (1970).

28. S. Lal and T.L. Sourkes, Potentiation and inhibition of the amphetamine stereotypy in rats by neuroleptics and other agents. Arch. Int. Pharmacodyn. 199, 289 (1972).

29. H. Ozawa and T. Miyauchi, Potentiating effect of lithium chloride on methylamphetamine-induced stereotypy in mice. Eur. J. Pharmacol. 41, 213 (1977).

30. D.S. Segal, M. Callaghan and A.J. Mandell, Alterations in behaviour and catecholamine biosynthesis induced by lithium. Nature 254, 58 (1975).

31. C. Cox, P.E. Harrison-Read, H. Steinberg and M. Tomkiewicz, Lithium attenuates "manic" activity in rats. Nature 232, 336 (1971).

32. D.C. U'Prichard and H. Steinberg, Selective effects of lithium on two forms of spontaneous activity. Br. J. Pharmacol. 44, 349 (1972).

33. J.R. Boissier and P. Simon, La reaction d'exploration chez la souris. Therapie 17, 1225 (1962).

34. P. Poitou, R. Boulu and C. Bohoun, Effect of lithium and other drugs on the amphetamine chlordiazepoxide hyperactivity in mice. Experientia 31, 99 (1975).

35. B.J. Carroll and P.T. Sharp, Rubidium and lithium: Opposite effects on amine-mediated excitement. Science 172, 1355 (1971).

36. I. Sanghvi and S. Gershon, Rubidium and lithium: Evaluation as antidepressant and anti-manic agents. Res. Comm. Chem. Path. Pharmacol. 6, 293 (1973).

37. M. Tomkiewicz and H. Steinberg, Lithium treatment reduces morphine self-administration in addict rats. Nature 252, 227 (1974).

38. J.M. Liebman and D.S. Segal, Lithium differentially antagonizes self-stimulation facilitated by morphine and (+)-amphetamine. Nature 260, 161 (1976).

39. L. Saarnivaara and M.L. Mattila, Comparison of tricyclic antidepressants in rabbits: antinocioception and potentiation of the noradrenaline pressor responses. Psychopharmacologia 35, 221 (1974).

40. J.D. Sinclair, Lithium-induced suppression of alcohol drinking by rats. Med. Biol. 52, 133 (1974).

41. M.H. Sheard, Behavioral effects of p-chlorophenylalanine; inhibition by lithium. Commun. Behav. Biol. Part A, 5, 71 (1970).

42. H. Geyer, I. Sanghvi and S. Gershon, The aggressive monoamines. Psychopharmacologia 28, 107 (1973).

43. I.P. Kiseleva and I.P. Lapin, Antagonistic effect on lithium carbonate in 5-hydroxy-tryptophan-induced head-twitches in mice. Pharmacol. Res. Commun. 1, 108 (1969).

44. D.G. Grahame-Smith and A.R. Green, The role of brain 5-hydroxytryptamine in the hyper-activity produced in rats by lithium and monoamine oxidase inhibition. Br. J. Pharmacol. 52, 19 (1974).

45. A. Judd, J. Parker and F.A. Jenner, The role of noradrenaline, dopamine and 5-hydroxy-tryptamine in the hyperactivity response resulting from the administration of tranylcypromine to rats pretreated with lithium or rubidium. Psychopharmacologia 42, 73 (1975).

46. J.B. Lassen and R.F. Squires, Potentiation of nialamide-induced hypermotility in mice by lithium and the 5-HT uptake inhibitors chlorimipramine and FG 4963. Neuro-pharmacology 15, 665 (1976).

47. D.F. Smith, Biogenic amines and the effect of short term lithium administration on open field activity in rats. Psychopharmacologia 41, 295 (1975).

48. J.F.J. Cade, Lithium salts in the treatment of psychotic excitement. Med. J. Aust. 36, 349 (1949).

49. T.A. Ramsey, J. Mendels, C. Hamilton and A. Frazer, The effect of lithium carbonate on self-stimulating behavior in the rat. Life Sci. 11, 773 (1972).

50. F.N. Johnson and S. Wormington, Effects of lithium on rearing activity in rats. Nature 235, 159 (1972).

51. M.H. Sheard, Effect of lithium on foot shock aggression in rats. Nature 228, 284 (1970).

52. L.A. Syme and G.J. Syme, Effects of lithium chloride on the activity of rats tested alone or in pairs. Psychopharmacologia 29, 85 (1973).

53. D.F. Smith and H.B. Smith, The effect of prolonged lithium administration on activity, reactivity, and endurance in the rat. Psychopharmacologia 30, 83 (1973).

54. M.L. Weischer, Uber die antigressive wirkung von lithium. Psychopharmacologia 15, 245 (1969).

55. M.H. Sheard, Lithium in the treatment of aggression. J. Nerv. Ment. Dis. 160, 108 (1975).

56. B.S. Eichelman, Jr. and N.B. Thoa, The aggressive monoamines. Biol. Psychiatry 6, 143 (1973).

57. J.H. Court, Manic-depressive psychosis: An alternative conceptual model. Br. J. Psychiatry 114, 1523 (1968).

58. S.J. Suomi, Factors affecting responses to social separation in rhesus monkeys. In Animal Models in Human Psychobiology, G. Serban and A. Kling, eds. (1976), p. 9, Plenum Press, New York.

59. A. Flemenbaum, Does lithium block the effects of amphetamine? A report of three cases. Am. J. Psychiatry 131, 7 (1974).

60. M.S. Buchsbaum, D.P. van Kammen and D.L. Murphy, Individual differences in average evoked responses to d- and l-amphetamine with and without lithium carbonate in depressed patients. Psychopharmacology 51, 129 (1977).

61. H. Beckmann, D.P. van Kammen, F.K. Goodwin and D.L. Murphy, Urinary excretion of 3-methoxy-4-hydroxyphenylglycol in depressed patients: Modifications by amphetamine and lithium. Biol. Psychiatry 11, 377 (1976).

62. J.C. Gillin, D.P. van Kammen, J. Graves and D.L. Murphy, Differential effects of d- and l-amphetamine on the sleep of depressed patients: Unaltered by lithium carbonate. Life Sci. 17, 1233 (1975).

63. M. Ebert, D.P. van Kammen and D.L. Murphy, Plasma levels of amphetamine and behavioral response. In Pharmacokinetics of Psychoactive Drugs: Blood Levels and Clinical Response, L. Gottschalk, ed. (1976) p. 157, Spectrum Pubs., New York.

64. L.-E. Jonsson, E. Anggard and L.-M. Gunne, Blockade of intravenous amphetamine euphoria in man. Clin. Pharmacol. Ther. 12, 889 (1971).

65. H.K.H. Brodie, D.L. Murphy, F.K. Goodwin and W.E. Bunney, Jr., Catecholamines and mania: The effect of alpha-methyl-para-tyrosine on manic behavior and catecholamine metabolism. Clin. Pharmacol. Ther. 12, 218 (1971).

66. D.R. Jasinski, J.G. Nutt, C.A. Haertzen and J.D. Griffith, Lithium: Effects on subjective functioning and morphine-induced euphoria. Science 195, 582 (1977).

67. A. Goldstein, Opioid peptides (endorphins) in pituitary and brain. Science 193, 1081 (1976).

68. F. Bloom, D. Segal, N. Ling, eand R. Guillemin, Endorphins: Profound effects in rats suggest new etiological factors in mental illness. Science 194, 630 (1976).

69. Y.F. Jacquet and N. Marks, The C-fragment of β-lipotropin: An endogenous neuroleptic or antipsychotogen? Science 194, 632 (1976).

70. R. Byck, Peptide transmitters: A unifying hypothesis for euphoria, respiration, sleep, and the action of lithium. Lancet 2, 72 (1976).

71. C.B. Pert and S.H. Snyder, Opiate receptor binding of agonists and antagonists affected differentially by sodium. Mol. Pharmacol. 10, 868 (1974).

72. L.L. Judd, R.B. Hubbard, L.Y. Huey, P.A. Attewell, D.S. Janowsky and K.I. Takahashi, Lithium carbonate and ethanol induced "highs" in normal subjects. Arch. Gen. Psychiatry 34, 463 (1977).

73. R.S. Ryback and R.S. Schwab, Manic response to levodopa therapy. Report of a case. New Eng. J. Med. 285, 788 (1971).

74. M. Schou, Lithium in psychiatric therapy and prophylaxis. J. Psychiatr. Res. 6, 67 (1967).

75. L.L. Judd, B. Hubbard, D.S. Janowsky, L.Y. Huey and P.A. Attewell, The effect of lithium carbonate on affect, mood, and personality of normal subjects. Arch. Gen. Psychiatry 34, 346 (1977).

76. D.L. Murphy and R.J. Wyatt, Enzyme studies in the major psychiatric disorders: I. Catechol-O-methyl-transferase, monoamine oxidase in the affective disorders, and factors affecting some behavior-related enzyme activities. In The Biology of the Major Psychoses: A Comparative Analysis, D.X. Freedman, ed. (1975) p. 277, Raven Press, New York.

77. D.L. Murphy, R. Belmaker and R.J. Wyatt, Monoamine oxidase in schizophrenia and other behavioral disorders. J. Psychiatr. Res. 11, 221 (1974).

78. M. Buchsbaum, R.D. Coursey and D.L. Murphy, The biochemical high-risk paradigm: Behavioral and familial correlates of low platelet monoamine oxidase activity. Science 194, 339 (1976).

79. D.L. Murphy, R.H. Belmaker, M. Buchsbaum, N.F. Martin, R. Ciaranello and R.J. Wyatt, Biogenic amine-related enzymes and personality variations in normals. Psychol. Med. 7, 149 (1977).

80. M. Zuckerman, The sensation-seeking motive. In Progress in Experimental Personality Research, B. Maher, ed. (1974), 7, 77, Academic Press, New York.

81. M. Buchsbaum, R.J. Haier and D.L. Murphy, Suicide attempts, platelet monoamine oxidase and the average evoked response. Acta Psychiatr. Scand. (in press).

82. M. Zuckerman, T. Murtaugh and J. Siegel, Sensation seeking and cortical augmenting-reducing. Psychopharmacology 11, 535 (1974).

83. D.E. Redmond, Jr. and D.L. Murphy, Behavioral correlates of platelet monoamine oxidase (MAO) activity in rhesus monkeys. Psychosom. Med. 37, 80 (1975).

84. F.N. Johnson, Chlorpromazine and lithium. Dis. Nerv. Syst. 33, 235 (1972).

85. H. Weingartner, H. Miller and D.L. Murphy, Affect state dependent learning. J. Abnormal Psychol. (in press).

86. H. Weingartner, B. Hall, D.L. Murphy and W. Weinstein, Imagery, affective arousal and memory consolidation. Nature 263, 311 (1976).

87. D.L. Murphy, Animal models for human psychopathology: Observations from the vantage point of clinical psychopharmacology. In Animal Models in Human Psychobiology, G. Serban and A. Kling, eds. (1976), p. 265, Plenum Press, New York.

88. W.E. Bunney, Jr. and D.L. Murphy, Switch processes in psychiatric illness. In Factors in Depression, N.S. Kline, ed. (1974), p. 139, Raven Press, New York.

89. W. Engelmann, A slowing down of circadian rhythms by lithium ions. J. Zeitschrift Fur Naturforschung 28, 733 (1973).

90. V. Seltzer and S.R. Tonge, Methylamphetamine withdrawal as a model for the depressive state: Antagonism of post-amphetamine depression by imipramine. J. Pharm. Pharmacol. 27, 16P (1975).

91. A. Kukopulos and D. Reginaldi, Does lithium prevent depressions by suppressing mania? Int. Pharmacopsychiatry 8, 152 (1973).

DISCUSSION

Dr. Stricker wondered how much of the decrease in activity of rats given lithium is the result of the animals becoming sick and thereby decreasing their activity. *Dr. Murphy* agreed that the toxic effects of lithium might be involved but he did not think it could be an overall explanation of the results. Other workers have shown that certain motor measures (e.g., atoxia, flinch-jump threshold) are not altered by lithium.

GENERAL DISCUSSION

Dr. Hanin opened the discussion by pointing out that some of the contributors (Drs. Sulser, Zigmond, Breese) used 6-hydroxydopamine as a specific destroyer of dopaminergic or adrenergic neurons whereas Dr. Butcher maintains that 6-hydroxydopamine is non-specific, that it will destroy anything it contacts.

Dr. Zigmond insisted that there is much reason to believe that when 6-hydroxydopamine is administered properly it is a very useful and relatively specific tool. He then offered various types of evidence for the specificity of 6-hydroxydopamine. Histological evidence shows via conventional microscopy a normal brain with only a cannula track. If the 6-hydroxydopamine is administered in too large a volume or too quickly, one does get a large hole. If a section is taken from the same brain and stained for histochemical fluorescence, there is a marked loss of fluorescent terminals - indicating rather marked specificity.

Dr. Zigmond continued with the comment that specificity is also shown biochemically: there is a depletion of norepinephrine and dopamine, but not of other neurotransmitters. He mentioned also Dr. Sulser's results where desmethylimipramine blocked the effects of 6-hydroxydopamine on norepinephrine terminals; this, he feels, is completely incompatible with the non-specific poison effect proposed by Butcher, et al.

With regard to behavioral effects Dr. Zigmond pointed out that 6-hydroxydopamine may be administered intraventricularly (as done by Dr. Stricker), or intracisternally (as done by Dr. Breese), or directly into the substantia nigra (as done by Drs. Ungerstedt, Marshall and others)- and the behavioral effects are more or less the same. Since dopamine depletion produced by each of these procedures is about the same and the behavioral effects are about the same, it seemed to Dr. Zigmond the behavioral changes are the result of dopamine depletion rather than non-specific effects.

Dr. Breese cautioned that one had to be very careful in the use of 6-hydroxydopamine. Dr. Zigmond had been talking about its use in rats; Dr. Breese had collaborated in primates. Here, intraventricular administration results in a great deal of non-specific damage to ventricular areas; this is not observed in the rat. Dr. Breese speculated that it might be due to the high dosage required in the monkey. In contrast, very small amounts of 6-hydroxydopamine administered to the monkey intracerebrally produces a moderate Parkinsonism.

Dr. Hanin expressed concern that in some histological sections immense, massive destruction may be seen after 6-hydroxydopamine. He wondered if some of the behavioral effects were due to this destruction rather than to neurotransmitter depletion.

Dr. Redmond indicated his disagreement with Drs. McKinney and Breese on 6-hydroxydopamine effects in primates. When he sacrifices animals acutely (3 to 4 weeks after 6-hydroxydopamine), he does not observe non-specific effects. A group of monkeys treated with 6-hydroxydopamine in Puerto Rico were sacrificed 2-3 years later. They still had more than 50% decrement in brain norepinephrine compared to controls and had almost no signs of non-specific damage. Dr. Redmond suggested that the difference between his results and others might be a consequence of concentration of 6-hydroxydopamine: if a high concentration of 6-hydroxydopamine is used, it is more likely to produce a non-specific effect; if a lower concentration is used and doses are repeated several times, one is less likely to obtain a non-specific effect.

Dr. McKinney felt that one might better talk about species, rather than primates in general. His studies had used the rhesus monkey (Macaca mulatta). Although Dr. McKinney had varied the concentration of 6-hydroxydopamine and the injection volume and had given it in minute doses with many injections with adequate time between injections, there always was considerable non-specific damage when 6-hydroxydopamine was given intraventricularly. Dr. McKinney concluded that he did not know whether the behavioral effects in the rhesus were due to the non-specific damage or to the amine depletion.

Dr. Meltzer questioned whether all of the deficits which Dr. Matthysse had proposed in his schizophrenia models are demonstrable in all schizophrenics. He thought that the strategy of drug testing and drug development should shift from the dopamine receptor model (which he felt would not result in any fundamental advances) to the type that Dr. Matthysse had proposed.

Dr. Matthysse stated that he felt that clinically we have the cognitive syndrome somewhat under control (i.e., the patient is able to think more clearly as his delusions diminish in magnitude), but we do not have any control over the social motivational aspects of the syndrome. It is a regular finding that the drug-treated schizophrenic is unmotivated and unhappy; he sits at home and does nothing and has no relation with people. We do not have models for social motivational withdrawal, but Dr. Matthysse thought it possible to design such an animal model. He felt that it would be necessary to use a domesticated species such as the dog, since defects of this kind would be fatal in the wild. He suspected that there must be dog models of social withdrawal and lack of motivated activity which could be used for testing drugs that affect those aspects of the schizophrenic syndrome.

Dr. Kornetsky argued that such a model would be used in the search for drugs to treat some of the symptomatic aspects of the disease. However, he thinks it is more logical to assume that some of the behavior is conditioned by the disease process and not the cause of the disease process, e.g., that the withdrawn schizophrenic isn't schizophrenic because he's withdrawn but, rather, that he is withdrawn because he has this particular disease.

Dr. Breese wondered if this implied that psychotropic drugs take time to work because people have to learn to be "uncrazy".

Dr. Kornetsky replied that since schizophrenics had had years of maladaptive behavior, even if a drug would completely reverse the presumed biochemical lesion, one would not expect immediate return to normalcy.

Dr. Friedhoff stated that although he was a principal advocate of the position that dopamine is involved in the pathogenesis of schizophrenia rather than as an etiological factor, he disagreed with some of Dr. Meltzer's comments. Dr. Friedhoff did not think we had any concept of what core schizophrenia is other than via the behavioral manifestations of schizophrenia. Since the only working hypothesis we have on such pathogenesis is related to dopamine, we should not drop this approach. He felt that Dr. Matthysse's models are extremely interesting and should be explored, but only as an alternative to other models.

Dr. Meltzer indicated that an excess of dopamine is characteristic of mania of certain types of psychotic depression, and of some forms of schizophrenia at some phases. There are other clearly schizophrenic behaviors which are not explainable by current dopaminergic theories. There are also problems with these theories in explaining the actions of some neuroleptics as well as our psychosis models. Thus, Dr. Meltzer sees the dopaminergic story as only a limited, not irrelevant portion of the story. There may be more use for the dopaminergic-antidopaminergic strategy of drug development, but there would be value to alternative approaches. Dr. Meltzer concluded by indicating that recent follow-up studies at Michael Reese had shown that about 60% of patients released on chronic neuroleptics still had severe cognitive impairment one to two years after discharge from the hospital even though these were acute schizophrenics with good prognosis.

Dr. Kornetsky cited other recent work in which discharged schizophrenics who were holding down jobs and were not on medication showed an attentional deficit when tested under conditions of sensory overload. Normal subjects, in contrast, tend to do as well or better in such tests when under sensory overload.

Dr. R. Vogel mentioned a recent paper by Czechoslovakian investigators who compared social effects in mice which were serotonin-depleted and mice given amphetamine acutely. Both the Chechoslovakians and Dr. Vogel observed in both groups agonistic behavior and social withdrawal.

Dr. Matthysse commented on the question of dopamine centrality. He stated that for years the treatment of choice in Parkinson's disease was anticholinergics, even though it is not an acetylcholine disease. Investigators should have recognized the possibility of dopa involvement because of the cholinergic-dopaminergic interactions in the striatum, but they didn't. In Huntington's disease, gabaminergic defects seem most critical, yet drugs of choice are dopaminergic blockers. Thus, clues may be derived from other than etiological mechanisms.

Dr. Stricker proposed that continuing emphasis should be placed on dopamine since he did not feel it was an arbitrary transmitter. He felt that the anatomy of brain catecholaminergic pathways would have predicted its significance even if nothing had been known of its clinical effects. The fact that schizophrenics have a hard time dealing under conditions of sensory overload is exactly what one would expect if a catecholaminergic pathway mediated brain function as it is generally assumed to do. A similar argument can be made with regard to Parkinson's disease; there is a non-specific inability to initiate movement. The non-specific contribution of aminergic neurons to arousal demands that they be involved in schizophrenia and various other general syndromes. Dr. Stricker concluded with the comment that the clinical data reinforce the predicted picture, which is entirely consistent with the general picture of function predicted by the anatomy of the aminergic neurons.

TAILS OF STRESS-RELATED BEHAVIOR: A NEUROPHARMACOLOGICAL MODEL

Seymour M. Antelman[*] and Anthony R. Caggiula[+]

[*]Department of Psychiatry, Western Psychiatric Institute and Clinic,
University of Pittsburgh School of Medicine, and Psychobiology Program,
[*][+]Departments of Psychology and [+]Pharmacology, Pittsburgh, Pa. 15260

CLINICAL AND NATURALISTIC OBSERVATIONS

Recent estimates of the prevalence of obesity in the U.S. range as high as 68% (1). Although the causes of this disorder are varied, and often involve genetic and/or metabolic factors, it has long been recognized that the stress associated with traumatic emotional events may also contribute to overeating in some obese individuals. Bruch, who perhaps more than any current-day investigator has focused our attention on the importance of emotional factors in overeating, has characterized the weight gain associated with such behavior as "reactive obesity" (2,3). This type of overeating and weight gain is frequently observed after a traumatic experience, such as the loss of a loved one. Although the precipitating stressors vary in kind from individual to individual, they often result in anxiety or depressive reactions which can be diminished by eating (2,3). The use of overeating to allay emotional states is not limited to extreme circumstances. It may also occur in response to everyday frustrating experiences, as is well illustrated in the following remarks of an unhappy young woman cited by Bruch (2): "Sometimes I think I'm not hungry at all. It is that I am just unhappy in certain things - things I cannot get. Food is the easiest thing to get that makes me feel nice and comfortable." Although this woman was able to control her eating during the day when surrounded and comforted by other people, she had considerable diffi- culty at night when she was alone and tense. In speaking of this she stated: "I think then that I am ravenously hungry and I do my utmost not to eat. My body becomes stiff in my effort to control my hunger. If I want to have any rest at all - I've got to get up and eat. Then I go to sleep like a newborn baby."

This type of eating pattern is very similar to the "Night Eating Syndrome" described by Stunkard (4) in some of his patients. This syndrome, which is characterized by hyperphagia and insomnia at night followed by morning anorexia, occurs during stressful periods in life and is diminished by their resolution. In addition to the foregoing clinical examples, stress-induced eating in humans has also been observed under controlled laboratory conditions (5).

Eating precipitated by stressful or activating situations is not restricted to humans but is also observed in lower animals. In natural surroundings, feeding is often associated with bouts of fighting or sexual behavior, and presumably is the result of the activation produced by these activities. In birds, where these effects are best documented, eating has been reported during boundary disputes in the prairie horned lark (6), both the great and blue tit (7,8), the avocet (9), the turkey (10), and the zebra finch (11), among others. Sporadic eating during a series of copulations has also been observed in the wild rat (12).

We have thus far alluded only to the influence of stress on eating in both lower animals and humans, but other behaviors can result, including aggression and sexual behavior. For example, many of the same types of stressful or activating stimuli which induce feeding also result in sexual behavior. Morris (11) noted in his description of the mounting of a nearby female by a male zebra finch during a fight with another male that "The first time displace- ment mounting was seen, it occurred at the point at which displacement feeding was seen on other occasions." Sexual activity associated with fighting has also been described in the cormorant (13) and the avocet (9). The wide variety of activating conditions which can precipitate sexual behavior in the wild is nicely illustrated by Bergman (14) who described such behavior in turnstones during the passing of a boat or airplane, a sudden hail storm or a fight with an enemy.

Relative to feeding and sexual behavior there have been few systematic field (in contrast to numerous laboratory) studies of aggression occurring as a result of activation. Rather, the stress associated with aggression more often serves as a trigger for other activation- related behaviors. Among the observations that have been reported are those of van Lawick- Goodall (15) on aggression which appeared to be caused by frustration. She noted, for instance, that chimpanzees often behaved aggressively when an artificially provided supply of bananas became depleted and also when a grooming partner stopped grooming.

In humans, when obese individuals are forced to adhere to strict dieting conditions and presumably thereby stressed or frustrated, there is an enhanced reactivity to goal objects other than food and a consequent increase in other oral activities and also in sexual behavior (16).

AN EXPERIMENTAL MODEL

Recently Antelman, Szechtman, Chin and Fisher (17) reported on a simple yet extremely powerful experimental manipulation, tail-pinch (TP), which permits us to duplicate in the laboratory, many of the stress-related behaviors seen in the wild. This manipulation may also provide a model system for studying some of the neurochemical mechanisms involved in stress-induced behavior in both lower animals and humans.

Tail-pinch-induced behavior (TPB) is an unusually reliable phenomenon. For instance, eating in food-sated animals (17) has now been seen within seconds of pinch application in over 95% of the approximately 4000 rats tested to date. This behavior appears normal and is readily observed in most animals without any squealing or other signs of aversion when a pressure of approximately 80-100 lbs per sq. in. is applied through a tail cuff. It is important to emphasize that the behavior produced by TP is not dependent on physical pain. In fact at the most effective TP pressures, the only behavioral sign of TP is the stimulus-bound nature of a behavior such as eating. Although TPB can also be produced at pressures that do result in behavioral signs of pain, the behavior appears to be fractionated and the animal is more easily distracted.

We believe that TP is simply a convenient way of applying an activating, mildly stressful stimulus to the animal. Indeed, the types of behavior induced by TP should and do occur in the presence of other activating stimuli. Feeding, for instance, has also been demonstrated with noise above a certain intensity (18) as well as with electric shock (19). Of course, we have already mentioned some of the stimuli which have been found to induce "tail-pinch behaviors" in nature.

The parallels between activation-induced behavior seen in the wild and TPB produced in our laboratory are not limited to feeding. In fact, virtually all of the biologically signif-icant consummatory behaviors displayed by the rat can be induced by TP. The particular behavior which occurs as a function of TP (or a similar stimulus) depends on the goal objects that are available and tends to be appropriate to those objects. Thus, maternal behavior is seen in the presence of rat pups (Sherman, Fisher and Antelman, submitted), male sexual behavior is seen in the presence of a receptive female (20,21), and aggression is often observed when pinch or shock is applied in the presence of another male (21). In a testing arena devoid of any goal objects, TP tends to induce "finger-nail biting" or stereotyped sniffing and/or gnawing behavior much as occurs with amphetamine.

In addition to the goal objects available in the environment, other factors, such as physiological predisposition, experience, and learning also serve to guide the outcome of TP. The role of physiological condition in determining the particular behavior seen during TP is clearly evident in our study of maternal behavior. Although rat mothers (1-10 days post-partum), virgin females and males all showed maternal behavior when tail-pinched in the presence of pups alone, pinch applied in a choice situation i.e., both food and pups available, resulted in the mothers preferring the pups and both virgins and males preferring food over pups. The influence of experience on the outcome of TP was demonstrated in a study which involved maintaining rats after weaning on either food pellets, wet mash or sweetened milk and then testing them for TP-induced feeding with each of these foods (Antelman and Rowland, unpublished observations). The results of this experiment indicated that the animals clearly preferred the familiar foods even though each food was presented singly rather than in a choice situation. The tendency of wild rats to avoid new foods (neophobia) (12) is another instance of the preference for the familiar during conditions of stress or activation.

Most of the TP data (or data derived from related techniques) can be grouped into three categories. The first category has already been alluded to and simply involves the induction of a particular behavior, e.g., eating, in the normal animal. A second grouping reflects the ability of TP, under appropriate conditions, to facilitate, and in some cases exaggerate, the behaviors which can be induced. Finally, in animals where a given behavior is depressed as a consequence of brain damage or pharmacological treatment, TP can often provide a temporary recovery of normal function.

Since we have already described examples of the induction of behavior by TP (i.e. the first category), we will now consider the second category listed above. Not only can TP induce perfectly normal-appearing feeding, it can also produce considerable obesity when applied for prolonged periods of time (22). Thus it was found that TP applied six times daily, for 10 min. periods each, produced a two-threefold increase in daily caloric intake and, over a five-day pinch period, a mean weight gain of 71 grams for pinched animals relative to only 17 grams for controls. As will be discussed below, this obesity has much in common with stress-induced overeating in humans.

We have already stated that shock applied to the tail or TP can increase the percentage of naive adult male rats copulating on their first sex test (23; Shaw, Caggiula and Antelman; unpublished observations). In addition, it has been known for a number of years that shock could also accelerate this behavior in experienced animals (24; 25 ; 26). Shock has also been shown to increase the number of prepuberal male rats that will copulate (27). Maternal behavior can likewise be potentiated by TP. Although virgin female rats typically require approximately six days of contact with rat pups before spontaneously displaying maternal behavior, this period can be significantly shortened by administering only eight TP trials on a single day, (Szechtman, Seigel, Rosenblatt and Komisaruk, submitted for publication).

The restorative or "therapeutic" function of TP or tail shock has been demonstrated for both feeding and sexual behavior under a variety of conditions in which these activities were impaired. Perhaps most dramatically, we (28) have found that TP can reverse the aphagia which occurs as a consequence of lateral hypothalamic lesions. Such animals will usually die unless tube-fed and, indeed, all but one of our unpinched controls did die in this study. In contrast, when lesioned animals were pinched several times each day, they ingested a sufficient amount of sweetened milk during this stress (and at no other time) to survive until they recovered the ability to eat spontaneously. Acute application of TP to cats with lateral hypothalamic lesions has similarly induced food ingestion (29). As we (28) have previously noted, these effects bear a striking resemblance to the paradoxical kinesia observed in Parkinsons' patients during periods of stress. Most recently, we have demonstrated that TP could also overcome the impairment of food intake produced by several pharmacological agents. These compounds, amphetamine, methylphenidate and apomorphine, are all capable of inducing substantial anorexia. After inducing anorexia in food-deprived animals with these agents we were able to show that TP could completely reverse this suppression of food intake (30; Antelman, Caggiula, Black, Eichler and Edwards, submitted for publication).

As was the case for feeding, stress can also serve a "therapeutic" function in temporarily restoring copulatory activity that had been suppressed by certain types of brain damage. In male rats not copulating, and in fact completely akinetic after intraventricular injections of the selective CA neurotoxin 6-hydroxydopamine (6-OHDA), TP almost invariably induced stimulation-bound copulation (31; Shaw, Caggiula, Antelman and Greenstone, unpublished data).

PHARMACOLOGY OF TAIL-PINCH BEHAVIOR

Involvement of Brain Catecholamines

When TPB was first observed by Antelman and Szechtman at the beginning of 1973 it was not immediately apparent that eating behavior could also be obtained. Rather, TP-induced gnawing and licking attracted our initial interest. This proved fortuitous, since their similarity to the behaviors observed following large doses of amphetamine guided our experimentation toward the suggestion that a common neurochemical element may help mediate both amphetamine stereotypy and TPB.

Since the brain catecholamines (CA), norepinephrine (NE) and dopamine (DA), are thought to be importantly involved in amphetamine-induced behaviors, with DA playing the central role in stereotypy, we proceeded to examine a possible role of these compounds in TPB. We began by determining the effects of the butyrophenone neuroleptic, haloperidol, in doses (0.1-0.4 mg/kg) which are thought to antagonize both NE and DA receptors (32). Within this dose range, haloperidol reduced the incidence of TP eating to approximately 50% of control without causing any apparent debilitation of the animals (17). A similar finding has recently been obtained with tail shock induced copulation (33). In order to decide if this finding was due to blockade of either NE or DA receptors we used receptor antagonists thought to be specific to each of these CA. Phentolamine and sotalol, presumed α -and β -NE antagonists respectively, failed completely to block either TP eating or tail shock copulation, even at high doses. Similar findings for TP eating have since been obtained with other NE antagonists, phenoxybenzamine (up to 10 mg/kg) and D, L-propranolol (up to 3 mg/kg) (Antelman and Black, unpublished observations). In contrast, when we tried the specific DA-receptor blocking agents, spiroperidol (0.062-0.250 mg/kg) and pimozide (0.50-2.0 mg/kg), significant attenuation was obtained for both TP eating (17) and tail shock induced copulation (33). These data have been confirmed by ourselves and others using a variety of neuroleptics, including the phenothiazines, chlorpromazine (Antelman and Szechtman, unpublished observations;34), and trifluoperazine (Antelman and Szechtman, unpublished observations) and α-flupenthixol, a potent thioxanthene (35). Promethazine, a phenothiazine with antihistamine rather than antidopaminergic properties also does not block TPB (Antelman and Szechtman, unpublished observations). The only DA antagonists tested thus far which have failed to attenuate TPB are the dibenzazepine clozapine, (10-40 mg/kg) and thioridazine (5-40 mg/kg), a piperidine phenothiazine (Antelman and Szechtman, unpublished observations). Although the reasons for this failure have not been empirically determined, there are at least two possibilities. The first, and most obvious alternative (though not necessarily the correct one) is that both of these compounds have strong anticholinergic properties which can mask their DA-receptor blocking effects (36). Second, as we have very recently proposed (3 7) and will discuss below, interference with NE function can counteract (or mask) the effects of DA-receptor blockade under stressful circumstances, and both clozapine and thioridazine have powerful adrenolytic properties (38). Finally, several important differences between these two drugs suggest that there may not be a single explanation for why they failed to block TPB. Thus, clozapine actually potentiated TP eating and also induced a significant amount of eating in non-stressed animals, (Antelman, Black and Rowland, submitted) while thioridazine did neither of these. Moreover, when each drug was administered in combination with the tyrosine hydroxylase inhibitor, α-methyl-p-tyrosine, thioridazine now effectively reduced TPB (as would be expected from the work of Ahlenius and Engel, 39 , 40 and Carlsson, et al., 41 , 42 , as well as from our own lab, 43), while clozapine remained ineffective (Antelman and Szechtman unpublished observations). Although our data are consistent with the hypothesis of DA involvement in TPB, caution must be observed in interpreting the effects of neuroleptics since it has been proposed that they may merely reflect the inability of an animal to perform a given task (44). It seems unlikely, however, that this stricture applies to our data. In fact, when the effects of neuroleptics are plotted as a function of individual TP trials a response sequence strongly reminiscent of the pattern which accompanies extinction of a rewarded response is seen (see Fig. 1). Thus, neuroleptics are least likely to produce an attenuation of TPB on the

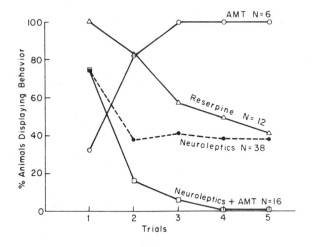

Fig. 1. Contrasting effects of AMT and neuroleptics on display of tail-pinch behavior on successive trials

Animals treated with AMT showed significantly less TPB on trial 1 as compared to trials 3-5 (p $<$ .062). Conversely, all animals treated with reserpine displayed TPB on trial 1 but showed less TPB on subsequent trials (1 vs. 3 p $<$ .04; 1 vs. 4 p $<$ .04; 1 vs. 5 p $<$ .008). This was also the case for neuroleptics (i.e., haloperidol, spiroperidol and pimozide) (trial 1 vs. 2 p $<$ .002; 1 vs. 3 p $<$ .02; 1 vs. 4 p $<$ .01; 1 vs. 5 p $<$ .006) as well as for the combination of ineffective doses of these neuroleptics and AMT (trial 1 vs. 2-5 p $<$ .001).

first trial, which actually occurs at a near-normal rate (17 ; 35). It is only on subsequent trials that the behavior is significantly depressed, a pattern of decline which cannot be ascribed to the time course of the drug. Although not reflected in Fig. 1 because it occurred on different trials for a given animal, there was actually a partial though transient reversal of the neuroleptic depression which tended to take place on trial three or four. Such a partial restoration of behavior is characteristic of the "spontaneous recovery" which is known to occur during extinction.

231

Despite our contention that the effects of neuroleptics on TPB are not explicable in terms
of a drug-induced performance deficit, the reader may yet be skeptical on this point. In
order to further satisfy our own apprehensions relative to this issue we determined the
effects of the anxiolytics, diazepam and chlordiazepoxide on TP-induced eating. When
acutely administered at high doses these drugs have considerable sedating properties.
Moreover, so far as is known they do not impair DA function, suggesting that they may serve
to reflect the extent to which TPB can be non-specifically interrupted. Our results
indicated that neither diazepam in doses up to 5 mg/kg or chlordiazepoxide in doses as high
as 50 mg/kg attenuated TPB to the slightest degree. Instead, they actually produced a
marked reduction in the pressure required to induce the response. Although prone and
markedly sedated, animals treated with these agents rose up at the slightest touch of the
tail, paradoxically showing intense emotional behavior (characterized by loud vocalizing).
and began to eat voraciously (Antelman and Szechtman, unpublished observations).
Most recently, Robbins, Phillips and Sahakian (46) have found an increase in amount eaten
during TP with chlordiazepoxide.

The data presented so far suggests that DA may play a critical role in the display of TPB.
In so doing they raise a whole host of related questions, such as whether newly synthesized
or stored material is involved, the particular DA system or systems which form the neuroanatomical substrate for stress-induced behavior, and the way in which other neurochemical
pathways interface with the key DA systems in mediating TPB.

To begin to answer some of these questions we sought to determine whether TPB depends on
uninterrupted synthesis of newly formed DA, or on its release from granular stores. To do
this we compared the effects of α-methyl-p-tyrosine (AMT), the tyrosine hydroxylase inhibitor
and FLA-63, the dopamine-β-hydroxylase inhibitor (DBHI) with those of reserpine, which is
known to release CA from their storage granules. Our results indicated that treatment with
the NE-synthesis inhibitor, FLA-63, not only failed to attenuate TPB, but actually significantly prolonged it. AMT, which additionally inhibited the synthesis of DA produced significant attenuation of TPB but only on the first trial (Fig. 1). A precisely complementary
result was obtained with reserpine, which had no influence whatever on the first trial but
produced significant and successively greater attenuating effects on subsequent TP trials
(17). Collectively, these results suggest that TPB initially relies on newly synthesized
DA, but continued induction over time is dependent on the release of DA stores. In addition,
the prolongation of the response to TP (after removal of the pinch itself) obtained with
FLA-63 suggests that NE (or some other β-hydroxylated amine) may normally exert a regulatory
influence on DA, at least during stress-related behavior. We will have more to say on this
topic when we deal with our theory of NE-DA interactions below.

We now approach the question of which DA pathway plays the dominant role in TPB. Data from
several sources suggest the nigrostriatal DA system as the most likely candidate. First,
lesioning of this pathway with 6-OHDA injected directly into the A9 region, resulted in
substantial and significant deficits in both the onset and maintenance of TPB (17). By
contrast, electrolytic lesions of the nucleus accumbens and olfactory tubercle, terminal
areas of the mesolimbic DA pathway, are completely devoid of any attenuating influence on
TPB (Antelman and Rowland, unpublished observations). Finally, when animals are made aphagic
following lesions of the lateral hypothalamus which transected (at least in part) the
nigrostriatal DA pathway, the onset of ingestion seen when these animals are tail pinched
corresponds to the time at which striatal DA receptors begin to manifest supersensitivity
(28).

Since the nigrostriatal DA system has been implicated in TPB it is logical to ask whether
this is reflected in a change in the activity of this system. We have examined this
question by studying the effects of TP on several indices of DA turnover. In these studies,
we have measured the decline in DA levels following treatment with AMT and the levels of the
DA metabolites, homovanillic acid (HVA) and dihydroxyphenylacetic acid (DOPAC). Although
the AMT method revealed a significant increase in cortical NE activity during TP, there was
no suggestion of any change in striatal DA activity either with this technique (17) or as
reflected in HVA levels (30). Indeed, to date, the only biochemical change of any kind
suggesting that TP affects DA activity in the nigrostriatal bundle is reflected in DOPAC
levels. This metabolite, which provides a measure of the deaminated products of DA, and is
thought to be a good index of its activity in the striatum (47), is consistently decreased
by TP (30; Antelman, Black and Edwards, unpublished observations). This finding may appear
odd in view of all the data presented above suggesting that TPB is dependent upon DA. These
data might have led to the prediction that TP would increase DA activity. In fact, the
finding of decreased DOPAC levels may not be inconsistent with this prediction at all. Thus,
York, et al (34), recording from single units in the substantia nigra, reported a preferential increase of cell discharge in the pars compacta (but not the pars reticulata) during
TP. Since the nigrostriatal bundle originates from the pars compacta these data strongly
suggest that TP does increase activity in striatal DA neurons. The apparent discrepancy
between our biochemical data and the electrophysiological work just cited is reminiscent of
similar findings with amphetamine. It therefore may be instructive to look at the

amphetamine data and the concepts which have been advanced to explain it.

Amphetamine produces an initial increase in the firing of striatal DA units followed by a marked depression of unit activity (48). Biochemically, although it increases CA release (49),it fails to accelerate the disappearance of DA after synthesis inhibition with AMT (50), and does not modify HVA activity except at exceedingly high doses (51; 30). Finally, to make the analogy with TP complete, amphetamine does produce a decrease in DOPAC levels (47; 30). Collectively, these data suggest that amphetamine, and perhaps also TP, by virtue of releasing DA,may activate compensatory mechanisms such as a neural feedback pathway and/or possibly presynaptic receptors (52). According to this notion the decrease in striatal DOPAC which occurs during TP may imply the activation of a homeostatic, compensatory mechanism.

Although a fair amount of evidence has been presented implicating DA as a key element in TPB, virtually all of it has been indirect, relying on the effects of receptor antagonists, synthesis inhibitors and lesions. It is only fair to ask whether DA agonists have any effects on TPB. Up to a point, a facilitation might be expected to occur. In fact, this is exactly what does occur. In an experiment referred to earlier and discussed in more detail below, we found that amphetamine (at 0.5 and 2.0 mg/kg) and methylphenidate, (at 5.6 mg/kg), both of which act to release DA, significantly increased the eating rate of tail-pinched animals (30; Antelman, Caggiula, Black and Edwards, submitted). This finding is all the more remarkable since we also found that the same doses of these compounds produced considerable, and in the case of methylphenidate, complete anorexia when administered without TP.

Involvement of Brain Serotonin

In response to a question about whether some phenomena which he had related to locus coeruleus NE neurons were due to regulation of these neurons by other pathways, Floyd Bloom (53) recently pointed out that "neurons exist by themselves only in the test tube and the fluorometer. In the behaving brain they are all tied to each other." This humorous but incisive truism bears on the work we have been describing here. Although we have so far emphasized the importance and perhaps primacy of the nigrostriatal DA bundle in relation to stress-induced behavior (using TPB as a model), this system does not exist in a vacuum. Several references have already been made to interactions with NE, and these will be discussed in more detail below. At this juncture, however, we would like to deal with the role played by serotonin (5HT) in TPB and to illustrate its possible interactions with DA. Our reasons for studying the possible influence of serotonin on TPB were that similar behaviors are known to be influenced by manipulating 5HT as can be induced by TP and related stimuli. For instance, pharmacologically increasing brain 5HT has been reported to decrease food intake (54), and to attenuate male copulatory behavior in rats (55; 56). Conversely, decreasing brain 5HT reportedly increases food intake (57; 58), and potentiates some features of male copulatory behavior (59).

In our studies (Kennedy, Antelman and Edwards, submitted) we observed the effects on TP-induced eating of three general types of pharmacological manipulations designed to influence brain 5HT activity: a) 5HT precursor loads, b) inhibition of monoamine oxidase (MAO) to prevent 5HT degradation, and c) inhibition of 5HT reuptake. We began our experiments by comparing the effects on TPB of the precursor amino acid, tryptophan, with those of 5-hydroxytryptophan, (5HTP) the immediate precursor of 5HT. The increase in total 5-hydroxyindoles [5HT plus 5-hydroxyindoleacetic acid (5-HIAA), a major metabolite of 5HT produced by the action of MAO and aldehyde oxidase] produced by tryptophan loading occurs without altering the regional pattern of brain 5HT metabolism found in normal animals. That is, those areas which have the highest 5-hydroxyindole levels in control animals also have the highest levels in tryptophan-treated animals (60). This is not the case after loading with 5HTP. Such a treatment bypasses the enzyme tryptophan-5-hydroxylase (which converts tryptophan to 5HTP), allowing 5HT to be synthesized in any neurons containing the ubiquitous L-aromatic-amino acid decarboxylases (61; 62), such as CA-containing neurons (63). Thus, tryptophan loading provides a more physiological means of increasing 5HT than does loading with 5HTP, since the former restricts the formation of 5HT to those neurons normally containing this transmitter (64; 65).

Despite the problems in interpreting the effects of 5HT loads, it is nevertheless informative to load rats with both tryptophan and 5HTP and to observe if either or both treatments alter the behavioral parameter being studied. If a given effect were found only after 5HTP, it might be due to non-specific formation of 5HT in brain areas not normally synthesizing 5HT, or to the DA displacement which occurs with this treatment (62). Conversely, if only tryptophan produced the observed effects, they could be due either to the amino acid itself or to a metabolite other than 5HT, such as tryptamine. On the other hand, if both manipulations produced similar effects, 5HT or a 5HT metabolite could be the responsible agent.

Our results indicated that 5HTP significantly reduced the incidence of TP eating at all doses tested (25, 50 and 75 mg/kg, i.p.) with the two higher doses producing approximately 50% attenuation. By contrast, tryptophan was much less effective. Although significant effects were obtained on both the incidence of, and the latency to begin the behavior, they were rather modest. Thus, 100 and 150 mg/kg of tryptophan, which have been reported to maximally elevate 5HT levels (66), only reduced the behavior to 88% and 73% of control levels, respectively.

What can explain the relatively small effect of tryptophan? One possibility is that most of the excess 5HT formed from tryptophan is rapidly metabolized by MAO without ever reaching the post-synaptic receptors (67,68). But, wouldn't the same argument apply to 5HTP? Not necessarily. Since 5HTP bypasses the hydroxylation step in the synthesis of 5HT, (which could become rate limiting when saturated with substrate), perhaps sufficient 5HT can be synthesized rapidly enough and spill over onto the receptors without being degraded by MAO (69; 65). On the other hand, tryptophan injections may allow 5HT to accumulate more slowly due to the regulating influence of tryptophan-5-hydroxylase, thereby making it subject both to binding by intraneuronal vesicles and to degradation by MAO (20). If the failure of tryptophan to produce a robust effect on TPB was due to its rapid metabolism by MAO, this should be remedied by combining it with MAO inhibitors (MAOI). Work in recent years has shown that there are two classes of MAO, referred to as type A and type B. While both of these can act to deaminate DA (71), type A selectively metabolizes 5HT (72) and NE (73) and type B selectively acts on phenylethylamine (72). Several drugs have been developed which preferentially inhibit either type A or type B MAO when given in low doses. Clorgyline (72, 71; 74) and Lilly 51641 (75; 74) specifically inhibit type A MAO at low doses, while low doses of deprenyl (72; 74) and pargyline (75) selectively interfere with type B MAO.

When clorgyline was administered alone it failed to affect TPB even at 20 mg/kg which produced 98% inhibition of type A MAO (using 5HT as the substrate) and only 26% inhibition of type B MAO (with phenylethylamine as the substrate). Similar results were seen with Lilly 51641 which produced only a marginal attenuation of behavior (to 83% of control) at 15 mg/kg. Quite different results were obtained, however, when these agents were given in conjunction with tryptophan. In each case, substantial and significant dose-dependent decreases in TPB were obtained when clorgyline was given in doses of 2-20 mg/kg and Lilly 51641 in doses of 5-15 mg/kg. Our assay data suggested that MAO A had to be inhibited by at least 75% before any meaningful behavioral attenuation could be expected. No signs of behavioral debilitation were seen at any dose level.

Although the findings with type A MAOI's might reflect serotonergic involvement in TPB, their interpretation is confounded by the fact that they also deaminate the CA. However, since DA is likewise metabolized by type B MAOI's, if these failed to replicate the results obtained with the A compounds, DA could effectively be ruled out in explaining these results. Moreover, in view of all the data already presented in favor of DA as an activator of TPB, type B MAOI's might actually be expected to facilitate this behavior.

Our results indicated that neither deprenyl nor pargyline produced any meaningful suppression of TPB, even when combined with tryptophan, so long as they were given in doses relatively selective in inhibiting type B MAO. Thus, 5 mg/kg of pargyline, which produced 95% inhibition of type B MAO, coupled with only 42% inhibition of the A enzyme, failed to attenuate TPB even when combined with 100 mg/kg of tryptophan. In fact, both pargyline and deprenyl significantly reduced the latency to begin TPB relative to tryptophan or saline-treated controls when given in low doses. However, when pargyline was combined with tryptophan at a dose (15 mg/kg) where it was no longer selective and also inhibited type A MAO by 76%, a significant attenuation of TPB to 67% of control was obtained.

While these data suggest that the findings with clorgyline and Lilly 51641 may have been due to 5HT, they do not yet rule out NE. However, two additional experiments argue against NE and strengthen the case for 5HT. In the first experiment we pretreated clorgyline-tryptophan injected animals with parachlorphenylalanine (PCPA), the tryptophan-5-hydroxylase inhibitor, in an attempt to determine whether the attenuation seen after these drugs might have been due to tryptamine, a non-serotonergic metabolite of tryptophan. Since tryptamine accumulation (but not 5HT) is actually accelerated in the presence of PCPA and tryptophan (70, and further enhanced by the addition of an MAOI which also increases CA levels (59), this combination treatment should also attenuate TPB if an increase in tryptamine or NE was the critical factor. Pretreatment with PCPA completely reversed the attenuating effects of clorgyline-tryptophan on TPB, strongly suggesting that it was 5HT, and not tryptamine or NE which was responsible for the original finding. A second experiment, using the selective 5HT reuptake blocker, Lilly 110140 (Fluoxetine)(77; 78), provided both an additional piece of evidence in support of a modulatory influence for 5HT in TPB and also another argument against explaining the behavioral effects of the type A MAOI's in terms of their action on NE neurons.

Lilly 110140 has been reported to enhance the effects of exogenously supplied 5HTP on plasma corticosterone (75). Behaviorally, it increases the depressive effects of tryptophan in pigeons performing a food-rewarded learned response (79). Moreover, it actually increases uptake of NE while blocking 5HT uptake (78). Thus, if as suggested by the previous experiment, 5HT rather than NE were involved in the effects of type A MAOI's and tryptophan, then a similar effect on TPB should be produced by Lilly 110140 (and tryptophan). Conversely, different results would be expected if the MAOI's were acting through NE. Since Lilly 110140 (10 mg/kg) produced a significant (though modest) attenuation of TPB to 83% of control which was potentiated (down to 53% of control) by the addition of 150 mg/kg of tryptophan, the results support a serotonergic rather than a noradrenergic interpretation of the MAOI data.

Taken in toto, the serotonergic data reported here would seem to make a strong case for a modulatory influence of 5HT on stress-induced behavior. These data also raise the question of whether evidence for a 5HT-DA interaction can be obtained with our TP paradigm. We have so far addressed this question in two ways. First, we have attempted to establish whether a synergism could be obtained by combining largely ineffective doses of a neuroleptic with doses of a tryptophan-clorgyline treatment which themselves were also ineffective in attenuating TPB. Conversely, we have also sought to determine whether the attenuation of TPB observed following the administration of a maximally effective dose of a DA-receptor antagonist could be significantly reversed by interfering with 5HT function.

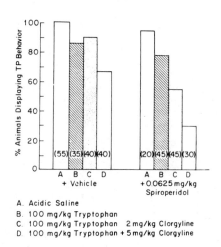

A. Acidic Saline
B. 100 mg/kg Tryptophan
C. 100 mg/kg Tryptophan   2 mg/kg Clorgyline
D. 100 mg/kg Tryptophan + 5 mg/kg Clorgyline

Fig. 2. DA-5HT interactions and tail-pinch behavior

Marginal attenuation of TPB by tryptophan and/or clorgyline, and its augmentation by a dose of spiroperidol previously shown to be ineffective when given alone. ( ) = trials; no. of trials/5 = N.

The data presented in Figs. 2 and 3 indicate that both effects can be obtained. In the first instance, we found that 0.062 mg/kg of spiroperidol, which failed to induce a significant blockade of TP eating (animals continued to respond on 95% of post-drug trials), reduced TP responding to 55% of control levels when combined with 100 mg/kg of tryptophan and 2 mg/kg of clorgyline. When administered alone, the tryptophan-clorgyline combination only marginally affected the response, reducing it to 90% of control levels.

The coupling of PCPA, the 5HT synthesis inhibitor (administered in a dose of 100 mg/kg per day for four consecutive days), with 0.25 mg/kg spiroperidol produced precisely the opposite results of the previous experiment (Fig. 3). Whereas spiroperidol given with the PCPA vehicle reduced responding down to 13% of control levels, the addition of PCPA significantly reinstated the behavior to 64% of control. By itself, PCPA significantly reduced the mean latency to begin TP-induced eating.

The results of these two experiments indicate that 5HT and DA interact reciprocally in the mediation of stress-induced behavior. Moreover, since PCPA did not completely reverse the influence of spiroperidol, this may suggest that the role of 5HT in TPB is secondary to that of DA. The recent indication from the work of Fibiger and Miller (80) that there may be projections from the dorsal raphe area to the zona compacta of the substantia nigra provides a possible anatomical basis for our findings.

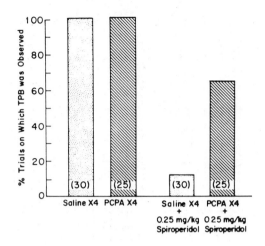

Fig. 3. DA-5HT interactions and tail-pinch behavior

Pretreatment with PCPA partially reverses the attenuating effects on TPB of
spiroperidol (p < .0003).

Possible Interactions Between NE and DA

Several times within the course of this manuscript we have made oblique reference to the
possibility of a stress-related interaction between NE and DA. We would now like to consider this hypothesis more deliberately. The suggestion of an interaction between NE and DA
first arose from some early observations on TP-induced feeding. Although all available
evidence indicates that this response depends on the nigrostriatal DA bundle, TP results in
an increased release of cortical NE (17). While this effect on NE could merely have been
an epiphenomenon, unrelated to the behavioral response, other data implied that this was
not the case. For instance, treatment with the NE-synthesis inhibitor, FLA-63, produced a
significant prolongation of the TP response (17). That is to say, although the TP response
is normally largely stimulation-bound, meaning that it begins shortly after pinch onset and
terminates with its cessation, FLA-63 caused the response to extend significantly beyond the
removal of the pinch. Viewed in another way, FLA-63 can be said to have retarded the
extinction of the tail-pinch response. The reader may recall that this is precisely the
opposite of what was found with DA-receptor antagonists, which actually seemed to precipitate extinction of the TP response. These results suggest the possibility that NE may
normally exert a regulatory influence on the nigrostriatal DA system, (and perhaps on other
DA pathways as well) during stressful situations, serving to keep responses related to this
system within certain limits. In the absence of this regulatory influence, (a situation
which occurs when a DBHI like FLA-63 is administered), DA is more likely to react to a
stressful situation by over responding in an uncontrolled fashion. According to this view,
interference with NE activity will, during sufficiently activating conditions, potentiate
"DA-dependent" behaviors while under relatively non-activating or even less stressful
circumstances, it will not affect, or may depress the same behaviors (37). A corollary of
this hypothesis is that depression of NE activity may either prevent or counteract the
effects of interfering with the function of DA-containing neurons. This suggests that
selective interference with, or damage to, DA-containing neurons may result in greater
functional impairment of the organism during stressful circumstances than similar damage
which is also accompanied by a substantial interference with brain NE. Much of the recent
work in our laboratories has dealt with the above proposition and it is to this that we now
turn.

Our initial approach to the question of whether interfering with NE activity could overcome
the effects of impairing DA function has involved mainly the coupling of DBHI's and neuroleptics during both stressful (tail-pinch) and non-stressful (ad-lib) feeding tests
(Antelman and Black, submitted). The DBHI's which have been used are FLA-63 (81) and
methimazole, an imidazole-2-thiol derivative, slightly less potent than FLA-63 but without
the peritoneal irritating effects of the latter (82). Haloperidol has been the primary
neuroleptic used to attenuate both ad-lib and TP-induced feeding behavior.

Fig. 4 illustrates the effect on TPB of combining FLA-63 (at 10 and 20 mg/kg) with
haloperidol (0.4 mg/kg). In the first experiment, haloperidol reduced TP responding to
28% of vehicle treated controls. By contrast, when 10 mg/kg of FLA-63 was combined with
haloperidol, eating behavior was significantly reinstated to 86%. As can be seen, similar

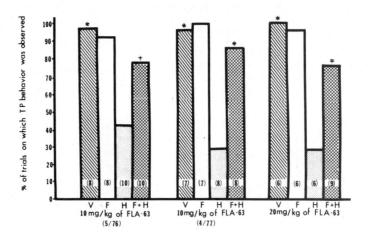

Fig. 4.   A DBHI (FLA-63) reverses the attenuation of TP
eating produced by haloperidol

The experiment with 10 mg/kg of FLA-63 was repeated in successive years. ( ) = N
All statistical comparisons are with haloperidol ∓ = p ⟨ .025 + = p ⟨ .005* =
p ⟨ .001   V = vehicle   H = Haloperidol (0.4mg/kg)   F = FLA-63.

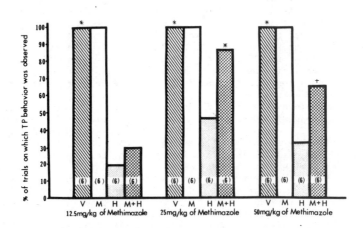

Fig. 5.   A DBHI (methimazole) reverses the attenuation of
TP eating produced by haloperidol

All statistical comparisons made with haloperidol. ( ) = N   * p ⟨ .001 + p ⟨ .005
V = vehicle   H = haloperidol   (0.4mg/kg)   M = methimazole.

results were obtained when 20 mg/kg of FLA-63 was combined with haloperidol. Fig. 5
depicts the results of the same type of experiment done with methimazole and haloperidol.
Again, similar results were obtained. At doses of 25 and 50 mg/kg, methimazole significantly reversed the attenuating influence of haloperidol on TPB when the two drugs were
combined. A lower dose of methimazole, 12.5 mg/kg, failed to modify the effect of
haloperidol. Since the equivalent of 12.5 mg/kg has been reported to produce a 50%
inhibition of DBH, while 25 and 50 mg/kg respectively resulted in approximately 60 and 80%
DBH inhibition (82), this may suggest that the enzyme must be inhibited by at least 60%

before stress can effectively reverse the behavioral concomitants of DA-receptor blockade.

The results of these experiments seem to suggest that inhibiting the synthesis of NE (or possibly some other β-hydroxylated amine) can indeed reverse the effects of DA-receptor blockade on TPB. However, there are other possible explanations of these data. For instance, the DBHI's could have affected the metabolism of haloperidol or, they might have acted directly on DA, enhancing its ability to compete with the neuroleptic for access to the post-synaptic receptor. If either of these explanations were valid we might expect that a DBHI would also counteract the suppressive effects of neuroleptics in a non-stress feeding situation. In order to test this possibility, we observed the effect of haloperidol or haloperidol plus either FLA-63 or methimazole on an ad-lib, four-hour feeding test in undeprived rats. In order to further minimize the activating properties of the situation, the animals (housed on a natural day-night cycle) were tested in the afternoon, prior to their period of maximal activity. Although haloperidol (0.4 mg/kg) significantly suppressed, (but did not completely eliminate), ad-lib feeding just as it had TP eating, neither FLA-63 (10 mg/kg) nor methimazole (50 mg/kg) showed even the slightest tendency toward reversing this suppression. Thus the ability of these compounds to counteract the influence of neuroleptics appears related to the activating or stressful properties of the environment. These data also provide support for our hypothesis that NE regulates, rather than inhibits DA, (37) since a simple disinhibitory influence of DBHI's on DA might have been expected to result in at least a partial tendency to counter the suppressive influence of haloperidol in the ad-lib feeding situation. The postulated regulatory influence of NE on DA may be the raison d'etre for the ability of DBHI's to reverse the behavioral deficits of neuroleptics. That is, although such an influence may be in the organism's best interests so long as the nigrostriatal DA system is intact, its continued presence when this system is impaired may actually be deleterious to the organism, by preventing a maximal response to stressful stimuli.

Recently, we have attempted to extend our DBHI-neuroleptic paradigm to stress-induced behaviors other than TP feeding. A more "traditional" stress-related behavior which we (Antelman, Caggiula and Black, in preparation) have examined in this way is shock induced aggression, which is typically thought to be mediated by NE (83). The results of this experiment, which paired FLA-63 and haloperidol, are displayed in Fig. 6. Once again the

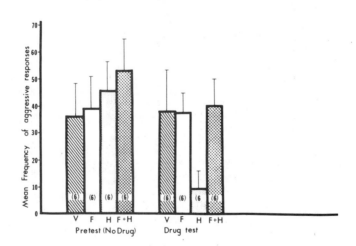

Fig. 6. A DBHI (FLA-63) reverses the attenuation of shock-
induced aggression produced by haloperidol

The substantial decrement in the frequency of aggressive responses (fighting, attack and defensive postures) produced by 0.4 mg/kg of haloperidol (relative to a pretest without drug) is significantly reversed (p < .01) by the addition of 10 mg/kg of FLA-63. See Fig. 4 for explanation of symbols.

data indicate that treatment with an NE-synthesis inhibitor can significantly overcome the attenuating effects of DA-receptor blockade during stress. These data appear to contradict any simple hypothesis ascribing irritable aggression to NE mediation. Instead, they support our previously stated notion that this behavior, in common with many other stress-related activities, is better understood when viewed from the standpoint of an interaction between

NE and DA (37). Unfortunately, it is not yet possible to claim that the NE-DA interaction hypothesis holds for still another stress-induced activity, TP-induced sexual behavior, since a recent initial test with this behavior proved unsuccessful. However, since the TP was applied in a punctate rather than sustained fashion, as it had been with eating, the differences between these results may turn out to be more apparent than real.

Finally, before leaving this section, we would like to briefly mention an extension of the DBHI neuroleptic-reversal experiment. There is a good deal of evidence suggesting a reciprocal relationship between DA and acetylcholine (Ach), at least so far as the striatum is concerned (84,85). Thus, it is well known that many of the striatal actions of neuroleptics can be modified by treatment with anticholinergic agents (36, 86). Indeed, as we mentioned earlier, this is a likely explanation for the failure of clozapine and thioridazine (neuroleptics with potent anticholinergic effects) to attenuate TPB. If we assume then, that the more traditional DA-receptor antagonists owe their effects on TPB and similar behaviors to an upset of the balance between DA and Ach in favor of the latter, two possibilities come to mind. The first, and more prosaic of the two, is that cholinergic agonists should attenuate TPB. Secondly, it should be possible to counteract the effects of these agents by the addition of DBHI's. We (Antelman and Black unpublished observations) have lately begun to test this hypothesis, and our preliminary findings, shown in Fig. 7,

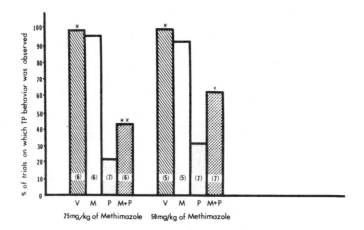

Fig. 7. A DBHI (methimazole) reverses the attenuation of TP eating produced by the cholinesterase inhibitor physostigmine

All statistical comparisons made with physostigmine (P = 0.5 mg/kg). * p < .001 ‡ p < .025. ** p < .05. See Fig. 4 for explanation of other symbols.

indicate that the anticholinesterase agent, physostigmine (at 0.5 mg/kg) does, in fact, reduce the incidence of TPB, while the addition of methimazole (in the same doses as were effective with haloperidol) significantly counteracts this effect.

In reflecting on all of the complex pharmacological relationships which have briefly been touched on in this section, it seems appropriate to conclude that the postulated interaction between NE and DA actually involves an interaction among NE, 5HT, DA, Ach and ?.

IS TPB A GOOD MODEL OF STRESS-RELATED BEHAVIOR?

Having now completed our litany of the wonders of the TP paradigm and its pharmacological substrates, we must now address ourselves more directly to the question implied by this chapter. That is, can the TP paradigm provide a reasonable model for studying stress-induced behaviors in humans?

One approach to this question has been suggested by Matthysse & Haber (87) in their recent paper on animal models of schizophrenia. These authors propose that the animal model be "isomorphic" with the clinical entity along several dimensions. That is, "Each formal characteristic of one system is required to have an exact counterpart in another, although the objects contained in the two systems may not be alike." While it may be premature to attempt to define a rigid set of criteria, since the formal characteristics of the two systems, which in this case are TPB and stress-related behavior in humans, have not been completely elucidated, there are several parallels which have clearly emerged.

It has frequently been observed that people who overeat and become obese are overly responsive to a wide range of stimuli, not just those associated directly with eating (16, 88).
Moreover, the eating behavior of these individuals tends to be stimulus-bound when presented with food (16,88), and they show an increased incidence of other oral behaviors or
sexual behavior when prevented from eating by strict dieting (89,16). These three characteristics, i.e. hyperreactivity to stimuli, stimulus-bound eating, and substitutability
of behavior, have all been previously described as prominent features of TPB.

It is also well known that obese individuals have trouble becoming satiated after the
ingestion of a given quantity of food (90), or after artificial expansion of the stomach
by inflating an intubated balloon (90). An exact parallel to this can be seen with TPB.
Animals will ingest significantly more sweetened milk in one "sitting" (i.e. up to 20cc)
when receiving continuous TP than would be expected during a period of spontaneous eating
(22).

Another parallel, and possibly the most interesting one, relates to the consequences of
stress-related or TP behavior. Bruck (3) and others (91, 90) have presented considerable
evidence to suggest that many stress-related overeaters tend to eat when emotionally tense,
or during other unpleasant states such as depression, and eating is reported to diminish or
prevent these states. Remarkably similar observations have been made in the course of experiments involving TP or related manipulations. For example, rats are largely unresponsive
to being pricked by a needle if the painful stimulus is applied during a TP-induced eating
bout, whereas the same stimulus produces vocalization and escape if the animal is prevented
from eating (Antelman and Rowland, unpublished observations). Similarly, the loud squealing and other signs of distress which normally accompany repeated tail shock of sexually
naive male rats will suddenly disappear when the shock results in copulation with a receptive female. Again, the vocalization can be reinstated by preventing the behavior
(Caggiula, unpublished observations).

The above examples suggest that one way of understanding the response to stress in both
humans and animals is within an adaptive or coping context. That is, the response may
actually function to reduce the adverse psychological and, as we shall see, biochemical
consequences of the stress. The reinstatement of biochemical homeostasis has been well
illustrated by experiments dealing with shock-induced aggression. For example, a number
of studies have found that animals shocked alone show changes in ACTH (92), blood pressure
(93), NE turnover (94), and cyclic AMP levels (95) which differ in either direction or
order of magnitude from those observed when the same shock occurs in the presence of a
partner against which aggression is possible.

Finally, stress-related overeating in humans and TP eating in rats show similar responses
to certain anorectic agents. Whereas amphetamine is purported to be ineffective against
emotionally-related overeating in humans (97), fenfluramine is thought to be the anorectic of
choice in this type of situation (96). As is illustrated in Figs. 8 and 9, we have analogous data in the TP situation. Here too, amphetamine (in otherwise anorectic doses)
fails completely to suppress TP eating. On the contrary, as we mentioned in an earlier
section, the rate of TP eating is actually significantly enhanced by amphetamine and its
congener, methylphenidate. In contrast to these results, TP was wholly ineffective in
counteracting fenfluramine anorexia.

The foregoing text illustrated parallels that are presently known to exist between TPB
in animals and stress-related eating in humans. However, the hallmark of a good model is
its ability to generate new information and to predict new ways of dealing with the
behavior being modeled. In this regard, the wealth of data, both behavioral and pharmacological, generated by the TP model may eventually prove to be invaluable. For instance,
one of the key features of TPB is the ability to interchange responses. In other words,
a particular stress will induce different behaviors depending on the cues in the enviorn-
ment. One might predict then, that the type of individual who uses food as a means of
alleviating stress (or, more properly, distress) might also be susceptible to the bland-
ishments of other "comforters" such as narcotics and alcohol. Conversely, the former
addict or alcoholic who originally used drugs as a means of dealing with stress may turn
to food or some other stress-related activity (89).When these seemingly disparate activities
are all subsumed under the rubric of stress -induced behaviors, the number of individuals
affected becomes legion and our tails of stress take on a new and added importance.

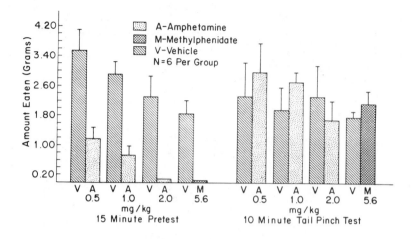

Fig. 8.   TP reverses amphetamine- and methylphenidate-anorexia

The mean (+ S.E.) amount of food eaten during a 15 minute pretest and a 10 minute tail-pinch period by 22-28h, food-deprived rats that had been presented with amphetamine, methylphenidate or the saline vehicle.   All drug-vehicle differences were statistically significant (smallest "t" = p < .01, two-tail) for the pretest but not significant (p > .10) for the tail-pinch period.

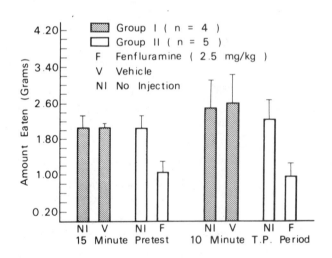

Fig. 9.   TP fails to reverse fenfluramine-anorexia

The mean (+ S.E.) amount of food eaten during 15 minute pretest and 10 minute tail-pinch period by two groups of 22-28h, food deprived rats on two successive weeks. On the first week, both groups were tested without any drug treatment.   On the second week, Group I received a saline vehicle injection while Group II received fenfluramine 15 minutes before the pre-test.   p < .05, two-tail for fenfluramine vs no injection on both the pre-test and tail-pinch comparisons.

We very much appreciate the invaluable assistance and heroic patience of Debra Greenstone, Cynthia Black, Dolores M. Saylor, Dr. Linda Johnson, Robert Scanlon, Claudia Kraft and George Lee.

Supported by PHS grants MH24114 (to S.M.A.) and MH16581 (to A.R.C.).

REFERENCES

1. U. S. Dept. of HEW (1967) <u>Obesity and Health</u>: <u>A sourcebook of current information for professional health personnel</u>, U. S. Public Health Service, Arlington, Va.

2. H. Bruch, (1973) <u>Eating Disorders,</u> Basic Books, New York.

3. H. Bruch, (1957) <u>The Importance of Overweight</u>, Norton, New York.

4. A. J. Stunkard, The night-eating syndrome, <u>Am. J. Med.</u> 19, 78 (1955).

5. J. E. Meyer and V. Pudel, Experimental studies on food-intake in obese and normal weight subjects, <u>J. Psychosom. Res.</u> 16, 305 (1972).

6. G. B. Pickwell, The Prairie Horned Lark, <u>Trans. Acad. Sci. St. Louis</u> 27, 1 (1931).

7. N. Tinbergen, Ueber das Verhalten Kamp fender Kohlmeisen (Parus m. major L.), <u>Ardea</u> 26, 22 (1937).

8. R. Hinde, The behaviour of the Great Tit (parus major) and other related species, <u>Behaviour Suppl.</u> 2, 1 (1952).

9. G.F. Makkink, An attempt at an ethogram of the European Avocet (Recurvirostra avosetta L.) with ethological and psychological remarks, <u>Ardea</u> 25, 1 (1936).

10. H. Raber, Analyse des Balzverhaltens eines domestizierten Truthahnes (Meleagris), <u>Behaviour</u> 1, 81 (1948).

11. D. Morris, The reproductive behaviour of the zebra finch (Pocphila guttata), with special reference to pseudo female behaviour and displacement activities, <u>Behaviour</u> 6, 271 (1954).

12. S. A. Barnett, Physiological effects of "social stress" in wild rats: I. Adrenal cortex, <u>J. psychosom. Res.</u> 3, 1 (1958).

13. N. Tinbergen (1951) <u>The Study of Instinct</u>, Claredon Press, Oxford.

14. G. Bergman, Der Steinwalzer, Arenaria i. interpres (L.) in seiner Beziehung zur Umwelt, <u>Acta Zool. fenn.</u> 47, 1 (1946).

15. J. van Lawick-Goodall, Behavior of free-living chimpanzees of the Gombe Stream area, <u>J. Exp. Psychol.</u> 40, 175 (1968).

16. M. L. Glucksman, Psychiatric observations on obesity, <u>Adv. psychosom Med.</u> 7, 194 (1972).

17. S. M. Antelman, H. Szechtman, P. Chin & A. E. Fisher, Tail pinch-induced eating, gnawing and licking behavior in rats: dependence on the nigrostriatal dopamine system <u>Brain Res.</u> 99, 319 (1975).

18. I. Kupfermann, Eating Behavior induced by sounds, <u>Nature</u> 201, 324 (1964).

19. G. M. Sterritt, Inhibition and facilitation of eating by electric shock, <u>J. Comp. Physiol. Psychol.</u> 55, 226 (1962).

20. A. R. Caggiula and R. Eibergen, Copulation of virgin male rats evoked by painful peripheral stimulation, paper presented Eastern Psychological Assn., 1969 Philadelphia, Pennsylvania.

21. A. R. Caggiula, Shock-elicited copulation and aggression in male rats, <u>J. Comp. Physiol.</u> 80, 393 (1972).

22. N. E. Rowland & S. M. Antelman, Stress-induced hyperphagia and obesity in rats. A possible model for understanding human obesity <u>Science</u> 191, 310 (1976).

23. A. R. Caggiula & R. Eibergen, Copulation of virgin male rats evoked by painful peripheral stimulation, <u>J. Comp. Physiol. Psychol.</u> 69, 414 (1969).

24. R. J. Barfield & B. D. Sachs, Sexual behavior: Stimulation by painful electric shock to the skin in male rats <u>Science</u> 161, 392 (1968).

25. A. R. Caggiula & M. Vlahoulis, Modifications in the copulatory performance of male rats by repeated peripheral shock, Behav. Biol. 11, 269 (1974).

26. B. D. Sachs & R. J. Barfield, Copulatory behavior of male rats given intermittent electric shocks: Theoretical implications. J. Comp. Physiol. Psychol. 86, 607 (1974)

27. D. A. Goldfoot & M. J. Baum, Initiation of mating behavior in developing male rats following peripheral electric shock Physiol. Behav. 8, 857 (1972).

28. S. M. Antelman, N. E. Rowland & A. E. Fisher, Stress-induced recovery from lateral hypothalamic aphagia, Brain Res. 102, 346 (1976).

29. P. Teitelbaum & D. Wolgin, Neurotransmitters and the regulation of food intake, In Gispen, W.H., van Wimersma Greldanus, Tj. B., Bohus, B. and deWied, D. (1975) Progress in Brain research. 235

30. S. M. Antelman, A. R. Caggiula, D. J. Edwards and N. E. Rowland, Tail pinch stress reverses amphetamine anorexia, Neuroscience Abstracts 2, 845 (1976).

31. A. R. Caggiula, D. H. Shaw, S. M. Antelman, D. J. Edwards, Interactive effects of brain catecholamines and variations in sexual and non-sexual arousal on copulatory behavior of male rats, Brain Res. 111, 321 (1976).

32. P. A. J. Janssen, The pharmacology of haloperidol, Int. J. Neuropsychiat. 3, 10 (1967).

33. S. M. Antelman, J. G. Herndon, A. R. Caggiula, and D. H. Shaw, Dopamine receptor blockade: prevention of shock-activated sexual behavior in naive rats Psychopharmacological Bulletin 11, 45 (1975).

34. D. H. York, S. Lentz and G. Love, Neural discharge associated with tail pinch induced stereotyped behavior, Fed. Proc. 35, 668 (1976).

35. B. J. Sahakian and T. W. Robbins, Isolation rearing enhances tail pinch induced oral behavior in rats, Physiology and Behavior 18, 53 (1977).

36. S. H. Snyder, S. P. Banerjee, H. I Yamamura, & D. Greenberg, Drugs, neurotransmitters and schizophrenia, Science 184, 1243 (1974).

37. S. M. Antelman & A. R. Caggiula, Norepinephrine-Dopamine interactions and behavior, Science 195, 646 (1977).

38. H. H. Keller, G. Bartholini, and A. Pletscher, Increase of 3-methoxy-4-hydroxyphenythyl-ene glycol in rat brain by neuroleptic drugs, Eur. J. Pharmacol. 23, 183 (1973).

39. S.Ahlenius and J. Engel, Behavioral effects of haloperidol after tyrosine hydroxylase inhibition, Eur. J. Pharmacol. 15, 187 (1971).

40. S. Ahlenius and J. Engel, On the interaction between pimozide and α -methyltyrosine, J. Pharm. Pharmac. 25, 172 (1973).

41. A. Carlsson, T. Persson, B. E. Roos, & J. Walinder, Potentiation of phenothiazines by α -methyltyrosine schizophrenia, Journal of Neural Transmission 33, 83 (1972).

42. A. Carlsson, B. E. Roos, J. Walinder, and A. Skott, Further studies on the mechanism of antipsychotic action: Potentiation by α -methyltyrosine of thioridazine effects in chronic schizophrenics, Journal of Neural Transmission 34, 125 (1973).

43. S. M. Antelman, H. Szechtman, P. Chin & A. E. Fisher, Inhibition of tyrosine hydroxylase but not dopamine- β - hydroxylase facilitate the action of behaviorally ineffective doses of neuroleptics, J. Pharm. Pharmac. 28, 66 (1976).

44. H. C. Fibiger, D. A. Carter and A. G. Phillips, Decreased intracranial self-stimulation after neuroleptics or 6-hydroxydopamine: Evidence for mediation by motor deficits rather than by reduced reward, Psychopharmac. 47, 21 (1976).

45. S. M. Antelman and H. Szechtman 1973, Unpublished observations.

46. T. W. Robbins, A. G. Phillips, & B. J. Sahakian, Pharmacol., Biochem. and Beh. In press.

47. R. H. Roth, J. R. Walters, G. K. Aghajanian, Effect of impulse flow on the release and synthesis of dopamine in the rat striatum, Usdin,E & Snyder, S. eds.(1973) Frontiers in Catecholamine Research Pergamon Press, Baltimore, Md. p 567.

48. G. V. Rebec and D.S. Segal, Amphetamine and Methylphenidate: Neurochemical, Electrophysiological and Behavioral Comparisons, Neuroscience Abstracts 2, 877(1976).

49. A. J. Azzaro and C. O. Rutledge, Selectivity of release of norepinephrine, dopamine and 5-hydroxytryptamine by amphetamine in various regions of rat brain, Biochem. Pharmacol. 22, 2801 (1973)

50. H. Corrodi, K. Fuxe, & T. Hokfelt, The effect of some psychoactive drugs on central monoamine neurons, Eur. J. Pharmacol. 1, 363 (1967).

51. A. Jori and E. Dolfinz, On the effect of anorectic drugs on striatum homovanillic acid in rats, Pharmacol. Res. Comm. 6, 175 (1974).

52. B. S. Bunney and G. K. Aghajanian, Dopaminergic Influence in the Basal Ganglia: Evidence for striatonigral Feedback Regulation In Yahr Ed. (1976) The Basal Ganglia Raven Press, New York 249.

53. F. E. Bloom, Norepinephrine: Central synaptic transmission and hypotheses of psychiatric disorders In Usdin, E., Hamburg, D.A., & Barchas, J.D. Eds. (1977) Neuroregulators and Psychiatric Disorders, Oxford University Press, N.Y. 19.

54. Z. L. Kruk, Dopamine and 5-hydroxytryptamine inhibit feeding in rats, Nature 246, 52 (1973).

55. G. L. Gessa & A. Tagliamonte, Possible role of brain serotonin and dopamine in controlling male sexual behavior, Adv. Biochem. Psychopharmacol. 11, 217 (1974).

56. C. O. Malmnas, Opposite effects of serotonin and dopamine on copulatory activity in castrated male rats, Adv. Biochem. Psychopharmacol. 11, 243 (1974).

57. S. T. Breisch, F.P. Zemlan & B. G. Hoebel, Hyperphagia and obesity following serotonin depletion by intraventricular p-chlorophenylalanine, Science 192, 382 (1976).

58. C. F. Saller and E. M. Stricker, Hyperphagia and increased growth in rats offer intraventricular injection of 5, 7-dihydroxytryptamine, Science 192, 385 (1976).

59. G. L. Gessa & A. Tagliamonte, Role of brain serotonin and dopamine in male sexual behavior, In Sandler, M. & Gessa, G.D. eds. (1975), Sexual Behavior: Pharmacology and Biochemistry, Raven Press, N.Y. p 117.

60. A.T.B. Moir and D. Eccleston, The effects of precursor loading in the cerebral metabolism of 5-hydroxyindoles, J. Neurochem. 15, 1093 (1968).

61. K. Fuxe, L. Butcher, & J. Engel, DL-5-hydroxytryptophan induced changes in central monoamine neurons after decarboxylase inhibition, J. Pharm. Pharmacol. 23, 420 (1971).

62. L.K.Y. Ng, T.N. Chase, R.W. Colburn, and I.J. Kopin, Release of ^{3}H-dopamine by 5-L-hydroxytryptophan, Brain Res. 45, 499 (1972).

63. L. M. Yunger and J. A. Harvey, Behavioral effects of L-5-hydroxytryptophan (L-5-HTP) following lesions in the medial forebrain bundle (MFB): effect of 6-hydroxydopamine (6-OHDA), Neuroscience Abstracts 1, 293 Society for Neurosciences, Bethesda Md.

64. G. K. Aghajanian and I. M. Asher, Histochemical fluorescence of raphe neurons: selective enhancement by tryptophan, Science 172, 1159 (1971).

65. G. K. Aghajanian, Discussion: Localization, uptake and metabolism of serotonin In Barchas, J. & Usdin, E. eds. (1973) Serotonin and Behavior, London: Academic Press p 162.

66. J. D. Fernstrom and R. J. Wurtman, Brain serotonin content: physiological dependence on plasma tryptophan levels, Science 173, 149 (1971).

67. D. G. Grahame-Smith, Metabolis compartmentation of brain monoamines, In Balazas, R. & Cremer, J. E. eds. (1976), Metabolic compartmentation in the brain, J. Wiley and Sons, N.Y. p 47.

68. D. G. Grahame-Smith, Studies in vivo on the relationship between brain tryptophan, brain 5HT synthesis and hyperactivity in rats treated with a monoamine oxidase inhibitor and L-tryptophan, J. Neurochem. 18, 1053 (1971b).

69. G. K. Aghajanian, A. W. Graham, and M. H. Sheard, Serotonin-containing neurons in brain: depression of firing by monoamine oxidase inhibitors, <u>Science</u> 169, 1100 (1970).

70. A. R. Green and D. G. Grahame-Smith, 5-hydroxytryptamine and other indoles in the CNS, In Iversen, L.L., Iversen, S.D., and Snyder, S.H. eds. (1975) <u>Handbook of Psychopharmacology,</u> Plenum Press, N.Y. p 169.

71. H. Y. T. Yang and N. H. Neff, The monoamine oxidase of brain: selective inhibition with drugs and the consequences for the metabolism of the biogenic amines, <u>J. Pharmacol. exp. Therap.</u> 187, 733 (1974).

72. H. Y. T. Yang and N. H. Neff, B-phenylethylamine: a specific substrate for type B monoamine oxidase, <u>J. Pharmacol. exp. Therap.</u> 187, 365 (1973).

73. C. Goridis & N. H. Neff, Monoamine oxidase in sympathetic nerves: a transmitter specific enzyme type, <u>Brit. J. Pharmacol.</u> 43, 814 (1971).

74. A. J. Christmas, C. J. Coulson, D. R. Maxwell, and D. Riddell, A comparison of pharmacological and biochemical properties of substrate-selective monoamine oxidase inhibitors, <u>Brit. J. Pharmacol.</u> 45, 490 (1972).

75. R. W. Fuller, Selective inhibition of monoamine oxidase, <u>Adv. Biochem. Pharmacol</u> 5, 339 (1972).

76. J. M. Saavedra and J. Axelrod, Effects of drugs on the tryptamine content of rat tissues, <u>J. Pharmacol. exp. Therap.</u> 185, 523 (1973).

77. R. W. Fuller, K. W. Perry, & B. B. Golloy, Effect of an uptake inhibitor on serotonin metabolism in rat brain: studies with 3-(p-trifluoromethylphenoxy)-N-methyl-3-phenylpropylamine (Lilly 110140) <u>Life Sciences</u> 15, 1161 (1975).

78. D. T. Wong, J. S. Horng, F. P. Bymaster, K. L. Hauser, and B. B. Golloy, A selective inhibitor of serotonin uptake: Lilly 110140, 3-(p-trifluoromethylphenoxy)-N-methyl-3-phenylpropylamine, <u>Life Sciences</u> 15, 471 (1974).

79. J. N. Hingtgen and M. H. Aprison, Behavioral depression in pigeons following L-tryptophan administration, <u>Life Sciences</u> 16, 1471 (1975).

80. H. C. Fibiger and J. J. Miller, Raphe Projections to the Substantia Nigra: A Possible Mechanism for Interaction between Dopaminergic and Serotonergic Systems, <u>Neuroscience Abstracts</u> 2, 487 (1976).

81. T. H. Svensson & B. Waldeck, On the significance of central noradrenaline for motor activity: experiments with a new dopamine β -hydroxylase inhibitor, <u>Eur. J. Pharmacol.</u>, 7, 278 (1969).

82. J. M. Stolk & D. P. Hanlon, Inhibition of brain dopamine- β -Hydroxylase activity by methimazole, <u>Life Sciences</u> 12, 417 (1973).

83. N. B. Thoa, B. Eichelman, J. S. Richardson, D. Jacobowitz, 6-Hydroxydopa depletion of brain norepinephrine and the facilitation of aggressive behavior, <u>Science</u> 178, 75 (1972).

84. P. G. Guyenet, F. Javoy, Y Agid, J. C. Beaujouan, and J. Glowinski, Dopamine receptors and cholinergic neurons in the rat neostriatum, <u>Adv. in Neurology</u> 9, 43 (1975).

85. J. T. Coyle and P. Campochiaro, Ontogenesis of Dopaminergic-Cholinergic Interactions in the rat striatum: a neurochemical study, <u>J. of Neurochem.</u> 27, 673 (1976).

86. H. C. Fibiger, A. P. Zis and A. G. Phillips, Haloperidol-induced disruption of conditioned avoidance responding: attenuation by prior training or by anticholinergic drugs, <u>Eur. J. Pharmacol.</u> 30, 309 (1975).

87. S. Matthysse and S. Haber, Animal Models of Schizophrenia In Ingle, D.J. and Shein, H. M. eds., (1975) <u>Model Systems in Biological Psychiatry</u>, MIT Press, Mass. 4.

88. S. Schacter, Obesity and eating, <u>Science</u> 161, 751 (1965).

89. J. LeMagnen, Stress et obesite, <u>La Recherche</u> 7, 777 (1976).

90. A. J. Stunkard, Obesity; M. Freedman and Kaplan, Comprehensive textbook of Psychiatry, Williams & Wilkins, Baltimore (1967).

91. W. W. Hamburger, Emotional aspects of obesity, Med. Clin. N. Amer. 35, 483 (1951)

92. R. L. Conner, J. Vernikos-Danellis, & S. Levine, Stress, fighting and neuroendocrine function, Nature 234, 564 (1971).

93. R. B. Williams & B. Eichelman, Social Setting: Influence on the physiological response to electric shock in the rat, Science 174, 613 (1971).

94. J. M. Stolk, R. L. Conner, S. Levine, & J. Barchas, Brain norepinephrine metabolism and shock-induced fighting behavior in rats: Differential effects of shock and fighting on the neurochemical response to a common footshock stimulus, J. Pharmacol. exp. Therap. 190, 193 (1974).

95. B. Eichelman, E. Orenberg, E. Seagraves, & J. Barchas, Influence on social setting on the induction of brain cyclic AMP in response to electric shock in the rat, Nature 263, 433 (1976).

96. T. Silverstone, Anorectic Drugs in Silverstone, J. T. Ed. (1975), Obesity: Pathogenesis and Management Publishing Sciences Group, Acton, Mass. p 193.

97. I. R. Innes and M. Nickerson, Norepinephrine, epinephrine, and the sympathomimetic amines, In Goodman, L.S. and Gilman, A. (1970), The Pharmacological Basis of Therapeutics, Macmillan Publishing Co., N.Y. p. 477.

ANIMAL MODELS OF AGGRESSIVE BEHAVIOR

Michael H. Sheard

Yale University Medical School and the
Connecticut Mental Health Center,
New Haven, Connecticut 06519, USA

INTRODUCTION

Aggressive behavior is behavior which inflicts or threatens to inflict harm or damage to some goal entity.

It is behavior of primary importance in biological adaptation and problems with aggressive behavior abound in many psychiatric syndromes. There is, for example, combativeness in psychoses, suicide and homicide, in affective illness, assaultiveness in alcoholism, violence associated with drug dependency, as well as the battered child and wife syndromes. On the other hand there are problems of over-control of appropriate aggressive behavior. Aggressive behavior can be classified into two main types - irritable or defensive, and predatory (1,2). Although Moyer (3) has attempted a more complex classification based on the stimulus which elicits the behavior, e.g., predatory - (cat - rat; rat - mouse); intermale; isolation induced in mice; fear induced; irritable aggression - (e.g., shock elicited fighting) territorial, maternal, and instrumental. The relevance of such animal models to aggression in man is disputed. At first sight it appears that there is little evidence for predatory aggression in man. His aggressive behavior can be classified as either irritable (e.g., fear or pain-induced - including "psychic pain") or instrumental. A case can be made, however, for assuming that the neural substrates for instrumental aggression may overlap with or actually be closely linked with those neural substrates which subserve predatory aggression (quiet biting attack) in other mammals. Animals can learn instrumental aggression. For instance, rats can learn to fight by presenting a positive stimulus contingent upon attack behavior (4) or by removing a negative stimulus contingent on attack (5).

In animals a fruitful approach to understanding aggressive behavior has been to investigate the genetic, neurophysiological and neurochemical substrates or correlates of aggressive behaviors. In this endeavor, it has become evident that it is imperative to consider different configurations of aggression. Evidence is accumulating that these different categories of aggressive behavior have different genetic, neurophysiological and neurochemical correlates, and are differentially affected by drugs (3,6,7,8). Moreover, to understand the action of a drug upon a specific aggressive behavior, it is necessary to measure its effect upon a number of other behaviors, such as general or specific reactivity, motor activity, learning, etc.

Two sets of experimental strategies are possible: 1) Behavioral aggression is elicited and the neurophysiological and neurochemical changes are correlated with changes in the behavior; or 2) The supposed neuroanatomical or neurochemical substrates are manipulated (e.g., by lesions, stimulation or drugs) and changes in behavioral aggression are monitored. These strategies are described in two models of the main categories of aggressive behavior; predatory (elicited behavior in the cat) and irritable (shock elicited fighting in the rat).

NEURAL BASIS OF ATTACK BEHAVIOR IN THE CAT

The neural basis of attack behavior in the cat is known in some detail from the work of Flynn and his colleagues at Yale (9). Many other workers have contributed to this area (10, 11,12). While the majority of the work has utilized electrical stimulation in the cat, the findings can be generalized to other species, for example, ring dove, rat (13,14), dog, opossum (15), and primates (16).

Different forms of aggressive behavior can be reproducibly elicited by electrical stimulation in the same animal. One form first reported by Hess (17) has been called affective defense (18), affective attack (19), defensive threat (15), and rage. This form of aggressive behavior is marked by intense autonomic arousal and vocalizations such as hissing, spitting and growling. When an attack occurs, it is frequently initiated by well-directed paw strikes with unsheathed claws. The stimulation which elicits this behavior is sometimes aversive and sometimes not (11,19,21).

The second form of attack behavior, first described by Wasman and Flynn (19) is called quiet biting attack since it occurs with little autonomic arousal, and with biting of the neck of the target animal. This behavior has also been called predatory, since the cat frequently shows a motor pattern characteristic of stalking a prey. However, the behavior can clearly be differentiated from eating behavior (20). It is interesting to note that the sites from which quiet biting attack may be elicited often demonstrate self-stimulation (13,14) suggesting the possibility that such aggressive behavior may be reinforcing and thus be a useful model for nonaffective aggression in man.

The neural substrate from which these two attack behaviors can be elicited includes sites in the hypothalamus (12,14,18,19), the midbrain (12,22,23,24), the preoptic region (25), the stria terminalis and its bed nucleus (22,25), and amygdala (22,26,27,28,29), thalamus (25, 30), as well as pons and medulla (9). Preoptic, hypothalamic and midbrain areas from which quiet biting attack can be elicited are lateral to those which elicit affective defense. In general these areas are separate from continuations of afferent systems subserving pain, which on stimulation can elicit affective displays and attack. Recent work in Flynn's laboratory has extended the neural substrate for attack caudally through the pons to the level of the brachium conjunctivum and the inferior olive, where the thresholds are lower than in the hypothalamus. It appears that limbic forebrain structures modulate the hypothalamus which in turn modulates the midbrain and pontine sites. Anatomical connections exist to underly such behavioral findings (9). Moreover, combined behavioral and reduced silver degeneration studies suggest separate pathways underlying quiet biting and affective defense attack (6,7). The former follows the descending medial forebrain bundle to the ventral tegmental area of the midbrain, whereas the latter travels in the dorsal longitudinal fasciculus (Schutz's bundle) in the ventrolateral periaqueductal region. Stimulation of these areas in the midbrain elicit respectively either quiet biting attack (31) or affective defense (32). The role of the hypothalamus appears to be a regulation of sites lower down the brainstem in midbrain and pons. The fact that these forms of aggressive behavior can be elicited by brain stimulation does not imply that such behavior is independent of environmental stimuli.

Concurrent stimulation and lesioning of limbic forebrain structures have been utilized to determine their role in aggressive behavior. For example, stimulation of the basal nuclei of the amygdala, the dorsal hippocampus, the rostral midline thalamus and medial piriform cortex suppresses hypothalamically elicited attack, while stimulation of the dorsolateral portion of the lateral nucleus of the amygdala, the ventral hippocampus, the posterior midline thalamic region and lateral piriform cortex facilitates hypothalamic attack (25,30, 33). In particular, Egger and Flynn (33) have shown that electrical stimulation in the magnocellular portion of the basal amygdaloid nucleus and in the anterior and medial portions of the lateral amygdaloid nucleus generally suppresses hypothalamically elicited attack. Stimulation in the dorsolateral portion of the lateral nucleus facilitates attack. Lesions in the region which produced suppression induced facilitation, especially when the lesion involved the border region between the magnocellular basal nucleus and the lateral nucleus. The precise neuroanatomical and neurophysiological basis for the modulatory behavior largely remains to be worked out.

In summary, there appears to be a hypothalamic-midbrain-pontine substrate which integrates and organizes aggressive behavior, as well as distinct regions which exercise a modulatory role, such as the amygdala. A third category of neural mechanisms resides in the sensory and motor systems through which the aggressive act is carried out. All three neural mechanisms present possible sites for drug action on aggressive behavior.

NEURAL BASIS OF IRRITABLE AGGRESSION IN RATS (SHOCK ELICITED FIGHTING)

The most commonly used animals are rats, although snakes and monkeys also show the phenomena (34). Animals are placed in a small enclosure and subjected to inescapable electric shock. They fight in behaviorally distinct patterns which are dependent upon the frequency and intensity of shock. The fighting behavior is not altered by deprivation of vision or olfactory bulb removal, but is reduced by removal of the vibrissae (35,36). Thus this type of aggression is clearly different from mouse killing (a form of predatory aggression in the rat), which does increase following olfactory bulb removal (37,38). Shock induced fighting is facilitated by lesions of the septum and ventromedial nucleus of the hypothalamus, while it is decreased with lesions of the amygdala and hippocampus (39,40,41,42). Moreover, with these lesions alteration in SEF does not correlate with changes in pain sensitivity or open field behavior. Neither lesions of the cingulate cortex nor dorsomedial nucleus of the thalamus have any effect on SEF (39).

ROLE OF NEUROTRANSMITTER SYSTEMS IN PREDATORY AGGRESSION

Catecholamines (CA)

In the cat, rage behavior can be induced with 1-dopa administration, and in surgically produced sham rage there is a decrease in brainstem norepinephrine (NE) which correlates with the number of attacks (43). In electrically stimulated rage there is also a decrease in brainstem NE associated with an increase in NE metabolites (44). Further evidence that catecholamines are implicated in this behavior comes from the finding that if NE is depleted from brain and its synthesis inhibited by a combination of reserpine and α-methyl-p-tyrosine (AMPT), surgically induced rage cannot be elicited (45). Also disulfiram - a dopamine-β-hydroxylase inhibitor, which inhibits production of NE from dopamine (DA), and haloperidol,a dopamine receptor as well as an adrenergic blocker, antagonize the rage behavior seen with l-dopa. On the other hand, the l-dopa rage is facilitated by protriptyline which blocks the reuptake of NE..

Previous experiments showed that facilitation of electrically-elicited attack occurred with several doses of d,l-amphetamine (46). These studies were repeated using d-amphetamine in doses of 0.125, 0.25, 0.5 and 1 mg/kg. A biphasic dose-response was found with facilitation occurring most consistently at a dose of 0.25 mg/kg and inhibition did not block the facilitating effect of a dose of 0.25 mg d-amphetamine on elicited attack. These studies point to a role for catecholamines in the regulation of elicited attack behavior and suggest further studies which would determine more accurately the separate roles of NE, DA and 5HT, as well as the interaction between catecholamine and serotonergic systems.

Serotonin (5HT)

A reduction of brain 5HT by p-chlorophenylalanine (PCPA) increases spontaneous predatory aggression in the cat (47) and lowers the threshold for hypothalamically elicited attack (48). Conversely, stimulation of the midbrain raphe region in cats at sites which cause a release of 5HT (49) produces an inhibition of elicited attack behavior (50). Further experiments to demonstrate a role for 5HT have used LSD. In five cats LSD was found to have a complicated effect on attack behavior. Low doses caused a period of inhibition which was followed by facilitation, which was again followed by a period of inhibition. P-Chloroamphetamine, which causes a rapid depletion of 5HT following a brief period of acute release, was found in three cats to produce an inhibition of attack followed by a prolonged facilitation (51). This agreed with the results found with PCA in SEF in rats. These studies implicate a role for 5HT in the regulation of elicited predatory aggression and affective defense.

Cholinergic System

Predatory aggression in the rat appears to be influenced primarily by cholinergic mechanisms. For example, cholinergic stimulation of the lateral hypothalamus (52,53) induces muricide, as does injection of pilocarpine. Early work on the elicitation of angry displays or affective attack by administering acetylcholine (ACh) or carbachol into the ventricles was complicated by the frequent presence of after discharges in the hippocampus or amygdala.

However, local injections into the brain itself have also produced aggressive behavior in the cat and rat (54,55). This has been shown to occur without seizure activity as monitored by EEG. However, the number of sites from which sham rage or affective attack could be involved by microinjection of cholinergic substances is very great and include septum, central grey, posterior hypothalamus and preoptic region (56). The best evidence for the role of the cholinergic system has been presented by Bandler (52,54). Frog killing was produced in the rat by injections of ACh or carbachol (in conjunction with a cholinesterase inhibitor) into the lateral hypothalamus, medial or midline thalamic nuclei or central midbrain tegmentum. It was demonstrated that these results were not due to osmotic effects, nonspecific neural activity or changes in general activity. The sites were similar to those from which electrical stimulation produces predatory aggression in the rat and cat. Neostigmine, a cholinesterase inhibitor, facilitates this behavior, while atropine has been reported to depress it. These results suggest a role for naturally occurring ACh in predatory aggression. There are conflicting reports on induction of mouse killing in non-killing rats with injection of carbachol into the brain. Some positive results have been reported which have not been confirmed by Bandler, who also could not suppress natural killing with atropine.

While ACh is important, both 5HT and CA affect predation. 5HT appears to play an inhibitory role since tryptophan and 5-hydroxytryptophan (5HTP) can inhibit muricide (57), while 5HT reduction by lesions (58) or PCPA (59,60) increases muricide. The catecholamine system is also involved since 6-hydroxydopamine (6OHDA), a drug which lowers NE and DA, increases the latency to killing (61). However, many of the changes in brain neurotransmitters occur both in rats which become muricidal and those which do not.

ROLE OF NEUROTRANSMITTER SYSTEMS IN SHOCK ELICITED FIGHTING (SEF)

Catecholamines

In general, drugs or environmental changes which increase central NE increase irritable aggression (62). For example, deprivation of rapid eye movement sleep or immobilization stress both increase SEF. Rubidium, which increases NE turnover (63,64,65), increases SEF, as do tricyclic antidepressants and monoamine oxidase inhibitors (MAOI) (66). However,

6OHDA, which lowers central NE and DA, or 6-hydroxydopa, which lowers NE alone, also increases SEF. This might be explained by the development of supersensitivity in postsynaptic CA receptors. Opposite effects on SEF have been described for intraventricular infusion of NE versus DA (67).

Serotonin

Evidence has been accumulating to suggest that 5HT has an inhibitory role in some aggressive behaviors, although this conclusion is still controversial. Depletion of 5HT by PCPA (68), raphe lesions (69,70) or the combination of raphe lesions and PCPA (71) increases mouse killing in rats, although increased muricide does not occur in every animal and may not be predicted by 5HT levels alone. Lesions of specific raphe nuclei increase SEF (72) whereas PCPA has been reported not to alter this form of behavior (62,73). The apparent lack of effect of PCPA on SEF was surprising in view of the fact that it has been consistently reported to reduce pain thresholds (74,75). It was recently shown that LSD augmented the acoustic startle reflex when relatively long interstimulus intervals (ISI) were used but not when relatively short intervals were used (76). This suggested that a critical factor in detecting behavioral changes induced by 5HT inhibition is the interval between eliciting stimuli. Interestingly, the one report in the literature showing an increase in SEF with PCPA did use a long ISI (77). Experiments were performed to test this idea and it was found that PCPA indeed enhances SEF when the ISI was long (15 sec) but not when short (3 sec). This difference could not be accounted for by a difference in pain threshold at different ISIs, since it was lowered by PCPA at both ISIs to a comparable degree. This finding resolved the apparent discrepancy between PCPA and lesion studies. The problem remains, however, as to why an increase in SEF could not be detected at short ISIs when 5HT is depleted with PCPA. The explanation appears to be that PCPA increases active avoidance behavior (78), and that this is incompatible with SEF but dissipates fast enough during a long ISI to allow fighting to appear.

Action of Indole Hallucinogens on SEF

Another category of drugs which predominantly affect brain 5HT systems are indole hallucino-gens. Basic studies on these compounds have shown that low doses of LSD specifically inhibit the firing rate of presynaptic neurons in the midbrain raphe region (79) and decrease the release of 5HT (80). High doses of LSD can also inhibit cells postsynaptic to the midbrain raphe (81) and thus mimic the action of 5HT on postsynaptic sites. The effects of LSD on single cells suggested that LSD could have a biphasic dose-response effect on behaviors modulated by the raphe 5HT system. Several such findings have in fact been reported (82,83, 84). In agreement with this notion it was found that LSD enhances SEF in rats when given in doses from 20-160 µg/kg (85). Since the effect of low doses is to specifically inhibit the presynaptic 5HT neurons and thereby decrease impulse-mediated release of 5HT, the results are consistent with the experiments which showed increased SEF when 5HT is lowered by raphe lesions of PCPA. At higher doses LSD did not affect SEF, suggesting that the postsynaptic agonistic effect may have effectively cancelled the presynaptic excitatory effect. A problem with high doses of LSD is that they can act like DA agonists. The DA agonist property may in part account for the lack of a depressive effect on SEF, and in other experiments it was found that in rats pretreated with the DA antagonist haloperidol, LSD does depress SEF. NN dimethyl tryptamine (DMT) has a biphasic dose response in acoustic startle, but the excitatory effect was relatively small and found at a low dose (0.25 mg/kg) (86). DMT has a weak and inconsistent excitatory effect on SEF between 0.125 and 1 mg/kg, whereas a signifi-cant depressive effect was found at 4 mg/kg and 8 mg/kg. Thus, both LSD and DMT had similar effects upon acoustic startle response and SEF. However, a surprising difference was found when 5-methoxydimethyl tryptamine (5MeODMT) was used. While 5MeODMT proved to be excitatory to startle, with response monotonically related to dose (0.12 - 8.0 mg/kg), with SEF it proved to have little or no effect at low doses and a significant depressant effect at higher doses. This finding suggests that effects on spinal cord receptors rather than on more rostral sites may be more important in the actions of 5MeODMT than in other indole hallucinogens. Indeed, 5MeODMT does not appear to be very potent judging from iontophoretic studies on forebrain areas receiving an identified 5HT input (87).

Effects of P-Chloroamphetamine (PCA) on SEF

The findings with 5HT depletion, together with those on indole hallucinogens, suggest that the raphe-5HT system has an inhibitory role in SEF. Thus release of 5HT can be expected to inhibit fighting, and 5HT depletion or depression of impulse-mediated 5HT release, to increase fighting. PCA shortly after administration markedly releases 5HT (88) and there-after causes a rapid depletion (89,90,91). PCA had, in fact, different effects upon SEF as a function of time after injection.

Fifteen minutes after injection PCA depressed fighting, at two hours fighting was somewhat greater than in controls, and at all later times fighting was greatly enhanced. Pretreatment of rats with AMPT did not block either the short- or long-time effect of PCA on fighting, whereas PCPA pretreatment completely blocked both the 15-minute depression and the 3-hour increase. These results strongly suggest that it is 5HT rather than CAs which mediates the changes. The results indicated that the effects of PCA were not mediated by newly synthe-sized CA, but it remains possible that PCA might influence SEF through release of stored CA (92,93,94). The fact that PCPA blocked the depressive effect of PCA on SEF also indicated that the inhibition did not simply result from animals being too sick to fight. Flinch-jump experiments were performed which suggest that the initial effect of PCA may be accounted for by changes in pain threshold, since at 15 minutes there was a marked increase in threshold followed by a decrease at 3 and 24 hours. However, pain threshold returned to normal at one and four weeks post-PCA when brain 5HT is still depleted, whereas fighting behavior remained enhanced. These long-term effects could not be explained on the basis that PCA-treated rats learned to fight more after the first few test sessions, since de novo testing at one and four weeks also showed augmentation. These results are consistent with the idea that inhibition of SEF is associated with enhanced release of 5HT, whereas enhancement of SEF is associated with 5HT inhibition or depletion (95).

Tricyclic Antidepressant Drugs: Antagonism of Effect of LSD on SEF

The tertiary amine antidepressant drugs depress the firing rate of presynaptic 5HT neurons in the midbrain raphe (96). However, this effect appears to be mediated indirectly via a block of reuptake, and moderate doses do not cause a net change in the release of 5HT (80). Therefore it could be expected that pretreatment with moderate doses of a tricyclic anti-depressant might block the action of LSD, since the presynaptic neurons are already inhibited. This was indeed found to be the case for acoustic startle (97). Experiments were performed to test the effect of chlorimipramine (CIMI) on SEF and on the excitatory effect of LSD. As a comparison another tricyclic drug, desipramine (DMI), was tested in a dose which has little or no effect on raphe firing (96). The results showed that CIMI and DMI in single doses (5 mg/kg) did not affect SEF. However, both CIMI and DMI antagonized the excitatory effect of LSD. DMI antagonism is inconsistent with the above rationale. DMI is more potent in blocking the reuptake of NE than in blocking the reuptake of 5HT (98). At the dose used, DMI does not block firing of raphe neurons but instead depresses the firing rate of neurons in the NE-containing locus coeruleus (99). This finding raises the possibility that a NE synapse is involved in the neural pathway to the final behavioral output. It also suggests the possibility that the tricyclic antidepressant drugs increase the sensitivity to LSD of neurons postsynaptic to the serotonergic raphe fibers (100).

Effect of Noise on SEF

There is anecdotal as well as some human experimental evidence on the effect of noise on aggression. Geen and O'Neal (101) found that exposure to a fight film increased the intensity and number of shocks delivered to a target when white noise was presented during the shock-delivering phase, but not when it was omitted. However, there are little or no experimental animal data. The effect of noise on SEF was examined by comparing levels of fighting at 75db versus 55 db (which was the ambient noise level with the cage fan on). Fighting was significantly increased at each of four shock intensities at 75 db. It is possible that noise could increase aggression by increasing arousal. For example, Berkowitz (102) has noticed that any arousal-enhancing stimulus in conjunction with aggres - sive cues may generate impulsive aggressive behavior in man. Many arousal mechanisms exhibit non-monotonic relationships to stimulus intensity, such that the behavioral effect is greatest at intermediate levels of stimulus intensity. For example, the relationship between startle response and background noise is non-monotonic (103,104). As background noise may be an effective means of manipulating arousal, it was of interest to determine whether shock elicited aggression would also bear a non-monotonic relationship to the level of background noise. An experiment confirmed that this was indeed the case. Rats fought more at 75 db, but less at 90 db, compared with 55 db. Varying the length of pre-exposure to noise before test shocks did not appear to have any effect. An experiment was run to measure how long the increase in aggression lasted after the noise intensity was reduced. It was found that there was a very rapid drop in the frequency of fighting, suggesting that the effect of noise on SEF is most dependent upon the level at the time of testing, with the effect developing and decaying very rapidly (105).

In summary, it is apparent that while the neural substrate for cat attack is well worked out, the pharmacology is not, whereas the pharmacology of shock elicited fighting is more advanced than a knowledge of its neural substrates. Moreover, it is clear that alterations in identified neurotransmitters can affect one or more types of aggressive behavior. Thus NE and DA have been implicated particularly in sham rage and affective attack as well as in irritable aggression. The cholinergic system is involved in predatory and affective attack and 5HT appears to play an inhibitory role in several forms of aggressive behavior.

ROLE OF DRUGS ON AGGRESSIVE BEHAVIOR AND NEUROTRANSMITTERS

Previous research has demonstrated that drugs and hormones can modify the latencies and thresholds of hypothalamically elicited aggressive behavior. For example, amphetamine administration has been shown to lower attack latencies (46) and lithium and imipramine administration has been shown to increase attack latencies (50,106). Similarly, both before and after gonadectomy, the administration of estrogen, testosterone and LH to male cats and

FSH to female cats has been found to decrease attack latencies, while the administration of FSH to male cats and estrogen, testosterone or LH to female cats has been shown to increase attack latencies (107).

In general, however, there is poor agreement between authors reporting on the effects of lithium on 5HT, due in part to species differences, toxicity and dose and chronicity differ-ences. A mechanism which may help to explain the disagreement has been presented by Knapp and Mandell (104,110). They found that chronic lithium administration gives rapid, and chronic, facilitation of tryptophan uptake with an acute increase in 5HT synthesis. This is followed by a drop of 5HT synthesis to normal, accompanied by a chronic decrease in the activity of tryptophan hydroxylase. Their observations offer a basis for rationalizing different results in studies of different duration (albeit not without difficulty) and point to a possible feedback mechanism regulating serotonin biosynthesis.

A problem with much of the previous work with lithium in animals is doubtful relevance to its clinical action in man. This is because it has either been studies in acute doses, or has been given chronically in doses which produce blood levels far higher than normally seen in clinical practice. Another factor has been the difficulty in administering lithium in such a way as to maintain a steady blood and brain level and avoid toxic complications.

It has been found that lithium caused a dose related increase in 5HIAA in the forebrain, particularly when synthesis was enhanced by electrical stimulation of the raphe region (108). This was not due to an action of lithium in blocking the exit of 5HIAA from brain since lithium still caused a time-dependent increase in concentration of 5HIAA after probenecid. Behavioral studies in rats depleted of brain 5HT following administration of PCPA showed an interesting exaggeration in sexual and aggressive behavior (68). It was thought therefore that lithium, by increasing the release of 5HT, might inhibit aggressive behavior. A series of experiments showed that lithium could indeed inhibit aggressive behavior in a wide range of situations. For example, shock-elicited aggression (112), territorial aggression (113), isolation-induced aggression (114), and PCPA-induced aggression (115) were all inhibited. Furthermore, studies of chronic assaultive behavior in the human showed that lithium could inhibit human aggression (116,117,118). In some of these human cases there was reduction of angry affect (116). The first study (116) was single blind but was recently confirmed in a large-scale study using a double-blind design (119). This was a three-year study of lithium versus placebo on impulsive aggressive behavior disorders in 16 through 22-year old incarcerated male delinquents. The success of this clinical study encourages studies of the action of lithium and other drugs on aggression in the animal models.

Effects of Lithium on SEF

A method of non-toxic lithium administration has been developed which ensures that rats gain weight the same as controls. This enables behavioral experiments to be carried out without serious risks of toxicity which have contaminated many previous studies of the effect of lithium in rats (111) and with maximum comparability of controls and treated animals. After 4-5 weeks on lithium, rats have been tested for SEF, general motor activity, and acoustic startle response. A consistent pattern of a decrease in SEF, an increase in general motor activity and a stabilization of the acoustic startle within a test session is observed. The latter finding consists of the lithium rats showing a lessened startle reflex to the first tone, and a smaller difference between the response to the first and last tone, than control rats. Preliminary results suggest that there is no change in sensitivity of the presynaptic 5HT receptor since low doses of LSD appear to have the same excitatory effect on SEF as in controls. Also single cell studies on midbrain raphe neurons showed no change in their responsiveness to low doses of LSD. However, a dose of DMT which had no effect on SEF in control rats was found to significantly depress fighting in lithium treated rats, suggesting that the postsynaptic 5HT receptor might have increased sensitivity.

Effects of Lithium on Elicited Attack

A dose-response study using oral administration of slow-release lithium determined the rela-tionship between non-toxic dose ranges and serum half life. This study was necessary in order to avoid the dangers of toxicity with lithium which would otherwise contaminate behavioral experiments. Using the knowledge of appropriate lithium dosage, the effect of lithium given in single daily oral doses was investigated on elicited attack behavior. Both quiet biting and affective defense sites were tested. Results showed consistent inhibition of attack behavior elicited from hypothalamic sites which was dose related. Moreover, at levels of serum lithium which inhibited hypothalamic site attack behavior, general motor patterns were not affected, nor were other elicited hypothalamic behaviors such as hissing and growling. Motor or attack behavior elicited from midbrain sites were not affected except in one cat which histology showed had a large lesion in the region of the electrode. The conclusion from these results was that lithium can inhibit electrically elicited attack behavior and also shows specificity as to the type of behavior affected and sites of action.

Supported by PHS Grant MH26446 and the State of Connecticut.

REFERENCES

1. D. Reis, Central neurotransmitters in aggression, Res. Publ. A.R.N.M.D. 52, 119 (1974).

2. M. H. Sheard, Effect of drugs in enhancing or inhibiting aggression, Psychopharm. Bull. (1977).

3. K. E. Moyer, Kinds of aggression and their physiological basis, Comm. Behav. Biol. 2A, 65 (1968).

4. R. Ulrich, M. Johnston, J. Richardson, and P. Wolff, The operant conditioning of fighting behavior in rats, Psychol. Rec. 13, 475 (1963).

5. A. J. Motshagen, and J. L. Slangen, Instrumental conditioning of aggressive behavior in rats, Aggress. Behav. 1, 1957 (1975).

6. C. C. Chi, and J. P. Flynn, Neural pathways associated with hypothalamically elicited attack behavior in cats, Science 171, 703 (1971).

7. C. C. Chi, and J. P. Flynn, Neuroanatomic projections related to biting attack elicited from hypothalamus in cats, Brain Res. 35, 49 (1971).

8. B. Eichelman, The catecholamines and aggressive behavior, Neurosci. Res. 5, 109 (1973).

9. J. P. Flynn, Neural basis of threat and attack, in R. G. Grenell, and S. Gabay, eds. Biological Foundations of Psychiatry, Raven Press, New York (1976).

10. R. R. Hutchinson, and J. W. Renfrew, Stalking attack and eating behaviors elicited from the same sites in the hypothalamus, J. Comp. Physiol. Psychol. 61, 360 (1966).

11. H. Nakao, Emotional behavior produced by hypothalamic stimulation, Am. J. Physiol. 104, 411 (1958).

12. F. M. Skultety, Stimulation of periaqueductal gray and hypothalamus, Arch. Neurol. 8, 608, (1963).

13. J. Panksepp, Aggression elicited by electrical stimulation of the hypothalamus in albino rat, Physiol. Behav. 6, 321 (1970).

14. C. Woodworth, Attack elicited in rats by electrical stimulation of the lateral hypothalamus, Physiol. Behav. 6, 345 (1970).

15. W. W. Roberts, M. L. Steinberg, and L. W. Means, Hypothalamic mechanisms for sexual aggressive and other motivational behaviors in the opossum, Didelphis Virginiana. J. Comp. Physiol. Psychol. 64, 1 (1967).

16. B. W. Robinson, M. Alexander, and G. Bowne, Dominance reversal resulting from aggressive responses evoked by brain telestimulation, Physiol. Behav. 4, 749 (1969).

17. W. R. Hess, Stammanglien reizzversuch. 10. Tagung der Deutschen Physiologischen Gesellschaft in Frankfurt, Berl. Ges. Physiol. 42, (1928).

18. W. R. Hess, and M. Brugger, Das subkortikale Zentrum der affektiven Abwehrreaktion, Helv. Physiol. Pharmacol. Acta 1, 33 (1943).

19. M. Wasman, and J. P. Flynn, Directed attack elicited from hypothalamus, Arch. Neurol. 6, 220 (1962).

20. M. D. Egger, and J. P. Flynn, Effects of electrical stimulation of the amygdala on hypothalamically elicited attack behavior in cats, J. Neurophysiol. 26, 705 (1963).

21. D. Adams, and J. P. Flynn, Transfer of an escape response from tail shock to brain-stimulated attack behavior, J. Exp. Anal. Behav. 14, 399 (1966).

22. A. Fernandez De Molina, and R. W. Hunsperger, Central representation of affective reactions in forebrain and brain stem: Electrical stimulation of amygdala, stria terminalis and adjacent structures, J. Physiol. (Lond.) 145, 251 (1959).

23. M. Sheard, and J. P. Flynn, Facilitation of attack by stimulation of the midbrain of cats, Brain Res. 27, 165 (1967).

24. R. W. Hunsperger, Affektreaktionen auf elektrische Reizung in Hirnstamm der Katze, Helv. Physiol. Pharmacol. Acta 14, 70 (1956).

25. M. MacDonnell, and J. P. Flynn, Attack elicited by stimulation of the thalamus in cats, Science 144, 1249 (1964).

26. E. Fonberg, The normalizing effect of lateral amygdalar lesions upon the dorsomedial amygdalar syndrome in dogs, Acta Neurobiol. Exp. 33, 449 (1973).

27. A. W. Zbrozyna, The organization of the defense reaction elicited from amygdala and its connections, in B. E. Eleftheriou, ed. The Neurobiology of the Amygdala, Plenum Press, New York (1972).

28. B. R. Kaada, Stimulation and regional ablation of the amygdaloid complex with reference to functional representations, in B. E. Eleftheriou, ed. The Neurobiology of the Amygdala, Plenum Press, New York - London (1972).

29. H. Ursin, and B. R. Kaada, Functional localization within the amygdaloid complex in the cat, Electroencephalogr. Clin. Neurophysio.. 12, 1 (1960).

30. A. Siegel, and J. P. Flynn, Differential effects of electrical stimulation and lesions of the hippocampus and adjacent regions upon attack behavior in cats, Brain Res. 7, 252 (1968).

31. R. J. Bandler, C. C. Chi., and J. P. Flynn, Biting attack elicited by stimulation of the ventral midbrain tegmentum in cats, Science 177, 364 (1972).

32. S. E. Edwards, and J. P. Flynn, Corticospinal control of striking in centrally elicited attack behavior, Brain Res. 41, 51 (1972).

33. M. D. Egger, and J. P. Flynn, Effects of electrical stimulation on hypothalamically elicited attack behavior in cats, J. Neurophysiol. 26, 705 (1963).

34. R. E. Ulrich, and N. H. Azrin, Reflexive fighting in response to aversive stimulation, J. Experimental Anal. Behav. 5, 511 (1962).

35. N. M. Bugbee, and B. Eichelman, Sensory alterations and aggressive behavior in the rat, Physiol. and Behav. 8, 981 (1972).

36. D. H. Thor, and W. B. Ghiselli, Suppression of mouse killing and apomorphine-induced social aggression in rats by local anesthesia of the mystacial vibrissae, J. Comp. and Physiol. Psychol. 88, 40 (1975).

37. P. Karli, M. Vergnes, and F. Didiergeorges, Rat-Mouse interspecific aggressive behavior and its manipulation by brain ablation and by brain stimulation, in S. Garattini, and E. B. Sigg, eds. Aggressive Behavior, John Wiley, New York (1969).

38. H. Bernstein, and K. E. Moyer, Aggressive behavior in the rat: effects of isolation, and olfactory bulb lesions, Brain Res. 20, 75 (1970).

39. B. Eichelman, Effect of subcortical lesions on shock-induced aggression in the rat, J. Comp. and Physiol. Psychol. 74, 331 (1971).

40. R. J. Blanchard, and D. C. Blanchard, Limbic lesions and reflexive fighting, J. Comp. and Physiol Psychol. 68, 603 (1968).

41. K. A. Miczek, and S. P. Grossman, Effects of septal lesions on inter- and intra-species aggression in rats, J. Comp. Physiol. Psychol. 79, 37 (1972).

42. A. B. Wetzel, R. L. Conner, and S. Levine, Shock-induced fighting in septal-lesioned rats, Psychonomic Science 9, 133 (1967).

43. D. J. Reis, and K. Fuxe, Brain norepinephrine: evidence that neuronal release is essential for sham rage behavior following brainstem transection in cat, Proc. Nat. Acad. Sci., U.S.A. 64, 108 (1969).

44. L. M. Gunne, and T. Lewander, Monoamines in brain and adrenal glands of cat after electrically induced defense reaction, Acta Physiol. Scand. 67, 405 (1966).

45. D. J. Reis, Brain monoamines in aggression and sleep, Clin. Neurosurg. 18, 471 (1971).

46. M. H. Sheard, The effects of amphetamine on attack behavior in the cat, Brain Res. 5, 330 (1967).

47. J. Ferguson, S. Henriksen, J. Cohen, G. Mitchell, J. Barchas, and W. Dement, "Hypersexuality" and behavioral changes in cats caused by administration of p-chlorophenylalanine, Science 168, 499 (1970).

48. M. F. MacDonnell, L. Fessok, and S. H. Brown, Aggression and associated neural events in cats, Quart. J. Stud. Alcohol. 32, 748 (1971).

49. M. H. Sheard, and A. J. Zolovick, Serotonin release in cat brain and cerebrospinal fluid on stimulation of midbrain raphe, Brain Res. 26, 455 (1971).

50. M. H. Sheard, The effect of stimulation of the raphe on hypothalamically elicited attack in cats, J. Psychiat. Res. 10, 151 (1974).

51. M. H. Sheard, The effect of p-chloroamphetamine on behavior, Psychopharm. Bull. 12, 59 (1976).

52. R. J. Bandler, Cholinergic synapses in the lateral hypothalamus for the control of predatory aggression in the rat, Brain Res. 20, 409 (1970).

53. R. J. Bandler, Direct chemical stimulation of the thalamus: effects on aggressive behavior in the rat, Brain Res. 26, 81 (1971).

54. R. J. Bandler, Cholinergic synapses in the lateral hypothalamus for the control of predatory aggression in the rat, Brain Res. 20, 409 (1970).

55. P. D. MacLean, Chemical and electrical stimulation of hippocampus in unrestrained animals. II. Behavioral findings, AMA Arch. Neurol. 78, 128 (1957).

56. R. D. Myers, Emotional and autonomic responses following hypothalamic chemical stimulation, Can. J. Psychol. 18, 6 (1964).

57. A. S. Kulkarni, Muricidal block produced by 5-hydroxytryptophan and various drugs, Life Science 7, 125 (1968).

58. L. D. Grant, D. V. Coscina, S. P. Grossman, and D. X. Freedman, Muricide after serotonin depleting lesions of midbrain raphe nuclei, Pharmacol. Biochem. Behav. 1, 77 (1973).

59. M. H. Sheard, The effect of p-chlorophenylalanine on behavior in rats: relation to 5-hydroxytryptamine and 5-hydroxyindoleacetic acid, Brain Res. 15, 524 (1969).

60. G. DiChiara, R. Camba, and P. F. Spano, Evidence for inhibition by brain serotonin of mouse killing behavior in rats, Nature 233, 272 (1971).

61. D. Jimmerson, and D. J. Reis, Effects of intrahypothalamic 6-hydroxydopamine on predatory aggression in rat, Brain Res. 61, 141 (1973).

62. B. Eichelman, and N. B. Thoa, The aggressive monoamines, Biol. Psych. 6, 143 (1973).

63. J. M. Stolk, W. J. Nowack, J. D. Barchas, and S. R. Platman, Brain norepinephrine: enhanced turnover after rubidium treatment, Science 168, 501 (1970).

64. J. M. Stolk, R. L. Conner, and J. D. Barchas, Rubidium-induced increase in shock-elicited aggression in rats, Psychopharma. 22, 250 (1971).

65. B. Eichelman, N. B. Thoa, and J. Perez-Cruit, Alkali metal cations: effects on aggression and adrenal enzymes, Pharmacol. Biochem. Behav. 1, 121 (1973).

66. B. Eichelman, and J. Barchas, Facilitated shock-induced aggression following anti-depressive medication in rat, Pharmacol. Biochem. Behav. 3, 601, (1975).

67. M. A. Geyer, and D. S. Segal, Shock induced aggression: opposite effects of intra-ventricularly infused dopamine and norepinephrine, J. Behav. Biol. 10, 99 (1974).

68. M. H. Sheard, The effect of p-chlorophenylalanine on behavior in rats: relation to 5-hydroxytryptamine and 5-hydroxyindoleacetic acid, Brain Res. 15, 524 (1969).

69. L. D. Grant, D. V. Coscina, S. P. Grossman, and D. X. Freedman, Muricide after serotonin depleting lesions of midbrain raphe nuclei, Pharmacol. Biochem. Behav. 1, 77 (1973).

70. M. Vergnes, G. Mack, et E. Kempf, Lésions du raphé et réaction d'agression inter-spécifique ratsouris. Effets comportementaux et biochimiques, Brain Res. 57, 67 (1973).

71. M. Vergnes, G. Mack, et E. Kempf, Contrôle inhibiteur du comportement d'agression interspécifique du rat: système serotoninergique du raphé et afférences olfactives, Brain Res. 70, 481 (1974).

72. B. L. Jacobs, and A. Cohen, Differential behavioral effects of lesions of the median or dorsal nuclei in rats: open field and pain elicited aggression, J. Comp. Physiol. Psychol. 90, 102 (1976).

73. R. L. Conner, J. M. Stolk, J. D. Barchas, W. C. Dement, and S. Levine, The effect of p-chlorophenylalanine (PCPA) on shock-elicited fighting behavior in rats, Physiol. Behav. 5, 1121 (1970).

74. J. A. Harvey, A. J. Schlosberg, and L. M. Yunger, Behavioral correlates of serotonin depletion, Fed. Proc. 34, 1796 (1975).

75. S. S. Tenen, The effects of p-chlorophenylalanine, a serotonin depletor, on avoidance acquisition pain sensitivity and related behavior in the rat, Psychopharma. (Berl.) 10, 204 (1967).

76. M. Davis, and M. H. Sheard, Effects of lysergic acid diethylamide (LSD) on temporal recovery (pre-pulse inhibition) of the acoustic startle response in the rat, Pharmacol. Biochem. Behav. 3, 861 (1975).

77. G. D. Ellison, and D. E. Bresler, Tests of emotional behavior in rats following depletion of norepinephrine or serotonin or both, Psychopharma. (Berl.) 34, 275 (1974).

78. S. S. Tenen, The effects of p-chlorophenylalanine, a serotonin depletor, on avoidance acquisition pain sensitivity and related behavior in the rat, Psychopharma. (Berl.) 10, 204 (1967).

79. G. K. Aghajanian, W. E. Foote, and M. H. Sheard, Lysergic acid diethylamide: sensitive neuronal units in the midbrain raphe, Science 161, 706 (1968).

80. D. W. Gallager, and G. K. Aghajanian, Effects of chlorimipramine and lysergic acid diethylamide on efflux of precursor-formed ^{3}H-serotonin, J. Pharmacol. Exp. Ther. 193, 785 (1975).

81. H. J. Haigler, and G. K. Aghajanian, Lysergic acid diethylamide and serotonin, a comparison of effects on serotonergic neurons and neurons receiving a serotonergic input, J. Pharmacol. Exp. Ther. 188, 688 (1974).

82. P. C. Dandiya, B. D. Gupta, M. L. Gupta, and S. K. Patni, Effects of LSD on open field performance in rats, Psychopharmacol. 15, 333 (1969).

83. L. E. Jarrard, Effects of d-lysergic acid diethylamide on operant behavior in the rat, Psychopharmacol. 5, 39 (1963).

84. H. A. Tilson, and S. B. Sparber, Similarities and differences between mescaline, d-lysergic acid diethylamide-25 (LSD) and d-amphetamine on various components of fixed interval responding in the rat, J. Pharm. Exp. Ther. 184, 376 (1973).

85. M. H. Sheard, D. I. Astrachan, and M. Davis, The effect of d-lysergic acid diethylamide (LSD) upon shock elicited fighting in rats, Life Sci. 20, 427 (1977).

86. M. Davis, and M. H. Sheard, Biphasic dose response effects of N-N-dimethyltryptamine on rat startle reflex, Pharmacol. Biochem. Behav. 2, 827 (1974).

87. C. de Montigney, and G. K. Aghajanian, personal communication.

88. E. Sanders-Bush, D. W. Gallager, and F. Sulser, On the mechanism of brain 5-hydroxy-tryptamine depletion by p-chloroamphetamine and related drugs and the specificity of their actions, in E. Costa, G. L. Gessa, and M. Sandler eds. Advances in Biochemical Psychopharmacology, Raven Press, New York (1974).

89. A. Pletscher, G. Bartholino, H. Bruderer, W. P. Burkard, and K. F. Gey, Chlorinated aralkylamines affecting the cerebral metabolism of 5-hydroxytryptamine, J. Pharmacol. Exptl. Therap. 145, 344 (1964).

90. R. W. Fuller, C. W. Hines, and J. Mills, Lowering of brain serotonin levels by chloroamphetamines, Biochem. Pharmacol. 14, 483 (1965).

91. F. P. Miller, R. H. Cox, W. R. Snodgrass, and R. P. Maickel, Comparative effects of p-chlorophenylalanine, p-chloroamphetamine and p-chloro-N-methyl-amphetamine on rat brain norepinephrine, serotonin and 5-hydroxyindole-3-acetic acid, Biochem. Pharmacol. 19, 435 (1970).

92. H. H. Frey, and M. P. Magnussen, Different central mediation of the stimulant effects of amphetamine and its p-chloro-analogue, Biochem. Pharmacol. 17, 1299 (1968).

93. A. K. Pfeifer, L. Gyorgy, and M. Fodor, Role of catecholamines in the central effects of amphetamine, Acta Med. Acad. Sci. (Hung.) 25, 441 (1968).

94. S. J. Strada, and F. Sulser, Comparative effects of p-chloroamphetamine and amphetamine on metabolism and in vivo release of ^{3}H-norepinephrine in the hypothalamus, Europ. J. Pharmacol. 15, 45 (1971).

95. M. H. Sheard, and M. Davis, Short and long term effects upon shock-elicited aggression, Europ. J. Pharmacol. 40, 295 (1976).

96. M. H. Sheard, A. Zolovick, and G. K. Aghajanian, Raphe neurons: effect of tricyclic antidepressant drugs, Brain Res. 43, 690 (1972).

97. M. Davis, D. W. Gallager, and G. K. Aghajanian, Tricyclic antidepressant drugs: attenuation of excitatory effects of d-lysergic acid diethylamide (LSD) on acoustic startle response, Life Sci. 20, 1249 (1977).

98. S. B. Ross, and A. L. Renyi, Inhibition of uptake of tritiated catecholamines by anti-depressants and related agents, Europ. J. Pharmacol. 2, 181 (1967).

99. H. S. Nyback, J. R. Walters, G. K. Aghajanian, and R. H. Roth, Tricyclic antidepressants: effects on the firing rate of brain noradrenergic neurons, Europ. J. Pharmacol. 32, 302 (1975).

100. M. H. Sheard, D. Astrachan, and M. Davis, Tricyclic antidepressant drugs: antagonism of effect of d-lysergic acid diethylamide (LSD) on shock elicited aggression, Commun. Psychopharmacol. 1, 167 (1977).

101. R. G. Geen, and E. C. O'Neal, Activation of cue-elicited aggression by general arousal, J. Pers. Soc. Psychol. 11, 289, (1969).

102. L. Berkowitz, Some determinants of impulse aggression: role of mediated associations with reinforcements for aggression, Psychol. Rev. 81, 165 (1974).

103. J. R. Ison, and G. R. Hammond, Modification of startle reflex in the RHT by changes in the auditory and visual environments, J. Comp. Physiol. Psychol. 75, 435 (1971).

104. M. Davis, Signal to noise ratio as a predictor if startle amplitude and habituation in the rat, J. Comp. Physiol. Psychol. 86, 812 (1974).

105. M. H. Sheard, D. Astrachan, and M. Davis, Effect of noise on shock elicited aggression in rats, Nature 257, 43 (1975).

106. B. Dubinsky and M. E. Goldberg, The effect of imipramine and selected drugs on attack elicited by hypothalamic stimulation in the cat, Neuropharmacol. 10, 537 (1971).

107. B. R. Inselman-Temkin, and J. P. Flynn, Sex dependent effects of gonadal and gonadotropic hormones on centrally elicited attack in cats, Brain Res. 60, 393 (1973).

108. M. H. Sheard, and G. K. Aghajanian, Neuronally activated metabolism of serotonin, effect of lithium, Life Sci. 9, 285 (1970).

109. S. Knapp, and A. J. Mandell, Short and long term lithium administration: effects on the brain's serotonergic biosynthetic systems, Science 180, 645 (1973).

110. S. Knapp, and A. J. Mandell, Effects of lithium chloride on parameters of biosynthetic capacity for 5-hydroxytryptamine in rat brain, J. Pharm. Exp. Ther. 193, 812 (1975).

111. G. R. Heninger, and M. H. Sheard, Lithium effects on the somatosensory cortical evoked response in the rat and cat, Life Sci. 19, 19 (1976).

112. M. H. Sheard, Effect of lithium on foot shock aggression in rats, Nature 228, 284 (1970).

113. M. H. Sheard, Aggressive behavior: modification by amphetamine p-chlorophenylalanine and lithium in rats, Aggressologie 14, 323 (1973).

114. M. H. Sheard, unpublished observations.

115. M. H. Sheard, Behavioral effects of p-chlorophenylalanine inhibition by lithium, Commun. Behav. Biol. 5, 71 (1970).

116. M. H. Sheard, Effect of lithium in human aggression, Nature 230, 113 (1971).

117. M. H. Sheard, Lithium in the treatment of aggression, J. Nerv. Ment. Dis. 160, 108 (1975).

118. J. P. Tupin, D. B. Smith, T. L. Classon, L. I. Kim, A. Nugent, and A. Groupe, Long term use of lithium in aggressive prisoners, Comp. Psychiat. 14, 311 (1973).

119. M. H. Sheard, J. L. Marini, C. I. Bridges, and E. Wagner, The effect of lithium on impulsive aggressive behavior in man, Am. J. Psychiat. 133, 1409 (1976).

DISCUSSION

Dr. Sanghvi inquired about the proposed mechanism of action of lithium and tricyclic anti-depressants. He also asked if 5-HTP had an effect on agressive behavior. Dr. Sheard suggested that the tricyclic antidepressants may block aggression by means of their anti-cholinergic activity since anticholinergics also block predator aggression. He indicated that both tryptophan and 5-HTP reduce predator agression as measured by muricidal behavior. He also stated that 5-HTP will abolish the enhancement of aggression produced by PCPA.

BIOGENIC AMINE METABOLISM IN THE AGGRESSIVE MOUSE-KILLING RAT

A. I. Salama and M. E. Goldberg

Department of Pharmacology, Warner-Lambert/Parke-Davis Pharmaceutical
Division, Ann Arbor, Michigan; and Department of Pharmacology, Squibb
Institute for Medical Research, Princeton, New Jersey

ABSTRACT

A commonly used model for pharmacological studies on aggressive behavior is the
mouse-killing or muricidal rat. These rats, show a spontaneous interspecific aggressiveness
towards mice, a behavior which is independent from the hunger state. This behavior can be
blocked selectively by antidepressants and stimulants. Biochemical correlates of this
behavior have been obtained. Mouse-killing rats have higher forebrain levels of norepinephrine (NE) than control, non-killer rats. They also have larger rate constants for the
decline of brain H^3-NE and consequently, have higher turnover rates for NE than controls
(non-killer). These differences have not been obtained in the hindbrain region. The effects
observed in the forebrain, occurred 2 hours after a killing episode and persisted for 24 hours
before returning to normal 48 hours after initial mouse-killing. Thus the change in turnover
rate appears to result from the killing episode and is not the etiology of this aggressive
response. No differences in the levels or rates of turnover of serotonin have been observed
in either of the brain regions studied. The effects of imipramine and amphetamine on NE
turnover in the forebrain of killer rats have been compared with those in drug-treated nonkiller rats. Both agents, given at doses which inhibit mouse-killing behavior, accelerate
the turnover rate of brain NE in non-killer rats. They do not, however, influence the existing
biochemical alterations obtained as a result of the killing episode. There is a greater
enhancement of spontaneous motor activity in the mouse-killing rat, compared to normal rats
after amphetamine. This increased sensitivity does not appear to be related to levels of NE
or dopamine, differences in levels of amphetamine in brain, or release and re-uptake of NE by
amphetamine. Relationships between NE metabolism and mouse-killing behavior are still not
entirely understood and require more extensive study.

INTRODUCTION

It has been shown that sensory deprivation, a condition similar to isolation, results in
behavioral disruption in humans. In the search for experimental models analogous to human
disorders, selected models of aggressive behavior have been used commonly in the laboratory.
The most widely used methods for studying aggression in rodents, have been those induced by
isolation, stress, brain lesioning and certain chemicals (see reviews of Valzelli, 1, 2).
Pharmacologists have used these models to evaluate centrally acting drugs with the aim of
finding selective 'antiaggressive agents' which might be applicable to certain clinical
disorders.

Extensive pharmacological studies using a diversity of psychotherapeutic agents have been
tested in a variety of aggressive models induced in rodents (1, 3, 4, 5, 6). In most cases
with the exception of the mouse-killing rat following certain antidepressants there was no
significant separation between doses capable of blocking aggression and those producing
neurotoxicity.

The mouse-killing response was first described by Karli (7), when he observed a spontaneous
interspecific aggressiveness in some rats toward mice, a behavior which was found to be
independent from hunger. Using disturbances of the rats ability to walk a rotating rod
(rotarod) as a measure of generalized behavioral depression, Horovitz et al (6) established
a ratio of rotarod to antimuricide activity, with ratios greater than unity, indicating a
selective effect on this interspecies aggressive response which these authors termed
'muricide'. Horovitz et al (8) were the first group to report that antidepressants, stimulants
and antihistamines selectively block the mouse-killing response in rats. Karli et al (9) and
Horovitz (10) related the specific site of action of antidepressants to the amygdala and
showed that amygdaloid lesions abolish mouse-killing in rats. Horovitz (11) also demonstrated that imipramine blocks electrical activity of the amygdala, and that direct

injections of imipramine into the amygdala, but not other limbic or hypothalamic regions blocked mouse-killing behavior. Electro-convulsive shock treatment, which is effective for the treatment of depression has also been observed to block mouse-killing behavior in rats (12). The question remains, as to whether the selective effect of the antidepressants on muricidal behavior represents a true relationship to their mechanism of action in man. Valzelli (1) and more recently Vogel (13) have reviewed the effects of many drugs including antidepressants on this muricidal response.

It is well known that emotional changes are accompanied by alterations in brain monoamines. The relationship of various central neurotransmitters to aggression has been extensively reviewed recently by Pradham (14), Eichelman (15, 16) and Welch (17). Although, many biochemical studies have appeared in recent years, there have been no common factors linking definite and consistent biochemical correlates to certain aggressive responses. The isolated fighting mouse has received the greatest attention, and at the same time has offered conflicting results from different laboratories. Welch and Welch (18) have demonstrated that in mice isolated for one week, there was a significantly higher level of brain norepinephrine (NE) compared to grouped animals, and that after α-methyl-p-tyrosine (αMT) the lowering of catecholamines was slower in isolated animals when compared to grouped animals. Valzelli (1) on the other hand, showed that mice rendered aggressive by prolonged isolation have normal brain serotonin (5HT) and NE levels, but demonstrated changes in the turnover rate of these amines, where 5HT synthesis occurred at a slower rate, and NE synthesis appeared to be faster when compared to non-aggressive mice.

Our investigations attempted to focus on the role of biogenic amines in the brain of mouse-killing rats, as well as the interactions of imipramine and amphetamine on biogenic amine metabolism in this aggressive model.

METHODS

In all of our studies we have housed the rats individually for 6 weeks and then tested for their ability to kill mice as described by Horovitz (6, 8). From a colony of Long Evans hooded rats subjected to the 6 week isolation and given a restricted food intake of 15 gm/day, rats were selected which had shown a positive mouse killing response within two minutes of challenge for 3 consecutive days, prior to biochemical studies, while those which failed to show a response (non-killers) served as controls. This method of treatment historically resulted in the selection of approximately 40 per cent of the colony as killer rats. It is important to note, and unless otherwise stated, that animals were sacrificed approximately 24 hours after the last presentation of mice. In all experiments, the rats were sacrificed by decapitation and the whole brain removed. The brain stem including the cerebellum (hindbrain) was separated from the forebrain by a cut above the corpora quadrigemina, and both regions were immediately frozen until ready for analysis.

Biochemical Studies

In agreement with earlier studies (19) we have found an increase of approximately 30 per cent in forebrain norepinephrine (NE) levels of the killer rat, when compared directly to isolated non-killer rats (20) (or indirectly to normal grouped non-killer rats). Since it is known that amygdaloid lesions block the muricidal response (10), it is quite possible that this type of behavior may be related to an excited state of the amygdala in animals predisposed to aggressive tendencies. Unfortunately, in these studies we were unable to measure directly NE levels of the amygdala. However, because of the substantial increase in NE levels in the forebrain of killer rats, it is reasonable to speculate that similar alterations occurred in this region. In support of this speculation, there were no changes in the steady state levels in the hindbrain of killer rats. Similarly, there were no significant differences between killer and non-killer rats on the levels of serotonin in forebrain or hindbrain regions (Table 1).

TABLE 1 Brain Norepinephrine and Serotonin Levels in Killer and Nonkiller Rats

Source	Type and (No.) of animals		Norepinephrine (µg/g ± S.E.)	Serotonin (µg/g ± S.E.)
Forebrain	Killer	(8)	0.69 ± 0.01*	0.38 ± 0.02
	Nonkiller	(10)	0.55 ± 0.01	0.32 ± 0.02
Hindbrain	Killer	(7)	0.53 ± 0.04	0.37 ± 0.01
	Nonkiller	(8)	0.54 ± 0.03	0.43 ± 0.04

* Significantly different from nonkiller rats (P < 0.01)

The levels of the amines should be considered only as an aspect of a dynamic process, and is determined by the relationship existing between the rate of formation and the rate of destruction of the amines. It is now accepted that changes in the tur over rate of biogenic amines provide a better indication of the functional state of neural activity than actual steady state levels.

In earlier studies (19) where measurement of NE turnover was obtained utilizing the methods and concepts of Neff and Costa (21), in which the rate of decline of NE levels after αMT is measured, we found a 52 per cent increase in turnover in the forebrain region of the killer rat. We reasoned that this higher rate of synthesis in the forebrain was required to maintain a higher steady state level. In these experiments there were no changes in the turnover rate of NE in the hindbrain region. In subsequent studies we noted that the doses of αMT used in turnover studies also blocked the muricidal response in most rats (unpublished observations). Thus the use of this agent, as a tool for studying turnover rates of NE could have confounded the earlier results. It should be mentioned that since αMT blocks the muricidal response, and inhibits tyrosine hydroxylase, the rate limiting enzyme in the biosynthesis of NE, there is a suggestion that the adrenergic system is implicated in the expression of muricidal behavior. To obtain a more reliable measure of NE turnover in this aggressive model the isotopic method described by Brodie et al (22), which measures the decline in specific activity of intraventricularly administered H^3NE was used. Furthermore, since it was intended to study the effects of amphetamine on NE turnover in the killer rats, we could not use the method of tyrosine hydroxylase inhibition with αMT, since it is well known that αMT antagonizes most pharmacologic effects of amphetamine (23).

Using the isotopic method, we obtained increases of 150 per cent in forebrain NE synthesis in the killer rat, and again no changes in turnover rate in the hindbrain region compared to nonkiller rats (Table 2). It appears that turnover results using αMT should not be considered absolute, possibly due to the inhibitory effect of the compound on this type of behavior. Thus, it appears that killer rats, which had been challenged with a mouse approximately 24 hours before sacrifice, maintain normally higher steady state levels of NE by a compensatory increase in synthesis rate. These differences may result from the innate aggressiveness of the animals. They do not appear to be due to the isolation period, since we found no changes in turnover rates between nonkillers (isolated) and grouped control animals. Similar observations were obtained by Valzelli (1) who has shown that increases in NE turnover rates in isolated fighting mice were due to aggressivity and not to isolation. Furthermore, that rate of NE utilization in the aggressive animals may be altered to explain the turnover results.

TABLE 2 Brain Norepinephrine Turnover in Nonkiller and killer Rats*

Source	Type of animal	Steady state level (µg/g ± S.E.)	Rate constant (k[hr⁻¹] ± S.E.)	Turnover rate (µg/g/hr)
Forebrain	Nonkiller	0.55 ± 0.01 (6)	0.15 ± 0.02 (18)	0.08
	Killer	0.70 ± 0.02 (12)†	0.28 ± 0.03 (36)†	0.20
Hindbrain	Nonkiller	0.50 ± 0.01 (6)	0.27 ± 0.05 (18)	0.14
	Killer	0.53 ± 0.02 (12)	0.34 ± 0.03 (36)	0.17

* Rats were given an intraventricular injection of ^{3}H-NE (5 µc/kg) and were sacrificed at various times later. Numbers in parentheses represent the number of animals.

†Significantly different from nonkiller rats (P < 0.01)

It was of interest to study the permanency of these neurochemical changes after a single mouse-killing episode. This information was required in order to determine whether these changes were indeed due to the stress of killing the mouse, or were due to the fact that these rats were aggressive subjects. From a rat colony, a number of killer and nonkillers were selected as described above in Methods. Ten days following their selection, both groups of rats were left undisturbed before the next presentation of a mouse. The animals were then challenged once more, and all pre-selected killer rats again exhibited their aggressive behavior, while the nonkillers failed to do so. Various times after this challenge, groups of killer and nonkiller rats were used to determine forebrain NE levels and turnover rate, using the isotopic method (22). Similar estimations were conducted in which the animals (referred to as normals) were housed in groups of five for 6 weeks and never challenged with a mouse. It is apparent from Table 3 that no differences in forebrain NE levels or turnover rates were observed between normal rats and nonkiller rats, thus suggesting an absence of effect by isolation. However, a significant elevation in steady state levels (20%) was observed in killer rats which had been challenged 2 hours prior to study. Together with this increase, there was 50 per cent increase in the rate constant, and consequently a 95 per cent increase in the turnover rate of NE compared with its nonkiller control. Twenty-four hours after the fighting episode, these parameters were still significantly elevated. However, by 48 hours as well as one week later both steady state levels and turnover rates had returned to control values. Thus it is conceivable, that the turnover changes observed within two hours after a killing episode are related to the stress induced by the killing episode, since they were not permanent and persisted for 24 to 48 hours. Since tyrosine hydroxylase activity can be altered by rapid changes in end-product inhibition or substrate availability, and gradual alterations in the amount of this enzyme occurs in response to prolonged neural stimulation (24), it is quite possible, that the dynamic changes occurring after the killing episode are related to an activation (or induction) of tyrosine hydroxylase. This would be consonant with results obtained in earlier stress studies.

TABLE 3 Turnover of Norepinephrine in the Forebrain of Killer Rats After a Single Killing Episode

Type of animal	Hr after killing episode	Steady state levels (μg/g $\pm$ S.E.M.)	Rate constant H^3NE decline [k(hr^{-1}) $\pm$ S.E.M.)	Turnover rate (μg/g/hr)
Normal	-	0.47 $\pm$ 0.01	0.16 $\pm$ 0.01	0.075
Nonkiller	2	0.50 $\pm$ 0.02	0.15 $\pm$ 0.02	0.075
Nonkiller	24	0.52 $\pm$ 0.01	0.17 $\pm$ 0.02	0.089
Killer	2	0.62 $\pm$ 0.01*	0.24 $\pm$ 0.02*	0.145
Killer	24	0.57 $\pm$ 0.01*	0.22 $\pm$ 0.02*	0.126
Killer	48	0.50 $\pm$ 0.01	0.18 $\pm$ 0.03	0.090
Killer	1 week	0.50 $\pm$ 0.02	0.15 $\pm$ 0.01	0.075

Rats were given an intraventricular injection of H^3NE and were sacrificed at various times later. Rate constants (k) were calculated from the method of least squares. Turnover rate is the product of k and the steady state levels. There were 10 rats in each group for steady state level determination.

* Significantly different from the respective non-killer group p < 0.01

Though there were no measurable differences in the steady state levels in either forebrain or hindbrain regions of 5HT in killer rats, (Table 1) it was still of interest to determine turnover rate of this amine. After inhibition of monoamine oxidase (MAO) by pargyline, we were unable to detect any differences in the rate of accumulation of 5HT in killer and non-killer rats, in either the forebrain or hindbrain region of the brain (Table 4). Thus, unlike the evidence presented for a decrease in the synthesis rate of this amine in the isolated fighting mouse (1) this model of aggression does not appear to be related to alterations in serotonin metabolism. However, it should be mentioned that Horovitz et al (8) have shown that iproniazid, another potent MAO inhibitor, had muricidal blocking properties. We have found, that pargyline also has some inhibitory effect on this type of behavior. Thus, it is conceivable that elevated brain 5HT levels resulting from MAO inhibition might cause muricidal blockade. Analogous to the decrease in NE after αMT and the resultant inhibition of mouse-killing, Kulkarni (25) has shown that high doses of 5-hydroxytryptophan also cause inhibition of muricidal behavior in some rats. Before concluding that 5HT turnover is not implicated in this model of aggression, we feel it should be studied by an isotopic procedure similar to that described for norepinephrine.

TABLE 4 Brain Serotonin Turnover in Nonkiller and Killer Rats*

Source	Type of animal	Steady state level (μg/g $\pm$ S.E.)	Rate constant (k[h^{-1}] $\pm$ S.E.)	Turnover rate (μg/g/hr)
Forebrain	Nonkiller	0.53 $\pm$ 0.01 (6)	0.40 $\pm$ 0.01 (24)	0.21
	Killer	0.53 $\pm$ 0.01 (6)	0.39 $\pm$ 0.01 (20)	0.21
Hindbrain	Nonkiller	0.40 $\pm$ 0.02 (6)	0.40 $\pm$ 0.01 (24)	0.16
	Killer	0.40 $\pm$ 0.02 (6)	0.36 $\pm$ 0.01 (20)	0.14

* Rats given 100 mg/kgi.p. of pargyline and sacrificed at various times later. Numbers in parentheses refer to the number of animals used.

Pharmacological Studies

As mentioned earlier, the tricyclic antidepressant imipramine and the stimulant amphetamine are representative agents which have been reported to block selectively the muricidal response, and are both used clinically in certain types of depression.

In order to select the proper dose and time parameters in which to evaluate the biochemical effects of these agents, they were studied for their inhibitory effects upon muricidal rats (20). For imipramine an ED_{50} of 13.4 mg/kg was obtained at the time of maximal effect, a value which compares favorably with that obtained by earlier workers (8). Based on these studies, a dose of 25 mg/kg i.p. of imipramine was selected for turnover studies, a dose which caused inhibition in most muricidal rats within a half hour, with a duration of about 4-5 hours. For amphetamine an ED_{50} of 1.8 mg/kg i.p. was obtained for blocking the muricidal response 30 minutes after administration. A 5 mg/kg i.p. dose of amphetamine was selected for study. This dose causes complete blockade initially and its effects persist for at least 4 hours in most animals.

After administration of imipramine or amphetamine to normal nonkiller rats the steady state levels of NE in the forebrain were unaltered; however, there was a significant increase in the rate constant and turnover rate of this amine (Table 5, 6). In killer rats there was a significant increase in NE forebrain levels after either drug treatment. However, the magnitude of this increase is similar to that observed in untreated killer rats as compared to nonkiller rats (Table 1). In addition, the rate constant of H^3-NE efflux and turnover rate in the forebrain of the killer rats after either imipramine or amphetamine is also elevated when compared to untreated nonkiller rats. These results obtained in the drug studies are difficult to interpret. Both imipramine and amphetamine, inhibit the re-uptake of NE, and have been reported to accelerated the turnover rate of NE in brain following acute administration (21, 27). Glowinski (27) has reported that stress markedly accelerates the increase in NE turnover rate obtained after 5 mg/kg of amphetamine. Clearly no such augmentation occurred in our studies when either drug was superimposed upon the already elevated turnover rate of NE in the forebrain of killer rats. These results suggest that some type of 'ceiling effect' exists between the factors involved.

TABLE 5 Effect of Imipramine on Forebrain Norepinephrine Turnover in Nonkiller and Killer Rats*

Treatment	Type of animal	Steady state level (µg/g ± S.E.)	Rate constant (k[hr^{-1}] ± S.E.)	Turnover rate (µg/g/hr)
Saline	Nonkiller	0.56 ± 0.01 (6)	0.15 ± 0.02 (18)	0.08
Imipramine	Nonkiller	0.55 ± 0.01 (10)	0.39 ± 0.01 (30)+	0.22
Imipramine	Killer	0.70 ± 0.02 (12)+	0.24 ± 0.02 (30)+	0.17

* Rats were given an intraventricular injection of ^{3}H-NE (5µc/kg) 30 min after saline or imipramine (25 mg/kg, i.p.) injection and were sacrificed at various times later. Number in parentheses represent the number of animals.

+ Significantly different from nonkiller (saline-treated) rats p < 0.01

TABLE 6 Effect of Amphetamine of Forebrain Norepinephrine Turnover in Nonkiller and Killer Rats*

Treatment	Type of animal	Steady state level (µg/g ± S.E.)	Rate constant (k[hr^{-1}] ± S.E.)	Turnover rate (µg/g/hr)
Saline	Nonkiller	0.52 ± 0.01 (6)	0.18 ± 0.03 (18)	0.09
Amphetamine	Nonkiller	0.60 ± 0.02 (6)	0.34 ± 0.04 (18)+	0.20
Saline	Killer	0.75 ± 0.02 (6)+	0.30 ± 0.04 (24)+	0.23
Amphetamine	Killer	0.79 ± 0.02 (6)+	0.28 ± 0.04 (18)+	0.22

* Rats were given intraventricular injection of ^{3}H-NE (5 µc/kg) 30 min after saline or amphetamine (5 mg/kg, i.p.) injection and were sacrificed at various times later. Numbers in parentheses represent the number of animals.

+ Significantly different from nonkiller (saline treated) rats p < 0.01

It is well known that animals submitted to prolonged isolation change their sensitivity to drugs. It has been demonstrated that mice submitted to prolonged isolation become more sensitive to the action of chlorpromazine. Similarly, after several types of stress, there is also increased sensitivity to the lethal effects of amphetamine (31).

Since the mouse-killing rat is obtained by isolation, we studied the involvement of brain NE and dopamine in locomotor activity and stereotype behavior after amphetamine in this aggressive model. The increased stereotype behavior (rearing) observed after 1 and 3 mg/kg of D-amphetamine was similar in nonkiller (normal) and mouse-killing (aggressive) rats. However, there was a much greater increase in spontaneous motor activity in the killer rat compared to normal rats following 3 mg/kg of amphetamine (32). This altered sensitivity to amphetamine suggested the possibility of biochemical changes in the brain of the killer rats. It is possible that the enhanced sensitivity after amphetamine could be attributable to higher brain levels of the stimulant. It is apparent from Table 7 that there were no significant differences in the uptake, peak level or disappearance of the stimulant in either killer or nonkiller rats. In earlier investigations (33) we showed that after amphetamine, peak brain levels occurred within 15 to 30 minutes, and then decreased rapidly. Thus, differences in the degree of enhanced locomotor activity between groups do not appear to be related to differences in brain levels of amphetamine.

TABLE 7   Amphetamine Levels in Forebrain of Nonkiller and Killer Rats

Time (min)	Dose (mg/kg)	Nonkiller	Killer
		Amphetamine Levels (μg/g $\pm$ S.E.)	
7.5	3	1.27 ± 0.1	1.66 ± 0.2
30	3	2.64 ± 0.2	2.72 ± 0.1
120	3	1.52 ± 0.1	1.46 ± 0.1

Groups of 10 animals were injected i.p. with amphetamine-C^{14}. They were sacrificed at the specified time intervals and amphetamine-C^{14} extracted as described in reference (32).

When interpreting the action of amphetamine on brain amines, one must account for the possible interference of the stimulant on uptake, storage and release of norepinephrine and dopamine. The influence of amphetamine on brain dopamine levels revealed no differences between killer and nonkiller rats after saline or drug administration (Table 8). However, in agreement with earlier observations (Table 1) there was about a 20 per cent increase in endogenous forebrain NE level in the killer rat under saline treatment, which was unaltered by amphetamine treatment.

TABLE 8   Norepinephrine and Dopamine Forebrain Levels in Nonkiller and Killer Rats

Type of Animal	Treatment	Dose (mg/kg)	Norepinephrine	Dopamine
			(μg/g $\pm$ S.E.)	
Nonkiller	Saline	-	0.42 ± 0.01	2.24 ± 0.10
Killer	Saline	-	0.52 ± 0.01*	2.78 ± 0.36
Nonkiller	Amphetamine	3	0.42 ± 0.01	2.78 ± 0.26
Killer	Amphetamine	3	0.54 ± 0.01**	2.47 ± 0.28

Groups of 5 animals each received saline or amphetamine and were sacrified 28 min later. The brain was quickly removed and dissected as described in reference (32).

* Significantly different from saline treated nonkiller rats p < .01
** Significantly different from amphetamine treated nonkiller rats p < .01

When amphetamine was administered after the intraventricular administration of H^3NE there were no differences in brain level of H^3NE in either killer or non-killer rats (Table 9).

TABLE 9   Release of H^3-Norepinephrine from Forebrain of Killer and Nonkiller Rats after Amphetamine

Type of Animal	Treatment	H^3-Norepinephrine in brain (dpm X 10^3/g $\pm$ S.E.)	Drug Induced Release
Nonkiller	Saline	149.6 ± 24	
Nonkiller	Amphetamine	147.0 ± 7	0
Killer	Saline	155.1 ± 10	
Killer	Amphetamine	147.2 ± 14	0

Groups of 5 animals each received H^3NE (5 μc/kg) into the lateral ventricle of the brain. One hr later, the animals were injected with saline or amphetamine (3 mg/kg) i.p. All animals were sacrificed 30 min later, the forebrain removed and analyzed for H^3NE.

Thus there appears to be no releasing effects on NE stores after 3 mg/kg of amphetamine in either groups of animals. Moore (34) has demonstrated that catecholamine release from brain by amphetamine was enhanced by conditions of aggregation. On the other hand, we have shown that several forms of stress do not enhance the depleting action of amphetamine (33). It is possible that differences in inhibition of neuronal catecholamine uptake by amphetamine could intensify the synaptic actions of released NE or DM, and explain the enhanced loco-motor activity observed in killer rats. Under non-drug conditions, there were no differences in the amount of H^{3}NE taken up by cortical brain slices in killer and nonkiller rats (32). Similarly, in the presence of amphetamine, there were no differences in the amount of uptake inhibition induced by the stimulant in cortical slices from either killer or nonkiller rats. It would appear, that the effect of amphetamine on locomotor activity is not correlated with inhibition of NE uptake between the two groups. In attempting to correlate enhanced stereo-typed behavior after amphetamine, Naylor and Costall (35) have concluded that there was no relationship between inhibition of dopamine uptake with stereotype enhancement. It is possible that killer rats are more sensitive to the central aspects of the stimulant properties of amphetamine.

CONCLUSION

From these biochemical studies, evidence is presented which indicates that NE metabolism may be involved in this model of aggressive behavior. Drugs such as antidepressants and stimulants which alter NE metabolism are selective in blocking mouse-killing in the rat. For example, the antidepressant desipramine, a more selective NE uptake inhibitor than imipramine, will inhibit muricidal behavior at lower doses, while chlorimipramine which is more selective toward inhibition of 5HT uptake, will require significantly higher doses. Also the tyrosine hydroxylase inhibitor αMT is effective in suppressing this aggressive behavior. In addition, NE metabolism is also involved in the stress accompanying mouse-killing, a condition apparently independent from this type of aggression. From the available biochemical information, it would appear that 5HT metabolism may not be involved in mouse killing. However, pharmacological evidence points to the contrary. For example, p-chlorophenylalanine (PCPA), a potent inhibitor of 5HT synthesis, not only increases mouse-killing in rats (36), but also potentiates the facilitatory effects on killing which are produced in that species by removal of the olfactory bulbs (37). This drug induced potentiation of muricide after bulbectomy seems to be due to the selective effects of PCPA on 5HT, since injections of pargyline or 5-hydroxytrypotophan which increase 5HT levels block muricide induced by olfactory injury. Kulkarni (25) has also reported that high doses of 5-hydroxytryptophan will influence muricidal tendencies.

The involvement of the cholinergic system in mouse-killing response should also be con-sidered. In support of this, McCarthy (38) showed that pilocarpine could induce mouse-killing in rats, as did carbachol and neostigmine (39), while atropine methyl sulfate injected in the lateral hypathalamus suppressed muricidal behavior. In addition, it appears that acetylcholine metabolism is different in mouse-killing rats, since these rats have a significantly higher choline acetyl transferase activity compared to nonkillers (40). Finally, anticholinergic agents are effective in inhibiting mouse-killing (6), shock elicited aggression (41) and predatory attack behavior in cats (42).

The interrelationships among these neurochemical systems, expression and/or inhibition of aggressive behavior, and the possible mechanism of action of antidepressants remain obscure. Obviously further studies are needed to clarify them.

REFERENCES

(1) L. Valzelli, Drugs and aggressiveness, Advances in Pharmacology 5, 79 Academic Press, New York (1967).

(2) L. Valzelli, Biological and pharmacological aspects of aggressiveness in mice, Neuro-Psycho-Pharmacology 5, 781, Excerp. Med. Fdn., Amsterdam (1967).

(3) R. D. Sofia, Effects of centrally active drugs on four models of experimentally-induced aggression in rodents, Life Sci 8, 705 (1969).

(4) R. E. Tedeschi, D. H. Tedeschi, A. Mucha, L. Cook, P. A. Mattis and F. S. Fellows, Effects of various centrally-acting drugs on fighting behavior of mice. J. Pharmacol. Exp. Therap 125, 28 (1959).

(5) J. B. Malick, R. D. Sofia, M. E. Goldberg, A comparative study of the effects of selected psychoactive agents upon three lesion-induced models of aggression in the rat, Arch. Int. Pharmacodyn 181, 459 (1969).

(6) Z. P. Horovitz, J. J. Piala, J. P. High, J. C. Burke, R. C. Leaf, Effects of drugs on drugs on the mouse killing (muricide) test and its relationship to amygdaloid function, Int. J. Neuropharmacol 5, 405 (1966).

(7) P. Karli, The Norway rat's killing response to the white mouse, an experimental analysis, Behavior (Leiden) 10, 81 (1956).

266

(8) Z. P. Horovitz, P. W. Raggozzino, R. C. Leaf, Selective block of rat mouse-killing by antidepressants, Life Sci 4, 1909 (1965).

(9) P. Karli, M. Vergnes, F. Didiergeorges, Rat-mouse interspecific aggressive behavior and its manipulation by brain ablation and by brain stimulation, Aggressive behavior 47, Excerp. Med Fdn. Amsterdam (1969).

(10) Z. P. Horovitz, Psychoactive drugs and limbic system of the brain, Psychosomatics 6, 281 (1965).

(11) Z. P. Horovitz, Relationship of the amygdala to the mechanism of action of two types of antidepressants (Thiazenone and Imipramine), Recent Advances in Biological Psychiatry 8, 21 Plenum Press, New York (1966).

(12) J. R. Vogel, D. R. Haubrich, Chronic administration of electroconvulsive shock: Effects on mouse-killing activity and brain monoamines in rats, Physiol. Behav 11, 725 (1973).

(13) J. R. Vogel, Antidepressants and mouse-killing (muricide) behavior, Antidepressants, Industrial Pharmacology 2, 99 Futura, Mt. Kisco, New York (1975).

(14) S. N. Pradhan, Aggression and central neurotransmitters, Int. Rev. Neurobiol 18, 213 (1975).

(15) B. Eichelman, The catecholamines and aggressive behavior: Chemical approaches to brain function. Neurosciences Research 5, 109 Academic Press, New York (1973).

(16) B. Eichelman, N. B. Thoa, The aggressive monoamines, Biol. Psychiat 6, 143 (1973).

(17) B. L. Welch, A. S. Welch, Aggression and the biogenic amine neurohumors: Aggressive behavior, Proceedings of the Symposium on the biology of aggressive behavior 188 Excerta. Medica, Fdn. Amsterdam (1969).

(18) B. L. Welch, A. S. Welch. Effect of grouping on the level of brain norepinephrine in white Swiss mice. Life Sci. 4, 1011 (1965).

(19) M. E. Goldberg, A. I. Salama, Norepinephrine turnover and brain monoamine levels in aggressive mouse-killing rats. Biochem. Pharmacol 18, 532 (1969).

(20) A. I. Salama, M. E. Goldberg, Neurochemical effects of imipramine and amphetamine in aggressive mouse-killing (muricidal) rats. Biochem. Pharmacol 19, 2023 (1970).

(21) N. H. Neff, E. Costa, Effect of tricyclic antidepressants and chlorpromazine on brain catecholamine synthesis. Antidepressant Drugs 28, Excerpta Med. Fdn. Amsterdam (1967).

(22) B. B. Brodie, E. Costa, A. Dlabac, N. H. Neff, H. H. Smookler, Application of steady state kinetics to the estimation of synthesis rate and turnover time of tissue catecholamines. J. Pharmacol. Exptl. Therap 154, 493 (1966).

(23) A. Weissman, B. K. Koe, S. S. Tenen, Antiamphetamine effects following inhibition of tyrosine hydroxylase. J. Pharmacol. Expt'l Therap 151, 339 (1966).

(24) R. A. Mueller, H. Thoenen, J. Axelrod, Increase in tyrosine hydroxylase activity after reserpine administration. J. Pharmacol Expt'l Therap 169, 74 (1969).

(25) A. S. Kulkarni, Muricidal block produced by 5-hydroxytryptophan and various drugs Life Sci 7, 125 (1968).

(26) J. Glowinski, J. Axelrod, Inhibition of uptake of tritiated noradrenaline in the intact rat brain by imipramine and structurally related compounds, Nature (Lond) 204, 1318 (1964).

(27) J. Glowinski, Effects of amphetamine on various aspects of catecholamine metabolism in the central nervous system of the rat. Int'l Symp. on amphetamines and related compounds 301, Raven press, New York (1970).

(28) S. Consolo, S. Garattini, L. Valzelli, Amphetamine toxicity in aggressive mice. J. Pharm. Pharmacol 17, 53 (1965).

(29) S. Consolo, S. Garattini, L. Valzelli, Sensitivity of aggressive mice to centrally acting drugs. J. Pharm Pharmacol 17, 594 (1965).

(30) B. L. Welch, A. S. Welch. Graded effect of social stimulation upon d-amphetamine toxicity, aggressiveness and heart and adrenal weight. J. Pharmacol Expt'l Therap 151, 331 (1966).

(31) M. E. Goldberg, A. I. Salama, Amphetamine toxicity and brain monoamines in three models of stress. Toxicol. Appl. Pharmacol 14, 447 (1969).

(32) A. I. Salama, M. E. Goldberg, Enhanced locomotor activity following amphetamine in mouse-killing rats. Arch. Int. Pharmacodyn 204, 162 (1973).

(33) A. I. Salama, M. E. Goldberg. Effects of several models of stress and amphetamine on brain levels of amphetamine and certain monoamines. Arch. Int. Pharmacodyn 181, 474 (1969).

(34) K. E. Moore, The role of endogenous norepinephrine in the toxicity of d-amphetamine in aggregated mice. J. Pharmacol Expt'l. Therap 144, 45 (1964).

(35) R. J. Naylor, B. Costall, The relationship between the inhibition of dopamine uptake and the enhancement of amphetamine stereotopy. Life Sci 10, 909 (1971).

(36) M. H. Sheard, The effect of p-chlorophenylalanine on behavior in rats: Relation to 5-hydroxytryptamine and 5-hydroxyindolacetic. Brain Res 15, 524 (1969).

(37) G. R. DiChiara, R. Camba, P. F. Spano. Evidence for inhibition by brain serotonin of mouse-killing behavior in rats. Nature 233, 272 (1971).

(38) D. McCarthy. Mouse-killing induced in rats treated with pilocarpine Fed. Proc 25, 385 (1966).

(39) D. E. Smith, B. M. King, B. G. Hoebel, Lateral hypothalamic control of killing: Evidence for a cholinoceptive mechanism. Science 169, 900 (1970).

(40) A. Ebel, G. Mack, V. Stefanovic, P. Mandel. Activity of choline acetyltransferase in the amygdala of spontaneous mouse-killer rats and in rats after olfactory lobe removal. Brain Res 57, 248 (1973).

(41) D. A. Powell, W. L. Milligan, K. Walters, The effects of muscarinic cholinergic blockade upon shock elicited aggression. Pharmacol. Biochem 1, 389 (1973).

(42) R. J. Katz, E. Thomas, Effects of Scopolamine and α-methylparatyrosine upon predatory attack in cats. Psychopharmacologia 42, 153 (1975).

THE GELLER CONFLICT TEST: A MODEL OF ANXIETY AND A SCREENING PROCEDURE FOR
ANXIOLYTICS*

James L. Howard and Gerald T. Pollard

Department of Pharmacology
Wellcome Research Laboratories
Research Triangle Park, N.C.

ABSTRACT

The Geller conflict test is presented as a good paradigm for predicting the antianxiety
action of drugs in humans and as a model which has some face validity with respect to some
aspects of human clinical anxiety. Problems in using the standard Geller are identified,
and data gathered in a procedure utilizing incremental shock rather than a fixed shock
level is presented to suggest that these problems can be overcome.

INTRODUCTION

Paradigms using response-contingent pairings of positive reinforcement and punishment have
proven to be uniquely useful in identifying drugs which have antianxiety effects in humans.
Although there are historical roots for approach-avoidance (conflict) paradigms (e.g.,
Masserman, 1943), the first satisfactory operant configuration was developed by Geller and
coworkers (Geller & Seifter, 1960; Geller, Kulak & Seifter, 1962). In the original Geller
paradigm (Geller & Seifter, 1960), rats were trained on a multiple VI 2-min/CRF** schedule
for milk reinforcement. Response rates in the CRF portion were suppressed by the addition
of footshock given concurrently with the positive reinforcement. Meprobamate and barbiturates increased the suppressed rate of responding whereas chlorpromazine did not.
Subsequently, the original Geller and many variations of it have been shown to correctly
identify clinically active anxiolytics, to predict their order of clinical potency, to be
insensitive (with minor exceptions) to stimulant, antipsychotic, antidepressant, or
analgesic drugs, to work in different species, and to be relatively independent of the
schedules of positive reinforcement or punishment. (For reviews of the literature, see
Cook & Davidson, 1973; Cook & Sepinwall, 1975; McMillan, 1975).

The purpose of this paper is twofold. First, some consideration will be given to the
response-contingent punishment paradigm as a model of human anxiety. Although speculative,
this is an attempt to confront a problem often ignored: Do animal models have any relation
to the human disease states for which they are used to develop drugs? Second, some problems
with the original Geller paradigm will be reviewed and a modification presented which
reduces these problems.

CONFLICT PARADIGMS AS MODELS OF HUMAN ANXIETY

Thompson and Schuster (1968; see also Brady, 1968) point out that preclinical drug evaluation
can proceed on three levels: the standard drug-correlational level, the analogous symptom
level, and the homologous disease level. Almost all preclinical work directed toward
finding new drugs useful in treating "mental" illness operates at the standard drug-
correlational level: if a procedure is differentially sensitive to known efficacious
drugs, then it can be used in screening for new drugs regardless of whether the procedure
has any face validity with respect to the disease state for which the drugs are to be
used. The Geller conflict test for anxiolytics clearly meets the requirements for a
standard drug-correlational model and is perhaps the single most useful operant paradigm
in screening for new drugs (Blackwell & Whitehead, 1975; Cook & Sepinwall, 1975).

---

*Appreciation for technical assistance is extended to Kenneth W. Rohrbach and Nancy E. Harto
 of Wellcome Research Laboratories.
**Multiple VI 2-min/CRF signifies a two-component operant behavior schedule. In the VI
 (variable interval) portion, signalled by one stimulus, bar-pressing is reinforced at
 variable intervals of time, with the mean interval being 2 min. In the CRF (continuous
 reinforcement) portion, signalled by a different stimulus, every bar-pressing is reinforced.

The danger in relying on models at the standard drug-correlational level is that they may only identify compounds which are like the standard drugs (Thompson & Schuster, 1968). At a higher level, the analogous symptom level in Thompson and Schuster's terminology, symptoms of the human disease state are reproduced in animals (Thompson & Schuster, 1968; Brady, 1968), and new drugs can be tested for their ability to alleviate these symptoms. Even more desirable are models at the homologous disease level, where not only are the salient features of the symptoms of the human disease state produced but the conditions producing the symptoms are like those producing the symptoms in human disease states (Thompson & Schuster, 1968; Brady, 1968). The more that the symptoms seen in an animal model resemble those seen in the human disease state, and the more that the conditions under which they are produced are homologous with those producing the human condition, then the more face validity the animal paradigm has for identifying drugs useful in treating the condition regardless of whether those drugs resemble known standard drugs.

The Geller conflict test appears to have some face validity as a model of human anxiety (Brady, 1968; Blackwell & Whitehead, 1975; Ray, 1965). However, in general there seems to have been little effort to consider how well or how poorly any animal models, including response contingent punishment models, correlate with the human disease states for which they are used to predict drug action. This reluctance to try to draw parallels in any absolute sense between human and animal behavior is understandable. As Hill and Tedeschi (1971) state regarding human psychopathology, "There may be valid counterparts of these manifestations in the behavior of animals, but the dangers of anthropomorphism are great, and inferences about the subjective state of an animal can be very misleading." Although it is easy to overinterpret and anthropomorphize, there would seem to be some value in examining the question of how closely an animal model, especially one in which a high correlation exists between drug action in human disease state and in animal model, resembles a human disorder.

In addressing the question of how closely the response-contingent punishment paradigms model human anxiety, one encounters a multifaceted problem. The term "anxiety" is used to cover a heterogeneous set of symptoms with varying etiologies (Klein, 1976; Boissier & Simon, 1969). Given the lack of diagnostic sophistication in human dysfunctions involving a primary complaint of anxiety and in the other psychopathological states in which anxiety is part of the symptomatology, one scarcely expects any animal model to be viewed as corresponding very well with a clinical entity as a whole (Blackwell & Whitehead, 1975; Boissier & Simon, 1969). Of course this problem is common to the whole enterprise of developing drugs to treat human mental disorders. Diagnostic imprecision and the lack of operational definitions of human psychopathological behaviors have been the greatest deterrents to the development of animal models of these human disease states (Cook & Sepinwall, 1975). However, although the problem cannot be easily resolved, it is possible to abstract from the clinical literature certain definitions and descriptions of anxiety which relate some aspects of the human clinical syndrome to some aspects of behavior observed in the response-contingent punishment paradigm.

A salient feature of the response-contingent punishment paradigm is that the approach-avoidance tendencies produce a conflict state in the animal, and conflict has been seen as a basic cause of psychoneurotic states in which anxiety is a prominent feature (Berger, 1968; Brady, 1968). Miller (1966) suggested that punishment of food-rewarded behavior was a useful means for evaluating in the animal laboratory the effects of drugs upon behaviors which would be clinically described as neurotic. From another point of view, the suppression of responding in the punishment paradigms can be considered as a passive avoidance response, and it has been suggested that a component of the human anxiety in psychoneurotic states arises from passive avoidance of social situations (Irwin, 1968). Blackwell and Whitehead (1975) argue that the response suppression produced by punishment which occurs in animal conflict tests can be related to the response suppression produced by punishment of appropriate responses in human childhood which leads to neuroticism and anxiety in later life.

From another perspective, it can be argued that fear and anxiety are closely related, distinguishable only in that fear is experienced under conditions of actual danger and anxiety under conditions of anticipation of danger (Klein, 1976; Boissier & Simon, 1969). Although fear is also a construct, it has often been inferred as a motivation experienced by animals when they engage in active or passive avoidance of an aversive event (Miller, 1957). If fear and anxiety are closely related constructs in humans, and given that fear is a term that has been used widely in explaining animal behavior, then it should not be unreasonable to suggest that an animal's behavior can be motivated by anxiety. Boissier and Simon (1969), after some consideration of this same question, conclude that "one can speak of anxiety in an animal when the animal is expecting an unpleasant sensation." Since the suppression of response rate occurs in the conflict paradigm in anticipation of the aversive event, maybe it is not such a great indulgence in anthropomorphizing to speak of experimentally-induced "fear" or "anxiety" (Ray, 1965).

271

Thus there seems to be some justification for the notion that some aspects of the heterogenous clinical state called anxiety are analogous to the motiviation experienced by animals in a conflict paradigm. Visual monitoring of the behavior of animals in a conflict situation reveals similarities to the behavior of some human psychoneurotic patients. Abortive bar presses, discontinuous or jerky movements, and other bizarre behaviors (personal observation; Hill & Tedeschi, 1971) are impressive, as is the return to normal, coordinated behavior under the influence of anxiolytics. It is also interesting that normal human subjects run in experimental conflict procedures display behavior qualitatively similar to that seen in animal paradigms (Beer & Migler, 1975). Finally, the usefulness of anxiolytics in reducing anxiety in patients anticipating surgery (Bookman & Randall, 1976) is suggestive, since such a situation bears a close superficial resemblance to an animal's anticipation of an aversive event in a conflict paradigm.

Although the above attempt to show some similarity between the response-contingent punishment paradigm and some aspects of what is labeled as anxiety in humans is neither rigorous nor exhaustive, it may be at least provocative. The usual emphasis on operational definitions and cautions about anthropomorphizing are certainly well taken; however, perhaps a willingness to indulge in speculation about the relationship between human and animal behavior can leaven what might otherwise become a precise but sterile exercise. And regardless of what opinion is held about the face validity of response-contingent punishment paradigms as homologous models of some forms of human anxiety, a strong case for the Geller conflict test as a model of anxiety in animals can be made on the basis of its ability to predict that certain drugs will show an anxiolytic activity in certain forms of human anxiety.

SOME PROBLEMS IN USING THE GELLER CONFLICT TEST

In our own employment of the Geller conflict test in screening new compounds for possible usefulness in treating human anxiety, we have encountered several problems:

1) The effectiveness of standard drugs in releasing suppressed rates of responding in conflict is inconsistent. This is manifested as absence of smooth dose response curves to anxiolytics in individual animals and in the instability over time of a given animal's response to a given dose of an anxiolytic.

2) Baseline response rates in conflict tend to wander. The amount of suppression of responding in conflict produced by a given shock level changes over sessions, often dramatically, requiring frequent readjustments of shock level in order to maintain a window of suppression from which good drug effects can be observed. Indeed, day-to-day variations often exceed in range the effect observed under drug. Sometimes baseline response rate in conflict is not recovered after treatment.

3) Training subjects is tedious. Careful and individually titrated increases in shock level are needed in order to achieve reduction of response rate in conflict without producing complete and sometimes permanent suppression of responding.

The Geller conflict test being used in our laboratory corresponds fairly closely to that originally proposed by Geller and Seifter (1960). Rats respond on a multiple VI 2-min/CRF (food plus shock) schedule—the CRF (food plus shock) portion constituting the conflict period. Four 12-min VI components alternate with four 3-min conflict periods. The conflict periods are signalled by a cue light over the bar. Positive reinforcement is a 45 mg Noyes pellet, and punishment is a 0.50 sec 60 Hz footshock delivered from a Grason-Stadler or Coulbourn shock generator with grid distributor. Male Long-Evans rats, 22 hr food-deprived, are used in daily tests 6 days per week. In most cases, drugs are administered on Tuesday and Friday. Figure 1 is a cumulative record showing a moderate rate of responding in the VI portions of the records with low response rates during conflict in the top panel (A) and the release of responding during conflict after the animal has been given an intraperitoneal injection of 10 mg/kg chlordiazepoxide (CDP) in the bottom panel (B).

Figure 2 illustrates one of the problems that we have observed in using the Geller paradigm for screening drugs. This animal, following training, was injected with 10 mg/kg of CDP approximately once every two weeks in order to demonstrate that a standard anxiolytic was capable of reliably producing release from suppression during conflict; VI rates of responding were virtually unchanged over the 3 months of time illustrated, but the drug-induced release from suppression was highly variable and required several readjustments of shock, always toward a lower level. Figure 3 illustrates the same kind of data from another animal showing variable responses to the injection of the same amount of drug over time and the need to readjust shock levels continually. Most papers reporting data gathered in the Geller paradigm provide only group means, not values for individual subjects or even variability estimates for group data; the few cases in which individual response rates in

Standard Geller (CPD 10)

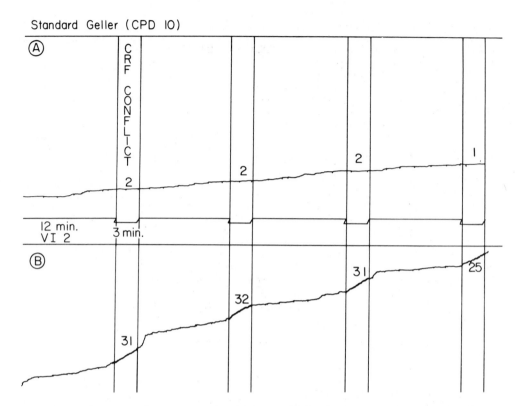

Fig. 1. Redrawn cumulative records showing a 22-hr food deprived rat responding on a multiple VI-2 minute/CRF (food plus shock) schedule during a 1-hr control session and, on the following day, during a session preceded 1 hr by an intra-peritoneal injection of 10 mg/kg chlordiazepoxide. In both sessions the shock level was 0.3 mA.

Rat # 47   Standard Geller

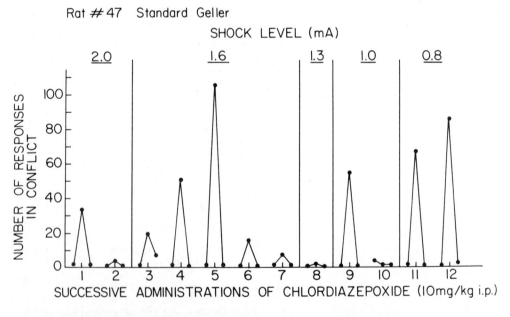

Fig. 2. Total number of responses made during a day's four conflict periods for the day preceding, day of, and day following intraperitoneal injection of 10 mg/kg chlordiazepoxide. The 12 injections of chlordiazepoxide were scattered over approximately a 3-month period during which the animal was run 6 days per week and occasionally received other drugs.

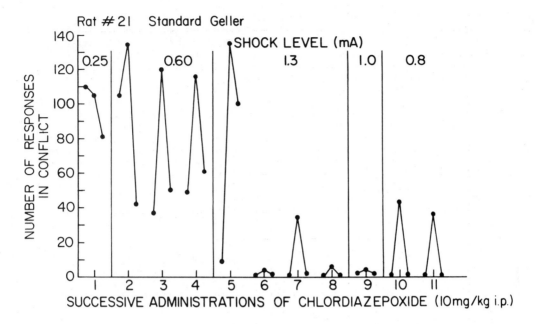

Fig. 3. Details same as for Fig. 2

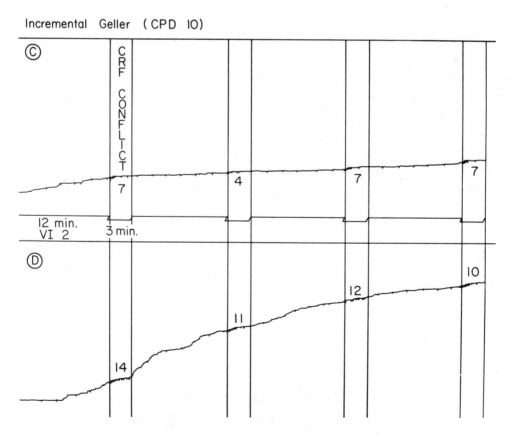

Fig. 4. Details same as for Fig. 1 except that incremental shock instead of fixed shock was used during conflict periods.

conflict have been reported show as much variability as we have typically found (e.g. Geller & Seifter, 1960). Davidson and Cook (1969) reported that a modified Geller configuration in which an FR-10 schedule replaced the CRF schedule in conflict resulted in more stable behavior which obviated the necessity for continuous titration of shock intensity (Davidson & Cook, 1969; Cook & Davidson, 1973); however, they too gave no estimates of variability of response, other than to indicate at what point group responses to drugs became significantly different from non-drug response rates in conflict.

In our laboratory we have employed a modification of the Geller which has reduced or eliminated some of the problems experienced with the standard paradigm. Figure 4 is a cumulative record of an animal responding on what we have chosen to call an incremental Geller conflict test. All features of the paradigm are identical to those described previously for the standard Geller we were using in our laboratory, with the exception that during the conflict portion of the schedule an incremental shock was used: following the first response in conflict, which is not punished by shock, the second response is punished by a .05 mA shock and each additional response increments the shock level by .05 mA; each conflict period restarts at a zero shock level. Two immediate advantages have been noted with this paradigm: it is extremely easy to train subjects, and the experimenter is freed from adjusting shock levels.

The next three figures present data from the first experiment run to compare the standard and incremental Geller paradigms in our laboratory. Seven male Long-Evans rats were run twice at four dose levels of CDP under both paradigms. VI rates were not different under the two paradigms. Figure 5 shows group dose response curves under both conditions in

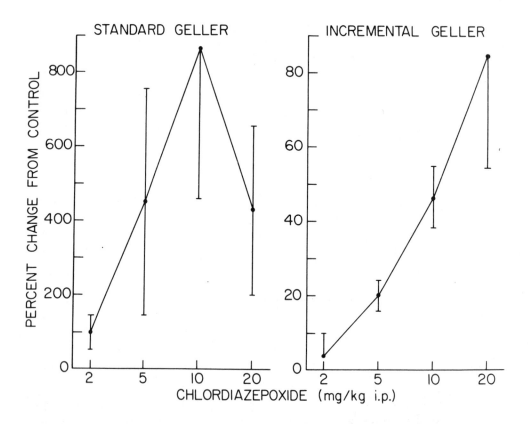

Fig. 5. Dose response curves gathered from seven rats run on both standard and incremental Geller paradigms. The percent change in responding during conflict after an intraperitoneal injection of 10 mg/kg chlordiazepoxide given 1 hr prior to the session was calculated based on the preceding control day. Two runs at each dose were averaged for each animal prior to calculating a group mean and a standard error of the mean.

conflict; a more orderly dose response curve with smaller standard errors is evident in the incremental Geller. The lower standard errors arise from two aspects of the data. First, the variation in control rate of responding from day to day is much less in the incremental Geller (average ratio of low control to high control day for the incremental paradigm is 1.63, for the standard paradigm is 13.96). Second, as illustrated in Fig. 6,

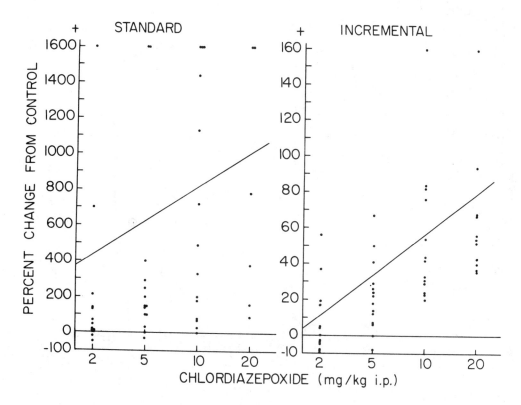

Fig. 6. Individual points used to calculate the group data presented in Fig. 5. Also shown is a regression line calculated from the points.

greater consistency was shown in terms of the response by individual animals to each increase in dose level. Figure 7 shows individual dose response curves for each of the seven animals; again, values for the incremental Geller appear to be more orderly.

Although the data presented comparing the two paradigms was from our first experiment and needs to be confirmed in a more systematic fashion, it does seem to indicate that the incremental Geller produces more consistent and orderly results. In addition, training subjects is easier, stable responding is attained more quickly, and baseline responding is more stable from day to day and over time without the necessity of adjusting shock level. Current studies in progress seem to be confirming these results.

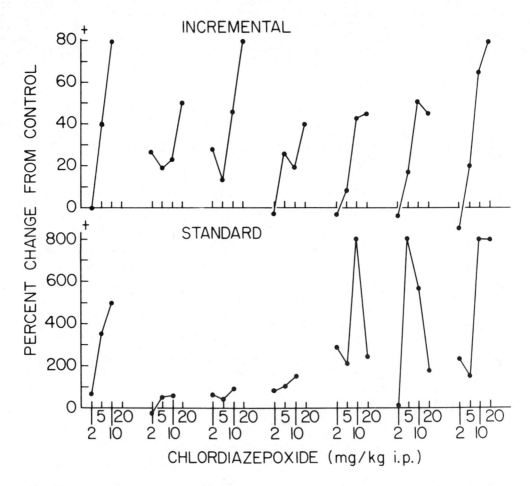

Fig. 7. Dose response curves for seven individual animals. Each point represents the mean of two determinations. Other data as in Fig. 5.

REFERENCES

Beer, B., & Migler, B. Effects of diazepam on galvanic skin response and conflict in monkeys and humans. In A. Sudilovsky, S. Gershon & B. Beer (Eds.)(1975), Predictability in Psychopharmacology: Preclinical and Clinical Correlations, Raven Press, New York, pp. 143-157.

Berger, F.M. The relation between the pharmacological properties of meprobamate and the clinical usefulness of the drug. In D.H. Efron (Ed.)(1968), Psychopharmacology, U.S. Government Printing Office, Washington, pp. 139-152.

Blackwell, B., & Whitehead, W. Behavioral evaluation of antianxiety drugs. In A. Sudilovsky, S. Gershon & B. Beer (Eds.)(1975), Predictability in Psychopharmacology: Preclinical and Clinical Correlations, Raven Press, New York, pp. 121-138.

Boissier, J., & Simon, P. Evaluation of experimental techniques in the psychopharmacology of emotion. Annals of the New York Academy of Sciences, 159, 898-914 (1969).

Bookman, P.H., & Randall, L.O. Therapeutic uses of the benzodiazepines. In L.L. Simpson (Ed.)(1976), Drug Treatment of Mental Disorders, Raven Press, New York, pp. 73-90.

Brady, J.P. Drugs in behavior therapy. In D.H. Efron (Ed.)(1968), Psychopharmacology, U.S. Government Printing Office, Washington, pp. 271-280.

Cook, L., & Davidson, A.B. Effects of behaviorally active drugs in a conflict-punishment procedure in rats. In S. Garattini, E. Mussini & L.O. Randall (Eds.)(1973), The Benzodiazepines, Raven Press, New York, pp. 327-345.

Cook, L., & Sepinwall, J. Behavioral analysis of the effects and mechanisms of action of benzodiazepines. In E. Costa & P. Greengard (Eds.)(1975), Mechanism of Action of Benzodiazepines, Raven Press, New York, 1975, pp. 1-28.

Cook, L., & Sepinwall, J. Psychopharmacological parameters of emotion. In. L. Levi (Ed.)(1975), Emotions—Their Parameters and Measurement, Raven Press, New York, pp. 379-403.

Cook, L., & Sepinwall, J. Reinforcement schedules and extrapolations to humans from animals in behavioral pharmacology. Federation Proceedings, 34, 1889-1897 (1975).

Davidson, A.B., & Cook, L. Effects of combined treatment with trifluoperazine-HCl and amobarbital on punished behavior in rats. Psychopharmacologia, 15, 159-168 (1969).

Geller, I., Kulak, J.T., & Seifter, J. The effects of chlordiazepoxide and chlorpromazine on a punishment discrimination. Psychopharmacologia, 3, 374-385 (1962).

Geller, I., & Seifter, J. The effects of meprobamate, barbituates, d-amphetamine and promazine on experimentally induced conflict in the rat. Psychopharmacologia, 1, 482-492 (1960).

Irwin, S. Anti-neurotics: Practical pharmacology of the sedative-hypnotics and minor tranquilizers. In D.H. Efron (Ed.)(1968), Psychopharmacology, U.S. Government Printing Office, Washington, pp. 185-204.

Hill, R.T., & Tedeschi, D.H. Animal testing and screening procedures in evaluating psychotropic drugs. In. R.H. Rech & K.E. Moore (Eds.)(1971), An Introduction to Psychopharmacology, Raven Press, New York, pp. 237-288.

Klein, D.F. Diagnosis of anxiety and differential use of antianxiety drugs. In L.L. Simpson (Ed.)(1976), Drug Treatment of Mental Disorders, Raven Press, New York, pp. 61-72.

Masserman, J.H. (1943) Behavior and Neurosis, Univ. of Chicago Press, Chicago.

McMillan, D.E. Determinants of drug effects on punished responding. Federation Proceedings, 34, 1870-1879 (1975).

Miller, N.E. Some animal experiments pertinent to the problem of combining psychotherapy with drug therapy. Comprehensive Psychiatry, 7, 1-12 (1966).

Miller, N.E. Experiments on motivation. Science, 126, 1271-1278 (1957).

Ray, O.S. Tranquilizer effects as a function of experimental anxiety procedures. Archives Internationales de Pharmacodynamie et de Thérapie, 153, 49-68 (1965).

DISCUSSION

Dr. Barry felt that a potential problem might be the difference between first trial results under chlordiazepoxide and those when the animal has a memory of the contingencies of increasing shocks. He thought it might be a particularly valuable procedure to test the same animals on several repeated occasions under a particular drug. *Dr. Howard* replied that he had tested animals on an incremental Geller test at the same dose administered about every third day for five or six times. There was pretty good consistency in terms of release of suppression. He had observed a learning or adaption by the test animals to the effects of anxiolytics.

Dr. R. Vogel mentioned another conflict procedure developed by John Vogel used in "analyzing anxiety mechanisms" which is somewhat simpler than the Geller; it involves a simple suppression of licking. *Dr. Howard* suggested that the major disadvantage of the conditioned suppression of drinking paradigm is the absence of a control for non-conflict behavior within the same experiment whereas in the original Geller one can look at the effects in a non-conflict situation interspersed with the effects in conflict behavior.

Dr. Tenen pointed out that the drug data obtained in the Vogel and Geller procedures are not identical. *Dr. Barry* agreed with Dr. Tenen and stated that sometimes the Geller procedure was sensitive to one-tenth the concentration required by the Vogel suppression technique. Furthermore, he felt the variability in the suppressed water-licking procedure was enormous. *Dr. Vogel* said that his experience with the simplified procedure had been limited to ethanol dose-response curves and that he had obtained similar results as with the Geller procedure. Although the variability in the procedure had required the use of an increased number of animals, this seemed preferable to him to the long period required in the Geller to establish a baseline.

THE USE OF ANIMAL MODELS FOR DELINEATING THE MECHANISMS OF ACTION OF ANXIOLYTIC AGENTS

Arnold S. Lippa, Eugene N. Greenblatt and Russell W. Pelham

Central Nervous System Research Dept., Lederle Laboratories, Pearl River, N.Y. 10965

INTRODUCTION

Investigations of the etiology of anxiety have benefited from the development of test systems which provide animal models for this mental state(1-4). Since it is not yet possible to identify animals with anxiety, most animal models depend on behavioral and/or pharmacological measures which are affected by psychotherapeutic drugs in apparent parallel with their clinical efficacy. These models conform to basic criteria concerning selectivity, sensitivity, relative potency and tolerance(1-4).

For instance, one of the most impressive properties of the drugs used in the treatment of anxiety (anxiolytics) is their ability to release both conditioned and unconditioned behaviors previously suppressed by punishment(1-4). In a series of now classic experiments, Geller and his co-workers(5-7) trained rats to bar press for food reinforcement; at discrete time intervals a tone signaled the onset of a conflict period during which responses were simultaneously reinforced with food and punished with electrical shock. Rats treated with different classes of anxiolytics accepted significantly more shocks during the conflict component than vehicle treated rats. Numerous studies utilizing a variety of psychotherapeutic drugs have demonstrated conflict behaviors to be highly selective for and sensitive to anxiolytics with a rank order of potency similar to that observed clinically(6,8-10). Since tolerance does not develop to the psychotherapeutic actions of the anxiolytics(11,12), it is significant that tolerance does not develop to the anti-conflict activity of these drugs(13,14). Similarly, other models have been generated which capitalize on the abilities of anxiolytics to stimulate food intake and to protect against the convulsions produced by pentylenetetrazol (see 1 for review).

In addition to predicting the psychotherapeutic actions of anxiolytics, animal models can also be used to investigate the mechanism(s) of action of these drugs(9,15). In fact, considerable interest has recently centered around the possibility that actions on brain serotonin (5HT), gamma aminobutyric acid (GABA) and glycine systems may be responsible for the pharmacological properties of the benzodiazepines, the most widely used class of anxiolytic drugs(16). Before implicating any neurotransmitter system as crucial in mediating the actions of the anxiolytics, multidisciplinary efforts are required to define by direct measurements the neuronal system(s) upon which the anxiolytics act and to relate these neuropharmacological actions to specific behavioral effects produced by these drugs. It is naive to assume that any one neuropharmacological property will be totally responsible for the therapeutic actions of the anxiolytics; rather, it is hoped that such studies will provide a point of origin for further investigations. The present paper is one of a series(1,17,18,19) in which we have utilized behavioral, pharmacological and electrophysiological techniques to address these problems.

GAMMA-AMINOBUTYRIC ACID (GABA)

GABA, a putative central nervous system neurotransmitter, has been postulated to play a role in the mechanism of action of benzodiazepines. Several studies have shown that benzodiazepines antagonize the convulsions produced by GABA receptor blockers (picrotoxin and bicuculline) and synthesis inhibitors(2,20,21). In addition, GABA agonists have been reported to antagonize the convulsions produced by pentylenetetrazole(22), a property which is believed characteristic of the anxiolytics(1,4). Neurochemical studies(20,21,23,24) have indicated that peripherally administered benzodiazepines and centrally administered GABA may affect the same receptor site since both of these treatments decrease cerebellar cyclic guanosine monophosphate (cyclic GMP) levels, an effect antagonized by GABA receptor blockers.

It should be pointed out, however, that the mode of action of any peripherally administered drug is complex and effects on various levels of integration must be considered. It is for

this reason that direct electrophysiological measurements are of crucial importance in
identifying the GABA-ergic properties of the benzodiazepines. To date, the electrophysiological results have been highly contradictory. On the one hand, Haefely and his colleagues
have shown that benzodiazepines mimic GABA in that they enhance presynaptic inhibition in the
spinal cord and in the cuneate nucleus(25,26,27) and enhance postsynaptic inhibition in the
cuneate nucleus and substantia nigra(25). Furthermore, these investigators have also reported
that benzodiazepines inhibit the firing rate of cerebellar Purkinje cells(25,28) which receive
a direct GABA-ergic inhibitory input from basket and stellate cells(29,30). However, all of
these results utilized peripheral administration of benzodiazepines and indirect effects cannot
be excluded. Conversely, Steiner and Felix(31) have reported that several benzodiazepines were
able to antagonize the effects of GABA on cerebellar Purkinje cells and cells in the lateral
vestibular nucleus [which receive a direct GABA inhibitory input from cerebellar Purkinje
cells(32,33)]. As in the experiments of Haefely, et al., the benzodiazepines were peripherally
administered.

We have also investigated the effects of both peripheral and microiontophoretic administration
of benzodiazepines on the firing of cerebellar Purkinje cells(18). Glass micropipettes were
stereotaxically placed into the cerebellar vermis of rats anesthetized with urethan (1 g/kg).
Single cerebellar Purkinje cells were identified by their amplitude, firing pattern and/or
ability to be inhibited by microiontophoretic administration of GABA(30,34). Intraperitoneal
injection of 12.5 mg/kg of chlordiazepoxide produced a suppression of Purkinje cell firing
rate with a latency of 3-15 minutes. No effect was produced by isovolumetric injections of
the vehicle solution (Fig. 1A and B). This inhibition lasted for as long as the recording
session(> 1 hour) (Fig. 1A). In 4 rats, 3 mg/kg of picrotoxin was intraperitoneally injected
20 minutes after chlordiazepoxide. Within 4-10 minutes, a reversal of the chlordiazepoxide-
induced suppression of firing rate was observed in all four animals (Fig. 1B). These results
indicate that a dose of chlordiazepoxide which had previously been found to produce marked
ataxia(1,17) also suppressed the firing of cerebellar Purkinje cells. The electrophysiological
effects of this dose of chlordiazepoxide were reversed by a dose of picrotoxin which had also
previously been found to reverse the ataxia produced by chlordiazepoxide(1,17). These data
support the hypothesis that peripheral administration of chlordiazepoxide can interfere with
cerebellar functioning through a GABA-ergic mechanism.

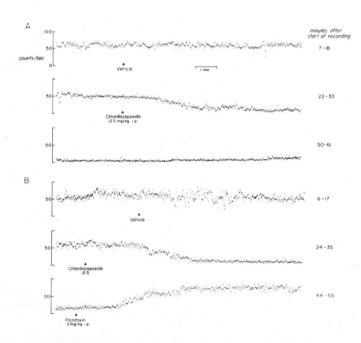

Fig. 1 The effects of chlordiazepoxide on single cerebellar Purkinje
cells. A. Ratemeter records of the effect of chlordiazepoxide
(12.5 mg/kg, i.p.) or vehicle on the firing (spikes/second) of
a single cerebellar Purkinje cell. B. Similar effects in
another animal - note the reversal of the depressant effects
after administration of picrotoxin (3 mg/kg, i.p.)

In additional studies, the microiontophoretic application of flurazepam, a soluble benzo-diazepine, inhibited the majority of Purkinje cells studied (19/23). Increasing amounts of inhibition were obtained by increasing ejection current (Fig. 3).

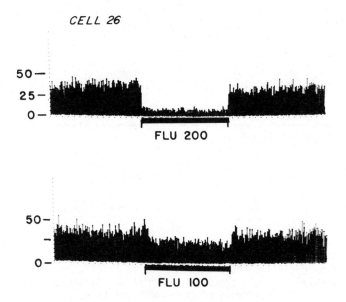

CELL 26

FLU 200

FLU 100

Fig. 2 The effect of flurazepam on a cerebellar Purkinje cell. Each computer generated histogram is comprised of the total number of counts per bin over 5 consecutive, triggered sweeps where bin time equals 128 msec. and length of sweep equals one min. Solid horizontal bars indicate ejection periods and numbers denote ejection current (nA) levels.

Two types of responses to flurazepam were obtained. In 13 cells, flurazepam produced an inhibition of firing with a short latency to onset and recovery (Fig. 3), an effect similar to that seen with GABA (Fig. 4). The concurrent iontophoretic administration of picrotoxin antagonized the depressant effects of flurazepam and GABA on all cells tested (7/7) (Figs. 3 and 4). In an additional 6 cells, flurazepam produced an inhibition with a long latency to onset (usually > 30 seconds) and the inhibition persisted well beyond the termination of ejection (usually > 1 minute). This long-lasting inhibition could be reversed by subsequent application of picrotoxin (see Fig. 5). Dray and Straughan(35) have recently reported similar effects on unidentified rat medullary neurons.

While the present results provide direct evidence for the GABA-ergic properties of at least one benzodiazepine, flurazepam, they do not offer any insight into whether the actions of flurazepam are due to a direct effect on the post-synaptic receptor or to a releasing action on presynaptic GABA stores. Because the benzodiazepines did not affect 3H-GABA binding(36)or GABA levels(21), Costa et al.(20,21) have suggested that these drugs may mobilize GABA from a presynaptic storage site. The fact that chlordiazepoxide-induced ataxia(1,17) and depression of cyclic GMP levels(21) can be antagonized by inhibiting GABA synthesis supports this hypo-thesis.

Having established direct neurophysiological evidence for the GABA-ergic property of at least one benzodiazepine, flurazepam, it is now possible to discuss the role that this neuropharma-cological property may play in mediating the actions of the benzodiazepines. Cook and Sepinwall(9) have reported that aminooxyacetic acid was ineffective in a conflict procedure when tested either at the peak time for anticonvulsant activity (40-50 min.) or 3-4 hours after administration when GABA levels should be elevated. On the basis of these data, the authors minimized the importance of the GABA-ergic properties of the benzodiazepines in mediating anti-conflict activity. In contrast, Stein et al.(15) reported that picrotoxin selectively antagonized the anti-conflict activity of oxazepam and emphasized the GABA-ergic properties as important for anti-conflict activity.

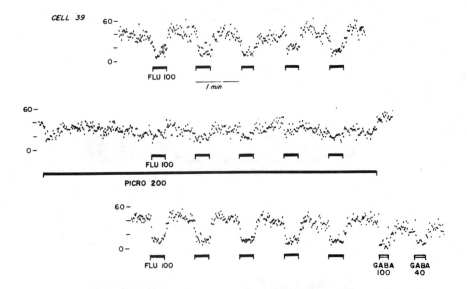

Fig. 3 Effects of iontophoretically applied flurazepam on a cerebellar
Purkinje cell. Ratemeter records of the firing rate of a
cerebellar Purkinje cell in counts/sec. First and third trace
indicate the effects of flurazepam before and after concurrent
iontophoretic administration of picrotoxin. Middle trace
indicated antagonism of flurazepam effect by concurrent ejection
of picrotoxin. Horizontal bars indicate ejection periods and
numbers denote ejection current levels (nA).

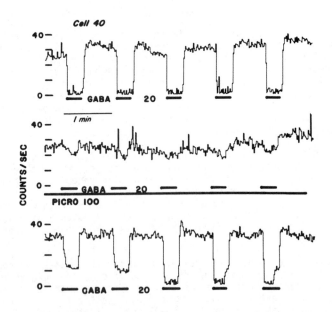

Fig. 4 Effects of GABA and picrotoxin on a cerebellar Purkinje cell.
Details similar to those in Fig. 3.

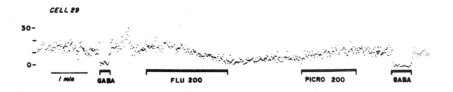

Fig. 5 Long latency effects of flurazepam. Details similar to Fig. 3.

In order to resolve this issue, we have developed a procedure for measuring both anti-conflict activity and ataxia in the same animals(1,17). Anti-conflict activity was measured in food (24 hours) and water (48 hours) deprived rats using an unconditioned thirsty rat conflict procedure(1). Rats were placed into a small darkened chamber containing a spout which provided access to a 10% dextrose solution. Once the animals had begun drinking, electric shock was applied on a 5 seconds on - 5 seconds off schedule. Ataxia was measured immediately before conflict testing by determining the ability of the same rats to traverse a 5/8" diameter inclined rod. Chlordiazepoxide (12.5 mg/kg, p.o., 30 minutes before testing) produced a significant (p < .05, Mann-Whitney U Test) increase in the number of shocks received as well as a significant (p < .05, Chi square test) increase in the number of animals unable to walk the rod (Fig. 6). Non-convulsant doses of picrotoxin (0.75, 1.5 and 3.0 mg/kg, i.p., 20 minutes after chlordiazepoxide) produced a dose-related reversal of the ataxia produced by chlor-diazepoxide without affecting the anti-conflict activity (Fig. 6). A similar effect in chlordiazepoxide-treated rats was observed after pretreatment with a non-convulsant dose of isoniazid (100 mg/kg, s.c., 30 minutes prior to chlordiazepoxide) (Fig. 6).

In view of these data and those reported by Cook and Sepinwall(9), it is unlikely that a GABA-ergic mechanism is responsible for the anti-conflict activity of the benzodiazepines. Rather, these data support the hypothesis(1,17) that the GABA-ergic properties of the benzodiazepines are responsible for the ataxia produced by these drugs. Since the benzodiazepines can produce GABA-mediated effects on the cerebellum(20,21) [a region involved in the maintenance of postural reflexes and motor coordination(37,38)], it appears reasonable to emphasize this neuronal substrate as a site of action for the production of ataxia by the benzodiazepines.

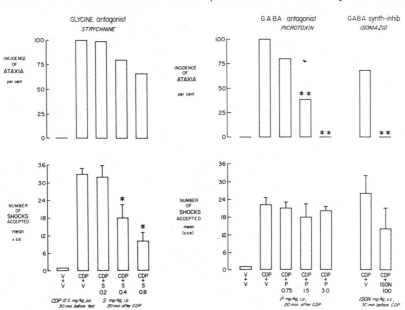

Fig. 6 The effect of drugs on the ataxia and anti-conflict activity of chlordiazepoxide. * p < .05, Mann-Whitney U test. ** p < .05, Chi square test.

GLYCINE

Snyder and his colleagues(36,39,40) have recently proposed that the benzodiazepines exert their pharmacological actions by directly stimulating glycine receptors. The glycine-ergic properties of the benzodiazepines were measured by their ability to displace 3H-strychnine [a purported glycine receptor blocker] from its binding sites on synaptic membranes. These authors also reported a significant correlation between the activity of the benzodiazepines in the binding assay and their activity in various animal and human tests. Conversely, Cook and Sepinwall(9) have reported no significant correlations between the potency of ten selected benzodiazepines in a rat conflict procedure and their reported activity in the binding assay. This approach may not be totally justified, however, since the receptor binding assay does not take into account the drug metabolism and absorption which take place in the experimental animal, as well as in man; as such, it may not accurately reflect the relative potencies observed in the in vivo situation. Supportive evidence for an important role for glycine in the mediation of the therapeutic action of anxiolytics comes from the report that the potencies of diazepam and chlordiazepoxide in antagonizing strychnine-induced convulsions are similar to their potencies in antagonizing pentylenetetrazole-induced convulsions(21), a procedure useful for the identification of anxiolytic drugs(1,4).

We have also reported(1,17) that non-convulsant doses of strychnine (0.2,0.4 and 0.8 mg/kg, i.p., 20 minutes after chlordiazepoxide) decreased the anti-conflict activity of chlordiazepoxide in a dose dependent manner (Fig. 6). The ataxia produced by chlordiazepoxide was unaffected by strychnine. Recently, Stein et al.(15) have reported that strychnine depressed unpunished responding at doses lower than those which attenuated the anti-conflict effects of oxazepam. These latter results suggest that our original observation concerning the attenuation of the anti-conflict effects of chlordiazepoxide may have been due to a non-specific depression produced by the interaction of strychnine and chlordiazepoxide.

In order to clarify these contradictions, studies were conducted to determine the effect of chronic diazepam treatment on the subsequent ability of diazepam to protect against pentylene-tetrazole-, bicuculline-, and strychnine-induced convulsions. Since tolerance is known to develop to the side effects but not to the therapeutic effects produced by the benzodiazepines (11,12), the use of chronic studies can provide further clarification of the importance of glycine-ergic as well as GABA-ergic mechanisms in mediating the therapeutic actions of benzodiazepines.

For the acute studies, groups of 8-10 mice were orally treated with graded doses of diazepam and subsequently challenged with a CD_{99} (dose estimated to produce convulsions in 99% of vehicle or untreated controls) of pentylenetetrazole (30 mg/kg, i.v.), bicuculline (0.6 mg/kg, i.v.) or strychnine sulfate (1.2 mg/kg, s.c.). Diazepam was administered 60 minutes prior to the injection of pentylenetetrazole or bicuculline and 30 minutes prior to strychnine. For the chronic studies, mice were treated with diazepam (2 mg/kg, p.o.) or isovolumetric amounts of the vehicle for 7 consecutive days. On the eighth day, groups of 8-10 mice were orally treated with graded doses of diazepam and subsequently challenged with a convulsant agent in the same manner as animals in the acute studies.

After treatment with the convulsant agents, all animals were isolated and observed for the presence of $\geq$ 5 seconds of uninterrupted clonic or tonic seizures. Antagonism of pentylene-tetrazole- and strychnine-induced convulsions was defined as the complete absence of seizures. Antagonism of bicuculline-induced seizures was defined as the absence of tonic seizures. Observations were made for 5 minutes after administration of bicuculline or pentylenetetrazole and for 30 minutes after administration of strychnine.

As can be seen in Fig. 7, the ability of diazepam to antagonize pentylenetetrazole-induced convulsions was similar after either acute or chronic administration [acute ED_{50}=0.82 (0.63-1.05)mg/kg; chronic ED_{50}=0.89 (0.62-1.29)mg/kg]. In contrast, following 7 days of diazepam treatment, the anti-strychnine activity of diazepam administered on the eighth day was less than 50% of that observed in acutely treated mice (p < .05 - probit analysis, see 41) [acute ED_{50}=1.03 (0.77-1.39)mg/kg, chronic ED_{50}=2.73 (1.06-7.00)mg/kg]. Likewise, the ability of diazepam to antagonize the convulsions produced by bicuculline was also reduced after chronic administration of diazepam [acute ED_{50}=1.60 (0.99-2.60)mg/kg/ chronic ED_{50}=3.05 (2.15-4.33)mg/kg].

Since tolerance to the anti-pentylenetetrazole activity of diazepam did not develop after chronic administration of this drug, these data support the hypothesis that the ability to antagonize convulsions produced by pentylenetetrazole may reflect the anxiolytic properties of psychoactive drugs(1,4). By contrast, tolerance to the anti-strychnine activity of diazepam did develop after chronic administration. These data minimize the importance of any glycine-ergic actions of diazepam in the mediation of its anxiolytic effects and support the interpretation of Stein et al.(15) that anti-conflict reversals after strychnine represent some general response-depressant effect of strychnine. Furthermore, the failure of in vivo electrophysiological experiments to detect any glycine-ergic actions on the part of diazepam (42), or flurazepam(35) seriously questions the existence of this property. Although benzodiazepines can interfere with 3H-strychnine binding(36,39), neither strychnine nor glycine affect 3H-diazepam binding(43). Nevertheless, whatever pharmacological proterties of diazepam are responsible for protecting against the convulsions produced by strychnine, they do not appear to be involved in the anxiolytic actions of this drug.

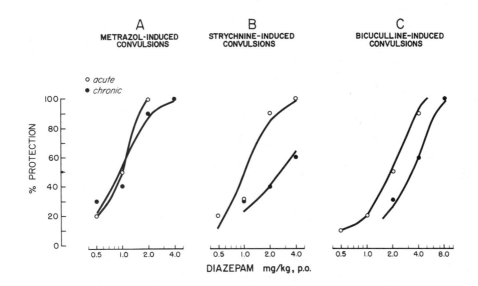

Fig. 7 The effects of acute and chronic diazepam treatment on the convulsions produced by pentylenetetrazole (Metrazol®), strychnine and bicuculline.

Finally, tolerance also developed to the anti-bicuculline activity of diazepam. These results confirm the original report of Juhasz and Dairman(44) and minimize the importance of GABA-ergic mechanisms in mediating the anxiolytic actions of diazepam. We have also reported that picrotoxin and isoniazed antagonized the ataxia but not the anti-conflict effects of chlordiazepoxide and suggested that the GABA-ergic properties of chlordiazepoxide were responsible for the production of this side effect(1,17). The present results are compatible with such a hypothesis.

SEROTONIN

Some of the earliest investigations into the mechanism of action of the anxiolytics suggested an important role for brain serotonin (5-HT) systems in the mediation of their therapeutic action. Specifically, it was reported that p-chlorophenylalanine (p-CPA), a 5-HT synthesis inhibitor, produced an anti-conflict effect similar to that seen with standard anxiolytics(14, 45,46). This effect was antagonized by restoring 5-HT levels with 5-hydroxytryptophan (5-HTP), a 5-HT precursor. Similar anti-conflict effects were also observed with the 5-HT antagonists, methysergide(15,47) and cinanserin(9,48). These results suggested that the anxiolytics produce their therapeutic effects by somehow inhibiting brain 5-HT systems. In fact, Stein et al.(13,14) have reported that although acute administration of oxazepam reduced both 5-HT and norepinephrine turnover, chronic administration only reduced 5-HT turnover. These data support a role for brain 5-HT in mediating anxiolytic action.

More recently, however, two different laboratories have been unable to obtain consistent anti-conflict effects with p-CPA(49,50). Since p-CPA has peripheral as well as central effects on 5-HT synthesis, we(50) re-investigated the role of 5-HT with the use of 5,6-dihydroxy-tryptamine (5,6-DHT), a neurotoxin which produces a destruction of brain serotonin fibers(61). The basic procedure utilized the thirsty rat conflict test described in the previous section. When rats were treated with two intraventricular injections of 5,6-DHT (75 µg of free base, separated by 48 hours), a significant anti-conflict effect was observed. This effect was apparent two days after the final injection and lasted as long as two weeks, with recovery occurring by three weeks. This anti-conflict effect paralleled the depletion and recovery of 5-HT measured in the rat brain(51). Similar but less long-lasting effects were also reported by Stein et al.(15).

TABLE 1   Effect of Test and Diazepam on Response Rate and Plasma Corticosterone

Treatment	Test Conditions	Response Rate (Shocks/Rat/Minute)	Plasma Corticosterone (µg%)
Vehicle	Deprived Controls	-	3.9 ± 1.5^a
Vehicle	Conflict	$1.2\pm.4^a$	28.0 ± 3.0^o
Diazepam (6 mg/kg, p.o.)	"	$4.5\pm1.2^*$	32.8 ± 2.9
(12 mg/kg, p.o.)	"	$5.0\pm1.0^*$	$13.5\pm4.2^{**}$
(24 mg/kg, p.o.)	"	$8.0\pm1.2^*$	$9.5\pm3.3^{**}$

a Mean + Standard Error
o Significantly different from Deprived Controls group, $p < 0.01$
* Significantly different from Vehicle group, $p < 0.01$
** Significantly different from both Deprived Controls and Vehicle-Conflict group
    $p < 0.01$

TABLE 2   Effect of Adrenalectomy (ADRX) or Hypophysectomy (HYPOX) on
Response Rate: Interaction with Diazepam

Treatment	Operation	Response Rate (Shocks/Rat/Minute)
Vehicle	Sham ADRX or HYPOX	0.8 ± 0.38^a
Vehicle	ADRX	1.2 ± 0.60
Diazepam (6 mg/kg, p.o.)	Sham ADRX or HYPOX	$5.0\pm1.4^*$
Diazepam (6 mg/kg, p.o.)	ADRX	19.0 ± 1.2^o
Vehicle	HYPOX	2.0 ± 1.0
Diazepam (6 mg/kg, p.o.)	HYPOX	24.0 ± 3.0^o

a Mean + Standard Error
* Significantly different from Vehicle-Sham group, $p < 0.01$
o Significantly different from Vehicle-Sham and Vehicle-Operated
    groups, $p < 0.01$

For example, Fig. 8 illustrates the effects one week after 5,6-DHT. An increase in punished responses equivalent to that produced by 8 mg/kg, p.o. of diazepam was observed. Combined treatment with 5,6-DHT and diazepam produced additive effects. Administration of 6-hydroxy-dopamine (6-OHDA) (2 intraventricular injections of 200 µg free base, separated by 48 hours) neither produced an anti-conflict effect nor interfered with the anti-conflict effects of diazepam. It is clear from these data that interference with normal 5-HT functioning can produce an effect on conflict responding similar to that seen with anxiolytic drugs. It is still premature, however, to ascribe a direct action of the anxiolytics on brain 5-HT systems; indirect effects via interactions with other systems should not be excluded(1).

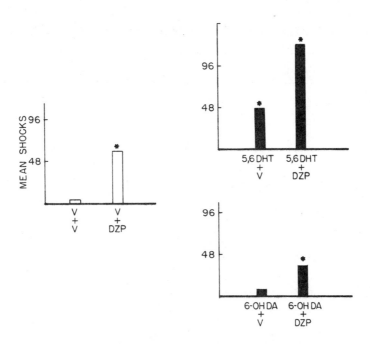

Fig. 8 The effects of 5,6-dihydroxytryptamine and 6-hydroxydopamine on
thirsty rat conflict responding. * p < .05, t test, compared
to appropriate vehicle treated controls.

One interaction which may prove to be of some importance involves the pituitary-adrenal
system. The potential importance of this hormonal system for the expression of anxiety may
be derived from the following. First, correlation between the degree of anxiety in humans
and the amount of cortisol production has been reported(52). Increases in circulating
cortisol, whether endogenously produced as in the case of Cushing's disease or exogenously
administered, can produce altered emotional states marked by anxiety, irritability and tension
(53-55). Second, glucocorticoids such as cortisol and corticosterone may act as required
factors in modulating 5-HT responses to specific regulatory signals. Adrenalectomy has been
reported to prevent the developmental rise of tryptophan hydroxylase activity in neonatal
rats as well as the elevation of enzyme activity in adult rats produced by stress, reserpine
or ethanol(see 56 for review). In adrenalectomized adult mice treated with ethanol or
reserpine, replacement with corticosterone did not enhance the drug effect but merely restored
the drug-induced increase of tryptophan hydroxylase to that achieved in normal animals(56).
Under certain conditions, tryptophan hydroxylase activity can be increased by corticosterone
administration(56,57), although negative results have also been reported(58,59). Finally,
anxiolytics have been reported to decrease corticosterone levels in rats(51,60).

For these reasons, it is significant that the conflict procedure increased plasma corticosterone
levels (Table 1). Diazepam produced a dose-related depression of plasma corticosterone
despite the fact that these rats received significantly more shocks than controls (Table 1).
Neither adrenalectomy nor hypophysectomy alone affected conflict responding, but both of
these treatments greatly potentiated the anti-conflict effect of diazepam (Table 2). Conversely, exogenous administration of corticosterone antagonized the anti-conflict actions
of diazepam (Table 3). Although these results might be interpreted to suggest that the
anti-conflict effects of diazepam were produced through a suppressive action on the production of corticosterone, additional studies have separated these two actions. Therefore,
the intraventricular administration of 5,6-DHT (as described earlier) produced an increase
in shocks without affecting plasma corticosterone levels (Table 4). In addition, low doses
of diazepam produced an increase in shocks without affecting plasma corticosterone levels
(Table 1).

These data emphasize the potential importance of brain serotonergic systems in mediating the
psychotherapeutic actions of anxiolytics and point to a pivotal role for this system in
integrating information concerning environmental contingencies. This idea is certainly not
new. The possibility that 5-HT might be involved in the modulation of sensory-motor processes
was first raised by Tenen(62) with the report that p-CPA produced a hyperalgesia to electrical
foot shock which was reversed by 5-HTP. Subsequent reports stated that 5-HT depleted animals
were more active than controls under conditions of novel environmental stimuli(63,64). Rats
exhibited increased reactivity to stimuli(65,66) and cats over-reacted to slight noises(67).

TABLE 3 Antagonism by Corticosterone of Diazepam Stimulation of Response Rate

Treatment		Response Rate (Shocks/Rat/Minute)
Vehicle + Oil	- -	0.7 ± 0.2^{a}
Vehicle + Corticosterone	- (0.5 mg/kg, i.p.)	0.6 ± 0.3
Vehicle + Corticosterone	- (1.0 mg/kg, i.p.)	0.7 ± 0.6
Diazepam + Oil	(12 mg/kg, p.o.) -	$15\pm2.0^{*}$
Diazepam + Corticosterone	(12 mg/kg, p.o.) (0.5 mg/kg, i.p.)	1.1 ± 0.5
Diazepam + Corticosterone	(12 mg/kg, p.o.) (1.0 mg/kg, i.p.)	0.6 ± 0.3

a Mean + Standard Error
* Significantly different from Vehicle + Oil group, p < 0.01

TABLE 4 Effect of Intraventricular (IVT) 5,6-DHT on Response Rate and Plasma Corticosterone

Treatment	Response Rate (Shocks/Rat/Minute)	Plasma Corticosterone (µg%)
Saline-Ivt + Vehicle	0.8 ± 0.5^{a}	26.8 ± 3.6^{a}
Saline-Ivt + Diazepam (8 mg/kg, p.o.)	$9.5\pm2.0^{*}$	$8.8\pm2.4^{*}$
5,6-DHT-Ivt + Vehicle	$8.3\pm3.0^{*}$	23.2 ± 3.7
5,6-DHT-Ivt + Diazepam (8 mg/kg, p.o.)	19.5 ± 3.0^{o}	$13.3\pm2.6^{*}$

a Mean + Standard Error
* Significantly different from Saline-Ivt + Vehicle Group, p < 0.01
o Significantly different from both Saline-Ivt + Vehicle and 5,6-DHT-Ivt + Diazepam group, p < 0.01

This increased responsiveness has been observed across all sensory modalities which have heretofore been investigated(68). Since the brain 5-HT system has also been reported to be affected by the pituitary-adrenal system(56), it may be of some relevance that glucocorticoids are primarily secreted during periods of uncertainty. These results have been obtained in humans as well as in experimental animals(see 69 for review). The relationships between pituitary-adrenal function and anxiety have already been discussed (see above). This hormonal system may thus provide affective information to be integrated with incoming sensory stimuli. The net result of this and other as yet unidentified neuro-hormonal interactions may determine the degree of behavioral suppression required by the complex contingencies of various environmental circumstances. For these reasons, it is significant that alterations in pituitary-adrenal functioning affected the anti-conflict activity of diazepam in a manner consistent with the reported effects of these alterations on brain 5-HT functioning. It should be pointed out, however, that suppression of the pituitary-adrenal system is neither necessary (see Table 4) nor sufficient (see Table 2) for obtaining anti-conflict activity. Certainly, pituitary-adrenal functioning can not be considered to be the sole modulatory influence on brain 5-HT systems. Further research must be directed toward identifying additional factors.

SUMMARY

Behavioral experiments have demonstrated the ability of picrotoxin (a GABA receptor blocker) and isoniazid (a GABA synthesis inhibitor) to antagonize the ataxia but not the anti-conflict effects produced by chlordiazepoxide. Tolerance developed to the GABA-ergic properties of diazepam, as measured by the ability of diazepam to protect against the convulsions produced by bicuculline (a GABA receptor blocker). Tolerance does not develop to the therapeutic actions of anxiolytics. Neurophysiological studies have demonstrated the ability of peripherally administered chlordiazepoxide or microiontophoretically administered flurazepam to inhibit the activity of cerebellar Purkinje cells, effects which are antagonized by picrotoxin. On the basis of these results we suggest that GABA mediated actions on cerebullum are responsible for the production of ataxia and minimize the importance of GABA-ergic actions in mediating the therapeutic effects of anxiolytics. Additional studies have reached similar conclusions concerning the proposed glycine-ergic properties of anxiolytics. In support of previous work, we have found that interfering with serotonergic functioning produces behavioral effects similar to those produced by anxiolytics. Endocrine interactions with brain serotonergic systems and anxiolytic drugs highlight the importance of these systems in mediating the actions of anxiolytics.

ACKNOWLEDGEMENTS

The authors wish to acknowledge the very capable technical contributions of Ms. B. Brooks, M. Sano and M. Wilfred and Messrs. D. Critchett and W. Smith without whom these experiments would not have been possible. The authors further wish to acknowledge the patience and care with which this manuscript was prepared by Ms. H. C. Gee.

REFERENCES

1. Lippa, A. S., P. A. Nash, and E. N. Greenblatt. (1977) Pre-clinical neuropsychopharmacological testing procedures for anxiolytic drugs. In: The Anxiolytics V. 3, ed. S. Fielding and H. Lal, Futura, New York.

2. Zbinden, G. and L. O. Randall, Pharmacology of benzodiazepines: Laboratory and clinical considerations, Adv. Pharmacol. 5, 213 (1967).

3. Randall, L. O. and B. Kappell (1973) Pharmacological study of some benzodiazepines and their metabolites, In: Garattini, S., Mussini, E., and Randall, L. O. The Benzodiazepines. Raven Press, New York.

4. Hill, R. T. and D. H. Tedeschi (1971) Animal testing and screening procedures in evaluating psychotropic drugs, In: Rech, R. H. and Moore, K. E., An Introduction to Psychopharmacology, Raven Press, New York.

5. Geller, I. and J. Seifter, The effects of meprobamate, barbiturates, d-amphetamine and promazine on experimentally induced conflict in the rat, Psychopharmacologia 1, 482 (1960).

6. Geller, I. (1962) Use of approach avoidance behavior (conflict) for evaluating depressant drugs, In: Nodine, J. H. and Moyer, J. H., Psychosomatic Medicine, Lea & Febiger, Philadelphia.

7. Geller, I., J. T. Kulak and J. Seifter, The effects of chlordiazepoxide and chlor-

promazine on a punishment discrimination, Psychopharmacologia 3, 374 (1962).

8. Cook, L. and A. B. Davidson (1973). Effects of behaviorally active drugs in a conflict-punishment procedure in rats, In: Garattini, S., Mussini, E., and Randall, L. O. The Benzodiazepines, Raven Press, New York.

9. Cook, L. and J. Sepinwall (1975) Behavioral analysis of the effects and mechanisms of action of benzodiazepines, In: Costa, E. and Greengard, P., Mechanism of Action of Benzodiazepines, Raven Press, New York.

10. Cook, L. and J. Sepinwall, Animal psychopharmacological procedures: Predictive value for drug effects in mental and emotional disorders, In: Airaksinen, M., CNS and Behavioral Pharmacology, Proceedings of the Sixth International Congress of Pharmacology, Pergamon Press, New York.

11. Goldberg, M. E., A. A. Manian and D. H. Efron, A. comparative study of certain pharmacological responses following acute and chronic administration of chlordiazepoxide. Life Sci. 6, 481 (1967).

12. Warner, R. S., Management of the office patient with anxiety and depression. Psychosomatics 6, 347 (1965).

13. Margules, D. L. and L. Stein, Increase of "anti-anxiety" activity and tolerance of behavioral depression during chronic administration of oxazepam, Psychopharmacologia 13, 74 (1968).

14. Stein, L., C. D. Wise, and B. D. Berger, (1973) Antianxiety action of benzodiazepines: Decrease in activity of serotonin neurons in the punishment system, In: Garattini, S., Mussini, E., and Randall, L. O., The Benzodiazepines, Raven Press, New York.

15. Stein, L., C. D. Wise, and J. D. Belluzzi, (1975) Effects of benzodiazepines on central serotonergic mechanisms, In: Costa, E. and Greengard, P., Mechanism of Action of Benzodiazepines, Raven Press, New York.

16. Costa, E. and P. Greengard (1975) Mechanism of Action of Benzodiazepines, Raven Press, New york.

17. Lippa, A. S., W. V. Smith and E. N. Greenblatt, Investigations into the mechanism of action of the benzodiazepines. Fed. Proc. 36, 1044 (1977).

18. Lippa, A. S. and D. J. Critchett, Neurophysiological evidence for the GABA-ergic properties of benzodiazepines, in preparation.

19. Lippa, A. S. and B. Brooks, Additional studies on the importance of glycine and GABA in mediating the actions of benzodiazepines, submitted for publication.

20. Costa, E., A. Guidotti and C. C. Mao (1975) Evidence for involvement of GABA in the action of benzodiazepines. In: Costa, E. and P. Greengard, Mechanism of Action of Benzodiazepines, Raven Press, New York.

21. Costa, E., A. Guidotti, C. C. Mao and A. Suria, New concepts on the mechanism of action of benzodiazepines. Life Sci. 17, 167 (1975).

22. DaVanzo, J. P., Grieg, M. E., and Cronin, M. A.: Anticonvulsant properties of aminooxyacetic acid, Am. J. Physiol. 201, 833 (1961).

23. Mao, C. C., A. Guidotti and E. Costa, The regulation of cyclic guanosine monophosphate in rat cerebellum: possible involvement of putative amino acid neurotransmitters. Brain Res. 79, 510 (1974).

24. Mao, C. C., A. Guidotti and E. Costa, Interactions between gamma amino butyric acid and guanosine cyclic 3', 5' - monophosphate in rat cerebellum. Molecular Pharmacol. 10, 735 (1974).

25. Haefely, W., A. Kulcsar, H. Mohler, L. Pieri, P. Polc and R. Schaffner (1975) Possible involvement of GABA in the central actions of benzodiazepines. In: E. Costa and P. Greengard, Mechanism of Action of Benzodiazepines, Raven Press, New York.

26. Stratten, W. P. and C. D. Barnes, Diazepam and presynaptic inhibition. Neuropharmacology 10, 685 (1971).

291

27. Polc, P., H. Mohler and W. Haefely, The effect of diazepam on spinal cord activities: possible sites and mechanism of action. Naunyn-Schmiedeberg Arch. Pharmacol. 284, 319 (1974).

28. Pieri, L. and W. Haefely, The effect of diphenylhydantoin, diazepam and clonazepam on the activity of Purkinje cells in the rat cerebellum, Naunyn-Schmiedeberg Arch. Pharmacol. 296, 1 (1976).

29. Roberts, E., T. N. Chase and D. B. Tower (Eds.) (1976) GABA in Nervous System Function Raven Press, New York.

30. Woodward, D. J., D. Rushmer, B. J. Hoffer, G. R. Siggins and A. P. Oliver, Evidence for the presence of stellate cell inhibition in frog cerebellum and for the mediation of this inhibition by GABA. Fed. Proc. 30, 318 (1971)

31. Steiner, F. A. and D. Felix, Antagonistic effects of GABA and benzodiazepines on vestibular and cerebellar neurones. Nature 260, 346 (1976)

32. Obata, K., M. Ito, R. Ochi and N. Sato, Further study on pharmacological properties of the cerebellar induced inhibition of Deiter's neurones. Exp. Brain Res. 11, 327 (1970).

33. Obata, K., M. Ito, R. Ochi and N. Sato, Pharmacological properties of the postsynaptic inhibition by Purkinje cell axons and the action of γ-aminobutyric acid on Deiter's neurones. Exp. Brain Res. 4, 43 (1967).

34. Eccles, J. C., M. Ito and J. Szentagothai, (1967) The Cerebellum as a Neuronal Machine. Springer-Verlag, Berlin.

35. Dray, A. and D. W. Straughan, Benzodiazepines: GABA and glycine receptors on single neurons in the rat medulla. J. Pharm. Pharmacol. 28, 314 (1976).

36. Snyder, S. H. and S. J. Enna, (1975) The role of central glycine receptors in the pharmacological actions of benzodiazepines. In: E. Costa and P. Greengard, Mechanism of Action of Benzodiazepines. Raven Press, New York.

37. Allen, G. I. and N. Tsukahava, Cerebrocerebellar communication systems. Physiol. Rev. 54, 957 (1974).

38. Guyton, A. C. (1976) Textbook of Medical Physiology, 5th edition, W. B. Saunders Co., Philadelphia, Penn.

39. Young, A. B., S. R. Zukin and S. H. Snyder, Interactions of benzodiazepines with central nervous system receptors: possible mechanism of action, Proc. Nat. Acad. Sci. 71, 2246 (1974).

40. Young, A. B. and S. H. Snyder, Strychnine binding associated with glycine receptors of the central nervous system, Proc. Nat. Acad. Sci. 70, 2832 (1973).

41. Finney, D. J. (1971) Probit Analysis, 3rd edition, Cambridge University Press.

42. Curtis, D. R., C. J. A. Game and D. Lodge, Benzodiazepines and central glycine receptors. Brit. J. Pharmacol. 56, 307 (1976).

43. Squires, R. F. and C. Braestrup, Benzodiazepine receptors in rat brain. Nature 266, 732 (1977).

44. Juhasz, L. and W. Dairman, Effect of subacute diazepam administration in mice on the subsequent ability of diazepam to protect against metrazol and bicuculline induced convulsions. Fed. Proc. 36, 377 (1977).

45. Geller, I. and K. Blum, The effect of 5-HTP on para-chlorophenylalanine (p-CPA) attenuation of conflict behavior, Eur. J. Pharmacol., 9,319 (1970).

46. Robichaud, R. C. and R. L. Sledge, The effects of p-chlorophenylalanine on experimentally induced conflict in the rat. Life Sci. 8, 965 (1969).

47. Graeff, F. G. and R. I. Schoenfeld, Tryptaminergic mechanisms in punished and non-punished behavior, J. Pharmacol. Exp. Ther. 173, 277 (1970).

48. Geller, I., R. J. Hartmann and D. J. Croy, Attenuation of conflict behavior with cinanserin, a serotonin antagonist: Reversal of the effect with 5-hydroxytryptophan and alpha-methyl-tryptamine, Res. Commun. Chem. Pathol. Pharmacol. 7, 165 (1974).

49. Blakely, T. A. and L. F. Parker, The effects of para-chlorophenylalanine on experimentally induced conflict behavior, Pharmacol. Biochem. Behav. 1, 609 (1973).

50. Cook, L. and J. Sepinwall (1975) Psychopharmacological parameters and methods, In: L. Levi, Emotions - Their Parameters and Measurement, Raven Press, New York.

51. Pelham, R. W., A. C. Osterberg, L. Thibault and T. Tanikella. Interactions between plasma corticosterone and anxiolytic drugs on conflict behavior in rats, Presented at the Fourth International Congress of the International Society of Psychoneuroendocrinology, Aspen, Colorado, 1975.

52. Sachar, E. J., Psychological factors relating to activation and inhibition of the adrenocortical response in man: a review. Prog. Brain Res. 32, 316 (1970).

53. Cleghorn, R. A. (1957) Steroid hormones in relation to neruopsychiatric disorders. In: Hormones, Brain Function and Behavior, Academic Press, New York.

54. Levitt, E. E., H. Persky, J. P. Brady and J. A. Fitzgerald. The effect of hydrocortisone infusion on hypnotically induced anxiety. Psychosom. Med. 25, 158 (1963).

55. Weiner, S., D. Dorman, H. Persky, T. W. Stach, J. Norton and E. E. Levitt, Effect on anxiety of increasing the plasma hydrocortisone level. Psychosom. Med. 25, 69 (1963).

56. Sze, P. Y. (1976) Glucocorticoid regulation of the serotonergic system of the brain. I In: Costa, E., E. Giacobini and R. Paoletti, Advances in Biochemical Psychopharmacology vol. 15, Raven Press, New York.

57. Azmitia, E. C. and B. S. McEwen, Corticosterone regulation of tryptophan hydroxylase in midbrain of the rat. Science 166, 1274 (1969).

58. Renson, J. (1973) Assays and properties of tryptophan-5-hydroxylase. In: Barchas, J. and E. Usdin, Serotonin and Behavior, Academic Press, New York.

59. Lovenberg, W., G. H. Besselar, R. E. Bensinger and R. Jackson (1973) Physiological and drug-induced regulation of serotonin synthesis. In: Barchas,J. and E. Usdin, Serotonin and Behavior, Academic Press, New York.

60. Lahti, R. A. and C. Barsuhn, The effects of minor tranquilizers on stress-induced increases in rat plasma corticosteroids. Psychopharmacologia 35, 215 (1974).

61. Baumgarten, H. G. and H. G. Schlossberger (1973) Effects of 5,6-dihydroxytryptamine on brain monoamine neurons in the rat. In: Barchas, J. and E. Usdin, Serotonin and Behavior, Academic Press, New York.

62. Tenen, S. S. The effects of p-chlorophenylalanine, a serotonin depletor, on active avoidance pain sensitivity, and related behavior in the rat. Psychopharmacologia 10, 204 (1967).

63. Brody, J. F. Behavioral effects of serotonin depletion and of p-chlorophenylalanine (a serotonin depletor) in rats. Psychopharmacologia 17, 14 (1970).

64. Persip, G. L. and L. W. Hamilton. Behavioral effects of serotonin or a blocking agent applied to the septum of the rat. Pharmacol. Biochem. Behav. 1, 139 (1973).

65. Mouret, J., P. Bobillier, and M. Jouvet. Insomnia following p-chlorophenylalanine in the rat. Eur. J. Pharmacol. 5,17 (1968).

66. Fibiger, H. C. and B. A. Campbell. The effect of para-chlorophenylalanine on spontaneous locomotor activity in the rat. Neuropharmacol. 10, 25 (1971).

67. Ferguson, J., S. Henriksen, H. Cohen, G. Mitchell, J. Barchas and W. Dement. Hypersexuality and behavioral changes in cats caused by administration of p-chlorophenylalanine. Science 168, 499 (1970).

68. Weissman, A. (1973) Behavioral pharmacology of p-chlorophenylalanine (PCPA). In: Barchas, J. and E. Usdin, Serotonin and Behavior, Academic Press, New York.

69. Warburton, D. M., Modern biochemical concepts of anxiety, Int. Pharmacopsychiat. 9, 189 (1974).

TION OF THE NUCLEUS LOCUS COERULEUS:
TUDIES OF ANXIETY

M.D.

of Medicine, 333 Cedar St., New Haven, Ct. 06510

ions in the function of the noradrenergic nucleus locus
ed a possible model for studies relevant to human anxiety.
de the similar behavioral effects of electrical stimulation
coeruleus-activating drug, piperoxane, and threatening but
ese behavioral effects, the effects of piperoxane, or elec-
oeruleus are diminished by clonidine, diazepam, morphine,
ions of the locus coeruleus. The relevance of these studies
pported by the known anti-anxiety properties of drugs which
function, and the previous reports that a drug which
s anxiety in humans. These data are felt to support an
locus coeruleus, of which human anxiety may be a part.

g, D.R. Snyder, J.W. Maas, and I have been studying the
eus locus coeruleus in nonhuman primates. These investiga-
c alterations in noradrenergic function in the brain than
pharmacologic studies (1-3). The locus coeruleus has the
highest density of norepinephrine-containing neurons in the central nervous system (4,5),
and its projections supply a major proportion of the total norepinephrine in the primate
brain (6). Numerous studies of the locus coeruleus in nonprimate mammals have failed to
establish a convincing function for the system, although effects on sleep, motor activity,
eating, memory, grooming, and gnawing have been reported (7-13). The remarkable finding
has been that such a diffuse and widely projecting system has so few effects on behavior.
We hoped that studies in nonhuman primates might provide information about possibly more
subtle functions of a specific neurotransmitter that has been related to a variety of human
emotions and psychiatric illnesses (14). Redmond, Huang, Snyder, and Maas recently reported
preliminary findings from these studies which noted the similarities of the behavioral
responses to low intensity electrical stimulation of the locus coeruleus system and the
responses to human threats, leading to the suggestion that the locus coeruleus system and
its projections may be related to human anxiety or fear (15). The questions most relevant
to this symposium are whether these effects were specific to the locus coeruleus system and
how one might determine that alterations in the function of the locus coeruleus were useful as a "model system" for studies of anxiety in humans.

Some recent pilot studies, undertaken with Y.H. Huang, M.S. Gold, and J. Baulu have helped
us to sharpen our hypotheses and perhaps to anticipate the answers to these questions. We
have attempted to test the specificity of a possible relationship between the locus coeruleus and threat-associated behaviors in monkeys by using pharmacologic methods to alter the
function of the locus coeruleus, and have combined these methods with electrical stimulation
or electrolytic lesions of the locus coeruleus.

Pharmacologic Methods to Alter Locus Coeruleus Function

Receptors on the cell bodies of the locus coeruleus have been described which, in addition
to post-synaptic connections from the cell, suggest a variety of pharmacologic methods
which might be expected to increase or decrease the function of the neurons of the locus
coeruleus. Cell body receptors have been characterized which bind norepinephrine (16,17),
epinephrine (18), γ-amino-butyric acid (GABA) (19), and the opiates (20). Aghajanian and
his co-workers, using physiologic recording of the electrical activity of single neurons in

the locus coeruleus, have demonstrated the functional effects of the systemic or direct applications of these and other compounds which are thought to affect these receptors: norepinephrine (21), clonidine (22), desmethylimipramine (23), morphine (24), epinephrine (21), and GABA (21) which decrease its activity, and piperoxane (21), which increases it. The interactions of the α-adrenergic agonist, clonidine, and the α-adrenergic antagonist, piperoxane, with each other and with the effects of NE, E, a β-adrenergic blocker, and GABA iontophoretically applied to locus coeruleus neurons have led to the characterization of these receptors as α-adrenergic and inhibitory to locus coeruleus function. Maas et al (25) have demonstrated the predicted effects of 10 μg/kg clonidine and 2.5 mg/kg of piperoxane on arterio-venous differences in the major metabolite of brain NE in Macaca arctoides. Although the specific effects of diazepam on locus coeruleus neuronal firing rates have not been reported in rats or monkeys, there is biochemical evidence that the benzodiazepines decrease norepinephrine turnover in locus coeruleus projection areas (26,27). Diazepam might be important to study in addition because of its anti-anxiety effects in humans. In contrast to the α-adrenergic receptors on the locus coeruleus cell bodies, the post-synaptic receptors of the locus coeruleus in the cerebellum (28) and hippocampus (29) have been characterized as β-adrenergic. Post-synaptic β-adrenergic blockade, therefore, might also be expected to block the effects of locus coeruleus activity on these and other, as yet unstudied, β-adrenergic projections. See Fig. 1 and Fig. 2.

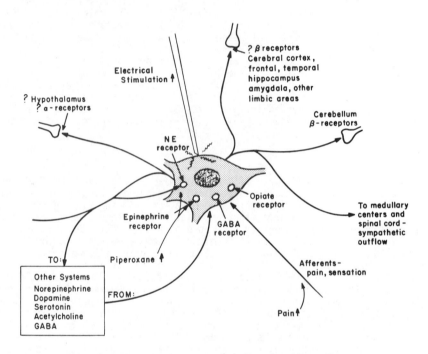

Fig. 1. Ways to increase (↑) the function of a
locus coeruleus neuron

The results of these pilot studies of alterations in locus coeruleus function will be described as tests of the hypothesis that the locus coeruleus is related to the expression of threat-associated behaviors.

METHODS

Measurement of Behaviors Associated with Threats in Monkeys

Certain behaviors in monkeys are increased by threatening situations. These have been noted in field investigations in situations of impending aggression, danger, conflict, or uncertainty (30). In the laboratory, these behaviors are quantitatively increased by such things as threatened attacks by humans, electrical shocks, novel conditions, or conflict. These behavioral manifestations are accompanied by the usual physiological signs of fear, such as increases in heart rate, blood pressure, urination, and defecation. Most of these behaviors can occur in chair-restrained monkeys and provide a preliminary method of comparing the effects of different ways of changing the function of the locus coeruleus.

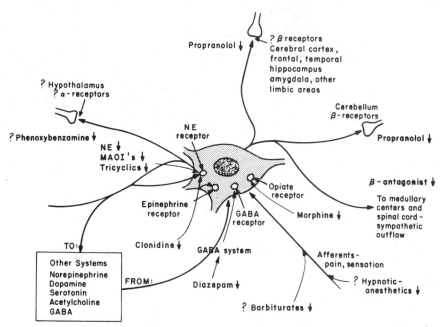

Fig. 2. Ways to decrease (↓) the function of a
locus coeruleus neuron

Each adult male or female <u>Macaca arctoides</u> studied was chair-trained for 1-2 months in a
sound dampened cubicle, with sham stimulation periods and saline intravenous infusions as
controls for electrical stimulation or drug infusion experiments. Drugs and electrical
stimulation were administered from outside the cubicle while the monkey was videotaped on
multiple view video cameras, with a computer-generated time base and coded event marking on
the tape. The tapes were later scored by three raters with replaying or frame-by-frame
analysis as necessary to achieve consensus. Sequential and simultaneous precisely defined
behaviors from an ethogram were entered into the computer for analysis (see Table 1 for the
list of categories). Specific behaviors, found to be associated with threats in previous
studies, were summed and plotted as one minute bin-width histograms (to allow comparison
with single neuronal activity in companion studies). All behavioral effects reported were
clear-cut, <u>i.e.</u> obvious increase or elimination of responses in association with a studied
variable.

Table 1   List of Behavioral Categories Scored from Videotape

Jump/startle*	Self groom	Lipsmack
Scratch*	Eye scan	Grimace
Handwring*	Attend/Stare	Threat
Hair/skin pull*	Lookout	Manipulate object
Self-mouth*	Brows up	Penile erection
Tongue movement	Open mouth/Brows up	Defecate
Chew	Head turn	Urinate
Grasp*	Body shake	Vocalize
Clutch*	Grind teeth	Arm drop
Struggle	Eat	Sit quietly
Yawn*	Drink	Drowsy
Freeze	Pout face/grunt purr	Eyes closed

*Behaviors most often associated with threatening situations

<u>Location of the Primate Locus Coeruleus in Vivo</u>

The locus coeruleus electrodes for recording, stimulation, or electrolytic lesions were
placed using the relationship between the locus coeruleus and the motor nucleus of the tri-
geminal nerve and the responses of single neurons in the locus coeruleus to physiologic
stimuli, as previously described (6). Electrodes were also placed 5 mm. rostral to the
locus coeruleus at a site which neurophysiologic, biochemical, and histological evidence
indicates is within the dorsal noradrenergic bundle (31). Histological and biochemical evidence for the location of electrodes or lesions within the locus coeruleus using these criteria has also been previously published (6,32).

RESULTS

The Effects of Increased Locus Coeruleus Function

We have found that the following procedures increase certain behaviors in monkeys so that it is virtually impossible to distinguish which stimulus is present: a fear stimulus, a frightening situation with no identifiable threat, a painful stimulus, intravenous piperoxane (1 mg/kg), and electrical stimulation of electrodes in the locus coeruleus or dorsal noradrenergic bundle (see Table 2). All of these procedures are also associated with some behavioral manifestations of increased alertness. The painful stimulus, piperoxane, and electrical stimulation have all been shown to produce increased function of the locus coeruleus in the rat (21,33,34).

Table 2 Procedures Increasing Threat-Associated Behaviors and their Effects on Alertness-Sedation Behaviors and Locus Coeruleus Function

Stimulus	Threat-Assoc. Behaviors	Alertness-Sedation Behaviors	L. Coeruleus Function (Species)
Fear	↑	↑	?
Frightening situation	↑	↑	?
Pain	↑	↑	↑(monkey)
Piperoxane (1 mg/kg)	↑	↑	↑(rat)
Electrical stimulation of locus coeruleus or dorsal NE bundle (biphasic, bipolar, 0.2-0.4 mA, 0.5 msec, and 5-50 Hz)	↑	↑	↑(monkey)

The Effects of Decreased Locus Coeruleus Function

Clonidine (10 μg/kg), diazepam (0.15 mg/kg), and morphine (0.2 mg/kg), administered intravenously, all decrease or eliminate the same specific behaviors which are increased by a fear-stimulus, a frightening situation, piperoxane, or electrical stimulation of the locus coeruleus (see Table 3).

Table 3 Procedures Decreasing Threat-Associated Behaviors and their Effects on Alertness-Sedation Behaviors and Locus Coeruleus Function

Stimulus	Threat-Assoc. Behaviors	Alertness-Sedation Behaviors	L. Coeruleus Function (Species)
Clonidine (10 μg/kg)	↓	↓	↓(monkey)
Diazepam (0.15 mg/kg)	↓	↓	?
Morphine (0.2 mg/kg)	↓	↓	↓
Propranolol (D,L-5 mg/kg)	↓	↓	↓
Lesions of the locus coeruleus	↓	No change	↓

Clonidine and morphine have been reported to decrease the functional activity of the locus coeruleus at these dose ranges in anesthetized rats (21,22,24), and this dose of clonidine has been shown to decrease arterio-venous norepinephrine metabolites in this species (25). The doses of morphine and diazepam were chosen because they are the mg/kg equivalents of amounts often used clinically in humans, and they were thought to be more relevant to their effects in humans than higher doses often studied with other animal "models" for anxiety (35). Propranolol (5 mg/kg) also decreased the same specific behaviors, most likely on the basis of post-synaptic (36,37) antagonism instead of changes in cell firing rates. All of the above procedures were also associated with some increases in "sedated" behaviors. However, each of the procedures eliminated the threat-associated behaviors without inducing sleep or drowsiness. Electrolytic lesions of the locus coeruleus, however, produced a most dramatic decrease in the threat-associated behaviors, not only with little detectable decrease in alertness, but increases in various social and non-social activities such as movement and aggression (38).

DISCUSSION

Criteria for a Central Neurophysiologic Model for Studies of Anxiety

These pilot studies support the hypothesis that specific threat-associated behaviors in chair-restrained monkeys are specifically affected by procedures which change the function of the locus coeruleus. These behavioral effects are consistent with an essential role for the locus coeruleus in mediating anxiety and fear, and can be confirmed to some extent from human reports of the effects of some of these drugs and procedures. The specific conditions which would allow alterations in the neurophysiology of the locus coeruleus to qualify as a model for anxiety should be quite rigorous. The reasoning process which might lead to an animal model of some human emotion, or for that matter, knowledge of the emotion of another human, is never more certain than analogy and inference, regardless of the apparent behavioral similarities to one's own emotions. It is particularly important, therefore, to be as precise as possible in specifying what types of inferences and conditions one is willing to accept and to formulate hypotheses to confirm or reject the model. A list of several such conditions and criteria is proposed in Table 4.

Table 4 Some criteria for a Central Neurophysiological Model of Anxiety

1. Quantifiable objective behavioral measures which meet these criteria:
    a. They are produced by the conditions which produce the emotion in humans
    b. They are quantitatively increased with increasing intensity of fear or anxiety producing stimuli
    c. They are selectively diminished, after identical stimuli, by pharmacologic agents which are anxiolytic in humans
    d. They correlate with the other peripheral and central manifestations of anxiety or fear in humans (pulse rate, tremor, skin conductance changes, cortisol secretion, etc.)
2. Behavioral and physiologic changes should be reproduced by alterations in central physiologic function, with these considerations:
    a. Central systems altered should be anatomically and neurochemically connected to other areas and systems known to regulate physiologic manifestations or correlates of anxiety
    b. Low intensity electrical stimulation should produce or increase the behavioral and physiologic manifestations and approximate qualitatively and quantitatively those resulting from fear-producing non-painful stimuli, painful stimuli, or the "memory" or anticipation of such stimuli.
    c. Specific electrolytic lesions should diminish or abolish the behavioral and physiologic manifestations in response to environmental stimuli effective in producing them in normal animals.
    d. "Specific" biochemical alterations in the function of the same areas should have effects consistent with the functional effects of low intensity electrical stimulation of the areas or lesions of the areas. Ideally, these same agents should also be anxiolytic or produce anxiety in humans
3. Central correlates of function in the proposed relevant areas or systems should exist for the following:
    a. Evidence of decreased function for all agents which are anxiolytic in humans. These changes should predict or correlate with anxiolytic potency, and the agents should block the effects of central physiologic activation of the areas.
    b. Evidence of increased function for all procedures, stimuli, or drugs which increase or produce anxiety or fear in humans
    c. Such peripheral evidence of function as is available from humans should not be inconsistent with the proposed central function

Do Behavioral Changes Associated with Locus Coeruleus Function Meet Criterion 1 (Table 4)?

The behavioral measures associated with locus coeruleus function also change with environmental stimuli associated with anxiety in humans. They are quantifiable and vary with the intensity of threatening stimuli. These behaviors are specifically diminished or eliminated by diazepam in clinically relevant and non-sedative doses. These behavioral changes correlate with several physiologic correlates of anxiety in humans. We are unable, however, to distinguish between the responses to acute frightening stimuli and the responses somehow perceived as threatening, but where no immediate threat is apparent to the animal. These characteristics have traditionally been used to distinguish fear from anxiety, which is sometimes defined to occur in a "free-floating" state without recognizable external precipitants (39).

A Speculative Anatomy of Locus Coeruleus Mediated Anxiety (Table 4, Criterion 2a)

The connections of the locus coeruleus system are adequate to place it in a central position in anxiety, although a number of the pathways necessary have not yet been identified anatomically. This highly speculative "anxiety" system involving the locus coeruleus is diagrammed in Fig. 3 to suggest that anxiety is a possible function of the locus coeruleus and some of its afferent and efferent connections.

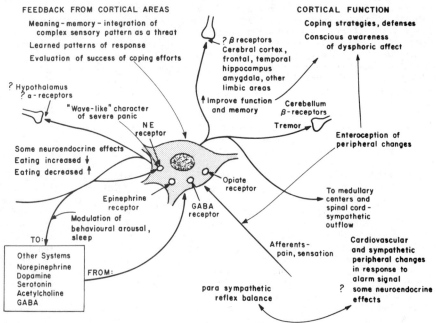

Fig. 3. A schematic Anatomy of Locus Coeruleus-Mediated Anxiety

The overall neuroanatomy.of the locus coeruleus system immediately suggests a generalized function. In the rat or cat the locus coeruleus has diffuse projections to the cerebral cortex, to the limbic system, to the medullary and spinal centers affecting cardiovascular sympathetic activity, bowel and sphincter activity, to the hypothalamus, and to the cerebellum (4,5,40,41). It has been suggested that a single locus coeruleus neuron may send efferents to all of these areas (42). The locus coeruleus receives numerous afferents from sensory modalities, other noradrenergic nuclei, the nucleus raphe pontis and magnus, the hypothalamus and substantia nigra (43), among those identified so far (40,43). In the normal individual, activation of the locus coeruleus by one of the afferent sensory or cortical pathways, hypothesized in Fig. 3, would produce the various manifestations of anxiety in the projection areas. Activation of the locus coeruleus-cortical or limbic efferent projection areas would lead to coping strategies, or defenses, to remove the threat. Cortical areas would register a disagreeable dysphoric affect is locus coeruleus activation were persistent or too strong. The cortex or limbic areas would provide feedback to the locus coeruleus after interpreting the meaning of sensory patterns, depending to some extent on learned responses (correctly associating the stimuli with the affective responses appropriate for danger, or incorrectly associating them with other stimuli, producing phobias), and an evaluation of the success of coping efforts, thereby increasing or decreasing locus coeruleus activity in a feedback loop for which we now have some behavioral evidence without known neuroanatomic pathways. The locus coeruleus may be involved in neuroendocrine changes associated with stress and anxiety; the principal endocrine effect of stress (increases in ACTH secretion [44]) has been reported in the cat after locus coeruleus electrical stimulation (45). We have recently presented evidence that the locus coeruleus system is inhibitory to eating (32), another known anxiety-fear effect which is consistent with an anxiety function for the locus coeruleus. The characteristic tremor of anxiety (46) may be associated with the cerebellar projections from the locus coeruleus. Modulation of arousal state may be an interactive function with other systems in the pontine reticular formation and elsewhere (47,48), and appears to be separable from the other behavioral effects, since locus coeruleus lesions are not visibly sedating. The caudal outflow may be responsible for cardiovascular and sympathetic peripheral changes associated both with anxiety and stimulation of the locus coeruleus (49,50). These effects are interactive with the parasympathetic system to produce the particular visceral manifestations of anxiety (51). Enteroception of these peripheral changes feeds back via the cortex and, perhaps, to the locus coeruleus directly, in a feedback loop, the blockade of which has been interpreted as the mode of anti-anxiety action of the β-adrenergic antagonist propranolol (52,53).

Effects of Alterations of Locus Coeruleus or Dorsal Bundle Activity (Table 4, Criteria 2b-d)

Locus coeruleus or dorsal NE bundle stimulation produces behaviors which can be quantitatively or qualitatively approximated to varying levels of response to environmental fear-producing stimuli by alteration of the rates of low intensity electrical stimulation. The behavioral effects of electrical stimulation are blocked by D,L-propranolol, a β-adrenergic blocker, indicating post synaptic β-adrenergic specificity of the stimulation, and by clonidine in doses that turn off the locus coeruleus system without activating post synaptic receptors (21,25). Identical behaviors are produced by piperoxane, which activates locus coeruleus single unit activity, and increases NE turnover in this species (25) by a biochemical means. Finally, the same behavioral responses noted to increase after procedures which increase locus coeruleus function are reduced by lesions of the locus coeruleus without behavioral sedation.

Central Correlates of Anxiety or Fear (Table 4, Criteria 3a,b)

The locus coeruleus activity correlates of anxiety or fear are not yet well demonstrated. Pain, one of the most potent of anxiety-producing stimuli, causes increased neuronal activity in the locus coeruleus of the rat (34) and the anesthetized monkey (6). There is some evidence already that drugs with known actions on the locus coeruleus decrease or increase anxiety in humans. Many drugs which are anxiolytic in humans block the action of the locus coeruleus system (see Fig. 2). Diazepam presumably blocks the effects of electrical stimulation of the locus coeruleus indirectly by way of the inhibitory GABA system (54,55) and the GABA receptors on the cell bodies of the locus coeruleus (19) which inhibit its firing (21). The benzodiazepines may act in the same way to produce their anxiolytic effects. The monoamine oxidase inhibitor drugs and the tricyclic anti-depressants have been reported to be of value in the treatment of panic and phobic disorders (56); several tricyclics diminish the neuronal function of the locus coeruleus (23). The barbiturate drugs have been used successfully in the treatment of anxiety (56) and turn off locus coeruleus activity (57). The opiates have also long been noted for their anti-anxiety effects (58); and morphine turns off the locus coeruleus in rats (24). The post synaptic β-adrenergic antagonist, propranolol, has also been reported to have anti-anxiety properties (59-61). Its effect in this model would be due to blockade of the effects of locus coeruleus β-adrenergic projections (28,29). Two drugs which activate the locus coeruleus (21) are reported to cause anxiety in humans: piperoxane (62,63) and yohimbine (64). Electrical stimulation in the region of the locus coeruleus in humans has been reported to produce feelings of fear and death, whereas a small lesion produces a marked calming effect (65).

It can be argued that none of these mechanisms or the procedures described in this study is entirely specific to the locus coeruleus. Each of the drugs studied or described affects other norepinephrine systems, other epinephrine synapses, or other neurotransmitter systems in addition to the locus coeruleus. Electrical stimulation might activate other passing systems in the area. The effects of all of them would not be expected to be virtually identical, however, unless the relevant effects were dependent on locus coeruleus function. Further evidence of specificity is the fact that the behavioral effects of electrical stimulation of the locus coeruleus are blocked by drugs which have relatively specific and previously described effects on locus coeruleus system function. It is still possible that many or most of the effects may result from effects of the locus coeruleus on other areas. These data suggest, however, that the locus coeruleus is necessary for these effects to occur.

Peripheral and Indirect Evidence from Humans for Locus Coeruleus Anxiety Function
Table 4, Criterion 3c

A central noradrenergic involvement in human anxiety is supported by the work of investigators who have noted correlations between epinephrine and fear (66), peripheral norepinephrine and anxiety (67; see ref. 14 for a review), and 24-hour urinary MHPG and anxiety in humans (68). Post et al. have very recently reported a strong correlation between CSF norepinephrine and anxiety but not activity in depressed patients (69). Other human and animal data has been reviewed implicating the "biogenic amines" in emotion (70), and the "catecholamines" in anxiety (71), "vigilance" (47), and "arousal" (48). These data are generally consistent with a role for the locus coeruleus-noradrenergic system in human anxiety.

CONCLUSION

These anatomical, theoretical, and speculative considerations about the data so far available suggest that although the locus coeruleus may be in the center of and essential to the behavioral and physiological expression of anxiety, it is not all there is to anxiety, and anxiety is not its only, or perhaps even its most important, role. To call the locus coeruleus an "anxiety" system, then, is at best only partly correct, since "anxiety" requires other areas and since anxiety is probably only a portion of the function of this primitive neurophysiologically inhibitory brain system. It is also not merely an "arousal" or "alerting" system, since these effects are separable from threat-associated behaviors pharmacologically and after locus coeruleus lesions. The evolutionary role and normal function of anxiety in a broad sense have been described to be to insure that threats to well-being are respected, and if possible, prevented (72). In a middle or normal range of function, such a system would be cautionary and associated with improved chances for survival, and its absence or diminished function would produce certain liabilities -- the failure to withdraw in the face of lethal danger or the inability to learn from certain kinds of experience or to inhibit certain responses. Perhaps "alarm" system is a better name for the locus coeruleus' apparent function, since "alarm system" conveys an adaptive nature, the deficit which would be present if the system were missing, and the unpleasantness of its ringing loudly or too long.

From a practical pharmacologic point of view, this "alarm system" may also be relevant to human anxiety, as suggested previously by the apparent convergence of possible mechanisms of anxiolytic activity on the locus coeruleus. This system has yet to predict a new anxiolytic drug, as has the Geller conflict procedure, although, if the model is correct, clonidine must have some heretofore undetected anxiolytic effects. The Geller procedure, however, reportedly does not detect the extremely potent anxiolytic activity of morphine (35), or the effects of D,L propranolol on anxiety (73), both of which have demonstrable behavioral and functional effects on the locus coeruleus system and threat-associated behaviors in monkeys.

This paper, due to length limitations, makes no attempt to do justice to the extensive body of animal studies which could be interpreted as supporting this alarm or anxiety model system, or to the many investigators who have been convinced of an association between the sympathetic nervous system and anxiety since Cannon's observations in 1915 (66). The implications of the findings reported here for the catecholamine hypothesis of the affective disorders (74,75) and other psychiatric illnesses must be discussed elsewhere. It is also perhaps best to let the critics of this interpretation of an alarm-fear-anxiety function for the locus coeruleus and others who have advocated the importance of other systems in mediating anxiety (76) speak for themselves. Our studies are only just beginning, and our findings must still be confirmed in every respect.

I hope that these preliminary observations and comments about the locus coeruleus and alarm, anxiety, or fear in monkeys will stimulate further studies which will improve the understanding of the effects and treatment of pathological human anxiety and its possible role in the etiologies of mental and psychosomatic illness.

ACKNOWLEDGEMENTS

These studies were undertaken in collaboration with Y.H. Huang, J.W. Maas, D.R. Snyder, M.S. Gold, and J. Baulu. G.K. Aghajanian, R.H. Roth, and J.P Flynn gave valuable advice. G. Hu, B.G. Erb, and S. Rek provided technical and editorial help. This work was supported by grants from the U.S.P.H.S., MH-24607, MH-25642, the State of Connecticut, and the H.F. Guggenheim Foundation. I thank these companies for generous gifts of drugs: Rhone-Poulenc (piperoxane) and Boehringer Ingelheim Ltd. (clonidine).

REFERENCES

(1) D.E. Redmond, Jr., J.W. Maas, A. Kling, C.W. Graham, H. Dekirmenjian, Social behavior of monkeys selectively depleted of monoamines, Science 174, 428 (1971).

(2) D.E. Redmond, Jr., A. Kling, J.W. Maas, and H. Dekirmenjian, Changes in primate social behavior after treatment with alpha-methyl-p-tyrosine, Psychosomatic Medicine 33, 97 (1971).

(3) D.E. Redmond, Jr., R.H. Hinrichs, J.W. Maas, and A. Kling, Behavior of free-ranging macaques after intraventricular 6-hydroxydopamine, Science 181, 1256 (1973).

(4) A. Dahlstrom and K. Fuxe, Evidence for the existence of monoamine containing neurons in the central nervous system 1. Demonstration of monoamines in the cell bodies of brain stem neurons, Acta Physiol. (Scand.) Suppl 232, 1 (1964).

(5) L.A. Loisou, Projections of the nucleus locus coeruleus in the albino rat, Brain Res. 15, 563 (1969).

(6) Y.H. Huang, D.E. Redmond, Jr., D.R. Snyder, and J.W. Maas, In vivo location and destruction of the locus coeruleus in the stumptail macaque (Macaca arctoides), Brain Res. 100, 157 (1975).

(7) N.-S. Chu and F.E. Bloom, Norepinephrine-containing neurons: changes in spontaneous discharge patterns during sleep and waking, Science 179, 908 (1973).

(8) N.-S. Chu and F.E. Bloom, Activity patterns of catecholamine-containing pontine neurons in the dorsal-lateral tegmentum of unrestrained cats, J. Neurobiol. 5, 527 (1974).

(9) P. Lidbrink, The effect of lesions of ascending noradrenaline pathways on sleep and waking in the rat, Brain Res. 74, 19 (1974).

(10) W. Ritter and L. Stein, Self stimulation of noradrenergic cell group (A6) in locus coeruleus of rats, J. Comp. Physiol. Psychol. 85, 443 (1973).

(11) D.G. Amaral and A. Routtenberg, Locus coeruleus and intracranial self-stimulation: a cautionary note, Behav. Biol. 13, 331 (1975).

(12) D.G. Amaral and J.A. Foss, Locus coeruleus lesions and learning, Science 188, 377 (1975).

(13) D.J. Micco, Complex behaviors elicited by stimulation of the dorsal pontine tegmentum in rats, Brain Res. 75, 172 (1974).

(14) D.L. Murphy and D.E. Redmond, Jr., (1975) Catecholamines in affect, mood, and emotional behavior in Catecholamines and Behavior, vol. 2, pp. 73-117, Plenum Press, New York.

(15) D.E. Redmond, Jr., Y.H. Huang, D.R. Snyder and J.W. Maas, Behavioral effects of stimulation of the locus coeruleus in the stumptail monkey (Macaca arctoides) Brain Res. 116, 502 (1976).

(16) T. Hokfelt, On the ultrastructural localization of noradrenaline in the central nervous system of the rat, Z. Zellforsch. 79, 110 (1967).

(17) M. Descarries, B. Droz, Incorporation de noradrenaline ^{3}HNA-^{3}H dan le system nerveux central du rat adulte. Etude radio-autographique en microscopie electronique, C.R. Acad. Sci. (Paris) 266, 2480 (1968).

(18) T. Hokfelt, K. Fuxe, M. Goldstein and O. Johnsson, Immunohistochemical evidence for the existence of adrenaline neurons in the rat brain, Brain Res. 66, 235 (1974).

(19) L.L. Iversen and F. Schon, The use of autoradiographic techniques for the identification and mapping of transmitter-specific neurons in CNS in New Concepts of Transmitter Regulation, ed. by A. Mandell and D. Segal, pp. 153-193, Plenum Press, New York (1973).

(20) M.J. Kuhar, C.B. Pert and S.H. Snyder, Regional distribution of opiate receptor binding in monkey and human brain, Nature 245, 447 (1973).

(21) J.M. Cedarbaum and G.K. Aghajanian, Noradrenergic neurons of the locus coeruleus: inhibition by epinephrine and activation by the alpha-antagonist piperoxane, Brain Res. 112, 413 (1976).

(22) T.H. Svensson, B.S. Bunney and G.K. Aghajanian, Inhibition of both noradrenergic and serotonergic neurons in brain by the alpha-adrenergic antagonist clonidine, Brain Res. 92, 291 (1975).

(23) H. Nyback, J.R. Walters, G.K. Aghajanian and R.H. Roth, Tricyclic antidepressants: effects on the firing rate of brain noradrenergic neurons, Europ. J. Pharmacol. 32, 302 (1975).

(24) J. Korf, B.S. Bunney and G.K. Aghajanian, Noradrenergic neurons: morphine inhibition of spontaneous activity, Europ. J. Pharmacol. 25, 165 (1974).

(25) J.W. Maas, S.E. Hattox, D.H. Landis and R.H. Roth, The determination of a brain arterio-venous difference for 3-methoxy-4-hydroxyphenethylene glycol (MHPG), Brain Res. 118, 167 (1976).

(26) K.M. Taylor and R. Laverty, The effect of chlordizepoxide, diazepam and nitrazepam on catecholamine metabolism in regions of the rat brain, Europ. J. Pharmacol. 8, 296 (1969).

(27) K. Fuxe, L.F. Agnati, P. Bolme, T. Hokfelt, P. Lidbrink, A. Ljungdahl, M.P. de la Mora, and S.-O Ogren, The possible involvement of GABA mechanisms in the action of benzodiazepines on central catecholamine neurons in Mechanisms of Action of Benzodiazepines, ed. by E. Costa and P. Greengard, Raven Press, New York (1975).

(28) B.J. Hoffer, G.R. Siggins, and F.E. Bloom, Studies on norepinephrine containing afferents to Purkinje cells of rat cerebellum. II. Sensitivity of Purkinje cells to norepinephrine and related substances administered by microiontophoresis, Brain Res. 25, 523 (1971).

(29) M. Segal and F.E. Bloom, The action of norepinephrine in the rat hippocampus. II. Activation of the input pathway, Brain Res. 72, 99 (1974).

(30) M. Bertrand (1969) The Behavioral Repertoire of the Stump Tail Macaque, pp. 95, 98-99, Karger, New York.

(31) Y.H. Huang, D.E. Redmond, Jr., D.R. Snyder, and J.W. Maas, Field and action potentials in the primate locus coeruleus following stimulation of the dorsal noradrenergic bundle, Brain Res. Bull., in press (1977).

(32) D.E. Redmond, Jr., Y.H. Huang, D.R. Snyder, J.W. Maas and J. Baulu, Hyperphagia and hyperdipsia after locus coeruleus lesions in the stumptailed monkey, Life Sci. 20, 1619 (1977).

(33) A.W. Graham and G.K. Aghajanian, Effects of amphetamine on single cell activity in a catecholamine nucleus, the locus coeruleus. Nature 234, 100 (1971).

(34) J. Korf, G.K. Aghajanian and R.H.Roth, Stimulation and destruction of the locus coeruleus: opposite effects on 3-methoxy-4-hydroxyphenylglycol sulfate levels in the rat cerebral cortex, Europ. J. Pharmacol. 21, 305 (1973).

(35) L. Cook and J. Sepinwall, Behavioral analysis of the effects and mechanisms of action of benzodiazepines, in Mechanism of Action of the Benzodiazepines, ed. by E. Costa and P. Greengard, Raven Press, New York, pp. 1-28 (1975).

(36) R. Laverty and K.M. Taylor, Propranolol uptake into the central nervous system, and the effect on rat behaviour and amine metabolism, J. Pharm. Pharmacol. 20, 605 (1968).

(37) N.-E. Anden and U. Strombom, Adrenergic receptor blocking agents: effects on central noradrenaline and dopamine receptors and on motor activity, Psychopharmacologia (Berl.) 38, 91 (1974).

(38) D.E. Redmond, Jr., Y.H. Huang, D.R. Snyder, J.W. Maas, and J. Baulu, Behavioral changes following lesions of the locus coeruleus in Macaca arctoides, Neurosci. Abs., 1, 472 (1976).

(39) H.I. Lief (1967) Psychoneurotic Disorders, I: Anxiety, conversion, dissociative, and phobic reactions in Comprehensive Textbook of Psychiatry, Chap. 23, 857, Williams and Wilkins Co., Baltimore.

(40) R. Freedman, S.L. Foote and F.E. Bloom, Histochemical characterization of neocortical projection of the nucleus locus coeruleus in the squirrel monkey. J. Comp. Neurol. 164, 209 (1975).

(41) R.M. Kobayashi, M. Palkovitz, D.M. Jacobowitz and I. Kopin, Biochemical mapping of the noradrenergic projection from the locus coeruleus, Neurology 25, 223 (1975).

(42) T. Maeda and N. Shimuzu, Projections ascendantes du locus coeruleus et d'autres neurones aminergique au niveau du prosencephale du rat, Brain Res. 36, 19 (1972).

(43) K. Sakai, M. Touret, D. Salvert, L. Leger, and M. Jouvet, Afferent projections to the cat locus coeruleus as visualized by the horseradish peroxidase technique, Brain Res. 119, 21 (1977).

(44) L. Levi, Neuro-endocrinology of anxiety, Chap. 7 in Studies of Anxiety, ed. by M.H. Lader, World Psychiatric Assoc. and Headley Bros., Ashford, Kent, pp. 40-52, (1969).

(45) D.G. Ward, W.E. Grizzle, and D.S. Gann, Inhibitory and facilitatory areas of the rostral pons mediating ACTH release in the cat, Endocrinology 99, 1220 (1976).

(46) C.D. Marsden, T.M.D. Gimlette, R.G. McAllister, D.A.L. Owen, and T.N. Miller, Effect of B-adrenergic blockade on finger tremor and achilles reflex time in anxious and thyrotoxic patients, Acta Endocrinologica 57, 353 (1968).

(47) P.J. Morgane and W.C. Stern, Interaction of amine systems in the central nervous system in the regulation of the states of vigilance. In Neurohumoral coding of brain

function, ed. by R.D. Myers and R.R. Drucker-Colin, Plenum, New York.

(48) D.W. Gallager and G.K. Aghajanian, Effect of antipsychotic drugs on the firing of dorsal raphe cells 1. Role of adrenergic system. Europ. J. Pharmacol. 39, 341 (1976).

(49) H. Przuntek and A. Phillipu, Reduced pressor responses to stimulation of the locus coeruleus after lesion of the posterior hypothalamus, Naunyn-Schmiedeberg's Arch. Pharmacol. 276, 119 (1973).

(50) M.E. Raichle and B.K. Hartman, J.O. Eichling and L.G. Sharpe, Central noradrenergic regulation of cerebral blood flow and vascular permeability, Proc. Nat. Acad. Sci., USA 72, 3726 (1975).

(51) E. Gellhorn, The neurophysiological basis of anxiety: a hypothesis, Perspect. Biol. Med. 8, 488 (1965).

(52) L.A. Gottschalk, W.N. Stone and G.C. Gleser, Peripheral versus central mechanisms accounting for antianxiety effects of propranolol, Psychosom. Med. 36(1), 47 (1974).

(53) J.A. Boon, P. Turner, D.C. Hicks, Beta-adrenergic receptor blockade with practolol in treatment of anxiety, Lancet 1, 814 (1972).

(54) H.H. Keller, R. Schaffner and W. Haefely, Interaction of benzodiazepines with neuroleptics at central dopamine neurons. Naunyn-Schmiedeberg's Arch. Pharmacol. 294, 1 (1976).

(55) A. Dray and D.W. Straughan, Benzodiazepines: GABA and glycine receptors on single neurons in the rat medulla, J. Pharm. Pharmacol. 28, 314 (1976).

(56) D.F. Klein and J.M. Davis (1969) Diagnosis and drug treatment of psychiatric disorders, Williams and Wilkins, Baltimore.

(57) Y.H. Huang, unpublished data.

(58) L.S. Goodman and A. Gilman (1965) The pharmacological basis of therapeutics, 3rd ed., Macmillan, New York.

(59) P. Turner, K.L. Granville-Grossman, J.V. Smart, Effect of adrenergic receptor blockade on the tachycardia of thyrotoxicosis and anxiety state, Lancet 2, 1316 (1965).

(60) K.L. Granville-Grossman, P. Turner, The effect of propranolol on anxiety. Lancet 1, 788 (1966).

(61) M.M. Suzman, Propranolol in the treatment of anxiety, Postgraduate Med. J. 52 (Suppl. 4), 168 (1976).

(62) M. Goldenberg, C.H. Snyder and H. Aranow, Jr., New test for hypertension due to circulating epinephrine, JAMA 135, 971 (1947).

(63) S. Soffer, Regitine and benodaine in the diagnosis of pheochromocytoma, Med. Clin. No. Amer. 38, 375 (1954).

(64) G. Holmberg and S. Gershon, Autonomic and psychic effects of yohimbine hydrochloride, Psychopharmacologia 2, 93 (1961).

(65) B.S. Nashold, Jr., W.P. Wilson and G. Slaughter (1974) Adv. in Neurology, Vol. IV, Raven Press, New York.

(66) W.B. Cannon (1915) Bodily Changes in Pain, Fear and Rage, Appleton, New York.

(67) R.J. Wyatt, B. Portnoy, D.J. Kupfer, F. Snyder and K. Engelman, Resting catecholamine concentrations in patients with depression and anxiety, Arch. Gen. Psychiat. 24, 65 (1971).

(68) D.R. Sweeney, J.W. Maas, and G.R. Heninger, State anxiety and urinary MHPG, Psychosom. Med. 39, 57 (1977).

(69) R.M. Post, C.R. Lake, D.C. Jimerson, and W.E. Bunney, Jr., CSF Norepinephrine in affective illness, New Research Abstracts, American Psychiatric Assoc., No. NR 24 (1977).

(70) J.J. Schildkraut and S.S. Kety, Biogenic amines and emotion, Science 156, 21 (1967).

(71) M. Lader, The peripheral and central role of the catecholamines in the mechanisms of anxiety, Int. Pharmacopsychiat. 9, 125 (1974).

(72) S. Freud (1936) The Problem of Anxiety, tr. by H.A. Bunker, M.D., The Psychoanalytic Quarterly Press and Norton, New York.

(73) L. Stein, C.D. Wise, and J.D. Belluzzi, Effects of benzodiazepines on central serotonergic mechanisms in Mechanism of action of benzodiazepines, ed. by E. Costa and P. Greengard, Raven Press, New York (1975) pp. 29-44.

(74) W.E. Bunney, Jr., and J.M. Davis, Norepineprine in depressive reactions: Review, Arch. Gen. Psychiat. 13, 483 (1965).

(75) J.J. Schildkraut, The catecholamine hypothesis of affective disorders: a review of supporting evidence, Amer. J. Psychiat. 122, 509 (1965).

(76) E. Costa and P. Greengard, eds. Mechanism of action of benzodiazepines, Raven Press, New York (1975).

DISCUSSION

Dr. Cooper expressed interest as to whether the monkeys would self-stimulate the locus coeruleus, which has been shown to be a "reward structure" in rats. *Dr. Redmond* said that they would. Although there seems to be some contradiction between the function proposed for the locus coeruleus and self-stimulation, he was not concerned. There is some controversy as to whether self-stimulation is mediated by the noradrenergic system. In addition, it has been shown that humans will choose to indulge in behaviors which are risky.

Dr. Kelleher indicated there is also ample evidence in monkeys that they will respond to produce what is generally conceded to be a very noxious electrical stimulation to the tail. Thus, it is not too surprising that they may respond to what would seem like a noxious stimulation in the CNS.

Dr. Redmond stated that there is evidence that animals can be shaped to stimulate areas in the brain that are thought to be pain-producing. Thus, he concluded, the fact that animals will self-stimulate is not evidence that the effect is pleasurable. *Dr. Kornetsky* cautioned that it would be better to talk about what is reinforcing and what is not reinforcing rather than to talk about pleasure and pain, since in one context something which would be defined as pain and a deterrent, may be painful yet still be reinforcing in another situation.

GENERAL DISCUSSION

Dr. Hanin wondered why Dr. Salama called his mouse-killing model one of aggression rather than paranoia or self-preservation, etc. He also asked if a non-aggressive rat placed in a cage with an isolated rat would be killed. *Dr. Salama* indicated that it was assumed to be a model of aggression since there was some aggressivity involved. When a non-aggressive rat is put in with an isolated rat, he has observed neither fighting nor biting. The "killer rat" seems to attack only smaller species, e.g., mice, frogs.

Dr. Lomax formulated a number of questions on the definition of "stress". When the crossed leg of a patient is hit with a hammer and the leg jerks, this is called a reflex arc. In a paper here, tail-pinching was said to cause stress; other experiments with rat tails include heating and cooling the tail. By stress to animals, does one mean activating the whole of the reticular formation or could it be a specific reflex in some parts of the system. For example, if a rat eats more after its tail is pinched, could it be merely a shifting of the threshold at which the animal adjusts its desire to eat? This would eliminate the necessity of introducing the stress concept.

Dr. Ginsburg made some additional comments on the model of increased motor activity as being inappropriate for hyperkinesis. He pointed out that it was not possible to identify the hyperkinetic children playing on a school ground. However, when these children are placed in an inhibitory learning situation where they are required to pay attention to a task, they are identifiable since they will be constantly getting up, etc. In the hybrid dogs of Paul Scott, it is not possible to note hypermotility, etc. until one tries to teach them an inhibitory task. Even though they may learn such commands as "Sit; Stay", they keep getting up. When these dogs are given amphetamine, some respond, some do not - their symptomotology may even be exacerbated by the amphetamine. Dr. Ginsburg did not know whether such result was caused by a number of different mechanisms.

Dr. Hanin noted that Dr. Antelman had shown that the tail-pinching behavior was dopaminergically mediated since it could be inhibited by dopamine antagonists and stimulated by dopamine agonists. He felt this was of relevance to the debate on the significance of dopamine.

On the question asked by *Dr. Hanin* as to whether lithium is an antimanic or an antidepressant or both, *Dr. Sanghvi* replied that the consensus at Bellevue was that it could not be classified as a typical antidepressant agent. He feels that it does not have antidepressant activity by itself but it may be useful in cutting down the required dosage of tricyclic thymoleptics. *Dr. Murphy* stated that the clinical data are fairly compelling that lithium does have acute antidepressant effect among bipolar patients. This is not yet an FDA-approved usage, but Dr. Murphy believes it will soon be.

Dr. Leith asked about the population who had shown interaction between lithium and amphetamine. *Dr. Murphy* replied that this combination had been used only on a depressed patient population; most of whom were not bipolar. Amphetamine had an acute antidepressant effect; even on those non-bipolar patients who did not respond to lithium.

Dr. Lippa mentioned Dr. Antelman's observation that agents which interfere with noradrenergic functioning seem to reverse the effects of haloperidol in his model. He then postulated the notion that those agents which have a low liability to produce extrapyramidal symptomatology are also potent adrenergic blockers and perhaps this is related to EPS liability rather than their anticholinergic property.

In answer to a final question as to whether he would classify amphetamine as an antidepressant, *Dr. Murphy* stated that it has acute antidepressant effects, but it does not have sustained antidepressant effects when given chronically. Thus, it is not of much practical use as an antidepressant.

VI. Models in Psychiatry: Genetic Factors

GENETIC MODELS OF BEHAVIOR DISORDERS

Benson E. Ginsburg
The University of Connecticut, Storrs, CT 06268

ABSTRACT

Most human behavioral syndromes express themselves in three dimensions: affective, ideational, and motoric. The science of syndromy, from the time of Claude Bernard, has recognized that there are often multiple etiologies involved within a syndrome and that there is no necessary 1:1 correspondence between a symptom complex and its underlying causes(s). The use of animal models makes it possible to control both genetic and environmental variables in relation to each other and thereby to separate a syndrome into its biological substituents. Analogies and homologies between humans and other forms are ascertainable in the affective and motoric realms, as in the case of monkeys deprived of mothering, and mice, dogs, rabbits, and baboons that show spontaneous seizures. Whether models for thought disturbances can be achieved below the human level remains moot, although Matthysse (this volume) has presented evidence that cognitive processing is amenable to research in animals.

With respect to therapy, the basic concepts and techniques for classical and instrumental conditioning were the results of animal research and have been successfully applied at the human level. Animal models in which there is a differential effect of the Y-chromosome (that may be autosomally mediated) on pubertal hormone levels and later aggressive behavior suggest that human XYY males are not necessarily at risk for such behavior and that it should be possible to identify and successfully treat those that are. Genetically mediated neuropathologies, endocrine anomalies, and variations in neurotransmitter mechanisms form the bases for testable hypotheses regarding human mental disease. The relatively new science of pharmacogenetics, for example, has already demonstrated that the same pharmacologic treatment administered to individuals with the "same" condition (e.g., phenothiazines to schizophrenics) will have ameliorative effects on some patients, but will not help others, and will make still others worse. The same is true for amphetamines and related substances in "hyperkinetic" children. Here, a model is available in which dogs of particular genetic origins are unable to perform in inhibitory training tasks and some of these are consistently helped by d-amphetamine, while others are not. This implies a variation in the mechanisms underlying the behavior and provides a research tool with which to investigate this variation, which is also encountered in children. Such results compel a new approach to the study of human behavioral conditions that are familial: namely, that of providing a genetic taxonomy for each such condition, using polymorphic linkage markers. Only in this way will it be possible to parse the biological variation within a syndrome such that the multiplicity of etiologies can be separated for meaningful study of causal mechanisms.

Animal studies also suggest that individuals such as identical twins or members of an inbred strain having the same encoded genotypes do not necessarily have the same effective genotypes. This is not only a matter of "reaction range" or how the same genes act under various environmental conditions, but also of the "genomic repertoire;" i.e., which of a number of genetically encoded alternatives will be activated during development to interact with environmental conditions in the determination of the phenotype. These events may be under the control of regulatory genes that interface with developmental events to selectively activate and/or suppress the expression of encoded genetic alternatives. These processes can be readily studied in animals and have powerful implications for the understanding and control of phenotypic variation, particularly in clinical states.

INTRODUCTION

The classical approaches to the use of animal models in behavioral and neurobiological research include the use of such models considered as simpler systems by comparison to the human; as systems where behavioral, pharmacological, biochemical, surgical, and other interventions may be performed; and as systems from which general principles can be derived that are applicable to our own species as well. Genetic analyses using animal models have demonstrated that almost every aspect of biological variation is genetically influenced, including neural structures, sensory mechanisms, emotionality, responses to drugs and alcohol, social behavior, gross teratology, and many others (1-7). In many instances, the same results can be brought about by environmental intervention during development (1-3). This should not be surprising because developmental processes subject to genetic variation are also subject to manipulation in other ways. For the most part, such interventions are deleterious. We can easily interfere with normal processes to create teratologies, but the reverse seldom occurs. That it can occur is demonstrated by the relative success in the management of PKU, cretinism and other developmental endocrine conditions, and by the creation of phenocopies of gene mutations in such organisms as Drosophila by means of manipulating aspects of the physical environment during sensitive periods in development.

The purpose of this presentation is to review some of the models we have been researching from the point of view of how best to establish and understand homologies of mechanism, how to apply such understanding to human behavior, and particularly to search for new principles and approaches that have potential for expanding our present frontiers and for providing both insight and the possibility of effective ameliorative intervention into behavioral problems.

THE HUMAN ANIMAL AS ITS OWN MODEL

Research involving genetic variables using our own species has been more than moderately successful. Twin studies, pedigree analyses, heritability studies, and related approaches to sibships, families and restricted gene pools have demonstrated that there are hereditary bases for a great many human variations, including psychiatric disorders (4, 8). These may vary from predispositions, as in schizophrenia, to seemingly strict determinism from which there is no escape, as in Down's syndrome. Among the former, the vogue has been to attempt to understand environmental-developmental factors as though these were controlling, and, therefore, must constitute the basis for therapy. Biological approaches have, for the most part, been associated with our increasing understanding of neurochemistry, and have sought to provide therapy by means of the modification of these processes through pharmacologic intervention.

The problem with the first of these approaches is that the investigators have been caught in the trap of their own contrivance: to be sure, major traumas suffered in childhood can effect later behavior, but Clarke and Clarke (36) have recently assembled and reviewed an impressive body of evidence in which children who had been severely deprived and mistreated during their early years, including some that had no language development by age 6 or 7, were restored to normal performance levels and mental health by the time of adolescence. Similar data exist for severely deprived animals (4, 9). The animal studies also show that the sensitive periods during which various environmental manipulations exert their maximal effects are a function of the genotype, and that the direction and magnitude of such effects are also a function of genotype (3). In some instances, the period of lability correlates with the source of the X-chromosome (10).

Two lessons are to be learned from these examples: one is that to focus on a particular approach often leads to dogmatic conclusions that are controverted in a broader context. The other is that to homogenize biological variation obscures an important area of investigation -- namely, that whether mouse or human, one organism is not exactly like another in its biological potential for response, and to treat biological variation as though it were statistical "error" deprives the investigator of a primary source of understanding (1). Our preconceptions are so powerful that when we find data that contradict them, we tend to rationalize the data as exceptions, or to misinterpret them in other ways. When, for example, we reported that wild animals reared as pets from infancy tended to revert if frightened later on, while those tamed later in life during a period of fearfulness that had to overcome this affect did not revert, we were reported as having shown the opposite, in accordance with the prevailing view. Similarly, when we excluded the dominant male of a wolf pack from possible paternity of a number of litters that might otherwise have been attributed to him, this was omitted from most reviews of the relationship between social organization, evolutionary adaptive advantage, and breeding structure, since it was assumed that male social dominance and paternity are associated. More recently, using genetic markers, the same lack of necessary relationship between dominance and paternity has been demonstrated for at least two species of macaques, but the incorporation of iconoclastic ideas into our prevailing scientific views are such that it will be awhile before the new facts, which arise out of asking the questions differently, will make themselves felt.

With respect to human behavioral syndromes, the possibility of multiple etiologies must be recognized, and the errors introduced by assuming that one mechanism is controlling, without adequately testing this view, stands in the way of further progress. The way out of this trap, since we cannot do genetic selection experiments with humans in which the presumed mechanism might be genetically dissociated from its putative effects, were it to be merely correlational (2, 11, 12), is to take advantage of the many known human genetic polymorphisms that can be used as linkage markers (13). By using these in conjunction with the familial distribution of a given syndrome, it becomes possible to demonstrate either that there is an invariable association with a particular chromosome or linkage group, suggesting that there may be only one controlling mechanism, or, that in some pedigrees one pattern of genetic association is involved, while others are involved when different families are analyzed. Such an approach would create a genetic taxonomy within a syndrome, making it possible to do research on causal mechanisms based on the knowledge that a number are involved, and on the further knowledge of which is which.

THE GENOMIC REPERTOIRE AS AN ALTERNATIVE MODE OF GENETIC ENGINEERING

Using our best estimates of what proportion of the encoded DNA is transcribed and translated in any given cell, we end up with an order of magnitude approximating 10% (14). The most notable exception to this is the brain, and particularly the higher cortical centers, where close to half the encoded genetic material is active (14). At any stage in the life cycle of an active, metabolizing cell, most of the encoded genetic material is merely physically, but not physiologically, present. It is only the physiologically effective portion of the genome that is available to interact with environmental conditions at any given time during the life cycle of the cell. An important consideration, therefore, becomes that of evaluating the silent part of the genotype. This undoubtedly contains structural genes that could have provided alternative metabolic and morphologic capacities. Where a deleterious gene comes to expression, it is not necessary to assume that the potential for a normal alternative does not exist. While this is undoubtedly the case in some instances, it is not necessarily the case in all, or even in the majority of such examples. It is entirely possible, and some of our work lends itself to this hypothesis (10, 12, 15), that segregation data following the Mendelian model may involve genes that interface with environmental events during development to determine whether still other (structural) genes become incorporated into the effective genotype. In this way, genetics provides an interface strategy between the environment and the metabolic potential of the cell, the tissue, or the organism. The study of how such genes act to selectively regulate the phenotypic expression of the genome should provide an effective clue to the means by which appropriate intervention can optimize the potential of the genotype in relation to its latent phenotypic capabilities.

Regulatory models have been offered for bacterial systems, but none have really been established for the mammalian nervous system. The conceptions summarized above apply at every stage of development from early gestation through puberty and senescence. While we know a great deal about how that portion of the genetic code of a cell that is activated is transcribed and translated into biochemical potential, and while we understand the lability that can be achieved in these processes, as, for example, by enzymatic inhibition or induction, we have very little understanding of the primary step; i.e., how particular sites on the chromosomes are differentially exposed in different cells and during different stages of development in order that the DNA at such sites may initiate the chain of events leading to some aspect of phenotypic expression. If, for example, the appropriate portion of the twenty-first chromosome in a Down's syndrome child could be induced never to divest itself of its histoprotein coat, it might, despite its physical presence, exert no physiological effect. Similarly, if the genetic capacity to manufacture insulin were found to be encoded but suppressed in certain individuals, the genetic problem would become not that of providing a potential that did not exist in the individual (as per the promise on the bright side of recombinant DNA research), but rather to reprogram the existing potential.

The questions that the theoretical possibilities of such an approach raise are: what is the evidence that latent normal capacities for phenotypic expression exist in individuals, who, under most circumstances, would not show these; how, if this is true, can one reprogram the genomic repertoire more optimally; do we have research tools for investigating these possibilities; and, finally, how do they apply to human behavioral conditions?

LATENT CAPACITIES

These are known to exist in the form of recessive genes, of genes whose activities are masked by non-allelic genes "epistatic" to them, by genes that are said not to be fully penetrant or to come to complete expressivity, and one could throw in other bits of genetic jargon, all indicating that we know from breeding experiments or genealogic histories that genetic systems capable of coming to phenotypic expression exist, but are not expressed in certain classes of individuals.

Our group has been one among many that has been working on the genetic substratem of sound-induced seizures in mice. This may be an exotic phenomenon of little practical clinical interest, since the seizures do not have those cortical components that would make them interesting as models of human epilepsy (16). There are, however, a number of independently occurring autosomal alleles that have been identified as contributing to seizure proneness, and the interactions of some of these have been analyzed (17, 18). There is also an inverse relationship between exposure to such seizures at weaning age and later tendencies towards intermale aggression (10, 19). Early work in our laboratory established that some aspect of glutamic acid metabolism was associated with seizure susceptibility in particular strains of mice (20-24). Glutamate exerted a protective effect against seizures in these genotypes, but not necessarily in others. Close structural analogues of glutamic acid had an exacerbating effect in these same genotypes. The continuation of this line of investigation eventually pointed to the GAD-GABA system as one directly involved in the seizure mechanism in these strains (15). Developmental GAD curves in brain differed among seizure prone and seizure resistant mice where this genetic system was a factor in the seizures. Eventually, mutants on each of these inbred strain backgrounds that changed the seizure incidence were discovered. The most interesting ones appeared to switch the GAD curves from one level of activity to another at about 18 days post partum with a concomitant change in the later seizure suscepti- bility. On the assumption, therefore, that both strains had both capacities encoded in their genotypes and that what the mutants were doing was to selectively activate (or suppress) one or another aspect of the encoded genotype, an experiment was attempted in which the activity of the mutant gene was mimicked by pharmacological means at the time when it would have acted in an individual not possessing this presumed regulatory mutant. The phenocopy was successful in producing the appropriate biochemical change, as well as the concomitant behavioral change, both of which persisted during the lifetime of the individual. These data are consistent with the notion that the genotype can be reprogrammed (in this case in either direction) in order to produce a permanent change in the phenotype of the treated individual.

The mechanism mediating the relationship between seizures and intermale aggression is not yet clear. However, we are now testing the hypothesis that a steroid alteration is involved that either lowers the levels of testosterone, or alters the binding efficacy of the target cells prepubertally (25-29).

ANIMAL MODELS FOR EPILEPSY

Spontaneous seizures that respond to the usual anti-epileptic drugs have been reported in dogs, rabbits, baboons (photic drive), gerbils (Lomax, this volume), and we have now found them in mice. These were first discovered by M. Fine in our laboratory in a line of C57BL/10 mice carrying a gene for susceptibility to audiogenic seizures. The spontaneous seizures typically occur between 50 and 80 days post partum. They are preceded by postural rigidity, followed by rhythmic waves of muscular contraction and relaxations along each flank, and by vocalizations (16). This is followed by clonic and clonic-tonic seizures which recur on Maxson's (16) interpretation. The spontaneous seizures may be a type of partial epilepsy or psychomotor epilepsy which is secondarily generalized. There is no epileptiform EEG pattern discernible prior to the age when the seizures begin. Since mice are unable to survive re- peated anoxic episodes such as those resulting from the seizures, lethality is a usual but secondary consequence. Koniecki and Friedrich (in 16) have provided evidence that this is not a demyelinating mutant, and Maxson et al. (16) have presented evidence that the model has potential for the study of genetics, development, and neural mechanisms in at least one type of epilepsy that can be precisely specified biologically by means of genetic control.

MECHANISMS OF AGGRESSIVE BEHAVIOR

Genetic selection for high degrees of aggression, or for its opposite, has been successful in animals (2, 30, 31). In some instances, there are specifiable interactions with conditions of rearing (1, 3). However, these are genotype-dependent in the sense that the same environ- mental manipulations will increase the aggressiveness of some genotypes, decrease that of others, and have no effect on still others. While correlations with testosterone levels and with the serotonergic system have been described, and the manipulation of these mechanisms have resulted in changes in aggression (25-29, 32), the total behavior involving neural centers, conditioning, transmitters, and hormones is still poorly understood.

Because of the equivocal association of the extra Y-chromosome with human male aggression, a number of us became interested in the role of the Y-chromosome in this phenomenon. Using mouse strains where the males differed in combativeness under specified conditions of rearing, we made reciprocal crosses such that the Y-chromosome from a highly aggressive strain could be combined with the X of a relatively non-aggressive strain and with the autosomal complement equally divided between the strains and constant for all F_1 individuals. The crosses were made reciprocally. Since the strains also differed in their responses to handling, it was possible to attempt to determine whether any aspect of this was heterosomally mediated. As it turned out, this aspect of the behavior is associated with the source of the X-chromosome, whereas the tendency of the males to be combative after sexual maturity when placed with other males is a function of the source of the Y-chromosome (10, 19). In order to control for conditions of rearing, animals were cross fostered shortly after birth. The natural experiments were perhaps the most compelling, since males bearing the Y of a high aggression strain grew up in the uterus of a low aggression mother and were nursed by her. The converse was true for those bearing the Y of the low aggression strain. We are now in the 20th generation of placing various Y-chromosomes on various autosomal complements (e.g., those of pure strains) in order to test the Y-autosomal interaction.

From the data we have gathered so far, it appears that the effect of the Y is to interact with particular autosomal constituents in order to regulate the pubertal testosterone level. One of the "high aggression" Y's is associated with a supranormal pubertal surge in testosterone, that is in turn regulated by the testosterone level itself, since it modulates back to normal by the time of complete sexual maturity, when the behavior manifests itself. We are now in process of mimicking the pubertal surge in a variety of genotypes that do not normally show it in order to determine whether this is sufficient to induce the behavior at maturity. We are also preventing the pubertal surge from occurring in those genotypes where it would normally do so by means of castration and silastic implants. This work will be supplemented by the use of anti-androgens that compete with testosterone at binding sites.

It may be that the pubertal surge is a sufficient precondition for the development of high aggression in most, if not all, genotypes of mice. It may also be that it only invokes such behavior through particular neural mechanisms in mice of some genetic constitutions but not in others. The experiments now in progress should answer these questions.

These researches have two-fold implications for the general strategy previously outlined. With respect to the human XYY condition, the genetic work with mice would suggest that only certain Y-chromosomes, if replicated (or even if not replicated), would be associated with a biological propensity for aggressive behavior, and, further, only in combination with certain autosomal genes. The experiments also suggest that diagnosis and amelioration of hyper-aggressive tendencies would be possible if the situation is homologous in man and mouse. With respect to diagnosis, monitoring hormone levels at the onset of puberty would indicate whether or not a supranormal surge of testosterone was beginning to occur. In such an event, the appropriate use of anti-androgens during this period would provide a physiological correction.

GENETIC PREDISPOSITION TO ALCOHOL AND DRUGS

Both individual and ethnic differences in the ability to metabolize ethanol have been reported for humans. In inbred strains of mice, there is a variation in alcohol preference that is associated with the strain and is, therefore, genetic. Selection for "sleeping time" as measured by the ability of the mouse to right itself in response to a given dose of ethanol has also been successful. Work with withdrawal seizures has demonstrated that genetic variation occurs and also that glucocorticoids are necessary in order for the normal withdrawal symptoms to appear (33). Low dosage alcohol administered transplacentally and/or via the mother's milk has been shown to have physical, behavioral, and biochemical effects, some of which are more pronounced in particular genotypes (5). The role of biological variation, while a complicating factor, is also a tool in that it separates some mechanisms from others and permits their individual analysis.

So far as differences in reactions to drugs are concerned, there has long been an interest in animal models that would help in the understanding of biochemical processes associated with behavioral conditions for which drug therapy would be useful. When it comes to problems of thought disturbances, it has been proposed that hallucinogenic substances given to animals, or, perhaps, amphetamines, which can induce schizophrenic-like episodes in people, can be used in order to determine the neural substrates of the behaviors in question, even though the symptoms and subjective content might be quite different in the animal model. It has also been suggested that natural models in which personality disorders similar to those found in humans can be identified in animals. One such situation, that of minimal brain dysfunction with hyperkinetic symptoms, has been described in several dogs by Corson and co-workers (34). More recently, through collaboration of Corson and J. P. Scott, F1 hybrids between beagles and Malaysian dogs imported from the Telom River area were found to be unusually resistant to restraint as exemplified by attempts to assess their behavior in a Pavlov stand, and to have difficulty with inhibitory training tasks. The present author, in a series of collaborative studies with R. Becker, had been looking for possible mouse models of this condition and had also investigated the possibility that crosses between wild and domestic Canids (beagle X coyote) might provide such a model. Because of the Corson study, we next investigated the Telomian-beagle hybrids provided by J. P. Scott and found them to have difficulty in restraining their motor movements if trained to an inhibitory task -- in this case, the standard obedience exercise of having to sit and stay upon a verbal command reinforced with a physical signal. While the dogs could be trained to sit, they would not retain this position as did the beagle controls. Experience has shown that this is a simple task easily mastered by most breeds or breed crosses. While other behavioral tests were also used, this was the simplest and most clear-cut. Each of 20 subjects was trained to the task in standard fashion, and all of the Telomian-beagle hybrids had difficulty in retaining the sit. The duration per sit per command was calculated for each dog, and the trained control level of performance was adjusted to a scale of 10. d-Amphetamine sulfate was orally administered using the range of doses reported by Corson (34). One of our findings was that the beagles were much more affected by the same oral dose on a per kilo basis than were the hybrids. Among the hybrids that could not sustain the sit, some were amphetamine responders in the sense that even when amphetamine was administered at levels that produced agitation, head wagging movements and other evidence of extreme stimulation, the hybrid responders were able to maintain the sit for as long as the controls. Other hybrids having exactly the same behavioral phenotype (e.g., they were unable to remain long in the sitting position without drugs) were not helped by amphetamines at any reasonable dose, and for some, the agitation and hyperkinesis was exaggerated (35).

This example illustrates the variable genetic bases for a behavioral phenotype. The fact of close genetic relationship does not mean genetic uniformity. The phenotype or syndrome is not a unitary one with respect to underlying mechanism, since the symptoms will be consistently alleviated by a particular pharmacological agent in some individuals but not in others. Since breeding experiments can be performed, the genetic bases for these differences can be clarified and can serve as a tool to identify and separate the several mechanisms involved in the behavior. Neurochemical investigations done with the collaboration of S. Bareggi (35) suggest that there are two mechanisms involved and that both are dopaminergic, but that one is primarily pre- and the other postsynaptic.

This example does not, however, lead to the frontier of attempting to reprogram the existing genotype. It is our hope that the developmental interactions exemplified in the Y-chromosome mechanism of hormonal control of behavior can be exploited to this end, since the Y-chromosome is one that has few active sites, and since the time when it is activated and de-activated with respect to the pubertal surge can be specified. We are hopeful that the intensive study of the steroid hormone-genetic feedback mechanism may provide a useful model for attacking the problem of genetic reprogramming and perhaps of even identifying the primary changes occurring at a chromosome surface that constitute the initial events involved in gene activation and expression.

ACKNOWLEDGMENT

These researches were supported by NIH grant MH 28021-01 and MH 27591-01, by The W. T. Grant Foundation, Inc., and by The University of Connecticut Research Foundation.

REFERENCES

1. B. E. Ginsburg, All mice are not created equal: recent findings on genes and behavior, Soc. Serv. Rev. 40, 121-134 (1966).

2. B. E. Ginsburg, Genetic parameters in behavioral research. In: J. Hirsch, ed. (1967) Behavior-Genetic Analysis, McGraw-Hill, New York, pp. 135-153.

3. B. E. Ginsburg, Genotypic factors in the ontogeny of behavior, Science and Psychoanalysis 12, 12-17 (1968).

4. B. E. Ginsburg, Anxiety: a behavioural legacy. In: Physiology, Emotion and Psychosomatic Illness (1972), Elsevier, Amsterdam, pp. 163-174.

5. B. E. Ginsburg, J. Yanai and P. Y. Sze, A developmental genetic study of the effects of alcohol consumed by parent mice on the behavior and development of their offspring. In: M. E. Chafetz, ed. (1975) Research, Treatment and Prevention, NIAAA, Washington, D. C., pp. 183-204.

6. B. Moisset and B. L. Welch, Effects of d-amphetamine upon open field behaviour in two inbred strains of mice, Experientia 29, 625 (1973).

7. T. H. Roderick, R. E. Wimer and C. C. Wimer, Genetic manipulation of neuroanatomical traits. In: L. Petrinovich and J. L. McGaugh, eds. (1976) Knowing, Thinking, and Believing, Plenum Press, New York.

8. R. R. Fieve, D. Rosenthal, and H. Bull (1975) Genetic Research in Psychiatry, Johns Hopkins University Press, Baltimore.

9. S. J. Suomi and H. F. Harlow, Social rehabilitation of isolate-reared monkeys, Develop. Psychol. 6, 487 (1972).

10. B. E. Ginsburg, The role of genic activity in the determination of sensitive periods in the development of aggressive behavior. In: J. Fawcett, ed. (1971) Dynamics of Violence, American Medical Association, Chicago, pp. 165-175.

11. B. E. Ginsburg and W. S. Laughlin, The distribution of genetic differences in behavioral potential in the human species. In: M. Mead, Th. Dobzhansky, E. Tobach and R. E. Light, eds. (1968) Science and the Concept of Race, Columbia University Press, New York, pp. 26-36.

12. B. E. Ginsburg and W. S. Laughlin, Race and intelligence, what do we really know? In: R. Cancro, ed. (1971) Intelligence, Genetic and Environmental Influences, Grune and Stratton, New York, pp. 77-87.

13. V. A. McKusick and F. H. Ruddle, The status of the gene map of the human chromosomes, Science 196, 390 (1977)

14. A. G. Motulsky and G. S. Omenn, Biochemical genetics and psychiatry. In: R. R. Fieve, D. Rosenthal and H. Brill, eds. (1975) Genetic Research in Psychiatry, Johns Hopkins University Press, Baltimore, pp. 3-14.

15. B. E. Ginsburg, J. S. Cowen, S. C. Maxson and P. Y. Sze, Neurochemical effects of gene mutations associated with audiogenic seizures. In: A. Barbeau and J. R. Brunette, eds. (1969) Progress in Neuro-Genetics, Excerpta Medica, Amsterdam, pp. 695-701.

16. S. C. Maxson, T. R. DeFanti and D. L. Koniecki, A spontaneous seizure mutant in C57BL/10Bg mice, paper presented at the 7th annual meeting of the Behavior Genetics Association, Louisville, Kentucky (1977).

17. B. E. Ginsburg, Causal mechanisms in audiogenic seizures. In: R. G. Busnel, ed. (1963) Psychophysiologie, Neuropharmacologie et Biochimie de la Crise Audiogene, Centre National de la Recherche Scientifique, Paris, pp. 227-240.

18. B. E. Ginsburg and D. S. Miller, Genetic factors in audiogenic seizures. In: R. G. Busnel, ed. (1963) Psychophysiologie, Neuropharmacologie et Biochimie de la Crise Audiogene, Centre National de la Recherche Scientifique, Paris, pp. 217-225.

19. B. E. Ginsburg, S. C. Maxson and M. K. Selmanoff, Genetics of aggressive behavior in mice. In: S. Genoves and J. Passy, eds. (1976) Comportamiento y Violencia, Editorial Diana, Mexico City, pp. 107-131.

20. B. E. Ginsburg, S. Ross, M. J. Zamis and A. Perkins, An assay method for the behavioral effects of L-glutamic acid, Science 112, 12-13 (1950).

21. B. E. Ginsburg, S. Ross, M. J. Zamis and A. Perkins, Some effects of l(+) glutamic acid on sound-induced seizures in mice, J. Comp. Physiol. Psychol. 44, 134-141 (1951).

22. B. E. Ginsburg, Genetics and the physiology of the nervous system. In: Genetics and the Inheritance of Integrated Neurological and Psychiatric Patterns (1954), Williams and Wilkins, Baltimore, pp. 39-56.

23. B. E. Ginsburg and J. L. Fuller, A comparison of chemical and mechanical alterations of seizure patterns in mice, J. Comp. Physiol. Psychol. 47, 344-348 (1954).

24. J. L. Fuller and B. E. Ginsburg, Effect of adrenalectomy on the anticonvulsant action of glutamic acid in mice, Amer. J. Physiol. 176, 367-370 (1954).

25. M. K. Selmanoff, S. C. Maxson and B. E. Ginsburg, Chromosomal determinants of intermale aggressive behavior in inbred mice, Behav. Genet. 5, 108 (1975).

26. M. K. Selmanoff, J. E. Jumonville, S. C. Maxson and B. E. Ginsburg, Evidence for a Y chromosomal contribution to an aggressive phenotype in inbred mice, Nature 253, 529-530 (1975).

27. M. K. Selmanoff, S. C. Maxson and B. E. Ginsburg, Chromosomal determinants of intermale aggressive behavior in inbred mice, Behav. Genet. 6, 53-69 (1976).

28. M. K. Selmanoff, B. D. Goldman, S. C. Maxson and B. E. Ginsburg, Correlated effects of the Y-chromosome of mice on developmental changes in testosterone levels and intermale aggression, Life Sciences 20, 359-366 (1977).

29. M. K. Selmanoff, B. D. Goldman and B. E. Ginsburg, Developmental changes in serum luteinizing hormone, follicle stimulating hormone and androgen levels in males of two inbred mouse strains, Endocrinology 100, 122-127 (1977).

30. B. E. Ginsburg, Genetics and personality. In: J. M. Wepman and R. W. Heine, eds. (1963) Concepts of Personality, Aldine, Chicago, pp. 63-78.

31. B. E. Ginsburg, Coaction of genetical and nongenetical factors influencing sexual behavior. In: F. Beach, ed. (1965) Sex and Behaviour, Wiley, New York, pp. 53-75.

32. B. E. Ginsburg and P. Y. Sze, Pharmacogenetic studies of the serotonergic system in association with convulsive seizures in mice. In: B. K. Bernard, ed. (1975) Aminergic Hypotheses of Behavior: Reality or Cliche? National Institute on Drug Abuse, Rockville, Maryland, pp. 85-95.

33. P. Y. Sze, J. Yanai and B. E. Ginsburg, Adrenal glucocorticoids as a required factor in the development of ethanol withdrawal seizures in mice, Brain Research 80, 155-159 (1974).

34. S. A. Corson, E. O. Corson, L. E. Arnold and W. Knapp, Animal models of violence and hyperkinesis. In: G. Serban and A. Kling, eds. (1976) Animal Models in Human Psychobiology, Plenum Press, New York.

35. B. E. Ginsburg, R. E. Becker, A. Trattner, J. Dutson and S. R. Bareggi, Genetic variation in drug responses in hybrid dogs: A possible model for the hyperkinetic syndrome, Behav. Genet. 6, 107 (1976).

36. A. M. Clarke and A.D.B. Clarke (1976) Early Experience, Myth and Evidence, Open Books, London.

A BIOCHEMICAL ANALYSIS OF STRAIN DIFFERENCES IN NARCOTIC
ACTION

G. Racagni, F. Cattabeni and R. Paoletti

Institute of Pharmacology and Pharmacognosy, University of
Milan, 20129 Milan, Italy

ABSTRACT

The biochemical effects of morphine have been investigated in two inbred
strains of mice in which analgesia and locomotor activity can be genetically
differentiated. The results suggest that striatal dopaminergic system seems
to be involved mainly in the running syndrome, whereas cholinergic neurons
might be implicated in morphine analgesia.

INTRODUCTION

In recent years, the development of the genetic approach to the laboratory
animals has resulted in many available lines of inbred strains of mice (1).
There are two important reasons for using such strains. The first is the more
satisfactory behavioural homogeneity of individuals belonging to the same
line, and the second is the difference between the behavioural patterns
characterizing each strain. The need of a behavioural homogeneity derives
from the large variability evident in the common laboratory animals which is
a critical point for studying many psychopharmacological agents and drug-
induced abnormalities in the central nervous system. A number of findings
indicate that central stimulating agents such as amphetamine may exert an
effect on exploratory activity in some animals, whereas the level of activity
remains unaffected or even depressed in other animals (2). In the case of
inbred mice, the different behavioural responses following the administration
of pharmacological agents are strictly strain dependent in their action (3,4,
5). The strain dependent action of some drugs has been correlated to quantitative and qualitative strain differences in biochemical parameters.
The use of inbred strains gives the possibility to establish if differences
in drug responses are related to differences in receptors or in the metabolic
rate of the drug under investigation. In the field of narcotic action, drugs
which induce analgesia or hyperkinesia in mice, the use of different strains
is potentially very useful in order to separate and characterize the single
effects. A genetic control of opiate-induced behaviour might be suggested by
the presence of wide variations among mice in their sensitivity to opiate-
induced running fit syndrome (6). Moreover it has been shown that strains of
inbred mice are characterized by different patterns of morphine addiction (7).
Recent findings obtained in two different strains of mice indicate, in fact,
that the effects of morphine on analgesia and running activity are two
distinct phenomena (8). Accordingly, C57 BL/6J mice (C57)showed the highest
locomotor response when running behaviour was considered whereas the same
strain was the less sensitive to the analgesic effect of morphine; DBA/2J
mice (DBA) were characterized by opposite response patterns. It was, therefore, of interest to study the biochemical differences in these strains which
probably might be responsible for the strain dependent effects of opiates and
for the dissociation between the effects of morphine on running and analgesia.

IMPLICATION OF THE STRIATAL DOPAMINERGIC NEURONS IN MORPHINE
INDUCED INCREASE OF LOCOMOTOR ACTIVITY IN INBRED STRAINS OF
MICE.

Narcotic analgesics act on monoaminergic neurons in brain (9,10,11). The administration of 6-hydroxydopamine antagonizes the analgesic response in naive
rats (12). Analgesic doses of morphine increase the turnover rate of dopamine
(DA) in rat striatum (10,11,12). Moreover, tolerance develops to the stimulatory effect of narcotic analgesics on catecholamine turnover during chronic

drug administration (14). Since the interpretation of these mechanisms is complicated when various behavioural patterns are present, such as analgesia and catalepsy in the rat, we have utilized the genetic specificity of behaviour offered by C57 and DBA mice to study the possible role of DA neurons in pharmacological effects of morphine. Table 1 shows that the increased motor activity after morphine in C57 strain is followed by a release of DA in the striatum.

TABLE 1 Effect of morphine on striatal 3-methoxy-tyramine (3-MT) content and running activity in two strains of mice

	C57		DBA	
	Running activity	3-MT nmoles/g	Running activity	3-MT nmoles/g
Saline	74 $\pm$ 9	0.53 $\pm$ 0.043	49 $\pm$ 2	0.48 $\pm$ 0.049
Morphine (10 mg/kg i.p.)	275 $\pm$ 25[+]	1.28 $\pm$ 0.20[+]	50 $\pm$ 4	0.28 $\pm$ 0.023[+]

The animals were sacrificed 30 min. after morphine administration.
Data on morphine-induced running activity were taken from Oliverio and Castellano (8).
Values are mean $\pm$ S.E.M. of five determinations.
[+]P < 0.01

In fact, the levels of 3-methoxy-tyramine (3-MT), which has been measured as an index of DA release (15,16) are doubled in this strain. 3-MT was detected by a mass fragmentographic technique recently described (17). Since the increase of DA release is associated with an activation of DA receptors, we have investigated in vivo the 3',5'-cyclic adenosine monophosphate (cAMP) accumulation in the striatum of C57 mice after morphine injection.
Table 2 shows that morphine elicits an increase of striatal cAMP concentrations. Moreover, haloperidol, a dopaminergic receptor blocker (11,15), antagonizes the increase of cAMP elicited by morphine (Table 2). No increase has been shown in the striatal cAMP concentration of the DBA strain.

TABLE 2 Striatal cAMP levels in C57 and DBA mice receiving morphine and haloperidol.

	C57	DBA
	cAMP (pmoles/mg protein)	
Saline	4.64 $\pm$ 0.38	1.08 $\pm$ 0.22
Morphine (10 mg/kg i.p.)	8.80 $\pm$ 0.66[+]	0.89 $\pm$ 0.19
Haloperidol (1 mg/kg i.p.)	5.27 $\pm$ 0.39	0.97 $\pm$ 0.10
Haloperidol + Morphine	5.30 $\pm$ 0.41	0.95 $\pm$ 0.11

Haloperidol and morphine were given 40 and 30 min. respectively before killing.
Each mean $\pm$ S.E.M. refers to at least five determinations.
[+]P < 0.01

IMPLICATION OF CHOLINERGIC NEURONS IN MORPHINE INDUCED
ANALGESIA IN INBRED STRAINS OF MICE.

It has been suggested that cholinergic mechanisms participate in the mediation
of cerebral analgesic action and in the expression of the symptoms of morphine
dependence. In fact, opiates have been shown to reduce the release of acetylcholine (ACh) from cortical neurons (18). Moreover, muscarinic receptor
blockers can suppress some manifestations of morphine withdrawal in mice and
rats (19,20). Recently, it has been shown that during morphine analgesia the
turnover rate of ACh in cortex of rat brain is decreased (21) and the ED 50
for analgesia of four analgetics with different chemical structures correlates
with the dose that inhibits by 50% the turnover rate of ACh in the cortex
(22). Table 3 shows that the turnover rate of ACh is markedly decreased in
cortex of DBA strain which is the most sensitive to morphine analgesia. No
change has been detected in the striatum. Moreover, no decrease has been
shown in the turnover rate of ACh in the cerebral cortex of C57 strain.
Turnover rate of ACh as been measured by infusing phosphoryl-deutered choline
at a constant rate through the tail vein for 6 minutes. The animals were
killed by microwave radiation which instantaneously inhibits brain choline
acetyltransferase and acetylcholinesterase and stabilizes brain choline and
ACh content (23). The ACh turnover rate was calculated according to Racagni
et al. (21).

TABLE 3   Effect of morphine (10 mg/kg i.p.) on acetylcholine
turnover rate in cerebral cortex of DBA and C57 mice.

	C57	DBA
	μmoles/g/h	
Saline	0.20 + 0.02	0.32 + 0.02
Morphine	0.30 + 0.01	0.13 + 0.01 [+]

The animals were sacrificed with microwave radiations 30
min. after morphine administration.
Values are mean + S.E.M. of five determinations.
[+] P < 0.01

DISCUSSION

The utilization of inbred strains is a new and promising approach in the
field of neuropharmacology. Several findings indicate that different biochemical mechanisms may account for the behavioural differences among inbred
strains of mice and for their sensitivity to the effects of drugs. Evidence
exists, for example, that large differences in sleeping-time are evident
between strains injected with the same dose of nembutal (24). Strain differences in a variety of behavioural responses related to alcohol consumption
have been reported (25). Clear differences in strain reactivity with regard
to the effects of psychoactive agents on learning have been found by Bovet
et al. (26). Genetic mechanisms seem to play an important role also when
the effects of morphine are assessed on the locomotor behaviour of mice.
In fact, the C57 strain is characterized by hyperactivity after morphine
administration (8), whereas no effect is evident when the opiate is injected
to the DBA mice. When the analgesic effects of morphine are analyzed the
patterns of reactivity to the opiate by these strains of mice are reversed
(8). On the bases of these behavioural differences, we have investigated in
both strains of mice two neuronal systems which have been postulated to be
involved in morphine action. In fact, DA and ACh have been studied extensively
with respect to their possible roles in the acute and chronic actions of
morphine (10,11,13). The data presented suggest that striatal dopaminergic
system seems to be involved mainly in the running syndrome produced by the
analgesic, since striatal dopaminergic neurons are activated (Table 1) and a
dopaminergic blocker, as haloperidol, antagonizes the increase of cAMP in the
C57 strain (Table 2). On the other hand, the decrease of turnover rate of ACh
in cholinergic structures of DBA mice (Table 3) proves that cholinergic
synapses might be involved in morphine analgesia. These biochemical results
are in agreement with the behavioural data. In fact, α-methyl-p-tyrosine, an
inhibitor of catecholamine synthesis, antagonizes the increased locomotor

activity elicited by morphine in C57 mice whereas it has no effect on analgesia in DBA strain (27). Moreover, the lesion of the septum, an area which is critical for eliciting analgesia, and which has been associated with opiate receptors and that is known to contain cholinergic cell bodies which project to hippocampus and cortex, antagonizes morphine analgesia whereas the running fit syndrome is unaffected (27). Therefore our biochemical analysis has confirmed the significance of using inbred strain in the evaluation of the mode of action of analgesic drugs. For the first time has been shown the very strict relation of central cholinergic system - and not of the dopaminergic - to the analgesic effect of morphine and the role of the dopaminergic system has been more correlated with hypermotility. This example can be now expanded to other biochemical parameters with an immediate correlation with single behavioural and pharmacological effects of the centrally active drugs.

REFERENCES

(1) J.S. Taats, Standard nomenclature for inbred strains of mice: fifth listing, Cancer Res. 32, 1609 (1972).
(2) A. Oliverio, B.E. Eleftheriou and D.W. Bailey, Exploratory activity: genetic analysis of its modification by scopolamine and amphetamine, Physiol. Behav. 10, 893 (1973).
(3) G. Lindzey, J. Loehlin, M. Manosevitz and D.D. Thiessen, Behavioral genetics, Ann. Rev. Psychol. 22, 39 (1971).
(4) D. Wahlsten, Genetic experiments with animal learning: a critical review, Behav. Biol. 7, 143 (1972).
(5) A. Oliverio, Genetic factors in the control of drug effects on the behaviour of mice, in The genetic of behaviour, ed. by J.H.F. Van Abeelen, p. 375, North-Holland, Amsterdam (1974).
(6) A. Goldstein, P. Sheehan, Tolerance to opioid narcotics. I. Tolerance to the "running fit" caused by levorphanol in the mouse, J. Pharmacol. Exp. Ther. 169, 175 (1969).
(7) K. Eriksson and K. Kiianmaa, Genetic analysis of susceptibility to morphine addiction in inbred mice, Ann. Med. Exp. Fenn. 49, 73 (1971).
(8) A. Oliverio and C. Castellano, Genotype dependent sensitivity and tolerance to morphine and heroine: dissociation between opiate-induced running and analgesia in the mouse, Psychopharmacologia 39, 13 (1974).
(9) E.L. Way, H.H. Loh and F.H. Shen, Morphine tolerance, physical dependence and synthesis of brain 5-hydroxytryptamine, Science 162, 1290 (1968).
(10) D.H. Clouet and M. Ratner, Catecholamine biosynthesis in brain of rats treated with morphine, Science 168, 854 (1970).
(11) E. Costa, D.L. Cheney, G. Racagni and G. Zsilla, An analysis at synaptic level of the morphine action in striatum and the accumbens: dopamine and acetylcholine interactions, Life Sciences 17, 1 (1975).
(12) J.H. Ayhan, Effect of 6-hydroxydopamine on morphine analgesia, Psychopharmacologia 25, 183 (1972).
(13) C. Gauchy, Y. Agid, J. Glowinski and A. Cheramy, Acute effects of morphine on dopamine synthesis and release and tyrosine metabolism in the rat striatum, European J. Pharmacol. 22, 311 (1973).
(14) C.B. Smith, M.I. Sheldon, H.J. Bednarczy and J.E. Villarreal, Morphine-induced increases in the incorporation of ^{14}C-tyrosine into ^{14}C-dopamine and ^{14}C-norepinephrine in the mous brain: antagonism by naloxone and tolerance, J. Pharmacol. Exp. Ther. 180, 547 (1972).
(15) A. Carlsson and M. Lindqvist, Effect of chlorpromazine and haloperidol on the formation of 3-methoxy-tyramine and normetanephrine in mous brain, Acta Pharmacol. 20, 140 (1963).
(16) A.M. Di Giulio, A. Groppetti, F. Cattabeni, A. Maggi and S. Algeri, Functional significance of striatal 3-methoxy-tyramine, Proceedings of the First Meeting of the European Society of Neurochemistry, Bath, England (1975).
(17) C.L. Galli, F. Cattabeni, T. Eros, P.F. Spano, S. Algeri, A.M. Di Giulio and A. Groppetti, A mass fragmentographic assay of 3-methoxy-tyramine in rat brain areas, J. Neurochem. 27, 795 (1976)
(18) K. Jhamandas, C. Pinsky and J.W. Phillis, Effects of morphine and its antagonists on release of cerebral cortical acetylcholine, Nature 228, 176 (1970).
(19) H.O.J. Collier, D.L. Francis and C. Schneider, Modification of morphine withdrawal by drugs interacting with humoral mechanism: some contradictions and their interpretation, Nature 237, 220 (1972).

(20) J. Crossland, Acetylcholine and the morphine abstinence syndrome, in
     Drugs and cholinergic mechanisms in the C.N.S., ed. by E. Heilbronn
     and A. Winter, p. 355, Försvarets Forskningsanstalt, Stockholm,
     Sweden (1970).
(21) G. Racagni, D.L. Cheney, G. Zsilla and E. Costa, The measurement of
     acetylcholine turnover rate in brain structures, Neuropharm. 15, 723
     (1976).
(22) G. Zsilla, D.L. Cheney, G. Racagni and E. Costa, Correlation between
     analgesia and the decrease of acetylcholine turnover rate in cortex
     and hypocampus elicited by morphine, meperidine, viminol R_2 and
     azidomorphine, J. Pharmacol. Exp. Ther. 199, 662 (1976).
(23) A. Guidotti, D.L. Cheney, M. Trabucchi, M. Doteuchi, C. Wang and R.R.
     Hawkins, Focused microwave radiation: a technique to minimize post
     mortem changes of cyclic nucleotides, dopa and choline and to preserve
     brain morphology, Neuropharm. 13, 1115 (1974).
(24) A.M. Brown, The investigation of specific responses in laboratory
     animals, Symposium on Laboratory Animals Centre, Royal Veterinary
     College, London, p. 9 (1959).
(25) J.L. Fuller, Measurement of alcohol preference in genetic experiments,
     J. Comp. Physiol. Psychol. 57, 85 (1964).
(26) D. Bovet, F. Bovet-Nitti and A. Oliverio, Genetic aspects of learning
     and memory in mice, Science 163, 139 (1969).
(27) C. Castellano, B.E. Llovera and A. Oliverio, Morphine-induced running
     and analgesia in two strains of mice following Septol lesions or
     modification of brain amines, Naunyn-Schmiedeberg's Arch. Pharmacol.
     288, 355 (1975).

EXTENSION OF ANIMAL MODELS IN CLINICAL EVALUATION OF NEW DRUGS

Louis Lemberger

Lilly Laboratory for Clinical Research
Wishard Memorial Hospital, and
The Departments of Pharmacology and Medicine
Indiana University School of Medicine
Indianapolis, Indiana 46202

The purpose of this symposium is to demonstrate that animal models for
psychiatric and neurologic diseases (1) serve a useful function in selecting
agents which might be suitable candidates as possible therapeutic agents in
man, and (2) elucidate the pathophysiology of diseases to ultimately achieve
a cure. Animal models have essentially two major functions in drug development: (a) screening new drugs to aid in finding agents with which to treat
diseases in man, and (b) studying the mechanism of action of potential drugs
or studying the mechanisms of adverse effects of drugs which occur in man.
The purpose of the present paper is to demonstrate that by using the same
pharmacologic principles that are applicable to animal models, and extending
them to human studies, one can accelerate the drug-development process.

Clinical pharmacologic investigation of new drugs is conducted in several
stages which have been designated as Phase I, Phase II, and Phase III evaluation. Phase I clinical pharmacologic evaluation consists of single-dose
administration to subjects in a dose-ranging study (these have also been
termed tolerance studies) to establish the safety of the compound in man. In
the recent past and in many instances at the present time, these studies were
carried out until the subjects presented with some adverse side effects.
These usually occurred when subjects could no longer tolerate the medication.
Multiple-dose administration of the new compound during Phase I studies was
directed towards further evaluation of the tolerance and safety of the compound. Thus it has been the practice in the past that during this early
period of drug evaluation in man, tolerance studies were carried out until
biochemical and physiologic responses to drugs' effects makes it possible to
obtain some indication of the proper dose that might be effective in treating
the diseases for which the drug is designed. Moreover, correlating blood-
level data and pharmacologic responsiveness in animal models can be applied
to the clinical situation in the case of certain drugs where a good correlation exists, e.g., drugs used for treating cardiac arrhythmias. In most
cases laboratory studies done in animals during the initial periods of pharmacologic screening of drugs are conducted in normal healthy animals. Similarly,
for studying their mechanisms of action, studies are also done in normal
animals. Therefore it appears feasible to conduct these same types of studies
in normal human subjects (those individuals who usually participate in Phase I
studies). This approach appears to provide a logical extension of the preclinical studies to the clinical pharmacologic studies necessary for the
evaluation of drugs.

This presentation will use two different drugs as examples of how one can
extend studies from animal models to human volunteers and demonstrate how the
information obtained by this approach may be of value in estimating and/or
determining the optimum dose of drug as well as in demonstrating its "efficacy"
in early Phase I studies, even before the compound goes into patients (i.e.,
Phase II). Again, this approach does not appear to be unreasonable at this
stage since, in the development of some drugs, much of the data which was
used to make the decision to continue its development was derived from data
generated in normal animals.

The first example to be considered involves studies with lergotrile, a dopamine agonist designed for the treatment of galactorrhea and Parkinsonism and
whose structure is shown in Fig. 1. The animal model utilized for screening
inhibitors of prolactin secretion involves the use of normal animals which
are pretreated with reserpine to affect the endogenous levels of prolactin
secretion. Reserpinization in these animals produces a relative depletion of

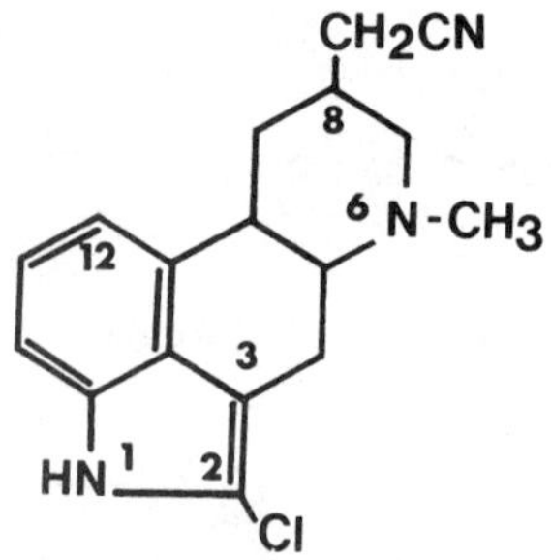

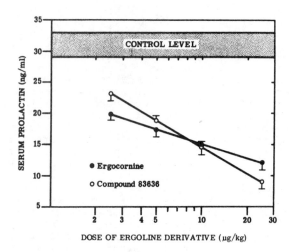

Fig. 1. Chemical structure of lergotrile (Lilly Compound 83636)

dopamine in the central nervous system, thereby decreasing the influence of prolactin inhibitory factor (PIF), and thus inducing an elevation in serum prolactin levels. This pharmacologic manipulation, when coupled with an assay which is sensitive for measuring changing levels of prolactin in the rat, has resulted in the use of this animal model as a screen for dopamine agonists.

The effects of the ergot derivatives ergocornine and lergotrile (Compound 83636) on prolactin levels in reserpinized rats are illustrated in Fig. 2 (Clemens et al., unpublished observations). Both decrease the serum prolactin levels in this animal model.

Fig. 2. Dose-response effect of ergocornine and lergotrile (Compound 83636) on serum prolactin in rats. Reserpine-primed rats (2 mg/rat) were treated with lergotrile or ergocornine at 2.5, 5, 10, and 25 μg/kg and compared to rats treated with vehicle alone (serum prolactin for control group = 32.1±4.0). Each value represents the mean of 10 rats ± SEM.

In an attempt to study the "efficacy" of lergotrile in normal subjects, the same principles that were used in the animal model were applied to man. With the knowledge that drugs such as the phenothiazines elevate serum prolactin in normal subjects by virtue of their dopamine receptor blockade, studies were designed to pretreat individuals with perphenazine, a potent phenothiazine derivative, and to investigate the effects of lergotrile on preventing or attenuating this perphenazine-induced elevation in serum prolactin.

The effect of lergotrile on perphenazine-induced elevation of serum prolactin is illustrated in Fig. 3. After the administration of perphenazine preceded one hour earlier by placebo medication, serum prolactin levels peaked at two hours. Pretreatment of subjects with a single dose of lergotrile (2 mg, 4 mg, or 6 mg) caused a dose-related delay in the onset of the perphenazine-induced

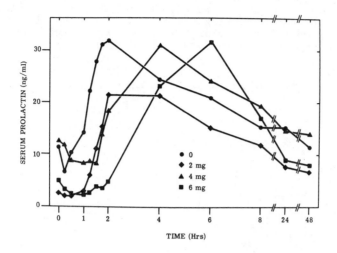

Fig. 3. Dose-response effect of lergotrile (Compound 83636) on perphenazine-induced elevation of serum prolactin in man. Either placebo or lergotrile (2, 4, or 6 mg) was orally administered 1 hr. prior to perphenazine (5 mg i.m.) administration. Each value represents the mean of 3 subjects. From Lemberger *et al*. (1).

elevation of serum prolactin. The effect of the single dose of 2 mg of lergotrile was similar to that after the combination of placebo and perphenazine. However, the administration of 4 mg or 6 mg of lergotrile resulted in a delayed onset in the perphenazine-induced elevation in serum prolactin, the peak occurring at 4 and 6 hr., respectively. In a similar fashion, attempts were made to study the effect of lergotrile on decreasing serum prolactin levels after they were already elevated secondary to perphenazine administration.

Lergotrile (4 or 6 mg) administered orally 1 hr. after perphenazine administration resulted in a decrease in serum prolactin levels towards baseline within 15 to 30 min. (Fig. 4). Likewise, the effect of chronic lergotrile administration was studied to determine if it could prevent the elevation of serum prolactin induced by perphenazine.

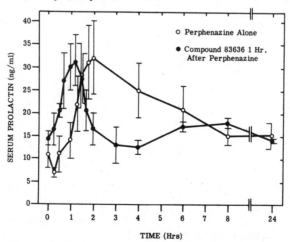

Fig. 4. Effect of lergotrile (Compound 83636) on perphenazine-induced elevation of serum prolactin in man. Lergotrile (4 or 6 mg) was orally administered 1 hr after administration of perphenazine (5 mg i.m.). The effect of perphenazine alone is illustrated for comparison purposes. Values represent mean and range for 2 subjects. From Lemberger *et al*. (1).

An increase in serum prolactin levels occurred after the administration of perphenazine to subjects who had been pretreated 1 hr. earlier with placebo medication (Fig. 5). The prolactin levels reached their maximum at 1 to 2 hrs. Pretreatment of subjects with lergotrile (2 mg 3 times daily for 7 days plus an additional 2-mg dose 1 hr. prior to perphenazine administration) attenuated the perphenazine-induced rise in serum prolactin.

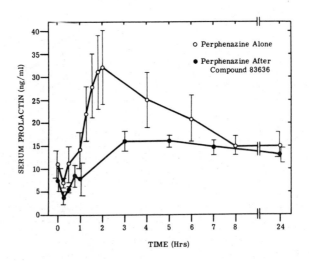

Fig. 5. Effect of chronic administration of lergotrile (Compound 83636) on perphenazine-induced elevation of serum prolactin in man. Either placebo or lergotrile (2 mg t.i.d.) was orally administered for 7 days plus 1 additional dose 1 hr. prior to the administration of perphenazine (5 mg i.m.). Each value represents the mean and range for 2 subjects. From Lemberger et al. (1).

Thus this study demonstrates that the dose necessary to lower serum prolactin and still ensure relative safety could be ascertained in man, using simple pharmacological techniques and manipulations in a similar fashion to those employed earlier in the animal studies.

The second example selected to illustrate the utility of extending animal models to early drug evaluation in humans involves studies with a reuptake inhibitor of norepinephrine.

dl-

$O-CH-(CH_2)_2-NH-CH_3$

$O-CH_3$

.HCl

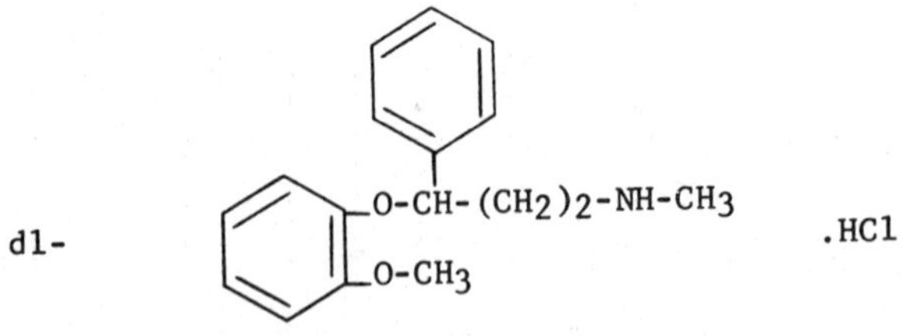

Fig. 6. Chemical structure of nisoxetine (Lilly Compound 94939)

Nisoxetine (Fig. 6), an inhibitor of norepinephrine re-uptake, was shown to potentiate the effects of norepinephrine on the guinea pig vas deferens in vitro (Aiken et al., unpublished observations). In addition, it blocked the effects of tyramine in this system. Moreover, when this compound was administered prior to the intravenous administration of norepinephrine, it potentiated the norepinephrine-induced contraction of the cat nictitating membrane. Similarly, the normal response to tyramine in this tissue preparation was attenuated by pretreatment with nisoxetine.

To further study the actions of this drug, another animal model was used which demonstrated that nisoxetine inhibited the reuptake of norepinephrine and serotonin in an in vitro system (3). This animal model utilized rat brain synaptosomes (pinched off nerve endings) which were incubated with radio-labeled biogenic amines and in the presence or absence of nisoxetine.

TABLE 1   Inhibition of Uptake into Synaptosomes (K_I values M)
Effect of Nisoxetine (Lilly 94939)

Biogenic Amine	$K_I(M)$
Norepinephrine	1.8×10^{-7}
Serotonin	1×10^{-6}

From Wong et al. (3)

In this system nisoxetine, at low concentration, prevented the accumulation of norepinephrine into the synaptosomal preparations (Table 1). However, a larger concentration was required to inhibit the accumulation of serotonin. To demonstrate the "effectiveness" of nisoxetine on the noradrenergic system, and to confirm its pharmacologic effects in normal subjects, infusions of norepinephrine (Levophed) were given over a 10-min. period. A dose of nor-epinephrine was selected during the period of placebo administration which was capable of elevating the blood pressure by approximately 25-30 mm of Hg. In addition, bolus injections of tyramine were given to subjects to determine what dose would produce similar pharmacologic effects, i.e., an elevation of 25-35 mm of Hg. Both of these pressor agents were also administered while the subjects were receiving nisoxetine at doses of between 10 to 20 mg twice daily for 7 days. The results obtained in a single individual are shown in Table 2.

TABLE 2   The Effect of Compound 94939 on Blood Pressure
Responses to Tyramine and Norepinephrine in a Typical Subject

DATE	DOSE OF 94939 MG	NO. OF DOSES	RESPONSE TO TYRAMINE			RESPONSE TO NE		
			DOSE TYR-AMINE MG	ΔBP mm/Hg	ΔBP* TYRAMINE DOSE	DOSE OF NE µg/MIN	ΔBP mm/Hg	ΔBP* NE DOSE
9-11-74	0		4.5	28	6.2	5.6	17	3.0
			4.5	29	6.4	7.8	37	4.9
			5.0	33	6.6			
9-13-74	0		4.5	33	7.3	5.6	36→52	6.4→9.3
			4.5	33	7.3	7.6	28→53	3.7→7.0
10-9-74	0		4.5	31	6.9	6.8	24	3.5
			4.0	29	7.3			
			4.0	35	8.8			
			4.0	32	8.0			
10-14-74	20	1st	4.0	13	3.3	4.0	>39	> 9.8
			6.0	24	4.0	1.8	40	22.2
			8.0	35	4.4			
10-17-74	20	7th	4.0	9	2.3	1.0	22	22.0
			6.0	15	2.5	1.43	32	22.4
			8.0	24	3.0			
			10.0	27	2.7			
10-21-74	20	15th	4.0	3	0.8	1.0	22	22
			8.0	22	2.8	1.4	28	20
			10.0	24	2.4			
			12.0	27	2.3			
			12.0	29	2.4			

*Represents mm/Hg increase per mg tyramine or per µg/min NE

For purposes of obtaining an expression of the difference in blood pressure sensitivity between the effects elicited during the placebo period and that during the nisoxetine period, a response ratio was calculated for each bio-genic amine (i.e., tyramine and norepinephrine).

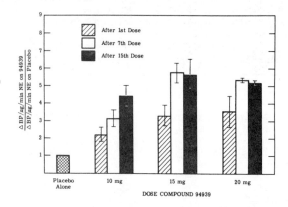

Fig. 7. The effect of nisoxetine after first dose, seventh dose, and fifteenth dose on the pressor response to intravenous infusion of norepinephrine. Each value represents the mean ± s.e. mean of two subjects given 10 mg b.i.d., two subjects given 15 mg b.i.d., and one subject given 20 mg b.i.d. The values were calculated from at least two independent observations per dose per subject. From Lemberger et al. (2).

As can be seen, nisoxetine prevented the reuptake of norepinephrine at the adrenergic nerve ending (Fig. 7), thus potentiating its effects on blood pressure to the extent that only about one-fifth (20%) of the dose of norepinephrine was required to produce the same elevation in blood pressure.

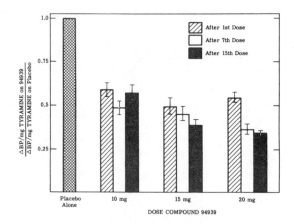

Fig. 8. The effect of nisoxetine after first dose, seventh dose, and fifteenth dose on pressor response to intravenous tyramine injection. Each value represents the mean ± s.e. mean of two subjects given 10 mg b.i.d., two subjects given 15 mg b.i.d., and one subject given 20 mg b.i.d. The values were calculated from at least three independent observations per dose per subject. From Lemberger et al. (2).

In contrast, nisoxetine, by preventing the uptake of tyramine into the adrenergic nerve ending, markedly attenuated tyramine's blood-pressure effect (Fig. 8). This necessitated a three-fold increase in the dose of tyramine to produce the same pharmacologic effect when tyramine was given during nisoxetine administration as compared to that produced by tyramine administration during the placebo period.

To simulate the animal studies done with the rat brain synaptosomes, studies were designed to investigate the effect of nisoxetine on the uptake of ^{3}H-serotonin into human platelets. It has been stated that the platelet possesses properties similar to synaptosomes with respect to its ability to accumulate biogenic amines (4). Thus, human platelets which are obtained by

a relatively benign procedure involving only the removal of blood by venipuncture and the subsequent harvesting of platelets were employed as a model
similar to the brain synaptosomes.

The effect of nisoxetine on the uptake of tritiated serotonin into platelets
could be used to determine if, in man, nisoxetine had specificity solely for
the noradrenergic system as had been earlier shown to exist in the animal
model.

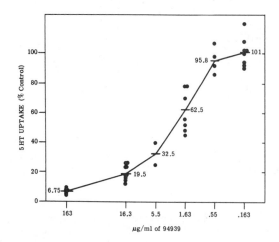

Fig. 9. The effect of nisoxetine on the uptake of (^{3}H)-5-HT
by human platelets. From Lemberger et al. (2).

The effects of the in vitro addition of nisoxetine on ^{3}H-serotonin uptake into
human platelets are shown in Fig. 9. It can be seen that at a concentration
of about 3 µg/ml nisoxetine inhibits the uptake of ^{3}H-serotonin by 50%. However, platelets which were isolated from subjects to whom nisoxetine had been
administered at a dose of 10-20 mg b.i.d. demonstrated no inhibition of
serotonin accumulation. This was indicative that nisoxetine would be ineffective on the serotonergic system at doses which were capable of producing
effects on the noradrenergic system. When nisoxetine blood levels were
determined after this dosage regimen, the concentration of drug present in
plasma was found to be several orders of magnitude less than that which would
be necessary to inhibit the uptake of serotonin in vitro. Thus, again it is
apparent that the major pharmacologic actions of this drug observed in animals
which made it interesting enough to support as a candidate for clinical trial,
could be confirmed in normal human volunteers at an early stage of the clinical
phase of the drug development process.

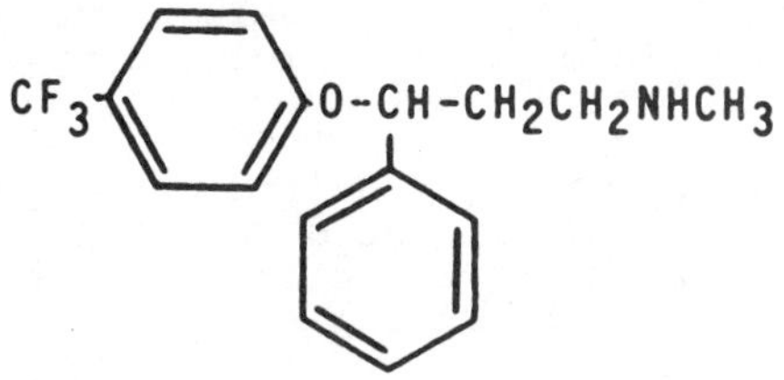

$$CF_3 - \bigcirc - O-CH-CH_2CH_2NHCH_3$$

Fig. 10. Chemical structure of fluoxetine (Lilly Compound 110140).

Clinical confirmation of data derived from animal studies has also been
obtained with fluoxetine, a compound similar in structure to nisoxetine (Fig.
10). In contrast to nisoxetine, this compound is a selective serotonin reuptake inhibitor. In the rat, Wong and his associates (5,6) showed that
fluoxetine blocked serotonin uptake into rat brain synaptosomes and platelets
at much lower doses than it affected norepinephrine or dopamine uptake.
Studies in man demonstrated that after multiple-dose administration of fluoxetine to man, this compound is capable of inhibiting the uptake of ^{3}H-serotonin
into platelets harvested from these individuals. The results are shown in
Fig. 11.

328

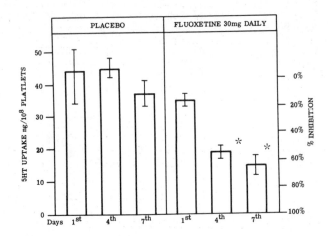

Fig. 11.   The effect of fluoxetine on ^{3}H-serotonin uptake
into human platelets.   *Significantly differs (P<0.005) from
placebo treatment for that day.

After four days of fluoxetine administration at a dose of 30 mg daily, the
uptake of ^{3}H-serotonin was inhibited by 55%.   Continued administration of
fluoxetine for a total of 7 days produced a 63% inhibition of ^{3}H-5HT uptake.
The degree of inhibition of ^{3}H-serotonin uptake correlated with the plasma
fluoxetine concentration.   To demonstrate the selectivity of fluoxetine on
the serotonergic system, studies using tyramine injections and norepinephrine
infusions were conducted in a manner previously described for the nisoxetine
studies.   In contrast to the nisoxetine studies, fluoxetine did not produce
any change in the sensitivity of the blood-pressure response resultant from
tyramine or norepinephrine administration.

In summary, the examples presented indicate that one can devise clinical
pharmacologic experiments in humans which could serve to confirm the animal
pharmacology.   The most important criteria which must be followed is that the
study design carefully consider the ethics involved in doing certain studies
and that careful, thoughtful planning be given to alternatives to the animal
model.   For example, if the animal model utilizes brain synaptosomes, it
goes without saying that a similar study in man would be unthinkable.   However, with the knowledge that platelets share similar properties to synaptosomes, one could conduct the study with minimal discomfort to the subject and
with a feeling of confidence that the results will be meaningful.   Thus when
biochemical or physiologic data are generated in animal models, it appears
possible that under certain circumstances similar data could be obtained in
humans, using techniques which are either noninvasive (e.g., monitoring blood
pressure, pulse, or EKG) or minimally invasive techniques such as utilizing
blood drawn by venipuncture.

As more knowledge is gained about the specific actions of drugs and as our
technology becomes more developed, it would seem that biochemical animal
models can be utilized in the future to better study the clinical pharmacologic
effects of drugs in normal subjects during Phase I studies.

References

(1)   L. Lemberger, R. Crabtree, J. Clemens, R. Dyke, and R. Woodburn,
        The inhibitory effect of an ergoline derivative (Lergotrile,
        Compound 83636) on prolactin secretion in man, J. Clin.
        Endocrinol. Metab. 39, 579 (1974).

(2)   L. Lemberger, S. Terman, H. Rowe, and R. Billings, The effect of
        nisoxetine (Lilly Compound 94939), a potential antidepressant,
        on biogenic amine uptake in man, Br. J. Clin. Pharmac. 3, 215 (1976).

(3) D.T. Wong, J.S. Horng, and F.P. Bymaster, dl-N-methyl-3-(o-methoxyphenoxy)-3-phenylpropylamine, hydrochloride, Lilly 94939, a potent inhibitor for reuptake of norepinephrine into rat brain synaptosomes and heart, Life Sci. 17, 755 (1975).

(4) W.B. Abrams and H.M. Solomon, The human platelet as a pharmacologic model for the adrenergic neuron, Clin. Pharmac. Ther. 10, 702 (1969).

(5) D.T. Wong, F.P. Bymaster, J.S. Horng, and B.B. Molloy, A new selective inhibitor for uptake of serotonin into synaptosomes of rat brain: 3-(p-trifluoromethylphenoxy)-N-methyl-3-phenylpropyl-amine, J. Pharmacol. Exp. Therap. 193, 804 (1975).

(6) J.S. Horng and D.T. Wong, Effects of serotonin uptake inhibitor, Lilly 110140, on transport of serotonin in rat and human blood platelets, Biochem. Pharmacol. 25, 865 (1976).

DISCUSSION

Dr. Usdin inquired whether the decrease in prolactin levels produced by lergotrile had an effect on the antipsychotic efficacy of phenothiazines in patients. *Dr. Lemberger* replied that his studies had been done with normals, not with schizophrenic patients. The level of perphenazine used (5 mg, i.m.) is the normal antiemetic dose. Dr. Small and he have given lergotrile to psychotic patients who were being treated with the slow-release drug, fluphenazine decanoate. In those individuals, a single dose of lergotrile (which is a potent dopamine agonist) was able to reverse the effects of the phenothiazine. He postulated that since fluphenazine decanoate is a very slow-release compound and yields very low blood levels, they were able to displace the phenothiazine from the receptor site and therefore exacerbated the schizophrenic response.

Dr. Tenen inquired whether 110140 had been tested in patients. *Dr. Lemberger* replied that this compound was still in Phase I clinical pharmacological studies. When the safety of the compound is determined, depressed patients will be given the drug. No behavioral effects have been seen in normal volunteers, even at doses up to 90 mg.

Dr. Cook suggested that often drugs act in an animal model similar to in the human on a mg/kg basis, probably because stimuli and environmental contingencies are similar. *Dr. Lemberger* felt this was not true in the case of such cannabinoids as Δ^9-THC or 11-hydroxy-Δ^9-THC. For example, a tachycardia or a "high" can be produced by a total i.v. dose in man of 1 mg, but effects are not seen in rats until after 10 mg. For fluoxetine, an effective dose for humans is about 30 mg per day, but Dr. Fuller in his animal studies on the blockade of ρ-chloroamphetamine depletion of serotonin uses a dose of 10 mg/kg in the rat. *Dr. Cook* wondered if the difference might be that the animal test has to be more severe since man is able to report symptoms but the animal requires measurement of something obvious. *Dr. Lemberger* pointed out that he measured biological or physiological responses, such as serum prolactin levels.

Dr. Salama asked what animal model Dr. Lemberger would use in the case of an uptake inhibitor and how he would project a therapeutic dose on the basis of animal experiments. *Dr. Lemberger* replied that he would not want to make such a speculation; that there are sex differences; species differences; that man metabolizes compounds in different pathways than some laboratory animals. Not only are there some pathways in humans different than in rats, but even within humans there are significant differences. As an example, Dr. Lemberger mentioned an individual who does not convert fluoxetine to norfluoxetine very well, even though the nor compound is the predominant one in the plasma of most people. This individual does not N-demethylate very well. Animal toxicology gives some indication of a relatively safe dose to start with in humans, but one must proceed cautiously from there.

THE USE OF AN ANIMAL MODEL FOR PARKINSONISM AND ITS EXTENSION TO HUMAN PARKINSONISM

Menek Goldstein,[1] Arthur F. Battista,[2] Abraham Lieberman,[3]
Jow Y. Lew[1] and Fumiaki Hata[1]

Departments of Psychiatry,[1] Neurosurgery[2] and Neurology[3]
Neurochemistry Laboratories
New York University Medical Center
New York, New York

ABSTRACT

Monkeys with unilateral lesions in the ventromedial areas of the brain stem which exhibit hypokinesia and tremor served as useful models for evaluating the antiparkinsonian efficacy of drugs. The administration of L-dopa or of putative dopamine agonists results in a relief of tremor with a concomitant occurrence of abnormal involuntary movements. The antitremorgenic activity of bromocriptine and of lergotrile is of longer duration than that of L-dopa. The abnormal involuntary movements evoked by the administration of bromocriptine or lergotrile are less severe than those evoked by administration of piribedil or of L-dopa. Clinical studies with bromocriptine and lergotrile indicate that these two drugs might be of therapeutic value in ameliorating parkinsonian symptoms. Adverse effects were similar to those observed with levodopa and carbidopa, except that in some patients abnormal involuntary movements and diurnal oscillations in performance were decreased while mental changes were increased. Thus, the animal model predicted correctly the antiparkinsonian efficacy and the severity of abnormal involuntary movements evoked by bromocriptine and lergotrile.

The interactions of bromocriptine and lergotrile with dopamine and with α receptors were investigated. Both drugs have mixed agonist-antagonist activities with respect to dopamine receptors. Lergotrile has a higher affinity for the agonist site than bromocriptine. Both drugs displace the binding of the α-adrenergic antagonist WB-4101. Bromocriptine displaces more effectively than lergotrile the binding of WB-4101 from the cerebral cortex membranes.

INTRODUCTION

Several animal models which mimic Parkinson disease have been described. The cardinal features of Parkinsons disease which have been reproduced in various animal models are based on the principal pathology of the disease, namely degeneration of the nigrostriatal dopamine pathway. The rotational behavior in rats following unilateral lesions of the nigrostriatal dopamine (DA) pathways (1) was extensively used as a model for biochemical and pharmacological studies of extrapyramidal dysfunctions. Monkeys with unilateral lesions in the ventromedial areas of the brain stem (VMT) exhibit some neurological and neurochemical deficits which are similar to those observed in Parkinsons disease (2,3). The data in Table I summarize the neurological and neurochemical deficits in Parkinsons disease and in monkeys with VMT lesions. While in human Parkinsonism the extrapyramidal symptomatology is dominated by rigidity, akinesia, tremor and gait disturbance, in monkeys with VMT lesions, only hypokinesia and tremor develop on the extremities contralateral to the lesion side. Thus, the neurological deficits in this animal model mimic partially the clinical symptoms of the human illness. Despite the obvious deficiencies to replicate all the cardinal features of Parkinsons disease the pharmacological studies in this animal model were useful in elucidation of the antiparkinsonian efficacy of various drugs. This paper describes the effects of L-dopa and of several putative DA receptor stimulating agents on the relief of tremor and on the occurrence of abnormal involuntary movements (AIM) in monkeys with VMT lesions. Furthermore, two ergot alkaloids, bromocriptine and lergotrile, were found to be effective antitremor agents in these animal models and therefore were tested as potential antiparkinsonian (AP) agents in clinical trials. To determine the interactions of ergot alkaloids with postsynaptic monoaminergic systems we have used binding assays with specific radioligands for measuring their affinities for the dopaminergic and α-adrenergic receptors.

TABLE 1

A Comparison of Pathological, Neurological and Neurochemical Deficits
in Parkinsons Disease and in Monkeys with VMT Lesions

Parkinsons Disease	Monkeys with VMT Lesions
Degeneration of the nerve cells in the pars compacta of the substantia nigra, lesions in other pigmental brainstem nuclei; (locus ceruleus, Lewy bodies)	Degeneration of the nerve cells in the pars compacta of the substantia nigra
Tremor, rigidity, akinesia, gait disturbance	Hypokinesia and tremor
Severe striatal dopamine deficiency; moderate decrease in cerebral 5-HT and NE levels. Decrease in the activities of DA synthesizing enzymes in the basal ganglia. Changes in the activity state of the DA receptor ?	Striatal DA and 5-HT deficiency. Decrease in the activities of DA synthesizing enzymes in the striatum. Changes in the activity states of the DA receptors ?

METHODS

Green monkeys (Ceropithecus sabaeus) were used and unilateral lesions were induced in the ventromedial tegmental region of the brain stem as previously described (2,3). Hypokinesia of the contralateral extremities appeared immediately afterward. In some monkeys a resting tremor (4 to 6 cycles per second) developed 5-10 days later. Recordings of tremor were obtained by means of a transducer attached to the extremities and were recorded on an electroencephalograph. The AIM were observed visually for a period of at least 6 hrs after each experiment, and some experiments were recorded cinematographically (4). Patients with Parkinsons disease were treated with bromocriptine or lergotrile as previously described (5,6). At weekly to monthly intervals a physician who was unaware of the patients medication assessed the patients using a Parkinsons disease evaluation form (7). The binding of H^3-haloperidol or H^3-apomorphine to postsynaptic striatal membranes was measured by the previously described procedures (8,9). The binding assay for H^3-DA to striatal membranes was modified by only using 6 mg of tissue per assay. Under these experimental conditions the specific binding to rat striatal membranes was found to be saturable. The binding of WB-4101 (a benzodioxane derivative, which is apparently a potent α-antagonist) (10) to postsynaptic cerebral cortical membranes was measured by the procedure of Greenberg et. al. (11).

RESULTS

The Effects of Drugs on Tremor Relief in Monkeys With VMT Lesions

DL-dopa alone or in combination with antiparkinsonian drugs. The effects of DL-dopa alone or in combination with other AP drugs on tremor in monkeys are summarized in Table 2. The intravenous administration of DL-dopa results in a transient disappearance of the tremor. The peripherally acting aromatic-L-amino acid decarboxylase (AADC) inhibitors such as L-methyldopa hydrazine (MK 486) or benserazide, (N-DL-seryl) N- (2,3,4 trihydroxy-benzyl)- hydrazine; (RO 4-4602) reduce the effective dose of L-dopa required for the relief of the tremor. Atropine, benztropine (cogentin) and diphenhydramine (benadryl) produce a dose dependent relief of the tremor. The combined administration of DL-dopa with each of these antiparkinsonian drugs have a much more pronounced effect on the relief of the tremor than their separate administration.

Putative DA Agonists. The administration of drugs which probably stimulate DA receptors (putative DA agonists) results in a transient relief of the tremor (Table 3). Apomorphine produces a transient relief of the tremor and potentiates the dopa induced relief of the tremor. Piribedil (Trivastal, ET 495), bromocriptine (CB 154) or lergotrile relieve the tremor for a longer time period than L-dopa. The tremor was relieved for the longest time period after administration of CB 154. Repeated administration of CB 154 enhances its antitremor effectiveness.

TABLE 2

Effects of DL-dopa Alone or in Combination with Other Antiparkinsonian
Drugs on Postural Tremor in Monkeys with VMT Lesions

Drug (mg/kg)	Pharmacologic responses on the activities of the contra-lateral limbs
None	Sustained hypokinesia and postural tremor
DL-dopa (15)	No effect
DL-dopa (30)	Tremor stopped for 0.5 - 2 hrs
DL-dopa (10) + MK-486 (10)	Tremor stopped for 0.5 - 2 hrs
Atropine (A) (0.2) or Benadryl (B) (10) or Cogentin (C) (0.25)	Tremor slightly diminished
DL-dopa (15) + (A) or (B) or (C)	Tremor stopped for 0.5 - 2 hrs

MK-486 was given i.p. one hr prior to the administration of DL-dopa
(i.v.).

A was given i.v. and B or C were given i.m. 30-60 min prior to the
administration of DL-dopa.

The Effects of L-dopa and of Putative DA Agonists on the Occurrence of AIM in Monkeys with VMT Lesions

A comparison of the effects of L-dopa and of putative DA agonists on the occurrence of AIM
is presented in Table 3. Type I AIM were evoked in normal monkeys either by administration
of L-dopa in combination with MK-486 or by administration of ET-495. The same dosages of
L-dopa plus MK-486 or of ET-495 given to monkeys with VMT lesions evoked AIM of type I and
II. Thus, monkeys with VMT lesions are more sensitive to the occurrence of AIM by DA
agonists than normal monkeys. It is noteworthy that bromocriptine and lergotrile evoked
initially sedation and less pronounced AIM than ET-495 or L-dopa. The blockade of DA
receptors with haloperidol abolishes the AIM evoked by trivastal or diminishes the tremor-
relieving activity of the drug. α-methyl-p-tyrosine, or inhibitor of tyrosine hydroxylase
diminishes tremor relieving and AIM-inducing activities of trivastal (4) and of bromocriptine. These results led us to suggest that presynaptic events might be partly involved in the AP efficacies of these drugs.

Clinical Studies: Treatment of Parkinsons Patients with Bromocriptine and Lergotrile

Bromocriptine. Fourteen patients with Parkinsons disease who were being treated with
levodopa combined with decarboxylase inhibitor (carbidopa) and who had signs of progression of their disease participated in this study. There was a statistically significant reduction in rigidity, tremor, bradykinesia, gait disturbance and total score after
treatment with bromocriptine (p < 0.01) (Table 4). Individually, ten of the 14 were improved by bromocriptine, in six the improvement was marked. In seven patients the combination of levodopa and decarboxylase inhibitor was completely eliminated. Two of these
patients were markedly improved two were moderately improved, two were unchanged and one
was worse. Adverse effects were similar to those observed with levodopa and carbidopa, except that in individual patients AIM and diurnal oscillations in performance ("on-off"
effect) were decreased while mental changes were increased (5).

Lergotrile mesylate. Twenty patients with advanced Parkinsons disease who were becoming
increasingly disabled on levodopa plus carbidopa and were able to tolerate 20 mg or more of
lergotrile participated in this study. Among the 20 patients completing a six month trial
there was a statistically significant reduction in rigidity, tremor, bradykinesia, total
score and AIM (p < 0.01) (Table 4). Individually, 11 of the 20 patients were considered
as "responders" showing a one stage or greater decrease in the severity of their disease.
In ten patients the duration and severity of the "on-off" effect was markedly decreased.
Among the 20 patients AIM were reduced in ten and increased in three when lergotrile was
added to levodopa plus carbidopa. Mental changes appeared in six patients, and dis-

TABLE 3

Effects of L-dopa and of Putative DA Agonists on
Postural Tremor and on the Occurrence of AIM

Surgical Lesion	Drug [a]	Motor Impairment	AIM [b] Type	Duration (Hrs)
None	MK 486 (10) + L-dopa (100)	None	I	1-2
None	ET 495 (3)	None	I	1-2
VMT	None	Hypokinesia and Tremor	None	---
VMT	MK 486 (10) + L-dopa (100)	Tremor stopped for 0.5 - 2 hrs	I, II	1-2
VMT	ET 495 (3)	Tremor stopped for 3-4 hrs	I, II	2-4
VMT	CB 154 (8)	Tremor stopped 2-6 hrs	I, II	1-2 hrs Initially sedation
VMT	CB 154 (8) [c]	Tremor stopped [c] 24-36 hrs	I, II	8-24 hrs [c]
VMT	Lergotrile (5)	Tremor stopped 2-5 hrs	Initially sedation and compulsive rotation	

[a] L-dopa and ET 495 were given i.v., the other drugs were given i.m. or i.p.

[b] AIM and other drug induced side effects

AIM, I: restlessness and aggressiveness
AIM, II: chorea-like movements, various types of stereotyped movements

[c] The drug was given every 24 hrs for 3 days. Tremor stopped the first day for 2-4 hrs, the second day for 24 hrs and the third day for 36 hrs. The AIM occurred the second and third day.

appeared when lergotrile was decreased or discontinued. Three patients developed persistant elevations in serum transaminase (SGOT, SGPT), after three to four months of treatment on a mean daily dose of 40 mg. of lergotrile. In these patients the drug was discontinued and transaminase returned to normal.

Displacement of Several Receptor Ligands by Bromocriptine and Lergotrile from Membrane Binding Sites

The binding of H^3-dopamine, H^3-apomorphine, and H^3-haloperidol to postsynaptic striatal membranes and of H^3-WB-4101 to cerebral cortical membranes was studied (12). It is evident from the results presented in Table 5 that lergotrile and bromocriptine displace the binding of H^3-DA, H^3-apomorphine and H^3-haloperidol. However, lergotrile displaces the binding of H^3-DA and H^3-apomorphine more effectively than bromocriptine. It should also be noted that bromocriptine displaces the binding of H^3-haloperidol more effectively than lergotrile. Both drugs displace the binding of H^3-WB-4101; bromocriptine has a higher affinity than lergotrile for the H^3-WB-4101 binding sites. The neuroleptics, haloperidol and clozapine, are potent displacers of H^3-WB-4101 binding, while (+) butaclamol is a less potent displacer.

DISCUSSION

In some aspects the experimental model produced in monkeys by placing VMT lesions replicates more closely human Parkinsonism than the other experimental models in rodents. Monkeys with VMT lesions exhibit hypokinesia and tremor of 4-6 cycles per sec which has a similar frequency as the Parkinsonian tremor. Although tremor is only one of the four major cardinal features of Parkinsons disease, the antitremor activity of various drugs in monkeys with VMT lesions is to some extent a reliable index for their antiparkinsonian

TABLE 4
Effect of Bromocriptine or of Lergotrile on Cardinal Parkinsonian Signs

Bromocriptine
Mean Scores

	Rigidity	Tremor	Bradykinesia	Gait	Totals
$\bar{a}$	38	12	46	69	43
$\bar{b}$	29	5	35	56	32
% change	24%	58%	24%	19%	26%

$\bar{a}$: before treatment with bromocriptine
$\bar{b}$: after treatment with bromocriptine

Lergotrile
Mean Scores

$\bar{c}$	32	16	50	37	35
$\bar{d}$	23	8	32	25	23
% change	28%	50%	36%	32%	34%

$\bar{c}$: before treatment with lergotrile
$\bar{d}$: after treatment with lergotrile

The results are the mean values obtained from 14 patients treated
with bromocriptine and from 20 patients treated with lergotrile.
The average dose of bromocriptine was 57 mg per day and the mean
reduction of levodopa + carbidopa was 31%. The average dose of
lergotrile was 52 mg per day and the mean reduction of levodopa
+ carbidopa was 15%. The cardinal signs of the disease were
evaluated as previously reported (5,6).

effectiveness in man. One of the major drawbacks in treatment of Parkinsonian patients
with L-dopa is the occurrence of AIM's. It is noteworthy that in this respect monkeys with
VMT lesions also replicate the clinical situation; L-dopa induced AIM is more severe in
monkeys with VMT lesions than in control monkeys and L-dopa induced dyskinesias occur in
parkinsonian patients but not in the control population (13). The results obtained in
experimental animal models (14,15) led us and others to evaluate clinically the therapeutic
efficacy of putative DA agonists in parkinsonian patients. Bromocriptine and lergotrile
were found to be effective AP agents (5,6,16) and the occurrence of AIM's following treatment with both drugs was less pronounced than following treatment with levodopa and carbidopa. Clinically, AIM's are less severe with lergotrile than with bromocriptine, and
similar results were obtained in monkeys with VMT lesions. The occurrence of less intense
AIM's following partial replacement of levodopa with bromocriptine and especially with
lergotrile might be related to the mixed agonist-antagonist properties of the latter two
drugs. Drugs which are mixed agonist-antagonist with respect to the DA receptors might have
some therapeutic advantages over pure agonists. The DA mixed agonist-antagonist may not
over-stimulate the DA receptors and therefore not produce the undesirable AIM's. Parkinsonian patients have so far not developed tolerance to the mixed agonist-antagonist, bromocriptine and lergotrile, while the development of tolerance to the DA agonist N-propylnor-
apomorphine has been reported (17). Preliminary studies in our laboratory indicate that
prolonged treatment of rats for 14 days with an ergot derivative which is a DA agonist (the
ergot derivative) does not displace the binding of H^3-haloperidol from striatal membranes)
produces a significant decrease in specific H^3-haloperidol binding. These results indicate
that prolonged treatment with DA agonists may produce a decrease in the DA receptor binding
sites or a change in the affinity of the receptor for the DA agonists. Thus, Parkinsonian
patients might not develop tolerance to bromocriptine and lergotrile because these drugs are
mixed DA agonist-antagonists. Furthermore, the unresponsiveness of a large number of
parkinsonian patients following prolonged treatment with levodopa may not solely be due, as
previously assumed, to the progressive degeneration of DA nigro-striatal neurons, but also
in part due to the desensitization of DA receptors induced by prolonged treatment with DA
agonists.

TABLE 5

Displacement of Several Receptor Ligands From Membrane Binding Sites
by Putative DA and α Agonists and Antagonists

Drug	K_i (nM)			
	H^3-DA [a]	H^3-Apomorphine [a]	H^3-Haloperidol [a]	H^3-WB-4101 [b]
Apomorphine	6.0	2.5	21.0	---
Dopamine	13.0	4.9	---	---
Bromocriptine	76.0	16.5	7.8	9.5
Lergotrile	36.0	13.0	48.0	57.0
Haloperidol	---	278.0	0.5	48.0
Clozapine	---	---	---	48.0
(+) Butaclamol	---	---	1.4	300.0

[a] rat striatal membranes

[b] rat cerebral cortex membranes

The concentration of radioligands were: 5 nM for H^3-DA, 2 nM for H^3-apomorphine, 4 nM for H^3-haloperidol and 0.2 nM for H^3-WB-4101.

Four to six concentrations of each drug were evaluated in triplicate to obtain IC_{50} values.

Data are the means from 3-5 determinations which varied less than 12%.

The K_D values for each radioactive ligand were determined from the corresponding Scatchard plot and the K_i was determined according to the equation $K_i = IC_{50}/1 + C/K_D$).

ACKNOWLEDGEMENTS

Supported by NINDS NS-06801 and NSF GB-27603.

REFERENCES

1. Ungerstedt, U., Mechanism of action of L-dopa studied in an experimental parkinson model, in Monoamines Noyaux Gris Centraux et Syndrome de Parkinson, 165 (1970), J. de Ajuriaguerra and G. Gauthier (Eds.), Bel-Air Symposium IV, Georg Cie S.A., Geneve.

2. Poirier, L.J. and Sourkes, T.L., Influence of the substantia nigra on the catecholamine content of the striatum, Brain 88, 181 (1965).

3. Goldstein, M., Anagnoste, B., Battista, A.F., Owen, W.S. and Nakatani, S., Studies of amines in the striatum in monkeys with nigral lesions, J. Neurochem. 16, 645 (1969).

4. Goldstein, M., Battista, A.F., Ohmoto, T., Anagnoste, B. and Fuxe, K., Tremor and involuntary movements in monkeys: Effect of L-dopa and of a dopamine receptor stimulating agent, Science 179, 816 (1973).

5. Lieberman, A., Kupersmith, M., Estey, E. and Goldstein, M., Treatment of parkinson's disease with bromocriptine, N. Eng. J. Med. 295, 1400 (1976).

6. Lieberman, A., Miyamoto, T., Battista, A.F. and Goldstein, M., Studies on the antiparkinsonian efficacy of lergotrile, Neurology 25(5), 459 (1975).

7. McDowell, F., Lee, J.E., Swift, T., Treatment of Parkinson's syndrome with I-dihydroxyphenylalanine (levodopa), Ann. Intern. Med. 72, 29 (1970).

8. Burt, D.R., Creese, I., Snyder, S.H., Properties of (^{3}H) haloperidol and (^{3}H) dopamine binding associated with dopamine receptors in calf membranes, Mol. Pharmacol. 12, 800 (1976).

9. Seeman, P., Lee, T., Chau-Wong, M., Tedesco, J., Wong, K., Dopamine receptors in human and calf brains using (^{3}H) apomorphine and antipsychotic drugs, Proc. Natl. Acad. Sci. 73, 4354 (1976).

10. Mottram, D.R. and Kapur, H., The α-adrenoreceptor blocking effects of a new benzo-dioxane, J. Pharm. Pharmacol. 27, 295 (1975).

11. Greenberg, D.A., U'Prichard, D.C., Snyder, S.H., Alpha-noradrenergic receptor binding in mammalian brain: Differential labeling of agonist and antagonist states, Life Sci. 19, 69 (1976).

12. Lew, J.Y., Hata, F., Ohashi, T. and Goldstein, M., The interactions of bromocriptine and lergotrile with dopamine and α-adrenergic receptors, J. Neural Transmission (In press).

13. Barbeau, A., L-dopa therapy in Parkinson's disease: A critical review of nine years' experience. Can. Med. Ass. J. 101, 59 (1969).

14. Miyamoto, T., Battista, A., Goldstein, M. and Fuxe, K., Long-lasting anti-tremor activity induced by 2-Br-α-ergocryptine in monkeys, J. Pharm. Pharmacol. 26, 452 (1974).

15. Goldstein, M., Battista, A.F., Nakatani, S. and Anagnoste, B., The effects of centrally acting drugs on tremor in monkeys with mesencephalic lesions, Proc. Natl. Acad. Sci. 63, 1113 (1969).

16. Lieberman, A., Zolfaghari, M., Boal, D., Hassouri, H., Vogel, B., Battista, A., Fuxe, K. and Goldstein, M., The antiparkinsonian efficacy of bromocriptine, Neurology 26(5), 405 (1976).

17. Cotzias, G.C., Papavasiliou, P.S. and Tolosa, E., Treatment of Parkinson's disease with aporphines: Possible role of growth hormone, N. Eng. J. Med. 294, 567 (1976).

DISCUSSION

Dr. Kling asked how long patients were kept on the combination therapy. *Dr. Goldstein* indicated that one patient was doing very well after one and a half years, that he was not showing any signs of tolerance developing.

Dr. Cook asked about the relative potency of the two isomers of butaclamol with respect to dopamine receptor blocking. *Dr. Goldstein* pointed out that there is about a 1000-fold difference between the two isomers with the L- being essentially inactive even at 10^{-6}M. There are similar differences with regard to behavioral effects.

EXTENSION OF ANIMAL MODELS TO CLINICAL EVALUATION OF ANTIANXIETY AGENTS*

Marian W. Fischman, Charles R. Schuster, and E. H. Uhlenhuth

Department of Psychiatry and Pharmacological and Physiological
Sciences, University of Chicago, 950 East 59th Street, Chicago,
Illinois 60637

It is tempting to develop animal models of human behavior, especially pathological behavior, which are based upon observations of similarities in response topography and maintaining events. It is the contention of this paper, however, that similarities in both overt behavior and the consequences of that behavior are insufficient to predict from non-humans to humans. It is essential that the effects of manipulations such as drug administration also be similar across the species in question. Since the discovery that the major and minor tranquilizers are useful in treating such clinical problems as anxiety, tension and agitation, scientists have attempted to reproduce these behaviors in the animal laboratory using a variety of behavioral procedures. The unifying concept in developing these behavioral procedures has been a motivational one. That is, the tranquilizers appear to be efficacious in relieving clinical problems in which aversive rather than appetitive stimulation is prepotent. The assumption here is that a generalized state of fear or anxiety is engendered when behavior is controlled by aversive stimulation, and, further, that the drugs which are useful in treating the behavioral syndrome called anxiety do so by modifying the underlying emotional state. Therefore, the argument continues, if one wishes to develop an animal model which distinguishes those drugs which should be efficacious in alleviating problems of anxiety, for example, from those drugs which are not, then aversive stimuli should be used to control behavior in the animal laboratory. Any changes in behavior which occur after the administration of the drug under study can be assumed to be due to changes in the underlying emotional state.

Early research on fear or anxiety was based on the two factor theory of avoidance advanced by Mowrer (28) and Miller (25). This hypothesis stated that there were two separate phases in the learning of an avoidance response. First the motive, fear or anxiety, is acquired on the basis of the pairing of a neutral and a noxious stimulus early in training. At this stage the organism learns to fear the conditioned stimulus and shows such symptoms of fear as urination, defecation, agitation and excitement. In the second phase, the previously neutral and now feared stimulus motivates escape responses which lead to its removal and the avoidance response is strengthened by a drive reduction mechanism of reinforcement. Observation of the organism shows an abatement of the signs of fear as the avoidance response is learned. Thus the concept of a generalized state of fear or anxiety was inextricably linked to the avoidance paradigm and to behavior maintained by postponement of or escape from aversive stimuli.

The validity of this motivational approach to animal models of behavior can be tested empirically (27). The hypothetical drive states such as fear or anxiety are dependent on environmental events such as presentation of a noxious stimulus. If the effects of specific classes of drugs are dependent on specific motivational states, then the behavioral effects of individual drugs should be determined by the environmental events which control the behavior being modified. Thus, for example, one would expect an anti-anxiety agent to specifically modify behavior maintained by shock avoidance at doses which have little or no effect on behavior maintained by presentation of an appetitive stimulus such as food. This prediction has not been substantiated when the relevant experimental studies were conducted.

It has been repeatedly shown that the behavioral effects of a number of psychotropic drugs are independent of the type of event maintaining behavior (5, 19, 20). In designing studies to illustrate this, it is important to compare behaviors with equivalent baselines. This

*This research was supported by MH-18611 and FDA 72-42 (E. H. Uhlenhuth, Principal Investigator), ADM-45-74-165 (C. R. Schuster, Principal Investigator) and a grant from Hoffmann-LaRoche Inc., Nutley, New Jersey.

means that the behaviors in question must be equated for rate and patterning under non-drug conditions. Thus, for example, Cook and Catania (5), using two different behavioral procedures, maintained comparable rates of lever pressing in two groups of squirrel monkeys. One group of monkeys received a train of low intensity electric shock pulses. The first lever press after 10 minutes (an escape response) turned off the electric shock for five minutes. The second group of monkeys was food deprived; the first lever press following a 10 minute fixed interval was followed by delivery of a food pellet. The schedule parameters were arranged so that the separation in time for food presentation or electric shock escape was roughly the same, and the general characteristics of the performances under the two schedules were similar. Rate of responding was low early in each interval and increased as the end of the interval approached. This similarity in performance under the two schedules of reinforcement continued when four different drugs were administered. Chlorpromazine and imipramine produced decreases in response rate under both schedules of reinforcement while meprobamate and chlordiazepoxide produced increases. It therefore seems that the pattern of baseline performance rather than the environmental stimulus maintaining the performance was of paramount importance in determining the drug effect in each case.

The actions of many drugs are determined by the baseline rate of responding. Thus, high rates of responding seen, for example, under many fixed ratio schedules of reinforcement can be decreased by doses of amphetamine which increase the lower rates maintained by fixed interval schedules (19, 32). Barbiturates, on the other hand, have been shown to decrease rates of responding maintained under some fixed interval schedules at the same doses that increase responding on fixed ratio schedules (11, 26). The rate dependency hypothesis, initially suggested by Dews (12) has been shown to hold for the amphetamines, barbiturates, minor tranquilizers and morphine (20, 29, 33).

Although the schedule-controlled rate and pattern of responding have been shown to be significant in determining the effects of drugs on behavior, there are a few studies in which the event maintaining behavior was the decisive determinant of the drug effect. Barrett (1), for example, has reported data in the squirrel monkey in which alcohol, chlordiazepoxide and pentobarbital were shown to increase responding maintained by food presentation but decrease that maintained by shock presentation. These data should not be interpreted as evidence that such differences are related to underlying motivational states. Rather, they indicate that more research is needed to adequately describe the factors determining the way in which a specific drug can modify behavior.

Regardless of the mechanisms involved in determining the effects of drugs on behavior, there is a growing body of evidence indicating that classes of drugs can be differentiated in the animal laboratory using carefully specified operant procedures. This has great relevance for the development of animal models of human pathology because correlations can be made between the effects of specific drugs in the animal laboratory and those found in the clinic. This does not mean that drug effects measured in infrahuman organisms are the same as their therapeutic effects in humans. We can only say, based on the data currently available, that it is possible to distinguish among psychomotor stimulants, barbiturates, neuroleptics and anti-anxiety agents using operant procedures in non-human research subjects. Comparable experimentation has not yet been carried out with human research subjects.

A particularly useful procedure for differentiating among classes of compounds is the conditioned avoidance paradigm. Substantial data have been collected since the initial report by Courvoisier (9) showing that chlorpromazine has a suppressant effect on responding by rats to avoid electric shock at doses which do not suppress escape from that stimulus. Avoidance procedures have been used extensively, and despite considerable differences in experimental design, they have repeatedly been shown to be effective in differentiating drugs with this selective suppressant effect (notably the phenothiazine compounds and narcotic analgesics) from those psychotropic agents which do not have this property (5).

The procedures used to study avoidance and escape performance have been of two general types. Early research utilized a discrete trial paradigm in which a conditioned stimulus (e.g., a tone or light) was presented and the conditioned avoidance response (e.g., pole climbing, wheel turning or lever pressing) was required within a specified period of time. In the absence of that response the aversive stimulus (e.g., electric shock) was presented and, if made, an escape response was measured. More recently, a free operant avoidance schedule originally described by Sidman (31) for use with electric shock stimuli has been used. In this procedure electric shock is delivered periodically (designated the shock-shock or S-S interval) in the absence of responding. Each response (usually a lever press) postpones the delivery of the electric shock for a specified period of time (designated the response-shock or R-S interval). The specificity of the phenothiazine effect on avoidance responding has been clearly demonstrated using either of these paradigms. Motivational interpretations of these results have assumed that if the presence of efficient avoidance responding indicates fear or anxiety towards the aversive stimulus, then the decrease in avoidance responding after administration of chlorpromazine can be interpreted as a reduction in anxiety of the organism being studied. This line of reasoning does not hold when one examines the effects of chlorpromazine on responding suppressed by punishment. The suppression should also reflect the common components of fear and anxiety. However, in this case, chlorpromazine is ineffective in increasing responding suppressed by immediate punishment. On the other hand minor tranquilizers such as meprobamate are relatively ineffective in decreasing responding maintained by the termination of an aversive stimulus while they are quite effective in increasing responding suppressed by immediate punishment (5, 15, 16, 20). This pharmacological separation of behaviors which might plausibly be considered to reflect the same underlying processes points out the necessity for careful analysis of behavioral processes.

An additional use of the conditioned avoidance paradigm as an animal model lies in the fact that effects obtained with phenothiazine drugs in non-human subjects are predictive of the clinical potency of these drugs when administered to humans with severe emotional and mental disorders. Table 1, taken from a paper by Cook and Catania (5), shows the correlation between the effects of a series of phenothiazines on discrete conditioned pole climbing avoidance performance and their effects in patients diagnosed as having severe mental and emotional disorders.

TABLE 1   Potency of phenothiazines on conditioned avoidance response (CAR) in rats compared with clinical potency*

Trade name	Generic name	CAR ED_{50} (mg/kg,[a] p.o.)	CAR rank	Clinical daily dose[b] (mg p.o.)	Clinical rank	Clinical rank by Schiele[c]
Stelazine	Trifluoperazine	0.9	1	3-40	1	1
Trilafon	Perphenazine	1.1	2	16-64	2	2
Dartal	Thiopropazate	1.5	3	20-80	3	3-4
Vesprin	Trifluopromazine	4.0	4	50-150	5	5
Compazine	Prochlorperazine	4.2	5	30-120	4	3-4
Thorazine	Chlorpromazine	9.9	6	75-1000	6	6
Sparine	Promazine	20.3	7	300-1000	7	7

The rank order of the relative potency of these drugs in the rat laboratory procedure correlates highly with the rank order of their relative clinical potency when this is defined as the average effective daily dose in the relevant patient population. The clinical dose estimates of Cook and Catania were supplemented with data collected by Schiele (30) on the relative potencies of this series of phenothiazines, and a similar rank order was obtained.

In developing an animal model for use in predicting drug effects in humans there are several problems. When generalizing from one species to another, there is always the possibility that the drug will affect behavior differently in the two species. For example, in most non-human species chlorpromazine has been found to cause a generally dose-dependent decrease in responding under a variety of experimental conditions (6, 17). However, Byrd (4) found that chlorpromazine enhanced response rates in chimpanzees responding under a multiple 10-minute fixed interval, 30-response fixed-ratio schedule of food delivery. On the other hand, d-amphetamine administered to these animals working on the same multiple schedule of food presentation (3) differentially affected response rates, enhancing low rates and causing decreases in high rates, similar to its effects in other species (12).

A second problem in using an animal model to predict behavior in clinical situations with humans is whether the behavior being studied in the non-human laboratory is analogous to the human behavior of interest. This means that behaviors must be studied in non-humans which are sensitive to the same manipulations as the relevant human behaviors being studied. Prediction across species is possible. The process of evolution indicates that there is a continuity in the fundamental mechanisms of behavior which extends across all vertebrates (34). The experimental analysis of behavior has shown that schedule-controlled responding is similar across species and, in general, the most relevant aspect of behavior when assessing a drug-behavior interaction is the rate and patterning of the behavior in question.

In the studies to be described here, the problem of generalizing from non-humans to humans has been approached by assuming that we are working with a new species whose behavioral

---

*Taken from Cook and Catania (5).

[a]Calculated as free base.

[b]Severe mental and emotional disorders in adults based on average use.

[c]Schiele (30).

repertoire has a continuity with those species already studied. We are attempting, as a first step, to use, with humans, procedures derived from the animal laboratory which have been shown to be sensitive and selective with respect to specific drug actions. Because the avoidance-escape procedure provides a means of differentiating classes of drugs in infra-human organisms, we extended the use of that procedure to humans in order to determine whether similar selectivity could be demonstrated using humans as the experimental organisms.

An initial pilot study was conducted in which the effects of chlorpromazine on electric shock avoidance and escape responding were determined using four human subjects button-pressing under a modified free operant avoidance schedule (14). Subjects drank a 10 ml chlorpromazine (in doses of 50, 75 or 100 mg) or placebo solution flavored with quinine and were tested in a one-hour experimental session one hour later. Surface skin electrodes were attached to two fingers of the right hand and subjects pressed a hand-held push button to avoid or escape electric shock (delivered in intensities of 0.35, 0.7, 1.5 or 3.0 mA). If the subject did not respond at all, a 10 sec chain of electric shocks, 0.5 sec in duration, each separated by 0.5 sec, was delivered every 30 sec (shock-shock interval). If the subject pressed the button during the 30 sec interval (an avoidance response) electric shock delivery was delayed for 30 sec (response-shock interval). Any response which occurred during the electric shock delivery (an escape response) terminated the electric shock and delayed its reoccurrence for 30 sec (response-shock interval). Eight experimental sessions were run for each of the electric shock intensities used. At each electric shock intensity the three chlorpromazine sessions (50, 75 and 100 mg) were interspersed with the five placebo sessions.

Data are presented for two subjects in Fig. 1.

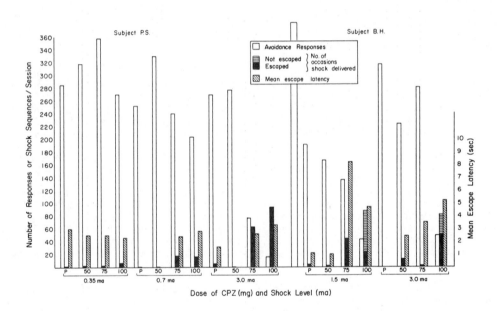

Fig. 1. The effects of varying dose of chlorpromazine and shock intensity level on behavior maintained by a free operant avoidance schedule in two humans. Subjects were given placebo (P) or 50, 75, or 100 mg of chlorpromazine and pressed a push button to avoid shock intensities of 0.35, 0.70, 1.5 or 3.0 mA. Baseline response rates were computed for each subject at each shock level using data from those days on which placebo was administered. The open bars represent the number of avoidance responses/session; closed bars represent the number of escape responses/session; horizontally striped bars indicate the number of occasions/session on which the shock was not escaped, and the diagonally striped bars show mean escape latency/session. Escape latency was adjusted to eliminate those occasions on which shock was not escaped. [Reprinted from (14)]

In general, as the dose of chlorpromazine was increased, the frequency of avoidance responding decreased. The effect on avoidance responding was relatively larger than the effect on escape responding. At the higher electric shock intensities, the dose effect relationship between chlorpromazine and its behavioral effect was more consistent. For the three subjects who showed some periods in which electric shock was not escaped when chlorpromazine was administered, the frequency of unescaped shocks was directly related to the dose.

The results of this initial study supported the frequently reported specificity of the suppressant effect of phenothiazines on avoidance behavior relative to escape behavior. The data collected suggested that in our new species under investigation, humans, there is a continuity in the maintenance of behavior by a free operant avoidance paradigm and the interaction of this behavior with chlorpromazine.

In a second study, five volunteer subjects were tested using the same modified free operant electric shock avoidance paradigm. In this case subjects were tested for a four hour experimental session in which they worked to avoid or escape electric shock for one hour, drank 30 ml of placebo or placebo solution with drug added, and then continued for three more hours during which the electric shock avoidance contingencies remained in effect. The electric shock intensity used was 3.0 mA. Chlorpromazine was administered in doses of 50, 75 and 100 mg, and pentobarbital in doses of 50, 100 and 150 mg. Doses were given in a mixed order interspersed with placebo sessions and all doses of one drug were completed before the other drug series was begun. Five subjects received chlorpromazine, four received pentobarbital.

Figure 2 presents summary data showing the mean percent change in avoidance response rate during the three hours after drug was administered as compared to the pre-drug avoidance response rate.

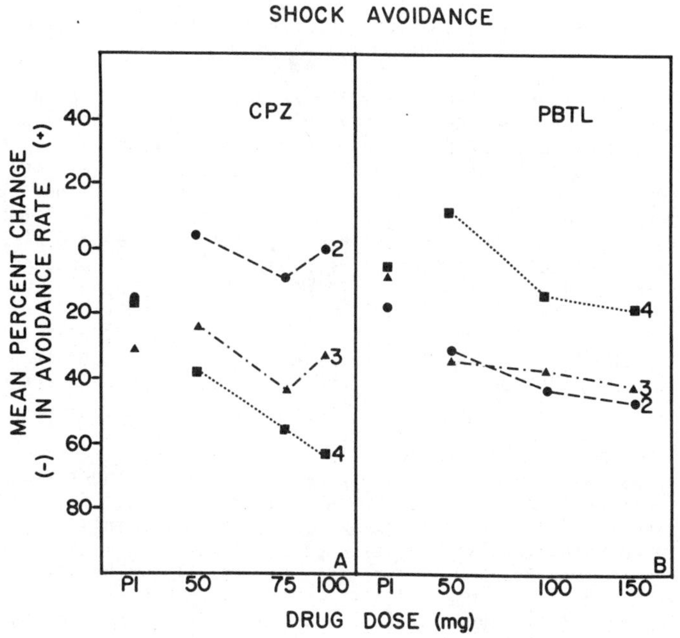

Fig. 2. Mean percent change in avoidance rate from the pre-drug experimental hour for subjects responding to avoid electric shock. The effects of chlorpromazine (CPZ) and pentobarbital (PBTL) are shown during the second (circles), third (triangles), and fourth (squares) hours of the experimental session. Data are also shown after ingestion of placebo (Pl).

344

Section A shows the effects of chlorpromazine and Section B the effects of pentobarbital. Although a considerable amount of inter-subject variability in response rate was seen, no consistent changes in rate were measured after placebo administration. Chlorpromazine caused a dose-related decrease in avoidance responding during the last hour of the experimental session while escape latencies during this period remained relatively unaffected. The mean escape latency calculated for all subjects over the three hours after chlorpromazine was administered was 1.75 seconds. There were virtually no periods when electric shock was not escaped during this drug series. Pentobarbital exerted its major suppressant effect during the first two hours after it was administered, although this effect was not dose related. The decrease in avoidance response rate was accompanied by an increase in escape latencies as well as an increase in the number of periods during which subjects sat through a chain of 10 electric shock pulses without making an escape response. The mean escape latency during the three hours after pentobarbital administration was 4.58 seconds, substantially higher than that seen after chlorpromazine. Thus the behavioral effects of these two drugs were differentiated using the modified free operant electric shock avoidance procedure. The interactions of these drugs with schedule controlled behavior have previously been shown to have generality over a wide range of experimental conditions in infrahuman organisms and this generality appears to extend to humans as well.

The use of electric shock stimuli with human research subjects is difficult because of the unwillingness of these subjects to place themselves in situations in which there is the possibility that they will receive an electric shock. An attempt was therefore made to use an alternative aversive stimulus in our avoidance procedure--one that was sufficiently aversive to maintain avoidance and escape behavior within the experimental session, but not so aversive that the subjects avoided the procedure entirely. A point loss procedure, with points convertible to money, was devised. Subjects began each four hour experimental session with 3600 points on the counter in front of them and in the absence of responding, points were subtracted at the rate of one per sec for 10 seconds every 30 seconds (R-S interval). This procedure was exactly the same as the electric shock avoidance procedure described above, with point loss substituted for the shock stimulus (see (13) for a full procedural description). The following drugs were tested in this procedure: d-amphetamine (5, 10, 15 and 20 mg), chlorpromazine (50, 75 and 100 mg), imipramine (75 and 150 mg), pentobarbital (50, 100, 150 and 200 mg) and diazepam (5, 10, 15 and 20 mg).

Although spacing of lever press responses varied between subjects, there was a pattern of responding for each subject that remained relatively constant across sessions. Inter-response time (IRT) distributions were calculated for each hour of every four-hour experimental session. An IRT is the time (in sec) between avoidance responses or between the last point lost and the next avoidance response. These were graphed as the percent of the total number of avoidance responses per hour and shown in Figs. 3-5 for one subject after placebo and one dose each of d-amphetamine, chlorpromazine, imipramine, pentobarbital and diazepam. This subject had a moderately high rate of avoidance responding; IRT distributions after placebo peaked at 5-6 seconds and were quite steep. There were rarely any points lost in placebo sessions, and a point loss sequence was always terminated after the first point was subtracted (an escape latency of 1.0 sec). d-Amphetamine (15 mg) caused an increase in avoidance responses and an abrupt shift of the IRT distributions to shorter latencies. No points were ever lost during the amphetamine series and responding did not return to pre-drug levels during the four-hour session.

One hundred mg of chlorpromazine (Fig. 4) severely disrupted the patterning of avoidance responding during the last two hours of the experimental session. There was a flattening of the IRT distributions with an increase in both long and short IRT's. The number of avoidance responses decreased over the four-hour session and the number of points lost increased. No periods occurred in which point loss was not escaped, and the average escape latency during the third and fourth hours were 2.3 and 4.1 seconds respectively. The effects of 150 mg of imipramine were somewhat similar to those seen after 100 mg of chlorpromazine. Avoidance responding decreased substantially over the four-hour session. Escape responding was less affected as shown by the fact that although there was an increase in the number of points lost per hour, there were very few periods of no escape--four in the third hour and two in the fourth hour. The mean escape latencies during successive hours of this session were 1.1, 2.1, 4.0 and 3.6 seconds.

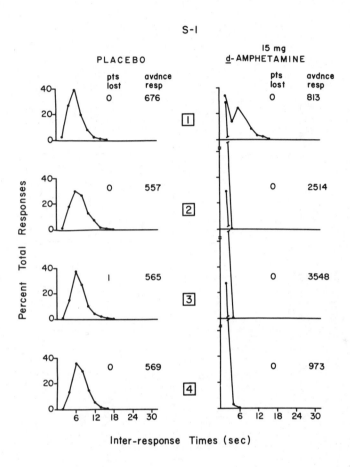

S-1

Fig. 3. Inter-response time distributions for subject 1, responding
to avoid point loss, after (A) placebo and (B) 15 mg d-amphetamine.
Session hours are indicated in the squares. The number of points lost
and the number of avoidance responses made are indicated for each hour
of the four-hour session. [Reprinted from (13)]

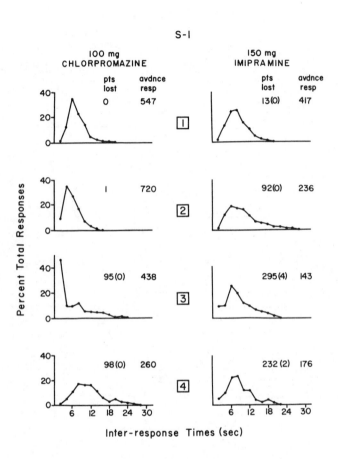

Fig. 4. Inter-response time distributions for subject 1, responding to avoid point loss, after (A) 100 mg chlorpromazine and (B) 150 mg imipramine. Session hours are indicated in the squares. The number of points lost and the number of avoidance responses made are indicated for each hour of the four-hour session. Numbers in parenthesis are the number of 10 sec periods in which no escape response was made. [Reprinted from (13)]

Diazepam and pentobarbital had suppressant effects on both avoidance and escape responding (Fig. 5).

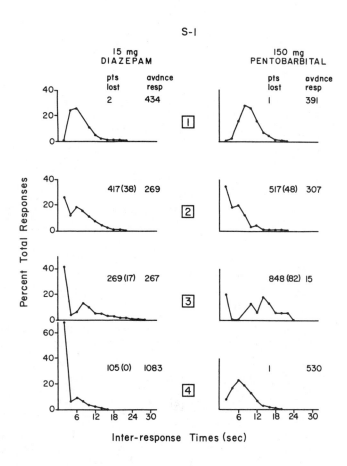

Fig. 5. Inter-response time distributions for subject 1, responding to avoid point loss, after (A) 15 mg diazepam and (B) 150 mg pentobarbital. Session hours are indicated in the squares. The number of points lost and the number of avoidance responses made are indicated for each hour of the four-hour session. Numbers in parenthesis are the number of 10 sec periods in which no escape response was made. [Reprinted from (13)]

Both drugs exerted their major suppressant effects on avoidance responding during the first or second hour after administration correlated with a substantial increase in the number of points lost as well as the number of periods when point loss was not escaped. By the last hour after administration of 150 mg of pentobarbital, the avoidance response rate increased and the IRT distribution was similar to that seen prior to drug. For this subject, there was an abrupt increase in avoidance responding during the last hour of the session when 15 mg of diazepam was administered. The increase was unlike that seen after d-amphetamine because concommittant with the increase in avoidance responding was the continued high level of point loss. Escape latencies increased substantially for this subject during the two hours after either of these drugs was ingested. After pentobarbital mean escape latencies for consecutive hours of the session were 1.0, 8.9, 10.4 and 1.0 seconds; after diazepam they were 1.0, 8.4, 5.0 and 4.2 seconds.

The data collected using the point loss avoidance paradigm show drug related differences. They replicate the selective suppressant effects of the phenothiazines found in non-human studies (8, 18) and clearly show the rate increasing effects of the amphetamines (10, 23) and the more general suppressant effects of pentobarbital and diazepam (2, 7, 23). We have thus been able to show that the avoidance paradigm is a useful model for differentiating the effects of different drugs. We have shown a continuity of behavior from non-human to human and have ruled out the possibility of any major species differences in avoidance-escape behavior and the interaction of that behavior with psychotropic drugs.

REFERENCES

(1) J. E. Barrett, Effects of alcohol, chlordiazepoxide, cocaine and pentobarbital on responding maintained under fixed-interval schedules of food or shock presentation, The Journal of Pharmacology and Experimental Therapeutics 196, 605 (1976).

(2) G. Bignami, L. diAcetis and G. L. Gatti, Facilitation and impairment of avoidance responding by phenobarbital sodium, chlordiazepoxide and diazepam--the role of performance baselines, Journal of Pharmacology and Experimental Therapeutics, 176, 725 (1971).

(3) L. D. Byrd, Effects of d-amphetamine on schedule-controlled key pressing and drinking in the chimpanzee, Journal of Pharmacology and Experimental Therapeutics, 185, 633 (1973).

(4) L. D. Byrd, Modification of the effects of chlorpromazine on behavior in the chimpanzee, Journal of Pharmacology and Experimental Therapeutics, 189, 24 (1974).

(5) L. Cook and A. C. Catania, Effects of drugs on avoidance and escape behavior, Federation Proceedings, 23, 818 (1964).

(6) L. Cook and R. T. Kelleher, Drug effects on the behavior of animals, Annals of the New York Academy of Science, 96, 315 (1962).

(7) L. Cook and J. Sepinwall, Psychopharmacological parameters of emotion. In, Levi, L. (1975) Emotions--Their Parameters and Measurement, Raven Press, New York.

(8) L. Cook and E. Weidley, Behavioral effects of some psychopharmacological agents, Annals of the New York Academy of Science, 66, 740 (1957).

(9) S. Courvoisier, J. Fournel, R. Ducrot, M. Kolsky and P. Koetschet, Propriétés pharmacodynamiques du chlorhydrate de chloro-3 (diméthylamino-3'propyl)-10 phenothiazine, Archives Internationale Pharmacodynamie, 92, 305 (1953).

(10) S. D. Dalrymple and R. Stretch, Effects of amphetamine and chlorpromazine on second-order escape behavior in squirrel monkeys, Psychopharmacologia (Berl.), 21, 268 (1971).

(11) P. B. Dews, Studies on behavior. I: Differential sensitivity to pentobarbital of pecking performance in pigeons depending on the schedule of reward, Journal of Pharmacology and Experimental Therapeutics, 113, 393 (1955).

(12) P. B. Dews, Studies on behavior. IV: Stimulant action of methamphetamine, Journal of Pharmacology and Experimental Therapeutics, 122, 137 (1958).

(13) M. W. Fischman, Evaluating the abuse potential of psychotropic drugs in man. In, Thompson, T. and Unna, K., eds. (1977) Predicting Dependence Liability of Stimulant and Depressant Drugs, University Park Press (in press).

(14) M. W. Fischman, R. C. Smith and C. R. Schuster, The effects of chlorpromazine on avoidance and escape responding in humans, Pharmacology, Biochemistry and Behavior, 4, 111 (1976).

(15) I. Geller and J. Seifter, The effects of meprobamate, barbiturates, d-amphetamine and promazine on experimentally induced conflict in the rat, Psychopharmacologia, 1, 482 (1960).

(16) I. Geller and J. Seifter, The effect of mono-urethans, di-urethans and barbiturates on a punishment discrimination, Journal of Pharmacology and Experimental Therapeutics, 136, 284 (1962).

(17) H. M. Hanson, J. J. Witoslawski and E. H. Campbell, Drug effects in squirrel monkeys trained on a multiple schedule with a punishment contingency, Journal of the Experimental Analysis of Behavior, 10, 565 (1967).

(18) G. A. Heise and E. Boff, Continuous avoidance as a base-line for measuring behavioral effects of drugs, Psychopharmacologia, 3, 264 (1962).

(19) R. T. Kelleher and W. H. Morse, Escape behavior and punished behavior, Federation Proceedings, 23, 808 (1964).

(20) R. T. Kelleher and W. H. Morse, Determinants of the specificity of behavioral effects of drugs, Ergebnesse der Physiologie, 60, 1 (1968).

(21) R. T. Kelleher, W. C. Riddle and L. Cook, Persistent behavior maintained by unavoidable shocks, Journal of the Experimental Analysis of Behavior, 6, 507 (1963).

(22) V. G. Laties and B. Weiss, Influence of drugs on behavior controlled by internal and external stimuli, Journal of Pharmacology and Experimental Therapeutics, 152, 388 (1966).

(23) A. Longoni, J. Mandelli and I. Pessotti, Study of antianxiety effects of drugs in the rat, with a multiple punishment and reward schedule. In, Garattini, S., Mussini, E. and Randall, L. O., eds. (1973) The Benzodiazepines, Raven Press, New York.

(24) M. J. Marr, Effects of chlorpromazine in the pigeon under a second-order schedule of food presentation, Journal of the Experimental Analysis of Behavior, 13, 291 (1970).

(25) N. E. Miller, Studies of fear as an acquirable drive. I: Fear as motivation and fear-reduction as reinforcement in the learning of new responses, Journal of Experimental Psychology, 38, 89 (1948).

(26) W. H. Morse, Use of operant conditioning techniques for evaluating the effects of barbiturates on behavior. In, Nodine, J. H. and Moyer, J. W., eds. (1962) Psychosomatic Medicine: The First Hannemann Symposium, Lea and Febiger, Philadelphia.

(27) W. H. Morse and R. T. Kelleher, Determinants of reinforcement and punishment. In, Honig, W. K. and Staddon, J. E. R., eds. (1977) Handbook of Operant Behavior, Prentice Hall, Englewood Cliffs, New Jersey.

(28) O. H. Mowrer, Anxiety-reduction and learning, Journal of Experimental Psychology, 27, 497 (1940).

(29) M. Richelle, B. Xhenseval, O. Fontaine and L. Thone, Action of chlordiazepoxide on two types of temporal conditioning in rats, International Journal of Neuropharmacology, 1, 381 (1962).

(30) B. C. Schiele, Newer drugs for mental illness, Journal of the American Medical Association, 181, 126 (1962).

(31) M. Sidman, Time discrimination and behavioral interaction in a free operant situation, Journal of Comparative and Physiological Psychology, 49, 469 (1956).

(32) C. B. Smith, Effects of d-amphetamine upon operant behavior of pigeons: Enhancement by reserpine, Journal of Pharmacology and Experimental Therapeutics, 146, 167 (1964).

(33) T. Thompson, J. Trombley, D. Luke and D. Lott, Effects of morphine on behavior maintained by four simple food-reinforcement schedules, Psychopharmacologia, 17, 182 (1970).

(34) B. Weiss and V. G. Laties, Comparative pharmacology of drugs affecting behavior, Federation Proceedings, 26, 1146 (1967).

ANIMAL MODELS OF DYSKINESIA

Harold L. Klawans, Ana Hitri, Paul A. Nausieda, and William J. Weiner
Department of Neurological Sciences, Rush-Presbyterian-St. Luke's Medical
Center, Chicago, Illinois

ABSTRACT

It has been suggested on clinical grounds that neuroleptic-induced tardive dyskinesias
result from denervation hypersensitivity of striatal dopamine receptors. Chronic pretreatment of laboratory animals results in behavioral evidence of striatal dopamine receptor site
hypersensitivity as well as biochemical evidence of altered dopamine receptors. Dopamine
agonist-induced dyskinesias are also felt to reflect drug-induced striatal dopamine receptor
hypersensitivity. Chronic animal studies have resulted in behavioral evidence of such
agonist-induced hypersensitivity.

INTRODUCTION

The term dyskinesia in relationship to human movement disorders is used in two separate ways.
Since dyskinesia merely means fragmentary, incomplete, or abnormal movement, the term is
sometimes used to encompass virtually all extrapyramidal hyperkinesias. At other times it
is used in a much more restricted sense and applied to a select group of iatrogenic movement
disorders which have not been given the benefit of the more precisely descriptive terms
such as chorea, athetosis, or dystonia. It is only in this more restricted role that the
term dyskinesia assumes any neurophysiologic and neuropharmacologic significance. This
review will not focus on all animal models of all dyskinesias in the broad sense but will
focus only on animal models of drug-induced dyskinesias. Animal models of two separate
classes of dykinesias will be explored: 1) neuroleptic-induced tardive dyskinesia and
2) dopamine agonist-induced dyskinesia.

TARDIVE DYSKINESIA

Phenomenologically, tardive dyskinesia is a choreiform movement disorder (1,2). Pharmacologically, it appears that the activity of dopamine at striatal dopamine receptors plays the
same primary role in the pathophysiology of the abnormal movements of tardive dyskinesia as
it does in other choreatic disorders such as Huntington's chorea (3,4) and levodopa-induced
dyskinesias (3,5).

Tardive dyskinesias occur in patients who have been treated with large doses of neuroleptics
for long periods of time (6,7). It is possible that brain damage may in some cases predispose a person to dyskinesia, but brain damage is not a necessary prerequisite. Tardive
dyskinesias usually appear late in the course of neuroleptic therapy. Lingual-facial-buccal
movements are the commonest hyperkinesia seen in this syndrome, but limb and truncal chorea
also occur (1,8). These movements may persist for years and may even become permanent
despite the discontinuation of the drug. Often these dyskinetic movements first appear
after a reduction in dosage of after discontinuation of phenothiazine therapy (1,7,9).

The neuroleptics that have been reported to produce tardive dyskinesias after prolonged
usage are the same agents that regularly result in drug-induced parkinsonism. Van Rossum
(10) has hypothesized that chlorpromazine and other neuroleptics act by blocking dopamine
receptors in the central nervous system. The competitive blockade by neuroleptics of dopaminergic receptors has been implicated as the basis of neuroleptic-induced parkinsonism (11).
Neuroleptics, by instituting a blockade of these receptors, may produce a "chemical denervation" of the dopamine receptors of the striatum. If the blockade is complete at any
receptor site, then this receptor is prevented from receiving its normal neurotransmitter
(dopamine) input.

It is possible that after such prolonged chemical denervation some receptors may develop
denervation hypersensitivity. Neurons whose dopamine receptors have developed this denervation hypersensitivity may respond in an abnormal manner to any dopamine that reaches these
receptors. It is quite possible that tardive dyskinesias may be the overt manifestation of
the abnormal response of such neurons.

If tardive dyskinesia is related to increased or altered activity of dopamine at striatal
dopamine receptors then an animal model of tardive dyskinesia should be based on striatal
dopaminergic function. Amphetamine- and apomorphine-induced stereotyped behavior appeared

to us to be a suitable starting point (12).

Stereotyped behavior has been produced in numerous animal species by the administration of amphetamine. This behavioral phenomenon has been studied most extensively in rats, and is characterized in that species by persistent sniffing, licking, and biting of the cage walls. Other animals show analogous kinds of movements. Guinea pigs gnaw persistently on the bars of the cage or the food dish when given large doses of amphetamines. The administration of amphetamines to a variety of animals results in some form of simple, repetitive, and persistent behavior in each species, but the specific form of this stereotyped behavior is characteristic of the species of animal (13). The concept that the exact form of the stereotyped behavior is species-specific is supported by the observation that other drugs which produce stereotyped behavior produce the same form of behavior in each species as does amphetamine. It appears also that, in all of the species investigated, the stereotyped movements consistently involve the branchiomeric musculature and more specifically the mouth parts and facial musculature.

Drugs which produce stereotyped behavior in animals include amphetamine, methamphetamine, apomorphine, cocaine, and levodopa (14,15). The common factor in the mechanism of action of these drugs is that they are all thought to increase the activity of dopamine or dopamine-like substances at the dopamine receptors. Amphetamine, methamphetamine, and cocaine are thought to cause the release of dopamine from nerve terminals in the caudate, thus making more dopamine available to act on the striatal dopamine receptors (16,17). Apomorphine seems to be able to activate these same receptor sites directly (14). Levodopa, the immediate precursor of dopamine, causes an increase in the synthesis of dopamine, which makes a greater quantity of dopamine available to act on the caudate. Whereas the form of stereotyped behavior is species-specific, the appearance of stereotyped behavior appears to be dependent upon increased activity of dopamine or dopamine analogs at dopamine receptors. The compulsive nature of stereotyped behavior which is characteristic of all species would appear to be a dopaminergic effect.

There is both pharmacological and anatomic evidence to suggest that both an intact caudate nucelus and a dopaminergic input to the caudate are necessary for the production of amine stereotyped behavior. Neuroleptics are known to strongly antagonize amphetamine-induced stereotyped behavior (13). The antagonism of this behavior as well as induction of a cataleptic state are characteristic properties of the neuroleptic phenothiazines and the butyrophenones. Reserpine and sedatives do not antagonize amphetamine-induced stereotyped behavior. Apomorphine, as mentioned above, produces stereotyped behavior, the nature of which is identical to that produced by amphetamine. Apomorphine, however, acts directly on the dopaminergic sites in the caudate. Its effect is not blocked by agents such as alpha-methyl-p-tyrosine which decreases the striatal content of dopamine (14). Haloperidol, however, by blocking the access to the receptor sites, blocks the apomorphine effect (13).

It has also been demonstrated that the caudate nucleus is the structure upon which dopamine or a dopamine analog must act in order to produce stereotyped behavior. Destruction of the caudate nucleus effectively prevents apomorphine-induced stereotyped behavior (18,19). Intracaudate injections of apomorphine, amphetamine, levodopa, and dopamine result in the characteristic behavior patterns, while injections of these drugs into other parts of the central nervous system do not (20-23). On the other hand, intracaudate injections of neuroleptics effectively prevent the appearance of drug-induced stereotyped behavior, but the injection of neuroleptics elsewhere does not (20). It appears that stereotyped behavior is a manifestation of dopaminergic function at striatal dopaminergic receptor sites and that as such this could be used as the basis of an animal model of tardive dyskinesia.

In an attempt to develop a workable animal model of tardive dyskinesia, we investigated the effect of prolonged neuroleptic pretreatment on stereotyped behavior. In the initial studies groups of guinea pigs were given daily subcutaneous injections of chlorpromazine (5 mg/kg for 3 weeks in one group, 10 mg/kg for 4 weeks in the other). Beginning 4 days after the last injection of chlorpromazine, amphetamine or apomorphine was given to the two groups of animals (24,25). It was found that doses of amphetamine which were too small to produce stereotyped behavior in control animals were effective in producing stereotyped behavior in the chlorpromazine pretreated animals. The animals were reevaluated weekly for 4 weeks, and it was found that the alteration in threshold for stereotyped behavior persisted. This implies that the neuroleptic pretreatment in some way altered the sensitivity of the dopaminergic receptors in the caudate to the dopamine released by the amphetamine. An alternative explanation may be that the absolute amount of dopamine available to act at the striatum is increased as a result of chlorpromazine pretreatment. It is known that chlorpromazine increases the synthesis and turnover of striatal dopamine by a feedback mechanism. However, studies have shown (26-29) that the concentration of dopamine in the brain after neuroleptic therapy is not elevated. Therefore, it is unlikely that amphetamine is releasing increased amounts of dopamine in chlorpromazine pretreated animals.

The studies in which apomorphine was given to these chlorpromazine pretreated animals also demonstrated a persistent decrease in the threshold for apomorphine-induced stereotyped behavior. Since apomorphine acts directly upon dopaminergic receptor sites in the striatum, these results support the hypothesis of a postsynaptic neuroleptic-induced alteration in sensitivity to dopaminergic agents.

It appears from this initial study that the neuroleptic pretreatment altered the sensitivity or the threshold of the dopaminergic receptors to dopamine and hence altered the threshold for the production of amine stereotyped behavior. Likewise, the caudate dopaminergic receptors in patients undergoing neuroleptic therapy may be altered in such a way that, upon termination of therapy, normal amounts of dopamine may trigger an abnormal response.

Since this original description of chlorpromazine-induced behavioral supersensitivity was reported, other investigators have reported similar findings with other neuroleptics in a variety of experimental animals (30-34). These studies all indicate that chronic administration of antipsychotic drugs followed by drug withdrawal produces a behavioral state of increased responsiveness to dopamine agonists, consistent with the development of supersensitivity that lasts for at least several weeks.

In our recent work with this behavioral model of neuroleptic-induced hypersensitivity, we have explored the question as to whether or not chronic neuroleptic pretreatment results in biochemically detectable hypersensitivity. In a recent attempt to provide biochemical evidence of hypersensitive dopamine receptors following chronic haloperidol, Burt et al. (35) reported an increase in the number of dopamine receptor sites when receptor binding was assayed with the dopamine antagonist ^{3}H-haloperidol. Our work investigated whether or not there is an alteration in dopamine receptor binding following chronic haloperidol in rats. Dopamine receptor binding in the striatum and nucleus accumbens is assayed with the naturally occurring agonist ^{3}H-dopamine. This approach has many advantages since haloperidol is not just a dopamine antagonist but also has some activity at norepinephrine and muscarinic receptor sites and since it is possible that haloperidol receptor sites may also exist (36).

In this work we compared white male Sprague Dawley rats who had received haloperidol (0.5 mg/kg IP) for 14 days followed by 7 days of saline injections with controls who had received saline for 21 days. The rats were sacrificed on day 21 and the whole brain was removed. The striatum and the nucleus accumbens were dissected from the fresh whole brain. The tissue was homogenized in 40 volumes of ice cold 50 nM Tris HCl buffer pH 7.4 containing 120 mM NaCl, 5 nM KCl, 2 mM CaCl$_2$, and 1 mM MgCl$_2$ according to the method of Creese (37). The homogenate was centrifuged at 50,000 x g for 10 minutes. The supernatant fluid was discarded and the pellet was rehomogenized in 50 volumes of the same buffer and recentrifuged. The supernatant fluid was discarded and the pellet was homogenized in 100 volumes of the same buffer. One ml aliquots corresponding to 10 mg of original wet tissue weight were incubated with various concentrations (2-255 mM) of ^{3}H-dopamine (SA 7.5 mg/µmole, New England Nuclear, Boston) in the same buffer as previous described for the tissue preparation with the addition of 0.1% ascorbic acid. The incubation was carried out at 4°C.

After incubation for 40 minutes, the samples were rapidly filtered under vacuum through Whatman GF/B filters. Each filter was rinsed with a single 5 ml saline aliquot. The filters were counted by liquid scintillation spectrometry. Specific binding of ^{3}H-dopamine was measured as the excess over blank tubes containing 100 µM dopamine or by extrapolation from the Scatchard curve according to the method of Chamness and McGuire (38).

It was found that equilibrium for ^{3}H-dopamine binding to striatal membranes was established at 40 minutes so that all binding studies where the amount of bound radioactive dopamine was monitored as a function of ^{3}H-dopamine concentration, the incubation time was 40 minutes to assure equilibrium.

Specific ^{3}H-dopamine binding was derived from graphical extrapolation and was found to be approximately 25% of total binding in the concentrations of dopamine used. Figure 1 shows the effect of chronic haloperidol (0.5 mg/kg) on specific ^{3}H-dopamine binding to striatal membranes (Michaelis-Menten curve). This illustrates that specific dopamine binding was increased after haloperidol pretreatment. In order to determine the number of binding sites and to accurately determine the affinity constants for the binding sites, the data is presented in a Scatchard plot in Fig. 2. The Scatchard plot demonstrates that after haloperidol pretreatment, there is a significant increase in ^{3}H-dopamine binding sites (1.5-fold) as well as an increase in binding affinity (nine-fold).

Figure 3 contrasts the results of chronic haloperidol on ^{3}H-dopamine specific binding in the nucleus accumbens (Scatchard plot). The slight increase in binding sites and in affinity after haloperidol was not significant (standard deviation overlap).

Table 1 is a summary of the effects of haloperidol pretreatment on ^{3}H-dopamine specific binding in caudate putamen and nucleus accumbens. Nucleus accumbens contains 60% more ^{3}H-dopamine binding sites than does the caudate putamen. Affinity constants are essentially the same in both anatomic areas and are in agreement with those reported earlier (37,39).

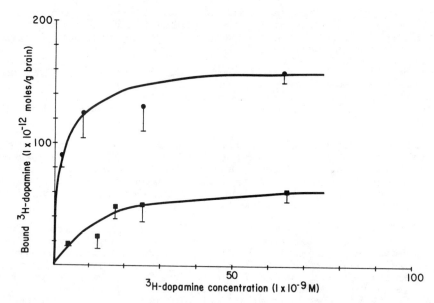

Fig. 1. Saturation curve of specific ^{3}H-dopamine binding to rat caudate putamen membranes. ■——■ controls; ●——● chronic haloperidol. Each point is the mean value of two separate experiments with the binding assays performed in duplicate.

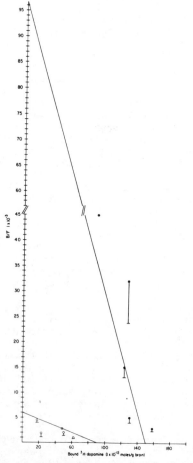

Fig. 2. Scatchard plot of ^{3}H-dopamine specific binding to caudate putamen membranes derived from control rat brains O——O and chronic haloperidol treated rat brains ●——●. The straight line was obtained by linear regression analysis with the slope representing the association constant.

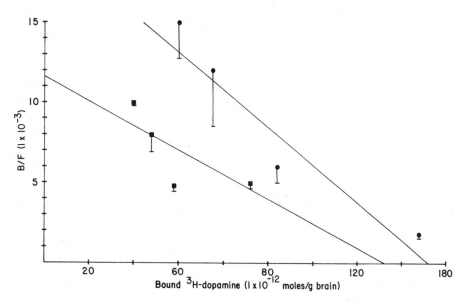

Fig. 3. Scatchard plot of ^{3}H-dopamine binding to nucleus accumbens membranes derived from control rat brains ■——■ and chronic haloperidol treated rat brains ●——● . Each point is the mean value of two independent experiments with the binding assays performed in duplicate.

TABLE 1   Effect of Chronic Haloperidol Pretreatment on Dopamine Receptor Site Binding

	Caudate Putamen		Nucleus Accumbens	
	n (picomoles/g brain)	Ka (M)	n (picomoles/g brain)	Ka (M)
Control	90	7×10^7	152	8×10^7
Haloperidol pretreatment	151	64×10^7	171	11×10^7

Haloperidol pretreatment resulted in a 60% increase in binding sites and a nine-fold increase in the affinity constant in the caudate putamen whereas the effect in the nucleus accumbens was not significant.

These results are direct biochemical evidence for the existence of supersensitive dopamine receptors in the rat striatum after chronic haloperidol pretreatment and as such support previous behavioral studies that have concluded that chronic neuroleptic treatment alters dopamine receptor site responsiveness. These results also support the study of Burt et al. (35) who demonstrated that dopamine receptor site binding is altered following chronic neuroleptic treatment. However, dopamine receptor site binding was assayed in that study by the use of ^{3}H-haloperidol, a dopamine antagonist. In this study ^{3}H-dopamine, the naturally occurring agonist, was employed to altered dopamine receptor site binding. Our studies demonstrate a 60% increase in striatal dopamine binding sites as compared to a 20% increase in the study of Burt et al. (35). We have also demonstrated a nine-fold change in binding affinity. The difference in magnitude of the altered dopamine receptor site binding characteristics following chronic haloperidol may be attributed, in part, to the use of the agonist instead of the antagonist to measure binding characteristics. The natural agonist may well be more specific for dopamine receptors and the binding of haloperidol to other sites (haloperidol, muscarinic and/or acceptor) could influence the results.

The fact that haloperidol induced altered binding characteristics of striatal dopamine receptors may explain in part at least the mechanism behind neuroleptic-induced tardive dyskinesia. These results then are consistent with the hypothesis that chronic neuroleptic therapy does induce striatal dopamine receptor hypersensitivity which could be the major factor in the pathophsyiology of tardive dyskinesia. The fact that chronic haloperidol treatment did not alter nucleus accumbens binding is consistent with the observation that tardive dyskinesia occurs without alteration of the underlying psychiatric state for which the patient is receiving neuroleptics. It has been suggested that dopamine acting at dopamine receptors plays a major role in the pathophysiology of schizophrenia (40). If this is true then neuroleptic-induced alterations in dopamine receptor site binding should worsen schizophrenia. The fact that these characteristics were unchanged in the nucleus accumbens may explain why the development of tardive dyskinesia is not related to any worsening of the underlying schizophrenia and is consistent with the hypothesis that these receptors play the major role in the pathophysiology of schizophrenia.

Because of the recent reports of the possible efficacy of lithium in the treatment of tardive dyskinesia (41-44), we have used this model to study the efficacy of lithium in the prevention and treatment of neuroleptic-induced alterations in the threshold for d-amphetamine- and apomorphine-induced stereotyped behavior (45).

The experimental groups consisted of the following: a) Control -- normal diet, saline injections for 3 weeks; b) Haloperidol pretreated -- normal diet, 0.5 mg/kg haloperidol for 3 weeks; c) Haloperidol and lithium treated -- lithium diet and 0.5 mg/kg haloperidol for 3 weeks. For all the above groups, the diet and injections were stopped one week prior to further testing. The fourth experimental group was d) Haloperidol pretreated, lithium treated -- normal diet and 0.5 mg/kg haloperidol for 3 weeks. After the 0.5 mg/kg haloperidol injections were stopped, the animals were placed on the lithium diet for 1 week and then given amphetamine or apomorphine.

The effect of lithium on haloperidol-induced decreased threshold for amphetamine-induced stereotyped behavior is summarized in Table 2 and Fig. 4 and 5. As shown in Table 2 3 mg/kg d-amphetamine does not induce stereotyped behavior in control animals but does in animals pretreated with haloperidol (p < 0.05). The simultaneous administration of lithium prevents this decreased threshold so that there is no significant difference between the control group and the group pretreated with both haloperidol and lithium but there is a significant difference (p < 0.05) between those receiving both haloperidol and lithium and those receiving haloperidol alone. Figure 4 shows the dose response curves for d-amphetamine in control animals, haloperidol pretreated animals, and haloperidol and lithium pretreated animals. Haloperidol pretreatment shifts the curve to the left indicating an increased response while animals pretreated with both haloperidol and lithium show no such shift. Figure 5 demonstrates that lithium given after 3 weeks of haloperidol pretreatment does not alter the haloperidol-induced shift of the dose response curve.

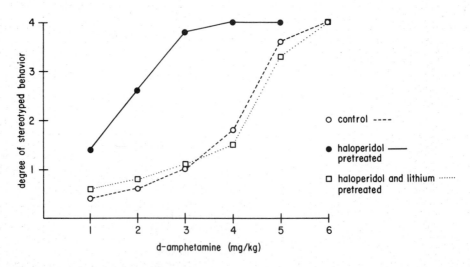

Fig. 4. Effect of concurrent lithium diet on haloperidol-induced alterations in amphetamine-induced stereotyped behavior.

The effects of lithium on haloperidol-induced decrease in the threshold for apomorphine-induced stereotyped behavior is summarized in Table 3. As above, haloperidol pretreatment decreased the threshold for apomorphine-induced stereotyped behavior and concurrent lithium pretreatment prevented this. Subsequent lithium treatment did not, however, prevent the haloperidol-induced decrease in threshold.

These results suggest that chronic haloperidol administration decreases the threshold to both amphetamine- and apomorphine-induced stereotyped behavior. Lithium given concurrently with haloperidol prevents this drug-induced alteration but the subsequent administration of lithium appears to have no effect.

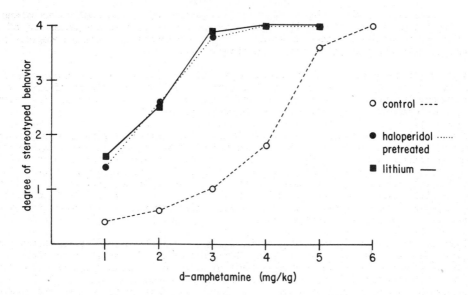

Fig. 5. Effect of subsequent lithium diet on haloperidol-induced alterations in amphetamine-
induced stereotyped behavior.

TABLE 2 Effect of Lithium Pretreatment and Treatment on Haloperidol-Induced
Decreased Threshold for Amphetamine-Induced Stereotyped Behavior

	Number	Dosage of d-amphetamine	Number developing stereotyped behavior	Significance
Control	12	3	0	
Haloperidol pretreated	6	3	6	$p < 0.05$
Haloperidol pretreated and lithium pretreated	6	3	0	vs. controls (not significant) vs. haloperidol pretreated ($p < 0.05$)
Haloperidol pretreated and lithium treated	6	3	6	$p < 0.05$

TABLE 3 Effect of Lithium Pretreatment and Treatment on Haloperidol-Induced
Decreased Threshold for Apomorphine-Induced Stereotyped Behavior

	Number	Dosage of apomorphine	Number developing stereotyped behavior	Significance
Control	12	.15	0	
Haloperidol pretreated	6	.15	6	$p < 0.05$
Haloperidol pretreated and lithium pretreated	6	.15	0	vs. controls (not significant) vs. haloperidol pretreated ($p < 0.05$)
Haloperidol pretreated and lithium treated	6	.15	6	$p < 0.05$

Interest in lithium in the treatment of human choreatic disorders began with several brief
reports of its efficacy in Huntington's chorea (43,46,47). Several preliminary reports
have claimed at least partial efficacy for lithium in ameliorating the abnormal movements

of tardive dyskinesia (41-44). Since neuroleptic-induced decreases in the threshold for stereotyped behavior appear to have the same physiologic basis as human tardive dyskinesia, the failure of lithium to reverse the former suggests that lithium will probably not be an effective treatment for tardive dyskinesia once the movements are present.

The successful prevention of haloperidol-induced hypersensitivity by the simultaneous ingestion of lithium suggests that lithium might be able to prevent the development of tardive dyskinesia. No human data is available on this issue.

Since tardive dyskinesia is a progressive disorder (48), it is possible that lithium may, however, have some role in the management of patients who are already manifesting some features of tardive dyskinesia. The concept that tardive dyskinesia is a progressive disorder is based on several clinical observations. First, there is the epidemologic evidence that the incidence of tardive dyskinesia is related to the duration of therapy. This suggests that the longer the pathogenesis persists the more frequently is the physiology disrupted to the point where clinical abnormalities appear. The observation that the disease is progressive in many individuals is more cogent. Tardive dyskinesia does not begin in most patients as a full-blown generalized choreatic disorder. It usually begins as an asymptomatic sign with mild lingual-facial-buccal dyskinesias and only later does it become more severe and more generalized. This increase in clinical manifestation represents two types of progression: 1) increase in the severity of the movements in those areas already involved, and 2) increase in the number of areas involved. The observation that lithium given concurrently with haloperidol prevents haloperidol-induced dopamine receptor hypersensitivity from developing, raises the possibility that lithium treatment might prevent the progression of tardive dyskinesia without necessarily decreasing the movements already present.

It has recently been suggested that the antipsychotic agent, clozapine, might not cause tardive dyskinesias. One study, using an analogous animal model, has shown that behavioral supersensitivity did not develop after withdrawal from 7 days of chronic clozapine administration to rats (49). In a more recent study, however, Smith and Davis (50) found that rats administered clozapine for 7 weeks did in fact exhibit behavioral supersensitivity as measured by apomorphine-induced stereotyped behavior.

DOPAMINE AGONIST-INDUCED DYSKINESIA

The chronic exposure to dopamine agonists in a variety of clinical settings has been shown to result in abnormal movements. The most common and best understood of these disorders is levodopa-induced dyskinesias in parkinsonism.

Neither the pathogenesis nor the pathophysiology of levodopa-induced dyskinesias in patients with parkinsonism is well understood. It is fairly well established that there is a direct relationship between the duration of levodopa therapy and the prevalence of dyskinesia (3,51,52). The basis of this direct relationship is not known, and aside from the relationship, little else about these dyskinesias is well understood or even generally accepted.

Various investigators have suggested that hypersensitivity of striatal dopamine receptor sites caused by prolonged denervation is related to the development of these dyskinesias (5,53,54). It is not, however, generally accepted that the pathophysiology of levodopa-induced dyskinesias is related to receptor site hypersensitivity (55) nor that the pathogenesis of the hypersensitivity is related to denervation.

Clinically the motor disorder which most closely resembles levodopa-induced dyskinesia in parkinsonism is another iatrogenic disorder -- neuroleptic-induced tardive dyskinesia (1,3). Both the character of the individual abnormal movements and the distribution and diversity of overall manifestations of these two disorders are so similar as to suggest that the pathophysiology (dopamine receptor site hypersensitivity) and pathogenesis (denervation) are similar in both states.

There are, however, a number of observations that are difficult to explain if levodopa-induced dyskinesias are due entirely to denervation hypersensitivity. These include the following:

1. The direct relationship between duration of levodopa therapy and prevalence of levodopa-induced dyskinesias

2. The direct relationship between duration of dyskinesias and the severity of the dyskinesias

3. The inverse relationship between duration of dyskiensias in the individual patient and the dosage of levodopa necessary to elicit dyskinesia (threshold dosage)

4. The change in time relationship between the dyskinesia and the individual dose of levodopa as the duration of dyskinesia increased.

Similar dyskinesias are also seen following chronic exposure to amphetamine and related agents.

Since we felt that the dyskinesias following chronic levodopa use in parkinsonism as well as the dyskinesias seen following chronic amphetamine abuse were related to the increased activity of dopamine at dopamine receptor sites within the striatum (3,56), it seems possible that supersensitivity due to excess innervation may well play a role in the pathogenesis of these two movement disorders.

Since amphetamine-induced stereotyped behavior and apomorphine-induced stereotyped behavior in animals have been intimately linked to the action of dopamine at dopamine receptor sites in the corpus striatum, it was felt that changes in stereotyped behavior following the chronic administration of amphetamine may act as a model of movement disorders in humans following the chronic use of amphetamine or levodopa.

In our initial experiments (57-59) we studied the effect of chronic amphetamine pretreatment on the subsequent response to both amphetamine and apomorphine. Recently, we have studied the effect of chronic pretreatment with levodopa (60), bromocriptine (61), and methylphenidate (62) on the behavioral responses to a variety of dopamine agonists including levodopa, amphetamine, and apomorphine.

In all of these studies, the exposure to the chronic dopamine agonists resulted in behavioral supersensitivity to subsequent exposure to dopamine agonists.

The initial studies showed that chronic pretreatment of guinea pigs with amphetamine produces an increased sensitivity to both amphetamine- and apomorphine-induced stereotyped behavior. These observations suggested that a postsynaptic mechanism was responsible for the hypersensitivity observed in these animals. The fact that catecholamine levels are decreased after chronic methamphetamine stimulation in rats (63) and guinea pigs (64), and that the pattern of amphetamine urinary metabolites in rats are unchanged by chronic amphetamine administration (65), adds additional evidence that the supersensitivity effects of amphetamine pretreatment are due to a receptor site change rather than an alteration in catecholamine or amphetamine metabolism.

These results were consistent with the hypothesis that a functional or structural change in the dopamine receptor site resulting in increased sensitivity can be caused by chronic administration of a dopamine agonist, viz., d-amphetamine. This drug-induced alteration could consist of a direct modification of the dopamine receptors themselves or a suppression of some other neuronal mechanism which normally antagonizes stereotyped behavior. Whichever mechanism is operating, the response of the cells upon which dopamine acts to elicit stereotypy is increased (supersensitive). A similar agonist-induced hypersensitivity might play a role in the development of amphetamine-induced dyskinesias.

Abnormalities associated with chronic amphetamine use include dyskinesias, complex stereotyped behavior, and a paranoid psychotic state which is often indistinguishable from acute or chronic paranoid schizophrenia (66,67). The involuntary dyskinetic movements are most often observed in the facial and masticatory musculature producing chewing, licking, teeth grinding, and protrusion of the tongue (66). Movements strikingly similar to these, known as lingual-facial-buccal dyskinesias, occur in Huntington's chorea and in a number of symptomatic choreas (56). Dyskinesias of the trunk and extremities are less common than lingual-facial-buccal movements in amphetamine-induced movement disorders (66) which is also the case in tardive and levodopa-induced dyskinesias (3).

Klawans and Weiner (56) have recently found that the administration of a single low dose of d-amphetamine does not produce dyskinesia in normal controls but is sufficient to exacerbate or uncover chorea in patients with Huntington's chorea, Sydenham's chorea, or chorea associated with systemic lupus erythematosus. Each of these disorders is believed to involve physiologic alterations within the basal ganglia and an increased sensitivity to dopamine.

In view of this information, the fact that amphetamine does not usually cause dyskinesias early in the course of abuse but does so after chronic abuse supports the hypothesis that chronic amphetamine administration produces supersensitivity to dopamine by altering the physiology of the basal ganglia. Supersensitivity following chronic agonism demonstrated in guinea pigs therefore appears to be an excellent model for the study of dyskinesias induced by chronic amphetamine abuse.

In a more recent study (62) we employed methylphenidate-induced stereotyped behavior in guinea pigs to assess the ability of chronic methylphenidate to induce receptor site hypersensitivity. After 3 weeks of treatment with methylphenidate, animals demonstrated a behavioral hypersensitive response to apomorphine. This suggests that chronic methylphenidate is capable of producing receptor site hypersensitivity.

In children with minimal brain dysfunction in whom chorea ensues shortly after initiation of methylphenidate therapy, chorea is probably related to the drug's central dopaminergic effect. In children who develop choreiform movements after chronic methylphenidate administration, it is suggested that methylphenidate may produce altered striatal dopamine receptor site responsiveness which results in chorea.

Other experiments have demonstrated the effect of long-term levodopa/carbidopa treatment and may serve as a model of levodopa-induced dyskinesias in humans. Chronic administration of doses of levodopa known to be subthreshold for acute induction of stereotypy produced definite stereotyped behavior after daily ingestion for 7 and 14 days. This alteration was also characterized by a decrease in the latent period between the ingestion of levodopa and the onset of stereotyped behavior. This decreased latency and threshold for stereotypy following chronic levodopa pretreatment may be viewed as a form of chronic agonist or "innervation-induced hypersensitivity."

These experiments may serve as a model for levodopa-induced dyskinesias in humans. Dyskinesias are seen in as high as 80% of the patients with Parkinson's disease receiving levodopa and are considered to be the major dose-limiting side effect of levodopa therapy (68). The appearance of the abnormal movements usually necessitates a decrease in the medication or prevents an increase in medication, thereby preventing maximum improvement of the parkinsonian disability (3).

The concept of chronic agonist-induced hypersensitivity helps to explain the clinical observations noted above which are not consistent with the hypothesis that levodopa-induced dyskinesias are due to denervation hypersensitivity. First, there appears to be a direct relationship between the duration of levodopa therapy and the prevalence of levodopa-induced dyskinesias. In our patient population, consisting of more than 200 patients, the prevalence of levodopa-induced dyskinesias increased from 37% at 6 months to 66% at 18 months to over 70% after 2 years of therapy. Similarly, in the experimental animals reported here, the prevalence of stereotypy increased from 40% after 7 days of levodopa ingestion to 74% after 14 days. In the same way that chronic subthreshold stimulation increased the prevalence of stereotyped behavior in animals, it is possible that chronic dopaminergic stimulation can contribute to the development of dopamine receptor site hypersensitivity in humans.

In many patients on long-term levodopa, there is a direct relationship between the duration of the dyskinesias and their severity. The dyskinesias are often quite mild at first and increase in severity as they persist. This increase often includes a wider distribution from solely lingual-facial-buccal movements to involvement of the trunk or limbs. There may also be an increase in the degree and severity of isolated, individual movements. The presence of these abnormal movements appears to indicate increased dopaminergic stimulation. While denervation hypersensitivity may play a pathogenic role in the initial excessive stimulation seen in these patients, these experiments suggest that continuation of levodopa stimulation by itself may further increase the sensitivity and account for the pathogenesis of these dyskinesias and their increasing severity with time.

The third pertinent observation is that there is often an inverse relationship between the duration of dyskinesias and the dosage of levodopa necessary to elicit the abnormal movements. It is possible that the chronic exposure to levodopa in patients with Parkinson's disease, like the guinea pigs exposed to chronic levodopa, has altered striatal dopaminergic sensitivity and thereby decreased the threshold dose of levodopa needed to elicit dyskinesias.

The final observation clarified by the concept of chronic agonist-induced hypersensitivity is the change in latency between time of levodopa ingestion and the appearance of dyskinesias. We have noted a large number of patients in whom dyskinesias were initially present 1 to 2 hours after levodopa ingestion. This time correlated fairly well with the maximal levodopa effect on the patient's parkinsonian signs and symptoms. After several months of chronic levodopa therapy, however, dyskinesias begin within minutes after levodopa ingestion and often precede all antiparkinsonian effects. The animal studies demonstrated a similar altered latency after chronic levodopa administration.

The clinical presentation and kourse of levodopa-induced dyskinesias are compatible with the concept of chronic agonist-induced hypersensitivity. Altered dopaminergic physiology may be induced by chronic levodopa stimulation in humans. The animal model of chronic agonist-induced hypersensitivity that is presented here may be of benefit in evaluating new therapeutic agents for levodopa-induced dyskinesias.

We have been using this approach to evaluate newer antiparkinsonian agents to see if these agents would cause dyskinesias. One such agent we have studied is bromocriptine (CB-154) a direct-acting dopamine agonist of proven clinical efficacy in parkinsonism. The capacity of bromocriptine to induce receptor site hypersensitivity was investigated utilizing this behavioral model in guinea pigs (61). Following 4 weeks of bromocriptine treatment, animals demonstrated a long-standing hypersensitivity to both amphetamine and apomorphine. This data suggests that chronic use of bromocriptine can induce receptor site hypersensitivity. These results may be an indication that such direct-acting dopamine agonists will offer no long-term advantage over current antiparkinsonian durgs because these agents should also induce dyskinesias due to agonist-induced hypersensitivity. The observed phenomena suggest that chronic dopaminergic agonism may not be an ideal therapy for parkinsonism.

ACKNOWLEDGMENTS

This work was supported in part by a grant from the United Parkinson Foundation, Chicago, Illinois.

REFERENCES

(1) H. L. Klawans, The pharmacology of tardive dyskinesia, Am. J. Psychiatry 130, 82 (1973).

(2) D. Tarsy and R. J. Baldessarini, The tardive dyskinesia syndrome, in Clinical Neuro-Pharmacology, H. L. Klawans (ed.), Raven Press, New York (1976).

(3) H. L. Klawans, Pharmacology of Extrapyramidal Movement Disorders, Karger, Basel (1973).

(4) H. L. Klawans, A pharmacologic analysis of Huntington's chorea, Eur. Neurol. 4, 148 (1970).

(5) H. L. Klawans, M. M. Ilahi, and D. Shenker, Theoretical implications of the use of L-dopa in parkinsonism. Acta Neurol. Scand. 46, 409 (1970).

(6) A. Faurbye, The structural and biochemical basis of movement disorders in treatment with neuroleptic drugs and in extrapyramidal diseases, Compr Psychiatry 11, 205 (1970).

(7) G. E. Crane, Tardive dyskinesias in patients treated with major neuroleptics: a review of the literature, Am. J. Psychiatry 124, 40 (1968).

(8) G. W. Paulson, "Permanent" or complex dyskinesias in the aged, Geriatrics 23, 105 (1968).

(9) R. Degkwitz, K. F. Binsack, and H. Herkert, Zum Problem der persistierenden extra-pyramidalen Hyperkinesen nach langfristiger Anwendung von Neuroleptika, Nervenartz 38, 170 (1967).

(10) J. M. Van Rossum, The significance of dopamine receptor blockade for the action of neuroleptic drugs, in Neuro-Psychopharmacology, H. Brill, J. O. Cole and P. Deniker (eds.), Excerpta Medica, Amsterdam (1966).

(11) H. L. Klawans, The pharmacology of parkinsonism, Dis. Nerv. Syst. 29, 805 (1968).

(12) R. Rubovits and H. L. Klawans, Implications of amphetamine-induced stereotyped behavior as a model for tardive dyskinesia, Arch. Gen. Psychiatry 27, 502 (1972).

(13) I. Munkvad, H. Pakkenberg, and A. Randrup, Aminergic systems in basal ganglia associated with stereotyped hyperactive behavior and catalepsy, Brain Behav. Evol. 1, 89 (1968).

(14) A. M. Ernst, Mode of action of apomorphine and dexamphetamine on gnawing compulsion in rats, Psychopharmacologia 10, 316 (1967).

(15) J. H. Wellner, M. Samach, and B. Angrist, Drug induced stereotyped behavior and its antagonism in dogs, Communications Behav. Biol. 5, 135 (1970).

(16) L. C. F. Hanson, Evidence that the central action of amphetamine is mediated via catecholamines, Psychopharmacologia 9, 78 (1966).

(17) J. Scheel-Kruger, Comparative studies of various amphetamine analogs demonstrating different interactions with the metabolism of the catecholamines in the brain, Eur. J. Pharmacol. 19, 47 (1971).

(18) C. Amsler, Beitrage zur Pharmakologie des Gehirns, Arch. Exp. Path. Pharmakol. 97, 1 (1923).

(19) A. Randrup and I. Munkvad, Dopa and other naturally occurring substances as causes of stereotypy and rage in rats. Acta Psychiatr. Scand. [Suppl.] 42, 193 (1966).

(20) R. L. Fog, A. Randrup, and H. Pakkenberg, Aminergic mechanisms in corpus striatum and amphetamine-induced stereotyped behavior, Psychopharmacologia 11, 179 (1967).

(21) R. Fog and H. Pakkenberg, Behavioral effects of dopamine and p-hydroxyamphetamine injected into corpus striatum of rats, Exp. Neurol. 31, 75 (1971).

(22) A. M. Ernst and P. G. Smelik, Site of action of dopamine and apomorphine on compulsive gnawing behavior in rats, Experientia 22, 837 (1966).

(23) R. L. Fog, A. Randrup, and H. Pakkenberg, Neuroleptic action of quaternary chlorpro-mazine and related drugs injected into various brain areas in rats. Psycho-pharmacologia 12, 428 (1968).

(24) H. L. Klawans and R. Rubovits, In experimental model of tardive dyskinesia, J. Neural. Transm. 33, 235 (1972).

(25) R. Rubovits, B. C. Patel, and H. L. Klawans, Effect of prolonged chlorpromazine pre-treatment on the threshold for amphetamine-induced stereotypy: a model for tar-dive dyskinesias, Adv. Neurol. 1, 671 (1973).

(26) K. F. Gey and A. Pletscher, Influence of chlorpromazine and chlorprothixene on the cerebral metabolism of 5-hydroxytryptamine, norepinephrine and dopamine,

J. Pharmacol. Exp. Ther. 133, 18 (1961).

(27) H. Nyback and G. Sedvall, Regional accumulation of catecholamines formed from tyrosine 14-C in rat brain: effect of chlorpromazine, Eur. J. Pharmacol. 5, 245 (1969).

(28) H. C. Guldberg and C. M. Yates, Effects of chlorpromazine on the metabolism of catecholamines in dog brain, Br. J. Pharmacol. 36, 535 (1969).

(29) H. Nyback and G. Sedvall, Effect of chlorpromazine on accumulation and disappearance of catecholamines formed from tyrosine C-14 in brain, J. Pharmacol. Exp. Ther. 162, 294 (1968).

(30) D. Tarsy and R. J. Baldessarini, Pharmacologically-induced behavioral supersensitivity to apomorphine, Nature [New Biol.] 245, 262 (1973).

(31) D. Tarsy and R. J. Baldessarini, Behavioral supersensitivity to apomorphine following chronic treatment with drugs which interfere with the synaptic function of catecholamines, Neuropharmacology 13, 927 (1974).

(32) G. Gianutsos and K. E. Moore, Dopaminergic supersensitivity in striatum and olfactory tubercle following chronic administration of haloperidol or clozapine, Life Sci. 20, 1585 (1977).

(33) B. Fjalland and I. Møller-Nielsen, Enhancement of methylphenidate-induced stereotypies by repeated administration of neuroleptics, Psychopharmacologia 34, 105 (1974).

(34) G. Gianutsos, R. B. Drawbaugh, M. D. Hynes, and H. Lal, Behavioral evidence for dopaminergic supersensitivity after chronic haloperidol, Life Sci. 14, 887 (1974).

(35) D. R. Burt, I. Creese, and S. H. Snyder, Antischizophrenic drugs: chronic treatment elevates dopamine receptor binding, Science 196, 326 (1977).

(36) A. Hitri, W. J. Weiner, R. L. Borison, B. I. Diamond, P. A. Nausieda, and H. L. Klawans, Dopamine receptor site binding in an animal model of tardive dyskinesia, Ann. Neurol. (submitted for publication).

(37) J. Creese, D. R. Burt, and S. H. Snyder, Dopamine receptor binding: differentiation of agonist and antagonist states with ^{3}H-dopamine and ^{3}H-haloperidol, Life Sci. 17, 993 (1975).

(38) C. G. Chamness and L. W. McGuire, Scatchard plots: common errors in correction and interpretation, Steroids 26, 538 (1975).

(39) P. Seeman, M. Chau-Wong, J. Tedesco, and K. Wong, Brain receptors for antipsychotic drugs and dopamine: direct binding assays, Proc. Natl. Acad. Sci. USA 72, 4376 (1975).

(40) H. L. Klawans, C. Goetz, and R. Westheimer, The pharmacology of schizophrenia, in Clinical Neuropharmacology, H. L. Klawans (ed.), Raven Press, New York (1975).

(41) G. M. Simpson, Tardive dyskinesia, Br. J. Psychiatry 122, 618 (1973).

(42) A. Prange, I. C. Wilson, C. E. Morris, and C. D. Hall, Preliminary experience with tryptophan and lithium in the treatment of tardive dyskinesia, Psychopharmacol. Bull. 9, 36 (1973).

(43) P. Dalen, Lithium therapy in Huntington's chorea and tardive dyskinesia, Lancet 1, 936 (1973).

(44) F. A. Reda, J. M. Scanlon, K. Kemp, and J. I. Escobar, Treatment of tardive dyskinesia with lithium carbonate, N. Engl. J. Med. 291, 850 (1974).

(45) H. L. Klawans, W. J. Weiner, and P. A. Nausieda, The effect of lithium on an animal model of tardive dyskinesia, Prog. Neuropsychopharmacology (in press).

(46) B. Mattson, Huntington's chorea and lithium therapy, Lancet 1, 718 (1973).

(47) M. J. Aminoff and J. Marshall, Treatment of Huntington's chorea with lithium carbonate, Lancet 1, 107 (1974).

(48) H. H. Klawans, Therapeutic approaches to neuroleptic-induced tardive dyskinesia, in The Basal Ganglia, M. D. Yahr (ed.), Raven Press, New York (1976).

(49) A. C. Sayers, H. R. Burki, W. Ruch, and H. Asper, Psychopharmacologia 41, 97 (1975).

(50) R. C. Smith and J. M. Davis, Behavioral evidence for supersensitivity after chronic administration of haloperidol, clozapine, and thioridazine, Life Sci. 19, 725 (1976).

(51) A. Barbeau, L-dopa therapy in Parkinson's disease: a critical review of nine years experience, Can. Med. Assoc. J. 101, 59 (1969).

(52) C. H. Markham, The choreoathetoid movement disorder induced by levodopa, Clin. Pharmacol. Ther. 12, 340 (1971).

(53) N. E. Anden, Pharmacological and anatomical implications of induced abnormal movements with L-dopa, in L-Dopa and Parkinsonism, A. Barbeau and F. H. McDowell (eds.), Davis, Philadelphia (1970).

(54) A. Carlsson, Biochemical implications of dopa-induced actions on the central nervous system, with particular reference to abnormal movements, in L-Dopa and Parkinsonism, A. Barbeau and F. H. McDowell (eds.), Davis, Philadelphia (1970).

(55) T. N. Chase, E. M. Holden, and J. A. Brody, Levodopa-induced dyskinesias, Arch. Neurol. 29, 328 (1973).

(56) H. L. Klawans and W. J. Weiner, The effect of d-amphetamine on choreiform movement disorders, Neurology (Minneap.) 24, 312 (1974).

(57) H. L. Klawans, P. Crosset, and N. Dana, The effect of chronic amphetamine exposure on stereotyped behavior: implications for the pathogenesis of L-dopa-induced dyskinesias, Adv. Neurol. 9, 105 (1975).

(58) H. L. Klawans, D. I. Margolin, N. Dana, and P. Crosset, Supersensitivity to d-amphetamine- and apomorphine-induced stereotyped behavior induced by chronic d-amphetamine administration, J. Neurol. Sci. 25, 283 (1975).

(59) H. L. Klawans and D. I. Margolin, Amphetamine-induced dopaminergic hypersensitivity in guinea pigs, Arch. Gen. Psychiatry 32, 725 (1975).

(60) H. L. Klawans, C. Goetz, P. A. Nausieda, and W. J. Weiner, Levodopa-induced dopamine receptor hypersensitivity, Ann. Neurol. (in press).

(61) P. A. Nausieda, W. J. Weiner, D. J. Kanapa, and H. L. Klawans, Bromocriptine-induced behavioral hypersensitivity: implications for the therapy of parkinsonism, Neurology (Minneap.), (in press).

(62) W. J. Weiner, P. A. Nausieda, and H. L. Klawans, Methylphenidate-induced chorea: case report and pharmacologic implications, Neurology (Minneap.), (submitted for publication).

(63) T. Lewander, Urinary excretion and tissue levels of catecholamines during chronic amphetamine intoxication, Psychopharmacologia 13, 394 (1968).

(64) T. Lewander, Effects of acute and chronic amphetamine intoxication on brain catecholamines in the guinea pig, Acta Pharmacol. (Kbh.) 29, 209 (1971).

(65) T. Lewander, Interference in the metabolism of amphetamine in the rat by psychoactive drugs, in Abuse of Central Stimulants, F. Sjoqvist and M. Tottie (eds.), Raven Press, New York (1969).

(66) P. H. Connell, Amphetamine Psychosis, Oxford University Press, London (1958).

(67) G. Rylander, Psychoses and the punding and choreiform syndromes in addiction to central stimulant drugs, Psychiatr. Neurol. Neurochir. 75, 203 (1972).

(68) A. Barbeau, H. Mars, and L. Gillo-Joffroy, Adverse clinical side effects of L-dopa therapy, in Recent Advances in Parkinson's Disease, F. A. McDowell and C. H. Markham (eds.), Davis, Philadelphia (1971).

DISCUSSION

Dr. Spiker suggested that although it is generally assumed that tardive dyskinesia is a secondary effect of administered drugs, it may, in fact, be secondary to the schizophrenia. He pointed out that in Kraepelin's classical book on dementia praecox, a section on motor movements has a description which would fit that which is currently described as tardive dyskinesia. Thus, Dr. Spiker felt that it should be borne in mind in looking for the etiology and/or treatment of tardive dyskinesias that some may be due to the illness rather than the drug. *Dr. Klawans* replied that the tardive dyskinesias are not limited to schizophrenics. Schizophrenics are the largest group at risk since they are the only ones who take the equivalent of a gram a day of chlorpromazine for twenty years. However, there are groups of non-psychotic patients, with no history of psychosis, who have been given neuroleptics (for recurrent nausea, for anxiety, etc.) and there are reports of tardive dyskensias developing among these patients.

Dr. Lomax commented that other drugs (e.g., morphine) result in an increased sensitivity to dopamine. He feels that this is a presynaptic phenomenon. Although the morphine-dependent rat is hypersensitive (e.g., with regard to stereotypic behavior or thermoregulation) to apomorphine, it is not supersensitive to intraventricular dopamine. He queried Dr. Klawans as to whether he had run similar tests, but *Dr. Klawans* indicated that he had not used the intraventricular route.

Dr. Friedhoff noted that his laboratory had some data which seemed at variance with Dr. Klawans' results with L-dopa. He found that after chronic L-dopa treatment there was a decrease in the amount of dopamine specifically bound to receptors as well as a decreased adenylate cyclase response. Dr. Friedhoff also observed an increase in total binding. *Dr. Klawans* agreed that there was a slight increase in total binding, but said that he also observed an increase in specific binding.

Dr. Sulser expressed difficulty in rationalizing Dr. Klawans' data with regard to hypersensitivity after exposure of receptors to agonists. In his work on norepinephrine receptors, beta receptors, etc., Dr. Sulser observed subsensitivity after the receptors were exposed to agonists. *Dr. Klawans* replied that he had less difficulty rationalizing agonist-induced sensitivity than he had in understanding why denervation produces hypersensitivity. There is the view that as the dopaminergic nigrostriatal neuron reaches the caudate and abuts on a caudate neuron, a dopamine-sensitive synapse is formed on the caudate neuron, indicating that the caudate neuron must have a membrane which can respond to some trophic influence by development of a receptor at that spot. It seems reasonable to him that when such caudate neurons are flooded with dopamine, they may become sensitive elsewhere. This seems a simple telelogical explanation. But he expressed a lack of understanding why multiple receptors are developed as the result of denervation. Finally, Dr. Klawans stated his feeling that if the agonist has a trophic function, an embryological function in determining the receptor site, it is easy to understand why agonists can induce increased numbers of receptors.

ANIMAL MODELS OF EPILEPSY

Peter Lomax and Joseph G. Bajorek
Department of Pharmacology, School of Medicine and the Brain Research
Institute, University of California, Los Angeles, California 90024,
U.S.A.

INTRODUCTION

The variety of models and experimental preparations for the study of the epilepsies exceeds the variety of expression of the epilepsies itself. A detailed survey of these models
is beyond the scope of the present discussion but the reader is referred to the excellent
review by Purpura *et al.* (1). In many instances these models have been developed primarily
for screening anticonvulsant drugs although many of them are also applicable to analysis of
epileptogenic mechanisms.

Many of these models may be classified as 'invasive' preparations in that the seizure
phenomenon is induced by some manipulation of an otherwise normal brain. Topical application of metals or convulsive drugs, electrical stimulation ('kindling') and freezing lesions
applied to limited brain areas can produce seizure activity and electroencephalographic
(EEG) evidence of a focus of epileptiform activity. Generalized convulsions may be produced by electroshock treatment or systemic injection of convulsant chemicals. *In vitro*
studies with tissue cultures or brain slices produce environmentally controlled micro-models
useful for neurophysiological and biochemical studies. Metabolic abnormalities, analagous
to certain human conditions resulting from physiological stress or trauma, can be produced
by changing the ionic or metabolic environment, e.g. oxygen and/or carbon dioxide tensions.

The present review will center around those models of epilepsy whose causality is believed to derive primarily from genetic and developmental abnormalities. Such models would
appear to offer greater scope in studying the etiology, ontogenesis and pathogenesis of
human idiopathic epilepsies, as distinct from the secondary epilepsies (following infection,
trauma, biochemical changes, etc.), than do the invasive models. It must be born in mind,
however, that each of the human epilepsies may well have a specific model best serving its
causality.

SPONTANEOUS EPILEPTOGENIC MODELS

Audiogenic Seizures

Auditory stimulation can precipitate a behavioral seizure state in mice, rats and rabbits that are selectively bred for susceptibility. Such naturally seizure prone animals,
especially DBA/2J mice and primable C57BL/6J mice, have been used extensively for anticonvulsant screening and for research on the mechanisms of epilepsy. Such animals offer the
advantage of extensive genetic control, relatively easy behavioral testing schemes and the
breeding proclivity needed for extensive biochemical exploration.

Any auditory stimulus of sufficient intensity and duration will induce the convulsive
state in the susceptible animal. Free running mice are exposed to the sound (e.g. an electric bell or buzzer, hand chimes); restraint appears to inhibit seizure induction. Following onset of the stimulus the animal exhibits a series of responses ranging from circling
behavior to loss of the righting reflex, tonic-clonic movements and respiratory arrest.
Behavioral scoring systems have been devised in order to assess the severity of the seizures quantitatively. Similarly, components of the tonic flexion or extension, running and
clonus have been timed and used as indices of drug efficacy.

Bizarre behavior patterns are easily mistaken for epileptiform attacks in both humans
and laboratory animals. The *sine qua non* for the diagnosis of an epileptic seizure is the
demonstration of an associated EEG abnormality. Such records are difficult to obtain in
the mouse as the calvarium is fragile and the small size of the animals renders recording
difficult. Recently cortical bipolar recordings have been obtained in free running mice

during seizures. The results have been disheartening. Cortical paroxysmal activity was minimal during audiogenic seizures and the proposal was advanced that the audiogenic seizure state reflects a subcortical abnormality (2). Abnormalities in cochlear hair cells and the auditory pathways have been reported in mice susceptible to audiogenic seizures (3,4). In humans epilepsy triggered by auditory stimuli has been reported (5). These observations suggest that although audiogenic seizure states respond to anticonvulsant therapy such models may be better suited to the study of auditory sensitivity and subcortical paroxysmal activity than to the investigation of the basic abnormalities underlying idiopathic epilepsy.

Photogenic Seizures

In 1966 Killam and his associates (6) described the induction of a seizure state in the baboon, *Papio papio*, using an intermittent light stimulus (ILS). Subsequently this animal has been intensively studied and various psychophysiological, neurophysiological and pharmacological parameters have been defined leading to the proposal that this baboon represents a natural animal model of human primary epilepsy. Indeed photogenic epilepsy is a characteristic of one form of the human disease.

Baboons from the Camanesce region of Senegal are especially susceptible to ILS induced seizures indicating a specific genetic inheritance. Although these animals can be raised and bred in captivity the associated costs and relatively long life span preclude extensive genetic controls (7).

Seizure phenomena are observed when the animals are subjected to close range light stimulation at about 25 Hz. The seizures are classified on a four point scale ranging from 0 (no seizures) to 3 (tonic/clonic convulsions). The response generally subsides after the stimulus is discontinued, but may be sustained in some animals (self sustained seizures). The latter are considered the most effective representatives of the epileptic model. The responsiveness has been maintained over 5 years of testing. Recent observations indicate that seizures can occur spontaneously (without ILS) or in response to stress or other external stimuli in some baboons but these effects have yet to be well delineated (8).

The implantation of electrodes for cortical or depth recordings are relatively simple in this species and electrophysiological data have been obtained from single unit studies and from acute and chronic EEG records. The EEG recordings show spike-wave abnormalities temporally related to the ILS stimuli. These progress and spread through cortical and subcortical structures, the records resembling those seen in human epilepsy (9).

Anticonvulsants effective in human epilepsies, such as diphenylhydantoin, phenobarbital and the benzodiazepines, are effective in suppressing the seizures in these baboons, although the responses to diphenylhydantoin tend to be inconsistent (8).

In attempts to relate this natural model to other chronic experimental models the induction of epileptogenic foci with alumina cream or chronic electrical stimulation (kindling) has been attempted (10,11). Foci induced by alumina cream tend to potentiate the photogenic seizures whereas kindling has little effect on the ILS threshold.

The close phylogenetic relationship of the primates, and their related neurophysiological complexity would favor their greater applicability to studies of human diseases. However, inasmuch as the basic mechanisms underlying the epilepsies may reside at the cellular or molecular level, the view has been expressed that this very complexity may confound investigations with currently available experimental techniques. Certainly the lethal methodologies dictated by biochemical studies limit adequate analysis in *Papio papio* due to cost limitations.

Reflex or Xenogenic Seizures

Certain animals have been noticed to develop spontaneous seizures when exposed to an unfamiliar, open-field environment. It has been suggested that these be referred to as "reflex" (12) or "xenogenic" (13) seizures.

McCarty and Southwick (14) have described such seizures in the Grasshopper mouse *(Onychomys torridus)* and proposed this animal as a model for the study of the epilepsies. Further investigations along these lines are yet to appear.

Such spontaneous motor seizures had earlier been reported to occur in the Mongolian gerbil *(Meriones unguiculatus)* when the animal was confronted with novel environmental stimuli (15). Several breeding pairs of this strain of gerbils were received at UCLA in 1967 and the occurrence of the seizure phenomenon was noted. It was felt that the animals might prove to be a useful model for the study of the epilepsies and so, in 1968, a selective breeding program, using a closed colony technique, was undertaken (16). Three distinct strains were developed:

a) Seizure sensitive animals (WJL/UC strain)

b) Seizure resistant animals (STR/UC strain)

c) Random bred animals (RB/UC strain)

The incidence of seizures in these animals was 98.3%, 15.7% and 64.7%, respectively.

The severity of the seizures varied from animal to animal. However, analysis of the motor behavior and its duration allowed rating of the seizures on a clearly defined scale (Table 1.) (17).

TABLE 1. Rating Scale for Severity of Seizures in Seizure Sensitive Gerbils from the UCLA Colony

Grade	Motor Characteristics
0	No seizure activity
1	Animal moving, vibrissae and pinnae twitching
2	Motor arrest, vibrissae and pinnae twitching
3	Motor arrest with myoclonic jerks
4	Clonic/tonic seizure
5	Clonic/tonic seizure with body rollover
6	Continued seizures progressing to death

The seizures can be induced by removing the animals from the home cage and placing them in a novel environment (open field). The latency to onset varies from 16-36 sec and the durations were 9.4 ± 1.1 sec (grade 1), 14.7 ± 1.4 sec (grade 2), 278 ± 20.0 sec (grade 3), 308.5 ± 17.7 sec (grade 4), 286 ± 5.5 sec (grade 5). No sex related differences in incidence, latency, duration or severity of the seizures were found. Following induction of a seizure the animals show refractoriness which declines over a 4 day period. Normally a 6 day intertrial period is recommended for reliable induction.

Animals are first tested on weaning (day 30) and the mean age of seizure onset is 54 ± 3 days (Fig. 1). The maximum seizure severity is reached by 6 months (Fig. 1) and is then maintained throughout life (approximately 3 years).

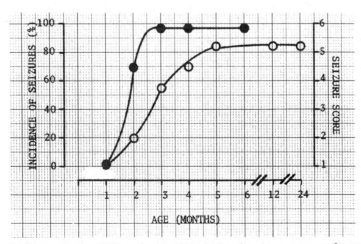

Fig. 1 Ontogenesis (●) and severity (○) of seizures as a function of age in gerbils of the WJL/UC strain (data derived from 381 and 581 animals respectively).

Fully grown gerbils are sufficiently large (70-100 g) to fit the head holder of a standard rat stereotaxic instrument. Using an atlas of the gerbil brain (18) multiple recording electrodes can be implanted and attached to the skull for chronic EEG recording. Cortical and depth electrodes reveal the epileptiform nature of the seizures. The occurrence of trigger foci and subsequent spread of seizure activity have been seen and the severity and spread of abnormal EEG activity has been correlated with the motor activity (19).

For testing anticonvulsant drugs statistically significant data can be generated using small groups of animals (e.g. 6) and recording the change in the mean seizure score. In order to administer drugs (or carry out other manouvers, such as hooking up recording electrodes) without precipitating a seizure the animal is anesthetized in its home cage with N_2O (80% with 20% O_2). The animal rapidly recovers after flushing the cage with O_2 and the severity of subsequent seizures is unaffected. Acute injection of diphenylhydantoin (25 mg.kg^{-1} i.p.) or phenobarbital (20 mg.kg^{-1} i.p.) significantly supresses the seizures whereas trimethadione (50 or 75 mg.kg^{-1} i.p.) is ineffective (20).

Preliminary studies of the role of brain amines, using various enzyme inhibitors, suggest that dopamine may be implicated in the genesis of the seizures: increasing brain dopamine levels suppressed seizure severity (21). Whether this finding indicates an underlying deficit in dopaminergic pathways awaits further elucidation.

Testing of the breeding data on several computer models, although not conclusive, strongly suggests that the etiology of the seizure disorder resides at a single autosomal locus with at least one dominant allele of variable penetrance (16). This conclusion is interesting as a similar mode of inheritance has been suggested for some forms of human epilepsy, most convinceingly in the case of infantile febrile convulsions (22). Indeed preliminary data (Bajorek and Lomax, unpublished data) indicate that febrile convulsions can be induced in immature gerbils.

CONCLUSIONS

Apart from anticonvulsant testing, natural models are needed for the study of the fundamental mechanisms of the epilepsies. Inasmuch as the human condition represents a spectrum of syndromes it is not to be expected that any single animal preparation would be appropriate for all of these. The discoveries of animals demonstrating genetically determined seizure traits encourages the hope that a fuller understanding of these perplexing diseases will emerge.

ACKNOWLEDGEMENTS

Supported by NSF grant GB-43531 and by US Navy ONR contract N00014-75-C-0506. J. G. Bajorek is supported by USPHS MH06415.

REFERENCES

(1) D. P. Purpura, J. K. Penry, D. B. Tower, D. M. Woodbury and R. D. Walters, eds., Experimental Models of Epilepsy, Raven Press, New York, 1972.

(2) S. C. Maxson and J. S. Cohen, Electroencephalographic correlates of the audiogenic seizure response of inbred mice, Physiol. Behav. 16, 623 (1976).

(3) C. H. Norris, Kanamycin priming for audiogenic seizures in mice, Pharmacologist 18, 183 (1976).

(4) J. F. Willott, Effects of unilateral spinal cordotomy and outer ear occlusion on audiogenic seizures in mice, Exptl. Neurol. 50, 30 (1976).

(5) R. G. Bickford and D. W. Glass, Sensory precipitation and reflex mechanisms, In: H. H. Jasper, A. A. Ward Jr. and A. Pope, eds., Basic Mechanisms of the Epilepsies, Little, Brown, Boston, 1969, p. 543.

(6) K. F. Killam, E. K. Killam and R. Naquet, Mise en evidence chez certains singes d'un syndrome photomyoclonique, Compt. Rendu. Acad. Scé (Paris) 262, 1010 (1966).

(7) R. Naquet and B. S. Meldrum, Photogenic seizures in baboons, In: D. P. Purpura, J. K. Penry, D. B. Tower, D. M. Woodbury and R. D. Walters, eds., Experimental Models of Epilepsy, Raven Press, New York, 1972, p. 373.

(8) E. K. Killam, Measurement of anticonvulsant activity in the *Papio papio* model of epilepsy, Fed. Proc., 35, 2264 (1976).

(9) R. Naquet, J. Catier and C. Menini, Neurophysiology of photically induced epilepsy in *Papio papio*, In: B. S. Meldrum and C. D. Marsden, eds., Advances in Neurology, vol. 10, Raven Press, New York, 1975, p. 107.

(10) S. Dimov and J. Lanoir, Chronic epileptogenic foci in the photosensitive baboon, *Papio papio*, Electroencephalogr. Clin. Neurophysiol. 34, 353 (1973).

(11) J. A. Wada and O. Takeshi, Spontaneous recurrent seizure state induced by daily electric amygdaloid stimulation in Senegalese baboons *(Papio papio)* Neurology 26, 273 (1976).

(12) W. J. Loskota, P. Lomax and S. T. Rich, The gerbil as a model for the study of the epilepsies: seizure habituation and seizure patterns, Proc. West. Pharmacol. Soc. 15, 109 (1972).

(13) D. Goldblatt, A. Konow, L. Shoulson and T. MacMath, Effect of anticonvulsants on seizures in gerbils, Neurology 21, 433 (1971).

(14) R. McCarty and C. H. Southwick, The development of convulsive seizure in the Grasshopper mouse *(Onychomys torridus)*, Develop. Psychobiol. 8, 547 (1975).

(15) D. D. Thiessen, G. Lindzey and H. C. Friend, Spontaneous seizures in the Mongolian gerbil, Psychonom. Sci. 11, 227 (1968).

(16) W. J. Loskota, The Mongolian gerbil *(Meriones unguiculatus)* for the study of the epilepsies and anticonvulsants, Ph.D. dissertation University of California at Los Angeles (1974).

(17) W. J. Loskota, P. Lomax and S. T. Rich, The gerbil as a model for the study of the epilepsies, Epilepsia 15, 109 (1974).

(18) W. J. Loskota, P. Lomax and M. A. Verity, A Stereotaxic Atlas of the Mongolian Gerbil Brain *(Meriones unguiculatus)* Ann Arbor Science, Ann Arbor, 1974.

(19) W. J. Loskota and P. Lomax, The Mongolian gerbil *(Meriones unguiculatus)* as a model for the study of the epilepsies: EEG records of seizures, Electroencephalog. Clin. Neurophysiol. 38, 597 (1975).

(20) W. J. Loskota and P. Lomax, The Mongolian gerbil *(Meriones unguiculatus)* as a model for the study of the epilepsies: anticonvulsant screening, Proc. West. Pharmacol. Soc. 17, 40 (1974).

(21) B. Cox and P. Lomax, Brain amines and spontaneous epileptic seizures in the Mongolian gerbil, Pharmacol. Biochem. Behav. 4, 263 (1976).

(22) E. Frantzen, M. Lennox-Buchthal, A. Nygaard and J. Stene, A genetic study of febrile convulsions, Neurology (Minneap.) 20, 909 (1970).

DISCUSSION

In reply to *Dr. Tedeschi's* question on anticonvulsants, *Dr. Lomax* indicated that he had found both phenobarbital and diphenylhydantoin to be effective at ca. 25 mg/kg in blocking seizures but that trimethadione was completely ineffective even at 75 mg/kg.

Dr. Kornetsky wondered if Dr. Lomax had done any learning experiments with his animals, especially prior to the time when they had maximum amount of seizures; such work could be a model with regard to epilepsy and antiepileptic drugs. *Dr. Lomax* said that although there have been studies of learning behavior in gerbils, he was unaware of any for the seizure-sensitive strain.

Dr. Reite wondered if seizures in implanted animals always started in limbic regions, but Dr. Lomax stated that it could start in any area of the brain.

ANIMAL MODELS OF HYPERACTIVITY

Alan M. Goldberg, Dept. of Environmental Health Sciences, The Johns
Hopkins University School of Hygiene and Public Health, Baltimore,
Maryland 21205

Ellen K. Silbergeld, Experimental Therapeutics Branch, Behavioral
Neuropharmacology Unit, N.I.N.C.D.S., National Institutes of Health,
Bethesda, Maryland 20014

INTRODUCTION

The childhood behavior and learning disorder varyingly known as minimal brain or minimal cerebral dysfunction (MBD) encompasses not only increased locomotor activity but also behavior
problems and learning disabilities. The description of children as "hyperkinetic" has been
used to denote inappropriate locomotion or disruptive movement rather than a indication of
the quantitative amount of locomotion expressed.

In approaching the development of animal models of this behavioral syndrome, most studies
have concentrated on the symptom of increased motility for homology with MBD. The other
symptoms of MBD have not been generally explored. These other aspects, problems in learning,
attention, and aggression, require further clinical definition and may prove important areas
of experimental research in the understanding of the syndrome. In addition, clinical studies
have not successfully defined the changes in whole body movement. Studies using various monitoring devices have failed to produce consistent data indicating clear increases in movement or response to medication in the MBD syndrome (1,2).

In the animal models, whole body locomotion is usually studied. To avoid confusion with the
connotions of the term "hyperkinesis" or "hyperactivity" as these words are used clinically,
the word "hypermotility" will therefore be used with reference to animal studies. This paper
will provide a review of studies dealing with attempts to produce hypermotile animals, chiefly rodents. In some cases, the studies have been undertaken with the intention of providing
information of potential use to the study and understanding of MBD hyperkinesis in children;
in other studies, the intent has been to elucidate neurochemical and neuroanatomical substrates involved in the control of mortility and pharmacological responses involving changes
in motility.

A distinctive feature of the clinical entity MBD is the use of stimulant medication (e.g. amphetamine and methylphenidate) to produce behavioral quieting and improved learning.

One of the presumed values for the development of animal models of neurologic and psychiatric
diseases is the possibility of developing useful nonhuman models for the development of novel
and safe therapies, particularly drugs. The investigations of the effects of drugs, particularly stimulants, on motility have been extensive and these papers will be included with
specific emphasis on neurochemistry as it applies to hypermotility.

Several reviews of MBD have recently appeared and the reader is referred to these for information on the clinical syndrome (3,4). Further, reviews of motility in animals have also
been published which provide further details on concepts, experiments, methods, and data to
which the reader is also referred (5).

II. MOTILITY

The behavioral entity of motility or locomotor activity in rodents is defined to a great extent by the methods used to measure or record its expression. One source of the differences
which exist in the experimental literature is certainly the diversity of methods and conditions used to detect, measure and define motility in rodents. The conditions and methods
themselves influence the expression of motor activity by animals and indeed different aspects
of behavior may be detected by different methods of measurement (6,7). The discussion in
this section is not intended as a critique or evaluation of the different methods in use,but
rather an attempt to indicate the diversity of measuring conditions and instrumentation and
to suggest, as a precaution when one attempts to compare studies, how these differences might
affect the results reported.

A. Conditions of Measurements

Many environmental conditions will affect the expressions of an animal motility. Most rodents are nocturnal in behavior, in that more activity is emitted during the animal's dark cycle (8). The provision of a regular light-dark cycle, length of the light-dark cycle and housing in groups or in isolation can influence the measured behavior. Locomotion has been shown to be affected by whether the animal's motility is measured individually or in a group; whether all animals within a group being measured have been pretreated in the same (9), whether the animal or group is measured in its home cage or in a novel environment. Measurement is also limited in time and spatial dimensions. Temporally, many studies select only the first few minutes for measurement; other studies choose only a later or longer period. Length of time and period selected influence the results. The longer the time span selected for observation or quantitation, the less variability is usually noted within a group (10). The time period selected also influences results; in a novel environment testing situation, locomotion may be more frequent during the first several minutes after placement in the situation as the animal presumably explores its new environment. After this early time period, activity, particularly during the light cycle for nocturnal animals, may be very low. Under conditions of environmental novelty, it may be difficult to distinguish reactivity from motility. One attempt to accustom animals to testing conditions has been the use of residential testing chambers in which groups of animals reside for long periods of time before and during testing procedures (11).

The spatial dimension of a detected activity is determined by the technology of activity measurement. In many cases, only two dimensional locomotion is exclusively or primarily detected; that is movement in the horizontal plane. Inclusion of "rearing" or use of other measuring techniques (see below) may add a third, vertical dimension to activity measurement. The integration of these three dimensions may be difficult, since it is not clear whether vertical movements are part of the same subset of locomotion as is horizontal movement (12). Also, the physical displacement of the body which is defined as "movement" varies across expreiments. With newer techniques, detection of movements as fine as vibrissa twitching or respiration is possible; it then becomes important for the experimenter to decide, and standardize, what constitutes a definable unit of locomotion.

B. Techniques of Measurement

This subsection will briefly describe some of the common methods of activity measurement used in animal studies. Again, the purpose of this section is not to critically evaluate the various approaches, but to caution that comparisons between studies must recognize the possible impact of different techniques upon the experimental results. Further, this discussion serves to exemplify the multitude of approaches currently used to define hypermotility. For a detailed description the reader is referred to (5).

The techniques can be subdivided into those which involve an observer to collect the data and those which rely upon automated devices which tend to eliminate observer bias and to quantitate the behavior. The observer-dependent methods usually involve limited time-sampling of continuous behavior over a discrete period. With the use of time-lapse photography, time-sampling has been significantly increased in frequency and precision over that possible with personal observation and manual recording (12, 13). Further refinement of this method has concentrated on reducing observer bias by rigorous definitions of behaviors and, very recently, by computer imaging of still photographs from time-lapse cameras automatically monitoring the animals (12).

An early automated device for motility measurement is the "jiggle" cage or stabilimeter. This device, usually employed in studies of groups of animals, detects movements by mechanical disturbance. In some cases, force displacement generated by the animals' movement is recorded by transduction through mechanical-electrical or pressure transducers; in other cases, movements have been detected by stylus and audio amplification.

Human observations and camera recordings have also been employed in quantitative approaches to the assessment of motility. The open field testing apparatus, which is a rectangular enclosure usually about one square meter for rats, provides a standard environment in which motility and other behaviors (rearing, sniffing, defecation, grooming) can be recorded over short continuous periods or by time sampling. The open field provides a spatial definition of locomotor behavior, since crossing of a line on its gridded floor is usually defined as a movement.

The basic principle of the open field—that some physical displacement of the animal's body in the horizontal (and sometimes vertical) plane is defined as a movement—serves as the basis for many of the automated approaches. The amount of displacement which constitutes a movement, and generates one unit of activity electronically, is determined by the design of the measuring device. In photocell-equipped cages, the distance between photocell emitter-detector pairs determines the amount of displacement the animal must move to produce counts. Recent devices using the principle of proximity plates or electromagnetic fields have attempted to increase the flexibility of the prototypic open field. These systems are designed to allow for adjustment by the experimenter of the amount of movement which defines an activity "unit". The more primitive of these systems are adjustable in terms of physical dimensions and speed of movement. The most recent electromagnetic devices are capable of "tracking" the animals in two or three dimensions. By computer processing, the data generated by all movements are separated into different categories based upon the extent and direction of displacement.

Some other methods measure activity emitted by the animal under more defined conditions. Activity in a running wheel, response to a hole board, passage through a tube or doorway connecting two cages, and inter-response time intervals in operant learning paradigms have all been used to provide data on motility. However, under such conditions, it becomes important (and difficult) to distinguish increases in motility from altered cue perception or motivation for reinforcement (5, 14).

Since all of these different measurement systems are complicated by a multitude of controlled and uncontrolled variables, activity must be carefully defined. As the literature now stands, activity is at best, a heterogenous collection of responses that all have as a common basis the physical movement of the animal. However, it should be noted, that measurement of the same animals in different situations can produce hypo, hyper or normal activity in relationship to other measures of activity.

III. Models

A. Spontaneous Motility

Motility has been studied in animals in which surgical, chemical or environmental interventions have been made. Before discussing these models it is appropriate to consider some apparently naturally occurring differences in motility. There are significant intra- and interstrain differences among laboratory rodents in terms of motility. In a study of ovariectomized rats of the same outbred strain, Sparber and Luther (15) found a significant bipolar distribution of door crossings as an index of motility. Segal et al. (16) compared inbred rat strains and found strain-dependent differences in mean motility in an open field.

Another aspect of "spontaneous" hypermotility is the observation that motility varies, roughly on an inverse basis,with age of rodents. Rats from age 5-20 days emit more locomotor activity (as measured in the open field than the same rats after weaning (17). Similar data have been found for mice measured in electromagnetic activity cages (18). This preweaning hypermotility has been called the "popcorn" phase (19), and probably represents a combination of hypermotility and hyperreactivity in young rodents. Hormonal status also affects motility; ovariectomy reduces motility in rats (15).

Motility can also be increased by manipulations of the environmental conditions under which animals are maintained between activity tests. The nutritional influences on motility have been discussed earlier in this conference (20). Increased motility, fighting, and reactivity have been described in rats on restricted postnatal nutrition (21, 22). Food deprivation immediately prior to measurement also tends to increase motility as measured by stabilimeter or running wheel (23). In addition, alteration of light-dark cycles in housing rooms influences motility. When light-dark alternations are reduced to 10 min periods instead of the usual 12 hour periods, the motility in electromagnetic activity cages of rats was significantly entrained during the dark periods and motility increased significantly over normally cycled (12 hr:12 hr) rats (24). Housing conditions also influence motility. When immature rats (3 weeks of age) are housed in isolation for 6-8 weeks, marked behavioral abnormalities are induced, including increases in motility, rearing and sniffing in the open field (25, 26).

B. Lesions

1. Anatomic lesions. Rodents frequently exhibit hypermotility as a result of destruction of several cortical areas. Amygdaloid, hippocampal, globus pallidus, and septal lesions have all produced hypermotile and hyperaggressive animals (7, 27). Changes in motility follow lesions affecting forebrain connections to the reticular formation (23, 28). However, in some cases the changes are detectable only when motility is measured in stabilimeter cages, while in other types of lesions, hypermotility is seen only when measured with running wheels (23). These results suggest that different pathways in the brain may control different aspects of motility which appear only under specific measuring paradigms.

2. Chemical lesions. The chemical agents 6-hydroxydopamine (6-OHDA) and 5,6-and 5,7-dihydroxytryptamine (DHT) have also been employed to lesion specific neurochemical pathways in the brain. Neonatal intracerebroventricular administration of 6-OHDA produces selective depletion of dopamine and norepinephrine-containing neurons (29). Subsequent to this treatment, the animal displays transitory changes in motility, including both hyper- and hypo-

motility (30, 31). When 6-OHDA is administered to mature animals by intraventricular injection, a selective depletion of norepinephrine occurs, and hypermotility in running wheels has been observed several weeks after treatment (32). Administration of 5,6-DHT to mature rats specifically depletes serotonin containing neurons and produces an initial hypermotility followed by a hypomotility (33). When 5,6-DHT is administered neonatally, or when serotonin is depleted by interference with its synthesis using p-chlorophenylalanine or p-chloramphetamine, increased motility has also been described (34, 35, 36). Contrary to the above, it has been shown that increased levels of serotonin (achieved by inhibition of monoamine oxidase and administration of tryptophan) or increased levels of tryptophan are associated with increased motility (37, 38, 39). However, this behavior, which has been proposed as being specifically serotonergic is also seen in rats after lesions of serotonergic terminals upon administration of tryptophan. This may actually represent a very complex behavior quite different from hypermotility (40).

The temporal nature of behavioral effects of the 6-OHDA and DHT is important to note; this characteristic has been cited as a point of similarity between 6-OHDA and childhood hyperkinesis (31). Changes in motility may be associated with time-dependent changes in neurotransmitter function after chemical lesioning. The first phase of behavioral alterations may be associated with release of transmitter from the damaged terminals; in the next phase, with loss of releasable transmitter from the destroyed terminals; the next phase with receptor supersensitivity subsequent to prolonged presynaptic denervation; and finally, with sufficient time, there may be even a final phase of behavior associated with neuronal regeneration which has been described for both 6-OHDA (41) and 5,6-DHT (42, 43). The temporal changes in the nature of neurochemical alterations consequent to chemical lesioning make it important to correlate observed changes in motility with the neurochemical status of the animal at the time of behavioral measurement.

C. Neuropharmacological Hypermotility: Amphetamine and Methylphenidate

Increases in motility have been produced by many drugs; examination of amphetamine and methylphenidate is of interest for several reasons. First, there has been substantial research on the anatomical and neurochemical substrates for their behavioral effects, and second these two drugs are currently the drugs of choice in treating hyperkinesis in children (44). Indeed, the therapeutic effect of these stimulants on hyperkinetic children constitutes one of the most interesting and provocative aspects of the syndrome (1).

Response of animals to amphetamine depends upon environment, conditions of measurement, dose, rate-dependence of behavior, and neurochemical integrity of the monoaminergic pathways of both mesolimbic and nigrostriatal areas. Rats habituated in a Y-maze show little or no stimulation of motility after amphetamine in distinction to naive rats (45). During the dark cycle, rats (in a residential maze with photocells) show less response to amphetamine at doses which produce marked hypermotility during the day (8). Also, the amphetamine dose-response curve of motility in rodents is not unimodal above a certain dose, relatively less hypermotility is elicited and stereotypic behavior begins to appear. In studies where specific behaviors have been monitored (46), high doses of amphetamine appear to disrupt patterning of behavior and to increase the incidence of certain behaviors over others.

Extensive research has been undertaken to elucidate the neurochemical substrates of amphetamine-induced hypermotility. The anatomical and synthetic integrity of presynaptic catecholamine pathways in the nigrostriatal system appears essential. Biochemical lesioning of these tracts with 6-OHDA and electrolytic lesions of the substantia nigra cell bodies suppress responses to amphetamine (47, 48, 49). Specifically, intact dopaminergic function appears important (50); in animals with noradrenergic depletion, potentiated response to amphetamine has been noted (32). Generally, it appears that increased cholinergic and GABA-ergic function antagonize the effects of amphetamine and methylphenidate (see below). Inhibition of serotonergic function potentiates response to methylphenidate, while MAO inhibition may partially antagonize stimulant response (36). The varying roles played by the monoaminergic neurotransmitters in the control of motility and response to amphetamine are difficult to distinguish. A recent review (51) has indicated the difficulties and suggested that the interactions of dopamine and norepinephrine vary with the behavioral state of the animal and with the relative activation of each pathway.

D. Hyperthyroidism

Administration of l-triiodothyronine to neonatal rats produces a hyperthyroid condition (52). Neonatal hyperthyroidism is associated with increased motility; conversely, chemical thyroidectomy with the antithyroid agent methimazole induces motility (in electromagnetic activity cages) following neonatal treatment (53). This interaction between altered thyroid function at critical periods of development and subsequent level of motility may be mediated by TRH which when injected intracerebrally produces increased motility (54, 55). Alternatively, these effects may be mediated by monoaminergic pathways (see below).

E. Anoxia and Carbon Monoxide

Deprivation of oxygen during perinatal development produces specific lesions of the CNS. Prolonged or preparturition barbiturate-induced anoxia are associated with hypomotility and
lethargy (56). Further, exposure of neonatal rats to carbon monoxide to the point of respiratory failure produces a hypermotility during the first few weeks of life. This hypermotility is completely reversed by 5 months of age (57).

F. Viral Exposure

The behavioral toxicology of neuroviruses is a field of recent study. Terzin (58) has suggested that altered motility may be a sensitive indicator of CNS damage by infection before
appearance of overt symptoms. Intracerebral innoculation of mice with hamster neurotrophic
measles virus produces an extreme hyperactivity accompanied by hypermotility (59). Hypermotility has also been reported following treatment with Newcastle Disease Virus. Altered
motility has been reported in rats innoculated with lymphocytic choriomeningitis (LCM) virus,
with both hypomotility (60) and hypermotility (61) reported. LCM rats showed significant
difference in acquisition of 2 learning tasks. The nature of the specific deficits suggested hypermotility (62). The alterations in behavior and the pathology depend upon age of
infection. However, changes in motility associated with LCM virus may represent a different
motor dysfunction, similar to cerebellar lesions or epilepsy.

G. Lead

Lead poisoning in children has a long history of associated neurological damage to both the
peripheral and central nervous system. Animal studies dealing with the behavioral and neurochemical effects of lead are very recent. Hypermotility has been reported in rodents exposed
to lead during the first few days of life. However, the animal data published to date have
not been consistent. The lack of consistency in all of the models discussed may have resulted from the many unknown or uncontrolled factors. In the case of lead, the route and
level of exposure have differed from study to study and the timing of lead exposure has not
been uniform. As a result, the internal dose at sites of toxic action in the CNS may vary
considerably. Further, the nutritional status has varied and the methods used to determine
activity have included most of the methods discussed previously. However, even with these
inconsistencies there has been a general tendency in the data that suggests that lead administration to the developing rodent produces hypermotility (18, 63, 64, 65). Further, it
should be noted that the altered response to amphetamine originally observed in hypermotile
mice (66) has been observed in the rat independent of the appearance of hypermotility (67).

At the levels of exposure used in most of the above studies, there have been no overt symptoms of lead encephalopathy such as cerebral edema or cerebellar hemorrhage. The cerebellar
hemorrhage generally associated with high level exposure in the rat has not been reported at
any dose used in the mouse (68, 69). However, a decrease in growth rate and delays in specific developmental landmarks have been observed (18, 70). It should be noted that the decrease in growth rate seen in the early studies can be prevented by limiting the number of
pups with each lactating dam to three of either rats or mice (71).

The pharmacology of lead-induced hypermotility has been studied mainly in the mouse (66, 72)
and these results have been summarized (73). Briefly, lead-induced hyperactivity is decreased by amphetamine, methylphenidate, dimethylaminoethanol and physostigmine; conversely
it is exacerbated by phenobarbital, I-dopa and benztropine.

It is interesting to note that in all of these studies one of the most consistent observations has been in the increased variability of the pharmacological reponse in the treated
group, i.e. the increase of the S.D. as a % of the mean. Further, this has been seen by
others (74). In a more general way, increased variability of pharmacological response has
been observed in many developmental studies of behavior following toxicological intervention
(75).

IV. CLINICAL CORRELATIONS

Several of the experimental models discussed above are intentional reproductions in animals
of conditions known to occur clinically. Indeed, some of these have been suggested as possibly etiologic in the production of MDB hyperkinesis in children (3). The observation that
the young in many species are more active than adults supports the hypothesis that hyperkinesis may be an aspect of neurodevelopmental immaturity rather than a result of pathology
or organic brain damage (76). Hyperthyroidism, as well as hypoglycemia, have also been suggested as underlying metabolic conditions producing behavior disorders in children.

The association of viral damage and anoxia with hyperkinesis was proposed early in the clinical literature on this condition (3). Many of the symptoms of MBD were first described in
children surviving the epidemic of von Economo's encephalitit in 1918. It should be noted,

however, that none of the paramyxoviruses used in the animal studies discussed above have been associated clinically with MBD. A relatively early observation of these symptoms was also noted in children suffering from anoxia trauma during birth (77). Finally, there have been descriptions of children surviving early exposure to lead which resemble the descriptions of hyperkinesis (78,79). A prospective study reported an increased frequency in the diagnosis of hyperkinesis and associated symptoms of behavioral abnormalities in groups of children with evidence of increased lead absorption but no overt symptoms of lead poisoning (80, 81). Also, retrospective evidence for increased lead absorption has been reported in children diagnosed as hyperkinetic for whom no certain etiologic factors are discernible (82). However, in this study, it may be difficult to separate cause and effect, since more active, less controllable children might be expected to exhibit more pica and thus consume more lead.

This discussion should not be construed to imply that any of these agents or conditions that produce hypermotility in animals is causally linked to the clinical syndrome of MBD. There is a great need for epidemologic studies dealing with the etiologies of MBD.

V. NEUROCHEMISTRY OF EXPERIMENTAL MODELS OF HYPERKINESIS

An analysis of the neurochemical basis of hypermotility is divided into a consideration on the neuropharmacology and the neurochemistry of various animal models. In this discussion, we will only consider acetylcholine, the monoamines and gamma aminobutryic acid. Again, as in the clinical correlates of the etiologies of MBD little information is available regarding the neurochemical changes seen in children (1).

A. Acetylcholine

Evidence for the possible involvement of cholinergic influences on paradigms of increased motility has been reviewed by Silbergeld and Goldberg (73). Cholinergic involvement in cases of spontaneous hypermotility has been suggested by the finding of decreased levels of acetylcholine (ACh) in the diencephalon of more active rats (83). Septal lesions resulting in hypermotility also have been associated with decreased cholinergic function, as reflected in lowered cortical ACh levels and decreased activity of choline acetyltransferase (84, 85, 86). Pharmacologically, enhancement of cholinergic function by inhibition of transmitter degradation with physostigmine was shown to suppress septal lesion induced hypermotility (87); moreover, an increased depressant response to arecoline was seen in animals hypermotile after lesions of the globus pallidus or substantia nigra (88). Lesion by x-irradiation also affects cholinergic function, reflecting possible compensation by presynaptic cholinergic terminals to loss of postsynaptic receptor sites through specific destruction following prenatal irradiation (89, 90).

Amphetamine and methylphenidate influence cholinergic function (91, 92). Increasing the functional activity of cholinergic pathways by physostigmine antagonizes the hypermotility and stereotypy induced by amphetamine and methyphenidate, while inhibition of cholinergic function by scopolamine potentiates the response to amphetamine (93, 94).

The possibility that hyperthyroid associated hypermotility involves altered cholinergic function is unclear and indirect. The available studies indicate only that in hypothyroid rats ACh levels are elevated by I^{131} destruction of the thyroid, and are unchanged by methimazole thyroidectomy (53, 95, 96).

Several studies have indicated that lead-induced hypermotility is associated with inhibition of cholinergic function. In vitro and in vivo exposure of the CNS to inorganic lead produces evidence of inhibited cholinergic function. Carroll et al.(92) have shown that cortical minces prepared from lead exposed mice release significantly less ACh upon potassium depolarization than minces prepared from age-matched control mice. Reductions in regional ACh levels have been reported in lead-treated rats (97). However, this has not been a consistent finding (98). The turnover of ACh in four brain regions--caudate, cortex, hippocampus and midbrain--is reduced by chronic lead treatment of rats (98). In vitro exposure of brain tissue (synaptosomes) to lead produces reduction in high affinity binding of the precursor choline (1, 72). However, this is complicated by the observation that minces from lead treated animals show no changes in either high or low affinity systems (92). Also, in vitro exposure of peripheral cholinergically innervated tissue to lead has been demonstrated to produce reductions in release of ACh consequent to electrical stimulation of the nerve trunk (99, 100, 101). Pharmacologically, the responses of lead-treated hypermotile mice suggest that decreases in cholinergic function play a role in hypermotility. Lead-induced hypermotility is reduced by physostigmine and oxotremorine, agents with cholinergic agonist properties and is exacerbated by the anticholinergics, atropine and benztropine. Dimethylaminoethanol also reduces hypermotility; however, the effects of this drug on cholinergic function are not yet clearly understood (102, 103).

In addition, it should be noted that alteration of the cholinergic system by direct pharma-

cological intervention is associated with changes in motility (104). A low dose of physostigmine intraperitoneally (50 ug/kg) produces hypermotility (105) while intracerebral injections
of atropine or the nicotinic blocker d-tubocurarine also produce hypermotility (106, 107).
The studies of unilateral injections of diisopropylflurophosphate, carbachol and ACh might
be interpreted to indicate a lateralized expression of increased motility following inhibition of cholinergic function such that the animal rotates contralateral to the site of injection (108).

B. Monoaminergic Transmitters - Dopamine, Norepinephrine and Serotonin

The separation and assignment of precise roles for each of the monoamine neurotransmitters
in control of behavior is very difficult (51). The use of chemical agents has helped to distinguish these processes, but investigation of these treatments has revealed that in some
cases the neurochemical depletions of the various transmitter compounds are not as specific
as originally supposed (29, 49). No clear interactive system prevails and indeed the relationship among these systems probably depends upon the level of functioning of these systems
in concert. However, the fact that the monoamines exert important effects on motility has
been indicated by many studies.

In spontaneous hypermotile animals, based on intra and interstrain separation of more and
less active rats, the more active groups have been shown to have indications of increased
monoaminergic receptor activity and possibly associated decreases in levels of dopamine and
activity of tyrosine hydroxylase (15, 16, 109, 110). The causal sequence of these effects--
presynaptic impairment and consequent postsynaptic receptor supersensitivity, or postsynaptic proliferation and consequent presynaptic inhibition--is unclear. Of relevance is the
production of hypermotility by chronic reserpinization, and the finding of an increased response of norepinephrine-stimulated adenylate cyclase (111). This study suggests that amplification of monoaminergic neurotransmission by supersensitive receptor-associated cyclase
may be involved with hypermotility. Similar results have been shown with response to l-dopa
in mice pretreated with reserpine and α-methyl-p-tyrosine, in which an altered receptor response has also been postulated (112).

In rodents made hypermotile by prolonged isolation, there is evidence that the hypermotility
is associated with an induction of increased catecholaminergic function. Isolation-induced
hypermotility is accompanied by increased urinary excretion of the norepinephrine metabolite
3-methoxy,4-hydroxyphenylglycol (MHPG) (113). Further, the beta-blocker propanolol reduces
both hypermotility and increased MHPG excretion (113). The hypothesis that increased noradrenergic function might be correlated with increased motility is further supported by the
report that intracerebral and intraventricular injections of norepinephrine stimulate motility (109, 114).

Use of chemical or anatomic lesions to produce hypermotility has shown that the monoamines
interact in control of motility. Septal-lesion hypermotility is associated with decreased
levels of norepinephrine and dopamine, although no change in turnover rates was observed
(115). Anatomic lesions of the substantia nigra, an area rich in cell bodies for dopamine-
containing tracts, also produce hypermotility (88). This would suggest that the removal of
the nigrostriatal dopamine pathway produces increased motility. However, increased motility
follows injection of dopamine or d-amphetamine into the nucleus accumbens and after injection of dopamine, d-amphetamine or apomorphine into the tuberculum olfactorium, both of
which are important constituents in the mesolimbic dopamine system (116). Since no stimulation of motility was found with intrastriatal injections of dopamine (117), this suggests
that control of motility by the monoamines may vary among different anatomic pathways.

Some evidence exists to suggest that behavioral changes consequent to viral infection of the
CNS involve alterations in monoaminergic systems. Guchhart and Monjan (118) have shown that
treatment of rats with lymphocytic choriomeningitis virus produces changes in noradrenergic
function, as reflected in increased COMT activity and decreased norepinephrine levels in the
hindbrain.

In hyperthyroid hypermotile rats, there appears to be an increased synthesis and turnover
of catecholamines (52) while hypothyroid, hypomotile rats appear to have decreased levels
and synthesis of norepinephrine and dopamine. (53).

In experimental and clinical lead poisoning, there is neurochemical evidence for increased
monoaminergic function. Increased steady state levels of brain norepinephrine have been reported in lead-treated mice and rats (72, 119). However, this has not been a consistent
finding. Similarly inconsidtent results have been reported for NE turnover (120, 121). Increased urinary excretion of the catecholamine metabolites have been reported in lead exposed mice and children (122). The effects of lead on serotonergic function have not been
extensively studied, although there is one report suggesting increased urinary excretion of
5-hydroxy indoleacetic acid in lead-exposed animals and children (123). In vitro, exposure
of brain tissue to lead has been shown to increase the activity of tyrosine hydroxylase

(124), and to increase calcium-dependent release of dopamine (125). Thus both norepinephrine and dopamine pathways appear to be activated by lead exposure. The neuropharmacologic response of lead-treated hypermotile rodents is less illuminating. Drugs thought to be monoaminergic agonists or to act through increasing monoamine release tend to reduce lead-induced hypermotility (d- and l-amphetamine, methylphenidate) while lead-induced hypermotility is exacerbated by l-dopa, apomorphine, and benztropine, drugs which are also thought to act by increasing monoamine receptor stimulation (72). An indication that monoaminergic function is involved in lead-induced hypermotility is the observation that inhibition of monoamine synthesis by α-methyl-p-tyrosine not only reduces lead-induced hypermotility but also restores a normal, stimulatory response to d-amphetamine in lead-treated mice (61, 72).

C. Gamma Amino Butyric Acid

Gamma amino butyric acid (GABA) has been proposed as a major inhibitory neurotransmitter in the central nervous system (126). It is not surprising that decreases in GABAergic function have been suggested as the biochemical basis for hyperkinesis (127). No studies of GABA function in spontaneously hypermotile animals have been conducted. However, in rodents where motility is stimulated by ethanol, this hypermotility is suppressed by increasing GABAergic function (128). Ethanol-induced motility is thought to be associated with the dopamine-releasing properties of ethanol (128, 129); the reduction of ethanol motility by GABAergic drugs suggests a dopamine-GABA interaction associated with the expression of motility. In another paradigm of drug-induced hypermotility (treatment of rats with tetrabenzine and iproniazid and training to a Sidman shock-avoidance schedule), steady state levels of GABA in excited rats were increased in telencephalon, diencephalon, mesencephalon, and cerebellum (130). In the absence of turnover studies, it is, however, difficult to interpret these results. They may suggest a connection between excitation-induced release of catecholamines and decreased release of GABA, which is reflected in increased neuronal levels. Further, it has been observed that increasing GABAergic function by sparing of released transmitter or receptor stimulation produces a sedated animal (131).

The direct effects of lead treatment on GABA have also been studied. The first investigations reported no changes in steady state whole brain levels of GABA (63) or synaptosomal uptake of GABA (72), but regional studies have uncovered significant changes in GABA levels (132, 133). The evidence taken from animal studies of steady state levels, enzyme activities of glutamic acid decarboxylase and GABA transaminase, GABA turnover, and uptake and release of GABA, all indicate a significant inhibition of GABAergic function in lead-treated animals, which is probably mediated at the receptor level (133).

VI. TOWARDS A NEUROCHEMICAL MODEL OF HYPERMOTILITY

This review has described some of the experimental paradigms producing hypermotility in rodents; some of the neuropharmacological and neurochemical characteristics of these paradigms have also been presented. An important question arises: considering the variety of means and conditions by which hypermotility is produced, are there consistent or common mechanism (s)? As a corollary to this question, is it possible to develop a general hypothesis of the underlying neurochemistry of increased motility in animals? Any discussion of these points is necessarily limited by the incomplete investigations done to date. Complete neurochemical studies of all experimental conditions producing hypermotility have not been conducted. Moreover, variations in treatment and measurement conditions make comparisons and cross-experimental generalizations difficult.

Many of the experimental conditions producing hypermotility appear to involve alterations in at least monoaminergic, cholinergic, and GABAergic function. In addition, there is considerable evidence suggesting anatomical and chemical interactions among these pathways in control of behavior and motility. Much of the data reviewed here are consistent with the hypothesis that hypermotility may be associated with increased functioning of one or more monoaminergic pathways and decreased functioning of cholinergic and GABAergic systems.

The substrate of increased function may be presynaptic, as in the increased release of dopamine and inhibition of reuptake inactivation mechanisms by amphetamine and methylphenidate. It may be associated with inhibition of enzymatic degradation, as in the hypermotility associated with trancylpramine-tryptophan loading; or it may be postsynaptic, as in the increased responsiveness of norepinephrine-stimulated adenylate cyclase seen in reserpinized hypermotile rats or spontaneously more active rat strains. However, some of the lesion studies do not appear consistent with this hypothesis. Chemical destruction of dopamine pathways by neonatal 6-OHDA and electrolytic lesioning of dopamine cell bodies in the substantia nigra have both been reported to produce at least a transitory hypermotility in animals. Interpretation of these results must recognize the possibility that the lesions are not neurochemically or anatomically specific. Further, there is a temporal dependence of postlesion behavior which varies with the development of neuronal damage, compensatory supersensitivity and possible regeneration.

A problem in accepting the hypothesis of increased monoaminergic neurotransmission as the only neurochemical substrate of hypermotility, and by extension of MBD hyperkinesis, is the apparent discrepancy between this hypothesis and the observed therapeutic effects of amphetamine and methylphenidate. These drugs are generally thought to act by increasing catecholaminergic function. However, amphetamine and methylphenidate also possess actions on cholinergic function since both release ACh (134). In addition, methylphenidate can affect the activity of neurons in the reticular formation by mechanisms which are similarly affected by cholinergic agonists and blocked by cholinergic antagonists (91). Thus, an explanation of this apparent discrepancy between monoaminergic dysfunction and pharmacologic response may be provided by examining the cholinergic system in hypermotility.

Many of the experimental models appear to show evidence of decreases in cholinergic function. This has been particularly well-described in the investigations of lead-induced hypermotility where consistent decreases in acetylcholine release and turnover have been reported. Cholinergic dysfunction has also been described in septally lesioned hypermotile rats. However, studies of pharmacological manipulation of central cholinergic function are less clear, with hypermotility reported consequent to treatment with both physostigmine and curare which are presumably producing opposite effects on cholinergic receptor activity. In addition, there appears to be an important interaction of cholinergic function with the effects of amphetamine and methylphenidate on motility. Enhancement of cholinergic transmission antagonizes stimulation by amphetamine and methylphenidate while cholinergic antagonism potentiates the effects of these two stimulants as well as other treatments which increase motility through monoaminergic hyperactivity (135). The status of the cholinergic nervous system consequent to administration of monoaminergic neurotoxins has not been fully examined. However, decreased AChE and increased ACh in the striatum after 6-OHDA has been reported (136).

Finally, the possibility that an important chemical substrate of inhibition GABA, has been altered has received relatively less investigation in paradigms of hypermotility. In ethanol-induced hypermotility, enhancement of GABAergic function reduced motility, while GABAergic enhancement itself is associated with behavioral sedation. Electrolytic lesioning of areas rich in cell bodies of GABAergic pathways -- the caudate and globus pallidus -- has been reported to produce hypermotility. In experimental lead poisoning, decreases in GABAergic activity at the receptor level appear to be associated with hypermotility.

This discussion of possible neurochemical bases for hypermotility has been simplified by an arbitrary restriction to only a few neurotransmitter candidates. Certainly, many other as yet unexplored chemical pathways may participate in the control of motility, and alterations in their functioning may be revealed as the fundamental neurochemical defect associated with both experimental and clinical behavioral alteration. Further, this discussion has proceeded on an assumption that the transmitter pathways behave invariantly in their influence on motility and in their interactions with each other. Actually, there is increasing evidence for anatomically determined differences in certain transmitters, particularly for their effects on motor behavior. The relative importance of mesolimbic versus nigrostriatal dopamine pathways is an example of this possible complexity. Chemical destruction by local injection of 6-OHDA or lesioning by dopaminergic neurons in the medial ventral tegmentum produces hypermotility (137). This is in contrast to the hypomotility usually found in animals with 6-OHDA destruction of the substantia nigra (138). Thus, anatomically determined differences in pathways utilizing the same neurotransmitters may help in explaining some of the apparent discrepancies noted above.

VII. CONCLUSIONS

Are there animal models of childhood hyperkinesis? All of the methods discussed have only considered the aspect of motility and have not examined the other symptoms associated with MBD. One of the reasons for this has been the lack of clear definition of this syndrome both clinically and in animal models in terms that can allow well controlled experimental studies. A second limitation of the available data is the absence of correlative studies of behavior, chemistry, and pharmacology of the models. A third consideration is our lack of understanding of those other, as yet undefined variables, that interact with the methods to produce hypermotility. In the experimental studies, the behavioral and neurochemical pharmacology has not been directed, in most cases, towards using amphetamine or methylphenidate responses as criteria for homology with MBD. Buchsbaum and Wender (139) have proposed that a nonstimulatory behavioral response or an altered visual or auditory evoked response to these drugs is diagnostic in MBD. Since only a limited percentage (60-80%) of MBD children are reported to respond "therapeutically" to these drugs (44), and the response of normal children is largely unknown, the identification of this syndrome with the pharmacological response may only be a first step in approaching this problem in experimental models.

Of the potential models for MBD, several offer unique opportunities to study both the chemistry and pharmacology of this syndrome. The seductiveness of some of these models derives from the clinical correlation, albeit limited, with the model system. The identification and development of appropriate animal models depends upon the correlation of these models with the neurochemistry of the human disease. For this reason further studies are needed on the precise behavioral alterations and associated biochemical characteristics of MBD, both in experimental animal studies and in the human.

VIII.   REFERENCES

1.   Silbergeld, E.K. (1977) Neuropharmacology of Hyperkinesis in Essman, W. and Valzelli, L. (Ed.) Current Developments in Psychopharmacology 4, Spectrum, New York.

2.   Werry, J. S. Pediat. Clin. N. Amer. 15:3, 581-599 (1968).

3.   de la Cruz, F., Fox, B.H. and Roberts, R. H. (Ed.) Minimal Brain Dysfunction Ann. N. Y. Acad. Sci. 205 (1973).

4.   Wender, P. H. (1971) MBD in Children, John Wiley, New York.

5.   Gross, C. G. (1968) Weiskrantz, L. (Ed.) Analysis of Behavioral Change, Harper and Row, New York.

6.   Iversen, S.D. and Iversen, L.L. (1975) Behavioral Pharmacology, Oxford University Press, Oxford.

7.   Norton, S. Brain Res. Bull 1, 193-202 (1976).

8.   Reiter, L. W.,Anderson, G. E., Laskey, J. W. and Cahill, D.F.,Envir. Health Persp. 12 119-124 (1975).

9.   Campbell, B.A. and Randall, P.J., Science 195, 888-891 (1977).

10.   Silbergeld, E. K. and Aylmer, C.G.G., unpublished observations.

11.   Norton, S., Culver, B. and Mullenix, P., Behav. Biol. 15, 317-331 (1975).

12.   Norton, S., this book.

13.   Norton, S., Mullenix, P. and Culver, B. Brain Res. 116, 49-67 (1976).

14.   Morrison, J., Olton, D.S., Goldberg, A.M. and Silbergeld, E.K., Dev. Psychobiol. 8(5), 389-396 (1975).

15.   Sparber, S.B. and Luther, I.G. Neuropharmacol. 9, 243-247 (1970).

16.   Segal, D.S., Kuczenski, R. and Mandell, A.J., Beh. Biol. 7, 75-81 (1972).

17.   Campbell, B.A., Lytle, L.D. and Fibiger, H.C., Science 166, 635-636 (1969).

18.   Silbergeld, E.K. and Goldberg, A.M., Life Sci. 13, 1273-1283 (1973).

19.   Makar, H.S., Lehrer, G.M. and Silides, D.J., Envir. Res. 10, 79-91 (1975).

20.   Michaelson, I.A., Bornschein, R.L., Loch, R.K., and Rafales, L.S., this book.

21.   Barnes, R.H., Neely, C.S., Kwong, E., Labadan, B.A. and Frankova, S., J. Nutr. 96, 467-476 (1968).

22.   Dobbing, J. and Smart, J.L.,Br. Med. Bull.30, 164-168, (1974).

23.   Lynch, G.S., J. Comp. Physiol. Psychol. 70, 48-49 (1970).

24.   Borbely, A., Brain Res. 114, 305-317 (1976).

25.   Weinstock, M., and Speiser, G., Psychopharmacol.30, 241-250 (1973).

26.   Denenberg, V.H., Morton, R.C., Haltmeyer, G.C., Animal Behav. 12, 205-208 (1964).

27.   Smythies, J.R., Brain Mechanisms and Behaviors. Academic Press, New York (1970).

28.   Millichap, J.G. (1974) in Vernadakis and Weiner (eds.), Drugs and the Developing Brain 475-488, Plenum, New York.

29. Jacobowitz, D.M. (1973) in Usdiv, E. and Synder, S. (Eds.) Frontiers in Catacholamine Research. 727-739, Pergamon Press, New York.

30. Iversen, S.D. and Creese, I.N. Adv. Neurol. 9, 81-92 (1975).

31. Shaywitz, B.A., Yager, R.D., and Klopper, J.H. Science 191, 305-307 (1976).

32. Sorensen, A., and Ellison, G.,Psychopharmacol. 32, 313-325 (1973).

33. Ellison, G. Brain Res. 103, 81-92 (1976).

34. Brase, D.A., and Loh, H.H. Life Sci. 16, 1005-1016 (1975).

35. Jacobs, B.L., Trimbach, C., Eubanks, E.E. and Trulson, M., Brain Res. 94, 253-261 (1975).

36. Breese, G.R., Cooper, B.R., and Hollister, A.S., Psychopharmacol. 44, 5-10 (1975).

37. Grahame-Smith, D.G., J. Neurochem. 18, 1053-1066 (1971).

38. Green, A.R., and Grahame-Smith, D.G., Neuropharmacol. 13, 949-959 (1974).

39. Foldes, A., and Costa, E., Biochem. Pharmacol. 24, 1617-1621 (1975).

40. Stewart, M., and Baldessarini, R., this book.

41. Bjorklund, A., Nobin, A., and Stenevi, U., Brain Res. 50, 214-220 (1973).

42. Nygren, L., Olson, L., and Sieger, A., Histochem. 28, 1-15 (1971).

43. Baumgarten, H.G., Lachenmayer, L., Bjorklund, A., Nobin, A. and Rosengren, E., Life Sci. 12, 357-364 (1973).

44. Millichap, J.G., Ann. N.Y. Acad. Sci. 205, 321-334 (1973).

45. Rushton, R., and Steinberg, H. (1964) in Steinberg (ed) Animal Behavior and Drug Action, 207-218, Churchill, London.

46. Norton, S., Physiol. Behav. 11, 181-186 (1973).

47. Creese, I., and Iversen, S.D., Brain Res. 55, 369-393 (1973).

48. Simpson, B.A., and Iversen, S.D., Nature 230, 30-32 (1971).

49. Breese, G.R. (1976) in L.L. Iversen, S.D. Iversen and Synder, S. (Ed.) Handbook of Psychopharmacol. Vol. I, 137-189, Plenun Press, New York.

50. Thornburg, J.E., and Moore, K.E., Neuropharmacol. 12, 853-866 (1973).

51. Antelman, S.M., and Caggiula, A.R., Science 195, 646-653 (1976).

52. Rastogi, S., and Singhal, R., J. Pharmacol. Exp. Therap. 198, 609-618 (1976).

53. Rastogi, S., Lapierre, Y., and Singhal, R., J. Neurochem. 26, 443-449 (1976).

54. Cohn, M. and Cohn, M.L., Neurosci. Abst., 669 (1976).

55. Breese, G. R., Cott, J.M., Cooper, B.R, Prange, A.J., Lipton, M.A. and Plotnikoff, N. J. Pharmacol. and Exp. Ther. 193, 11-22 (1975).

56. Windle, W.F., and Becker, R.F., Am. J. Obstet. Gynecol. 45, 183-200 (1943).

57. Culver, B., and Norton, S., Exper. Neurol. 50, 80-98 (1976).

58. Terzin, A.L., Brit. J. Exp. Pathol. 49, 107-118 (1968).

59. McFarland, H., and Silbergeld, E.K., unpublished observations.

60. Hotchin, J., and Seegal, R., Science 196, 671-674 (1977).

61. Monjan, A., personal communication.

62. Monjan, A., Gohl, L.S., Hudgens, G.A., Bull. World Health Organ. 52, 487-493 (1975).

63. Sauerhoff, M.G., and Michaelson, I.A., Science 182, 1022-1024 (1973).

64. Shih, T-M, Khachaturian, Z.S., Reisler, K.L., Rizk, M.M. and Hanin, I., Fed. Proc. 35, 307 (1976).

65. Domer, F., and Lhera, J-C, Fed. Proc. 36, 1045 (1977).

66. Silbergeld, E.K., and Goldberg, A.M., Exp. Neurol. 42, 146-157 (1974).

67. Sobotka, T., and Cook, M., Am. J. Ment. Defic. 79, 5-9 (1974).

68. Rosenblum, W., and Johnson, M., Arch. Path. 85, 640-652 (1968).

69. Silbergeld, E.K., and Goldberg, A. M., Environ. Health Perspec. 7, 227-232 (1974).

70. Maker, H.S., Lehrer, G.M., and Silides, D.J., Envir. Res. 10, 79-91 (1975).

71. Goldberg, A.M., unpublished observations.

72. Silbergeld, E.K., and Goldberg, A.M., Neuropharmacol. 14, 431-444 (1975).

73. Silbergeld, E.K., and Goldberg, A.M., (1976) in Hanin, I. and Goldberg, A.M. (Eds.) Biology of Cholinergic Function, 619-645, Raven Press, New York.

74. Breese, G.R., personal communication.

75. Sparber, S.B. and Shideman, F.E., Dev. Psychobiol. 2, 56-59 (1969).

76. Kinsbourne, M., Ann. N.Y. Acad. Sci. 205, 268-273 (1973).

77. Strauss, A.A., and Werner, H., Am. J. Psychiatry 97, 1194-1202 (1941).

78. Byers, R.K. and Lord, E., Am. J. Dis. Children 66, 471-494 (1943).

79. Waldron, H.A., and Stoten, D.S., (1974) Subclinical Lead Poisoning, Academic Press, New York.

80. De la Burde, B., and Choate, M.S., J. Pediat. 87, 638-642 (1975).

81. Baloh, R., Sturm, R., Green, B., and Gleser, G., Arch. Neurol. 32, 326-330 (1975).

82. David, O., Clark, J., and Voiller, K., Lancet II, 900-902 (1972).

83. Benesova, O., and Benes, V., Activa Nervosa Superior (Praha) 16, 816-827 (1974).

84. Pepeu, G., Mulas, A., Ruffi, A., and Sotgui, P., Life Sci. 10, 181-184 (1971).

85. Sorensen, J.P. and Harvey, J.A., Physiol. Behav. 6, 723-725 (1971).

86. Lewis, P.R., Shute, C.C.D., and Silver, A., J. Physiol. 191, 215-224 (1967).

87. Stark, P., and Hendersen, J.K., Neuropharmacol. 11, 839-847 (1972).

88. Costall, B., Naylor, R.J., and Olley, J.E., Neuropharmacol. 11, 317-330 (1972).

89. Valcana, T., Liao, C., and Timiras, P.S., Environ. Physiol. Biochem. 4, 47-63 (1974).

90. Valcana, T., Liao, C., and Timiras, P.S., Brain Res. 73, 105-120 (1974).

91. Shih, T-M, Khachaturian, Z.S., Barry, H., and Hanin, I., Neuropharmacol. 15, 55-60 (1976).

92. Carroll, P.T., Silbergeld, E.K. and Goldberg, A.M., Biochem. Pharmacol. 26, 397-402 (1977).

93. Fjalland, B., and Møller Nielsen, I., Psychopharmacol. 34, 111-118 (1974).

94. Klarvans, H.L., Rubovits, R., Patel, B.C., and Weiner, W.J., J. Neurol. Sci. 17, 303-308 (1972).

95. Singhal, R., Rastogi, S., and Hrdina, P., Life Sci. 7, 1617-1626 (1976).

96. Valcana, T. (1971) in Ford (Ed.) Influence of Hormones on the Nervous System, 174-184, S. Karger, New York.

97. Modak. A., Weintraub, S., and Stavinoha, W., Toxicol. Appl. Pharmacol. 34, 340-347 (1975).

98. Shih, T-M, and Hanin, I., Fed. Proc. 36, 977 (1977).

99. Kostial, K., and Vouk, V.W., Br. J. Pharmacol. 12, 219-222 (1957).

100. Silbergeld, E.K., Fales, J.T., and Goldberg, A.M., Neuropharmacol. 13, 795-801 (1974).

101. Manalis, R.S., and Cooper, G., Nature 243, 354-356 (1974).

102. Zahniser, N., Chou, D. and Hanin, I., J. Pharmacol. Exp. Ther. 200, 545-559 (1977).

103. Goldberg, A.M. (1977) Dis. Nervous System, In Press.

104. Hingtgen, J.N., and Aprison, M.H. (1976) in Goldberg, A.M. and Hanin, I. (Eds.) Biology of Cholinergic Function, 515-566, Raven Press, New York.

105. Wray, S.R., Neuropharmacol. 15, 269-271 (1976).

106. Leaton, R.N., and Rich, R.H., Physiol. Behav. 8, 839 (1972).

107. McKenzie, G.M., Gordon, R.J. and Viik, K., Brain Res. 47, 439-456 (1972).

108. Myers, R.D. (1975). Handbook of Drug and Chemical Stimulation of the Brain. Van Nostrand and Reinhold, New York.

109. Segal, D.S., Geyer, M.A. and Weiner, B.E., Science 189, 301-303 (1975).

110. Skolnick, P., and Daly, J.W., Science 184, 175-177 (1974).

111. Williams, B.J., and Pirch, J.H., Brain Res. 68, 227-234 (1974).

112. Ahlenius, S., Anden, N-E, and Engel, E., Brain Res. 62, 189-199 (1973).

113. Speiser, A., and Weinstock, M., Pharmacol. Biochem. Behav. 4, 531-534 (1976).

114. Benhert, O. Life Science 8, 943-948 (1969).

115. Bernard, B.K., Berchek, J.R., and Yutzey, D.A., Pharmacol. Biochem. Behav. 3, 121-126 (1975).

116. Pijnenburg, A.J., Honig, W.M.W., Van der Heyden, J.A.M., and van Rossum, J.M., Eur. J. Pharmacol. 35, 45-58 (1976).

117. Costall, B., Naylor, R.J., and Pinder, K., J. Pharm. Pharmacol 26, 753-757 (1974).

118. Guchhait, R., and Monjan, A', Neurosci. Abst, 1054 (1976).

119. Golter, M., and Michaelson, I.A., Science 187, 359-361 (1975).

120. Michaelson, I.A., Greenland, R.D., and Roth, W., Pharmacologist 16, 250 (1974).

121. Schumann, A'M., Dewey, W.L., Borzelleca, J.F., and Alphin, R.S., Fed. Proc. 36, 405 (1977)

122. Silbergeld, E.K., and Chisholm, J.J., Science 192, 153-155 (1976).

123. Stankovic, M., Dugandzic, B., Stankovic, B., Milovanovic, Lj., and Koricanac, Z.(1973) Arhiv. Hig. Rada. Toksikol, Vol. 24, 37-44.

124. Wince, L.C., Donavan, C.H. and Azzaro, A.J., Pharmacologist 18, 473 (1976).

125. Silbergeld, E.K., Life Sci. 20, 309-318 (1977).

126. Aprison, M.H.,and Werman, R., (1968) in Ehrenpris and Solnitzky (Eds.) Neuroscience

127. Roberts, E. (1977) in Yahr (Ed.) Basal Ganglia, 191-204, Raven Press, New York.

384

128. Cott, J.M., Carlsson, A., Engel, E. and Lindquist, N-S, _Arch. Pharmacol_. 295, 203-209 (1976).

129. Seeman, P., and Lee, T., J. _Pharmacol. Exp. Therap_. 190, 131-140 (1974).

130. McBride, W.J., Hingtgen, J.N., and Aprison, M.H., _Pharmacol. Biochem. Behav_. 4, 53-57 (1976).

131. Grimm, V., Gottesfeld, Z., Wassermann, I. and Samuel, D., _Pharmacol. Biochem. Behav_. 3, 573-578 (1971).

132. Piepho, R., Ryan, C. and Lacz, J., _Pharmacologist_ 18, 125 (1976).

133. Silbergeld, E.K., Miller, L.P., Kennedy, S., and Eng, N., _Neurosci. Abst_., In Press, (1977).

134. Pepeu, G., and Bartholini, A., _Eur. J. Pharmacol_. 4, 253-254 (1968).

135. Hingtgen, J.N., and Aprison, M.H., _Neuropharmacol_. 9, 417-425 (1970).

136. Kim, J.S., _Brain Res_. 55, 64-72 (1973).

137. LeMon, M., Galey, D., and Cardo, B., _Brain Res_. 88, 190-194 (1975).

138. Ungerstedt, U., _Acta Physiol. Scand_. 367, 95-121 (1971).

139. Buchsbaum, M., and Wender, P.H., Arch. Gen. Psychiat. 29, 764-770 (1973).

We would like to express our sincere appreciation to Gloria Rosal for the typing of this manuscript. The research conducted in the authors' laboratory cited in this chapter was supported by grants from the U.S. Public Health Service NIEHS00034 and 00454.

ANIMAL MODELS IN HEPATIC COMA

Josef E. Fischer, M.D.

Department of Surgery, Massachusetts General Hospital and Harvard Medical
School, Boston, Massachusetts 02114
Supported in part by PHS Grants No. AM15347 and AM19100

ABSTRACT

Animal models of hepatic encephalopathy in three species, rat, dog and monkey are described.
Each has advantages and all are useful in studying various aspects of altered cerebral
metabolism in liver disease.

INTRODUCTION

There are many aspects of hepatic disease one would want to mimic in animal models with
hepatic damage, including the peripheral (cardiovascular) manifestations, renal manifestations and lastly, but most importantly, the brain manifestations including the coma-like
state seen with profound liver disease. Our own interest in hepatic coma has involved
altered neurotransmission in the brain. Most of the models I will describe have as their
basis investigations of some of the neurochemical changes which occur within the brain.

In describing hepatic disease, it is important to define what type of hepatic disease one is
attempting to reproduce in an animal model. Acute and chronic hepatic damage and the brain
disorders sustained in acute and chronic hepatic damage may not be brought about by the
same mechanism. The two may, in fact, be entirely different diseases. Thus, one must take
care to define what it is one is trying to reproduce experimentally.

There are three traditional methods for producing hepatic damage.

1. The first involves the extirpation of the liver. The proponents of this form of "hepatic
damaged" animal model feel that in the absence of all hepatic function some of the metabolic
byproducts of liver failure will be reproduced without a necrotic liver within the experimental animal providing its other toxic products.

2. Those proponents of the vascular form of hepatic damaged model emphasize that it is the
presence of a sick, damaged liver which may in itself contribute to the symptomatology of
hepatic failure and it makes little sense to extirpate the liver. Thus, most of these
forms of hepatic damage involve total deprivation of hepatic blood supply, generally accomplished in staged fashion by an end-to-side portacaval shunt, thereby depriving the liver of
portal flow, followed approximately 48 hours later by hepatic artery ligation. Total vascular deprivation thus results with massive hepatic necrosis. This model will reproduce
acute hepatic necrosis. In the event one wishes to mimic the effects of chronic hepatic
damage, an end-to-side portacaval shunt in previously normal animals will result ultimately
in diminished hepatic function and an animal model which, for better or for worse, may mimic
to some extent, chronic human hepatic damage.

3. Poisoning, in graded fashion, using halogenated hydrocarbons, either carbon
tetrachloride or some other substance such as 2,4 dimethylnitrosimine in graded dosage may
be utilized either to give acute fulminant hepatic necrosis or chronic damage-simulating
cirrhosis.

At various times in our experimental approach to liver disease, we have found it necessary
to use different size animal models for different purposes. I will therefore describe
models for hepatic damage in three animals: the rat, the dog and the monkey.

I. EXPERIMENTAL HEPATIC DAMAGE IN THE RAT

Although the traditional form of inducing hepatic damage in the rat was by graded doses of
carbon tetrachloride or by the use of certain deficient diets, and while my initial experience in doing research was with these forms of hepatic damage, we have never utilized these

on a large scale in the experimental laboratory. The reason for this is that these models are very difficult to reproduce in the large numbers necessary for animal experiments. Thus, most of the work with rats has involved animals with portacaval shunt (1), either chronic animals with portacaval shunt alone which tend to mimic chronic liver disease or acute animals with total hepatic devascularization to mimic acute fulminant hepatitis (2).

A. Acute Fulminant Hepatitis in the Rat

This is generally produced by a two-stage ligation of hepatic blood supply. End-to-side portacaval shunt is carried out by a Teflon button technique (1) and followed 48 hours later by hepatic artery ligation. In order to support the animal during the acute period and to avoid artifact, hypoglycemia is prevented by glucose infusions and the animal is warmed. Animals generally become comatose at approximately 8 hours. Biochemical measurements in these animals have revealed remarkable similarities to human fulminant hepatic necrosis including a marked hyperaminoacidemia, marked hyperammonemia (much higher than seen in man), and a marked increase in brain glutamine (3). These animals tend to hyperventilate as patients do and do develop a respiratory alkalosis. Various studies in these animals have included increases in brain false or co-transmitters octopamine, phenylethanolamine, increases in brain serotonin, normal brain energy relationships, etc. (3,4,5).

B. Animals with Chronic End-to-Side Portacaval Shunts

The rat with an end-to-side portacaval shunt will undergo partial hepatic atrophy as judged by a decreased liver/body weight ratio (6). This animal will also manifest some of the striking biochemical findings associated with chronic human hepatic disease, including an altered neutral amino acid profile including increased aromatic amino acids, phenylalanine, tyrosine, and free tryptophan as well as decreased branched chain amino acids valine, leucine, and isoleucine (7). Plasma and brain octopamine are increased and seem to vary as does plasma tyrosine (6). Brain tryptophan and, in some parts serotonin are also increased and seem to vary as does the relationship between tryptophan and its competing neutral amino acids (8). More recently we have utilized these animals to demonstrate changes in the blood brain barrier.

II. DOG MODELS OF HEPATIC INSUFFICIENCY

In trying to develop a model for hepatic insufficiency, we have basically used two methods, a chemical and a vascular method for producing hepatic insufficiency.

A. Dimethylnitrosamine Poisoning as a Model for Acute Fulminant Necrosis

We did not originally intend to reproduce acute fulminant hepatic necrosis by treating animals with 2,4 dimethylnitrosamine. What we were hoping to reproduce was a chronic animal model for cirrhosis. What ultimately resulted, however, was a model for acute fulminant hepatitis in which animals died with prominent symptoms including fulminant hepatic coma and hemmorhage from the gastrointestinal tract (9). At autopsy, these animals manifested massive hepatic necrosis or acute yellow atrophy. On analysis of their amino acid patterns, which we have found to be good differentiator between acute fulminant hepatic necrosis and chronic hepatic insufficiency, these animals manifested amino acid patterns quite similar to that of acute fulminant hepatitis (9,10).

B. Chronic End-to-Side Portacaval Shunts in Dogs

This is the animal model with which we have the most experience. The normal dog with chronic end-to-side portacaval shunt undergoes gradual hepatic damage and ultimately dies with a finite life span of approximately two to three months following chronic portal flow diversion, presumably on the basis of substrate deprivation (11). Others have related the chronic hepatic atrophy and liver failure in these shunted dogs to lack of hepatotrophic substances such as insulin or glucagon. Whatever the etiology, these animals go through a rather characteristic pattern of hepatic insufficiency, including recognizable stages of hepatic encephalopathy, finally ending in coma and death. In addition, these animals manifest a rather characteristic amino acid pattern which we have associated with chronic hepatic insufficiency, including increased aromatic amino acids, phenylalanine, tyrosine and free tryptophan and decreased branched chain amino acids, valine, leucine and isoleucine. These animals have been used in experiments of hormonal changes in hepatic failure and nutritional support, as well as the treatment of hepatic encephalopathy, utilizing infusions of amino acids calculated to normalize a plasma amino acid pattern. (Fig. 1).

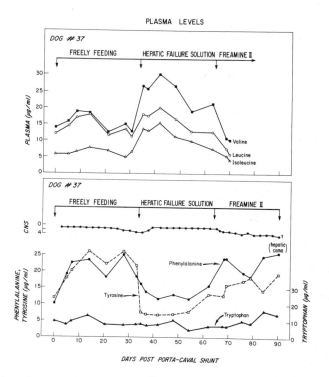

PLASMA LEVELS

Fig. 1. Changes in neutral amino acids in a normal dog
following end-to-side portacaval shunt.

When the animal becomes symptomatic with increased aromatic amino acids and decreased branched chain amino acids, infusional therapy to correct the amino acid pattern is begun with
improvement. With another mixture which does not correct amino acid pattern, animals die in
hepatic encephalopathy. (Reference 11).

More recently we have combined the ability to place indwelling CSF cannulas in these animals
and to withdraw amounts of CSF serially without anesthetizing the animal to correlate the
changes in the periphery and in the brain, at least as manifested by lateral ventricular
CSF. These findings, interestingly enough, as compared with cisternal CSF obtained by
repeated ventricular puncture, show some minor but significant alterations in differences
between CSF produced or at least harvested from these two regions. Their significance is
not as yet known.

It is of interest that the amino acid patterns obtained in dogs subjected to forms of
hepatic damage are exactly similar to the amino acid patterns obtained in man, a more marked
hyperaminoacidemia being seen in fulminant viral hepatitis as opposed to other characteristic
alterations, increased aromatic amino acids, and decreased branched chain amino acids as
seen in patients with chronic hepatic insufficiency (9,10). (Fig. 2-4).

388

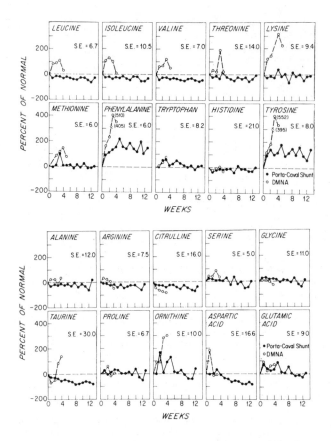

Fig. 2. Amino acid patterns in dogs with two
types of liver damage – portacaval shunt, dimethylnitrosamine
Similarity is seen to patterns in man. (Fig. 3,4) (Reference 9)

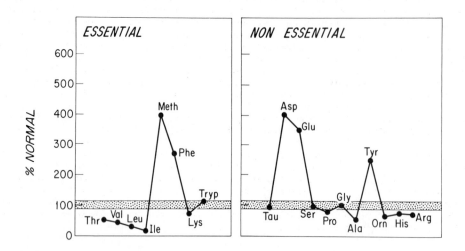

Fig. 3.

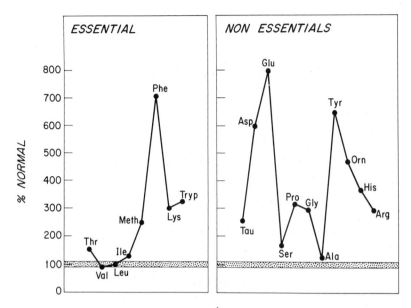

Fig. 3 and 4. Amino acid patterns in patients with fulminant hepatitis
(Fig. 3) with marked hyperaminoacidemia and chronic liver disease
(Fig. 4) with decreased branched chain amino acids and increased
aromatic amino acids. (Reference 10).

III. NON-HUMAN PRIMATE MODELS OF HEPATIC ENCEPHALOPATHY

Because of difficulty in obtaining serial samples of CSF in an awake, restrained dog, because
of the calvarium, it was initially decided to place lateral ventricular CSF cannulas in
monkeys. The procedure carried out is as follows: Using a special cannula, a lateral
ventricular needle is placed down to just above the ventricle. This is allowed to heal for
two weeks and after two weeks, ventricular puncture, which may be carried out repeatedly,
is begun. During these periods, the animal, which is neurologically normal and awake, is
placed in a restraining chair, the needle placed, and using a slow pump, lateral ventricular
CSF is harvested and used for analysis. After a baseline period of about 4 weeks, the
animal is then subjected to chronic end-to-side portacaval shunt. In the normal primate,
as in man, end-to-side portacaval shunt is somewhat better tolerated than in the dog.
Hepatic encephalopathy, nonetheless, will develop with a characteristic amino acid pattern
similar to that of patients with chronic hepatic insufficiency, only rather than in
appearing in two months, it generally takes four or five months to develop. During this
period of time, these animals will manifest various grades of hepatic encephalopathy,
including ataxia, asterixis, tremor, drowsiness, stupor and frank coma (13).

These animals have been utilized for the study of various therapeutic maneuvers in hepatic
encephalopathy, including amino acid infusion, intestinal sterilization, and lactulose.

The significance of these studies for this meeting and those questions to which they are
addressed is obvious. We have been studying the brain. We have been studying rather
gross alterations produced by rather drastic means. However, the experimental techniques
are applicable to a number of different situations, including more subtle alterations
brought about by drugs, changes in diet, and changes in environment. It is hopeful that
some of the techniques we have developed may be utilized more widely in investigation of
neurological and psychiatric disorders.

REFERENCES

1. Funovics, J.M., Cummings, M.G., Shuman, L., James, J.H. and Fischer, J.E., An improved non-suture method for portacaval anastomosis in the rat, Surgery 77, 661 (1975).

2. Dodsworth, J.M., Cummings, M.G., James, J.H. and Fischer, J.E., Depletion of brain norepinephrine in acute hepatic coma, Surgery 75, 811 (1974).

3. Biebuyck, J.F., Funovics, J.M., Dedrick, D.F., Scherer, Y.D. and Fischer, J.E., Neuorchemistry of hepatic coma: alterations in putative transmitter amino acids, Proceedings of an International Symposium on Artificial Support Systems for Acute Hepatic Failure. Artificial Liver Support. Roger Williams and Iain M. Murray-Lyon (Eds.), (1974).

4. Fischer, J.E. and Baldessarini, R.J., False neurotransmitters and hepatic failure, Lancet 2, 75 (1971).

5. Fischer, J.E. and James, J.H., Treatment of hepatic coma and hepatorenal syndrome: mechanism of action of L-Dopa and Aramine, Am. J. Surgery 123, 222 (1972).

6. James, J.H., Hodgman, J.M., Funovics, J.M. and Fischer, J.E., Alterations in brain octopamine and brain tyrosine following portacaval anastomosis in rats, J. of Neurochem. 27, 223 (1976).

7. James, J.H., Hodgman, J.M., Funovics, J.M., Yoshimura, N. and Fischer, J.E., Brain tryptophan, plasma free tryptophan and distribution of plasma neutral amino acids, Metabolism 25, 471 (1976).

8. Cummings, M.G., Soeters, P.B., James, J.H., Keane, J.M. and Fischer, J.E., Regional brain study of indoleamine metabolism following chronic portacaval anastomosis in the rat, J. of Neurochem. 27, 501 (1976).

9. Aguirre, A., Yoshimura, N., Westman, T. and Fischer, J.E., Plasma amino acids in dogs with two experimental forms of liver damage, J. Surg. Res. 16, 339 (1974).

10. Rosen, H.M., Yoshimura, N., Hodgman, J.M. and Fischer, J.E., Plasma amino acid patterns in hepatic encephalopathy of differing etiology, Gastroenterology 72, 483 (1977).

11. Fischer, J.E., Funovics, J.M., Aguirre, A., James, J.H., Keane, J.M., Wesdorp, R.I.C., Yoshimura, N. and Westman, T., The role of plasma amino acids in hepatic encephalopathy, Surgery 78, 276 (1975).

12. Starzl, T.E., Francavilla, A. and Halgrimson, C.G., The origin, hormonal nature and action of hepatotrophic substances in portal venous blood, Surg. Gyn. Obstet. 137, 179 (1973).

13. Smith, A.R., Rossi-Fanelli, F., James, J.H. and Fischer, J.E., Serial study of biochemical changes in ventricular CSF and plasma in hepatic coma in the monkey. Submitted for publication.

NARCOTIC MECHANISMS AND DYSKINESIAS

Kristin R. Carlson

Department of Pharmacology, School of Medicine
University of Pittsburgh, Pittsburgh, PA 15261, U.S.A.

ABSTRACT

A syndrome resembling tardive dyskinesia in humans is produced in rhesus
monkey and guinea pig by chronic narcotic treatment. Following narcotic
administration, both species are highly sensitive to dopaminergic agonists
such as methamphetamine and apomorphine; in the monkey, the behavior elicited by these agonists consists of oral dyskinesias identical to those seen
in tardive dyskinesia. Stress exacerbates the symptoms, as is the case in
tardive dyskinesia. The dyskinesias can be blocked by dopaminergic antagonists but are not affected by sedatives. As in humans with well-established symptomatology, the condition appears to be relatively permanent.
It is suggested that the same neurochemical mechanism is responsible for
narcotic-induced dyskinesias and neuroleptic-induced tardive dyskinesia,
i.e., blockade of nigrostriatal dopamine receptors by the chronic drugs
leads to receptor hypersensitivity to dopaminergic agonists after the
chronic drug is terminated.

INTRODUCTION

Tardive dyskinesia is an iatrogenic disease resulting from long-term use of
neuroleptics in the treatment of schizophrenia (for review see ref. 1).
The primary symptoms consist of repetitive and involuntary movements of the
oral and facial musculature; for example, patients roll and protrude their
tongues (2-4). Although the symptoms may be present during neuroleptic
therapy, they typically appear when drug dosage is reduced or treatment is
terminated (3,5,6). Since the mechanism of action common to all clinically-
useful neuroleptic agents is blockade of dopamine receptors (7,8), it is
thought that prolonged blockade renders the receptors supersensitive (9-11)
or increases their number (12). Thus, when the blockade is lifted by
halting neuroleptic administration, dopamine activates supersensitive receptors in the nigrostriatal system, thereby producing oral dyskinesias (10,
11,65).

Several other clinical observations support this hypothesis. Administration of dopaminergic agonists can precipitate oral dyskinesias in symptom-
free patients (13) and can exacerbate present symptoms (13-15). Although
there is no universally accepted treatment for tardive dyskinesia (16),
reinstitution of dopaminergic blockade by resuming neuroleptic therapy can
often alleviate the symptoms (13,15,17). Other characteristics of the
syndrome to which we will return in developing animal models are the
exacerbation of symptoms by emotional stress (2,15,18,19) and the apparently
permanent nature of the condition once it is established (20,21).

Klawans and his coworkers (e.g., this volume) have investigated the development of hypersensitivity in the nigrostriatal system of the guinea pig as a
consequence of chronic neuroleptic administration, and he and others have
proposed this preparation as an animal model of tardive dyskinesia (22-24).
Following neuroleptic treatment, lower doses of dopaminergic agonists were
sufficient to elicit the stereotyped gnawing behaviors which are dependent
on the nigrostriatal system (25,26) and are analogous to oral dyskinesias.

In recent years evidence has accumulated that an identical syndrome follows chronic narcotic administration in experimental animals, and that the same mechanism, the development of dopaminergic supersensitivity, is responsible. Narcotics such as morphine and methadone act as post-synaptic dopamine receptor blockers; for example, biochemical changes indicative of receptor blockade are found in rat striatum following treatment (27-34). Behaviorally, the classic symptoms of dopamine receptor blockade, catalepsy and antagonism of agonist-induced stereotyped behavior (24,33,35,36), are produced by morphine and methadone (33,34,37,38). The purpose of the present paper is to review evidence, mainly from our laboratory (39-42), that a model of tardive dyskinesia can be produced in rhesus monkey and guinea pig by chronic narcotic treatment, in that the characteristics of the narcotic-induced syndrome match those of the neuroleptic-induced condition in humans. Specifically, we will examine the elicitation of oral dyskinesias by dopaminergic agonists, the exacerbation of symptoms by stress, the elimination of symptoms by dopaminergic antagonists, and the permanence of the phenomenon.

PRIMATE MODEL

In the normal rhesus monkey, oral dyskinesias are elicited only after several months of treatment with doses of methamphetamine (MA) in the 10-20 mg/kg range (43,44). Thus, supersensitivity of the striatal dopamine system would be evident in the elicitation of oral dyskinesias by lower doses or a shorter period of administration.

Effect of Chronic Oral Methadone Treatment

Eight male rhesus monkeys (Macaca mulatta) were maintained on low doses of methadone hydrochloride (Dolophine; Lilly; 10 mg/ml) by making available for 1 hr. daily a fixed amount of methadone (MD) mixed with 100 ml Tang orange drink. They consumed 0.5-2.6 mg/kg/day for 10-22 months, and had been withdrawn from MD for 2-17 months at the beginning of this study. Each monkey's drug history is available in Carlson and Eibergen (39). These MD doses were not sufficient to produce physical dependence, as the monkeys did not respond to a naloxone challenge nor to abrupt MD withdrawal. Eleven male monkeys with no experience with MD served as controls.

Using a squeeze-back observation cage, all monkeys were injected i.m. with methamphetamine hydrochloride (Sigma; 50 mg/ml as the salt) on a cycle mimicking human "spree" abuse: once daily for 4 days followed by 3 rest days. Each monkey's MA-elicited behavior was recorded regularly for a minimum of 6 hr. post-injection and periodically in his home cage thereafter. Table 1 presents the results of this experiment, in which it can be seen that all the former-methadone (M) monkeys developed oral dyskinesias; the remarkable aspect of these results is that 7 of them exhibited dyskinesias at the very low dose of 2.0 mg/kg after no more than 4 injection days, and in 4 cases on the very first day. An example of the most common oral dyskinesia is illustrated in Fig. 1. Once elicited, oral dyskinesias were the characteristic response to MA on all subsequent injection days. Dyskinesias appeared 5-10 min. post-injection, occurred at rates of 25-60 events/min., and, depending on MA dose, persisted for 36-72 hr.

With two exceptions, the control (C) monkeys failed to develop dyskinesias even during prolonged administration of increasing doses of MA, but rather showed the types of stereotyped behaviors typical of normal monkeys (43,44). C8's dyskinesia was atypical in that it lacked the highly predictable rhythmicity and continuity characteristic of the M monkeys' dyskinesias. C6 eventually developed an oral dyskinesia which was indistinguishable from those of the M monkeys.

Thus, prior treatment with MD sensitized the striatal dopamine system to such an extent that very low doses of MA were immediately effective in eliciting oral dyskinesias. We suggest that this phenomenon represents a state of latent supersensitivity, unmasked by administration of a dopaminergic agonist, which corresponds to the precipitation of symptoms of tardive dyskinesia by such agonists (13).

TABLE 1   Results of MA Administration

Monkey	Dyskinesia	Day*	MA Dose	Behavior
M1	Yes	1	2.0	Tongue protrusion
M2	Yes	1	2.0	Tongue protrusion
M3	Yes	1	2.0	Jaw displacement
M4	Yes	1	2.0	Sucking cheek
M5	Yes	4	2.0	Wide mouth opening
M6	Yes	4	2.0	Tongue protrusion
M7	Yes	59	6.0	Tongue protrusion
M8	Yes	2	2.0	Tongue protrusion
C1	No	8	1.0-2.0	Mild chewing
C2	No	8	1.0-2.0	Body jerk
C3	No	31	2.0-5.0	Shuffling movements
C4	No	45	2.0-5.0	Body jerk
C5	No	40	2.0-5.0	Writhing
C6	Yes	47	5.0	Tongue protrusion
C7	No	48	2.0-5.0	Body jerk
C8	Yes	2	1.0	Tongue protrusion
C9	No	8	1.0-2.0	Mild chewing
C10	No	8	1.0-2.0	No response
C11	No	8	1.0-2.0	Circling

*Day indicates the injection day using the dose (mg/kg) specified on which dyskinesias were first observed, or the total number of injection days covering the dosage range specified for those monkeys not exhibiting dyskinesias.

Fig. 1. A typical tongue protrusion in M1 following 4.0 mg/kg MA.

394

Effects of Stress

The severity of symptoms of tardive dyskinesia is influenced by the patient's state of arousal. For example, symptoms may be absent during sleep, and are often more intense when the patient is under emotional stress (2,15, 18,19). We examined this phenomenon in several of our dyskinetic monkeys by stressing them and measuring the rate of their tongue protrusions, the most common and easily quantified dyskinesia. Each monkey received a dose of MA which was itself subthreshold for eliciting tongue protrusions (42), and was subjected to a stressor, the loud alarm buzzer of a Gra-Lab Timer. A twenty-sec. sampling interval during each minute was signalled by a recycling clock, and the number of tongue protrusions during each sampling interval was counted. The buzzer was sounded during approximately half the sampling intervals, buzzer and non-buzzer intervals occurring in pseudo-random order. Data were recorded for the 15 min. immediately preceding injection, and for various periods post-injection.

The results are shown in Fig. 2. The buzzer itself had no effect on the rate of tongue protrusions before MA administration, nor did MA elicit tongue protrusions in the absence of the stressor. However, after MA the buzzer dramatically increased the rate of tongue protrusions in the M monkeys. When the buzzer was sounded they immediately began to extend their tongues, and ceased shortly after the buzzer was turned off. C8, who as noted earlier showed a different pattern of tongue protrusions than the M monkeys, was not affected by the buzzer. Further, in identical test sessions the rate of other C monkeys' characteristic behaviors (e.g., body jerks) was not altered by the buzzer. Thus, the M monkeys' results were not due to a generalized augmentation of stereotyped behaviors, but to a stress-induced exacerbation of one type, oral dyskinesias, which resulted originally from sensitization of the striatal dopamine system by MD. These results further substantiate the correspondence of this model with the condition of tardive dyskinesia.

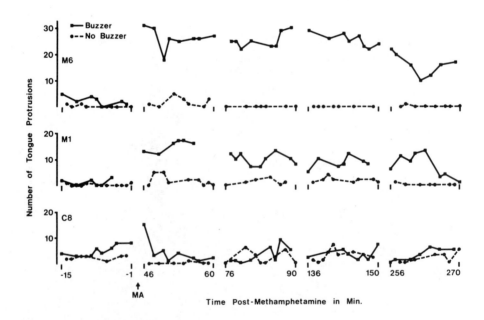

Fig. 2. Effect of stress on the rate of oral dyskinesias in 3 monkeys. Number of tongue protrusions in the 20-sec. observation period during each min. are plotted as a function of time before and after MA administration. Stress periods were those in which a buzzer was sounded (Buzzer), non-stress periods were quiet (No Buzzer). M6 illustrates the largest effect and M1 the smallest effect observed in former methadone monkeys. Reprinted with permission from ref. 42.

Effect of Acute Dopaminergic Blockade

The symptoms of tardive dyskinesia are often ameliorated by reinstitution of therapy with the agent originally responsible for the condition, i.e., a dopaminergic blocking drug (13,15,17). We tested the ability
of various dopaminergic blockers and other agents to suppress oral dyskinesias and other stereotyped behaviors when administered after MA and to
prevent their appearance when given before MA. The rate/min. of behaviors
was determined for 30 min. preceding MA administration and for at least
4 hr. after MA.

The complete results of this study, including time-action curves, are
available in Eibergen and Carlson (42); Table 2 presents the aspects most
relevant to the present discussion. The dopaminergic blockers chlorpromazine and spiroperidol immediately suppressed ongoing stereotyped behaviors
when administered after MA, and prevented the onset of stereotyped behaviors when given before MA. Clozapine also reduced the intensity of oral
dyskinesias in rather low oral doses, and completely suppressed dyskinesias at a higher parenteral dose. The above results were not due to a general depression of activity, since sedative doses of phenobarbital and
diazepam had no effect on the rate of oral dyskinesias. These data support
the role of a hypersensitive striatal dopamine system in this phenomenon,
and are consistent with the resemblence of this condition to tardive
dyskinesia.

TABLE 2   Effect of Acute Blockade

Blocking Drug and Dose	Before/After MA Dose	Monkey	Effect of Blocking Drug on Behavior
Chlorpromazine			
0.5	After 3.5	M7	Completely suppressed
1.0	After 2.0	M8	Completely suppressed
1.0	Before 2.0	C4	Completely blocked onset
Spiroperidol			
0.025	After 5.0	M1	Completely suppressed
0.05	After 5.0	M7	Completely suppressed
0.025	Before 5.0	M7	Completely blocked onset
Clozapine			
p.o.0.1	Before 5.0	M2	Cut number in half
0.5	Before 6.0	M2	Cut number in half
i.m.2.5	After 5.0	M4	Completely suppressed
Phenobarbital			
30.0	Before 6.0	M7	No effect
10.0	Before 2.5	M8	No effect
Diazepam			
1.0	After 4.0	M4	No effect

A Preliminary Test With Apomorphine and Stress

We were struck by the fact that one of the monkeys (M6) showing tongue
protrusions in response to MA had been withdrawn from MD 17 months
previously. This suggested a prolonged effect of MD and a relatively
permanent alteration in the sensitivity of the striatal dopamine system.
In order to specifically test for dopaminergic supersensitivity, one can
use as an agonist apomorphine hydrochloride (APO), which is generally
considered to directly stimulate post-synaptic dopamine receptors
(38,45-50). One monkey (M9) with a MD history and no experience
with MA was available; he had consumed an average of 2.0 mg/kg/day
MD for 24 months and had been withdrawn from MD 26 months previously.
A control monkey (C12) had consumed Tang during the same period. Both

monkeys were injected s.c. with 0.5 mg/kg APO using the squeeze-back
observation cage, rated for the intensity of elicited behavior every 5 min.,
and at 24 and 48 hr. post-injection were subjected to the presumed stress
of being returned to the squeeze-cage for observation.

APO elicited an oral dyskinesia in M9 consisting of intense, very loud
bruxism in which the mandible was moved laterally at a very rapid rate. At
peak effect he stood immobile while engaging in bruxism, and as the effect
waned the bruxism alternated with bouts of compulsive chewing on parts of
the cage. C12, on the other hand, exhibited stereotyped body movements in
which he laid on his left side, slowly raising his right limbs in the air
and returning them to a resting position. This activity alternated with
periods of normal behavior. Thus, the most striking effect was the immediate elicitation of an oral dyskinesia in M9; there were also quantitative
differences between the monkeys in the intensity and duration of their
behaviors, as shown in Fig. 3 A&B. Although these results are preliminary,
they do indicate at the very least that M9's striatal dopamine system continued to be supersensitive over 2 years after MD was discontinued, a considerable portion of a monkey's life-span.

Only M9 reacted to the stress test (Fig. 3C). At 24 hr. post-APO he exhibited intense bruxism immediately upon placement in the squeeze-cage, and
at 48 hr. the bruxism was still present although diminished in intensity
and duration. Recalling that in a prior experiment we found that the
buzzer alone did not trigger dyskinesias, but that the buzzer in conjunction with a sub-threshold dose of MA was effective, the present results
suggest that the residual effects of APO can persist for days and that the
imposition of a stressor during that time can reactivate oral dyskinesias.
This occurred only in a monkey sensitized by prior MD treatment, and thus is
consistent with other animal experimental data and the clinical picture of
tardive dyskinesia.

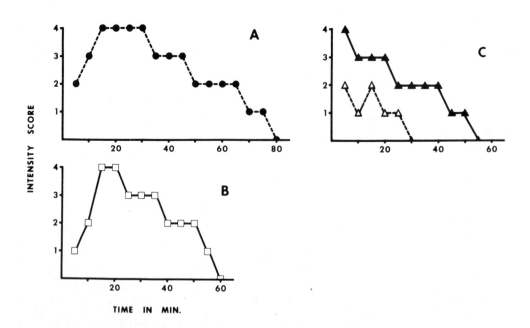

Fig. 3. Score rating the intensity of stereotyped behavior
as a function of time. A: Intensity of bruxism in M9.
B: Intensity of body movements in C12. C: Intensity of
bruxism in M9 following placement in the squeeze-cage 24 hr.
(▲) and 48 hr. (△) post-APO.

GUINEA PIG MODEL

In this species the behavior equivalent to oral dyskinesias is stereotyped
chewing and gnawing. This is the rodents' characteristic response to adequate doses of dopaminergic agonists (25,26,45), and the dependence of this
behavior on the striatal system has been solidly documented (25,26). The
difference between primates (including man) and rodents lies in the ability
of the drug to elicit gnawing upon first administration in a normal
animal; in rodents this occurs regularly. Thus, supersensitivity of the
striatal dopamine system is evidenced in the guinea pig by a shift to the
left of a dose-response curve, when measuring the intensity of stereotyped
behavior in response to various doses of an agonist.

It is well-established in rats and guinea pigs that dopaminergic supersensitivity results from chronic neuroleptic administration (22-24,41,51-53);
this phenomenon has been proposed as an animal model of tardive dyskinesia
(22-24). In addition, there is evidence of the potentiation of agonist-
elicited locomotion and aggression after chronic narcotic treatment (54-57).
This section will summarize some of the evidence from our laboratory that
stereotyped oral behaviors in response to agonists and stress are also
enhanced by prior narcotic treatment.

Supersensitivity to Methamphetamine After Chronic Methadone

Ten male guinea pigs were injected s.c. twice daily for 35 days with MD,
receiving 10 mg/kg/day for 21 days, 15 mg/kg/day for 9 days, and regularly
decreasing doses for the final 5 days. Ten control subjects received
equivalent volumes of saline (SAL). Beginning on the eighth day following
treatment they were challenged with 0.5, 1.0, 2.0, and 3.0 mg/kg MA. At
30 and 60 min. post-injection the amount of time they spent in stereotyped gnawing during a 5-min. observation period was measured. As shown
in Fig. 4, there was a significant (p<.01) shift to the left of the dose-
response curve of the MD animals at 30 min; identical results were
obtained at 60 min. Thus, prior MD treatment altered those animals'
response to MA in a direction indicating supersensitivity of the striatal
dopamine system.

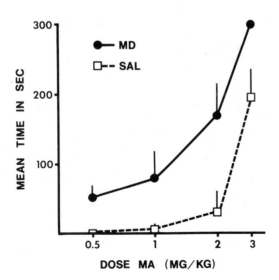

Fig. 4. Time spent in stereotyped
oral behaviors (Mean ± SEM) as a
function of MA dose during a 5-min.
period beginning at 30 min. post-
injection. Reprinted with permission from ref. 41.

Sensitivity to Stress During and After Chronic Methadone

A stressful stimulus, such as tail-pinch in rats, can elicit the oral behaviors which are dependent on the functioning of the striatal dopamine system (58-59). Thus, we tested for supersensitivity of that system in the above guinea pigs by stressing them with various intensities of aversive foot shock and measuring the amount of time during the 60 sec. period following each shock that they spent in stereotyped mouth movements.

As can be seen in Fig. 5, the "dose-response" curve of the MD animals was shifted significantly (p<.01) to the left both during and following treatment, i.e., they were hypersensitive to the oral behavior-eliciting effects of this stressor. The results during MD treatment confirm scattered reports that morphine-dependent rhesus monkeys (60) and methadone maintenance patients (61) are more susceptible to the effects of stress. We suggest that the results following MD are analogous to the exacerbation of symptoms found in tardive dyskinesia patients under emotional stress.

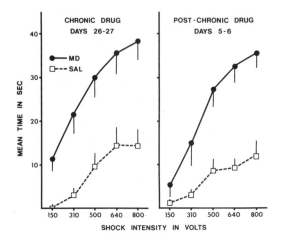

Fig. 5. Time spent in stereotyped oral behaviors (Mean ± SEM) as a function of foot shock intensity during and following chronic drug treatment. Reprinted with permission from ref. 41.

Supersensitivity to Apomorphine After Chronic Narcotics

In an extension of the above work we treated chronically with MD, morphine sulfate (MS), or saline (SAL) for a shorter period, 3 weeks. The MD and MS guinea pigs were injected s.c. twice daily with 20 mg/kg/day the first week, 30 mg/kg/day the second week, and 40 mg/kg/day the third week, while the SAL animals received equivalent volumes of saline. On days 5-7 and 33-35 after the termination of treatment, all animals were challenged with 0.1, 0.2, and 0.4 mg/kg APO, the order of doses determined by a Latin Square. They were rated for the intensity of stereotyped behavior using an 8-point scale at 5-min. intervals post-injection; mean scores across the 7 rating times are reported here.

Fig. 6 shows that the MD and MS dose-response curves were significantly (p<.01) displaced from that of the SAL group one week after chronic treatment. By week 5, however, the MS group's supersensitivity had dissipated but the MD group continued to exhibit a highly significant (p<.001) degree of supersensitivity. Thus, both narcotics are capable of inducing dopaminergic hypersensitivity, but MD appears to produce a more persistent effect.

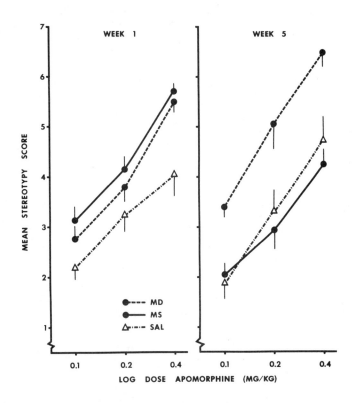

Fig. 6. Stereotypy score (Mean ± SEM) as a
function of APO dose, 1 week and 5 weeks
following chronic drug treatment.

DISCUSSION

Evidence has been presented suggesting that narcotic-induced dyskinesias
in two species of experimental animals can be used as a model of tardive
dyskinesia in humans. The symptomatology is virtually the same in the
two conditions, and the factors which ameliorate or exacerbate the
condition in humans have the same effect in these animals. Further, at
least in the case of methadone-induced dyskinesias, the condition
appears to be very long-lasting.

The major difference between tardive dyskinesia and the proposed model
is that in tardive dyskinesia the symptoms often arise spontaneously
when neuroleptic therapy is halted. Although agonists do exacerbate
the symptoms, their administration is not usually a condition for observing dyskinesias. We observed no spontaneous dyskinesias following
termination of narcotic treatment in either monkey or guinea pig,
and had to precipitate them with agonists, albeit in very low doses.
It is possible that there is a fundamental difference between humans
and other species in this respect. Several studies of chronic neuroleptic administration in monkeys have shown that oral dyskinesias may
be seen in some animals during drug treatment, but that they rapidly
disappear upon drug discontinuation (62,63). Further, in no such
study in monkey (39,42,62-64) or rodent (22-24,41,51-53) has there been
any evidence of the spontaneous appearance of dyskinesias or other
stereotyped behavior following termination of neuroleptic treatment.
Thus, it is perhaps not surprising that narcotic-treated animals did
not spontaneously display these behaviors.

The most commonly accepted explanation of tardive dyskinesia involves overactivity of nigrostriatal dopamine function resulting from prior chronic blockade by neuroleptics (for review see ref. 65). In view of the known activity of narcotics as dopaminergic blocking agents, it is reasonable to propose that their chronic administration results in a state of latent dopaminergic hypersensitivity which is responsible for the agonist-elicited symptoms resembling those of tardive dyskinesia.

This narcotic-induced phenomenon attains clinical relevance when one considers the thousands of individuals in long-term methadone maintenance therapy. Proper medical follow-up when these patients leave treatment is strongly suggested, in order to assess the degree to which they are at risk to the development of extrapyramidal dyskinesias.

ACKNOWLEDGEMENTS

This research was supported by USPHS grants MH20121 and DA00883, and by a grant from the Benevolent Foundation of Scottish Rite Freemasonry, Northern Jurisdiction, U.S.A. The collaboration of Dr. Robert D. Eibergen and the skilled technical assistance of John Almasi are gratefully acknowledged.

REFERENCES

1. D. Tarsy & R. J. Baldessarini, The tardive dyskinesia syndrome. In Clinical Neuropharmacology, Vol. 1, H. L. Klawans, Jr., (Ed.), Raven Press, New York (1976).

2. G. E. Crane, Tardive dyskinesia in patients treated with major neuro-leptics: A review of the literature, Am. J. Psychiat. 124, 40, Suppl. (1968).

3. G. E. Crane & G. Paulson, Involuntary movements in a sample of chronic mental patients and their relation to the treatment with neuroleptics, Int. J. Neuropsychiat. 3, 286 (1967).

4. L. Uhrbrand & A. Faurbye, Reversible and irreversible dyskinesia after treatment with perphenazine, chlorpromazine, reserpine and electro-convulsive therapy, Psychopharmacologia 1, 408 (1960).

5. D. E. Casey & D. Denney, Deanol in the treatment of tardive dyskinesia, Am. J. Psychiat. 132, 864 (1975).

6. J. P. Curran, Tardive dyskinesia: Side effect or not? Am. J. Psychiat. 130, 406 (1973).

7. I. Creese, D.R. Burt, & S. H. Snyder, Dopamine receptor binding pre-dicts clinical and pharmacological potencies of antischisophrenic drugs, Science 192, 481 (1976).

8. P. Seeman, T. Lee, M. Chau-Wong, & K. Wong, Antipsychotic drug doses and neuroleptic/dopamine receptors, Nature 261, 717 (1976).

9. N.-E. Anden, S.G. Butcher, H. Corrodi, K. Fuxe & U. Ungerstedt, Receptor activity and turnover of dopamine and noradrenaline after neuroleptics, Eur. J. Pharmac. 11, 303 (1970).

10. H. L. Klawans, Jr., The pharmacology of tardive dyskinesia, Am. J. Psychiat. 130, 82 (1973).

11. H. L. Klawans, Jr., The pharmacology of tardive dyskinesias. In The Pharmacology of Extrapyramidal Movement Disorders, Monographs in Neural Sciences, Vol. 2. S. Karger, Basel (1973).

12. D. R. Burt, I. Creese & S. H. Snyder, Antischizophrenic drugs: Chronic treatment elevates dopamine receptor binding in brain, Science 196, 326 (1977).

13. J. Gerlach, N. Reisby & A. Randrup, Dopaminergic hypersensitivity and cholinergic hypofunction in the pathophysiology of tardive dyskinesia, Psychopharmacologia 34, 21 (1974).

14. W. E. Fann, J. M. Davis & I.C. Wilson, Methylphenidate in tardive dyskinesia, Am. J. Psychiat. 130, 922 (1973).

15. N.S. Kline, On the rarity of "irreversible" oral dyskinesias following phenothiazines, Am. J. Psychiat. 124, 48, Suppl., (1968).

16. H. Kazamatsuri, C. Chien, & J. O. Cole, Therapeutic approaches to tardive dyskinesia: A review of the literature, Arch. Gen. Psychiat. 27, 491 (1972).

17. G.M. Simpson & E. Varga, Clozapine - a new antipsychotic agent, Curr. Therap. Res. 16, 679 (1974).

18. R. Druckman, D. Seelinger & B. Thulin, Chronic involuntary movements induced by phenothiazines, J. nerv. ment. Dis. 135, 69 (1962).

19. G. Jacobson, R. J. Baldessarini & T. Manschreck, Tardive and with-drawal dyskinesia associated with haloperidol, Am. J. Psychiat. 131, 910 (1974).

20. G.E. Crane, Persistent dyskinesia, Br. J. Psychiat. 122, 395 (1973).

21. American College of Neuropsychopharmacology, Neurologic syndromes associated with antipsychotic drug use, N. Engl. J. Med. 289, 20 (1973).

22. H.L. Klawans, Jr. & R. Rubovits, An experimental model of tardive dyskinesia, J. neural Trans. 33, 235 (1972).

23. R. Rubovits, B.C. Patel & H. L. Klawans, Jr., Effect of prolonged chlorpromazine pretreatment on the threshold for amphetamine stereo-typy: A model for tardive dyskinesias, In Advances in Neurology, Vol. 1, A. Barbeau et al. (Eds.), Raven Press, New York (1973).

24. A.C. Sayers, H.R. Burki, W. Ruch & H. Asper, Neuroleptic-induced hypersensitivity of striatal dopamine receptors in the rat as a model of tardive dyskinesias: Effects of clozapine, haloperidol, loxapine and chlorpromazine, Psychopharmacologia 41, 97 (1975).

25. A. Randrup & I. Munkvad, Stereotyped activities produced by amphetamine in several animal species and man. Psychopharmacologia 11, 300 (1967).

26. R. Fog, On stereotypy and catalepsy: Studies on the effect of amphetamines and neuroleptics in rats, Acta neurol. Scand. (Suppl. 50) 48, 11 (1972).

27. L. Ahtee, Catalepsy and stereotyped behavior in rats treated chronically with methadone: relation to brain homovanillic acid content, J. Pharm. Pharmac. 25, 649 (1973).

28. L. Ahtee, Catalepsy and stereotypies in rats treated with methadone; Relation to striatal dopamine, Eur. J. Pharmac. 27, 221 (1974).

29. L. Ahtee & I. Kaariainen, The effect of narcotic analgesics on the homovanillic acid content of rat nucleus caudatus, Eur. J. Pharmac. 22, 206 (1973).

30. K. Kuschinsky & O. Hornykiewicz, Morphine catalepsy in the rat: Relation to striatal dopamine metabolism, Eur. J. Pharmac. 19, 119 (1972).

31. Z. Merali, P.K. Ghosh, P.D. Hrdina, R.L. Singhal, & G.M. Ling, Alterations in striatal acetylcholine, acetylcholine esterase and dopamine after methadone replacement in morphine-dependent rats, Eur. J. Pharmac. 26, 375 (1974).

32. J. Perez-Cruet, G. DiChiara, & G.L. Gessa, Accelerated synthesis of dopamine in the rat brain after methadone, *Experientia* 28, 926 (1972).

33. S. K. Puri, C. Reddy, & H. Lal, Blockade of central dopaminergic receptors by morphine: Effect of haloperidol, apomorphine or benztropine, *Res. Commun. Chem. Path. & Pharmac.* 5, 389 (1973).

34. H. A. Sasame, J. Perez-Cruet, G. DiChiara, A. Tagliamonte, P. Tagliamonte, & G. Gessa, Evidence that methadone blocks dopamine receptors in the brain, *J. Neurochem.* 19, 1953 (1972).

35. B. Costall & R. J. Naylor, On catalepsy and catatonia and the predictability of the catalepsy test for neuroleptic activity, *Psychopharmacologia* 34, 233 (1974).

36. C. Ezrin-Waters, P. Muller, & P. Seeman, Catalepsy induced by morphine or haloperidol: Effects of apomorphine and anticholinergic drugs, *Can. J. Physiol. Pharmac.* 54, 516 (1976).

37. R. Fog, Behavioral effects in rats of morphine and amphetamine and of a combination of the two drugs, *Psychopharmacologia* 16, 305 (1970).

38. R. C. Srimal & B.N. Dhawan, An analysis of methylphenidate induced gnawing in guinea pigs, *Psychopharmacologia* 18, 99 (1970).

39. K. R. Carlson & R. D. Eibergen, Susceptibility to amphetamine-elicited dyskinesias following chronic methadone treatment in monkeys, *Ann. N.Y. Acad. Sci.* 281, 336 (1976).

40. R.D. Eibergen & K. R. Carlson, Dyskinesias elicited by methamphetamine: Susceptibility of former methadone-consuming monkeys, *Science* 190, 588 (1975).

41. R. D. Eibergen & K.R. Carlson, Behavioral evidence for dopaminergic supersensitivity following chronic treatment with methadone or chlorpromazine in the guinea pig, *Psychopharmacology* 48, 139 (1976).

42. R.D. Eibergen & K.R. Carlson, Dyskinesias in monkeys: Interaction of methamphetamine with prior methadone treatment. *Pharmac. Biochem. Behav.* 5, 175 (1976).

43. E. H. Ellinwood, Jr., Effect of chronic methamphetamine intoxication in rhesus monkeys, *Biol. Psychiat.* 3, 25 (1971).

44. E. H. Ellinwood, Jr. & O. Duarte-Escalante, Chronic methamphetamine intoxication in three species of experimental animals, In *Current Concepts on Amphetamine Abuse*, E. H. Ellinwood, Jr. and S. C. Cohen (Eds.), U.S. Gov't. Printing Office, Washington (1970).

45. T.L. Sourkes & S. Lal, Apomorphine and its relation to dopamine in the nervous system. In *Advances in Neurochemistry*, Vol. 1, B.W. Agranoff & M. H. Aprison (Eds.), Plenum, New York (1975).

46. N.-E. Anden, A. Rubenson, K. Fuxe & T. Hokfelt, Evidence for dopamine receptor stimulation by apomorphine, *J. pharm. Pharmac.* 19, 627 (1967).

47. A. M. Ernst, Mode of action of apomorphine and dexamphetamine on gnawing compulsion in rats, *Psychopharmacologia* 10, 316 (1967).

48. A. M. Ernst & P. G. Smelik, Site of action of dopamine and apomorphine on compulsive gnawing behavior in rats, *Experientia* 22, 837 (1966).

49. U. Ungerstedt, L.L. Butcher, S.G. Butcher, N.-E. Anden, & K. Fuxe, Direct chemical stimulation of dopaminergic mechanisms in the neostriatum of the rat, *Brain Res.* 14, 461 (1969).

50. **A.** Weissman, B.K. Koe & S.S. Tenen, Antiamphetamine effects following inhibition of tyrosine hydroxylase, *J. pharmac. exp. Therap.* 151, 339 (1966).

51. B. Fjalland & I. Moller-Nielsen, Enhancement of methylphenidate-induced stereotypies by repeated administration of neuroleptics, Psychopharmacologia 34, 105 (1974).

52. D. Tarsy & R. J. Baldessarini, Behavioral supersensitivity to apomorphine following chronic treatment with drugs which interfere with the synaptic function of catecholamines, Neuropharmacology 13, 927 (1974).

53. J. E. Thornburg & K. E. Moore, Supersensitivity to dopaminergic agonists induced by haloperidol. In Aminergic Hypotheses of Behavior: Reality or Cliche?, B. K. Bernard (Ed.), National Institute on Drug Abuse, Rockville, Md. (1975).

54. E. Eidelberg & R. Erspamer, Dopaminergic mechanisms of opiate actions in brain, J. Pharmac. exp. Therap. 192, 50 (1975).

55. H. Lal, J. O'Brien, & S. K. Puri, Morphine-withdrawal aggression: Sensitization by amphetamines, Psychopharmacologia 22, 217 (1971).

56. S. K. Puri & H. Lal, Effect of dopaminergic stimulation or blockade on morphine-withdrawal aggression, Psychopharmacologia 32, 113 (1973).

57. G. Gianutsos, M. D. Hynes, S. K. Puri, R. B. Drawbaugh & H. Lal, Effect of apomorphine and nigrostriatal lesions on aggression and striatal dopamine turnover during morphine withdrawal: Evidence for dopaminergic supersensitivity in protracted abstinence, Psychopharma-cologia 34, 37 (1974).

58. S. M. Antelman & H. Szechtman, Tail pinch induces eating in sated rats which appears to depend on nigrostriatal dopamine, Science 189, 731 (1975).

59. S. M. Antelman, H. Szechtman, P. Chin & A. E. Fisher, Tail pinch-induced eating, gnawing and licking behavior in rats: Dependence on the nigrostriatal dopamine system, Brain Res. 99, 319 (1975).

60. S. Holtzman & J. Villarreal, Morphine dependence and body temperature in rhesus monkeys, J. Pharmac. exp. Therap. 166, 125 (1969).

61. P. Renault, C. R. Schuster, R. Heinrich & B. van der Kolk, Altered plasma cortisol response in patients on methadone maintenance, Clin. Pharmac. Ther. 13, 269 (1972).

62. G. A. Deneau & G. E. Crane, Dyskinesias in rhesus monkeys tested with high doses of chlorpromazine. In Psychotropic Drugs and Dysfunctions of the Basal Ganglia, G. E. Crane & R. Gardner (Eds.), U. S. P. H. S., Washington (1969).

63. G. Paulson, Effects of chronic administration of neuroleptics: Dyskinesias in monkeys, Pharmac. Ther. B. 2, 167 (1976).

64. F. S. Messiha, The relationship of dopamine excretion to chlorpro-mazine-induced dyskinesia in monkeys. Arch. int. Pharmacodyn. 209, 5 (1974).

65. R. J. Baldessarini & D. Tarsy, Mechanisms underlying tardive dyskinesia. In The Basal Ganglia, M. D. Yahr (Ed.), Raven Press, New York (1976).

DISCUSSION

Dr. Lemberger inquired about the minimal doses used by Dr. Carlson and how she determined that the treated animals did not have physical dependence. *Dr. Carlson* replied that the monkeys were given 0.5-2.5 mg/kg p.o. of methadone and that they were challenged with naloxone to show lack of physical dependence. In the later guinea pig studies, animals were given 20 mg/kg/day the first week; 30, the second week; and 40, the third week. When withdrawal was abrupt, the animals were very hyperactive, irritable, etc.

Dr. Lippa indicated that when one generates an animal model, it is critical to show that those drugs that are active in man are active in animals. Equally critical is showing that drugs that are inactive in man are inactive in the animal model. In light of this, he asked Dr. Carlson to comment on the clinical relevance of her model; e.g., do narcotics produce tardive dyskinesia in man? *Dr. Carlson* stated that she had not read of any cases of narcotics producing tardive dyskinesia. However, she noted that there are many significant differences (e.g., in age sex) between the populations of individuals who develop tardive dyskinesia after neuroleptic therapy and the populations of methadone maintenance clinics. The elderly female patient most susceptible to tardive dyskinesia is not the young male population found in methadone maintenance clinics. Although Dr. Carlson has not read any case reports of tardive dyskinesia following withdrawal of heroin or methadone, she pointed out that methadone has been around only about ten or twelve years and this was the period of clinical use of neuroleptics before tardive dyskinesia became a well-recognized clinical entity.

Dr. Bunney mentioned an electrophysiological system in use in his lab in which an effect on firing rate is noted after acute administration of haloperidol or an antipsychotic or a dopamine agonist. The firing rate of nigrostriatal dopamine cells is increased by morphine, and this is blocked by tiny doses of naloxone. Neither morphine nor naloxone has effect on the firing rates of these cells after haloperidol or dopamine agonists. Thus, Dr. Bunney wondered whether this is really a dopamine receptor effect or some other system interacting, but not at the dopamine receptor. *Dr. Carlson* queried whether morphine antagonized an apomorphine effect on firing rate; *Dr. Bunney* replied that it did not, that the presence of morphine or naloxone had no effect on what dopamine agonists or antagonists did to the firing rate of these cells.

In response to a question of *Dr. R. Vogel*, *Dr. Carlson* stated that at the apomorphine dose used in her study (0.5 mg/kg), it was not an emetic.

BRAIN NEUROTRANSMITTER RECEPTORS AND CHRONIC ANTIPSYCHOTIC DRUG TREATMENT: A MODEL FOR TARDIVE DYSKINESIA.

R.M. Kobayashi[1], J.Z. Fields[2], R.E. Hruska[2], K. Beaumont[2] and H.I. Yamamura[2].

[1]VA Hospital and Dept. Neurosciences, UCSD, La Jolla, CA 92037.
[2]Dept. Pharmacology, Univ. of Arizona, Tucson, AZ 85724.

ABSTRACT

The model that tardive dyskinesia is related to denervation supersensitivity of dopamine receptors was evaluated by measurement of the effects of chronic treatment (21 days) with the antipsychotic drugs, haloperidol (Haldol, 5 mg/kg) and clozapine (30 mg/kg), on brain dopamine, muscarinic and GABA receptors. Chronic haloperidol treatment significantly increased the number of dopamine receptor binding sites in the caudate nucleus for up to 7 days after cessation of chronic treatment and was back to control levels by 14 days. By contrast, acute Haldol treatment did not alter the number of dopamine receptors. Acute injection of high dose (100 mg/kg) clozapine markedly reduced muscarinic receptor binding; this was not observed after acute or chronic treatment with 30 mg/kg.

BACKGROUND

Chronic treatment with antipsychotic drugs is complicated by the emergence of tardive dyskinesias, which are abnormal involuntary movements particularly affecting orofacial structures (1-3). Since a prominent effect of the antipsychotic drugs is blockade of brain dopamine receptors, it has been hypothesized that tardive dyskinesia is related to supersensitivity of dopamine receptors after prolonged antagonism by chronic drug treatment (3). This speculation is consistent with the clinical observation that dyskinesia may worsen or initially appear after reduction of the dose, while increasing the dose may reduce the symptoms, at least temporarily. Two effective antipsychotic drugs, thioridazine and clozapine, are associated with a low incidence of extrapyramidal side effects and exhibit affinity for the muscarinic cholinergic receptor in brain (4). Increasing affinity for the muscarinic receptor has been linked with a lowered incidence of motor abnormalities (4). Thus, it has been proposed that antipsychotic efficacy is directly related to dopamine receptor blockade while extrapyramidal side effects are related inversely to muscarinic cholinergic receptor blockade (4). We examined these hypotheses by measuring the effects of chronic in vivo administration of haloperidol and clozapine on dopamine, muscarinic and gamma-aminobutyric acid (GABA) receptor binding in the caudate and hippocampus of the rat brain. We also determined the effects of acute in vivo administration of a series of dopaminergic and cholinergic drugs on muscarinic binding in seven brain regions.

METHODS

Male Sprague-Dawley rats weighing an average of 185 gm were housed under diurnal lighting conditions and fed standard laboratory chow and water ad lib. Drugs and dosages used in the acute studies were atropine sulfate (1 and 5 mg/kg), oxotremorine (0.5 and 2.5 mg/kg), physostigmine (0.5 and 1.0 mg/kg), haloperidol (kindly supplied by McNeil Laboratories, 0.5 and 5.0 mg/kg), apomorphine (0.5 and 5.0 mg/kg), clozapine (kindly supplied by Sandoz Pharmaceuticals, 30 and 100 mg/kg dissolved in 10% citric acid), spiroperidol (kindly supplied by Janssen Pharmaceutica, 0.5 and 5.0 mg/kg dissolved in dilute acetic acid) and deanol (20 and 200 mg/kg). Animals were sacrificed by decapitation 30 mins after i.p. injection. Controls received 0.3 cc saline 30 mins before sacrifice. For the chronic drug studies, animals were injected i.p. with saline or haloperidol (5 mg/kg) or clozapine (30 mg/kg) daily for 21 days. Groups of chronically treated rats were sacrificed at intervals of 1 day, 7 days and 14 days following the end of the 21 days of haloperidol or saline injection and at 1 day and 7 days after termination of clozapine treatment.

After decapitation, the brains were rapidly removed and sectioned in the coronal plane at intervals of approximately 5 mm on chilled plates. With a stereomicroscope, brain regions were removed with a needle of 3.5 mm inner diameter or with a microknife. For the acute drug studies, the regions removed were nucleus accumbens, caudate-putamen, septum, cingulate cortex, amygdala, hippocampus and hypothalamus. Samples from individual rats were used to measure muscarinic cholinergic receptor binding. In the chronic drug study, the caudate nucleus and hippocampus were removed; corresponding regions from two animals were pooled and dopamine, muscarinic and GABA receptor binding was determined in aliquots of the same sample. Tissue was frozen until assayed, usually within one week.

Brain regions were homogenized in 10 volumes of ice-cold 0.05 M sodium-potassium phosphate buffer, pH 7.4, using a Brinkman Polytron PT-10 (setting no. 5, 15 secs). For dopamine receptor binding, the homogenate was diluted 4-fold and centrifuged twice at 18,000 rpm for 20 mins (Sorvall RC 2-B) with resuspension of the intermediate pellet in fresh buffer. The final pellet was resuspended to achieve a 5% homogenate. The tissue (250 µl of caudate and 1,000 µl of hippocampus) was incubated with ^{3}H-spiroperidol (1 Ci/mmole; 0.8 nM) in cold, freshly prepared 50 mM Tris buffer containing ions as follows: 120 mM NaCl, 5 mM KCl, 2 mM CaCl$_2$ and 1 mM MgCl$_2$, to give a final pH of 7.1 at 37°C. Total volume of the incubation mixture was 10 ml. After incubation at 37°C for 30 mins, the reaction was terminated by rapid filtration under vacuum through Whatman GF/B filters and rinsed 3 times with 5 ml of ice cold 50 mM Tris buffer. Radioactivity in the filter was extracted after solubilization for approximately 12 hours in 10 ml of scintillation cocktail (consisting of 1 l of Triton X-100, 2 l of toluene and 16 gm of Omnifluor). Parallel tubes to assess non-specific binding included the above plus 0.1 µM (+) butaclamol. Specific binding was calculated as the total minus non-specific binding. This method is similar to that reported by Burt et al. (5).

Muscarinic receptor binding was determined using the specific ligand ^{3}H-quinuclidinylbenzilate (QNB) as reported by Yamamura and Snyder (6). Aliquots of the homogenate were centrifuged at 48,000 xg for 10 mins and the supernatant discarded. The samples were rehomogenized, centrifuged and the supernatant discarded. The final pellet was resuspended in 19 volumes of 0.05 M sodium-potassium phosphate buffer. The samples were analyzed in a final total volume of 0.2 ml, containing 1 nM ^{3}H-QNB. A parallel set of tubes to determine non-specific binding contained 1 µM atropine. After incubation at 37°C for 1 hr, the reaction was terminated by rapid filtration under vacuum through Whatman GF/B filters. The incubation tubes were twice washed with 5 ml of ice-cold phosphate buffer and the filters were then twice washed with 5 ml of the phosphate buffer. The filters were placed in vials containing 12 ml of Triton X-100:toluene phosphor for approximately 12 hrs and the radioactivity measured by liquid scintillation spectrometry.

GABA receptor binding was measured using ^{3}H-GABA as described (7). The homogenate was washed, centrifuged at 18,000 rpm for 20 mins and resuspended three times to remove endogenous GABA. The pellets were resuspended in 20 volumes (for hippocampus) or 40 volumes (for caudate) of 0.05 M Tris-citrate buffer, pH 7.1. Duplicate 200 µl aliquots of homogenate were added to 2 ml of 0.05 M Tris-citrate buffer containing 10 nM ^{3}H-GABA (35 Ci/mmole) alone or in the presence of 0.5 mM unlabeled GABA to measure non-specific binding. The reaction mixture was incubated at 4°C for 5 min and the reaction was terminated by centrifugation for 10 min at 48,000 xg. The supernatant fluid was discarded and the pellet rinsed rapidly and superficially with 5 ml of the Tris-citrate buffer and then by 10 ml of ice-cold distilled water. The radioactivity was extracted with 10 ml of Triton-toluene-Omnifluor and the radioactivity was counted by liquid scintillation spectrometry at an efficiency of 46%. Specific ^{3}H-GABA binding was calculated by subtracting non-specific binding from the total bound radioactivity.

Protein content of the aliquots was determined by the method of Lowry et al. (8). Receptor binding is expressed as femtomoles per mg protein. Statistical analysis was conducted using the t-test.

RESULTS

Acute administration of drugs active on dopaminergic and muscarinic systems failed to consistently alter brain muscarinic receptor binding, with the exception of clozapine (Fig. 1). At the high dose of 100 mg/kg but not at 30 mg/kg, clozapine reduced the number of ^{3}H-QNB binding sites in the caudate-putamen, cingulate cortex, amygdala, hippocampus and hypothalamus but not in the nucleus accumbens or septum. Injection of two different doses of haloperidol and the related drug spiroperidol did not prominently alter ^{3}H-QNB binding.

Chronic treatment with haloperidol (5 mg/kg) significantly increased the number of dopamine receptor binding sites in the caudate nucleus when measured at intervals of 1 day and 7 days after chronic treatment and was similar to control values by 14 days after treatment (Table 1). By contrast, haloperidol did not alter dopamine receptor binding in the hippocampus or ^{3}H-QNB binding in the caudate or GABA binding in the caudate or hippocampus. However, ^{3}H-QNB binding in the hippocampus was significantly reduced 14 days after chronic haloperidol treatment. Clozapine did not alter binding to any of the three ligands in either caudate or hippocampus when measured at 1 day or 7 days after chronic treatment.

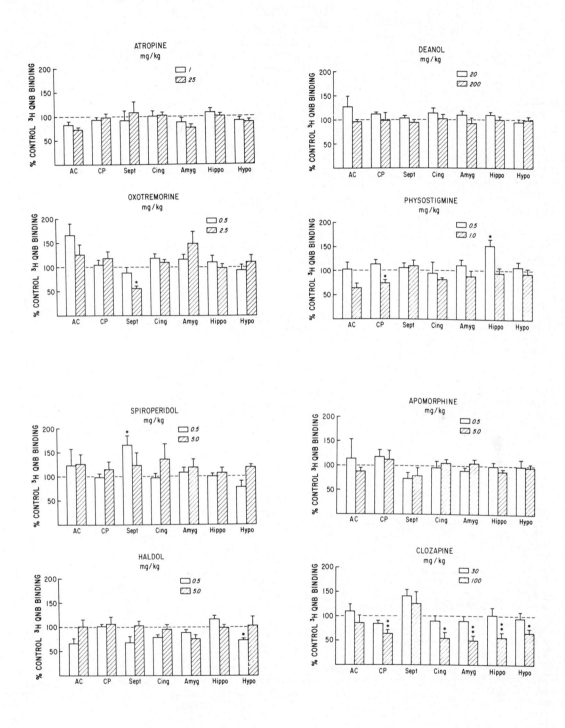

Fig. 1. Effects of acute administration of drugs on
^{3}H-QNB binding in rat brain

Drugs were injected i.p. 30 min before sacrifice. Brain regions were dissected and samples frozen until assayed for ^{3}H-QNB binding. Abbreviations are: AC, nucleus accumbens; CP, caudate-putamen; Sept, septum; Cing, cingulate cortex; Amyg, amygdala; Hipp, hippocampus; Hypo, hypothalamus. Significance was determined by t-test (*p<.01, **p<.001).

408

TABLE 1 Time Course and Effects of Chronic Haloperidol and
Clozapine Treatment on Brain Receptor Binding

Receptor (Region)	Treatment	Receptor Binding (fmoles/mg protein) Interval after stopping treatment		
		1 day	7 days	14 days
DA (Caudate)	Control	189 ± 16	184 ± 7	203 ± 15
	Haloperidol	251 ± 13*	250 ± 11***	201 ± 14
	Clozapine	216 ± 6	139 ± 23	—
DA (Hippocampus)	Control	7.4 ± 1.8	8.7 ± 1.9	13.0 ± 2.9
	Haloperidol	8.0 ± 2.1	12.3 ± 6.1	9.9 ± 1.3
	Clozapine	2.9 ± 2.3	10.8 ± 2.3	—
Muscarinic (Caudate)	Control	535 ± 19	506 ± 45	538 ± 44
	Haloperidol	607 ± 32	576 ± 29	466 ± 42
	Clozapine	591 ± 37	396 ± 71	—
Muscarinic (Hippocampus)	Control	545 ± 25	709 ± 62	680 ± 41
	Haloperidol	561 ± 23	615 ± 50	466 ± 34**
	Clozapine	812 ± 119	531 ± 63	—
GABA (Caudate)	Control	243 ± 56	287 ± 45	250 ± 63
	Haloperidol	274 ± 58	254 ± 16	277 ± 52
	Clozapine	318 ± 66	401 ± 92	—
GABA (Hippocampus)	Control	131 ± 27	143 ± 29	112 ± 20
	Haloperidol	161 ± 42	188 ± 43	114 ± 22
	Clozapine	199 ± 45	124 ± 8	—

Rats received daily i.p. injections of saline, haloperidol (5 mg/kg) or clozapine (30 mg/kg) for 3 weeks and were killed 1, 7 or 14 days later. Caudate nucleus and hippocampus were immediately removed and frozen until assayed for receptor binding. Each value is the mean ± S.E.M. for 4-6 rats in each group. Significance determined by t-test (*p<.02, **p<.01, ***p<.001).

DISCUSSION

The hypothesis that links tardive dyskinesia to denervation supersensitivity of brain dopamine receptors is consistent with much of the clinical and laboratory observations. However, it is largely inferential and direct substantive evidence has been lacking. Our observation that chronic haloperidol administration increased dopamine receptor binding sites in the caudate nucleus by 33-36% provides evidence which is more direct. That this is selective is suggested by the lack of effect of haloperidol on dopamine receptor binding in the hippocampus and on muscarinic and GABA receptor binding as well. Similar increases in the range of 18-50%, in dopamine receptor binding in the caudate by chronic haloperidol has also been reported by others (9,10) but these studies did not examine regions other than the striatum and did not examine receptors other than dopamine. The time course of this effect on dopamine receptors is similar to that reported by Burt et al. (9), who reported an increase in dopamine receptor binding at 5 days after stopping chronic treatment, with a return to control levels by 17 days. Increased dopamine receptor binding was entirely attributable to an increase in the number of binding sites rather than to a change in affinity (9). This increase in the number of dopamine receptor sites in the caudate may underlie the behavioral supersensitivity to apomorphine observed after chronic treatment with antipsychotic drugs (11-13). The reason for the reduced 3H-QNB binding in the hippocampus 14 days after treatment is unclear, particularly since there was no change during the earlier intervals or in other receptors.

Compared to haloperidol, clozapine has a lower affinity for the dopamine receptor but a higher affinity for the muscarinic receptor (5,6). In addition, clozapine has been associated with a low incidence of tardive dyskinesia. Thus, it was of particular interest to examine the effects of clozapine and to compare these to haloperidol. Marked reduction of 3H-QNB binding to 50-66% of control values in 5 of 7 regions was obtained after acute injection of 100 mg/kg but not following the lower dose of 30 mg/kg. Persistent binding of the drug to receptor sites is a possible explanation for this reduction in 3H-QNB binding. The tissue samples in both the acute and chronic study were washed twice in an effort to remove any remaining drug. To further evaluate for persistent binding, striatum was washed 4 times after acute injection of 100 mg/kg of clozapine. However, even after this additional washing, 3H-QNB binding was still reduced to 55% of control values (p<.05) (unpublished observation by the authors). Atropine, a potent and specific muscarinic antagonist, did not alter 3H-QNB binding, even after acute administration of up to 306 mmoles/kg. Since the IC_{50} for atropine (ie., concentration of the drug which displaced specific ^{3}H-QNB binding by 50%) is 13-26 x greater than for clozapine (4,6), high affinity alone does not provide sufficient explanation. No alteration of dopamine, muscarinic or GABA receptor binding was produced by clozapine when examined 1 and 7 days after terminating chronic drug treatment. This is possibly dose related, since the dose of 30 mg/kg used in the chronic study failed to alter muscarinic binding in the acute drug study. In the chronic study, the higher dose of 100 mg/kg was lethal in a large number of animals and could not be used.

Haloperidol has been shown to reduce GABA content in the striatum by 37% but produced no change in either glutamic acid decarboxylase or GABA-transaminase (14) and in the present study, no change in caudate or hippocampal ^{3}H-GABA binding.

Our data indicate that chronic treatment with the antipsychotic drug haloperidol is associated with an increase in ^{3}H-spiroperidol binding sites in the caudate nucleus. This increase in dopamine receptors may provide a suitable laboratory model for tardive dyskinesia.

RMK is supported by a Clinical Investigator award, VA and MH 26072. HIY is supported by RSDA MH 00095, MH 27257 and CCHD.

REFERENCES

1. R.M. Kobayashi, Orofacial dyskinesia. Clinical features, mechanisms and drug therapy, West. J. Med. 125, 277 (1976).
2. R.M. Kobayashi, Drug therapy of tardive dyskinesia, New Eng. J. Med. 296, 257 (1977).
3. H.L. Klawans, The pharmacology of tardive dyskinesias, Am. J. Psychiatry 130, 82 (1973).
4. S.H. Snyder, D. Greenberg and H.I. Yamamura, Antischizophrenic drugs and brain cholinergic receptors, Arch. Gen. Psychiatry 31, 58 (1974).
5. D.R. Burt, I. Creese and S.H. Snyder, Properties of ^{3}H-dopamine binding associated with dopamine receptors in calf brain membranes, Mol. Pharmacol. 12, 800 (1976).
6. H.I. Yamamura and S.H. Snyder, Muscarinic cholinergic binding in rat brain, Proc. Nat. Acad. Sci. 71, 1725 (1974).
7. S.J. Enna and S.H. Snyder, Properties of gamma-aminobutyric acid (GABA) receptor binding in rat brain synaptic membrane fractions, Brain Res. 100, 81 (1975).
8. O.H. Lowry, N.J. Rosebrough, A.L. Farr and R.J. Randall, Protein measurement with the Folin phenol reagent, J. Biol. Chem., 193, 265 (1951).
9. D.R. Burt, I. Creese and S.H. Snyder, Antischizophrenic drugs: Chronic treatment elevates dopamine receptor binding in brain. Science 196, 326, (1977).
10. P. Muller and P. Seeman, Increased specific neuroleptic binding after chronic haloperidol in rats, Neuroscience Abstracts 1, 874 (1976).
11. H.L. Klawans, Jr. and R. Rubovits, An experimental model of tardive dyskinesia, J. Neural Transm. 33, 235 (1972).
12. D. Tarsy and R.J. Baldessarini, Behavioral supersensitivity to apomorphine following chronic treatment with drugs which interfere with the synaptic function of catecholamines, Neuropharmacol. 13, 927 (1974).
13. P.F. VonVoightlander, E.G. Losey and H.J. Triezenberg, Increased sensitivity to dopaminergic agents after chronic neuroleptic treatment, J. Pharmacol. Exp. Therap. 193, 88 (1975).
14. J-S. Kim and R. Hassler, Effects of acute haloperidol on the gamma-aminobutyric acid system in rat striatum and substantia nigra, Brain Res. 88, 150 (1975).

EFFECT OF STRIATAL KAINIC ACID LESIONS ON MUSCARINIC CHOLINERGIC RECEPTOR BINDING: CORRELATION WITH HUNTINGTON'S DISEASE

Henry I. Yamamura[1], Robert E. Hruska,[1] Robert Schwarcz[2] and Joseph Coyle[2]

[1]Department of Pharmacology, College of Medicine, University of Arizona Health Sciences Center, Tucson, Arizona 85724 and [2]Department of Pharmacology and Experimental Therapeutics and Psychiatry and the Behavioral Sciences, The Johns Hopkins University, School of Medicine, Baltimore, Maryland 21205.

ABSTRACT

Muscarinic cholinergic receptor alterations in kainic acid-induced lesions in the rat caudate nuclei have been examined. We find significant decreases in muscarinic cholinergic receptor binding in lesioned caudate nuclei when compared to contralateral non-lesioned caudate nuclei. Kainic acid injection induces biochemical changes in the rat caudate nuclei that resemble those seen in Huntington's disease and suggest that kainic acid-induced lesions of the caudate nuclei may serve as a useful animal model for Huntington's disease.

INTRODUCTION

The degeneration of neurons in the rat caudate nuclei after kainic acid injections appears to resemble the neurochemical findings in the neostriata of postmortem samples from patients with Huntington's Disease (HD). (1).

In patients dying of HD, an autosomal dominant neurological disorder, there are selective decreases in glutamic acid decarboxylase (GAD) and choline acetyltransferase (ChAc) activities within the neostriatum (2-5). Since these enzymes have been used as markers for the presence of GABAergic and cholinergic neurons respectively, the decrease in their activities tend to support a selective loss of GABA and acetylcholine containing neurons in the caudate and putamen of HD brains.

Neurotransmitter receptor binding studies have been performed to investigate alterations in receptor densities in the neostriatum of postmortem HD brains (6-11). In these studies significant decreases in dopamine, serotonin and muscarinic cholinergic receptor binding have been shown to occur while no alterations are seen in GABAergic and β-adrenergic receptor densities (8). No significant deviation in receptor affinity for these binding agents were observed suggesting that the decrease in dopaminergic, serotonergic and muscarinic cholinergic receptor densities is associated with striatal neuronal degeneration.

Recently, Coyle and his associates, have reported that microinjection of kainic acid, a rigid analogue of glutamate, into the rat caudate nuclei causes degeneration of 70-85% of the neuronal cell bodies while leaving axons and nerve terminals intact (12). Neurochemical examination of rat caudate nuclei after kainic acid lesions showed that the activities of both ChAc and GAD are reduced about 70% while the activity of tyrosine hydroxylase is elevated approximately 50% (13).

Since the striatal injections of kainic acid produce profound decreases in the presynaptic neurochemical markers in the rat caudate nuclei, we have examined for neurotransmitter receptor alterations in the kainic acid-induced caudate nuclei. To evaluate the muscarinic cholinergic receptor alteration in kainic acid lesions, we have used the potent and specific antagonist, 3-quinuclidinyl benzilate (QNB) radiolabeled to a high specific activity (14).

METHODS

2 ug of kainic acid (Sigma Chemical Co.) in 1 μl of sodium phosphate buffered saline were microinjected into the left corpus striatum (coordinates = 7.9A; 2.6L; 4.8V) according to the procedure of Coyle and Schwarcz (12). The rats were allowed to recover for 5 days after kainic acid injection at which time they were decapitated and their brains rapidly removed and dissected in the cold room. The muscarinic cholinergic receptor was measured in striatal homogenates by the method of Yamamura and Snyder (14) with [^{3}H]-3-quinuclidinyl benzilate (QNB; 13 Ci/mmol; Amersham/Searle Corp.) as the ligand.

Briefly, ^{3}H-QNB binding was studied by incubating aliquots of washed rat caudate nuclei homogenate in 2 ml of 0.05 M sodium-potassium phosphate buffer (pH 7.4) for 60 min at 37° with varying concentrations of ^{3}H-QNB in the presence and absence of 1 uM atropine. The amount of ^{3}H-QNB displaced by the atropine is termed specifically bound ^{3}H-QNB. The reaction was terminated by rapid filtration through Whatman GF/B glass fiber filters, followed by four 5 ml rinses with ice cold buffer. Bound ^{3}H-QNB retained on the filter was extracted into 8 ml of a toluene based scintillation cocktail and the radioactivity was monitored in a Searle Mark II liquid scintillation counter with 45% efficiency.

Protein contents were determined by the method of Lowry using bovine serum albumin as a standard (15).

The significance of the difference between groups was analyzed using a two-tailed students "t" test.

RESULTS AND DISCUSSION

Specific binding of ^{3}H-QNB to membrane preparations from kainic acid lesioned caudate nuclei is reduced 43% below values for the contralateral non-lesioned caudate nuclei (Table 1). The decline of ^{3}H-QNB binding induced by kainic acid lesions of rat caudate nuclei parallels the reduction in binding of ^{3}H-QNB in HD caudate nuclei (Table 1).

TABLE 1 Muscarinic Cholinergic Receptor Binding in Caudate Nuclei from Kainic Acid Lesioned Rats and Huntington's Disease Patients

Kainic Acid Lesions			HD
Lesioned	Non-Lesioned	Δ%	Δ%
^{3}H-QNB Bound (fmol/mg tissue)			
43 $\pm$ 5.1 (14)	77.0 $\pm$ 11.6 (14)	-43**	-52;-64* a,b

Rats received a unilateral injection of 2 ug in 1 ul of kainic acid and sacrificed after 5 days. Data are the means $\pm$ S.E.M. from Scatchard analysis. Data are the averages of duplicate determinations for each individual sample with the number of samples in parentheses. For Huntington's Disease caudate, the values are taken from [a]Wastek et al (1976); [b]Enna et al (1976).

* P < 0.01

** P < 0.025

We evaluated the effect of kainic acid lesions on the apparent dissociation constant (K_D) as well as on the total number of binding sites (Bmax) (data is not shown). The K_D values were 11 to 13 pM for the lesioned and non-lesioned rat caudate nuclei, respectively, as determined from the rate of association (7 X 10^8 M^{-1} min^{-1} for both lesioned and non-lesioned caudate nuclei)and from the rate of dissociation (9 X 10^{-1} min $^{-1}$ for the lesioned and 8 X 10^{-1} min $^{-1}$ for the non-lesioned caudate nuclei).

The apparent K_D was also determined from saturation isotherms (data is not shown). The analyses of the saturation curves by Scatchard plots revealed a K_D value of 20 to 50 pM for both the lesioned and non-lesioned caudate nuclei. Therefore, the kainic acid-induced lesions had no significant effect on the apparent dissociation constant. We have reported similar findings in postmortem HD brain samples (7). However, kainic acid lesions do appear to decrease markedly the density of specific ^{3}H-QNB binding sites as determined from Scatchard analysis.

Since the alterations in muscarinic cholinergic receptor binding effected by kainic acid lesions of the rat caudate nuclei parallels the cholinergic receptor binding alterations found in HD caudate nuclei, these findings strengthen the usefulness of the kainic acid-induced lesions as an animal model of Huntington's disease.

REFERENCES

(1) J.T. Coyle, R. Schwarcz, J.P. Bennett and P. Campochiaro, Clinical, Neuropathologic and Pharmacologic Aspects of Huntington's Disease: Correlates with a New Animal Model. Neuropsychopharm. (in press).

(2) E.D. Bird, A.V.P. Mackay, K.N. Rayner and L.L. Iversen, Reduced Glutamic Acid Decarboxylase Activity of Postmortem Brain in Huntington's Chorea, Lancet 1,1090 (1973)

(3) E.D. Bird and L.L. Iversen, Huntington's Chorea: Postmortem Measurement of Glutamic
     Acid Decarboxylase, Choline Acetyltransferase and Dopamine in Basal Ganglia,
     Brain 97, 457 (1974).

(4) T.L. Perry, S. Hansen and M. Kloster, Huntington's Chorea: Deficiency of Gamma Amino-
     butyric Acid in Brain, N. Eng. J. Med. 288, 337 (1973).

(5) W.L. Stahl and P.D. Swanson, Biochemical Abnormalities in Huntington's Chorea Brains,
     Neurology 24, 813 (1974).

(6) T.D. Reisine, J.Z. Fields, G.J. Wastek, P.C. Johnson and H.I. Yamamura, Dopaminergic
     Receptor Alterations in Huntington's Disease (submitted).

(7) H.I. Yamamura, G.J. Wastek, P.C. Johnson and L.Z. Stern, Biochemical Characterization
     of Muscarinic Cholinergic Receptors in Huntington's Disease, Symposium on
     Cholinergic Mechanisms and Psychopharmacology, 1977.

(8) S.J. Enna, E.D. Bird, J.P. Bennett, D.B. Bylund, H.I. Yamamura, L.L. Iversen and
     S. H. Snyder, Huntington's Chorea: Changes in Neurotransmitter Receptors in the
     Brain, N. Engl. J. Med. 294, 1305 (1976).

(9) G.J. Wastek, L.Z. Stern, P.C. Johnson and H.I. Yamamura, Huntington's Disease: Regional
     Alterations in Muscarinic Cholinergic Receptor Binding in Human Brain, Life Sci.
     19, 1033 (1976).

(10) S.J. Enna, L.Z. Stern, G.J. Wastek and H.I. Yamamura, Neurobiology and Pharmacology of
     Huntington's Disease, Life Sci. 20, 205 (1977).

(11) C.R. Hiley and E.D. Bird, Decreased Muscarinic Receptor Concentration in Postmortem
     Brain in Huntington's Chorea, Brain Res. 80, 355 (1974).

(12) J.T. Coyle and R. Schwarcz, Lesion of Striatal Neurones with Kainic Acid Provides
     a Model for Huntington's Chorea, Nature 263, 244 (1976).

(13) R. Schwarcz and J.T. Coyle, Striatal Lesions with Kainic Acid: Neurochemical Character-
     istics. Brain Res. (in press).

(14) H.I. Yamamura and S.H. Snyder, Muscarinic Cholinergic Binding in Rat Brain, Proc.Nat'l.
     Acad. Sci. U.S.A. 71, 1725 (1974).

(15) O.H. Lowry, N.J. Rosebrough, A.L. Fair and R.J. Randell, Protein Measurement with the
     Folin Phenol Reagent, J. Biol. Chem. 193, 265 (1951).

DISCUSSION

Dr. Cooper commented that in his own work on kainic acid, he had observed a very marked reduction in striatal mass when animals were sacrificed two weeks after kainic acid injection. There was about a 70% reduction of striatal mass after 3 μg and about 20-25% reduction in animals injected with 1 μg. *Dr. Yamamura* stated that he had given his animals 2 μg of kainic acid in 1 μl of buffer. His striata assays were both on a per mg basis and on mg protein. At 5 days, there was a marked reduction in the protein content of striata. *Dr. Cooper* suggested that his results indicated an initial edema of the striata, followed by a weight loss as the striata were phagocytized. As far as tyrosine hydroxylase is concerned, on a per unit weight basis there is little change. However, if one looks at activity per striatum, there is great depletion since the mass is only about 20% of that before kainic acid. Dr. Cooper concluded with the comment that there is also a reduction in striatal mass in Huntington's Disease.

BEHAVIORAL AND NEUROCHEMICAL EFFECTS
OF CENTRAL CATECHOLAMINE DEPLETION:
A POSSIBLE MODEL FOR "SUBCLINICAL" BRAIN DAMAGE*

Michael J. Zigmond and Edward M. Stricker

Departments of Life Sciences and Psychology
University of Pittsburgh, Pittsburgh, Pennsylvania 15260

INTRODUCTION

The contribution of brain catecholamines (CAs) to behavior was first suggested in 1956, when Holzbauer and Vogt (1) demonstrated that the concentration of norepinephrine (NE) in brain was lowered after administration of the antipsychotic sedative agent, reserpine. Since that time, considerable psychopharmacological evidence has accumulated indicating that NE and dopamine (DA) both have important roles in the mediation of central arousal during the behavioral response to diverse sensory stimuli (2,3). For example, both electrocortical activation and behavioral responsiveness are decreased by drugs that depress activity at central catecholaminergic synapses, whereas both responses are increased by drugs that augment activity at those synapses.

Given the apparent importance of CAs in brain function, it is not surprising that dysfunctions in central catecholaminergic pathways have been implicated in a variety of neurological and psychiatric disorders, such as schizophrenia, Minimal Brain Dysfunction, and Parkinsonism. In many cases, the connection between CAs and the dysfunction is based largely on the therapeutic effectiveness of drugs such as amphetamine and chlorpromazine, which are known to affect central catecholaminergic neurotransmission. In fact, such disorders would be classified as "subclinical" from the standpoint of the neurologist because they are accompanied by few, if any, classical or "hard" neurological symptoms of brain damage, a finding that has led some to deny the existence of an underlying organic pathology. Moreover, in other disease states, such as Parkinsonism, neurological symptoms do not begin to appear until long after the presumed onset of damage to the central nervous system. Thus, even in a circumstance where one can be sure of neurological damage, it is possible to identify a lengthy preclinical period.

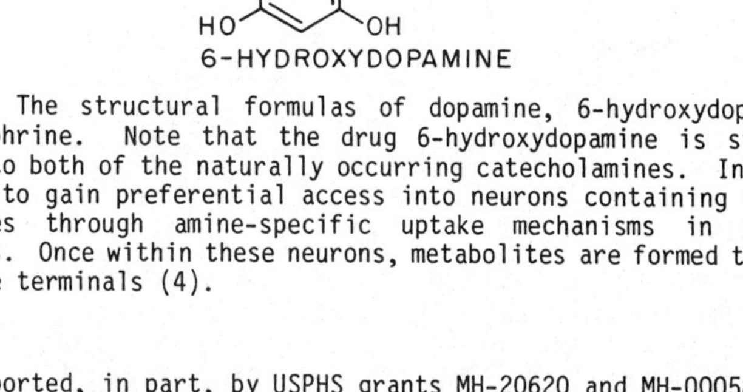

Fig. 1. The structural formulas of dopamine, 6-hydroxydopamine, and norepinephrine. Note that the drug 6-hydroxydopamine is structurally similar to both of the naturally occurring catecholamines. Indeed, it is believed to gain preferential access into neurons containing these catecholamines through amine-specific uptake mechanisms in their axon terminals. Once within these neurons, metabolites are formed that destroy the nerve terminals (4).

*This work was supported, in part, by USPHS grants MH-20620 and MH-00058. Some of this work was carried out with the collaboration of M. Friedman, T. Heffner, G. Kapatos, J. Marshall, C. Volz, and M. Zimmerman. Technical assistance was provided by L. Howdyshell, D. McKeag, S. Wuerthele, and J.-S. Yen.

For the past several years we have been investigating the behavioral effects of selective destruction to central noradrenergic and dopaminergic projections in the adult male albino rat. We believe our findings may help to explain the above paradoxes of subclinical brain damage; that is to say, the appearance of behavioral disorders in the absence of conspicuous organic pathology, and the lack of prominent neurological symptoms despite known brain lesions.

In the studies to be reported here, 100-200 μg 6-hydroxydopamine (6-HDA) was injected into the cerebrospinal fluid of rats by way of the lateral ventricle to produce permanent depletions of CAs throughout the brain (Fig. 1). This procedure results in little detectable tissue damage and no loss of other known neurotransmitters (5-7). In most cases, we have further limited the lesion by pretreating animals with desmethylimipramine (25 mg/kg, i.p.) and either pargyline (50 mg/kg, i.p.) or tranylcypromine (5 mg/kg, i.p.), thereby blocking the effects of 6-HDA on noradrenergic neurons while potentiating the loss of DA (Table 1). While some nonspecific effects of this treatment undoubtedly occur, results obtained with this procedure are consistent with the effects that follow injections of 6-HDA into the cisterna magna or the placement of 6-HDA or electrolytic lesions along the ascending DA projections, methods which cause similar depletions of DA but which produce different patterns of nonspecific damage (8-12).

TABLE I

EFFECT OF 6-HYDROXYDOPAMINE TREATMENTS ON BRAIN CATECHOLAMINES AND INGESTIVE BEHAVIORS

| Group[a] | Norepinephrine (μg/gm)[b] | | | Dopamine (μg/gm)[b] | | Ingestive Behavior |
	Dien-cephalon	Brain stem	Telen-cephalon	Striatum	Telen-cephalon	
Control[c]	0.76+0.04	0.38+0.02	0.28+0.01	8.27+0.49	0.80+0.11	Normal
6-HDA	0.14+0.02	0.13+0.02	<0.02[d]	3.44+0.16	0.19+0.03	Mild anorexia and hypodipsia
Pargyline + 6-HDA	0.09+0.01	0.14+0.01	<0.02[d]	<0.04[d]	<0.10[d]	Aphagia and adipsia
DMI, pargyline + 6-HDA	--[e]	--	0.25+0.02	<0.04[d]	--	Aphagia and adipsia

[a]Rats received 6-HDA (200 μg, intraventricularly), alone or 30 minutes after pretreatment with pargyline (50 mg/kg, i.p.), or pargyline and desmethylimipramine (DMI) (25 mg/kg, i.p.). Rats given 6-HDA alone or after pargyline pretreatment received two treatments, 48 hours apart; rats given 6-HDA after pargyline and DMI received a single treatment. Animals were killed at least 2 weeks after treatment, and tissues were prepared and analyzed spectrophotofluoro-metrically.

[b]Values are mean + standard error of the mean for 4-10 animals, expressed as micrograms of catecholamine per gram of fresh brain weight.

[c]There was no significant effect on catecholamine concentrations of any vehicle injections or sham lesions, and control values have been pooled.

[d]Below the sensitivity of the assay.

[e]No data available.

Central CA-containing neurons originate in the brain stem and project throughout the brain. Their thin, unmyelinated fibers are difficult to detect by standard histological techniques; indeed, until their visualization by histochemical fluorescence, their existence was not generally recognized. Accordingly, we measure destruction of the catecholaminergic neurons by biochemical determination of CA levels in the telencephalon, an area where normally a high concentration is found.

SOME EFFECTS OF CENTRAL DOPAMINE-DEPLETING LESIONS

Damage to central catecholaminergic projections produces few, if any prominent behavioral dysfunctions. For example, using 6-HDA, we have produced DA depletions of up to 80-90% and have found that in most obvious respects the lesioned rats are indistinguishable from vehicle-injected controls despite their extensive brain damage. A more careful examination of these animals, however, reveals significant deficits in their ability to behave appropriately in response to a variety of stimuli. For example, unlike control animals, the lesioned rats do not increase their food intakes when given large doses of insulin or 2-deoxyglucose, do not build nests when placed in a cold environment, do not increase their water intakes when made hypotensive, and do not perform a learned response in order to avoid an electric shock (13-19).

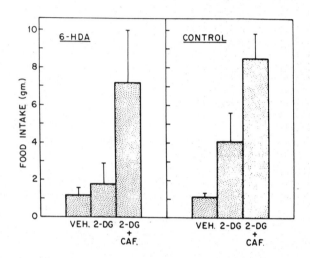

Fig. 2. The effect of caffeine on feeding induced by 2-deoxyglucose (2-DG) in 6-HDA-treated rats or unlesioned controls. Lesioned animals were given two ivt. injections of 200 µg 6-HDA. Caffeine (25 mg/kg, i.p.) was given concurrently with 2-DG (375 mg/kg, i.p.) just before the 4-hr feeding test. Intakes were compared with the mean food consumption during three days of pretesting when the vehicle solution (0.15 M NaCl) was injected instead. Values are means + S.E.M. for 4-6 rats. Rats given 6-HDA showed a significantly greater food intake when pretreated with caffeine than when 2-DG alone was given (P < .001) (from Ref. 24).

Although it is possible that the individual behavioral dysfunctions are unrelated, it seems more likely that they reflect a common inability of the lesioned animal to perform well during a period of pronounced physiological or psychological stress.* However, there are several illuminating instances in which rats with central CA-depleting lesions become able to behave appropriately despite acute and profound stress. First, these animals drink much more readily in response to dehydration at night (20,21), when the capacity for CA release may be greater than in the day (22,23). Second, they eat and drink more normally when they are pretreated with caffeine (Fig. 2) or apomorphine (24,25), drugs that should produce a temporary increase in central dopaminergic activity. Finally, they eat in response to severe glucoprivation after receiving moderate doses of insulin each day for several weeks, and this improvement is paralleled by an increase in tyrosine hydroxylase activity in brain (Fig. 3) (17,26).

In contrast with the subtle behavioral effects of the above lesions, specific destruction of 90% or more of the ascending dopaminergic neurons leads to marked functional impairments in rats (14,16). These animals are akinetic and cataleptic, and will not respond to normal sensory stimulation. For example, they will not feed themselves and will die of starvation unless maintained by intragastric intubation of liquid nutrients. Nevertheless, it appears

---

*A parallel may be drawn between these findings and those from studies of damage to the sympathetic nervous system. Damage to the peripheral neurons, like damage to the central ones, is not usually accompanied by obvious dysfunctions under controlled laboratory conditions. However, when exposed to more stressful situations, animals with central lesions display behavioral inadequacies, while sympathectomized animals show marked physiological impairments.

that the lesioned animals have not lost the capacity to behave, since normal responses can be temporarily reinstated in such animals by a variety of procedures. For example, although they will not eat dry laboratory chow or drink water, they may consume palatable foods and fluids such as chocolate milk and mixed cereal (16,27). These ingestive behaviors are potentiated by the administration of amphetamine or other nonspecific activators (e.g., tail pinch) (12,18,28). In addition, rats that are akinetic in their individual home cages show normal motor activity and exploratory behavior when placed in a group cage with other rats or in a shallow ice bath, and swim normally when placed in a large tank of tepid water (29). In each case, a higher threshold of arousal appears to be required in order to initiate the appropriate behavior. However, the therapeutic effects dissipate rapidly when the animals are removed from the activating situation, and the rats then respond as poorly as they had prior to activation (Fig. 4).

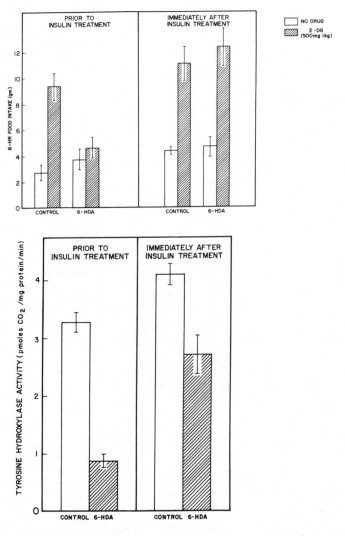

Fig. 3. The effects of chronic treatment with insulin on (upper) food intake induced by 2-deoxyglucose (2-DG) and (lower) tyrosine hydroxylase activity in hippocampus, in 6-HDA-treated rats or unlesioned controls. Lesioned animals were given two intraventricular injections of 200 µg 6-HDA. Two weeks later, food intake was measured for 6 hr following 2-DG (500 mg/kg, i.p.) and was compared with the mean food consumption during three days of pretesting when no drug was given. Half of the animals were then treated for 16 days with protamine-zinc insulin (1-8 U/day, s.c.). Four days after the last administration of insulin, feeding to 2-DG was again determined. Representative animals were sacrificed 24 hr after the 2-DG tests for analysis of tyrosine hydroxylase. Values are means ± S.E.M. for 4-8 rats (from Ref. 26).

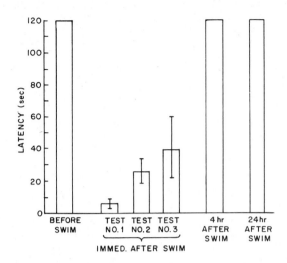

Fig. 4. Latencies for 6-HDA-treated rats to turn around on an inclined
cage before 5 min of swimming, and at various times afterwards. Lesioned
animals were given four intraventricular injections of 200 µg 6-HDA, each
preceded by an injection of desmethylimipramine (25 mg/kg, i.p.) and the
fourth also preceded by an injection of tranylcypromine (5 mg/kg). Values
are means + S.E.M. for 6 rats. Rats were found to be greatly improved
during the first 2 min after they were removed from the activating
situation (P < .02), but no residual effect remained 4 hr or 24 hr after
swimming (from Ref. 29). Copyright 1976 by the American Psychological
Association. Reprinted by permission.

Many of the debilitated, brain-damaged rats, when maintained by intragastric intubation of
nutrients, eventually will recover voluntary feeding and drinking behaviors and otherwise
appear normal (Fig. 5) (16,27). However, these animals are even more intolerant of
glucoprivation, cold stress, and hypotension than are rats with somewhat smaller depletions of
DA (18). Indeed, they occasionally revert to their prior unresponsive state when exposed to a
prolonged and intense stimulus, and it may be many weeks, if ever, before they "recover" again.

If destruction of central dopaminergic neurons provides the basis for the severe behavioral
impairments that initially are observed after the brain lesions, then it is not immediately
apparent how to account for recovery of function since the DA depletions are permanent. One
possibility is that resumption of feeding, drinking, and other behaviors results from
functional recovery of the damaged pathway that is not revealed by measurements of amine
concentration. In support of this hypothesis, we have observed that, in animals which had
recovered from the initial impairments following 6-HDA-induced brain lesions, these
dysfunctions can be reinstated by further disruption of central dopaminergic neurons; in fact,
the potency of agents such as α-methyltyrosine and spiroperidol is increased (Fig. 6)
(27,30,31). Moreover, animals with no detectable DA in telencephalon (i.e., less than 2% of
control values) do not recover, even if they are maintained for one year (18). These results

Fig. 5. Food and water intakes
and body weight of a rat given two
intraventricular injections of
200 µg 6-HDA, each 30 min after
pargyline (50 mg/kg, i.p.).
Intragastric (IG) feedings are
shown. The bottom line indicates
access to highly palatable foods
(from Ref. 27). Copyright 1973 by
the American Association for the
Advancement of Science.
Reprinted by permission.

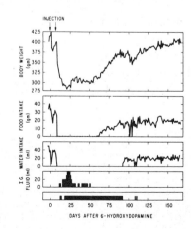

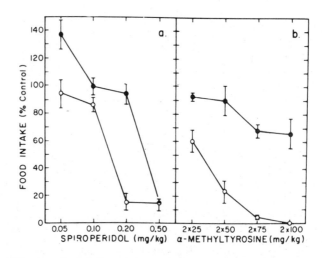

Fig. 6. Effect of 6-HDA on anorexia caused by sprioperidol or α-methyltyrosine. Rats given two intraventricular injections of 200 μg 6-HDA, following pretreatments with desmethylimipramine (25 mg/kg, i.p.) and pargyline (50 mg/kg, i.p.), or the control treatment, later received i.p. injections of (a) spiroperidol or (b) α-methyltyrosine. Each point represents the mean ± S.E.M. for 6-8 animals. Rats given 6-HDA (o) showed a significantly greater reduction in feeding than control animals (●) after either spiroperidol or α-methyltyrosine (both Ps < .01) (from Ref. 31).

would not be expected if resumption of the various behaviors had resulted from transfer of functions formerly served by the amines to another neurochemical pathway. Instead, it seems probable that the dopaminergic neurons continue to serve an important function in the lesioned animals.

COMPENSATORY CHANGES AFTER CATECHOLAMINE-DEPLETING LESIONS

We believe that most of the characteristics of animals sustaining damage to catecholaminergic projections can be explained by considering the role of these neurons in behavior and the compensatory neurochemical changes that occur at residual catecholaminergic synapses after lesions.

The low, stable firing rate of NE- and DA-containing cells under most conditions suggests the presence of a central catecholaminergic tone similar to that observed at many peripheral sites (32,33). Pharmacological studies support the presence of several feedback systems that appear designed to promote such basal tone. For example, the systemic administration of an agonist such as amphetamine is followed by a decrease in firing rate and in CA turnover, while haloperidol, an antagonist, produces an increase in these indices of catecholaminergic activity (32,33). Under conditions where such compensatory mechanisms would be inadequate to restore basal activity, a change in the responsiveness of the postsynaptic cell also has been observed. Thus, prolonged depletion of CAs with reserpine, or chronic receptor blockade with haloperidol, is followed by an enhanced behavioral response to agonists, an increase in the response of adenylate cyclase to NE or DA, and an increase in the number of apparent receptors (34-36).

Assuming that receptors are activated by CAs from many adjacent fibers, and that the communications within individual neurons are equivalent, it is possible that alterations of turnover, synthesis, and receptor activity at residual catecholaminergic synapses provide the basis for recovery of function following 6-HDA-induced brain lesions.* We will now consider

---

*This model, which proposes that amine from one terminal acts meaningfully on a relatively distant receptor as a sort of "neurohormone", is inconsistent with the traditional concept of neurotransmission. However, that concept was derived from studies of the cholinergic neuromuscular junction, a 200 Å gap, and it may not be applicable either to the noradrenergic sympathetic neuroeffector junction, which can exceed 1000 Å, or to the catecholaminergic "synapse" in the central nervous system, whose size is unknown. In this regard, it seems noteworthy that electron microscopic visualizations of central catecholaminergic neurons do not reveal thickenings at the presynaptic terminal or at the postsynaptic membrane that are characteristic of a peripheral cholinergic synapse (37).

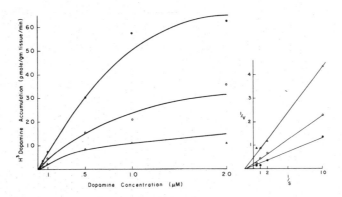

Fig. 7. The accumulation of H^{3}DA by synaptosome-rich fractions of striatum prepared from a control rat (●) and two animals treated 2 months previously with desmethylimipramine (25 mg/kg, i.p.) and 6-HDA (200 µg, ivt.). The lesioned animals showed 53% (o) and 77% (▲) loss of DA. A decrease in the Vmax for DA transport was observed in the lesioned animals which was proportional to the loss of DA.

some evidence that adaptive changes in transmitter availability and membrane sensitivity do, in fact, occur after damage to catecholaminergic projections. We will then propose a schema for recovery of function based on such events, and discuss the implications of this schema for subclinical brain damage.

Increased Turnover in Residual Catecholaminergic Neurons. In 1965, it was observed that patients who had suffered from Parkinson's disease had large depletions of DA and of homovanillic acid (one of the principal metabolites of DA) in basal ganglia, but that the ratio of homovanillic acid to DA was elevated (38). These observations suggest that an increased turnover of DA had occurred in remaining terminals. Similarly, turnover of DA in residual neurons has been shown to increase after damage to the dopaminergic neurons of the nigrostriatal bundle in experimental animals (39,40). This change may explain the enhanced sensitivity of lesioned animals to α-methyltyrosine (Fig. 6a), an effect that is proportional to the loss of DA (30), since after inhibition of synthesis an increase in turnover would increase the rate of DA depletion.

Increased Efficacy of Released Transmitter. Damage to central catecholaminergic projections is also accompanied by an increased responsiveness to CA agonists. As in the peripheral nervous system, this probably is the result of two separate processes. First, lesions produce a rapid degeneration of nerve terminals, the principal site of CA inactivation. This is reflected in a decrease in the uptake of labeled CA (Fig. 7) (41,42), and an increase in the electrophysiological response to CA (43,44). Such findings suggest that in lesioned animals

Fig. 8. The effect of L-dopa on motor activity in control rats (▲) or in rats that had been given desmethylimipramine (25 mg/kg, i.p.) and 75 µg (o), 150 µg (▲), or 200 µg (●) of 6-HDA intraventricularly. (These treatments had little effect on brain NE, but produced striatal DA depletions of 11, 42, and 86%, respectively.) Dopa was administered i.p. 5 min prior to testing and 30 min after pretreatment with RO4-4602 (50 mg/kg, i.p.), a peripheral decarboxylase inhibitor. Motor activity was measured as the number of photobeam crossings during the first hour of testing. Each point represents the mean + S.E.M. for 4-6 rats. Dopa-induced motor activity was significantly elevated by the DA-depleting lesion (Ps < .05) (from Ref. 50).

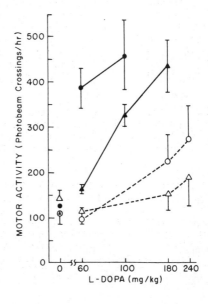

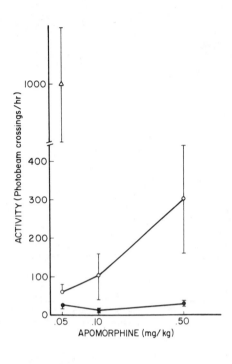

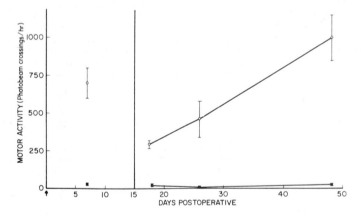

Fig. 9. The effect of apomorphine on motor activity in control rats (●) or rats that had been given desmethyl-imipramine (25 mg/kg, i.p.) and 150 µg or 200 µg 6-HDA. Apomorphine was administered i.p. 5 min prior to testing. Testing procedures as in Fig. 8. Each point represents the mean + S.E.M. for 4-7 rats.

(Upper) Dose-effect relationship 48 days after 150 µg (o) or 200 µg (Δ) 6-HDA treatment. (These treatments had little effect on NE, but depleted striatal DA by 70 and 91%, respectively.) Apomorphine-induced motor activity was significantly elevated by the larger DA-depleting lesion (Ps < .001).

(Lower) Apomorphine-induced motor activity in rats with large DA depletions (87-95%), as a function of days since treatment (arrow). By the seventh postoperative day, the response to apomorphine (.5 mg/kg, i.p.) was significantly elevated in the lesioned animals (o) but not in the controls (●). Thereafter the response to apomorphine (.05 mg/kg, i.p.) continued to increase for at least 48 days postoperatively (from Ref. 50).

released transmitter will be taken up less rapidly by surrounding terminals, making it more available to the postsynaptic membrane. Such "denervation supersensitivity" would provide an explanation for the enhanced responsiveness of lesioned rats to dopa (45,46), which, like the decrease in uptake, appears within 24 hours of 6-HDA treatment and is roughly proportional to the loss of DA (Fig. 8).

Second, lesions can produce an increase in the sensitivity of the postsynaptic membrane to direct-acting agonists. This can be seen in the enhanced responsiveness of adenylate cyclase to NE and DA in 6-HDA-treated rats (Table 2) (47-50), an increase in the number of apparent catecholaminergic receptors (36,51), and an increase in the amount of motor activity that is elicited in DA-depleted rats by apomorphine (52,53). In contrast to the enhanced sensitivity to dopa, sensitivity to apomorphine develops slowly and is readily apparent only in rats with lesions that deplete DA by at least 80% (Fig. 9) (46,50).

The Development of Additional Catecholaminergic Nerve Terminals. Long-term recovery from damage to peripheral adrenergic fibers is probably a result of regenerative sprouting from the lesioned axons. Although collateral sprouting from residual terminals has also been observed in the brain (54,55), such growth probably is not of much significance in recovery of function following damage to central catecholaminergic neurons since it is not detectable by histo-chemical fluorescence and does not result in the restoration of CA levels.

TABLE II

INCREASE IN ACTIVITY OF STRIATAL ADENYLATE CYCLASE FOLLOWING INCUBATION WITH DOPAMINE[a]

DA (μM)	Control Animals	6-Hydroxydopamine Lesioned Animals
	(pmoles/mg protein/min)[b]	
1	26.5 ± 3.0	58.0 ± 4.6
10	82.5 ± 7.3	112.0 ± 4.9
100	122.7 ± 8.4	130.2 ± 6.0

[a]Rats received desmethylimipramine (25 mg/kg, i.p.) and 6-hydroxydopamine (200 μg, ivt.) (n=6) or vehicle (n=6) at least 2 months prior to the experiment.

[b]Adenylate cyclase was measured as the rate of conversion of ^{14}C-ATP to ^{14}C-cAMP in homogenates of striatum using a Tris maleate buffer (80 mM, pH 7.4) in the presence of ATP (1 mM) and $MgCl_2$ (4 mM). No change in basal enzyme activity was detected. Values are the mean increase above this baseline $\pm$ S.E.M. Enzyme from lesioned animals was significantly more responsive to DA (p < .01).

Summary. The following description summarizes our conception of the specific mechanisms in the recovery process (Fig. 10). Soon after subtotal damage to a bundle of CA-containing fibers, CA release from the residual neurons will represent only a fraction of that which occurred before the lesion, and consequently the rate of receptor activation will be lower than normal. These decreases in CA release and receptor stimulation lead to immediate increases in CA synthesis and release from residual neurons. Together with increased efficacy of released CA due to the loss of uptake sites, these compensatory processes may provide for recovery of function. With somewhat larger lesions, the above alterations may not be sufficient to restore basal receptor activation; nevertheless, recovery of function occurs as the sensitivity of the postsynaptic membrane to CA is enhanced. Finally, it seems likely that functional adjustments within interrelated pathways utilizing other neurotransmitters also may be of importance in promoting recovery of function.

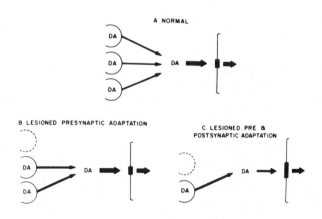

Fig. 10. A schema for recovery of function following subtotal damage to central dopamine (DA)-containing neurons. In the animal with relatively small lesions (B), increased release from residual dopaminergic terminals together with decreased reuptake of DA serve to maintain synaptic DA levels at near-normal levels (A). In the animal with relatively large lesions (C), these changes are inadequate to maintain synaptic DA levels but increases in the number of postsynaptic receptors serve to maintain synaptic function.

GENERAL IMPLICATIONS OF THE SCHEMA

There are three features inherent in this formulation that should be specified because they provide a framework for considering the behavioral impairments observed both in experimental animals after brain lesions and in human patients with subclinical brain damage.

Rapid Recovery Should Occur Despite Large Lesions. The schema provides an explanation for the observation that extensive damage to catecholaminergic neurons in laboratory animals can occur without major behavioral deficits. Following even relatively large lesions, increased CA turnover, together with decreased transmitter inactivation and adjustments in other, inter-related pathways, may restore function so rapidly that obvious behavioral deficits never become apparent. In the case of still larger lesions, more gradual processes, such as the synthesis of new membrane receptors, also are involved and recovery should be delayed. Finally, with near-total CA depletions, the residual capacity for transmitter biosynthesis should be too small to support function, even when the efficacy of released amine is enhanced, and recovery may not be possible. These predictions are supported by our findings that prominent behavioral deficits do not occur until DA has been depleted by about 90%, and that the time required for recovery is increased as depletions are increased above that level (18,27).

Lesioned Animals Should Be Intolerant of Intense Stimuli. The schema may also explain why lesioned animals are intolerant of intense stimuli, such as severe glucoprivation, cold stress, or hypovolemia. Catecholaminergic neurons normally synthesize transmitter at a rate well below their maximal potential as defined by the availability of the rate-limiting enzyme, tyrosine hydroxylase (56). This permits intact animals to increase turnover rapidly in response to stimuli that increase the demand for catecholaminergic neurotransmission. To the extent that lesioned animals already have utilized a significant portion of this reserve capacity to elevate basal turnover in residual neurons, they will have compromised their ability to respond to increased demand.

Note that this formulation implies that more normal behaviors should occur when the stimulus intensity is reduced, the efficacy of availab.. transmitter is increased, or the capacity for transmitter biosynthesis is increased. Our observations are consistent with each prediction: lesioned animals which do not increase their feeding when exposed to severe cold will increase intake when placed in less severe conditions (in order to decrease stimulus intensity), and lesioned animals which do not increase their feeding in response to 2-deoxyglucose will eat when pretreated with caffeine (to increase transmitter efficacy) or chronic insulin (to increase transmitter biosynthesis) (17,24,26).

Lesioned Animals Should Be Unresponsive to Weak Stimuli. In addition to their intolerance of intense stimuli, lesioned animals appear relatively unresponsive to weak ones. For example, during the early phases of recovery they may not eat lab chow or drink tap water but they will eat a palatable liquid diet. Later, when the initial deficits have disappeared, the animals show an abnormally delayed drinking response to a progressive loss of plasma volume, increasing their water intake only when the hypovolemic stimulus becomes substantial. This may result from the fact that fewer nerve terminals are available to respond to changes in input and that a larger increase in stimulus intensity will therefore be required to produce a given increment in catecholaminergic activity.

Taken together with the relative intolerance of lesioned rats to excessive stimulation, these findings suggest that damage to central CA-containing neurons truncates the range of stimuli that can be accommodated. Therapy may consist of pharmacological and environmental inter-ventions which expose the animal to a narrow range of inputs, enough to elicit a behavioral response but not so much that the capacity of the residual neurons is exceeded.

IMPLICATIONS OF THE SCHEMA: SUBCLINICAL BRAIN DAMAGE

We have proposed that certain neurochemical events, which occur following subtotal destruction of central catecholaminergic nerve terminals, underlie many aspects of the behavior of the lesioned animals. We also believe that these events may shed some light on the etiology, detection, and treatment of certain related dysfunctions in human patients.

Parkinson's Disease.   The best documented example of a human neurological dysfunction involving the loss of central catecholaminergic neurons is Parkinsonism. Upon autopsy, Parkinsonian patients invariably have a marked loss of DA-containing cells in the substantia nigra, a more moderate loss of NE-containing cells in the locus coeruleus, and extensive CA depletions in several brain regions, particularly the basal ganglia (57). Large DA-depleting lesions lead to most of the cardinal symptoms of Parkinsonism: hypokinesia, rigidity, and postural disorders (i.e., all but tremor). In addition, as indicated above, we have found that lesioned rats show (i) no prominent symptoms until DA · depletions exceed 90%, (ii) "paradoxical kinesia" when exposed to various brief activating situations, and

(iii) increased symptoms when confronted with physiological or psychological stress, all characteristics which parallel phenomena that are well documented in patients with Parkinson's Disease. Thus, significant symptomatology in Parkinsonism is invariably associated with DA depletions of 70% or greater (57). Moreover, a brief period of marked improvement sometimes occurs when Parkinsonian patients are presented with certain emergencies. For example, patients who have been akinetic for years have been reported to run from a burning building, and to return to their previous unresponsive condition when they reach safety moments later (58). Paradoxically, stress commonly exacerbates otherwise mild symptoms, as in the case of patients whose motor performance worsens when they become angry (58). Such effects typically are short-lived, but "traumatic Parkinson's Disease" has been observed in which permanent symptoms apparently are precipitated by physiological or emotional stress (59).

We believe that the lack of neurological deficits despite significant neuronal damage can be explained in terms of the compensatory changes which occur in residual neurons, and that these changes have implications for the effects of stress, as well. For a brief period an intense stimulus might be expected to increase transmitter release sufficiently to restore function. However, if DA release exceeds the capacity for synthesis, stores eventually will be depleted and function should be once again impaired. Moreover, it seems possible that under certain conditions stress-induced exhaustion of residual DA and the subsequent loss of all input to the postsynaptic receptors can lead to long-term impairment of those receptors, a phenomenon that may be analogous to the transsynaptic atrophy which occurs in certain sensory systems.

This perspective also has clear implications for the detection of Parkinson's Disease in individuals with few neurological symptoms. The performance of these subjects on various tasks should be unusually susceptible to disruption, such as by increasing task complexity or distracting stimulation, especially when damage to dopaminergic neurons is relatively large. Similarly, the behavioral responses of such subjects should be more severely affected following treatment with drugs that decrease dopaminergic activity (such as α-methyltyrosine or spiroperidol; see Fig. 6), while unusual increases in behavioral arousal might be expected following treatment with drugs that increase dopaminergic activity (such as dopa or apomorphine; see Figs. 8 and 9).

Minimal Brain Dysfunction. The Minimal Brain Dysfunction syndrome includes hyperkinesis, attentional deficits and learning disabilities. The neural correlates of this condition are unknown since such children typically fail to show classical signs of organic brain damage and even soft neurological signs may be absent. However, we believe that, like Parkinsonian patients, these children have several important similarities to the animal model we have described. First, the disability is often associated with birth trauma, including hypoxia, a condition found to damage central catecholaminergic neurons (60). Second, the behavioral deficits seen in Minimal Brain Dysfunction can be improved by using catecholaminergic agonists, such as amphetamine, methylphenidate and caffeine, or by increasing the stimulus value of the educational materials. Third, such children are unusually sensitive to the disruptive effects of stressful environments, and fatigue easily (61-64).

While the hyperkinetic symptoms associated with Minimal Brain Dysfunction have been interpreted as a sign of overarousal, the parallels with animals sustaining DA-depleting lesions, and particularly the efficacy of environmental and pharmacological stimulation, suggest the converse. In fact, psychophysiological signs of underarousal have been reported (65). Thus, hyperactivity may represent a behavioral compensation which serves to heighten arousal levels and which becomes unnecessary under proper environmental conditions or following stimulant drug medication.

SUMMARY AND CONCLUSIONS

Observations from this and other laboratories indicate that damage to CA-containing projections is not associated with obvious behavioral impairments until the loss of amine is considerable. This appears due, in part, to compensatory adjustments at residual catecholaminergic synapses and in other, interrelated pathways. However, while these adaptive changes permit apparently normal behavior under neutral laboratory conditions, the range of stimuli to which the animals will respond is reduced. Such animals are relatively unresponsive to mild stimuli but intolerant of intense ones. This may result from the small number of residual CA-containing terminals, and the inability of those terminals to increase significantly their rate of transmitter release, respectively.

Animals with CA-depleting brain lesions have several features which are reminiscent of human patients with Parkinson's Disease or Minimal Brain Dysfunction. These include a lack of correspondence between presumed central damage and neurological deficits, a temporary improvement in performance with increased stimulus intensity or catecholaminergic agonists, and an increased susceptibility to the disruptive effects of stress. The latter characteristic can be used to detect the dysfunctions in human subjects with subclinical brain damage and to assess the severity of the problem. Conversely, because the neurochemical adaptations contract the range of stimuli which elicit an optimal performance, in therapy the human subject should be alerted to the subtle differences in task complexity, emotional upset, and environmental stimulation that will promote or hinder his behavioral capabilities.

REFERENCES

(1) N. Holzbauer and M. Vogt, Depression by reserpine of the noradrenaline concentration in the hypothalamus of the cat, Journal of Neurochemistry 1, 8 (1956).

(2) P. Bolme, K. Fuxe, and P. Lidbrink, On the function of central catecholamine neurons-- their role in cardiovascular and arousal mechanisms, Research Communications in Chemical Pathology & Pharmacology 4, 657 (1972).

(3) M. Jouvet, The role of monoamines and acetylcholine-containing neurons in the regulation of the sleep-waking cycle, Ergebnisse der Physiologie, Biologischen Chemie und Experimentellen Pharmakologie 64, 166 (1972).

(4) C. Sachs and G. Jonsson, Mechanisms of action of 6-hydroxydopamine, Biochemical Pharmacology 24, 1 (1975).

(5) F. E. Bloom, Fine structural changes in rat brain after intracisternal injection of 6-hydroxydopamine. In: Malmfors, T. and Thoenen, H., eds. (1971), 6-Hydroxydopamine and Catecholamine Neurons, North-Holland, Amsterdam.

(6) J. C. Hedreen and J. P. Chalmers, Neuronal degeneration in rat brain induced by 6-hydroxydopamine; a histological and biochemical study, Brain Research 47, 1 (1972).

(7) B. R. Jacks, J. de Champlain, and J. P. Cordeau, Effects of 6-hydroxydopamine on putative transmitter substances in the central nervous system, European Journal of Pharmacology 18, 353 (1972).

(8) P. Teitelbaum and A. N. Epstein, The lateral hypothalamic syndrome: recovery of feeding and drinking after lateral hypothalamic lesions, Psychological Review 69, 74 (1962).

(9) U. Ungerstedt, Adipsia and aphagia after 6-hydroxydopamine induced degeneration of the nigro-striatal dopamine system, Acta Physiologica Scandinavica Supplementum 367, 95 (1971).

(10) H. C. Fibiger, A. P. Zis, and E. G. McGeer, Feeding and drinking deficits after 6-hydroxydopamine administration in the rat: similarities to the lateral hypothalamic syndrome, Brain Research 55, 135 (1973).

(11) G. R. Breese, R. D. Smith, B. R. Cooper, and L. D. Grant, Alterations in consumatory behavior following intracisternal injection of 6-hydroxydopamine, Pharmacology, Biochemistry and Behavior 1, 319 (1973).

(12) J. F. Marshall, J. S. Richardson, and P. Teitelbaum, Nigrostriatal bundle damage and the lateral hypothalamic syndrome, Journal of Comparative and Physiological Psychology 87, 808 (1974).

(13) K. M. Taylor and R. Laverty, The effects of drugs on the behavioral and biochemical actions of intraventricular 6-hydroxydopamine, European Journal of Pharmacology 17, 16 (1972).

(14) M. J. Zigmond and E. M. Stricker, Deficits in feeding behavior after intraventricular injection of 6-hydroxydopamine in rats, Science 177, 1211 (1972).

(15) B. R. Cooper, G. R. Breese, J. L. Howard, and L. D. Grant, Effect of central catecholamine alterations by 6-hydroxydopamine on shuttle box avoidance acquisition, Physiology and Behavior 9, 727 (1972).

(16) E. M. Stricker and M. J. Zigmond, Effects on homeostasis of intraventricular injection of 6-hydroxydopamine in rats, Journal of Comparative and Physiological Psychology 86, 973 (1974).

(17) E. M. Stricker, M. I. Friedman, and M. J. Zigmond, Glucoregulatory feeding by rats after intraventricular 6-hydroxydopamine or lateral hypothalamic lesions, Science 189, 895 (1975).

(18) E. M. Stricker and M. J. Zigmond, Recovery of function following damage to central catecholamine-containing neurons: a neurochemical model for the lateral hypothalamic syndrome. In: Sprague, J. M. and Epstein, A. N., eds. (1976), Progress in Psychobiology and Physiological Psychology, vol. 6, Academic Press, New York.

(19) J. G. Van Zoeren and E. M. Stricker, Effects of preoptic, lateral hypothalamic, or dopamine-depleting lesions on behavioral thermoregulation in rats exposed to the cold, Journal of Comparative and Physiological Psychology in press (1977).

(20) E. M. Stricker, Drinking by rats after lateral hypothalamic lesions: a new look at the lateral hypothalamic syndrome, Journal of Comparative and Physiological Psychology 90, 127 (1976).

(21) N. Rowland, Circadian rhythms and the partial recovery of regulatory drinking in rats after lateral hypothalamic lesions, Journal of Comparative and Physiological Psychology 90, 382 (1976).

(22) M. J. Zigmond and R. J. Wurtman, Daily rhythm in the accumulation of brain catecholamines synthesized from circulating H^3-tyrosine, Journal of Pharmacology and Experimental Therapeutics 172, 416 (1970).

(23) J. DiRaddo and C. Kellogg, In vivo rates of tyrosine and tryptophan hydroxylation in regions of rat brain at four times during the light-dark cycle, Naunyn-Schmiedeberg's Archives of Pharmacology 286, 389 (1975).

(24) E. M. Stricker, M. B. Zimmerman, M. I. Friedman, and M. J. Zigmond, Caffeine restores feeding response to 2-deoxy-D-glucose in 6-hydroxydopamine-treated rats, Nature 267, 174 (1977).

(25) J. F. Marshall and U. Ungerstedt, Apomorphine-induced restoration of drinking to thirst challenges in 6-hydroxydopamine-treated rats, Physiology and Behavior 17, 817 (1976).

(26) M. J. Zigmond, C. D. Volz, E. M. Stricker, and M. I. Friedman, Effects of chronic insulin on brain tyrosine hydroxylase and glucoprivic feeding in 6-hydroxydopamine-treated rats, Neuroscience Abstracts 2, 842 (1977). (Abstract)

(27) M. J. Zigmond and E. M. Stricker, Recovery of feeding and drinking by rats after intraventricular 6-hydroxydopamine or lateral hypothalamic lesions, Science 182, 717 (1973).

(28) S. M. Antelman, N. E. Rowland, and A. E. Fisher, Stress related recovery from lateral hypothalamic aphagia, Brain Research 102, 346 (1976).

(29) J. F. Marshall, D. Levitan, and E. M. Stricker, Activation-induced restoration of sensorimotor functions in rats with dopamine-depleting brain lesions. Journal of Comparative and Physiological Psychology 90, 536 (1976).

(30) M. J. Zigmond and E. M. Stricker, Ingestive behavior following damage to central dopamine neurons: implications for homeostasis and recovery of function. In: Usdin, E., ed. (1974), Neuropsychopharmacology of Monoamines and Their Regulatory Enzymes, Raven Press, New York.

(31) T. G. Heffner, M. J. Zigmond, and E. M. Stricker, Effects of dopaminergic agonists and antagonists on feeding in intact and 6-hydroxydopamine-treated rats, Journal of Pharmacology and Experimental Therapeutics 201, 386 (1977).

(32) A. W. Graham and G. K. Aghajanian, Effects of amphetamine on single cell activity in a catecholamine nucleus, the locus coeruleus, Nature 234, 100 (1971).

(33) B. S. Bunney, J. R. Walters, R. H. Roth, and G. K. Aghajanian, Dopaminergic neurons: effect of antipsychotic drugs and amphetamine on single cell activity, Journal of Pharmacology and Experimental Therapeutics 185, 560 (1973).

(34) G. C. Palmer, F. Sulser, and G. A. Robison, Effects of neurohumoral and adrenergic agents on cyclic AMP levels in various areas of the rat brain in vitro, Neuropharmacology 12, 327 (1973).

(35) D. Tarsy and R. J. Baldessarini, Behavioural supersensitivity to apomorphine following chronic treatment with drugs which interfere with the synaptic function of catecholamines, Neuropharmacology 13, 927 (1974).

(36) D. R. Burt, I. Creese, and S. H. Snyder, Antischizophrenic drugs: chronic treatment elevates dopamine receptor binding in brain, Science 196, 326 (1977).

(37) V. M. Tennyson, R. Heikkila, C. Mytilineou, L. Cote, and G. Cohen, 5-Hydroxydopamine 'tagged' neuronal boutons in rabbit neostriatum: interrelationship between vesicles and axonal membrane, Brain Research 32, 341 (1974).

(38) H. Bernheimer and O. Hornykiewicz, Herabgesetzte konzentration der homovanillinasure im gehirn von parkinsonkranken menschen als ausdruck der storung des zentralen dopaminstoffwechsels, Klinische Wochenschrift 43, 711 (1965).

(39) D. F. Sharman, L. J. Poirier, G. F. Murphy, and T. L. Sourkes, Homovanillic acid and dihydroxyphenylacetic acid in the striatum of monkeys with brain lesions, Canadian Journal of Physiology and Pharmacology 45, 57 (1967).

(40) Y. Agid, F. Javoy, and J. Glowinski, Hyperactivity of remaining dopaminergic neurones after partial destruction of the nigro-striatal dopaminergic system in the rat, Nature, New Biology 245, 150 (1973).

(41) L. L. Iversen and N. J. Uretsky, Regional effects of 6-hydroxydopamine on catecholamine containing neurons in rat brain and spinal cord, Brain Research 24, 364 (1970).

(42) M. J. Zigmond, J. P. Chalmers, J. R. Simpson, and R. J. Wurtman, Effect of lateral hypothalamic lesions on uptake of norepinephrine by brain homogenates, Journal of Pharmacology and Experimental Therapeutics 179, 20 (1971).

(43) P. Feltz and J. de Champlain, Enhanced sensitivity of caudate neurons to microionto-phoretic injections of dopamine in 6-hydroxydopamine treated cats, Brain Research 43, 601 (1972).

(44) U. Ungerstedt, T. Ljungberg, B. Hoffer, and G. Siggins, Dopaminergic supersensitivity in the striatum. In: Calne, D. G., Chase, T. N., and Barbeau, A., Advances in Neurology 9, 57 (1975).

(45) R. I. Schoenfeld and N. J. Uretsky, Enhancement by 6-hydroxydopamine of the effects of dopa upon the motor activity of rats, Journal of Pharmacology and Experimental Therapeutics 186, 616 (1973).

(46) J. E. Thornburg and K. E. Moore, Supersensitivity to dopamine agonists following unilateral, 6-hydroxydopamine-induced striatal lesions in mice, Journal of Pharmacology and Experimental Therapeutics 192, 42 (1975).

(47) G. C. Palmer, Increased cyclic AMP response to norepinephrine in the rat brain following 6-hydroxydopamine, Neuropharmacology 11, 145 (1972).

(48) A. Kalisker, C. O. Rutledge, and J. P. Perkins, Effect of nerve degeneration by 6-hydroxydopamine on catecholamine-stimulated adenosine 3',5'-monophosphate formation in rat cerebral cortex, Molecular Pharmacology 9, 619 (1973).

(49) R. K. Mishra, E. L. Gardner, R. Katzman, and M. H. Makman, Enhancement of dopamine-stimulated adenylate cyclase activity in rat caudate after lesions in substantia nigra: evidence for denervation supersensitivity, Proceedings of the National Academy of Sciences, United States 71, 3883 (1974).

(50) M. J. Zigmond and E. M. Stricker, Compensatory changes after intraventricular administration of 6-hydroxydopamine: a neurochemical model for recovery of function. In: Jonsson, G., Malmfors, T., and Sachs, C., eds. (1975), Chemical Tools in Catecholamine Research, vol. 1, North-Holland, Amsterdam.

(51) J. R. Sporn, T. K. Harden, B. B. Wolfe, and P. B. Molinoff, β-adrenergic receptor involvement in 6-hydroxydopamine-induced supersensitivity in rat cerebral cortex, Science 194, 624 (1976).

(52) U. Ungerstedt, Postsynaptic supersensitivity after 6-hydroxydopamine induced degeneration of the nigro-striatal dopamine system. Acta Physiologica Scandinavica Supplementum 367, 69 (1971)

(53) R. I. Schoenfeld and N. J. Uretsky, Altered response to apomorphine in 6-hydroxydopamine-treated rats, European Journal of Pharmacology 19, 115 (1972).

(54) A. Björklund, R. Katzman, U. Stenevi, and K. A. West, Development and growth of axonal sprouts from noradrenaline and 5-hydroxytryptamine neurons in the rat spinal cord. Brain Research 31, 21 (1971).

(55) R. Katzman, A. Björklund, C. Owman, U. Stenevi, and K. A. West, Evidence for regenerative axon sprouting of central catecholamine neurons in the rat mesencephalon following electrolytic lesions, Brain Research 25, 579 (1971).

(56) W. Lovenberg and E. A. Bruckwick, Mechanisms of receptor mediated regulation of catecholamine synthesis in brain. In: Usdin, E. and Bunney, W. E. Jr., eds. (1975), Pre-and Postsynaptic Receptors, Dekker, New York.

(57) O. Hornykiewicz, Parkinson's disease: from brain homogenate to treatment, Federation Proceedings, Federation of American Societies for Experimental Biology 32, 183 (1973).

(58) R. S. Schwab and I. Zieper, Effects of mood, stress and alertness on the performance in Parkinson's disease, Psychiatria et Neurologia 150, 345 (1965).

(59) R. S. Schwab and A. C. England, Parkinson syndromes due to various specific causes. In: Vinken, P. J. and Bruyn, G. W., eds. (1968), Handbook of Clinical Neurology, Wiley, New York.

(60) N. T. Zervas, H. Hori, M. Negora, R. J. Wurtman, F. Larin, and M. H. Lavyne, Reduction in brain dopamine following experimental cerebral ischaemia, Nature 247, 283 (1974).

(61) Cruickshank, W. N., Bentzen, F. A., Ratzeburg, F. H., and Tannhauser, M. T. (1961), A Teaching Method for Brain-injured and Hyperactive Children, Syracuse University Press, Syracuse.

(62) P. H. Wender, The minimal brain dysfunction syndrome in children. I. The syndrome and its relevance for psychiatry. II. A psychological and biochemical model for the syndrome. The Journal of Nervous and Mental Disease 155, 55 (1972).

(63) B. D. Garfinkel, C. D. Webster, and L. Sloman, Methylphenidate and caffeine in the treatment of children with minimal brain dysfunction, The American Journal of Psychiatry 132, 723 (1975)

(64) S. Zentall, Optimal stimulation as theoretical basis of hyperactivity, American Journal of Orthopsychiatry 45, 549 (1975).

(65) J. Satterfield and M. Dawson, Electrodermal correlates of hyperactivity in children, Psychophysiology 8, 191 (1971).

AN ANIMAL MODEL OF MYOCLONUS RELATED TO CENTRAL SEROTONERGIC NEURONS

R. Malcolm Stewart and Ross J. Baldessarini
Departments of Neurology and Psychiatry, Harvard Medical School
Massachusetts General Hospital and the Mailman Research Center
McLean Hospital, Belmont, MA 02178

INTRODUCTION

Myoclonus refers to several types of rapid involuntary arrhythmic jerking movements (1). This activity may be seen in varying clinical situations differing in etiology, pathological process and anatomic site (2). In this sense myoclonus is a symptom of many motor disorders rather than a single clinical pathological entity. Nevertheless, a growing association is developing between abnormalities of serotonin (5-HT) metabolism (Fig. 1) in the central nervous system (CNS) and some forms of myoclonus (3-8), based on clinical studies of the endogenous level of the serotonin metabolite 5-hydroxyindole acetic acid (5-HIAA) in the cerebro-spinal fluid (CSF) and on clinical responses to administration of the serotonin precursor 5-hydroxytryptophan (5-HTP). This relationship between human myoclonus and CNS serotonin metabolism has been further strengthened by studies of various animal models in which repetitive myoclonic jerking movements have been produced either by a serotonin precursor (9), receptor agonist (10), or by pharmacological manipulations which increase the availability of 5-HT in the CNS (11-12). In this report we will review some of these animal behavioral syndromes and evaluate how they may be relevant to some clinical forms of myoclonus and to the laboratory and clinical investigation of the pharmacology and pathophysiology of this group of disorders.

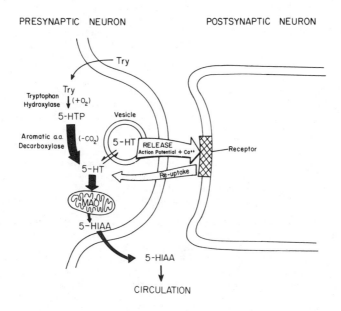

Fig. 1. Serotonergic synapse

Figure 1 is a diagrammatic representation of the principal biochemical processes occuring at a serotonergic synapse. Synthesis of serotonin (5-HT) from tryptophan (Try) via 5-hydroxytryptophan (5-HTP) utilizes two enzymes, tryptophan hydroxylase, the rate limiting step (small arrow), and L-aromatic amino acid decarboxylase. Serotonin is taken up into storage vesicles for subsequent release and action at the postsynaptic receptor. Presynaptic reuptake of serotonin is a major inactivating mechanism which leads either to storage in vesicles or catabolism by monoamine oxidase (MAO) to 5-hydroxyindoleacetic acid (5HIAA) which is found in the body fluids.

CLINICAL ASPECTS OF MYOCLONUS

Symptomatology

Myoclonus is a quick arrhythmic repetitive or non-repetitive contraction of a muscle or even a group of skeletal muscles, typically at 10-50 contractions per minute. These contractions are diffuse and usually involve the limbs and the trunk. Myoclonus can be activated by various sensory stimuli including touch, noise and light (1). Willed movement may evoke a special form of the disorder called "intention" myoclonus (2). Myoclonus may be present during activity or rest but generally disappears with sleep. With strong stimuli the myoclonic jerks may progress to a generalized seizure. The term myoclonus has also been used when describing a more restricted form of repetitious rhythmic clonus of the branchiomeric musculature involving palate, pharynx, vocal cords, face and jaw (1). The rhythmic type probably represents a separate category of neurological diseases and we restrict our concerns to the more diffuse type. In addition myoclonus must be distinguished from other involuntary movements such as tremor, chorea and restricted forms of epilepsy (epilepsia partialis continua).

Clinical Disease Spectrum

Myoclonus is not restricted to a single clinical entity and has been seen in a variety of conditions (1). Intention myoclonus frequently follows post-anoxic encephalopathy (2). A rapidly evolving dementia associated myoclonus in adults is sometimes seen with Creutzfeldt-Jacob disease. A similar symptom complex in children may be seen in subacute sclerosing panencephalitis (SSPE). Herpes simplex encephalitis may produce myoclonus, dementia and a confusional state in both adults and children. Myoclonus is also noted in degenerative and metabolic diseases such as Hallervorden-Spatz disease, Wilson's disease and the cerebral lipidoses, including the familial myoclonic epilepsy of Unverricht-Lundborg-Lafora, and myoclonus in combination with progressive cerebellar degeneration (Ramsy-Hunt Syndrome) (1). Myoclonus is sometimes evident in children with neuroblastomas (9). In addition myoclonus has been seen following various drugs and medications including Metrazol, overdoses of tricyclic antidepressants (13-14) and after 5-HTP given to infants with Down's syndrome (4).

Neuropathology

The clinical pathological correlations for human myoclonus remain imprecise. Myoclonus has been noted with lesions of cerebral cortical gray matter, thalamus, substantia nigra, basal ganglia, dentate nucleus of the cerebellum, lesions of the brain stem and spinal cord (1). Usually however, the pathological changes are so diffuse that an exact anatomical localization of the areas responsible for myoclonus is impossible. The smallest pathological lesions which appear to cause myoclonus are located in the thalamus and cerebellum (2). The pathological processes involved have also been varied and include acute and chronic infections, toxic and metabolic disorders, vascular and post-anoxic states, as well as degenerative and neoplastic conditions.

Clinical Pharmacology

Although in humans there is an increasing association between the metabolism of serotonin (5-HT) in the CNS and some forms of myoclonus, there is as yet no consistent or unifying pattern in this relationship. For instance there is a differential response of the myoclonus to the administration of the serotonin precursor 5-hydroxytryptophan (5-HTP). When infants with Down's syndrome who had relatively low platelet serotonin levels were given 5-HTP, they developed myoclonus which ceased when this agent was stopped (4). Moreover, in some infants with a massive myoclonic seizure disorder called the infantile-spasm syndrome, about half have elevated blood platelet 5-HT levels (14-16); furthermore, the use of a low tryptophan diet has been reported to be beneficial in some cases of this syndrome (3). 5-HTP has aggravated myoclonic jerking in patients with Tay-Sachs disease (7), or a ceramide lactosidase deficiency (7,17). On the other hand 5-HTP may have either no effect or may improve some forms of myoclonus, especially those of the post-anoxic type which have been shown to have low CSF levels of 5-HIAA, the main metabolite of 5-HT (5-8). In addition myoclonic jerks sometimes follow L-dopa therapy in Parkinson's disease and have been treated successfully with the serotonin receptor blocker methysergide (18). Myoclonus has also been reported following intoxication by overdoses of the amine reuptake blocker imipramine (13), and its effects suggest that this condition may inflict an imbalance of CNS neurotransmitters, with a relative excess of serotonin over acetylcholine (14). The benzodiazepine derivative clonazepam has also been reported to ameliorate the post-anoxic type of intention myoclonus (19). Although the exact pharmacological mechanism is unknown, benzodiazepines have been reported to reduce the turnover of 5-HT (20), and to increase brain levels of intracisternally administered labeled 5-HT in animals (21).

EXPERIMENTAL ASPECTS OF MYOCLONUS

Central Serotonergic Behavioral Syndrome

One difficulty in studying myoclonus in the laboratory has been the lack of specific animal models. Progress in this regard has been hampered by the lack of a precise understanding of the pathological anatomy or biochemistry of myoclonus on which to build a model. This situation arises in part because many different motor disorders are included under the rubric "myoclonus" which really stands for a syndrome or a class of diseases rather than a single clinical entity (I).

Although it has not been entirely possible to dissect the least common denominator from the clinical cases of myoclonus, several leads on which to develop animal models are known. For instance, it appears that the serotonin system is involved in, if not causally related to myoclonus, especially since low values of the serotonin metabolite 5HIAA have been reported in the CSF in the post anoxic

Table I - Animal models with myoclonus, twitching or jerking

Challenge	Pretreatment	Species	References
1. Serotonin and its precursors			
Tryptophan	MAOI	rat	II
"	"	mouse	23
5-hydroxytryptophan (5-HTP)	---	Guinea pig	9
"	MAOI	rat	II,31,33,36
"	MAOI	mouse	17
"	Reuptake blockade	guinea pig	13
"	"	rat	39,40
"	"	mouse	39
"	Chronic receptor blockade	guinea pig	44
"	Serotonin neurotoxin	rat	38,41
"	" "	mouse	40
"	Decarboxylase inhibitor	mouse	35-37,42
5-hydroxytryptamine (5-HT) *	Serotonin neurotoxin	rat	40
2. Serotonin receptor agonists			
Tryptamines			
Tryptamine	MAOI	rat	26
Dimethyl-tryptamine (DMT)	---	mouse rat	53
5-methoxy-N-N-dimethyl tryptamine (5MeODMT)	---	rat	10,27
"	MAOI	rat	10,25
Phenylethylamines			
2,5-dimethoxy-4-methyl amphetamine (DOM)	---	rat	24
p-methoxyamphetamine (pMA)	---	rat	24
p-methoxyphenylethylamine (pMPEA)	---	rat	24
Other			
Quipazine	---	rat	30
3. Serotonin releasing or depleting agents			
p-chloroamphetamine (pCA)	---	rat	29
fenfluramine	---	rat	28

MAOI = Monoamine oxidase inhibitor; (*) given intraventricularly. All other drugs were given peritoneally.

form of myoclonus (22), and blood platelet 5-HT may be low in Down's syndrome which is often associated with myoclonus when treated with 5-HTP (4). The differential response of myoclonus to 5-HTP in the various clinical conditions is another important lead (7). This relationship of clinical myoclonus to serotonin is one of the strongest links to the various animal models which might be considered appropriate to the question of myoclonus (Table I).

In laboratory animals, repetitive myoclonic jerking movements and twitches have been produced by serotonin precursors (9,23), serotonin receptor agonists, (10, 24-28), and serotonin releasing agents (29,30). This stereotyped behavior may sometimes occur when a serotonin precursor or agonist is given alone but is much more likely following pretreatment with drugs which increase the availability of serotonin in the CNS, such as inhibitors of monoamine oxidase (MAO) (11, 31-33,34) or of peripheral aromatic amino acid decarboxylase (35-37), blockers of serotonin reuptake (13, 38, 39), or the dihydroxy-lated tryptamines - neurotoxins which can be used to destroy serotonin nerve terminals selectively (38,40,41).

A striking feature of this behavioral syndrome following various manipulations of the serotonin system is the arrhythmic repetitive jerking movements of limbs, head and trunk which can progress to a generalized clonic-tonic seizure (11, 38,40,41). These jerking movements are often accompanied by ataxia and fine head and body tremor, as well as postural changes of rigidity and immobility of limbs or the tail. Autonomic nervous system manifestations such as flushing of the skin, salivation, diarrhea and ejaculation may accompany this syndrome. This behavior has been also described as a "hyperactivity syndrome" (11), "reciprocal forepaw treading" (12,41), "myoclonus" (9, 13, 38, 39) and "the piano playing syndrome" (42). Although other manifestations may be present to varying degree, the jerking or twitching movements appear to be a common element of the various behavioral responses, which seem to represent a discrete syndrome associated with excessive availability of central 5HT neurotransmission (11, 12, 38).

This behavior can be quantified using a number of different methods. Animals with myoclonus may be evaluated using a behavioral rating scale, established by observing responses over a wide range of doses of the agent inducing the myoclonus. For rats one useful behavioral scale, developed with 5-HTP in rats pretreated with 5,7-dihydroxytryptamine, is as follows: 0 = no change in behavior; 1 = unsteady gait, head tremor; 2 = intermittent myoclonus; 3 = continuous myoclonus; 4 = generalized seizures (38). One must be aware of species differences in the behavioral response, however, and the rating scale must be adjusted accordingly. Following a challenge dose of 5-HTP, mice also develop jerking movements but show much less tremor and autonomic signs than rats. Myoclonus has also been recorded objectively, quantitatively and sensitively using an electronic monitor and these results correlate closely with observations made using a behavioral rating scale (40). Although electrome-chanical recording of myoclonus may be more sensitive, simultaneous visual inspection of the behavior is recommended to provide assurance that artifacts such as scratching, grooming and locomotion are accounted for. Myoclonus may also be recorded using electromyographic techniques employing intramuscular electrodes in either waking (38) or anesthetized animals (43). An example of such a recording is given in Fig. 2.

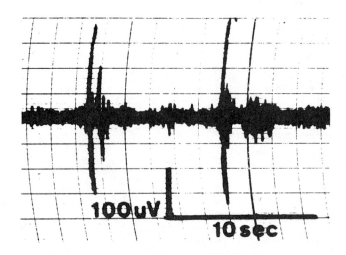

Fig. 2. Electromyographic recording of myoclonus

Figure 2 shows the electromyographic recording of myoclonic jerking from the posterior cervical region of a rat. This muscle activity was induced by 5-hydroxytryptophan (65 mg/kg, i.p.) 10 days after giving the serotonin neurotoxin 5,7-dihydroxytryptamine (200 μg, intracisternally) with desmethylimipramine (DMI, 25 mg/kg, i.p.).

Under certain conditions, this behavioral syndrome can be produced by precursors of serotonin (Table I). Although tryptophan alone does not produce the jerking movements, following pretreatment with the MAO inhibitor tranylcypromine, a characteristic behavioral syndrome of hyperactivity including tremor, clonic movements, forepaw treading, erect and retroflexed (Straub) tail, and salivation has been noted (11, 23). A similar but more rapid response has been noted when 5-hydroxytryptophan (5-HTP), the immediate precursor of serotonin was given with an MAO inhibitor to rats or mice (11,31-33). Interestingly 5-HTP alone has been reported to produce myoclonic jerking in young guinea pigs (9), but not in the young rat (38). The same precursor in combination with a blocker of the reuptake of 5-HT has produced myoclonus or twitching movements (39); this is particularly striking in the guinea pig (13). Prolonged presumptive blockade of serotonin receptors by methysergide has been reported to enhance this response to 5-HTP in the guinea pig, possibly by a mechanism of disuse super-sensitivity of serotonin receptors in the CNS (44). Attempts to produce such an effect by repeated treatment with methysergide in the rat, which normally does not develop myoclonus in response to 5-HTP alone, were unsuccessful (R.M. Stewart, and R.J. Baldessarini, unpublished observations). Myoclonus has however been produced following 5-HTP in rats treated with both 5,6- and 5,7-dihydroxy-tryptamine (DHT) (38, 41), although tryptophan was not effective under the same conditions (38). In contrast to other acute pharmacological manipulations, such as pretreatment with an MAO inhibitor or a blocker of the reuptake of serotonin, lesions induced with a serotonin neurotoxin (DHT) are chronic and animals will respond over days to weeks to repeated injections of 5-HTP. This response was blocked by serotonin receptor blockers and high, centrally effective, doses of a decarboxylase inhibitor (38, 41). Because of chronic denervation, postsynaptic mechanisms involving serotonin supersensitivity may be involved in addition to presynaptic ones, such as the loss of 5-HT uptake (38). Serotonin given alone intraperitoneally or even intraventricularly in high doses to normal rats failed to produce myoclonus (38), although striking myoclonus did occur following intraventricular infusion of 5-HT into the lateral ventricle of rats previously lesioned with 5,7-DHT (J. Warbritton, R.M. Stewart and R.J. Baldessarini unpublished observations).

In laboratory animals, this behavioral syndrome marked by repetitive myoclonic jerking or twitching has been produced by a number of agents which are considered to be agonists of serotonin receptors in the CNS. Conversely, this response appears to be selective for the serotonin system in that agonists of dopamine such as apomorphine and L-dopa do not produce this syndrome, and physostigmine, which enhances central cholinergic function, produces tremor but not myoclonus (38). Among suspected serotonin agonists, this syndrome has been noted to occur in normal animals treated with certain methylated tryptamines including 5-methoxy-N,N-dimethyltryptamine (5MeODMT) (10, 25-27). In addition, certain of the hallucinogenic phenylethylamines are believed to stimulate 5-HT receptors and produce this syndrome when given to normal animals; these include 2,5-dimethoxy-4-methyl-amphetamine (DOM), p-methoxyamphetamine (pMA) and p-methoxyphenethylamine (pMPEA). This effect of these phenylethylamines does not seem to be mediated simply by the release of serotonin, since it occurs in animals pretreated with the tryptophan hydroxylase inhibitors p-chlorophenylalanine or H22/54 to deplete the CNS of serotonin (24), although release of 5-HT by these compounds possibly may also occur and contribute to their myoclonus-inducing effects (24). Quipazine (2-Cl-piperazinyl-quinoline maleate) is also reported to be a 5-HT agonist and to produce a similar behavioral syndrome (30). Head twitches have also been induced by the benzodiazepine clonazepam, the effects of which were blocked by the antiserotonin drug cyproheptadine (45). Finally a syndrome of tremor, myoclonic jerking and piloerection can be produced acutely by agents which cause rapid presynaptic release of endogenously stored serotonin and include p-chloramphetamine (29) and fenfluramine (30).

Monoamine oxidase (MAO) inhibitors have a special role in the pharmacology of this behavioral syndrome. This enzyme is responsible for the degradation of biogenic amines to their deaminated metabolites and may have an especially important role in regulating the physiologic availability of serotonin. MAO type A is believed to catabolize serotonin relatively selectively, while MAO type B selectively oxidizes several phenylethylamine derivatives. The treatment of rats with tranyl-cypromine, a nonselective inhibitor of both forms of MAO, followed by a challenge with tryptophan produces a "hyperactivity syndrome" well described by Grahame-Smith (11). As noted previously this syndrome appears to be remarkably similar to the myoclonic jerking syndrome produced by other serotonin-enhancing or mimicking agents. Curiously, no such response to tryptophan was obtained after pretreatment with selective inhibitors of MAO, either clorgyline which inhibits type A MAO or deprenyl which inhibits type B MAO. However when both clorgyline and deprenyl were administered together to inhibit both forms of MAO, the myoclonic behavioral response to tryptophan did develop (31). This study suggests that although serotonin is normally metabolized selectively by type A MAO, type B may also be important and the hyperactivity syndrome following an MAO-inhibitor will not be produced unless both forms of MAO are inhibited (31). In addition MAO inhibitors can enhance the behavioral response to putative serotonin receptor agonists such as 5MeODMT or quipazine although the significance of these effects is not clear.

Neuroanatomical Substrate

The behavioral syndrome of limb jerking, tremor, and postural change appears to be mediated by structures in the brain stem and spinal cord (40, 45). For instance, this syndrome when induced by tryptophan and an MAO inhibitor was still elicited in adult rats prepared by transection of the neuroaxis caudal to the red nucleus of the brain stem, and removal of the cerebellum (46). The syndrome can also be produced by 5-HTP in animals pretreated by a serotonin neurotoxin: either

5,7- or 5,6-DHT. While 5,7-DHT produces lesions throughout the neuroaxis, 5,6-DHT more selectively lesions serotonin neurons in the lower brain stem and spinal cord (40). After 5,6-DHT jerking movements are prominent but tremor and autonomic signs are less noticeable than with 5,7 DHT (40). Focal myoclonus has also been reported after injection of picrotoxin into the caudate nucleus (47), suggesting that effects at sites other than the lower brain stem or spinal cord may also be capable of producing such effects.

Experimental Pharmacology

Agents (Table 2) which decrease the myoclonic response include the proposed serotonin receptor blockers methysergide, d-lysergic acid diethylamide (LSD), 2-bromolysergic acid diethylamide (BOL), methiothepin, and cyproheptadine (38). Recently the β-adrenergic blocker, propranolol has been shown to inhibit the hyperactivity induced by serotonin agonists or precursors (48-50); the significance of this observation remains unclear. In addition, while α-adrenergic blocking agents were ineffective, high centrally effective doses of an inhibitor of L-aromatic amino acid decarboxylase greatly attenuated the response to 5-HTP in the 5,7-DHT induced myoclonus (38).

Although we have stressed the relationship of this behavioral syndrome to the metabolism of serotonin in the CNS (Fig. I), other proposed neurotransmitters may also be important. For instance a dopamine pathway is thought to be involved in some way between the point of 5-HT receptor stimulation and the final motor response (51). Interesting, however, the L-dopa induced myoclonic response in man can be blocked by methysergide, a serotonin receptor antagonist (18). A gamma-aminobutyric acid (GABA) mediated system may also be involved since treatment with amino-oxyacetic acid (AOAA) increases GABA in the CNS and can attenuate this syndrome, while picrotoxin, a putative GABA-antagonist, can enhance it (52). Furthermore local injection of picrotoxin into the caudate nucleus of the rat will produce focal myoclonus which can be stopped by a subsequent injection of GABA into the caudate (52).

Table 2 - Pharmacology of the central serotonergic behavioral syndrome

AGONISTIC

5-HT receptor agonists
   tryptamines
   phenylethylamines
5-HT and precursors
5-HT releasing agents

POTENTIATED BY

MAO inhibitors
GABA antagonists
5-HT reuptake blockade
Prolonged 5-HT receptor blockade

NOT AGONISTIC

Catecholamine - agonists
   (apomorphine, L-Dopa, amphetamine)

ACh -agonists (physostigmine)

ANTAGONISTIC

Ergot alkaloid derivatives
   (methysergide, LSD)
β-adrenergic blockers
   (e.g., propranolol)
DA-synthesis inhibitors (e.g. AMPT)*
GABA-agonists

Prolonged 5-HT agonism or precursors

NOT ANTAGONISTIC

Butyrophenones
ACh-antagonists (ATROPINE)
α-adrenergic antagonists
   (phentolamine, phenoxybenzamine)

*AMPT (α-methyl-p-tyrosine does not have this effect in all models.

In laboratory animals repetitive jerking movements have been produced by both serotonin precursors and proposed serotonin agonists including the methylated tryptamines (10,25,27,53). This myoclonus may occur whenever a serotonin agonist or precursor is given alone, but the behavioral response is greatly enhanced by drugs which are thought to increase the availability of serotonin in the CNS, such as inhibitors of monoamine oxidase or of peripheral aromatic amino acid decarboxylase, and blockers of serotonin reuptake (Table 2). The myoclonus which follows neurochemical lesioning by the serotonin neurotoxins 5,7-DHT or 5,6-DHT may be more complex and reflect both presynaptic as well as post synaptically mediated supersensitivity to serotonin. Putative serotonin receptor antagonists block this central serotonergic syndrome which appears to be mediated by brain stem and spinal cord mechanisms. Neurotransmitter systems other than those using serotonin may modulate this behavior as well.

IMPLICATIONS OF THE ANIMAL MODELS

These animal behavioral syndromes may be relevant to some clinical forms of myoclonus, especially pediatric types, and those associated with tricyclic antidepressant drug toxicity. The relationship of these models to those forms of clinical myoclonus seen in adults which are ameliorated by 5-HTP or clonazepam is less clear. Nevertheless, these models should be useful in evaluating pharmacological agents that act as agonists and antagonists at serotonin receptors in the CNS. In addition the functional interrelationships between the serotonin system and other neurotransmitter systems may be evaluated.

ACKNOWLEDGEMENTS

We acknowledge the excellent secretarial skills of Elizabeth Stein. This study was supported in part by USPHS (NIMH) Grants MH-16674 and MH-25515 to the Department of Psychiatry. R.M.S. received support from BRSG Research Grant B-75-39 to the Massachusetts General Hospital. R.J.B. is a recipient of a National Institute of Mental Health Career Research Scientist Award, MH-74370.

REFERENCES

1.   DeJong, R.N. (1967) The Neurologic Examination, Harper and Row, New York, p 543.

2.   J.W. Lance and R.D. Adams, The syndrome of intention or action myoclonus as a sequel to hypoxic encephalopathy Brain 86, 111 (1963).

3.   N.N. Lowe, J.F. Bosma, M.D. Armstrong, Infantile spasms with mental retardation. I. Clinical observations and dietary experiments Pediatrics 22, 1153 (1958).

4.   M. Coleman, Infantile spasms associated with 5-hydroxytryptophan administration in patients with Down's syndrome Neurology (Minneap) 21, 911 (1971).

5.   D. Chadwick, R. Harris, P. Jenner, E.H. Reynolds and C.D. Marsden, Manipulation of brain serotonin in the treatment of myoclonus Lancet 434 (1975).

6.   M.H. Vanwoert and V.H. Sethy, Therapy of intention myoclonus with L-5-hydroxytryptophan and a peripheral decarboxylase inhibitor, MK 486 Neurology (Minneap) 25, 135 (1975).

7.   J.H. Growdon, R.R. Young and B.T. Shahani, L-5- hydroxytryptophan in treatment of several different syndromes in which myoclonus is prominent Neurology (Minneap) 26, 1135 (1976).

8.   M.H. Van Woert, D. Rosenbaum, J. Howieson and M.B. Bowers, Long term therapy of myoclonus and other neurologic disorders with L-5-hydroxytryptophan and carbidopa New Eng J. Med 296, 70 (1977).

9.   H.L. Klawans, C. Goetz, and W. Weiner, 5-hydroxytryptophan-induced myoclonus in guinea pigs and the possible role of serotonin in infantile myoclonus Neurology (Minneap), 23, 1234 (1973).

10. K. Fuxe, B. Holmsted and G. Jonsson, Effects of 5-methoxy-N,N-dimethyltryptamine on central monoamine neurons Eur. J. Pharmacol. 19, 25 (1972).

11. D.G. Grahame-Smith, Studies in vivo on the relationship between brain tryptophan, brain 5-HT synthesis and hyperactivity in rats treated with a monoamine oxidase inhibitor and L-tryptophan J. Neurochem. 18, 1053 (1971).

12. B.L. Jacobs, An animal behavioral model for studying central serotonergic synapses Life Sciences 19, 777 (1976).

13. R. Westheimer and H.L. Klawans, The role of serotonin in the pathophysiology of myoclonic seizures associated with acute imipramine toxicity Neurology (Minneap) 24, 1175 (1974).

14. S. Lippman, R. Moskovitz and L. O'Tuama, Tricyclic-induced myoclonus Am. J. Psychiatry 134, 1 (1977).

15. S. Ota, Study of serotonin metabolism in pediatrics. 2. Blood serotonin levels in various diseases in children Acta Paediatr. Jpn. 73, 61 (1969).

16. M. Coleman, D.J. Boullin and M. Davis, Serotonin abnormalities in the infantile spasm syndrome Neurology (Minneap) 21, 421 (1971).

17. M.L. Goldstein, E.H. Kolodny, G.G. Gascon, et al., Macular cherry-red spot, myoclonic epilepsy and neurovisceral storage in a 17-year-old girl Trans. Am. Neurol. Assoc. 99, 110 (1974).

18. H.L. Klawans, C. Goetz, D. Bergen, Levodopa-induced myoclonus Arch. Neurol. 32, 331 (1975).

19. M.A. Goldberg and J.D. Dorman, Intention myoclonus: successful treatment with clonazepam Neurology (Minneap) 26, 24 (1976).

20. C.D. Wise, B.D. Berger and L. Stein, Benzodiazepines: anxiety-reducing activity by reduction of serotonin turnover in the brain Science 177, 180 (1972).

21. T.N. Chase, R.I. Kutz and I.J. Kopin, Effect of diazepam on fate of intracisternally injected serotonin-C^{14} Neuropharmacology 9, 103 (1970).

22. F. L'hermitte, M. Petertalvi, R. Marteau, et al., Analyse pharmacologigue d'un cas de myoclonies d'intention et d'action post-anoxiques Rev. Neurol. (Paris) 124, 21 (1971).

23. S.J. Coine, R.W. Pickering and B.T. Warner, A method for assessing the effects of drugs on the central actions of 5-hydroxytryptamine Brit. J. Pharmacol. 20, 106 (1963).

24. N. Anden, Hans Corrodi, K. Fuxe and L. Meek, Hallucinogenic phenylethylamines: interactions with serotonin turnover and receptors Eur. J. Pharmacol. 25, 176 (1974).

25. D.G. Grahame-Smith, Inhibitory effect of chlorpromazine on the syndrome of hyperactivity produced by L-tryptophan or 5-methoxy-N,N-dimethyltryptamine in rats treated with a monoamine oxicase inhibitor Br. J. Pharmacol. 43, 856 (1971).

26. D.H. Tedeschi, R.E. Tedeschi and E.J. Fellows, The effects of tryptamine on the central nervous system, including a pharmacological procedure for the evaluation of iproniazide-like drugs J. Pharmacol. Exp. Ther. 126, 223 (1959).

27. R.F. Squires, Evidence that 5-methoxy-N,N-dimethyl tryptamine is a specific substrate for MAO-A in the rat: implication for the indoleamine dependent behavioral syndrome J. Neurochem. 24, (1975).

28. A.R. Green, M.B.H. Youdim, and D.G. Grahame-Smith, Quipazine: its effects on rat brain 5-hydroxytryptamine, monoamine oxidase activity and behavior Neuropharmacology (1976).

29. M.E. Trulson and B.L. Jacobs, Behavioral evidence for the rapid release of CNS serotonin by PCA and florfluoramine Eur. J. Pharmacol. 36, 149 (1976).

30. J.H. Growdon, Postural changes, tremor and myoclonus in the rat immediately following injections of p-chloroamphetamine Neurology (Minneap) In press (1977).

31. A.R. Green and M.B.H. Youdin, (1976) Use of a behavioral model to study the action of mono-amine oxidase inhibition in vivo In Monoamine Oxidase and its Inhibition, Ciba Foundation Symposium 39, Elsevier, Amsterdam, p. 231.

32. R.F. Squires and J.B. Lassen, The inhibition of A and B forms of MAO in the production of a characteristic behavioral syndrome in rats after L-tryptophan loading Psychopharmacologia (Berl.) 41, 145 (1975).

33. M. Nakamura, H. Fukushima and S. Kitagawa, Effects of amitriptyline and isocarboxazid on 5-hydroxytryptophan induced head twitches in mice Psychopharmacology 48, 101 (1976).

34. S.M. Hess and W. Doepener. Behavioral effects of brain amine contents in rats Arch. Int. Pharmacodyn. 134, 89 (1961).

35. K. Modigh, Effects of chlorimipramine and protriptyline on the hyperactivity induced by 5-hydroxytryptophan after peripheral decarboxylase inhibition in mice J. Neural. Transmission 34, 101 (1973).

36. K. Modigh, Central and peripheral effects of 5-hydroxytryptophan on motor activity in mice Psychopharmacologia (Berl.) 23, 48 (1972).

37. L.L. Butcher, J. Engel and K. Fuxe, Behavioral, biochemical and histochemical analyses of the central effects of monoamine precursors after peripheral decarboxylase inhibition

Brain Research 41, 387 (1972).

38.　R.M. Stewart, J.H. Growdon, D. Cancian, and R.J. Baldessarini, 5-hydroxytryptophan-induced myoclonus: increased sensitivity to serotonin after intracranial 5,7-dihydroxytryptamine in the adult rat Neuropharmacology 15, 449 (1976).

39.　A.R. Beabien, D.C. Carpenter, L.F. Mathieu, M. MacConaill and P.D. Hrdina　Antagonism of imipramine poisoning by anticonvulsants in the rat　Toxicol. Applied Pharmacol. 38, 1 (1976).

40.　R.M. Stewart, S.C. Gerson, G. Sperk, A. Campbell and R.J. Baldessarini, Biochemical and behavioral studies of the effects of dihydroxylated tryptamines in the rodent brain New York Acad. Sci. In press (1977).

41.　M.E. Trulson, E.E. Eubanks and B.L. Jacobs, Behavioral evidence for supersensitivity following destruction of central serotonergic nerve terminals by 5,7-dihydroxytryptamine J. Pharmacol. Exp. Therap. 198, 23 (1976).

42.　J. Maj and L. Pawlowski, The effect of L-5-hydroxytryptophan (5-HTP) on locomotor activity in mice Pol. J. Pharmacol. Pharm. Supp 127, 145 (1975).

43.　D. Bieger, L. Larochelle and O. Hornykiewicz, A model for the quantitative study of central dopaminergic and serotonergic activity Eur. J. Pharmacol. 18, 128 (1972).

44.　H.L. Klawans, D.J. D'Amico and B.C. Patel, Behavioral supersensitivity to 5-hydroxytryptophan induced by chronic methysergide pretreatment Psychopharmacologia (Berl.) 44, 297 (1975).

45.　M. Nakamura and H. Fukushima, Head twitches induced by benzodiazepines and the role of biogenic amines Psychopharmacology 49, 259 (1976).

46.　B.L. Jacobs and H.Klemfuss, Brainstem and spinal cord mediation of a serotonergic behavioral syndrome Brain Research 100, 450 (1975).

47.　C.D. Marsden, B.S. Maldrum, C. Rycock and D. Tarsy, Focal myoclonus produced by injection of picrotoxin into the caudate nucleus of the rat. J. Physiology 246, 96 (1974).

48.　A.R. Green and D.G. Grahame-Smith, (-)-Propranolol inhibits the behavioral responses of rats to increased 5-hydroxytryptamine in the central nervous system Nature 262, 594 (1976).

49.　D.N. Middlemiss, L. Blakeborough, and S.R. Leather, Direct evidence for an interaction of β-adrenergic blockers with the 5-HT receptor Nature 267, 290 (1977).

50.　W. Weinstock, C. Weiss and S. Gitter, Blockade of 5-hydroxytryptamine receptors in the central nervous system by β adrenergic antagonists Neuropharmacology 16, 273 (1977).

51.　A.R. Green and D.G. Grahame-Smith, The role of brain dopamine in the hyperactivity syndrome produced by increased 5-hydroxytryptamine synthesis in rats Neuropharmacology 13, 949 (1974).

52.　A.R. Green, A.F.C. Tordoff and M.R. Bloomfield, Elevation of brain GABA concentrations with amino-osyacetic acid; effect on the hyperactivity syndrome produced by increased 5-hydroxytryptamine synthesis in rats J. Neural Transmission 39, 103 (1976).

53.　N. Anden, H. Corrodi and K. Fuxe, Hallucinogenic drugs of the indolealkylamine type and central monoamine neurons J. Pharmacol Exp. Ther. 179, 236 (1971).

DISCUSSION

In reply to a question of *Dr. Norton, Dr. Stewart* indicated that the 5,7-DHT treated animals (prior to administration of 5-HTP) had fairly normal locomotor activity. There was, possibly, somewhat less total 24-hour activity, but none had jerking movements or any signs of autonomic nervous system discharge. *Dr. Norton* asked about observations on actual mechanisms of loco-motion. *Dr. Stewart* replied that he saw no evidence of paralysis, that for the first few post-injection days, the animals seemed slightly hyperactive; there was no change in tone, no spacticity or rigidity.

Dr. Stricker inquired about the injection site as well as norepinephrine levels in the treated animals. *Dr. Stewart* said that intracisternal injection had been used rather than intraventricular to minimize effects on other systems. Norepinephrine levels were found to be normal, as were those of dopamine. Uptake of GABA, glutamate, dopamine, and norepinephrine were normal, but serotonin uptake was decreased.

In reply to a comment of *Dr. Sanghvi, Dr. Stewart* emphasized that pretreatment with desmethyl-imipramine was essential to block uptake into catecholaminergic neurons.

GENERAL DISCUSSION

Dr. W. Vogel had a question on the dyskinesia model. Animals pretreated with antipsychotic or with a narcotic antagonist are then challenged with amphetamine or methamphetamine and stereotyped behavior is measured. However, the antipsychotic-treated patient develops dyskinesia without an amphetamine or methamphetamine challenge. If such patients have the antipsychotic medication discontinued, there are reports that they have decreased response to amphetamine. Dr. Vogel felt that these two items are not consistent.

Dr. Carlson agreed with the accuracy of Dr. Vogel's comment. She stated that she was unaware of any experimental animal preparation in which there was the spontaneous appearance of dyskinesia after drug (chlorpromazine, haloperidol, narcotics, etc.) discontinuation. She could only conclude that there was something unusual about humans in this respect.

Dr. Friedhoff also commented on Dr. Vogel's question. He speculated that when supersensitivity was produced by blockade (e.g., with haloperidol or other antipsychotic drug), there was an adjustment on the presynaptic side upon withdrawal of the drug. There is a decrease of dopamine release and a decrease in dopamine turnover, which compensates for the increased sensitivity. In humans, the adjustment seems only transient and then the release of dopamine tends to normalize itself, giving rise to the spontaneous production of symptoms.

NEUROLEPTIC-STIMULATED PROLACTIN SECRETION IN THE RAT

AS AN ANIMAL MODEL FOR BIOLOGICAL PSYCHIATRY:

I. COMPARISON WITH ANTI-PSYCHOTIC ACTIVITY

Herbert Y. Meltzer,[1,2] Victor S. Fang,[3] Richard Fessler,[2]
Miljana Simonovic[1,4] and Dusanka Stanisic[1]

Departments of Psychiatry,[1] Medicine,[3] and Pharmacology
and Physiological Sciences,[4] University of Chicago Pritzker
School of Medicine and the Illinois State Psychiatric
Institute[2]

ABSTRACT

Prolactin secretion from the anterior pituitary gland in the rat is under tonic inhibition by dopamine. The effect of dopamine is most likely a direct one at the pituitary. Blockade of dopamine receptors by neuroleptics leads to a marked stimulation of prolactin secretion. The ability of neuroleptics to disinhibit prolactin secretion in the rat is parallel to their anti-psychotic properties in man in the following ways: 1) those phenothiazines, butyro-phenones, thioxanthines, substituted benzamides, dibenzodiazepines and related compounds which are anti-psychotic increase plasma prolactin in the rat but those which are not anti-psychotic do not increase plasma prolactin; the possible exceptions are metoclopramide and perlapine, both of which have been reported to lack anti-psychotic activity, but on the basis of limited clinical study only; 2) metabolites of chlorpromazine which do not inhibit amphetamine-induced stereotypy do not increase plasma prolactin levels; 3) the anti-psychotic effects of butaclamol and flupenthixol stereoisomers parallel their potency as stimulators of prolactin secretion; 4) the ability of a wide variety of anti-psychotics to stimulate prolactin secretion is highly correlated with clinical anti-psychotic potency; 5) no tolerance develops to the anti-psychotic action of neuroleptics or their ability to stimulate rat or human prolactin secretion. These results suggest that the pituitary dopamine receptors which regulate prolactin secretion can serve, with due caution, as an animal model for further study of selected features of the dopamine receptors which mediate the anti-psychotic action of neuroleptics in man. Although prolactin has been implicated in mammary carcinogenesis in the rodent, there is no substantial evidence that the increased prolactin produced by neuroleptics is a risk factor for carcinogenesis in man. Thus, the effect of neuroleptics on prolactin in the rat is not likely to be useful as an animal model for mammary carcinoma in man.

DIRECT DOPAMINERGIC INHIBITION OF PROLACTIN SECRETION

Prolactin secretion from the anterior pituitary gland in primates and laboratory rodents is under tonic inhibition by a dopamine-dependent mechanism (1,2). This may be due to an action of dopamine directly at the pituitary, or indirectly at the hypothalamic level, to release a so-called prolactin inhibitory factor (PIF) (1,2). There are a variety of studies which demonstrate a direct effect of dopamine itself on the pituitary gland to suppress prolactin release. Thus, L-DOPA, the precursor of dopamine, can inhibit prolactin secretion in stalk-sectioned monkeys (3); intravenous dopamine can decrease serum prolactin levels in man (4,5) and dopamine, in minute concentrations, can inhibit prolactin secretion from anterior pituitary glands in vitro (2,6).

We have found that dopamine itself can decrease plasma prolactin levels in rats. Alpha-methylparatyrosine methyl ester (AMPT), an inhibitor of dopamine synthesis, was administered to increase rat plasma prolactin levels. As can be seen in Table 1, the AMPT-induced increase in prolactin was inhibited by dopamine in a dose-dependent manner. The method used to obtain blood samples and measure rat prolactin by radioimmunoassay has been described elsewhere (7). The ability of dopamine to reverse the increase in plasma prolactin produced by AMPT could be due to a direct effect at the pituitary or median eminence to inhibit prolactin release, or it could be due to increased clearance of prolactin from blood. Evidence has been recently presented that dopamine promotes the clearance of prolactin from the blood after it has been secreted, a factor which must be considered in interpreting in vivo experiments (8).

TABLE 1 Effect of Dopamine on Rat Plasma Prolactin

Drug 1	Dose (mg/kg)	Drug 2	Dose (mg/kg)	Plasma Prolactin[*] (ng/ml)	p<
Saline	-	Saline	-	8.2 ± 3.5	-
AMPT	50	Saline	-	49.4 ± 38.9	.05
AMPT	50	Dopamine	1	68.7 ± 41.4	.05
AMPT	50	Dopamine	10	3.0 ± 2.2	NS

[*] Mean ± S.D.
The interval between drug 1 and drug 2 was 30 min; the interval between drug 2 and sacrifice was 30 min. All groups consisted of 5 rats.

In addition to these studies demonstrating the effect of dopamine itself on prolactin secretion, corrobative evidence for the dopaminergic control of prolactin secretion is provided by the demonstration that known dopamine agonists such as apomorphine, ergot alkaloids and piribedil also markedly suppress prolactin secretion in man and laboratory animals (9-11). Dopamine agonists also inhibit prolactin secretion in vitro (2,9,12).

INFLUENCES OTHER THAN DOPAMINE ON PROLACTIN SECRETION

There is a general acceptance that the dopaminergic influence on prolactin secretion is the most potent factor which regulates prolactin secretion from the pituitary, regardless of whether it is solely or majorly via a direct influence on the pituitary, or via release of PIF (1,2). However, there are a number of other factors which can be shown to affect prolactin secretion in vivo in man or laboratory animals or in vitro. They include serotonin (13,14), thyrotropic releasing hormone (15,16), GABA (17,18), histamine (19), acetylcholine (20), prostaglandins (21,22), endorphins, including met-encephalin (23,24), neutral amino acids (25), estrogens (26-28), stress (29,30), the menstrual cycle (31), sleep (32), pineal hormones (33) and circadian rhythms (34). This is not an exhaustive list. Whether any neurotransmitters other than dopamine participate in the regulation of prolactin secretion through the 24 hr cycle, or under such special conditions as suckling and stress, has not been definitively demonstrated. Each factor that is identified as affecting prolactin secretion in vivo or in vitro, leads to speculation that it has a role in regulating prolactin secretion but there is no conclusive evidence for any of these candidates, except possibly estrogen. Nevertheless, the possibility that non-dopaminergic influences participate in the regulation of prolactin secretion must be considered in the interpretation of all experimental data.

DECREASED DOPAMINERGIC ACTIVITY INCREASES PROLACTIN SECRETION

The ability of dopamine or dopamine agonists to inhibit prolactin secretion leads to the expectation, which has been abundantly confirmed, that dopamine antagonists, such as the neuroleptics, markedly promote prolactin secretion in vivo (35,36) and in vitro (2). Other means of decreasing the availability of dopamine at the pituitary, such as inhibiting its synthesis with AMPT (37), decreasing impulse flow in tubero-infundibular dopamine neurons with gamma-hydroxybutyrolactone (38) and preventing intracellular dopamine storage with reserpine (39) markedly increases prolactin secretion. In this paper, we will review selected studies on the effect of anti-psychotic drugs on plasma or serum prolactin levels in man and laboratory animals. The focus will be to attempt to determine whether the effect of anti-psychotic drugs on prolactin secretion is an effective model for their action at neo-striatal dopamine receptors, which is relevant to the capacity of these drugs to induce parkinsonian side effects and tardive dyskinesia or their capacity to act at meso-limbic or meso-cortical dopamine receptors, which is believed to be relevant to their anti-psychotic activity (40-42).

EFFECT OF NEUROLEPTICS ON PROLACTIN SECRETION IN VIVO AND IN VITRO

Ben-David, Dikstein and Sulman (43) studied the effect of phenothiazines on the histologic maturation of breast tissues in estrogenized adult female rats and determined a so-called mammotrophic index (MTI). No relationship was found between the MTI and anti-psychotic properties. Chlorpromazine sulfoxide and perphenazine sulfoxide, which are not anti-psychotic, were mammotrophic. It was subsequently observed that anti-psychotic agents such as chlorpromazine and reserpine can cause persistent lactation in the rat (44). Ben-David (45) reported that perphenazine increased rat plasma prolactin in vivo using a bioassay method. Extensive in vitro studies have lead to the conclusion that the action of the neuroleptics to increase prolactin secretion is directly at the pituitary itself, and is due to antagonism of the action of dopamine (2,12,46). This is in agreement with evidence we have presented elsewhere that intravenous dopamine, which does not cross the blood-brain-barrier, can inhibit the fluphenazine-induced increase in prolactin secretion in man (47). This indicates the action of dopamine at the pituitary or median eminence can overcome the neuroleptic action which leads to disinhibition of prolaction secretion. It does not rule out additional effects of neuroleptics at dopaminergic receptors other than the median eminence or pituitary.

CORRELATION BETWEEN ANTI-PSYCHOTIC ACTION AND DISINHIBITION OF PROLACTIN SECRETION

Clemens, Smalstig and Sawyer (48) reported that the anti-psychotic agents they studied: pimozide, haloperidol, sulpiride, chlorpromazine, thioridazine and fluphenazine increased male rat serum prolactin levels. Promethazine, pyrathiazine, methdilazine and ethopropazine, non-anti-psychotic drugs, did not raise serum prolactin levels. Thiethylperazine, which was believed to be a non-anti-psychotic phenothiazine, did raise serum prolactin levels in rats. Sachar et al. (49) subsequently demonstrated it did the same in man. Further clinical investigation has now demonstrated thiethylperazine is anti-psychotic (Rotrosen, J. and Angrist, B., in preparation).

We have provided additional information in support of the relationship between disinhibition of prolactin secretion and anti-psychotic action. Four members of the dibenzodiazepine series of drugs have been studied for anti-psychotic efficacy: loxapine, clothiapine, clozapine, and perlapine. Loxapine, clothiapine and clozapine have proven anti-psychotic action (50-52) but perlapine is believed to be mainly a sedative (53,54). Clozapine has been reported to have relatively weak dopamine receptor blocking effects in vivo (55-58). It produces no extrapyramidal side effects in man (59-61). Seeman et al. (62) and Creese, Burt and Snyder (63) have reported that clozapine can displace ^{3}H-haloperidol in an in vitro dopamine receptor binding assay, and that the magnitude of this effect is consistent with its average clinical dose in relation to other anti-psychotic drugs.

We have found that all four dibenzodiazepines can increase rat plasma prolactin (64,65). We have determined the ED_{500}, the dose which raises plasma prolactin five-fold, using a log probit plot. Of the four drugs, clozapine was the least potent (ED_{500}, 4.33 mg/kg), loxapine the most potent (ED_{500}, 0.20 mg/kg), and perlapine and clothiapine, approximately equally potent (ED_{500}, 1.27 mg/kg and 1.81 mg/kg, respectively). To rule out that the increase in prolactin might be due to a serotonergic agonist effect, which would also account for an increase in rat plasma prolactin levels, we pretreated rats with the serotonin antagonist methysergide, 10 mg/kg, 15 min prior to perlapine, 10 mg/kg, and sacrificed the rats 30 min later. Methysergide did not significantly effect the increase in prolactin due to perlapine (65). Similar results have been obtained with clozapine (Meltzer and Fang, unpublished data).

In so far as loxapine, clothiapine and clozapine increase rat plasma prolactin levels and are anti-psychotic, our results are consistent with the hypothesis that the dopamine receptors which influence anti-psychotic action and prolactin secretion have a similar response to neuroleptic drugs. Perlapine appears to be an exception. However, Wilk and Stanley (66) have recently demonstrated that perlapine produces effects on rat limbic dopamine metabolism that are identical to those produced by classical anti-psychotic neuroleptics. They have proposed that perlapine will prove to be anti-psychotic with further testing, a prediction with which we concur.

There is one more compound, metoclopramide, which has been reported to elevate plasma prolactin levels in the rat (67) and man (68) but to lack significant anti-psychotic properties (69,70). Since metoclopramide is chemically related to sulpiride, which is a highly effective anti-psychotic drug (71), it is possible that further clinical testing with metoclopramide will reverse this apparent inconsistency. Should perlapine or metoclopramide, or both, prove not to be anti-psychotic, then the fact that these drugs increase prolactin secretion in rats and man would indicate that ability of an apparent dopamine blocking agent to stimulate prolactin secretion is not an infallible guide to anti-psychotic activity. It should be pointed out that recent studies have suggested metoclopramide may stimulate prolactin secretion directly rather than by interfering with the dopaminergic inhibition of prolactin secretion (Fang, V.S., in preparation).

We have determined the ED_{200}, the dose which doubles plasma prolactin, from log probit plots, for 10 anti-psychotic drugs of known clinical potency. The Spearman rank order correlation between ability to stimulate prolactin secretion and potency as an anti-psychotic was 0.86 (p<.05) (48). Further evidence of the correlation between anti-psychotic activity and ability to disinhibit prolactin secretion is provided by our finding that 7-hydroxy-chlorpromazine was as potent as chlorpromazine in elevating rat plasma prolactin (72). 7-hydroxychlorpromazine is as effective as chlorpromazine in increasing brain dopamine turnover in mice (73) and nearly as effective in inhibiting amphetamine-induced stereotypy in the rat (74). On the other hand, the following derivatives of chlorpromazine have no effect on brain dopamine turnover or amphetamine-induced stereotypy or rat prolactin secretion: chlorpromazine sulfoxide, 8-hydroxychlorpromazine, 7,8-dihydroxychlorpromazine, 8-hydroxy-7-methoxychlorpromazine, and 7-methoxychlorpromazine (72).

There is still additional evidence supporting the similarity of the dopamine receptors which mediate neuroleptic action as anti-psychotic agents, as anti-amphetamine and anti-apomorphine agents, and as agents which can disinhibit prolactin secretion. Meltzer, Paul and Fang (7) have reported that the cis (α)-flupenthixol and (+)-butaclamol can markedly stimulate rat prolactin secretion whereas their stereoisomers (trans (β)-flupenthixol and (-)-butaclamol) are over 100-fold weaker in increasing prolactin secretion. β-flupenthixol and (-)-butaclamol have much less efficacy in reversing amphetamine- or apomorphine-stereotypy than do their stereoisomers (see Ref. 7). Confirmatory data was recently reported by Caron et al. (75) who studied the ability of butaclamol, flupenthixol and thioxanthine stereoisomers to compete for binding to pituitary dopamine receptors in vitro with ^{3}H-dihydroergocryptine. The stereospecificity found was identical to that we observed in vivo. They also found the same stereospecificity in blocking the ability of dopamine to inhibit prolactin release from rat anterior pituitary cells in culture.

TOLERANCE AND DISINHIBITION OF PROLACTIN SECRETION

An important contrast between the response of dopamine receptors in the striatum and those in the limbic regions to neuroleptics is the development of tolerance. There is evidence from studies of the effect of acute and chronic treatment with neuroleptics on rat brain dopamine metabolism and anti-stereotypy that tolerance to the effect of these drugs develops in the striatum (76,77). It is well-established that the parkinsonian side effects of neuroleptics diminish within the first few weeks of treatment in most patients, so that anti-cholinergic therapy need not be continued. Further, cerebrospinal fluid (CSF) levels of homovanillac acid (HVA), the major metabolite of dopamine, are initially increased in man after neuroleptic treatment, but return to normal within 4-10 weeks of treatment (78). Since CSF HVA largely reflects the rate of dopamine metabolism in the striatum (79), these findings suggest the development of tolerance in the striatum to the adaptive mechanisms which increase dopamine synthesis and metabolism following dopamine receptor blockade. On the other hand, tolerance does not usually develop to the anti-psychotic action of the neuroleptics as evidenced by their continued clinical efficacy over many years, usually at lower doses during the maintenance phase of treatment than the acute phase.

We have found that no tolerance develops to the neuroleptic-induced increase in serum prolactin levels in man (36) or plasma prolactin levels in rats. As illustrated in Fig. 1, for patients receiving at least 200 mg chlorpromazine orally, b.i.d., serum prolactin levels increase within a day after neuroleptic treatment is started and remain persistently elevated in morning blood samples obtained just prior to the next oral dose. This pattern has been present in all 125 patients we have studied. These hospitalized patients have generally been studied for up to 3 months. With R. Moline and A. Christopolous, we have measured serum prolactin levels in 12 out-patients who had received neuroleptics for 5-108 months (median 25 months). All had increased serum prolactin levels. Other investigators have reported normal serum prolactin levels in a few patients treated with neuroleptics for a period of years (80-83). It would be important to study such patients intensively to determine if a genuine tolerance did develop, whether they failed to ingest or absorb drugs, or whether there could be a primary failure of the pituitary to synthesize and secrete prolactin. We have treated rats with chlorpromazine 5 mg/kg at 9 AM and 4 PM chronically for 4 weeks. No decrease in the ability of chlorpromazine to increase rat plasma prolactin was noted (Meltzer and Fang, unpublished data).

Thus, the dopamine receptors which mediate the effect of neuroleptics on prolactin secretion, tho which mediate the anti-psychotic action of neuroleptics and those which mediate anti-amphetamine and anti-apomorphine stereotypy have many parallel responses to a large series of phenothiazines, thioxanthenes, butyrophenones, substituted benzamides, dibenzodiazepines and anti-psychotics of other structures. The possible exceptions, metoclopramide and perlapine, will require further study before they can be accepted as drugs which increase prolactin secretion but are not anti-psychotic. The minimal implication of these findings is that disinhibition of prolactin secretion, in vivo or in vitro, can be used as a screening method for anti-psychotic drugs. This method of screening would probably produce results qualitatively similar to those yielded by some of the other methods currently in use: blocking amphetamine- and apomorphine-induced stereotypy and competition with ^{3}H-haloperidol for dopamine receptor binding in vitro (63) but the prolactin response may be a more reliable procedure than the other methods and is readily quantitated. To the extent that the prolactin response is correlated with average clinical anti-psychotic dose in man, the ability to increase plasma prolactin in the rat can be used as an indication of expected clinical dose in man, which would be of substantial value in early clinical trials. As we

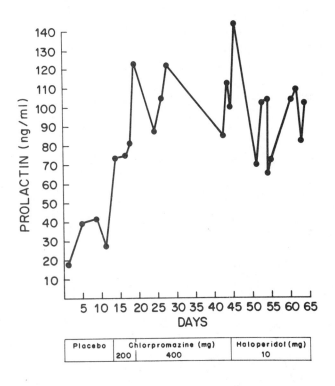

Placebo	Chlorpromazine (mg)		Haloperidol (mg)
	200	400	10

Fig. 1 Effect of chronic treatment with chlorpromazine
or haloperidol on serum prolactin levels

have stated elsewhere, we believe it is preferable to emphasize the search for anti-psychotic
drugs which are not dopamine receptor blockers (84).

EFFECT OF ESTROGENS ON INCREASE IN PLASMA PROLACTIN PRODUCED BY NEUROLEPTICS

The studies we have reviewed have for the most part been carried out in adult male rats.
Investigations utilizing the prolactin response to anti-psychotics as a means of investigating
the effects of these drugs on dopaminergic mechanisms, must be cognizant of the potent effect
of estrogens on prolactin secretion (26,27) and in particular, the strong interaction between
estrogens and reserpine or phenothiazines to disinhibit prolactin secretion. Thus, Van der
Gugten et al. (85) reported that reserpine was significantly more effective in increasing
plasma prolactin levels in rats given exogenous estrogen than in control rats. Ojeda and
McCann (27) studied the increase in plasma prolactin produced in female rats by pimozide on
the day of birth and at 3, 6, 10, 25 and 35 days. No effect was noted at day 0 but the
pimozide-induced stimulation of prolactin secretion increased up until day 35. They proposed
that plasma prolactin levels in the female rat depend on the balance between the stimulatory
effect of estrogen on prolactin secretion and the inhibitory control exerted by the hypothalamic dopaminergic system. Clemens, Smalstig and Sawyer (49) found that pimozide and chlorpromazine produced much greater increases in prolactin secretion in female rats than male
rats. Bohnet, Aragona and Friesen (86) also found larger increases in plasma prolactin in
female rats given chlorpromazine or the long-acting neuroleptic, fluphenazine decanoate.

As can be seen in Fig. 2, we have found that chronic administration of neuroleptics produces

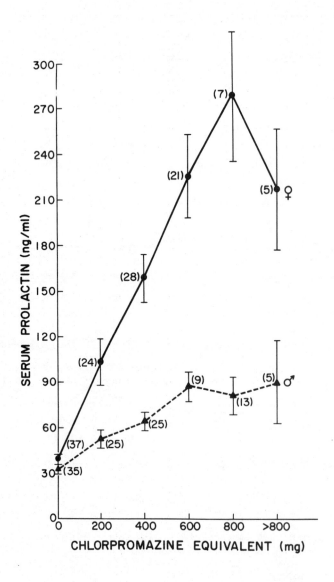

Fig. 2 Effect of neuroleptic treatment on serum prolactin levels

Patients were treated with chlorpromazine, thioridazine, fluphenazine
or haloperidol on a b.i.d. schedule. Blood samples were obtained after
one week at each dose level, 12 hrs after the last dose. All dosages
were converted to chlorpromazine equivalent according to Davis (101).
No adjustments were made for differences in subject's body weight.

markedly greater increases in serum prolactin levels in female schizophrenics than in male
schizophrenics. Buckman and Peake (28) have reported that perphenazine, 8 mg orally,
produced significantly larger increases in serum prolactin levels in normal females than
males. Estrogen potentiated the effect of perphenazine on serum prolactin levels in both
groups, but more so in females than in males. Buckman, Peake and Srivastava (87) subsequently
reported that perphenazine, 8 mg orally, produced a larger integrated increase in prolactin
secretion in females given perphenazine during the late follicular phase of the menstrual
cycle, when estrogen levels are high, compared to the early follicular phase when estrogen
levels are low.

FACTORS OTHER THAN ESTROGEN AND SEX WHICH AFFECT PROLACTIN RESPONSE TO NEUROLEPTICS

Bohnet, Aragona and Friesen (86) have reported that there are variations in the duration of the chlorpromazine-induced plasma prolactin increase and the rate of its return to normal in male and female rats of different ages. The magnitude of the prolactin response was greater in female rats 40 or 60 days old, compared to those 20 days old. Difference in the prolactin response to perphenazine in several strains of mice have been reported (88). It was speculated that these strain-specific differences may be an important factor in the development of mammary tumors. Augustine and MacLeod (89) found that thyroid ablation did not affect the prolactin response to perphenazine in male rats. However, l-thyroxine, 200 µg/kg, inhibited the stimulation of prolactin synthesis by perphenazine.

PROLACTIN AND BREAST CANCER

There is very strong evidence that prolactin is a factor in mammary tumors in rats and mice (90). Fortunately, there is no solid evidence that the same holds true for humans (91-92), although many investigators still believe the question is moot (93-95). Since neuroleptics, including reserpine, all produce persistent elevations in plasma prolactin levels, were prolactin to either initiate or promote mammary (or other) carcinomas in humans, the risk/ benefit ratio of these drugs in susceptible persons would be dramatically affected in an adverse way. Careful epidemiologic studies of patients chronically treated with neuroleptics are indicated. The few studies to date do not indicate an increased risk (96-99). Indeed, there is evidence that they may have some protective effects (96,97,100). In any event, there is no solid evidence that the rat or mouse represents a good animal model for the possible mammary carcinogenicity of the neuroleptics.

CONCLUSION

This paper has focused on several possible ways in which the prolactin response to anti-psychotic drugs in rats can have significance for psychopharmacology. The best use appears to be as a system for exploring effects at a dopamine receptor which may have a number of features in common with the limbic dopamine receptor postulated to be relevant to the anti-psychotic action of these drugs. There appears to be limited usefulness to explore extrapyramidal effects via the prolactin response although there is one report in man of a correlation between extrapyramidal response to neuroleptics and serum prolactin levels (82).

It must be kept in mind that the dopamine receptors at the pituitary receive their dopamine via the pituitary portal circulation, not across a synapse, as in the limbic region. This probably accounts for our inability to demonstrate an effect of dopamine uptake blockers such as benztropine on the prolactin response to L-DOPA in rats (Meltzer, Simonovic and Fang, unpublished data). Further, the pituitary gland does not have the complex neuronal circuitry that the striatal and limbic regions possess, so that many of the neurotransmitter and pharmacologic interactions present in these latter dopaminoceptive regions will not pertain in the pituitary. The fact that the pituitary and median eminence are outside the blood-brain-barrier will also contribute to differences in drug response in the two regions. Despite these differences, we believe the studies reported here indicate that the rat pituitary prolactin response to neuroleptics is a useful animal model for the anti-psychotic action of neuroleptics.

ACKNOWLEDGEMENTS

Supported in part by USPHS MH 25116, USPHS MH 29206 and State of Illinois 710-21. HYM is recipient of USPHS RSA MH 47,808. MS is supported by USPHS MH 14274.

REFERENCES

1. Meites, J. and Clemens, J.A., Hypothalamic control of prolactin secretion, Vitam. and Horm. 30, 165 (1972).

2. MacLeod, R.M., Regulation of prolactin secretion in Frontiers in Neuroendocrinology 4, 169 (1975), L. Martini and W.F. Ganong (Eds.), Raven Press, New York.

3. Diefenbach, W.P., Carmel, P.W., Frantz, A.G. and Ferin, M., Suppression of prolactin secretion by L-DOPA in the stalk-sectioned monkey, J. Clin. Endocrinol. Metab. 43, 638 (1976).

4. Massara, F., Cammani, F., Belforte, L. and Molinatti, G.M., Dopamine and inhibition of prolactin and growth-hormone secretion, Lancet 1, 913 (1976).

5. Leblanc, H., Bachelin, C.L., Abu-Fadel, S. and Yen, S.S.C., Effect of dopamine infusion on pituitary hormone secretion in humans, J. Clin. Endocrinol. Metab. 43, 668 (1976).

6. Shaar, C.J. and Clemens, J.A., The role of catecholamines in the release of anterior pituitary prolactin in vitro, Endocrinology 95, 1202 (1974).

7. Meltzer, H.Y., Paul, S.M. and Fang, V.S., Effect of flupenthixol and butaclamol isomers on prolactin secretion in rats, Psychopharmacology 51, 181 (1977).

8. Van der Gugten, A.A., Sahuleka, P.C., Van Galen, G.H. and Kwa, H.G., Prolactin inhibition test with L-DOPA: Decrease and restoration of plasma prolactin levels in the rat by a peripheral process, J. Endocrinol. 68, 369 (1976).

9. Smalstig, E.B., Sawyer, B.D. and Clemens, J.A., Inhibition of rat prolactin release by apomorphine in vivo and in vitro, Endocrinology 95, 123 (1974).

10. Nagasawa, H. and Meites, J., Suppression by ergocornine and iproniazid of carcinogen-induced mammary tumors in rats: Effects on serum and pituitary prolactin levels, Proc. Soc. Exp. Biol. Med. 135, 469 (1970).

11. Chase, T.N. and Shoulson, I., Dopaminergic mechanisms in patients with extrapyramidal disease, Adv. in Neurol. 9, 359 (1975).

12. MacLeod, R.M. and Lehmeyer, J.E., Pituitary gland alpha-adrenergic receptors and their function in prolactin secretion, Endocrinology 92, A-50 (1973).

13. Kamberi, I.A., Mical, R.S. and Porter, J.C., Effects of melatonin and serotonin on the release of FSH and prolactin, Endocrinology 88, 1288 (1971).

14. Lu, K-H and Meites, J., Effect of serotonin precursors and melatonin on serum prolactin release in rats, Endocrinology 93, 152 (1973).

15. Noel, G.L., Dimond, R.C., Wartofsky, L, Earll, J.M. and Frantz, A.G., Studies of prolactin and TSH secretion by continuous infusion of small amounts of thyrotropin-releasing hormone, J. Clin. Endocrinol. Metab. 39, 6 (1974).

16. Lester, R.C., Underwood, L.E., Marshall, R.N., Friesen, H.G. and Van Wyck, J.J., Evidence for a direct effect of thyrotropin-releasing hormone (TRH) on prolactin release in humans, J. Clin. Endocrinol. Metab. 36, 1148 (1974).

17. Ondo, J.G. and Pass, K.A., The effect of neurally active amino acids on prolactin secretion, Endocrinology 98, 1248 (1976).

18. Mioduszewski, R., Grandison, L. and Meites, J., Stimulation of prolactin release in rats by GABA, Proc. Soc. Exp. Biol. Med. 151, 44 (1976).

19. Donoso, A.O. and Bannza, A.M., Acute effects of histamine on plasma prolactin and luteinizing hormone levels in male rats, J. Neural Transmission 39, 95 (1976).

20. Libertun, C. and McCann, S.M., Blockade of the release of gonadotropins and prolactin by subcutaneous or intraventricular injection of atropine in male and female rats, Endocrinology 92, 1714 (1973).

21. Ojeda, S.R., Harms, P.G. and McCann, S.M., Central effect of prostaglandin E, (PGE), on prolactin release, Endocrinology 95, 613 (1974).

22. Warberg, J., Eskay, R.L. and Porter, J.C., Prostaglandin-induced release of anterior pituitary hormones: Structure activity relationships, Endocrinology 98, 1135 (1976).

23. Rivier, C., Vale, W., Ling, N., Brown, M. and Guillemin, R., Stimulation in vivo of the secretion of prolactin and growth hormone by β-endorphin, Endocrinology 100, 238 (1977).

24. Rivier, C., Brown, M. and Vale, W., Effect of neurotensin, substance P and morphine sulfate on the secretion of prolactin and growth hormone in the rat, Endocrinology 100, 751 (1977).

25. Davis, S.L., Plasma levels of prolactin, growth hormone, and insulin in sleep following the infusion of arginine, leucine and phenylalanine, Endocrinology 92, 1256 (1972).

26. Chen, C.L. and Meites, J., Effects of estrogen and progesterone on serum and pituitary prolactin levels in ovariectomized rats, Endocrinology 86, 503 (1970).

27. Kalra, P.S., Fawcett, C.P., Krulich, L. and McCann, S.M., The effects of gonadal steroids on plasma gonadotropins and prolactin in the rat, Endocrinology 92, 1256 (1973).

28. Buckman, M.T. and Peake, G.T., Estrogen potentiation of phenothiazine-induced prolactin secretion in man, J. Clin. Endocrinol. Metab. 37, 977 (1973).

29. Noel, G.L., Suh, H.K., Stone, G.J., and Frantz, A.G., Human prolactin and growth hormone release during surgery and other conditions of stress, J. Clin. Endocrinol. Metab. 35, 840 (1972).

30. Ajika, K., Kalra, S.P., Fawcett, C.P., Krulich, L and McCann, S.M., The effect of stress and nembutal on plasma levels of gonadotropins and prolactin in ovariectomized rats, Endocrinology 90, 707 (1972).

31. Perez-Lopez, F.R. and Robyn, C., Studies on human prolactin regulation, Life Sci. 15, 599 (1974).

32. Sassin, J.R., Frantz, A.G., Kapen, S. and Weitzman, E.D., The nocturnal rise of human prolactin is dependent on sleep, J. Clin. Endocrinol. Metab. 37, 436 (1973).

33. Rønneklew, O.K., Krulich, L. and McCann, S.M., An early morning surge of prolactin in the male rat and its abolition by pinealectomy, Endocrinology 92, 1339 (1973).

34. Dunn, J.D., Arimura, A. and Scheving, L.E., Effect of stress on circadian periodicity in serum LH and prolactin concentration, Endocrinology 90, 29 (1972).

35. Kleinberg, D.L., Noel, G.L. and Frantz, A.G., Chlorpromazine stimulation and L-DOPA suppression of plasma prolactin in man, J. Clin. Endocrinol. Metab. 33, 873 (1971).

36. Meltzer, H.Y. and Fang, V.S., The effect of neuroleptics on serum prolactin levels in schizophrenic patients, Arch. Gen. Psychiatry 33, 279 (1976).

37. Lu, K.-H. and Meites, J., Inhibition by L-DOPA and monoamine oxidase inhibitors of pituitary prolactin release. Stimulation by methyldopa and d-amphetamine. Proc. Soc. Exp. Biol. 137, 480 (1971).

38. Meltzer, H.Y. and Fang, V.S., Effect of gamma-butryolacton and baclofen on plasma prolactin in male rats, Biochem. Pharmacol. 26, 645 (1977).

39. Lu, K.-H., Amenomori, Y., Chen, C.L., and Meites, J., Effects of central acting drugs on serum and pituitary prolactin levels in rats, Endocrinology 87, 667 (1970).

40. Snyder, S.H., Banerjee, S.P., Yamamura, H.I. and Greenberg, D., Drugs, neurotransmitters and schizophrenia, Science 184, 1243 (1974).

41. Stevens, J.R., An anatomy of schizophrenia? Arch. Gen. Psychiatry 29, 177 (1973).

42. Meltzer, H.Y. and Stahl, S.M., The dopamine hypothesis of schizophrenia, Schizophrenia Bulletin 2, 19 (1976).

43. Ben-David, M., Dikstein, S. and Sulman, F.G., Production of lactation by non-sedative phenothiazine derivatives, Proc. Soc. Exp. Biol. Med. 118, 265 (1965).

44. Talwalker, P.K., Meites, J., Nicoll, C.S. and Hopkins, T.F., Effects of chlorpromazine on mammary glands of rats, Amer. J. Physiol. 199, 1073 (1960).

45. Ben-David, M., Mechanism of induction of mammary differentiation in Sprague-Dawley female rats by perphenazine, Endocrinology 83, 1217 (1968).

46. Quijada, M., Illner, P., Krulich, L. and McCann, S.M., The effect of catecholamines on hormone release from anterior pituitaries and ventral hypothalami incubated in vitro, Neuroendocrinology 13, 151 (1973).

47. Meltzer, H.Y., Goode, D.J. and Fang, V.S., The effect of psychotropic drugs on endocrine function. I. Neuroleptics, precursors and agonists in A Review of Progress, A. Dimascio and M. Lipton (Eds.), Raven Press, New York.

48. Clemens, J.A., Smalstig, E.B. and Sawyer, B.D., Anti-psychotic drugs stimulate prolactin release, Psychopharmacologia (Berl.) 40, 123 (1974).

49. Sachar, E.J., Gruen, P.H., Altman, N., Halpern, F.S. and Frantz, A.G., Use of Neuroendocrine techniques in psychopharmacological research in Hormones, Behavior and Psychopathology, E. Sachar (Ed.), Raven Press, New York.

50. Gerlach, I., Koppelhus, P., Helweg, E. and Monrad, A., Clozapine and haloperidol in a single blind cross-over trial: Therapeutic and biochemical aspects in the treatment of schizophrenia, Acta Psych. Scand. 50, 410 (1974).

51. Jacobsson, L., Noren, M.-B., Perris, C. and Rapp, W., A controlled trial of clothiapine and chlorpromazine in acute schizophrenia syndromes, Acta Psych. Scand., Supp. 255, 53 (1974).

52. Bishop, M.P. and Gallant, D.M., Loxapine: A controlled evaluation in chronic schizophrenic patients, Curr. Therap. Res. 12, 594 (1970).

53. Take, Y., Nikoda, T., Nakagima, R., Chiba, S., Saji, Y. and Nagawa, Y., Pharmacological studies of 6-(4-methyl-1-piperazinyl) morphanthridine (MP-11), J. Takeda Res. Lab. 29, 416 (1970).

54. Stille, G., Sayers, A.C., Lavener, H. and Eichenberger, E., 6-(4-methyl-1-piperazinyl)-morphanthridine (perlapine), a new tricyclic compound with sedative and sleep promoting properties, Psychopharmacologia (Berl.) 28, 325 (1973).

55. Bartholini, G., Haefely, W., Jalfre, M., Keller, H.H. and Pletscher, A., Effects of clozapine on cerebral catecholaminergic neurone systems, Br. J. Pharmacol. 46, 736 (1972).

56. Andén, N.-E. and Stock, G., Effect of clozapine on the turnover of dopamine in the corpus striatum and in the limbic system, J. Pharm. Pharmacol. 25, 346 (1973).

57. Burki, H.R., Ruch, W., Asper, H., Baggiolini, M. and Stille, G., Effect of single and repeated administration of clozapine on the metabolism of dopamine and noradrenaline in the brain, Eur. J. Pharmacol. 27, 180 (1974).

58. Hyttel, J., Effect of neuroleptics on the disappearance rate of ^{14}C labelled catecholamines formed from ^{14}C tyrosine in mouse brain, J. Pharm. Pharmacol. 26, 588 (1974).

59. Gerlach, J., Thorsen, K. and Fog, R., Extrapyramidal reaction and amine metabolites in cerebrospinal fluid during haloperidol and clozapine treatment of schizophrenic patients, Psychopharmacologia (Berl.) 40, 341 (1975).

60. Matz, R., Rick, W., Oh,D., Thompson, H. and Gershon, S., Clozapine - a potential antipsychotic agent without extrapyradmidal manifestations, Cur. Ther. Res. 16, 687 (1974).

61. Simpson, G.M. and Varga, E., Clozapine - a new antipsychotic agent, Cur. Ther. Res. 16, 679 (1974).

62. Seeman, P., Lee, T., Chan-Wong, M. and Wong, K., Antipsychotic drug doses and neuroleptic/dopamine receptors, Nature 261, 717 (1976).

63. Creese, I., Burt, D.R. and Snyder, S.H., Dopamine receptor binding predicts clinical and pharmacological potencies of antischizophrenic drugs, Science 192, 481 (1976).

64. Meltzer, H.Y., Daniels, S. and Fang, V.S., Clozapine increases rat serum prolactin levels, Life Sci. 17, 339 (1975).

65. Meltzer, H.Y., Fessler, R.G. and Fang, V.S., Perlapine: Relationship between stimulation of prolactin secretion and anti-psychotic activity, Psychopharmacology (In Press).

66. Wilk, S. and Stanley, M., Perlapine and dopamine metabolism: Prediction of antipsychotic efficacy, Eur. J. Pharmacol. 41, 65 (1977).

67. Yamauchi, J., Takahara, J. and Ofuji, T., Stimulation of prolactin secretion by metoclopramide in man and rats, Abs. Fifth Int. Congress of Endocrinol., Hamburg, Germany (1976).

68. Judd, S.J., Lazaries, L. and Smythe, G., Prolactin secretion by metoclopramide in man, J. Clin. Endocrinol. Metabol. 43, 313 (1976).

69. Borenstein, P. and Bles, G., Effets cliniques et electroencephalographiques du metoclopramide en psychiatrie, Therapie 30, 975 (1965).

70. Nakra, B.R.S., Bond, A.J. and Lader, M.H., Comparative psychotropic effects of metoclopramide and prochlorperazine in normal subjects, J. Clin. Pharmacol. 15, 449 (1975).

71. Haase, H.J., Florie, L. and Ulrich, F., Klinisch-neuroleptische untersuchung des N-[1-Athyl-pyrrolidur-2-yl)-methyl]-2-methoxy-5-sulfamoyl-benzamid-Neuroleptikums sulpirid (Dogmatil(R)) an akut erkrankten schizophrenen, Int. Pharmacopsychiat. 9, 77 (1974).

72. Meltzer, H.Y., Fang, V.S., Simonovic, M. and Paul, S.M., Effect of metabolites of chlorpromazine on plasma prolactin levels in male rats, Eur. J. Pharmacol. 41, 431 (1977).

73. Nyback, H. and Sedvall, G., Effect of chlorpromazine and some of its metabolites on synthesis and turnover of catecholamines formed from C^{14}-tyrosine in mouse brain, Psychopharmacologia (Berl.) 26, 155 (1972).

74. Lal, S. and Sourkes, T.L., Effect of various chlorpromazine metabolites on amphetamine-induced stereotyped behavior in the rat, Eur. J. Pharmacol. 17, 283 (1972).

75. Caron, M.G., Raymond, V., Lefkowitz, R.J. and Labrie, F., Identification of dopaminergic receptors in anterior pituitary: Correlation with the dopaminergic control of prolactin release, Fed. Proc. 36, 278 (1977).

76. Scatton, B., Garret, C., Glowinski, J. and Julow, L., Effect of a long-acting injectable neuroleptic, the palmitic ester of pipotiazine, on dopamine metabolism in the rat neostriatum in Pharmacology and Biochemistry of Nigro-Striatal System. Dopaminergic and Non-Dopaminergic Systems. CINP Congress, Paris (1974) (In Press).

77. Møller-Nielsen, I., Fjalland, B., Pedersen, V. and Nymark, M., Pharmacology of neuroleptics upon repeated administration, Psychopharmacologia (Berl.) 34, 95 (1974).

78. Post, R.M. and Goodwin, F.K., Time-dependent effects of phenothiazines on dopamine turnover in psychiatric patients, Science 190, 488 (1975).

79. Sourkes, T.L., On the origin of homovanillac acid (HVA) in the spinal fluid, J. Neural Transmission 34, 153 (1973).

80. Beumont, P.J.V., Corker, C.S., Friesen, H.G., Kolakowska, T., Mandelbrote, B.M., Marshall, J., Murray, M.A.F., and Wiles, D.H., The effects of phenothiazines on endocrine function: II. Effects in men and post-menopausal women, Br. J. Psychiatry 124, 420 (1974).

81. Wilson, R.G., Hamilton, J.R., Boyd, W.D., Forrest, A.P.M., Cole, E.N., Boyns, A.R. and Griffiths, K., The effect of long term phenothiazine therapy on plasma prolactin, Br. J. Psychiatry 127, 71 (1975).

82. Kolakowska, T., Wiles, D.H., McNeilly, A.S. and Gelder, M.G., Correlation between plasma levels of prolactin and chlorpromazine in psychiatric patients, Psychol. Med. 5, 214 (1975).

83. De Rivera, I.L., Lal, S., Ettigi, P., Hontela, S., Muller, H.F. and Friesen, H.G. Effect of acute and chronic neuroleptic therapy on serum prolactin levels in men and women of different age groups, Clin. Endocrin. 5, 273 (1976).

84. Meltzer, H.Y., Dopamine receptors and average clinical dose, Science 194, 545 (1976).

85. Van der Gugten, A.A., Verhofstad, F., Sala, M. and Kwa, H.G., Effect of reserpine on plasma prolactin dependent on the presence of oestrogens, Acta Endocrinologica 65, 309 (1970).

86. Bohnet, H.G., Aragona, C. and Friesen, H.G., Age and sex differences in prolactin response to phenothiazines in rats, Biol. Reprod. 15, 168 (1976).

87. Buckman, M.T., Peake, G.T. and Srivastava, L.S., Endogenous estrogen modulates phenothiazine stimulated prolactin secretion, J. Clin. Endocrinol. Metab. 43, 901 (1976).

88. Sinha, Y.N., Salocks, C.B. and Vanderlaan, W.P., Prolactin and growth hormone levels in different inbred strains of mice: Patterns in association with estrous cycle, time of day, and perphenazine stimulation, Endocrinology 97, 1112 (1975).

89. Augustine, E.C. and MacLeod, R.M., Prolactin and growth hormone synthesis: Effect of perphenazine, α-methyltyrosine and estrogen in different thyroid states, Proc. Soc. Exp. Biol. Med. 150, 557 (1975).

90. Smithline, F., Sherman, L. and Kolodny, H.D., Prolactin and breast carcinomas, N. Eng. J. Med. 292, 784 (1975).

91. Mack, T.M., Henderson, B.E., Gerkins, V.R., Arthur, M., Baptistas, J. and Pike, M.C., Reserpine and breast cancer in a retirement community, N. Eng. J. Med. 292, 1366 (1975).

92. Laska, E.M. Siegel, C., Meisner, M., Fischer, S. and Wanderlung, J., Matched-pairs study of reserpine use and breast cancer, Lancet 2, 296 (1975).

93. Jick, H., Reserpine and breast cancer: A perspective, JAMA 233, 896 (1975).

94. Kwa, H.G., De Jong-Bakker, M., Engelsman, E. and Cleton, F.J., Plasma prolactin in human breast cancer, Lancet 1, 433 (1974).

95. Henderson, B.E., Gerkins, V., Rosario, I., Casagrande, J. and Pike, M.C., Elevated serum level of estrogen and prolactin in daughters of patients with breast cancer, N. Eng. J. Med. 293, 790 (1975).

96. Katz, J., Kunofsky, S., Patton, R.E. and Allaway, N.C., Cancer mortality among patients in New York mental hospitals, Cancer 20, 2194 (1967).

97. Rassidakis, N.C., Kelepouris, M., Goulis, K. and Karaioseffidis, K., Malignant neoplasms as a cause of death among psychiatric patients, II. Inter. Mental Health Res. Newsletter 14 (1972).

98. Rassidakis, N.C., Kelepouris, M. and Fox, S., Malignant neoplasms as a cause of death among psychiatric patients, Inter. Mental. Health Res. Newsletter 13 (1971).

99. Rassidakis, N.C., Goulis, C., Kotroutsos, E. and Callarou, T., Malignant neoplasms as a cause of death among psychiatric patients. III. (Relatives of patients and malignancy), Inter. Mental Health Res. Newsletter 14 (1973).

100. Eicke, W.J., Gunstage verlaufe bei karzinompatienten unter zusatzlicker behand ung mit phenothiazin derivaten, Med. Klein. 68, 1015 (1973).

101. Davis, J.M., Dose equivalence of the anti-psychotic drugs, J. Psychiat. Res. 11, 65 (1974).

DISCUSSION

Dr. Salama asked if Dr. Meltzer had tried a combination of a neuroleptic with an anticholinergic to alter prolactin levels. He also asked whether after prolactin levels had been increased by a neuroleptic, a GABA agonist could be used to lower them. *Dr. Meltzer* replied that neither anticholinergics nor cholinergics had an effect on prolactin secretion. The literature on GABA effects on prolactin is conflicting, indicating that GABA both stimulates and inhibits prolactin secretion. In his own work with muscimol and diazepam, Dr. Meltzer has not seen significant prolactin effects. He has not tried to reduce prolactin levels with a GABA agonist.

Dr. Ursillo wondered if other antipsychotic agents showed the prolonged response elicited by clozapine. *Dr. Meltzer* indicated that the agents studied by Dr. Sachar and himself gave about the same response - 6 to 8 hours after i.m. administration. After oral administration, there are problems with differences in absorption.

TARDIVE DYSKINESIA AND ANTIPSYCHOTICS

W.E. Fann, M.D. and J.R. Stafford, M.D. and Jeanine Wheless
Baylor College of Medicine, Houston, Texas 77030

INTRODUCTION

It has generally been assumed that the antipsychotic properties of the neuroleptics are correlated with disturbances in extrapyramidal function. Neuroleptics are thought to render their
antipsychotic properties by blockade of dopamine receptor sites accompanied by enhanced turnover of striatal dopamine, but the resulting relative cholinergic dominance in the striatal
portion of the brain is probably the origin of extrapyramidal side effects (1,2). Consequently, antischizophrenic effects and extrapyramidal symptoms seem to be inextricably linked.

In a study of 64 psychotic patients treated with haloperidol for up to 2 years, Man noted
that the sooner the patient developed extrapyramidal symptoms, the sooner psychotic symptoms
improved. If the patient did not develop side effects, even at a high dose, the chances were
that he would not improve or would show only minimal improvement. The author stated that, by
watching for the development of extrapyramidal symptoms, one apparently could predict whether
or not a patient will improve. Extrapyramidal symptoms occurred in 76.5% of the sample; 78%
showed improvement in their psychosis. All symptoms were controlled with antiparkinson agents
or reduction in dosage (3).

Haase stated, "The more affinity a drug possesses for the extrapyramidal system, as shown by
the production of extrapyramidal psychokinetic inhibitions, the higher is its neuroleptic
potency (4)." Ayd also found a correlation between the absolute frequency of drug-induced
extrapyramidal reactions and the milligram potency of the phenothiazines -- the more potent
the drug, the more frequent the occurrence of striopallidal symptoms. Thirty-five percent of
patients treated with chlorpromazine, 44% of patients treated with thiopropazate, and 52% of
patients prescribed fluphenazine developed extrapyramidal reactions (5).

Among the extrapyramidal reactions which might occur are akinesia, acute dystonia, akathisia,
and parkinsonism. These side effects generally occur early in treatment and usually respond
to a reduction in dosage or addition of a corrective medication (5,6,7,). More recently,
another syndrome, tardive dyskinesia, has been associated with antipsychotic drug use. Tardive
dyskinesia generally occurs late in the course of treatment and often after discontinuation
of neuroleptic administration; symptoms may persist for months or years and sometimes are
permanent and irreversible; the condition shows poor response to any type of therapy (6,8,9).
Tardive dyskinesia should be distinguished from the other drug-induced extrapyramidal side
effects; therefore, a brief description of each reaction follows.

The most frequent extrapyramidal reaction to neuroleptics is akinesia, a condition characterized by weakness, muscular fatigue, and apathy. The patient is constantly aware of fatigue
in limbs used for ordinary motor acts such as walking or writing and, in advanced form, may
complain of aches and pains in the muscles of the affected limbs. Because of these symptoms,
there is often a reduction in voluntary activity (5).

Acute dystonic reactions are of abrupt onset and consist of bizarre muscular spasms, facial
grimacing and distortions, retrocollis, torticollis, dysarthria, and labored breathing.
Oculogyric crises -- spasm of the external ocular muscles with painful upward gaze persisting
for minutes of hours -- may also occur. Because of the unusual symptoms, dystonic reactions
have been misdiagnosed as tetany, seizure, or hysteria (5,6,7,). They occur more often in
males than females and in young rather than old patients; occurrence appears to be a matter
of individual sensitivity as well as dose and type of antipsychotic (6). Dystonic reactions
can be relieved promptly by parenteral administration of a variety of agents such as antiparkinsonian agents, barbiturates, and antihistamines (5,6,7).

Akathisia or motor restlessness refers to a subjective desire to be in constant motion; patients
complain of an inability to sit or stand still. Akathisia may be mistaken for psychotic

agitation leading to a further increase in dose which invariably worsens the patient's condition. Parenteral anti-Parkinson agents may produce immediate response. Oral doses may also allay the restlessness, often without lowering the dose of neuroleptic. However, akathisia is generally the most difficult to manage extrapyramidal reaction and the condition may not respond adequately to such medication. It is frequently necessary to lower dosage of the neuroleptic or add a sedative agent such as diazepam, diphenhydramine or small doses of a barbiturate (5,6).

Drug-induced Parkinsonism may be clinically indistinguishable from idiopathic or post-encephalic types. Cases of moderate severity will display mild rigidity, tremor, and a slowing of movement (bradykinesia). More severe cases will involve stooped posture, a marked pill-rolling tremor , a mask-like fixed facies, increased salivation with drooling, cog-wheeling rigidity, and an involuntary tendency to shorten steps and increase speed in walking (marche petit pas) (6,10,11). Treatment includes reduction of dose or addition of a conventional anti-Parkinsonian agent. It is generally preferable to add oral anti-Parkinson agents since this permits uninterrupted neuroleptic therapy and avoids the risk of loss of therapeutic benefit. The clinician should be aware, however, that while anticholinergic agents mask drug-induced parkinsonism, they may also increase the intensity and duration of tardive dyskinesia in the patient (11,12,13).

INCIDENCE OF EXTRAPYRAMIDAL REACTIONS

Extrapyramidal reactions occur in about 1% of general medical patients treated with phenothiazines or tricyclic antidepressants. Depending upon dosage and duration of treatment, between 21% and 79% of psychiatric patients receiving long-term phenothiazine therapy have been reported to develop extrapyramidal symptoms (14).

Among a survey of 3,775 psychiatric patients treated with phenothiazines, Ayd reported that 39.9% developed extrapyramidal symptoms -- 2.3% developed varied dystonias and dyskinesia; 15.4% developed parkinsonism; 21.2% developed akathisia. Of the patients exhibiting these side effects, 63% were women. Akathisia and parkinsonism occurred twice as often in women as men; dyskinesia, however, occurred twice as often in man (5).

Patients who have blood relatives with naturally-occurring Parkinsonism are more apt to have a neuroleptic-induced extrapyramidal reaction. Young patients are more likely to develop dystonic reactions or oculogyric crises. The older the patient, the more likely the development of parkinsonian symptoms (15).

Studies have suggested that older patients (6,8,16), females (5,6,8,15,17), and individuals with organic brain disorders (6,8), are more susceptible to develop extrapyramidal symptoms. Ayd, however, suggests that striopallidal symptoms due to neuroleptics are a matter of individual susceptibility even more than a matter of chemical structure, milligram potencey, dosage, or duration of treatment, pointing to the fact that 61.9% of the patients in his survey never had a neurologic reaction to neuroleptics (5).

TARDIVE DYSKINESIA

Tardive dyskinesia (TD) has been observed with increasing frequency since it was first reported in the late 1950's. The syndrome was initially controversial. In 1968 Kline challenged the fact that the syndrome exists at all, stating that the incidence of side effects has been misrepresented and that TD should not be regarded as a significant danger in therapy with neuroleptics (18). As recently as 1973 the relationship between the administration of neuroleptics and the occurrence of tardive dyskinesia has been questioned with the suggestion that acceptance of a cause-effect relationship might be premature (19). Two studies in 1968 attempted to ascertain the prevalence of TD in institutionalized patients who never received neuroleptic drugs. Crane studied chronic psychiatric patients in the US and Turkey, the two sample groups being comparable except for their exposure to drugs. Whereas no symptoms of dyskinesia were observed in the 97 Turkish patients, only 4 of whom had received neuroleptics, a considerable number of patients in the US exhibited TD, the difference reaching statistical significance (20). Greenblatt et al. conducted a survey of geriatric patients in a nursing home. Only 2 of 100 untreated patients exhibited dyskinesia; 20 of 52 patients receiving neuroleptics were afflicted (21). In 1973 the American College of Neuropsychopharmacology took a firm stand in the controversy, declaring that the etiologic relationship between neuroleptic administration and tardive dyskinesia was sufficiently probable to urge caution in the routine prolonged use of these drugs (22). Tardive dyskinesia is now considered a well defined neurologic entity and a major iatrogenic disease. The practice of maintaining literally millions of chronically psychotic patients on neuroleptics has currently been challenged (23) and clinicians are urged to use caution when prescribing these drugs on a long-term basis.

At the present time the pathogenesis of TD remains a matter of conjecture, although is is probable that a variety of pathophysiological factors are involved. Because TD is similar

to the hyperdopaminergic disease, Huntington's Disease (both conditions are improved by dopamine blocking agents and both are exacerbated by drugs that enhance CNS catecholamine action
and by anticholinergic compounds such as antiparkinsonian agents) (10,13,24,25), it has been
speculated that TD is also due to a hyperdopaminergic state resulting from long-term neuroleptic administration. It is postulated that prolonged dopaminergic blockade eventually
induces a chemical denervation of dopaminergic receptors, rendering the receptors supersensitive to dopamine. If dopamine blockade is diminished by reduction in dosage or discontinuation of the neuroleptic, the system responds in an exaggerated manner to what would
ordinarily be a normal level of dopamine input. This relative excess of dopaminergic tone is
thought to underly the clinical manifestations of tardive dyskinesia (2,6,9,11,13,24,25).

Clinical features of the syndrome include the following: 1) symptoms become manifest after
neuroleptics are significantly reduced or withdrawn; 2) symptoms are masked or disappear
when neuroleptics are reinstated or dose is significantly increased; 3) anticholinergic drugs
do not relieve and often worsen the symptoms of TD (6,8,9,13).

The classic symptom of TD is a bucco-facial-lingual triad consisting of involuntary movements of the lips, jaws, and tongue. Characteristic manifestations include smacking and
sucking movements of the lips, thrusting, rolling, and "fly-catching" movements of the
tongue, lateral jaw movements, and puffing of the checks. These symptoms often worsen under
emotional tension and disappear during sleep (6,9).

TD is identified by sight alone and unlike most disorders, the symptoms are best observed
when the examiner does not appear to be examining. The dyskinesias may be inhibited by the
patient when he is alerted but can be seen while the patient is walking, sitting or convers-
ing (7). Tongue and mouth movements, generally the most common manifestations, seem to be
less altered by voluntary control (24).

In addition to oral-facial dyskinesias,the extremeties may show choreiform movements (variable, purposeless, involuntary quick movements) or athetoid movements (continuous, arrhythmic,
wormlike slow movements) in the distal parts of the limbs (6,9). Circular movements of the
big toe are frequently noted and in resting patients in whom the features of TD occur in-
cessantly, both toes and shoes may be traumatized (7). Severe dystonia involving muscles
controlling balance of the body may be painful and greatly reduce the patient's activity.
These symptoms also disappear during sleep (6,9).

It has been suggested but not established that fine vermicular (wormlike) movements of the
tongue may be one of the earliest signs of the development of tardive dyskinesia. Periodic
examination of the tongue provides a simple and convenient early detection technique (9).

Symptoms of Parkinsonism may coexist with tardive dyskinesia making diagnosis difficult.
Because it appears that these two conditions are reciprocal in their pathophysiology, treatment for one condition may aggravate the other (25).

Since drug-induced parkinsonism is symptomatically equivalent to the idiopathic or post-encephalitic types of Parkinson's Disease, clinicians have reasoned that the atropine-like agents
that are effective in the treatment of Parkinson's Disease might also reverse the drug-
induced phenomena and it had become standard practice to prescribe antiparkinsonian drugs
prophylactically at the onset of neuroleptic therapy. However, not all patients started on
neuroleptics develop drug-induced parkinsonism and these atropine-like agents may increase the
patient's risk of developing TD. Furthermore, in patients with tardive dyskinesia the insti-
tution of an anticholinergic regimen has been found to worsen the condition. It is important
to note this because many patients with extrapyramidal symptoms are treated with anticholinergic
agents although the efficacy of anticholinergic therapy has been demonstrated only with parkinsonian syndrome. It is recommended that physicians prescribe antiparkinsonian drugs only
after the development of extrapyramidal symptoms rather than prophylactically (25,26,27).

INCIDENCE AND EPIDEMIOLOGY OF TD

Fann et al. surveyed a VA and a state mental hospital population and found that 36% of chronic
patients manifested the tardive dyskinesia syndrome. The reaction was slightly more prevalent
in females, occurring in 44% as compared to 33% of the males, but the difference was not
statistically significant (28).

In a sample of 332 chronic patients, TD was present in 56%. There was a highly significant
difference between age groups: prevalence of TD was 46% among patients under 49 years of
age; 60% in patients between 50 and 70 years; and 75% in patients older than 70 years. The
appearance of TD is probably related to individual susceptibility which might possibly increase with the changes of dopamine metabolism that occur with aging. No significant difference by sex was reported except that significantly more choreoathetoid movements occurred
in females than in males (16).

Twelve psychiatric patients with tardive dyskinesia were observed from the inception of the syndrome. The authors reported that first symptoms rarely persisted if antipsychotics were discontinued. They concluded that length of time symptoms have persisted prior to drug discontinuation, not age at onset, may be a major variable determining reversibility of the syndrome (29).

There has been no clear evidence that TD is specifically related to any particular drug (30), dosage level (28,30), duration of treatment (30), or the presence of brain damage (23,28). Neuroleptics are capable of evoking TD in non-psychotic as well as psychotic patients in the absence of obvious organic, metabolic, or neurological factors (23,31) and the continued use of neuroleptics in non-psychotic patients has been deemed inappropriate by some physicians (32).

Crane suggests that patients who do not exhibit parkinsonian features while on neuroleptics are less likely to develop TD upon withdrawal of the drug (33); another study did not confirm this finding (16). Good (34) and Fann and Lake (25), on the other hand, suggest that persons who develop acute neurological symptoms such as dystonia, akathisia, and parkinsonism, may be at high risk to develop TD.

An epidemiological study of tardive dyskinesia reported two interesting findings: 1) TD was significantly more frequent in schizophrenic patients with an insidious beginning than in patients with acute onset. 2) Prevalence of TD was lower if treatment begins at a younger age; the mean age at the beginning of neuroleptic treatment was significantly higher in patients who subsequently developed TD than in those who did not. Paradoxically, the mean duration of treatment and total amount of drugs are less in patients who begin treatment at a more advanced age. This study did not show a simple correlation between mean total amount of administered neuroleptic and TD (30) in disagreement with Crane who reported that risk of TD rose with total quantity of drug and duration of treatment (8).

ATYPICAL CASES OF TARDIVE DYSKINESIA

Classically, TD develops after prolonged use of neuroleptics, usually more than 2 years, and especially in patients prescribed high doses. However, there are reports in the literature of patients developing TD following short-term, low-dose therapy with neuroleptic drugs. TD symptomatology developed in a patient treated for 30 weeks with haloperidol never exceeding 10 mg/day. After medication was discontinued, symptoms lessened significantly within 4 weeks and had completely abated by 12.5 weeks (35). A 17 year old woman maintained for 2 years on haloperidol began experiencing tremulousness in the hands unresponsive to benztropine mesylate. Symptoms worsened to include facial tics, grimacing and head bobbing so severe that the patient had to use her hands to support her head. Haloperidol was discontinued and the patient was prescribed 100 mg thioridazine. Dyskinesias lessened considerably with disappearance of facial tics and grimacing but a slight head bobbing remained (36). Crane described the case of a 41 year old female treated with neuroleptics for less than one year who developed TD. TD subsided in a matter of days after withdrawal of all medication (37). Jacobsen et al. reported 4 cases of TD associated with haloperidol at doses of 4-20 mg/daily. Two of the cases involved temporary oral-facial dyskinesias and the other two exhibited a more persistent complex mixture of neurological features (38). A 19 year old male given relatively low doses of neuroleptic for less than 6 months developed tardive dyskinesia. The patient exhibited severe oral-facial dyskinesia including protrusion of tongue, chewing movement, movement of the chin, and contraction of the upper lip exposing upper teeth. Mouth movements greatly improved 4 months after the discontinuation of medication and by the sixth drug-free month, symptoms had cleared completely (39). Crane and Smeets studied 39 geriatric patients who had had no previous exposure to neuroleptics. Moderate to moderately severe dyskinesia was detected only in persons treated for at least 7 months with a minimum daily dose of at least 72 mg chlorpromazine or equivalent and a total dose of at least 14,000 mg (17). Marcott reported that a patient who received 900 mg of mesoridazine in 72 hours became dysarthric, restless and agitated with gross tremor. After medication was discontinued, the patient developed lip smacking, protrusion of the tongue, and grimacing with persisted 5 days after cessation of mesoridazine (40).

DRUG COMBINATIONS OR DISEASES INFLUENCING THE DEVELOPMENT OF EXTRAPYRAMIDAL SIDE EFFECTS

Although the high incidence of extrapyramidal side effects from moderate to high doses are well documented, complications occurring with low doses of neuroleptics necessitate the study of precipitating pathophysiological circumstances. One such factor may be the use of alcohol. Seven young patients treated with neuroleptics had a sudden occurrence of drug-induced akathisia and dystonia after imbibing ethyl alcohol. In 6 cases, symptoms subsided promptly with the administration of benztropine mesylate, biperiden HCl or biperiden lactate. One untreated case persisted for 36 hours. The author suggests that, "it is possible that so-called neuroleptic nonreactors precipitate extrapyramidal side effects through brain dysfunction induced by alcohol or other means (41) Lutz also described 4 cases of short-lasting akathisia during combined ECT and phenothiazine therapy (42).

A toxic neurological reaction following the combined use of lithium and haloperidol occurred in 7 patients with bipolar mania. None of the patients had a prior indication of TD or any other state of receptor hypersensitivity and in all patients the extrapyramidal symptoms persisted longer than would be expected with haloperidol alone. Antiparkinsonian agents had a limited effect and even exacerbated the condition of one patient. In all patients, symptoms were reversible (43).

The rate of extrapyramidal reactions in patients taking neuroleptics or antidepressants increases when prednisone is given concomitantly. However, the finding is complicated by the fact that of 8 patients who received both trifluoperazine and prednisone, 4 had systemic lupus erythematosus (SLE). Three patients with SLE developed extrapyramidal reactions whereas none without SLE developed these side effects (14).

Good described a patient who presented with facial grimacing, myoclonus, and catatonia-like symptoms associated with gluthemide discontinuance and antihistamine use. It is likely that additive or interactive effects of gluthemide and antihistamine led to significant alteration in the metabolism of dopamine and reduction in its intracellular concentration. This activity would account for the neuromotor signs. The catatonia-like phenomena may be related to prior inhibition of dopamine reuptake by striatal neurons with subsequent receptor hypersensitivity or hyperstimulation (34).

Altered thyroid states may influence the effect of psychotropic drugs. Lake and Fann report a case of neurotoxic rigidity apparently due to haloperidol in a hyperthyroid patient. The reaction was extreme, far greater than would be expected from the dose, and was temporally related to the waxing and waning of the patient's hyperthyroid state; haloperidol was well tolerated with no unusual response in the euthyroid state (44).

A patient with untreated hypoparathyroidism developed a severe dystonic reaction following IM prochlorperazine for nausea. Subsequently, the authors studied the effect of prochlorperazine in 5 untreated hypothyroid patients. All 5 had a severe dystonic reaction within 5 to 31 hours, indicating a striking phenothiazine sensitivity in the disease. The authors speculate that the extrapyramidal reaction may be related to the vascular and perivascular calcification of the basal ganglions that frequently occurs in hypoparathyroidism or to the generalized increased excitability of the nervous system caused by hypocalcemia (45).

OTHER DRUGS CAUSING EXTRAPYRAMIDAL SIDE EFFECTS

Although the neuroleptics are the most common drug type to cause extrapyramidal reactions, other drug classes have also been implicated in these neurologic conditions. Sympathomimetic vasoconstrictive agents in the antihistamines, which are structurally related to the phenothiazines, have been related to persistent involuntary movements of the face and mouth. Two females with prolonged use of antihistamic decongestants developed oral-facial dyskinesias which improved with haloperidol and worsened with placebo. Both patients had reduced homovanillic acid accumulation during probenecid loading, indicating functional or structural derangement of central dopamine pathways as in patients with neuroleptic-induced dyskinesia(46). Favis reports a case of facial dyskinesia following the use of a proprietary cough syrup in recommended dose for 2 days. The condition was alleviated by IM diazepam and reoccurred when the patient took more of the medication (47). Sovner stated that the mechanism of action in antihistamine-induced dyskinesia might be alteration of central dopaminergic pathways by blocking reuptake of dopamine by individual neurons (48).

Dystonic reactions to the halogenated phenothiazines are well documented. Sananman reported a case of dyskinesia occurring with a halogenated phenethylamine. A 43 year old female took a single fenfluramine tablet and within 2 hours experienced retrocollic movements of the head and neck with tongue and throat muscle spasms which made it difficult to breathe. The reaction subsided within 90 seconds following IV diphenhydramine (49). The occurrence of dyskinesia after sympathomimetic drugs has not been widely recognized but it is important that physicians be aware of the possibility of these side effects because diet medications and other sympathomimetic agents are prescribed frequently.

Fann et al. report 2 cases of buccal-facial-lingual dyskinesia occurring in patients treated with tricyclic antidepressants (50). Although the tricyclic antidepressants have little effect on striatal dopamine, they possess potent anticholinergic properties, similar to the neuroleptics. The appearance of dyskinesia in these patients lends support to the hypothesis that drug-induced hyperkinetic disorders are related to a diminution of CNS acethlcholine activity as well as an increase in dopamine activity.

Two young patients treated with depot injections of flupenthixol (not currently marketed in the US) developed chorei-form movements. In one, a 40 year old male, symptoms slowly and completely disappeared; in the other, a 28 year old male, however, the condition failed to respond to treatment and only when lithium was administered was there some amelioration of symptoms (51).

Palatucci reports a case of a 19 year old psychiatric patient who developed a severe dyskinetic reaction following parenteral administration of methylphenidate. Although methylphenidate has been used successfully to reverse the acute neurotoxic side effects of phenothiazines (64), it caused them in this instance. The reaction persisted up to 30 hours (52).

Kirschberg described an acute extrapyramidal reaction involving dyskinetic movements of the face, arms, and legs in a 15 year old female given two 250 mg doses of ethosuximide for petit mal seizures. Symptoms were abruptly and completely relieved by IV injection of diphen-hydramine HCl (53).

Amodiaquine HCl, an effective and relatively non-toxic anti-malarial drug, was implicated in 4 cases of involuntary movements. Dyskinesias included protrusion of the tongue, difficulty in speaking, excessive salivation, fasciculation of tongue and facial muscles, and intention tremor. Symptoms cleared within 3 hours in 3 patients prescribed IV benztropine mesylate; symptoms improved within 24 hours in one patient prescribed oral benzhexol (54).

Diazoxide, a non-diuretic benzothiadiazine vasodilator, is used in the long-term treatment of severe hypertension. Extrapyramidal symptoms were noted in 15% of acute and chronic hypertensive patients treated with diazoxide. Symptoms included the range of extrapyramidal reactions such as tremulousness, restlessness, coarse rhymthmical tremor, cogwheel rigidity, fixed facies, tightness of jaw preventing opening of the mouth, and oculogyric crises. Dosage adjustment or the use of diazepam or procyclidine usually controlled symptoms (55).

Reserpine depletes brain catecholamines and, like the phenothiazines, can cause a parkinsonian syndrome. A 54 year old male treated with reserpine for hypertension developed an oral-facial dyskinesia which persisted for 6 months (56).

TREATMENT OF TARDIVE DYSKINESIA

Drug therapy of TD has been approached along 3 main lines, based on the hypothesis that the condition results from a relative excess of brain dopamine. These techniques include modification of neuroleptic medication, depletion or blockade of dopamine,and cholinergic alteration (9,57,58).

Neuroleptics, dopamine-blocking agents, paradoxically can ameliorate the manifestation of the condition which they cause and some clinicians recommend prescribing, continuing, or increasing dosage of phenothiazines in patients with TD (57,58). Turek et al. studied oral-facial dyskinesia in 56 chronic inpatients with TD during various sequences of drug cessation and doubling of original dose of neuroleptics alone or with antiparkinsonian agents. Results indicated a tendency for dyskinetic symptom severity to be lessened during neuroleptic treatment phases and increased during subsequent drug-free intervals. Antiparkinsonian medication failed to reduce symptom manifestations (59).

Although effective in masking symptoms, the neuroleptics are not curative and their use in treating tardive dyskinesia is likely to aggravate in the underlying pathology. The reinstatement of antipsychotic therapy should be avoided except in patients requiring neuroleptics for control of otherwise unmanageable psychotic behavior (6,9,). Crane goes a step farther, stating that ". . .patients who are very sensitive to the neurotoxic effects of neuroleptics should not receive such compounds, regardless of their mental condition (37)."

Reserpine, which depletes dopamine by preventing intraneuronal storage, has been used effectively to treat TD, although it paradoxically can cause the syndrome (56,57). Tetrabenazine, another dopamine-depleting agent, has demonstrated effective, though short-lived suppression of symptoms. This drug is not currently marketed in the US and, because its effects are not lasting, it is doubtful that tetrabenazine will be prominent in the treatment of TD (9,57).

If tardive dyskinesia is due to an imbalance between the dopaminergic and cholinergic systems, drugs which would increase acetylcholine content in the striatum should restore balance and thereby control TD (60). Deanol, a putative acetylcholine precursor, has been investigated in TD with varying results. Widroe and Heisler report good results to deanol in doses of 100-500 mg daily (58). Fann et al. administered deanol to 10 patients with TD and all 10 exhibited partial or complete relief of symptoms after 5 days (10,61). Crane, however, reported marginal improvement in only 18% of his patients treated with deanol (62).

Physostigmine, a centrally active anticholinesterase, has also shown contradictory results. All of 7 patients with TD prescribed the drug in one study showed significant suppression of movement at 24 hours (24). A prior study by Fann et al. reports an absence of clinical changes in 10 patients whose tardive dyskinesia was treated with physostigmine (63).

Methylphenidate has been used successfully to reverse the acute neurotoxic side effects of phenothiazines (64). In patients with TD, however, it tends to aggravate symptoms. Of 17 patients with TD prescribed methylphenidate, 6 became worse, 8 showed no change, and 3 improved (63,65).

Our more recent work has involed the following phases: 1) A thirty day trial of deanol dosage to 1200 mg/daily in 10 patients diagnosed as suffering from TD (66). 2) A sequential administration of 14 days each deanol 1200 mg/daily, methylphenidate 40 mg/daily and cypro-heptadine 16 mg/daily, with 7 day placebo intervals. This study included measurement of movements with triaxial vector accelerometry as well as videotape recordings, and the deter-mination of plasma choline levels (67), before and after the 14 day deanol phase. An interesting finding was the presence of higher than normal baseline values of plasma choline and the failure of choline increase 12 hours after the last dose of 400 mg deanol. It should be noted that 7 of the 10 subjects were on continous neuroleptic treatment (68,69).

A more recent review of literature concerning deanol treatment of TD notes some 69 cases reported. Improvement was observed in approximately 50% although the reviewer notes the paucity of carefully controlled double-blind studies (69).

Our experience includes 10 of the cases included in Casey's review plus an additional nine-teen cases in the two studies noted in the preceding paragraphs. Clinical response to deanol was seen as pronounced in seven patients, moderate but significant in nine patients, and slight or insignificant in thirteen others.

Patient movements during the five minute videotape recording and the one hour quantitative accelerometry profile appear to correlate poorly with more lengthy informal ward observation. The variability in severity of the dyskinesia within each patient seems to correlate with levels of anxiety and/or voluntary suppression of movements.

Our experience with deanol suggest that it is a safe and often useful compound in the treat-ment of TD, but by no means a panacea. More precise selection of agent may be made when a series of short-acting agents (e.g., physostigmine, haloperidol, benztropine, baclofen, apomorphine, methylphenidate, and L-dopa) are administered to construct a profile of neurotransmitter responsivity. Correlation of such a profile's prediction with subsequent longer-term treatment of dyskinesia will require large scale and carefully controlled double-blind evaluation. Long term L-dopa therapy may result in relief of TD, in contrast to its reported initial exacerbation (70).

Agents which suppress acetylcholine have been most widely used clinically to treat TD. How-ever, the possibility that a cholinergic deficit exists in tardive dyskinesia suggests that long-term suppression of acetylcholine-dependent mechanisms may contribute to this condition. If this hypothesis is true, then a compound which elevates striatal dopamine levels would be preferable to one which suppresses the acetylcholine system in order to restore balance (71).

PREVENTION

Because of the lack of the adequate substitutes for the treatment of psychoses, tardive dyskinesia has generally been accepted as an undesirable but occasionally unavoidable con-sequence of the benefits of prolonged drug therapy. Physicians can minimize the risk in long-term patients by titrating dose to the lowest possible levels that provide adequate control. Many patients can be satisfactorially maintained for long period without anti-psychotic drugs and "drug holidays" are advised. These drug-free intervals help protect the patient from potential hazards of chronic administration and can allow the physician to ascertain if there is evidence of tardive dyskinesia (9,57).

The most important tool to diminish TD remains early detection of symptoms. Often early and insidious manifestations are over-looked and the syndrome is not diagnosed until it is well established, making prognosis poorer (8,67,72). Patients receiving neuroleptics should be examined regularly for the appearance of early signs of emerging TD. These signs include fine vermicular movements of the tongue, circular movements of the big toe, tics in the facial region, ill-defined abnormal mouth or eye movements, mild mouthing or chewing move-ments, the presence of rocking or swaying movements, or the occurrence of restless limb movements in the absence of the subjective discomfort associated with akathisia (6,9). Symptoms of slight intensity, such as those listed above, are often reversible when anti-psychotic medication is withdrawn. Symptoms of moderate or marked intensity are difficult to ameliorate and can become irreversible and permanent.

REFERENCES

1. S. Snyder, D. Greenberg, H. Yamamura, Antischizophrenic drugs and brain cholinergic receptors, Arch. Gen. Psych. 31, 58 (1974).

2. J.G. Bruch, A brief review of drug related neurologic disorders in the elderly. In: Drug Issues in Georpsychiatry, Williams & Wilkins (1974).

3. P.L. Man, Long-term effects of haloperidol, Dis. Nerv. Sys. 34, 113 (1973).

4. H.G. Hasse, et al., (1964) The Action of Neuroleptic Drugs: A Psychiatric, Neurological and Pharmacological Investigation, YearBook Medical Publishers, Inc.

5. F.J. Ayd, A survey of drug-induced extrapyramidal rractions, JAMA 175(12), 1054 (1961).

6. American College of Neuropsychopharmacology-FDA Task Force, Neurologic syndromes assoc. with antipsychotic drug use, New England J. Med. 289(1), 20 (1973).

7. G.W. Paulson, Tardive Dyskinesia, Ann Rev Med 26, 75 (1975).

8. G.E. Crane, Tardive dyskinesia in patients treated with major neuroleptics: a review of the literature, Amer. J. Psych. 124(8), 40 (1968).

9. K.E. Clyne and P.P. Juhl, Tardive Dyskinesia, Amer. J. Hosp. Pharm. 33, 481 (1976).

10. W.E. Fann, C.R. Lake, J.L. Sullivan, R.D. Miller, Neuroleptic-induced movement disorders: pharmacology and treatment. In: Psychopharmacogenetics, Plenum Press (1975).

11. W.E. Fann and C.R. Lake, Drug-induced movement disorders in the elderly: an appraisal of treatment. In: Drug Issues in Geropsychiatry, Williams & Wilkins (1974).

12. W.E. Fann and C.R. Lake, Amantadine versus trihexyphenidyl in the treatment of neuro- leptic-induced parkinsonism, Amer. J. Psych. 133(8),940 (1976).

13. H.L. Klawans, The pharmacology of tardive dyskinesias, Amer. J. Psych. 130(1), 82 (1973).

14. Boston Collaborative Drug Surveillance Program, Drug-induced extrapyramidal symptoms, JAMA 224(6), 889 (1973).

15. F.J. Ayd, Haloperidol: fifteen years of clinical experience, Dis. Ner. Sys. 33, 459 (1972).

16. A. Jus, R. Pineau, R. Lachance, et al., Epidemiology of tardive dyskinesia, Part I, Dis. Ner. Sys. 37, 210 (1976).

17. G.E. Crane and R.A. Smeets, Tardive dyskinesia and drug therapy in geriatric patients, Arch. Gen. Psych. 30, 341 (1974).

18. N.S. Kline, On the rarity of irreversible oral diskinesias following phenothiazines, Amer. J. Psych. 124(8), 48 (1968).

19. J.P. Curren, Tardive dyskinesia -- side effect or not? Amer. J. Psych. 130(4), 406 (1973).

20. G.E. Crane, Dyskinesia and neuroleptics, Arch. Gen. Psych. 19, 700 (1968).

21. D.L. Greenblatt, J.R. Dominick, B.A. Stotsky, A. DiMascio, Phenothiazine-induced dyskinesia in nursing home patients, J. Amer. Ger. Soc. 16, 27 (1968).

22. American College of Neuropsychopharmacology-FDA Task Force, Neurological syndromes assoc. with antipsychotic drug use. A special editorial report, Arch. Gen. Psych. 28, 463 (1973).

23. R.J. Baldessarini, Tardive dyskinesia: an evaluation of the etiologic association with neuroleptic therapy, Can. Psych. Ass. J. 19(6), 551 (1974).

24. W.E. Fann, C.R. Lake, C.J. Gerber, G.M. McKenzie, Cholinergic suppression of tardive dyskinesia, Psychopharmacologia 37, 101 (1974).

25. W.E. Fann and C.R. Lake, On the coexistence of parkinsonism and tardive dyskinesia, Dis. Ner. Sys. 35, 324 (1974).

26. J.C. Pecknold, J.V. Ananth, T.A. Ban, H.E. Lehmann, Lack of indication for use of anti-parkinson medication. A follow-up study, Dis. Ner. Sys. 32, 538 (1971).

27. W.E. Fann, C.R. Lake, B.W. Richman, Drug-induced parkinsonism. A re-evaluation, Dis. Ner. Sys. 36, 91 (1974).

28. W.E. Fann, J.M. Davis, D.S. Janowsky, The prevalence of tardive dyskinesia in mental hospital patients, Dis. Ner. Sys. 33, 182 (1972).

29. F. Quitkin, A. Rifkin, L. Gochfeld, D.F. Klein, Tardive dyskinesia; are first signs reversible? Amer. J. Psych. 134(1), 184 (1977).

30. A. Jus, R. Pineau, R. Lachance, et al., Epidemiology of tardive dyskinesia. Part II, Dis. Nerv. Sys. 37, 257 (1976).

31. H. Hussey, Tardive dyskinesia. An editoral, JAMA 228(8), 1030 (1974).

32. W.E. Thornton and B.P. Thornton, Tardive dyskinesia and low dosage, Amer. J. Psych. 130 (12), 1401 (1973).

33. G.E. Crane, Pseudoparkinsonism and tardive dyskinesia, Arch Neurol 27, 426 (1972).

34. M.I. Good, Catatonialike symptomatology and withdrawal dyskinesias, Amer. J. Psych. 133 (12), 1454 (1976).

35. G.L. Stimmel, Tardive dyskinesia with low-dose, short-term neuroleptic therapy, Amer J. Hosp. Pharm. 33, 961 (1976).

36. M. Hale, Reversible dyskinesia caused by haloperidol, Amer. J. Psych. 131(12), 1413 (1974).

37. G.E. Crane, Rapid reversal of tardive dyskinesia, Amer. J. Psych. 130, 1159 (1973).

38. G. Jacobson, R. Baldessarini, T. Manschreck, Tardive and withdrawal dyskinesia associated with haloperidol, Amer. J. Psych. 131(8), 910 (1974).

39. R. Moline, Atypical tardive dyskinesia, Amer. J. Psych 132(5), 534 (1975).

40. D.B. Marcotte, Neuroleptics and neurologic reactions, S. Med. J. 66(3), 321 (1973).

41. E. Lutz, Neuroleptic-induced akathisia and dystonia triggered by alcohol, JAMA 236(21), 2422 (1976).

42. E.G. Lutz, Short-lasting akathisia during combined electro-convulsive and phenothiazine therapy, Dis. Ner. Sys. 29, 259 (1968).

43. J.B. Loudon and H. Waring, Toxic reactions to lithium and haloperidol, Lancet II, 1088 (1976).

44. C.R. Lake and W.E. Fann, Possible potentiation of haloperidol neurotoxicity in acute hyperthyroidism, Br. J. Psych. 123, 523 (1973).

45. M. Schaaf and C.A. Payne, Dystonic reactions to prochlorperazine in hypoparathyroidism, New Eng. J. Med. 275(18), 991 (1966).

46. B.T. Thach, T.N. Chase, J.F. Basma, Oral facial dyskinesia associated with prolonged use of antihistamic decongestants, New Eng. J. Med. 293(10), 486 (1975).

47. G. Favis, Facial dyskinesia related to antihistamines? New Eng. J. Med. 294(13), 730 (1976).

48. R. Sovner, Dyskinesia associated with chronic antihistamine use, New Eng. J. Med. 294(2), 113 (1976).

49. M.L. Sananman, Dyskinesia after fenfluramine, New Eng. J. Med. 291(8), 422 (1974).

50. W.E. Fann, J.L. Sullivan, B.W. Richman, Dyskinesia associated with tricyclic antidepressants, British J. Psych. 128, 490 (1976).

51. A. Gibson, Choreiform movements after depot injections of flupenthixol, British J. Psych. 125, 111 (1974).

52. D.M. Palatucci, Iatrogenic dyskinesia -- a unique reaction to parenteral methylphenidate, J. Nerv. Men Dis. 159(1), 73 (1974).

53. G.J. Kirschberg, Dyskinesia -- an unusual reaction to ethosuximide, Arch. Neurol. 32, 137 (1975).

54. M.O. Akindele and A.O. Odejide, Amodiaquine-induced involuntary movements, British Med. J. 2, 214 (1976).

55. D. Neary, H. Thurston, J.E.F. Pohl, Development of extrapyramidal symptoms in hypertensive patients treated with diazoxide, British Med. J. 3, 474 (1973).

56. S.M. Wolf, Reserpine: cause and treatment of oral-facial dyskinesia, Bull. LA. Neuro. Soc. 38, 80 (1973).

57. R.M. Kobayashi, Drug therapy of tardive dyskinesia, New Eng. J. Med. 296(5), 257 (1977).

58. H.J. Widroe and S. Heisler, Treatment of tardive dyskinesia, Dis. Ner. Sys. 37, 162 (1976).

59. I. Turek, A. Kurland, T. Hanlon, M. Bohm, Tardive dyskinesia: its relation to neuroleptic and antiparkinson drugs, British J. Psych. 121, 605 (1972).

60. L. DeSilva and C.Y. Huang, Deanol in taridve dyskinesia, British Med. J. 3, 466 (1975).

61. W.E. Fann, J.L. Sullivan, R.D. Miller, G.M. McKenzie, Deanol in tardive dyskinesia: a preliminary report, Psychopharmcologia 42, 135 (1975).

62. G.E. Crane, Deanol for tardive dyskinesia, New Eng. J. Med. 292, 926 (1975).

63. W.E. Fann, J.M. Davis, I.C. Wilson, C.R. Lake, Attempts at pharmacological management of tardive dyskinesia. In: Psychopharmaoclogy and Aging, Plenum Press (1973).

64. W.E. Fann, Use of methylphenidate to counteract the acute dystonic reactions to phenothiazines, Amer. J. Psych. 122, 1293 (1966).

65. W.E. Fann, J.M. Davis, I.C. Wilson, Methylphenidate in tardive dyskinesia, Amer. J. Psych. 130(8), 922 (1973).

66. W.E. Fann, J.R. Stafford, J.I. Thornby, R. Morrell, D. Burgum, Long term deanol treatment in tardive dyskinesia, in preparation.

67. W.E. Fann, JR. Stafford, J.L. Malone, J.D. Frost, B.W. Richman, Clinical research techniques in tardive dyskinesia, Amer. J. Psych. in press.

68. E. Domino, W.E. Fann, J.R. Stafford, Failure of deanol to elevate plasma choline, Dis. Ner. Sys. in preparation.

69. D.E. Casey, Deanol in tardive dyskinesia, Dis. Ner. Sys. in press.

70. J.R. Stafford, W.E. Fann, Deanol in tardive dyskinesia, Dis. Ner. Sys. in press.

71. W.E. Fann, C.R. Lake, G.M. McKenzie, Adrenergic and cholinergic factors in extrapyramidal disorders. In: Neurotransmitter Balances Regulating Behavior (1975).

72. G.E. Crane, Prevention and management of tardive dyskinesia, Amer. J. Psych. 129(4), 466 (1972).

DISCUSSION

Dr. Zahniser reported elevation in plasma and red blood cell choline after acute administration of a high dose of deanol to rodents or humans. However, when H^2-labeled choline was used as a tracer, no H^2-labeled choline could be found in the brains of mice after administration of deanol, even though blood levels were very high. This seems to indicate that there may be a lipid-bound form of choline which is converted, probably in the liver, from deanol but does not pass the blood-brain barrier - at least at short times after administration. She also indicated that she has not seen an increase in brain levels of choline even after 300 mg/kg of deanol nor has she found evidence to indicate that deanol is an acute precursor of acetylcholine. Finally, *Dr. Zahniser* asked Dr. Stafford how many of his patients responded to deanol. *Dr. Stafford* replied that this depended on the rating system. On a clinical global basis, about 50% improved significantly. This could not be substantiated when blind video tape ratings were used or by sensitive quanititative physiological techniques.

Dr. Salama asked about the effect of lithium. *Dr. Stafford* replied that it has been reported that lithium gives at least transient improvement of dyskinesia.

Dr. Barasch reported that Dr. Yen Koo had observed the development of tardive dyskinesia in mice which had been kept on ethanol. Dr. Stafford agreed that acute alcohol administration could precipitate dyskinetic syndromes.

THE CARDIOVASCULAR TOXICITY OF TRICYCLIC ANTIDEPRESSANTS

Duane G. Spiker, M.D.
University of Pittsburgh School of Medicine and
Western Psychiatric Institute and Clinic
Pittsburgh, PA 15261

ABSTRACT

The human and animal literature concerning the cardiovascular effects of the tricyclic anti-depressants is reviewed. With mild to moderate intoxication the sympathomimetic and atropinic effects predominate while a "quinidine-like" effect characterizes severe toxicity. Animal studies show that cardiac failure associated with this latter effect is probably the cause of death in patients who die from tricyclic antidepressant overdose. The implications of this for clinical management of these patients are discussed.

INTRODUCTION

Taken in large amounts, the tricyclic antidepressants (TCA) have toxic effects on the heart (1, 2, 3, 4, 5). In the 453 patients described in these articles, there were 18 fatalities (4.0%), usually from cardiovascular collapse. In view of this, it seemed appropriate to review the cardiovascular toxicity of the TCA. Recent articles have discussed the cardiac effects of therapeutic doses of these drugs in normal patients (6, 7, 8, 9) and in patients with heart disease (9, 10). These will not be reviewed again.

PHARMACOLOGICAL EFFECTS

The TCA appear to have at least three pharmacological effects on the heart. They have some sympathomimetic effect in that they potentiate the pressor response to norepinephrine (NE) in man (11, 12), dogs (13, 14, 15), and cats (16, 17, 18). This effect can be reversed with beta adrenergic blockage (13). Beta adrenergic blocking agents can also prevent (19) and reverse (20) the usual increase in heart rate seen following administration of a TCA. The TCA appear to act on the sympathetic system by blocking the reuptake of NE by the nerve ending (9, 21, 22). They diminish the usual pressor response of tyramine (17, 18) and prevent the uptake of radioisotope labeled NE into the heart (23, 24) and aortic strips (25). They do not significantly increase its efflux from the heart (23).

The TCA also have atropine-like anticholinergic properties (18, 26). Both amitriptyline (15) and protriptyline (18) antagonize the pressor response of acetylcholine and the effect of peripheral vagal stimulation on the heart in anesthetized dogs. Imipramine diminishes the bradycardia and hypotension produced by peripheral vagal stimulation (16). The increased heart rate seen in dogs following the administration of amitriptyline or protriptyline can be partially antagonized by both physostigmine and pyridostigmine (20). However, one report showed that small doses of imipramine increased the hypotensive response to acetylcholine and higher doses had no effect (14). They felt this may be due to a cholinesterase inhibiting effect of imipramine.

The TCA also have another effect on the heart which cannot be readily explained by their effect on the sympathetic or parasympathetic systems. Different authors have used various names to describe it. It has been called a "membrane-stabilizing" (27), "quinidine-like" (28, 29), or "anti-arrhythmic" (30) effect. The common factor in these reports is the decreased conduction velocity in the heart. This appears to be dose dependent (27, 29, 31, 32). Decreased contractility and decreased coronary flow also appear to be associated with this effect (27). The relationship of this effect with the previously described adrenergic effect and atropine-like effect of the TCA is unclear. The decreased conduction velocity is not affected by propranolol (28, 32) or practolol (29).

HUMAN TOXICITY

As noted earlier, there are several reports in the literature describing large series of patients who allegedly overdosed with tricyclic antidepressants (1, 2, 3, 4, 5). It is obviously impossible to know precisely how much was actually taken by each patient. Even if this were known, a better measure would be the plasma concentration of the TCA. This would take into account the amount actually absorbed by the patient, his rate of metabolism of the drug, and to some extent the amount of drug already in his body before he overdosed. To date, only one group has reported the plasma TCA levels in a large number of patients who overdosed with them (3, 31, 33). The most consistent and reliable clinical finding correlating with total plasma TCA levels in their studies was the duration of QRS interval on a routine EKG (r=.75) (31). All patients with a plasma total TCA >1000 ng/ml had a QRS >100 milliseconds. As the plasma TCA level fell, the QRS reverted to normal. As a point of reference, the therapeutic range of nortriptyline is generally considered to be approximately 50-175 ng/ml (34, 35, 36, 37). They also reported in a later paper that death, cardiac arrest, need for supportive respiration, unconsciousness, grand mal seizure, ventricular rate >120/min, cardiac arrhythmias, and bundle branch block also correlate with plasma total TCA level (3). In addition, they noted that some of their patients still had high tricyclic levels several days after their overdose (31). They felt this might explain the sudden unexpected deaths in some patients who had apparently recovered from their overdose (38, 39, 40).

Accepting this finding that a QRS >100 milliseconds indicates a significant overdose of a TCA, other studies which did not measure plasma TCA levels can be reviewed. Forty-two percent of Thorstrand's patients had a QRS >110 milliseconds (4). He reported that 26% of his patients had a systolic blood pressure <100 mm Hg. Seventy-three percent had a heart rate >90 beats/min. In another paper, he reports on 10 patients who all had a QRS >100 milliseconds (41). In this elegant study, he measured cardiac output while these patients were comatose and after they were awake. While they were comatose they were in a hyperkinetic state with increased cardiac output. When he gave these patients propranolol, their cardiac output dropped but the QRS was not affected. He felt that this decrease in cardiac output following propranolol indicated his patients' hearts were still under the adrenergic stimulation of the TCA and not showing much myocardial depression.

Based on EKG changes, most of Gaultier's patients would also be considered to have taken significant overdoses of TCA (1). His report focuses primarily on EKG changes, but he does note that bradycardia and hypotension are ominous signs which usually precede death. Noble's (5) and Goel's (2) patients presumably did not ingest large amounts of TCA as judged by their reports that their patients had little problem with conduction disturbances on their EKGs. Their patients probably should be considered as showing mild to moderate TCA intoxication. The most prominent cardiac disturbance in their patients was tachycardia.

TOXICITY STUDIES IN ANIMALS

There have been numerous animal studies of the cardiovascular toxicity of the TCA (1, 13, 14, 16, 17, 19, 21, 23, 27, 28, 30, 32, 42, 43, 44, 45, 46, 47, 48). Fortunately, there is some agreement on the results. With low doses, there is an increased heart rate (13, 19, 30, 45, 46, 47), variable or increased blood pressure (13, 14, 19, 42), increased inotropy (13, 23, 45), and little effect on cardiac output (19). At high doses, there is decreased heart rate (13, 17, 19, 30, 45, 47), decreased blood pressure (14, 19, 42), decreased inotropy (13, 23, 42), and decreased cardiac output (19, 44, 45, 49). Thus for heart rate, blood pressure, and inotropy there is a dual effect of the TCA with increased effect at low doses and decreased effect at high doses (44).

The effect on coronary flow and peripheral resistence is less clear. Coronary flow seems to be variable or increased at low doses and decreased at high doses (30, 44). Peripheral resistance is variable at low doses and increased at high doses (19, 43, 44, 49). There appears to be a linear relationship between increasing TCA dose and decreased conduction velocity (27, 28, 30, 32, 45, 47). Arrhythmias are rare and occur only at very high doses (19, 30, 45, 50).

Based on human and animal studies, the following series of events seem to occur with increasing intoxication with TCA. Initially, there is an increased heart rate and variable blood pressure changes. Cardiac contractility and output then begin to decrease. Peripheral resistance begins to increase and conduction velocity slows. Various arrhythmias may begin to occur as cardiac output worsens. Decreasing heart rate and blood pressure are usually a prelude to death from cardiovascular collapse.

Thus there seems to be a dual effect of the TCA on the heart. The "adrenergic" and "atropine-like anticholinergic" effects predominate at low or moderate doses and the "membrane-stabilizing" or "quinidine-like" effect predominates at very high doses. This has obvious clinical implications. For example, as noted earlier, most deaths in humans are the result of cardiovascular collapse (1, 2, 3, 4. 5, 39). The animal literature strongly suggests that the critical malfunction of the heart is the decreased cardiac output associated with the "quinidine-like" effect (13, 23, 42, 43, 45, 47, 49) and not arrhythmias or conduction problems as stressed in the clinical literature (1, 38, 39, 51, 52). Animal studies clearly show that this occurs even before the blood pressure begins to fall (42, 49). Consequently, patients who overdose with TCA should be monitored for signs of heart failure. If this occurs, the animal literature shows that the drug of choice is probably a cardiac glycoside

(17, 42). The dose needed to increase contractility has only minimal effect on decreasing conduction velocity (42). Drugs which suppress contractility, such as propranolol or lido-caine, are probably contraindicated (27, 28, 41, 42, 53).

CONCLUSION

The TCA have clearly been shown to be effective in the treatment of depressed patients. Unfortunately, this is the very group of patients most at risk to take their medication in an overdose. The effect of this on the cardiovascular system is complex and depends on the amount of the TCA taken. Animal studies have been very important in elucidating these effects. These studies strongly suggest that the most critical effect is a direct myocardial depression which results in decreased cardiac contractility and output. This is rarely mentioned in the clinical literature, even though this probably is the cause of death in those patients who die from TCA overdoses. Although much has been done, more work in animals on the cardiovascular toxicity of the TCA and its treatment is needed.

REFERENCES

(1) M. Gaultier, J. R. Boissier, A. Gorceix, et al, The cardiotoxicity of imipramine in man and animals, Proc. Eur. Soc. Drug Toxicity 6, 171 (1965).

(2) K. M. Goel, R. A. Shanks, Amitriptyline and imipramine poisoning in children, Br. Med. J. 1, 261 (1974).

(3) J. M. Petit, D. G. Spiker, J. F. Ruwitch, et al, Tricyclic antidepressant plasma levels and adverse effects after overdose, Clin. Pharmacol. Ther. 21, 47 (1977).

(4) C. Thorstrand, Clinical features in poisonings by tricyclic antidepressants with special reference to the ECG, Acta. Med. Scan. 199, 337 (1976).

(5) J. Noble, H. Matthew, Acute poisonings by tricyclic antidepressants: clinical features and management of 100 patients, Clin. Toxicol. 2, 403 (1969).

(6) V. E. Ziegler, B. T. Co, J. T. Biggs, Plasma Nortriptyline and ECG findings, Am. J. Psychiatr. 134(4), 441 (1977).

(7) B. Davies, (1975) Effects on the heart of different tricyclic antidepressants, in J. Mendels (ed): Sinequan (Doxepin HCl), Excerpta Medica.

(8) J. Vohra, G. D. Burrows, G. Sloman, Assessment of cardiovascular side effects of therapeutic doses of tricyclic antidepressant drugs, Aust. NZ J. Med. 5, 7 (1975).

(9) J. W. Jefferson, A review of the cardiovascular effects and toxicity of tricyclic antidepressants, Psychosom. Med. 37, 160 (1975).

(10) S. J. Kantor, J. T. Bigger, A. H. Glassman, et al, Imipramine - induced heart block a longitudinal case study, JAMA 231, 1364 (1975).

(11) A. J. Prange, R. L. McCurdy, C. M. Cochrane, The systolic blood pressure response of depressed patients to infused norepinephrine, J. Psychiatr. Res. 5, 1 (1967).

(12) A. J. Prange, E. Pustrom, C. M. Cochrane, Imipramine enhancement of norepinephrine in normal humans, Psychiatr.Dig. 25, 27 (1964).

(13) A. Kaufmann, N. Basso, P. Aramendia, The cardiovascular effects of N-(γ methylamino-propyl-imino-dibenzyl) - HCL (desmethylimipramine) and Guanethidien, J. Pharm. Exp. Ther. 147, 54 (1965).

(14) M. Osborne, E.B. Sigg, Effects of imipramine on the peripheral autonomic system, Arch. Int. Pharmacodyn. 129, 273 (1960).

(15) K. D. Cairncross, On the peripheral pharmacology of amitriptyline, Arch. Int. Pharmacodyn. 154, 438 (1965).

(16) E. B. Sigg, Pharmacological studies with tofranil, Can. Psychiatr. Assoc. J. (suppl) 4, 575   (1959).

(17) K. Greef, J. Wagner, Animal experiments on the cardiotoxic action of antidepressants, Germ. Med. 1, 92-94 (1971).

(18) K. D. Cairncross, M. W. McCulloch, F. Mitchelson, The action of protriptyline on peripheral autonomic function, J. Pharmacol. Exp, Ther. 149, 365 (1965).

(19) H. Brunner, P. R. Hedwall, M. Meier et al, Cardiovascular effects of preparation CIBA 34, 276-BA and imipramine, Agents and Actions 2, 69 (1971).

(20) M. L. Torchiana, H. C. Wenger, B. Lagerquist, et al, Pharmacological antagonism of the toxic manifestations of amitriptyline and protriptyline in dogs, Toxicol. Appl. Pharmacol. 21, 383 (1972).

(21) N. Barth, E. Muscholl, The effects of the tricyclic antidepressants desipramine, doxepin and iprindole on the isolated perfused rabbit heart, Nauyn Schmiedelbergs Arch. Pharmacol. 284, 215 (1974).

(22) A. Carlsson, Modification of sympathetic function, Pharmacol. Rev. 18, 541 (1966).

(23) R. S. Davis, J. H. McNeil, The cardiac effects of cocaine and certain antihistamines and antidepressants, <u>Arch. Int. Pharmacodyn.</u> 201, 262 (1973).

(24) L. L. Iversen, Inhibition of noradrenaline uptake by drugs, <u>J. Pharm. Phamacol.</u> 17, 62 (1965).

(25) J. F. Connelly, A. W. Vernables, A case of poisoning with "tofranil", <u>Med. J. Aust.</u> 1, 108 (1961).

(26) K. D. Cairncross, S. Gershon, I. D. Gust, Some aspects of the mode of action of imipramine, <u>J. Neuropsychiatr.</u> 4, 224 (1963).

(27) A. Langsler, W. G. Johansen, M. Ryg, et al, Effects of dibenzpine and imipramine on the isolated rat heart, <u>Eur. J. Pharmacol.</u> 14, 333 (1971).

(28) C. Thorstrand, Cardiovascular effects of poisoning by hypnotic and tricyclic antidepressant drugs, <u>Acta. Med. Scand.</u> (suppl) 583, 1   (1975).

(29) J. Vohra, D. Hunt, G. Burrows, et al, Intracardiac conduction defects following overdose of tricyclic antidepressant drugs, <u>Eur. J. Cardiol.</u> 2, 453   (1975).

(30) E. Marmo, L. Coscia, S. Cataldi, Cardiac effects of antidepressants, <u>Jpn. J. Pharmacol.</u> 22, 283 (1972).

(31) D. G. Spiker, A. N. Weiss, S. S. Chang, et al, Tricyclic antidepressant overdose: clinical presentation and plasma levels , <u>Clin. Pharmacol. Ther.</u> 18, 539 (1975).

(32) C. Thorstrand, J. Bergstrom, J. Castenfors, Cardiac effects of amitriptyline in rats, <u>Scand. J. Clin. Lab. Invest.</u> 36, 7 (1976).

(33) D. G. Spiker, J. T. Biggs, Tricyclics antidepressants prolonged plasma levels after overdose, <u>JAMA</u> 236, 1711 (1976).

(34) Kragh-Sorensen, Asberg, Marie, et al, Plasma-nortriptyline levels in endogenous depression, <u>Lancet</u> 1, 113 (1973).

(35) M. Asberg, Plasma Mortriptyline levels- relationship to clinical effects, <u>Clin. Pharmacol. Ther.</u> 16, 215 (1974).

(36) M. Asberg, B. Cronholm, F. Sjoqvist, et al, Relationship between plasma level and therapeutic effect of nortriptyline, <u>Br. Med.J.</u>   3, 331 (1971).

(37) V. E. Ziegler, P. J. Clayton, J. R. Taylor, Nortriptyline plasma levels and therapeutic response, <u>Clin. Pharmacol. Ther.</u> 20, 458 (1976).

(38) R. J. Barnes, S. M. Kong, R. W. Y. Yu, Electrodardiographic changes in amitriptyline poisoning, <u>Brit. Med. J.</u> 3, 222 (1968).

(39) J. W. Freeman, G. R. Mundy, R. R. Beattie, et al, Cardiac abnormalities in poisoning with tricyclic antidepressants, <u>Br. Med. J.</u> ii 610 (1969)

(40) L. Sedal, M. G. Korman, P. O. Williams, et al, Overdosage of tricyclic antidepressants a report of two deaths and a prospective study of 24 patients, <u>Med. J. Aust.</u> 2, 74 (1972).

(41) C. Thorstrand, Cardiovascular effects of poisoning with tricyclic antidepressants, <u>Acta. Med. Scand.</u> 195, 505 (1974).

(42) A. R. Laddu, P. Somani, Desipramine toxicity and its treatment, <u>Toxicol. Appl. Pharmacol.</u> 15, 287 (1969).

(43) K. D. Cairncross, S. Gershon, A pharmacological basis for the cardiovascular complications of imipramine medication, <u>Med. J. Aust.</u> ii 372 (1962).

(44) E. B. Sigg, M. Osborne, B. Korol, Cardiovascular effects of imipramine, <u>J. Pharm. and Exp Therap.</u> 141, 237 (1963).

(45) J. C. Evreux, V. Vincent, G. Faucon, Cardiac disorders caused by amitriptyline, <u>Proc. Eur. Soc. Drug Toxicity.</u> 9, 58 (1968).

(46) E. Elonen, Effect of β-adrenoceptor blocking drugs, physostigmine, and atropine on the toxicity of doxepin in mice, Med. Biol. 53, 231 (1975).

(47) H. Kato, Y. Noguchi, K. Takagi, Comparison of cardiovascular toxicities induced by dimetacrine, imipramine and amitriptyline in isolated guinea pig atria and anesthetized dogs, Jpn. J. Pharmacol. 24, 885, (1974).

(48) T. Baum, A. T. Shropshire, G. Rowles, M. I. Gluckman, Antidepressants and cardiac conduction: iprindole and imipramine, Eur. J. Pharmacol. 13, 287 (1971).

(49) J. K. Van De Ree, N. E. Zimmerman, A. N. P. Van Heijst, Intoxication with tricyclic antidepressants, J. Eur. Toxicol. 5, 302 (1974).

(50) N. Barth, Arrhythmias evoked by infusion of tricyclic antidepressants and noradrenaline on isolated rabbit hearts, Naunyn Schmiedebergs Arch. Pharmacol. 282, R3 (1974).

(51) T. L. Slovis, J. E. Ott, D. T. Teitelbaum, W. Lipscomb, Physostigmine therapy in acute tricyclic antidepressant poisoning, Clin. Toxicol. 4, 451 (1971).

(52) W. K. Hong, P. Mauer, R. Hochman, J. G. Caslowitz, J. A. Paraskos, Amitriptyline cardiotoxicity, Chest 66, 304 (1974).

(53) J. A. Young, W. H. Galloway, Treatment of severe imipramine poisoning, Arch. Dis. Child. 46, 353 (1971).

tensive effects. From a conservative viewpoint, it has been argued that, since the neuroleptics are far more potent displacers of antagonist ligands and the binding/pharmacological
correlations are based on antagonist ligand potencies, the issue of the relevance of binding
studies is predicated on the assumption that binding sites of drugs which are, after all
not the endogenous transmitters, do represent the physiological neurotransmitter receptor.
Thus, do ^{3}H-haloperidol binding sites have any real physiological significance (104)? We
feel that the biochemical characteristics of antagonist binding sites certainly fulfill the
criteria expected of receptors, and more importantly that the existence of the same relative
pharmacological specificity, in each system, for the binding of an antagonist and the
transmitter itself, viz. ^{3}H-haloperidol and ^{3}H-dopamine (16); ^{3}H-LSD and ^{3}H-serotonin (23);
^{3}H-WB-4101 and ^{3}H-norepinephrine (49), implies that the binding of these antagonists can be
equated with interactions at physiological receptors. Indeed, as mentioned in the neuroleptics section above, there is reason to believe that antagonist binding is to the true physiological receptor whereas agonist binding is to a non-physiological, desensitized state.

Since there is a fund of evidence to suggest that the main therapeutic mechanism of tricyclic
antidepressants is presynaptic, receptor studies may only be of importance in determining
postsynaptic modulatory influences, and indeed it may be the case that tricyclic/α-receptor
interactions determine the amount of anti-agitation potential for each drug. The uniqueness
of the ^{3}H-diazepam binding site, and the therapeutic correlations of benzodiazepines acting
potently at that site, illustrate a major advantage of receptor binding techniques; namely,
an avenue towards the possible discovery of a new, selective transmitter system. The ergot
alkaloids represent more of a "miss" with respect to receptor binding, where studies present
the negative implication that the therapeutic actions of a drug like bromocriptine, based on
activation of the dopaminergic system, are probably presynaptic, not postsynaptic, in origin.

Nevertheless, when allowance is made for inherent difficulties of interpretation, receptor
binding remains a valuable heuristic tool for the elucidation of mechanisms of therapeutic
and side effects of psychotropic drugs, and for the design of more potent and selective
agents.

REFERENCES

1. F. E. Bloom, To spritz or not to spritz; the doubtful value of aimless iontophoresis, Life Sci. 14, 1819 (1974).

2. C.P. Pert and S.H. Snyder, Opiate receptor: demonstration in nervous tissue, Science 179, 1011 (1973).

3. L. Terenius, Characteristics of the receptor for narcotic analgesics in synaptic mem-brane fractions from rat brain, Acta Pharmacol. Toxicol. 33, 377 (1973).

4. E.J. Simon, J.M. Hiller and I. Edelman, Stereospecific binding of the potent narcotic analgesic [^{3}H]etorphine to rat brain homogenate, Proc. Nat. Acad. Sci., USA 70, 1947 (1973).

5. S.H. Snyder and J.P. Bennett, Jr., Neurotransmitter receptors in the brain: biochemical identification, Ann. Rev. Physiol. 38, 153 (1976).

6. S.H. Snyder, D.C. U'Prichard and D.A. Greenberg, Neurotransmitter receptor binding in the brain. Proc. Am. Coll. Neuropsychopharmacology Raven Press, New York, in press.

7. U. Ungerstedt, Stereotaxic mapping of monoamine pathways in the rat brain, Acta. Physiol. Scand. 82, Suppl. 367, 1 (1971).

8. D.B. Bylund and S.H. Snyder, Beta adrenergic receptor binding in membrane preparations from mammalian brain, Mol. Pharmacol. 12, 568 (1976).

9. D.C. U'Prichard, D.A. Greenberg, P. Sheehan and S.H. Snyder, Regional distribution of α-noradrenergic receptor binding in calf brain, Brain Res., in press.

10. S.J. Enna, J.P. Bennett, Jr., D.R. Burt, I. Creese and S.H. Snyder, Stereospecificity of interaction of neuroleptic drugs with neurotransmitters and correlation with clinical potency, Nature 263, 338 (1976).

11. S.J. Enna, J.P. Bennett, Jr., D.R. Burt, I. Creese, D.C. U'Prichard, D.A. Greenberg and S.H. Snyder, Stereospecificity and clinical potency of neuroleptics, Nature 267, 184 (1977).

12. K. Voith and J.R. Cummings, Behavioral studies on the enantiomers of butaclamol demonstrating absolute optical specificity for neuroleptic activity, Can. J. Physiol Pharmacol. 54, 551 (1976).

13. S.H. Snyder and H.I. Yamamura, Antidepressants and the muscarinic acetycholine receptor, Arch. Gen. Psychiat. 34, 236 (1977).

14. D.C. U'Prichard, D.A. Greenberg, P. Sheehan and S.H. Snyder, Tricyclic antidepressants: correlation of therapeutic properties with affinity for alpha noradrenergic receptor binding sites in the brain, Submitted to Science.

15. J.W. Maas, Biogenic amines and depression, Arch. Gen. Psychiat. 32, 1357 (1975).

16. D.R. Burt, I. Creese and S.H. Snyder, Properties of [3H]haloperidol and [3H]dopamine binding associated with dopamine receptors in calf brain membranes, Mol. Pharmacol. 12, 800 (1976).

17. P. Seeman, M. Chau-Wong, J. Tedesco and K. Wong, Brain receptors for antipsychotic drugs and dopamine: direct binding assays, Proc. Nat. Acad. Sci., USA 72, 4376 (1975)

18. P. Seeman, T. Lee, M. Chau-Wong, J. Tedesco and K. Wong, Dopamine receptors in human and calf brains, using [3H]apomorphine and an antipsychotic drug, Proc. Nat. Acad Sci., USA 73, 4354 (1976).

19. D.A. Greenberg, D.C. U'Prichard and S.H. Snyder, Alpha noradrenergic receptor binding in mammalian brain: differential labeling of agonist and antagonist states, Life Sci. 19, 69 (1976).

20. D.C. U'Prichard and S.H. Snyder, [3H]Epinephrine and [3H]norepinephrine binding to α-noradrenergic receptors in calf brain membranes, Life Sci. 20, 527 (1977).

21. H.I. Yamamura and S.H. Snyder, Muscarinic cholinergic binding in rat brain. Proc. Nat. Acad. Sci., USA 71, 1725 (1974).

22. A.S.V. Burgen, C.R. Hiley and J.M. Young, The properties of muscarinic receptors in mammalian cerebral cortex, Brit. J. Pharmacol. 51, 279 (1974).

23. J.P. Bennett, Jr. and S.H. Snyder, Serotonin and lysergic acid diethylamide binding in rat brain membranes: relationship to postsynaptic serotonin receptors, Mol. Pharmacol. 12, 373 (1976).

24. S.J. Enna and S.H. Snyder, Properties of γ-aminobutyric acid (GABA) receptor binding in rat brain synaptic membrane fractions, Brain Res. 100, 81 (1975).

25. H. Möhler and T. Okada, GABA receptor binding with 3H(+)bicuculline-methiodide in rat CNS, Nature 267, 65 (1977).

26. A.B. Young and S.H. Snyder, Strychnine binding associated with glycine receptors of the central nervous system, Proc. Nat. Acad. Sci., USA 70, 2832 (1973).

27. H. Kosterlitz (1976) Opiates and Endogenous Opioid Peptides, Elsevier, North Holland, Amsterdam.

28. S.H. Snyder, S.P. Banerjee, H.I. Yamamura and D. Greenberg, Drugs, neurotransmitters and schizophrenia, Science 184, 1243 (1974).

29. A.P. Feinberg and S.H. Snyder, Phenothiazine drugs: structure-activity relationships explained by a conformation that mimics dopamine, Proc. Nat. Acad. Sci., USA 72, 1899 (1975).

30. Janssen,P.A.J. and Van Bever,W.F (1975) in Current Developments in Psychopharmacology, vol. 2 (W.B. Essman and L. Valzelli, Eds.) Spectrum, New York, p. 165.

31. S.H. Snyder, I. Creese and D.R. Burt, The brain's dopamine receptor: labeling with [3H]dopamine and [3H]haloperidol, Psychopharm. Commun. 1, 663 (1975).

32. L.L. Iversen, M.A. Rogawski and R.J. Miller, Comparison of the effects of neuroleptic drugs on pre- and postsynaptic dopaminergic mechanisms in the rat striatum, Mol. Pharmacol. 12, 251 (1976).

33. Creese, I. and Snyder, S.H., Behaviroal and biochemical properties of the dopamine receptor, Proc. Am. Coll. Neuropsychopharmacology, Raven Press, New York, in press.

34. I. Creese., D.R. Burt and S.H. Snyder, Dopamine receptor binding predicts clinical and pharmacological properties of antischizophrenic drugs, Science 192, 481 (1976).

35. P. Seeman, T. Lee, M. Chau-Wong and K. Wong, Antipsychotic drug doses and neuroleptic/dopamine receptors, Nature 261, 717 (1976).

36. I. Creese, D.R. Burt and S.H. Snyder, Dopamine receptors and average clinical doses, Science 194, 546 (1976).

37. I. Creese, A.P. Feinberg and S.H. Snyder, Butyrophenone influences on the opiate receptor, Eur. J. Pharmacol. 36, 231 (1976).

38. D.R. Burt, S.J. Enna, I. Creese and S.H. Snyder, Dopamine receptor binding in the corpus striatum of mammalian brain, Proc. Nat. Acad. Sci., USA 72, 4655 (1975).

39. S.H. Snyder, The opiate receptor in normal and drug altered brain function, Nature 257, 185 (1975).

40. Hulme, E.C., Burgen, A.S.V. and Birdsall, N.J.M. (1976) in Smooth Muscle Pharmacology and Physiology, INSERM 50, 49.

41. S.H. Snyder, The glycine synaptic receptor in the mammalian central nervous system, Brit. J. Pharmacol. 53, 473 (1975).

42. S.H. Snyder, Neurotransmitter and drug receptors in the brain, Biochem. Pharmacol. 24, 1371 (1975).

43. D.C. U'prichard, D.A. Greenberg and S.H. Snyder, Binding characteristics of a radiolabeled agonist and antagonist at central nervous system alpha noradrenergic receptors, Mol. Pharmacol. 13, 454 (1977).

44. D.R. Burt, I. Creese and S. H. Snyder, Binding interactions of lysergic acid diethylamide and related agents with dopamine receptors in the brain. Mol. Pharmacol. 12, 631 (1976).

45. J. Mickey, R. Tate and R.J. Lefkowitz, Subsensitivity of adenylate cyclase and decreased β-adrenergic receptor binding after chronic exposure to (-)-isoproterenol in vitro, J. Biol. Chem. 250, 5727 (1975).

46. C. Mukherjee and R.J. Lefkowitz, Desensitization of β-adrenergic receptors by β-adrenergic agonists in a cell-free system: resensitization by guanosine 5^1-(β, γ-imino)triphosphate and other purine nucleotides, Proc. Nat. Acad. Sci., USA 73, 1494 (1976).

47. M. Weber, T. David-Pfeuty and J.P. Changeux, Regulation of binding properties of the nicotinic receptor protein by cholinergic ligands in membrane fragments from Torpedo marmorata, Proc. Nat. Acad. Sci., USA 72, 3443 (1975).

48. R.J. Lefkowitz and L.T. Williams, Catecholamine binding to the β-adrenergic receptor, Proc. Nat. Acad. Sci., USA 74, 515 (1977).

49. D.C. U'Prichard and S.H. Snyder, Binding of [^{3}H]catecholamines to α-noradrenergic receptor sites in calf brain, J. Biol. Chem. in press.

50. U. Ungerstedt, Postsynaptic supersensitivity after 6-hydroxydopamine induced degeneration of the nigro-striatal dopamine system, Acta Physiol. Scand. suppl. 367, 69 (1971).

51. Creese,I. and Iversen,S.D. (1975) in Pre and Postsynaptic Receptors (E. Usdin and W.E. Bunney, Eds.) Marcel Dekker, New York, p. 171.

52. I. Creese, D.R. Burt and S.H. Snyder, Dopamine receptor binding enhancement in lesion-induced behaviorally supersensitive rats, Science in press.

53. D.R. Burt, I. Creese and S.H. Snyder, Antischizophrenic drugs: chronic treatment elevates dopamine receptor binding in the brain, Science 196, 326 (1977).

54. G.E. Crane, Tardive dyskinesia in patients treated with major neuroleptics: a review of the literature, Am. J. Psychiat. 124, 40 (1968).

55. S.J. Peroutka, D.C. U'Prichard, D.A. Greenberg and S.H. Snyder, Neuroleptic drug interactions with norepinephrine alpha receptor binding sites in rat brain, Neuropharmacology in press.

56. J.M. van Rossum, Different types of sympathomimetic α-receptors, J. Pharm. Pharmacol. 17, 202 (1965).

57. D.A. Greenberg and S.H. Snyder, Selective labeling of α-noradrenergic receptors in rat brain with [³H]dihydroergokryptine, Life Sci. 20, 927 (1977).

58. D.A. Greenberg and S.H. Snyder, Pharmacologic properties of [³H]dihydroergokryptine binding sites associated with α-noradrenergic receptors in rat brain membranes, Mol. Pharmacol. in press.

59. A.S. Horn and O.T. Phillipson, A noradrenaline sensitive adenylate cyclase in the rat limbic forebrain: preparation, properties, and the effects of agonists, adrenolytics and neuroleptic drugs, Eur. J. Pharmacol. 37, 1 (1976).

60. J.B. Blumberg, J. Vetulani, R.J. Stawarj and F. Sulser, The noradrenergic cyclic AMP generating system in the limbic forebrain: pharmacological characterization and possible role of limbic noradrenergic mechanisms in the mode of action of anti-psychotics. Eur. J. Pharmacol. 37,357 (1976).

61. P.A. Janssen, C.J.E. Niemegeers and K.H.L. Schellekens, Is it possible to predict the clinical effects of neuroleptic drugs (major tranquillizers) from animal data? Part I. "Neuroleptic activity spectra" for rats, Arznei Forsch 15, 104 (1965).

62. A.S. Horn, A.C. Cuello and R.J. Miller, Dopamine in the mesolimbic system of the rat brain: endogenous levels and the effects of drugs on the uptake mechanism and stimulation of adenylate cyclase activity, J. Neurochem. 22, 265 (1974).

63. S.H. Snyder, D. Greenberg, and H.I. Yamamura, Antischizophrenic drugs and brain cholinergic receptors, Arch. Gen. Psychiat. 31, 58 (1974).

64. R.J. Miller and C.R. Hiley, Anti-muscarinic properties of neuroleptics and drug-induced Parkinsonism, Nature 248, 596 (1974).

65. B. Bopp and J.H. Biel, Antidepressant drugs, Life Sci. 14, 415 (1974).

66. Poeldinger, W.J. (1976), in Current Developments in Psychopharmacology, vol. 3 (W.B. Essman and L. Valzelli, Eds.). Spectrum, New York, p. 181.

67. S.B. Ross and A.L. Renyi, Inhibition of the uptake of tritiated 5-hydroxy-tryptamine in brain tissue. Eur. J. Pharmacol. 7, 270 (1969).

68. A. Carlsson, H. Corrodi, K. Fuxe and T. Hökfelt, Effect of antidepressant drugs on the depletion of intraneuronal brain 5-hydroxytryptamine stores caused by 4-methyl-α-ethyl-meta-tyramine, Eur. J. Pharmacol. 5, 357 (1969).

69. S.B. Ross and A.L. Renyi, Inhibition of the uptake of tritiated catecholamines by antidepressant and related agents, Eur. J. Pharmacol. 2, 181 (1967).

70. A.S. Horn, J.T. Coyle and S.H. Snyder, Catecholamine uptake by synaptosomes from rat brain. Structure-activity relationships of drugs with differential effects on dopamine and norepinephrine neurons, Mol. Pharmacol. 7, 66 (1971).

71. J.J. Schildkraut, Norepinephrine metabolites as biochemical criteria for classifying depressive disorders and predicting responses to treatment: preliminary findings, Am. J. Psychiat. 130, 695 (1973).

72. Byck, R. (1975) in The Pharmacological Basis of Therapeutics (L.S. Goodman and A. Gilman, Eds.), MacMillan, New York, p.152.

73. L. E. Hollister, Complications from psychotherapeutic drugs, Clin. Pharmacol Ther. 5, 322 (1964).

74. A.F. Poussaint, K.S. Ditman and R. Greenfields, Amitriptyline in childhood enuresis, Clin. Pharmacol. Ther. 7, 21 (1966).

75. P.E. Gordon, Meprobamate and benactyzine (Deprol) in the treatment of depression in general practice (a controlled study), Dis. Nerv. Syst. 28, 234 (1967).

76. S.H. Snyder, S.J. Enna and A.B. Young, Brain mechanisms associated with therapeutic actions of benzodiazepines: focus on neurotransmitters, Am. J. Psychiat. 134, 662 (1977).

77. Margules, D.L. and Stein, L. (1966) in Neuro-Psycho-Pharmacology (H. Brill, Ed.), Excerpta Medica, Amsterdam, p. 108.

78. Greenblatt, D.J. and Shader, R.I. (1974) Benzodiazepines in Clinical Practice, Raven Press, New York.

79. R.F. Schmidt, Presynaptic inhibition in the vertebrate central nervous system. Ergebn. Physiol. exp. Pharmakol. 63, 20 (1971).

80. Snyder, S.H. and Enna, S.J. (1975) in Mechanism of Action of Benzodiazepines, (E. Costa and P. Greengard, Eds.) Raven Press, New York, p. 81.

81. A.B. Young, S.R. Zukin and S.H. Snyder, Interaction of benzodiazepines with central nervous system glycine receptors: possible mechanism of action, Proc. Nat. Acad. Sci., USA 71, 2246 (1974).

82. D.R. Curtis, C.J.A. Game and D. Lodge, Benzodiazepines and central glycine receptors, Brit. J. Pharmacol. 56, 307 (1976).

83. A. Dray and D.W. Straughan, Benzodiazepines: GABA and glycine receptors on single neurons in the rat medulla, J. Pharm. Pharmacol. 28, 314 (1976).

84. Stein, L., Wise, C.D. and Belluzi, J.D. (1975) in Mechanism of Action of Benzodiazepines, (E. Costa and P. Greengard, Eds.) Raven Press, New York, p. 29.

85. J.A. Gray, J.D.T. McNaughton and P.H. Kelly, Effect of minor tranquilizers on hippocampal θ rhythm mimicked by depletion of forebrain noradrenaline, Nature 258, 424 (1975).

86. R.F. Squires and C. Braestrup, Benzodiazepine receptors in rat brain, Nature 266, 732 (1977).

87. C. Braestrup and R.F. Squires, Brain specific benzodiazepine receptors in rats characterized by high affinity ^{3}H-diazepam binding, Proc. Nat. Acad. Sci., USA, in press.

88. H. Möhler and T. Okada, Benzodiazepine receptor: demonstration in the central nervous system, Science, in press.

89. H. Möhler and T. Okada, Properties of ^{3}H-diazepam binding to benzodiazepine receptors in rat cerebral cortex, Life Sci., in press.

90. G.K. Aghajanian, LSD and CNS transmission, Ann. Rev. Pharmacol. 12, 157 (1972).

91. H. Corrodi, K. Fuxe, T. Hökfelt, P. Lidbrink and U. Ungerstedt, Effect of ergot drugs on central catecholamine neurons: evidence for a stimulation of central dopamine neurons, J. Pharm. Pharmacol. 25, 409 (1973).

92. G. Anlezark, C. Pycock and B. Meldrum, Ergot alkaloids as dopamine agonists: comparison in two rodent models, Eur. J. Pharmacol. 37, 295 (1976).

93. A.M. Johnson, D.M. Loew and J.M. Vigouret, Stimulant properties of bromocriptine on central dopamine receptors in comparison to apomorphine, (+)-amphetamine and L-DOPA, Brit. J. Pharmacol. 56,59 (1976).

94. H.G. Floss, J.M. Cassady and J.E. Robbers, Influence of ergot alkaloids on pituitary prolactin and prolactin-dependent processes, J. Pharm. Sci. 62, 699 (1973).

95. S.W.J. Lamberts and J.C. Birkenhäger, Bromocriptine in Nelson's syndrome and Cushing's disease, Lancet 2, 811 (1973).

96. G. Langer, E.J. Sachar, P.H. Gruen and F.S. Halpern, Human prolactin responses to neuroleptic drugs correlate with antischizophrenic potency, Nature 266, 639 (1977).

97. R. Kartzinel, I. Shoulson and D.B. Calne, Studies with bromocriptine, part 2. Double-blind comparison with levodopa in idiopathic Parkinsonism, Neurology 26, 511 (1976).

98. M. Trabucchi, P.F. Spano, G.C. Tonon and L. Frattola, Effects of bromocriptine on central dopaminergic receptors, Life Sci. 19, 225 (1976).

99. L.T. Williams, D. Mullikin and R.J. Lefkowitz, Identification of α-adrenergic receptors in uterine smooth muscle membranes by [^{3}H]dihydroergocryptine binding, J. Biol. Chem. 251, 6915 (1976).

100. W.J. Strittmatter, J.N. Davis and R.J. Lefkowitz, α-Adrenergic receptors in rat parotid cells I. Correlation of [^{3}H]dihydroergocryptine binding and catecholamine stimulated potassium efflux, J. Biol. Chem. in press

101. J.N. Davis, W. Strittmatter, E. Hoyler and R.J. Lefkowitz, [^{3}H]dihydroergocryptine binding in rat brain, Brain Res. in press.

102. I. Creese, R. Schneider and S.H. Snyder, manuscript in preparation.

103. A. Closse and D. Hauser, Dihydroergotamine binding to rat brain membranes, Life Sci. 19, 1851 (1976).

104. O. Hornykiewicz, Psychopharmacological implications of dopamine and dopamine antagonists: a critical evaluation of the evidence. Ann. Rev. Pharmacol. Toxicol. 17, 545 (1977).

DISCUSSION

Dr. Tenen asked if Dr. U'Prichard had tried to displace WB-4101 with benzodiadepines. *Dr. U'Prichard* said the benzodiazepines were ineffective even at 10^{-5}M.

Dr. Lomax asked if the level of endogenous dopamine could affect the determination of halo-peridol binding; *Dr. U'Prichard* thought this very unlikely since the membranes were washed and re-suspended 3 to 4 times.

X. Animal Models: Future Directions

FUTURE DIRECTIONS

A panel discussion on future directions in Animal Models research was chaired by
Roger R. Kelleher. The other panel members were Herbert Barry, III, Thomas N. Chase,
Leonard Cook and William T. McKinney, Jr. The panelists did not prepare manuscripts; they
were asked to give a short informal talk before the floor was open to all attending the
session.

Dr. Kelleher proposed that three areas of possible future use of animal models were the
following: 1. To study behavioral deficits in geriatric patients and to screen drugs of
potential utility in improving what is loosely called "memory and learning" in old people;
2. To study behaviorally-induced cardiovascular or gastric changes and the role of such
changes in psychosomatic medicine as well as screening for drugs effective in this area;
3. To study drug dependence.
Dr. Kelleher expressed the hope that investigators would not get so preoccupied with animal models that they neglected slight things which did not seem to have relevance or to fit
in well with the popular models of the moment. This has been a mistake in the past and undoubtedly will occur in the future. Dr. Kelleher cited an example: It is well known that
under appropriate conditions animals will repeatedly press a lever to obtain presentation of
food or water after the animals have been deprived. It is not as well known that under appropriate conditions monkeys can be trained to repeatedly press a lever to receive severe
electric shocks to their tails, under somewhat similar conditions to those in the food or
water reward paradigm. Similarly, it probably is not surprising that animals will repeatedly
press a lever to obtain injections of cocaine or morphine, but it seems surprising that not
only monkeys, but rats also will lever press to receive injections of apomorphine - a dopamine agonist.
Chlorpromazine presents an entirely different picture. Even well engendered behaviors can
be suppressed by injection of chlorpromazine. Furthermore monkeys will actively avoid or
postpone the injection of chlorpromazine. Dr. Kelleher felt that this difference of behavior
towards chlorpromazine (as opposed to morphine, shock, apomorphine, etc.) must have the
seeds of significance, but he did not know what they were; that possibly it may help explain
how seemingly noxious events may reinforce behavior. Perhaps it is even related to the dopamine picture.

The next panelist, *Dr. Barry*, proposed that in looking for animal models of human psychiatric conditions from the point of view of the behavioral pharmacologist, the place to look is
in models of chronic conflict, since unresolved chronic conflict underlies various types of
mental illness. Among others, there already is the Geller Technique of food approach which
conflicts with receiving painful shock; this is discussed in Dr. Howard's paper. Dr. Kelleher
has mentioned the training of monkeys to lever press to receive painful electric shocks -
which must involve a chronic conflict in these animals. Dr. Barry emphasized that a good
model of human psychiatric illness must involve continuous conflict with as much prolongation
as possible. He felt that the McKinney socially deprived model fits these criteria.

Dr. Chase stated his opinion that the Federal government is very concerned about the use of
animal models in research, that it is estimated that NINCDS, e.g., is spending seven million
dollars this year in support of animal model research. Dr. Chase pointed out that the Federal government was getting more and more into the regulation of not only clinical research
but also preclinical research. He forecast that such research would become not only more
centrally organized in the next decade, but also more bureaucratized, more socially responsive, more conservative - and much duller. He was afraid that as a result not only of such
regulation but also the increasing fear of the clinical researcher to be sued and the fear
of the preclinical researcher of not being funded, research would become increasingly conservative, that few would dare to run any experiment until every last detail has been checked
out.
Another trend commented upon by Dr. Chase was that of increased outside intervention by
non-scientists; this might be summed up by the statement that it is thought that science is
too important, too valuable, too costly and even too dangerous to be managed by scientists.
Dr. Chase felt there will be increasing supervision of science by non-scientists, there will
be increasing regulation of all types of research. Primate research is already under heavy
control.
Economics is also getting to be more significant in research. Cost of equipment, supplies
and facilities is escalating as is the per capita cost to do clinical trials. This latter is
related not only to inflation but also to increasing safety requirements. However, according
to Dr. Chase, the public attitude towards science has cooled. Some of this may be related
to oversell - both to the public and to Congress - e.g., indicating a cure just around the
corner. Scientists have not been very successful in educating the public on realistic expectations from science.

There is an increasing haste for results. Although the public may be overly skeptical about the scientist playing his own intellectual games for his own pleasure, there may be some truth in this. Dr. Chase feels there will be increasing pressure for applied research. Although the pressures may make research a lot more difficult and less interesting and less rewarding, he feels that we shall have to struggle more just to stay where we are. Specifically with regard to animal models, Dr. Chase feels there will be increasing support for these not only as a means to predict safety factors but also as means to decrease the huge investment required before potential new drugs can be introduced.

Dr. Cook, the next member of the panel to make comments, indicated that he felt there was too much emphasis in animal model work on neuroleptics. He feels that drugs for the improvement of learning and memory in the aged are definitely on the way. He recalled the opposition to the concept of tranquilizers in 1953 and 1954 and anticipated even greater opposition to the introduction of drugs enhancing performance.

Another type of drug which Dr. Cook suggested was needed was one to take care of severe conflict. The benzodiazepines and the meprobamates take care of mild to moderate conflict.

Dr. Cook concluded with the suggestion that we have sufficient screening procedures in animal models, that what is lacking is an understanding of the pathology of mental illnesses and application of this in animal models.

Dr. McKinney, the last of the panelists, started his comments with a discussion of the imbalance in budget support; he felt the support given to mental health research was very depressing. He then went on to point out the need to educate not only outsiders but even people within the fields of psychiatry and neurology of the potential value of animal models.

Although Dr. McKinney agreed that such a statement was probably overstated he did express the thought that most current preclinical screening tests are almost useless - in terms of human psychopathology, in a social context. He then turned to the very difficult problems currently being faced by primate research: a world-wide shortage of primates; an attack on use of primates in research as opposed to studies on rats or on man. Dr. McKinney feels one weakness in primate research is that the animals used in heart disease studies and a number of other studies are at least partial social isolates, and very little is currently known concerning possible neurobiological effects of partial social isolation.

Dr. McKinney concluded with some comments on the reverse of animal models, the application of some ethological techniques to human studies. He hoped there would be increasing use of such items as improved observational techniques to human clinical studies: He hoped that there would be more of a focus on animal models for understanding basic mechanisms of disease rather than an excessive concern with what clinical syndrome the animals represent.

Dr. Lomax related an anecdote about a skin rash which developed in his dog when his son left for prolonged periods. This rash could be prevented by Valium treatment; leading Dr. Lomax to speculate that the dog-human interaction had led to a model of clinical depression. He wondered if in the future it might be profitable to look for animal/human models in addition to or in place of animal/animal models.

In a reply to a question by *Dr. Usdin, Dr. Chase* suggested that it might be appropriate to have more organization in letting Congress and the public know what we are trying to accomplish with animal models.

Dr. Cook expressed disagreement with the statement that we do not have adequate animal models. He felt that current existing models were good enough to allow almost sure predictability of the efficacy of a drug in treating schizophrenia, reasonably good predictability of efficacy in treating depression, almost perfect predictability of efficacy in treating neuroses. Our current procedures predict such side effects as extrapyramidal symptoms and hypotension: Dr. Cook felt the major problem was one of not understanding what is available.

Dr. Kornetsky disagreed with Dr. Cook; he felt that we had many screens, but few good animal models of human diseases. He also expressed the thought that we may have oversold science to the public. As in art or music, there will always be much that is mundane in addition to the great symphonies.

Dr. Chase made the analogy between the Lindbergh flight and flight to the moon. Lindbergh's flight was designed by one man; the moon flight required an enormous team. Whether or not we like it, we are in an age of big science and the individual brilliant entreprenurial scientist has greater difficulty to succeed. Perhaps we can try to turn this around.

Dr. Cook emphasized the progress that has been made. Not so long ago, patients in mental institutions were handled by chemical strait jackets, chloral hydrate, barbital, etc. Today's drugs, retrospectively, are magic bullets. However, ten or fifteen years from now, chlorpromazine, the benzodiazepines, etc. may be seen as horribly non-specific. As far as the future goes, Dr. Cook predicted that drugs that will be wanted will be those having specific effects; our current drugs will then be referred to as non-specific CNS depressants.

Dr. Usdin agreed that it would be good to have antipsychotic drugs which did not affect the dopaminergic system. He asked which screening tests in current use did not involve the dopaminergic system. *Dr. Cook* mentioned such models as punished behavior, discrete conditioned avoidance. He also mentioned the work of Drs. Aghajanian and Bunney separating mesolimbic effects from effects on systems producing extrapyramidal effects.

Dr. Goldstein asked Dr. Cook to describe his ideal drug of the future. *Dr. Cook* suggested that this might be one that modulated behavior without affecting the dopaminergic system.

Dr. Kornetsky proposed that our aim should not be to find better and better drugs but rather to find the cause of diseases and how to prevent them.

Dr. Cook expressed the philosophy that there are very real areas of medical need which are approached very ethically by making drugs available. Obviously, if the same benefits to the patient can be achieved without supplying the drugs, this should be done.

Dr. Antelman was not sure whether the preferable alternative was schizophrenia or tardive dyskinesia. Although one could not make out well in the world with schizophrenia, neither can one with the protruding tongue, etc., of dyskinesia.

Dr. McKinney expressed the point of view that the primary purpose of animal models' studies is to understand diseases, not to develop drugs. He did not think that we could ever create a foolproof system in terms of animal models for human diseases, but he does not think it is vital that a system be foolproof before it is of value.

Dr. Barry suggested the need for more chronic, long-term studies. *Dr. Kelleher* agreed with the need for long-term studies and pointed out the difficulties in getting support for these.

The final statements in the session were made by *Dr. Kobayashi*. He felt that the field of "animal models" should be broadened to "conceptual models" - to include not only animal models, but also pharmacological models, surgical models, etc., not just pharmacological animal models. Dr. Kobayashi then pointed out the difference between current work and that of Copernicus who had a concept, designed and produced his own instrument, did the work and published his results rapidly.

Finally, Dr. Kobayashi mentioned an article he had read on imagination and creativity. The author of that article commented on creativity in art, creativity in music, creativity in journalism. The author's point of view was that perhaps the most creative people in recent generations have been scientists; that in order to be a real contributor to science, one has to be creative and imaginative; that science is where the height of creativity and imagination can be found. Dr. Kobayashi thought perhaps this could be found among animal models; that they could serve as the base needed to be able to understand how human biology functions.